SECOND EDITION

Simulation of Dynamic Systems

with MATLAB® and Simulink®

Harold Klee
Randal Allen

CRC Press
Taylor & Francis Group
Boca Raton   London   New York

CRC Press is an imprint of the
Taylor & Francis Group, an **informa** business

CRC Press
Taylor & Francis Group
6000 Broken Sound Parkway NW, Suite 300
Boca Raton, FL 33487-2742

© 2011 by Taylor and Francis Group, LLC
CRC Press is an imprint of Taylor & Francis Group, an Informa business

No claim to original U.S. Government works

Printed in the United States of America on acid-free paper
10 9 8 7 6 5 4 3 2 1

International Standard Book Number: 978-1-4398-3673-6 (Hardback)

Library of Congress Cataloging-in-Publication Data

Klee, Harold.
   Simulation of dynamic systems with MATLAB and Simulink. -- 2nd ed. / Harold Klee, Randal Allen.
      p. cm.
   Includes bibliographical references and index.
   ISBN 978-1-4398-3673-6 (hardback)
   1. Computer simulation. 2. SIMULINK. 3. MATLAB. I. Allen, Randal, 1964- II. Title.

QA76.9.C65K585 2011
003'.3--dc22   2010035461

Visit the Taylor & Francis Web site at
http://www.taylorandfrancis.com

and the CRC Press Web site at
http://www.crcpress.com

To Andrew, Cassie and in loving memory
of their mother and devoted wife, Laura.

Harold Klee

To Dave Lundquist and Steve Roemerman who believed in me.

Randal Allen

Contents

Foreword

As the authors point out in the preface, there is not yet extant a universally accepted definition of the term *simulation*. Another approach to defining the field would be "the art of reproducing the behavior of a system for analysis without actually operating that system." The authors have written a seminal text covering the simulation design and analysis of a broad variety of systems using two of the most modern software packages available today. The material is presented in a particularly adept fashion enabling students new to the field to gain a thorough understanding of the basics of continuous simulation in a single semester and providing, at the same time, a more advanced treatment of the subject for researchers and simulation professionals. The authors' extensive treatment of continuous and discrete linear system fundamentals opens the door to simulation for individuals without formal education in a traditional engineering curriculum.

However defined, simulation is becoming an increasingly important component of curricula in engineering, business administration, the sciences, applied mathematics, and the like. This text will be a valuable resource for study in courses using simulation as a tool for understanding processes that are not amenable to study in other ways.

Chris Bauer, PhD, PE, CMSP
Orlando, Florida

Simulation has come a long way since the days analog computers filled entire rooms. Yet, it is more important than ever that simulations be constructed with care, knowledge, and a little wisdom, lest the results be gibberish or, worse, reasonable but misleading. Used properly, simulations can give us extraordinary insights into the processes and states of a physical system. Constructed with care, simulations can save time and money in today's competitive marketplace.

One major application of simulation is the simulator, which provides interaction between a model and a person through some interface. The earliest simulator, Ed Link's Pilot Maker aircraft trainer, did not use any of the simulation techniques described in this book. Modern simulators, however, such as the National Advanced Driving Simulator (NADS), cannot be fully understood without them.

The mission of the NADS is a lofty one: to save lives on U.S. highways through safety research using realistic human-in-the-loop simulation. This is an example of the importance simulation has attained in our generation. The pervasiveness of simulation tools in our society will only increase over time; it will be more important than ever that future scientists and engineers be familiar with their theory and application.

The content for *Simulation of Dynamic Systems with MATLAB® and Simulink®* is arranged to give the student a gradual and natural progression through the important topics in simulation. Advanced concepts are added only after complete examples have been constructed using fundamental methods. The use of MATLAB and Simulink provides experience with tools that are widely adopted in industry and allow easy construction of simulation models.

May your experience with simulation be enjoyable and fruitful and extend throughout your careers.

Chris Schwarz, PhD
Iowa City, Iowa

Preface

In the first article of *SIMULATION* magazine in the Fall of 1963, the editor John McLeod proclaimed simulation to mean "the act of representing some aspects of the real world by numbers or symbols which may be easily manipulated to facilitate their study." Two years later, it was modified to "the development and use of models for the study of the dynamics of existing or hypothesized systems." More than 40 years later, the simulation community has yet to converge upon a universally accepted definition. Either of the two cited definitions or others that followed convey a basic notion, namely, that simulation is intended to reinforce or supplement one's understanding of a system. The definitions vary in their description of tools and methods to accomplish this.

The field of simulation is experiencing explosive growth in importance because of its ability to improve the way systems and people perform, in a safe and controllable environment, at a reduced cost. Understanding the behavior of complex systems with the latest technological innovations in fields such as transportation, communication, medicine, aerospace, meteorology, etc., is a daunting task. It requires an assimilation of the underlying natural laws and scientific principles that govern the individual subsystems and components. A multifaceted approach is required, one in which simulation can play a prominent role, both in validation of a system's design and in training of personnel to become proficient in its operation.

Simulation is a subject that cuts across traditional academic disciplines. Airplane crews spend hours flying simulated missions in aircraft simulators to become proficient in the use of onboard subsystems during normal flight and possible emergency conditions. Astronauts spend years training in shuttle and orbiter simulators to prepare for future missions in space. Power plant and petrochemical process operators are exposed to simulation to obtain peak system performance. Economists resort to simulation models to predict economic conditions of municipalities and countries for policymakers. Simulations of natural disasters aid in preparation and planning to mitigate the possibility of catastrophic events.

While the mathematical models created by aircraft designers, nuclear engineers, and economists are application specific, many of the equations are analogous in form despite the markedly different phenomena described by each model. Simulation offers practitioners from each of these fields the tools to explore solutions of the models as an alternative to experimenting with the real system.

This book is meant to serve as an introduction to the fundamental concepts of continuous system simulation, a branch of simulation applied to dynamic systems whose signals change over a continuum of points in time or space. Our concern is with mathematical models of continuous-time systems (electric circuits, thermal processes, population dynamics, vehicle suspension, human physiology, etc.) and the discrete-time system models created to simulate them. The continuous system mathematical models consist of a combination of algebraic and ordinary differential equations. The discrete-time system models are a mix of algebraic and difference equations.

Systems that transition between states at randomly occurring times are called discrete-event systems. Discrete-event simulation is a complementary branch of simulation, separate from continuous system simulation, with a mathematical foundation rooted in probability theory. Examples of discrete-event systems are facilities such as a bank, a tollbooth, a supermarket, or a hospital emergency room, where customers arrive and are then serviced in some way. A manufacturing plant involving multiple production stages of uncertain duration to generate a finished product is another candidate for discrete-event simulation.

Discrete-event simulation is an important tool for optimizing the performance of systems that change internally at unpredictable times due to the influence of random events. Industrial engineering programs typically include a basic course at the undergraduate level in discrete-event simulation.

Not surprisingly, a number of excellent textbooks in the area have emerged for use by the academic community and professionals.

In academia, continuous simulation has evolved differently than discrete-event simulation. Topics in continuous simulation such as dynamic system response, mathematical modeling, differential equations, difference equations, and numerical integration are dispersed over several courses from engineering, mathematics, and the natural sciences. In the past, the majority of courses in modeling and simulation of continuous systems were restricted to a specific field like mechanical, electrical, and chemical engineering or scientific areas like biology, ecology, and physics.

A transformation in simulation education is underway. More universities are beginning to offer undergraduate and beginning graduate courses in the area of continuous system simulation designed for an interdisciplinary audience. Several institutions now offer master's and PhD programs in simulation that include a number of courses in both continuous and discrete-event simulation. A critical mass of students are now enrolled in continuous simulation–related courses and there is a need for an introductory unifying text.

The essential ingredient needed to make simulation both interesting and challenging is the inclusion of real-world examples. Without models of real-world systems, a first class in simulation is little more than a sterile exposition of numerical integration applied to differential equations.

Modeling and simulation are inextricably related. While the thrust of this text is continuous simulation, mathematical models are the starting point in the evolution of simulation models. Analytical solutions of differential equation models are presented, when appropriate, as an alternative to simulation and a simple way of demonstrating the accuracy of a simulated solution. For the most part, derivations of the mathematical models are omitted and references to appropriate texts are included for those interested in learning more about the origin of the model's equations.

Simulation is best learned by doing. Accordingly, the material is presented in a way that permits the reader to begin exploring simulation, starting with a mathematical model in Chapter 1. A detailed derivation of the mathematical model of a tank with liquid flowing in and out leads to a simulation model in the form of a simple difference equation. The simulation model serves as the vehicle for predicting the tank's response to various inputs and initial conditions. Additionally, the derivation illustrates the process of obtaining a mathematical model based on the natural laws of science.

Chapters 2 and 4 present a condensed treatment of linear, continuous-time, and discrete-time dynamic systems, normally covered in an introductory linear systems course. Coverage is limited to basic topics that should be familiar to a simulation practitioner. Section 2.7 is extended to include a discussion of additional common nonlinear elements, namely, dead zone, quantization, relay, and saturation. The instructor can skip some or all of the material in these chapters if the students' background includes a course in signals and systems or linear control theory.

Numerical integration is at the very core of continuous system simulation. Instead of treating the subject in one exhaustive chapter, coverage is distributed over three chapters. Elementary numerical integration in Chapter 3 is an informal introduction to the subject, which includes discussion of several elementary methods for approximating the solutions of first-order differential equations. The material in Chapters 2 through 4 is a prerequisite for understanding general purpose, continuous simulation programs that are popular in the engineering and scientific community.

Simulink®, from The MathWorks, is the featured simulation program because of its tight integration with MATLAB®, the de facto standard for scientific and engineering analysis, and data visualization software. Chapter 5 takes the reader through the basic steps of creating and running Simulink models. Section 5.5 includes new material related to simulation implementation of nonlinear systems using specific blocks from the Simulink library. Due to the popularity of the Kalman filter, a case study has been added in Section 5.12 on this topic. The continuous-time Kalman filter equations are developed and modeled in Simulink, including simulated output. Subsequently, the steady-state continuous-time Kalman filter equations are developed and modeled in Simulink. The steady-state results are compared with the continuous-time results. Finally, the discrete-time Kalman filter equations are developed and modeled in Simulink. The discrete-time results are compared with the continuous-time results.

Chapter 6 delves into intermediate-level topics of numerical integration, including a formal presentation of One-Step (Runge–Kutta) and multistep methods, adaptive techniques, truncation errors, and a brief mention of stability.

Chapter 7 highlights some advanced features of Simulink useful in more in-depth simulation studies. A new section (Section 7.5) on S-blocks is introduced and an example is presented showing how to make the discrete-time Kalman filter available for drag-and-drop from the Simulink library. Other simulation programs offer similar features and the transition from Simulink to other simulation software is straightforward.

Chapter 8 is for those interested in more advanced topics on continuous simulation. Coverage includes a discussion of dynamic errors, stability, real-time compatible numerical integration, and multi-rate integration algorithms for simulation of systems with fast and slow components. Due to the popularity of Lego's Mindstorms™ NXT, a case study has been added in Section 8.7 on this topic.

All but two chapters conclude with a case study illustrating one or more of the topics discussed in that chapter. The featured text examples and case studies are analyzed using MATLAB script files and Simulink model files, all of which are available from CRC Press.

The text has been field-tested in the classroom for several years in a two-semester sequence of continuous simulation courses. Despite numerous revisions based on the scrutiny and suggestions of students and colleagues, it is naïve to think the final product is free of errors. Further suggestions for improvement and revelations of inaccuracies can be brought to the attention of the authors at rallen397@cfl.rr.com and klee@mail.ucf.edu.

Numerous individuals deserve our thanks and appreciation for helping to make this book possible. In particular, a sincere "thank you" to Nora Konopka at Taylor & Francis/CRC Press for committing to the second edition and seeing it through to fruition.

For MATLAB® and Simulink® product information, please contact:

The MathWorks, Inc.
3 Apple Hill Drive
Natick, MA, 01760-2098 USA
Tel: 508-647-7000
Fax: 508-647-7001
E-mail: info@mathworks.com
Web: www.mathworks.com

Authors

Dr. Harold Klee received his PhD in systems science from Polytechnic Institute of Brooklyn in 1972, his MS in systems engineering from Case Institute of Technology in 1968, and his BSME from The Cooper Union in 1965.

Dr. Klee has been a faculty member in the College of Engineering at the University of Central Florida (UCF) since 1972. During his tenure at UCF, he has been a five-time recipient of the college's Outstanding Teacher Award. He has been instrumental in the development of simulation courses in both the undergraduate and graduate curricula. He is a charter member of the Core Faculty, which is responsible for developing the interdisciplinary MS and PhD programs in simulation at UCF. Dr. Klee served as graduate coordinator in the Department of Computer Engineering from 2003 to 2006. Two of his PhD students received the prestigious Link Foundation Fellowship in Advanced Simulation and Training. Both are currently enjoying successful careers in academia.

Dr. Klee has served as the director of the UCF Driving Simulation Lab for more than 15 years. Under the auspices of the UCF Center for Advanced Transportation Systems Simulation, the lab operates a high-fidelity motion-based driving simulator for conducting traffic engineering–related research. He also served as editor-in-chief for the *Modeling and Simulation* magazine for three years, a publication for members of the Society for Modeling and Simulation International.

Dr. Randal Allen is an aerospace and defense consultant working under contract to provide 6DOF aerodynamic simulation modeling, analysis, and design of navigation, guidance, and control systems. His previous experience includes launch systems integration and flight operations for West Coast Titan-IV missions, propulsion modeling for the Iridium satellite constellation, and field applications engineering for MATRIXx. He also chairs the Central Florida Section of the American Institute of Aeronautics and Astronautics (AIAA).

Dr. Allen is certified as a modeling and simulation professional (CMSP) by the Modeling and Simulation Professional Certification Commission (M&SPCC) under the auspices of the National Training and Simulation Association (NTSA). He is also certified to deliver FranklinCovey's Focus and Execution track, which provides training on achieving your highest priorities.

Dr. Allen's academic background includes a PhD in mechanical engineering from the University of Central Florida, an engineer's degree in aeronautical and astronautical engineering from Stanford University, an MS in applied mathematics, and a BS in engineering physics from the University of Illinois (Urbana-Champaign). He also serves as an adjunct professor at the University of Central Florida in Orlando, Florida.

1 Mathematical Modeling

1.1 INTRODUCTION

1.1.1 Importance of Models

Models are an essential component of simulation. Before a new prototype design for an automobile braking system or a multimillion dollar aircraft is tested in the field, it is commonplace to "test drive" the separate components and the overall system in a simulated environment based on some form of model. A meteorologist predicts the expected path of a tropical storm using weather models that incorporate the relevant climatic variables and their effect on the storm's trajectory. An economist issues a quantitative forecast of the U.S. economy predicated based on key economic variables and their interrelationships with the help of computer models. Before a nuclear power plant operator is "turned loose" at the controls, extensive training is conducted in a model-based simulator where the individual becomes familiar with the plant's dynamics under routine and emergency conditions. Health care professionals have access to a human patient simulator to receive training in the recognition and diagnosis of disease. Public safety organizations can plan for emergency evacuations of civilians from low-lying areas using traffic models to simulate vehicle movements along major access roads.

The word "model" is a generic term referring to a conceptual or physical entity that resembles, mimics, describes, predicts, or conveys information about the behavior of some process or system. The benefit of having a model is to be able to explore the intrinsic behavior of a system in an economical and safe manner. The physical system being modeled may be inaccessible or even nonexistent as in the case of a new design for an aircraft or automotive component.

Physical models are often scaled-down versions of a larger system of interconnected components as in the case of a model airplane. Aerodynamic properties of airframe and car body designs for high-performance airplanes and automobiles are evaluated using physical models in wind tunnels. In the past, model boards with roads, terrain, miniaturized models of buildings, and landscape, along with tiny cameras secured to the frame of ground vehicles or aircraft, were prevalent for simulator visualization. Current technology relies almost exclusively on computer-generated imagery.

In principle, the behavior of dynamic systems can be explained by mathematical equations and formulae, which embody either scientific principles or empirical observations, or both, related to the system. When the system parameters and variables change continuously over time or space, the models consist of coupled algebraic and differential equations. In some cases, lookup tables containing empirical data are employed to compute the parameters. Equations may be supplemented by mathematical inequalities, which constrain the variation of one or more dependent variables. The aggregation of equations and numerical data employed to describe the dynamic behavior of a system in quantitative terms is collectively referred to as a mathematical model of the system.

Partial differential equation models appear when a dependent variable is a function of two or more independent variables. For example, electrical parameters such as resistance and capacitance are distributed along the length of conductors carrying electrical signals (currents and voltages). These signals are attenuated over long distances of cabling. The voltage at some location x measured from an arbitrary reference is written $v(x, t)$ instead of simply $v(t)$, and the circuit is modeled accordingly.

A mathematical model for the temperature in a room would necessitate equations to predict $T(x, y, z, t)$ if a temperature probe placed at various points inside the room reveals significant variations in temperature with respect to x, y, z in addition to temporal variations. Partial differential equations describing the cable voltage $v(x, t)$ and room temperature $T(x, y, z, t)$ are referred to as "distributed parameter" models.

The mathematical models of dynamic systems where the single independent variable is "time" comprise ordinary differential equations. The same applies to systems with a single spatial independent variable; however, these are not commonly referred to as dynamic systems since variations of the dependent variables are spatial as opposed to temporal in nature. Ordinary differential equation models of dynamic systems are called "lumped parameter" models because the spatial variation of the system parameters is negligible or else it is being approximated by lumped sections with constant parameter values. In the room temperature example, if the entire contents of the room can be represented by a single or lumped thermal capacitance, then a single temperature $T(t)$ is sufficient to describe the room. We focus exclusively on dynamic systems with lumped parameter models, hereafter referred to simply as mathematical models.

A system with a lumped parameter model is illustrated in Figure 1.1. The key elements are the system inputs $u_1(t), u_2(t), \ldots, u_r(t)$, which make up the system input vector $\underline{u}(t)$, the system outputs $y_1(t), y_2(t), \ldots, y_p(t)$, which form the output vector $\underline{y}(t)$, and the parameters $p_1, p_2, \ldots, p_m$ constituting the parameter vector $\underline{p}$. The parameters are shown as constants; however, they may also vary with time.

Our interest is in mathematical models of systems consisting of coupled algebraic and differential equations relating the outputs and inputs with coefficients expressed in terms of the system parameters. For steady-state analyses, transient responses are irrelevant, and the mathematical models consist of purely algebraic equations relating the system variables.

An example of a mathematical model for a system with two inputs, three outputs, and several parameters is

$$p_1 \frac{d^2}{dt^2} y_1(t) + p_2 p_3 \frac{d}{dt} y_1(t) + p_4 y_1(t) + p_5 \frac{d}{dt} y_2(t) + p_6 y_2(t) = p_7 u_1(t) \tag{1.1}$$

$$p_8 \frac{d}{dt} y_2(t) + \frac{p_9}{p_{10}} y_2(t) + p_{11} y_1(t) y_2(t) = p_{12} \frac{d}{dt} u_1(t) + p_{13} u_1(t) + p_{14} u_2(t) \tag{1.2}$$

$$p_{15} y_3(t) = \frac{p_{16} y_1^{p_{17}}(t)}{y_2(t)} \tag{1.3}$$

The order of a model is equal to the sum of the highest derivatives of each of the dependent variables, in this case $y_1(t), y_2(t), y_3(t)$, and the order is therefore $2 + 1 + 0 = 3$. Equation 1.1 is a linear differential equation. Equation 1.2 is a nonlinear differential equation because of the term involving the product of $y_1(t)$ and $y_2(t)$. The mathematical model is nonlinear due to the presence

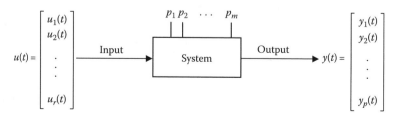

FIGURE 1.1 A system with a lumped parameter model.

of the nonlinear differential equation and the nonlinear algebraic equation (Equation 1.3). It is to be borne in mind that it is the nature of the equations that determines whether a math model is linear or nonlinear. An adjective such as linear or nonlinear applies to the mathematical model as opposed to the actual system.

It is important to distinguish between the system being modeled and the model itself. The former is unique, even though it may exist only at the design stage, while the mathematical model may assume different forms. For example, a team of modelers may be convinced that the lead term in Equation 1.1 is likely to be insignificant under normal operating conditions. Consequently, two distinct models of the system exist, one third order and the other second order. The third-order model includes the second derivative term to accurately reflect system behavior under unusual or nontypical conditions (e.g., an aircraft exceeding its flight envelope or a ground vehicle performing an extreme maneuver). The simpler second-order model ignores what are commonly referred to as higher-order effects. Indeed, there may be a multitude of mathematical models to represent the same system under different sets of restricted operating conditions. Regardless of the detail inherent in a mathematical model, it nevertheless represents an incomplete and inexact depiction of the system.

A model's intended use will normally dictate its level of complexity. For example, models for predicting vehicle handling and responsiveness are different from those intended to predict ride comfort. In the first case, accurate equations describing lateral and longitudinal tire forces are paramount in importance, whereas passenger comfort relies more on vertical tire forces and suspension system characteristics.

Mathematical modeling is an inexact science, relying on a combination of intuition, experience, empiricism, and the application of scientific laws of nature. Trade-offs between model complexity and usefulness are routine. Highly accurate microclimatic weather models that use current atmospheric conditions to predict the following day's weather are of limited value if they require 48 h on a massively parallel or supercomputer system to produce results. At the extreme opposite, overly simplified models can be grossly inaccurate if significant effects are overlooked.

The difference between a mathematical model and a simulation model is open to interpretation. Some in the simulation community view the two as one and the same. Their belief is that a mathematical model embodies the attributes of the actual system and simulation refers to solutions of the model equations, albeit generally approximate in nature. Exact analytical solutions of mathematical model equations are nonexistent in all but the simplest cases.

Others maintain a distinction between the two and express the view that simulation model(s) originate from the mathematical model. According to this line of thinking, simulating the dynamics of a system requires a simulation model that is different in nature from a mathematical model. A reliable simulation model must be capable of producing numerical solutions in reasonably close agreement with the actual (unknown) solutions to the math model. Simulation models are commonly obtained from discrete-time approximations of continuous-time mathematical models. Much of this book is devoted to the process of obtaining simulation models in this way. More than one simulation model can be developed from a single mathematical model of a system.

Stochastic models are important when dealing with systems whose inputs and parameters are best modeled using statistical methods. Discrete event models are used to describe processes that transit from one state to another at randomly spaced points in time. Probability theory plays a significant role in the formulation of discrete event models for describing the movement of products and service times at different stages in manufacturing processes, queuing systems, and the like. In fact, the two pillars of simulation are continuous system simulation, the subject of this book, and discrete event simulation.

There is a great deal more to be said about modeling. Entire books are devoted to properly identifying model structure and parameter values for deterministic and stochastic systems. Others concentrate more on derivation of mathematical models from diverse fields and methods of obtaining solutions under different circumstances. The reader is encouraged to check the references section at the end of this book for additional sources of material related to modeling.

Modeling is essential to the field of simulation. Indeed, it is the starting point of any simulation study. The emphasis, however, in this book is on the presentation of simulation fundamentals. Accordingly, derivation of mathematical models is not a prominent component. For the most part, the math models are taken from documented sources listed in the references section, some of which include step-by-step derivations of the model equations. The derivation is secondary to a complete understanding of the model, that is, its variables, parameters, and knowledge of conditions that may impose restrictions on its suitability for a specific application.

Simulation of complex systems requires a team effort. The modeler is a subject expert responsible for providing the math model and interpreting the simulation results. The simulationist produces the simulation model and performs the simulation study. For example, an aerodynamicist applies principles of boundary layer theory to obtain a mathematical model for the performance of a new airfoil design. Starting with the math model, simulation skills are required to produce a simulation model capable of verifying the efficacy of the design based on numerical results. Individuals with expert knowledge in a particular field are oftentimes well versed in the practice of simulation and may be responsible for formulation of alternative mathematical models of the system in addition to developing and running simulations.

A simple physical system is introduced in the next section, and the steps involved in deriving an idealized math model are presented. In addition to benefiting from seeing the process from start to finish, the ingredients for creating a simulation model are introduced. Hence, by the end of this chapter, the reader will be able to perform rudimentary simulation.

1.2 DERIVATION OF A MATHEMATICAL MODEL

We begin our discussion of mathematical modeling with a simple derivation of the mathematical model representing the dynamic behavior of an open tank containing a liquid that flows in the top and is discharged from the bottom. Referring to Figure 1.2, the primary input is the liquid flow rate $F_1(t)$, an independent variable measured in appropriate units such as cubic feet per minute (volumetric flow rate) or pounds per hour (mass flow rate). Responding to changes in the input are dependent variables $H(t)$ and $F_0(t)$ the fluid level, and flow rate from the tank.

Once the derivation is completed, we can use the model to predict the outflow and fluid level response to a specific input flow rate $F_1(t)$, $t \geq 0$. Note that we have restricted the set of possible inputs to $F_1(t)$ and in the process relegated the remaining independent variables, that is, other variables which affect $F_0(t)$ and $H(t)$, to second-order importance. Our assumption is that the eventual model will be suitable for its intended application. It must be borne in mind that if extremely accurate predictions of the level $H(t)$ are required, it may be necessary to include second-order effects such as evaporation and hence introduce additional inputs related to ambient conditions, namely, temperature, humidity, air pressure, wind speed, and so forth.

The derivation is based on conditions of the tank at two discrete points in time, as if snapshots of the tank were available at times "t" and "$t + \Delta t$," as shown in Figure 1.3.

The following notation is used with representative units given for clarity:

$F_1(t)$: Input flow at time t, ft³/min
$H(t)$: Liquid level at time t, ft
$F_0(t)$: Output flow at time t, ft³/min
A: Cross-sectional area of tank, ft²

FIGURE 1.2 Tank as a dynamic system with input and outputs.

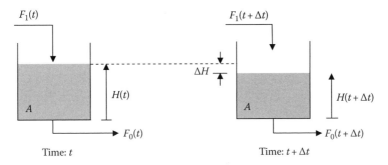

FIGURE 1.3 A liquid tank at two points in time.

At time $t + \Delta t$, from the physical law of conservation of volume,

$$V(t + \Delta t) = V(t) + \Delta V \qquad (1.4)$$

where
$V(t)$ is the volume of liquid in the tank at time t
ΔV is the change in volume from time t to $t + \Delta t$

The volume of liquid in the tank at times t and $t + \Delta t$ is given by

$$V(t) = AH(t) \qquad (1.5)$$

$$V(t + \Delta t) = AH(t + \Delta t) \qquad (1.6)$$

Equations 1.5 and 1.6 assume constant cross-sectional area of the tank, that is, A is independent of H.

The change in volume from t to $t + \Delta t$ is equal to the volume of liquid flowing in during the interval t to $t + \Delta t$ minus the volume of liquid flowing out during the same period of time. The liquid volumes are the areas under the input and output volume flow rates from t to $t + \Delta t$ as shown in Figure 1.4.

Expressing these areas in terms of integrals,

$$\Delta V = \int_{t}^{t+\Delta t} F_1(t)dt - \int_{t}^{t+\Delta t} F_0(t)dt \qquad (1.7)$$

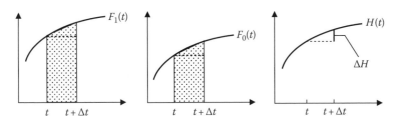

FIGURE 1.4 Volumes of liquid flowing in and out of tank from t to $t + \Delta t$.

The integrals in Equation 1.7 can be approximated by assuming $F_1(t)$ and $F_0(t)$ are constant over the interval t to $t + \Delta t$ (see Figure 1.4). Hence,

$$\int_t^{t+\Delta t} F_1(t)\mathrm{d}t \approx F_1(t)\Delta t \tag{1.8}$$

$$\int_t^{t+\Delta t} F_0(t)\mathrm{d}t \approx F_0(t)\Delta t \tag{1.9}$$

Equations 1.8 and 1.9 are reasonable approximations provided Δt is small. Substituting Equations 1.8 and 1.9 into Equation 1.7 yields

$$\Delta V \approx F_1(t)\Delta t - F_0(t)\Delta t \tag{1.10}$$

Substituting Equations 1.5, 1.6, and 1.10 into Equation 1.4 gives

$$AH(t + \Delta t) \approx AH(t) + [F_1(t) - F_0(t)]\Delta t \tag{1.11}$$

$$\Rightarrow A[H(t + \Delta t) - H(t)] \approx [F_1(t) - F_0(t)]\Delta t \tag{1.12}$$

$$\Rightarrow A\left[\frac{\Delta H}{\Delta t}\right] \approx F_1(t) - F_0(t) \tag{1.13}$$

where ΔH is the change in liquid level over the interval $(t, t + \Delta t)$. Note that $\Delta H / \Delta t$ is the average rate of change in the level H over the interval $(t, t + \Delta t)$. It is the slope of the secant line from pt A to pt B in Figure 1.5.

Consider what happens as pt B gets closer to pt A; that is, Δt gets smaller.

$$\text{End pt } \Delta H/\Delta t$$

$$B \quad \frac{H(t + \Delta t) - H(t)}{\Delta t} : \text{Slope of line } AB$$

$$B' \quad \frac{H(t + \Delta t') - H(t)}{\Delta t'} : \text{Slope of line } AB'$$

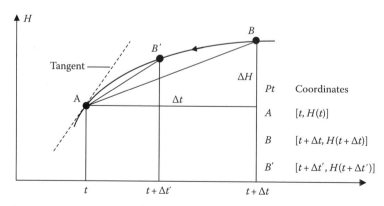

FIGURE 1.5   Average rate of change $\Delta H/\Delta t$ as Δt gets smaller.

In the limit as Δt approaches zero, pt B approaches pt A, and the average rate of change in H over the interval $(t, t + \Delta t)$ becomes the instantaneous rate of change in H at time t, that is,

$$\lim_{\Delta t \to 0} \frac{\Delta H}{\Delta t} = \frac{dH}{dt} \tag{1.14}$$

where dH/dt is the first derivative of $H(t)$. From the graph, it can be seen that dH/dt is equal to the slope of the tangent line of the function $H(t)$ at t (pt A).

Taking the limit as Δt approaches zero in Equation 1.13 and using the definition of the derivative in Equation 1.14 give

$$\lim_{\Delta t \to 0} A \left[\frac{\Delta H}{\Delta t} \right] = \lim_{\Delta t \to 0} [F_1(t) - F_0(t)] \tag{1.15}$$

$$\Rightarrow A \left[\frac{dH}{dt} \right] = F_1(t) - F_0(t) \tag{1.16}$$

Since there are two dependent variables, a second equation or constraint relating F_0 and H is required in order to solve for either one given the input function $F_1(t)$. It is convenient at this point to assume that F_0 is proportional to H, that is, $F_0 = \text{constant} \times H$ (see Figure 1.6). The constant of proportionality is expressed as $1/R$ where R is called the fluid resistance of the tank. At a later point, we will revisit this assumption.

$$F_0 = \frac{1}{R} H \tag{1.17}$$

Equations 1.16 and 1.17 constitute the mathematical model of the liquid tank, namely,

$$A \frac{dH}{dt} + F_0 = F_1 \quad \text{and} \quad F_0 = \frac{1}{R} H$$

where F_1, F_0, H, and dH/dt are short for $F_1(t)$, $F_0(t)$, $H(t)$, and $(d/dt)H(t)$.

In this example, the model is a coupled set of equations. One is a linear differential equation and the other is an algebraic equation, also linear. The differential equation is first order since only the first derivative appears in the equation and the tank dynamics are said to be first order.

The outflow F_0 can be eliminated from the model equations by substituting Equation 1.17 into Equation 1.16 resulting in

$$A \frac{dH}{dt} + \frac{1}{R} H = F_1 \tag{1.18}$$

Before a particular solution to Equation 1.18 for some $F_1(t)$, $t \geq 0$ can be obtained, the initial tank level $H(0)$ must be known.

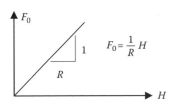

FIGURE 1.6 A tank with outflow proportional to fluid level.

There are several reasons why an analytical approach to solving Equation 1.18 may not be the preferred method. Even when the analytical solution is readily obtainable, for example, when the system model is linear, as in the present example, the solution may be required for a number of different inputs or forcing functions. Recall from studying differential equations what happens when the right-hand side of the equation changes. A new particular solution is required that can be time-consuming, especially if the process is repeated for a number of nontrivial forcing functions.

Second, the input $F_1(t)$ may not even be available in analytical form. Suppose the input function $F_1(t)$ is unknown except as a sequence of measured values at regularly spaced points in time. An exact solution to the differential equation model is out of the question since the input is not expressible as an analytic function of time.

EXERCISES

1.1 A system consists of two tanks in series in which the outflow from the first tank is the inflow to the second tank as shown in Figure E1.1:

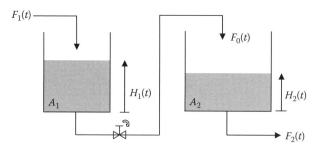

FIGURE E1.1

(a) Find the algebraic and differential equations that form the mathematical model of the two-tank system. Assume both tanks are linear, that is, the outflows are proportional to the liquid levels, and R_1 and R_2 are the fluid resistances of the tanks.

(b) Eliminate the flows $F_0(t)$ and $F_2(t)$ from the model to obtain a model in the form of two differential equations involving the system input $F_1(t)$ and the tank levels $H_1(t)$ and $H_2(t)$.

(c) Obtain the model differential equations when $F_0(t)$ and $F_2(t)$ are present instead of $H_1(t)$ and $H_2(t)$.

(d) The initial fluid levels in the tanks are $H_1(0)$ and $H_2(0)$. Suppose that the flow into the first tank is constant, $F_1(t) = \overline{F}_1$, $t \geq 0$. Obtain expressions for $H_1(\infty)$ and $H_2(\infty)$, the eventual fluid levels in Tanks 1 and 2, respectively. Do $H_1(\infty)$ and $H_2(\infty)$ depend on the initial fluid levels? Explain.

(e) Find the ratio of tank resistances R_1/R_2 if $H_1(\infty) = 2H_2(\infty)$.

(f) Suppose the flow between the two tanks is reduced to zero by closing the valve in the line. Show that this is equivalent to $R_1 = \infty$ and determine the values of $H_1(\infty)$ and $H_2(\infty)$ assuming the inflow to the first tank is still constant.

1.2 The two tanks in Exercise 1.1 are said to be noninteracting because the flow rate from the first tank only depends on the fluid level in the first tank and is independent of the fluid level in the second tank. Suppose the discharged fluid from the first tank enters the second tank at the bottom instead of the top as shown in Figure E1.2.

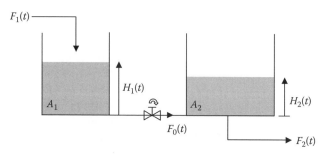

FIGURE E1.2

The flow between the tanks is now a function of the fluid levels in both tanks. The driving force for the intertank flow is the difference in fluid levels, and, for the time being, we can assume that the two quantities are proportional. That is,

$$F_0(t) \propto [H_1(t) - H_2(t)] \Rightarrow F_0(t) = \frac{H_1(t) - H_2(t)}{R_{12}}$$

where R_{12} represents a fluid resistance involving both tanks. The fluid resistance of the second tank is still R_2.

(a) The general form of the differential equation model for the system of interacting tanks is

$$\frac{dH_1}{dt} + a_{11}H_1 + a_{12}H_2 = b_1 F_1$$

$$\frac{dH_2}{dt} + a_{21}H_1 + a_{22}H_2 = b_2 F_1$$

   Note: H_1, H_2, and F_1 are short for $H_1(t)$, $H_2(t)$, and $F_1(t)$.

   Find expressions for a_{11}, a_{12}, a_{21}, a_{22}, b_1, and b_2 in terms of the system parameters A_1, A_2, R_{12}, and R_2.

(b) The tanks are initially empty, $H_1(0) = 0$ and $H_2(0) = 0$. The flow into the first tank is constant, $F_1(t) = \overline{F}_1$, $t \geq 0$. Show that the final fluid levels in both tanks after a sufficient period of time has elapsed, $H_1(\infty)$ and $H_2(\infty)$, can be obtained from the solution of the following system of equations:

$$a_{11}H_1(\infty) + a_{12}H_2(\infty) = b_1\overline{F}_1$$

$$a_{21}H_1(\infty) + a_{22}H_2(\infty) = b_2\overline{F}_1$$

(c) Solve for $H_1(\infty)$ and $H_2(\infty)$ in terms of the system parameters A_1, A_2, R_{12}, and R_2 and the constant inflow $\overline{F}_1$. Are the results different if the tanks are not initially empty? Explain.

(d) Using the following baseline values unless otherwise stated:

$$A_1 = A_2 = 25 \text{ ft}^2, \quad R_{12} = 3 \text{ ft per ft}^3/\text{min}$$

$$R_2 = 1 \text{ ft per ft}^3/\text{min}, \quad \overline{F}_1 = 5 \text{ ft}^3/\text{min}$$

   find the eventual fluid levels $H_1(\infty)$ and $H_2(\infty)$ and flows $F_0(\infty)$ and $F_2(\infty)$.

(e) Repeat part (d) with $A_2 = 75 \text{ ft}^2$.

(f) The valve between the tanks is opened, some resulting in $R_{12} = 2$ ft per ft^3/min. The remaining baseline values remain the same. Find $H_1(\infty)$, $H_2(\infty)$, and flows $F_0(\infty)$ and $F_2(\infty)$.

(g) Suppose Tank 1 initially holds 10 ft of liquid and Tank 2 has 4 ft. Find the initial rates of change in level for both tanks.

(h) Is it possible for the fluid level in Tank 2 to exceed the level in Tank 1? Explain.

(i) How does the model change if there is a separate flow, say $F_3(t)$, directly into the top of Tank 2?

1.3 Consider a cone-shaped tank with circular cross-sectional area like the one shown in Figure E1.3.

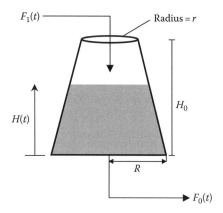

FIGURE E1.3

(a) How does this affect the derivation of the mathematical model?
(b) Find the math model for this case.

1.3   DIFFERENCE EQUATIONS

Looking back at Figures 1.3 and 1.4, recall that the level of fluid in the tank at time $t + \Delta t$ is equal to the level at time t plus the change in liquid level over the interval $(t, t + \Delta t)$. Thus,

$$H(t + \Delta t) = H(t) + \Delta H \tag{1.19}$$

From Figure 1.5, it is apparent that the change in level ΔH is simply the product of the average rate of change $\Delta H/\Delta t$ and the time interval Δt, that is,

$$H(t + \Delta t) = H(t) + \frac{\Delta H}{\Delta t} \Delta t \tag{1.20}$$

Solving for $\Delta H/\Delta t$ in Equation 1.13 and substituting the result into Equation 1.20 give

$$H(t + \Delta t) = H(t) + \frac{1}{A}[F_1(t) - F_0(t)]\Delta t \tag{1.21}$$

Keep in mind that Equation 1.21 is approximate because of the approximations to the integrals in Equations 1.8 and 1.9. Assuming the output flow F_0 is proportional to the level H, as we did to obtain Equation 1.17, gives

$$H(t + \Delta t) = H(t) + \frac{1}{A}\left[F_1(t) - \frac{1}{R}H(t)\right]\Delta t \tag{1.22}$$

$$\Rightarrow H(t + \Delta t) = \left(1 - \frac{\Delta t}{AR}\right)H(t) + \frac{\Delta t}{A}F_1(t) \tag{1.23}$$

Since Equation 1.23 is only approximate, $H(t)$ is replaced by $H_A(t)$ to distinguish it from the actual solution $H(t)$. From Equation 1.23 with $H(t)$ replaced by $H_A(t)$,

$$H_A(t + \Delta t) = \left(1 - \frac{\Delta t}{AR}\right)H_A(t) + \frac{\Delta t}{A}F_1(t) \tag{1.24}$$

Equation 1.24 is a difference equation that can be solved for the approximate solution $H_A(t)$ when $F_1(t)$, $t \geq 0$ and $H_A(0)$ are known. As we shall see, the approximate solution $H_A(t)$ can only be determined at discrete times, namely, $t = 0$, Δt, $2\Delta t$, $3\Delta t, \ldots$.

1.3.1 RECURSIVE SOLUTIONS

Difference equations are easily solved because of their inherent structure. The solution values $H_A(n\Delta t)$, $n = 1, 2, 3, \ldots$ are obtained in a sequential fashion by repeated application of the difference equation. The process begins with initial conditions $H_A(0)$ and $F_1(0)$ and proceeds as follows.

Starting with $t = 0$, from Equation 1.24

$$H_A(\Delta t) = \left(1 - \frac{\Delta t}{AR}\right)H_A(0) + \frac{\Delta t}{A}F_1(0) \tag{1.25}$$

Choosing $H_A(0)$ in Equation 1.25 equal to the known initial level $H(0)$ produces the first computed value for the approximate level, namely,

$$H_A(\Delta t) = \left(1 - \frac{\Delta t}{AR}\right)H(0) + \frac{\Delta t}{A}F_1(0) \tag{1.26}$$

The process can be repeated to obtain $H_A(2\Delta t)$ by letting $t = \Delta t$ in Equation 1.24, resulting in the following equation:

$$H_A(2\Delta t) = \left(1 - \frac{\Delta t}{AR}\right)H_A(\Delta t) + \frac{\Delta t}{A}F_1(\Delta t) \tag{1.27}$$

Substituting $H_A(\Delta t)$ from Equation 1.26 into the right-hand side of Equation 1.27 yields

$$H_A(2\Delta t) = \left(1 - \frac{\Delta t}{AR}\right)\left[\left(1 - \frac{\Delta t}{AR}\right)H(0) + \frac{\Delta t}{A}F_1(0)\right] + \frac{\Delta t}{A}F_1(\Delta t) \tag{1.28}$$

Expanding Equation 1.28 gives

$$H_A(2\Delta t) = \left(1 - \frac{\Delta t}{AR}\right)^2 H(0) + \left(1 - \frac{\Delta t}{AR}\right)\frac{\Delta t}{A}F_1(0) + \frac{\Delta t}{A}F_1(\Delta t) \tag{1.29}$$

Another iteration of Equation 1.24 with $t = 2\Delta t$ and $H_A(2\Delta t)$ from Equation 1.29 leads to $H_A(3\Delta t)$. The result is

$$H_A(3\Delta t) = \left(1 - \frac{\Delta t}{AR}\right)^3 H(0) + \left(1 - \frac{\Delta t}{AR}\right)^2\frac{\Delta t}{A}F_1(0) + \left(1 - \frac{\Delta t}{AR}\right)\frac{\Delta t}{A}F_1(\Delta t) + \frac{\Delta t}{A}F_1(2\Delta t) \tag{1.30}$$

Figure 1.7 illustrates the process up to this point.

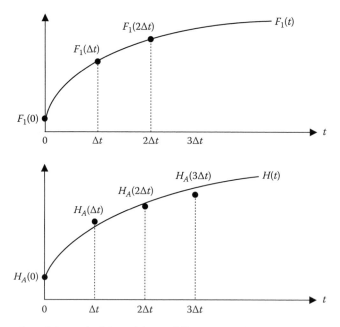

FIGURE 1.7   Illustration of the method for solving a difference equation.

Following the nth iteration, $H_A(n\Delta t)$ is known. The next iteration uses it and $F_1(n\Delta t)$ to generate a numerical value for $H_A([n+1]\Delta t)$ from

$$H_A([n+1]\Delta t) = \left(1 - \frac{\Delta t}{AR}\right)H_A(n\Delta t) + \frac{\Delta t}{A}F_1(n\Delta t) \qquad (1.31)$$

Equation 1.31 is a more common form of the difference equation than Equation 1.24. The solution to the difference equation is $H_A(n\Delta t)$, $n = 0, 1, 2, 3, \ldots$. It represents an approximation to the actual level $H(t)$ only at discrete times $t = 0, \Delta t, 2\Delta t, 3\Delta t, \ldots$. The accuracy of the approximate solution depends on the size of Δt because the difference equation is based on the use of $\Delta H/\Delta t$ as a suitable approximation for the first derivative dH/dt. As the step size Δt is reduced, the approximation is improved at the expense of more computations required to approximate $H(t)$ for a fixed period of time.

In order to find a solution $H(t)$, $t \geq 0$ to the mathematical model of the tank, $F_1(t)$ is required for $t \geq 0$. However, the solution $H_A(n\Delta t)$, $n = 0, 1, 2, 3, \ldots$ to Equation 1.31 requires knowledge of the input $F_1(t)$ only at the discrete times $t = 0, \Delta t, 2\Delta t, 3\Delta t, \ldots$. Similarly, calculating a single value of the approximate solution, for example, $H_A(25\Delta t)$, requires only a finite number of discrete inputs, namely, $F_1(n\Delta t)$, $n = 0, 1, 2, \ldots, 24$.

EXERCISES

1.4 Find the difference equation, similar to Equation 1.31, relating $F_{0,A}([n+1]\Delta t)$ to $F_{0,A}(n\Delta t)$ and $F_1(n\Delta t)$.

1.5 A tank with cross-sectional area $A = 5$ ft^2 is initially filled to a level of 10 ft. The flow out is given by $F_0 = H/R$, $R = 1$ ft per ft^3/min. There is no flow into the tank.
   (a) Find $H_A(n\Delta t)$, $n = 0, 1, 2, \ldots, 10$ when $\Delta t = 2.5$ min.
   (b) Find $H_A(n\Delta t)$, $n = 0, 1, 2, \ldots, 25$ when $\Delta t = 1$ min.
   (c) Find $H_A(n\Delta t)$, $n = 0, 1, 2, \ldots, 100$ when $\Delta t = 0.25$ min.
   (d) Plot the results and comment on the differences.

1.6 Repeat Exercise 1.5 for the case where the outflow is described by $F_0 = cH^{1/2}$, $c = 3$ ft^3/min per ft$^{1/2}$.

1.4 FIRST LOOK AT DISCRETE-TIME SYSTEMS

The variables $F_1(t)$, $F_0(t)$, and $H(t)$ in the liquid tank shown in Figure 1.3 are referred to as continuous-time (or simply continuous) signals. The reason is because there is a continuum of values between any two points along the t-axis where the variables are defined. Equation 1.18 is a continuous-time model and the system is a continuous-time system because it involves only continuous-time variables.

In contrast to the continuous-time signals $F_1(t)$, $F_0(t)$, and $H(t)$, the sequence of sampled input flow values, $F_1(n\Delta t), n = 0, 1, 2, \ldots$ and the sequence of approximate tank levels $H_A(n\Delta t), n = 0, 1, 2, \ldots$ are classified as discrete-time (discrete for short) signals because the independent variable "n" is discrete in nature. The difference equation (Equation 1.31) is classified as a discrete-time model, and the underlying system with purely discrete-time input and output signals is likewise a discrete-time system.

Figure 1.8 portrays the liquid tank continuous-time system with dependent variable $H(t)$ considered as the output.

A complete description of the system includes the following:

System: Continuous time
Independent variable: $t \geq 0$
Input: $F_1(t)$, $t \geq 0$
Dependent variables: $H(t)$, $F_0(t)$, $t \geq 0$
Output: $H(t)$, $t \geq 0$
Model: $A\dfrac{dH}{dt} + F_0(H) = F_1(t)$

The differential equation model is shown with a term $F_0(H)$ representing an algebraic function relating the outflow $F_0(t)$ to the fluid level $H(t)$. We have assumed this function to be linear; however, a more accurate description will be introduced later.

Figure 1.9 is a comparable diagram of the liquid tank discrete-time system with discrete-time input $F_1(n)$ and output $H_A(n)$. $F_1(n)$ is short for $F_1(n\Delta t)$, $n = 0, 1, 2, \ldots$ the sampled values of the input flow. $H_A(n)$ is short for $H_A(n\Delta t)$, $n = 0, 1, 2, \ldots$, the values computed from the difference equation in Equation 1.31. Note that $H_A(n)$, $n = 0, 1, 2, \ldots$ differs from $H(n\Delta t)$, $n = 0, 1, 2, \ldots$, the sampled values of the continuous-time level $H(t)$.

A complete description of the system includes the following:

System: Discrete time
Independent variable: $n = 0, 1, 2, \ldots$
Input: $F_1(n)$, $n = 0, 1, 2, \ldots$
Dependent variables: $H_A(n)$, $F_{0,A}(n)$, $n = 0, 1, 2, \ldots$
Output: $H_A(n)$, $n = 0, 1, 2, \ldots$
Model: $H_A(n + 1) = \left(1 - \dfrac{\Delta t}{AR}\right) H_A(n) + \dfrac{\Delta t}{A} F_1(n)$

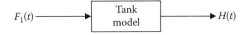

FIGURE 1.8 Liquid tank continuous-time system.

FIGURE 1.9 Liquid tank discrete-time system.

Difference equations can always be solved recursively. Expressions for the first three values $H_A(n)$, $n = 1, 2, 3$ are given in Equations 1.26, 1.29, and 1.30. It is sometimes possible to recognize a general pattern for $H_A(n)$ from results of the first several iterations. In this example, $H_A(n)$ is

$$H_A(n) = \left(1 - \frac{\Delta t}{AR}\right)^n H_A(0) + \frac{\Delta t}{A}\left(1 - \frac{\Delta t}{AR}\right)^{n-1} F_1(0) + \frac{\Delta t}{A}\left(1 - \frac{\Delta t}{AR}\right)^{n-2} F_1(1)$$

$$+ \frac{\Delta t}{A}\left(1 - \frac{\Delta t}{AR}\right)^{n-3} F_1(2) + \cdots + \frac{\Delta t}{A}\left(1 - \frac{\Delta t}{AR}\right) F_1(n-2) + \frac{\Delta t}{A} F_1(n-1), \quad n = 1, 2, 3, \ldots$$

$$(1.32)$$

Using summation notation, the general solution with $H_A(0)$ replaced by $H(0)$ is

$$H_A(n) = \left(1 - \frac{\Delta t}{AR}\right)^n H(0) + \frac{\Delta t}{A} \sum_{k=0}^{n-1} \left(1 - \frac{\Delta t}{AR}\right)^{n-k-1} F_1(k), \quad n = 1, 2, 3, \ldots \quad (1.33)$$

Equation 1.33 is the general solution to the difference equation model in Equation 1.31. When specific values of $H_A(n)$ are required, say $H_A(100)$, it eliminates the need for recursive solution of the previous 99 values $H_A(n)$, $n = 1, 2, 3, \ldots, 99$. The summation in Equation 1.33 requires some effort; however, the z-transform introduced in Chapter 4 provides a way to avoid the sum altogether.

Examination of Equation 1.33 reveals several important features of the approximate solution. First of all, notice the influence of the initial tank level $H(0)$ on the estimated level $H_A(n)$ at the current time n. The first term on the right-hand side of Equation 1.33 is the only term in the expression involving $H(0)$. Furthermore, the effect of $H(0)$ on $H_A(n)$ is reduced as the discrete-time variable n increases, provided the term in parenthesis, $1 - \Delta t/AR$, is less than 1 in magnitude. This appears reasonable if we ask ourselves, "How important is the initial tank level with respect to the current level after a significant amount of time has elapsed?" Clearly, the answer is "not very significant at all," and so we should not be surprised to see the only term containing $H(0)$ in the expression for $H_A(n)$ monotonically decreasing as n increases.

The second point of interest relates to the "memory" inherent in the system. By this, we mean how far back in discrete-time inputs must we go when calculating the current discrete-time output $H_A(n)$. Based on Equation 1.32, the answer is "all the way back" to the initial input $F_1(0)$. As a result, the discrete-time system is said to have infinite memory because the current discrete-time output $H_A(n)$ depends on all past values of the discrete-time input $F_1(k)$, $k = 0, 1, 2, \ldots, n - 1$. The nature of this dependency is a weighted sum with the most recent inputs receiving the higher weights, as expected.

The following example illustrates the use of Equation 1.33 to obtain an approximate solution to the level in the tank when the input flow is constant.

Example 1.1

A tank with cross-sectional area of 10 ft² receives a constant input flow of 5 ft³/min. The fluid resistance of the tank is 2 ft/(ft³/min), and the tank is initially filled to a level of 4 ft.

(a) Find the difference equation for obtaining an approximate solution for the level $H(t)$ using a time step $\Delta t = 0.25$ min.
(b) Solve the difference equation recursively to obtain the approximate fluid level $H_A(n)$, $n = 1, 2, 3$.
(c) Use the general solution to find $H_A(3)$ directly and compare your answer with the result from part (b).

(a)
$$\frac{\Delta t}{A} = \frac{0.25}{10} = 0.025, \quad 1 - \frac{\Delta t}{AR} = 1 - \frac{0.25}{10(2)} = 0.9875$$

$$H_A(0) = H(0) = 4, \quad F_1(n) = 5, \quad n = 0, 1, 2, 3, \ldots$$

The difference equation (Equation 1.31) (with Δt omitted) becomes

$$H_A(n+1) = 0.9875 H_A(n) + (0.025)5, \quad n = 0, 1, 2, 3, \ldots$$

(b) $H_A(n)$, $n = 1, 2, 3$ are easily computed.

$$
\begin{aligned}
n = 0 \Rightarrow H_A(1) &= 0.9875 H_A(0) + 0.025(5) \\
&= 0.9875(4) + 0.125 \\
&= 4.0750 \\
n = 1 \Rightarrow H_A(2) &= 0.9875 H_A(1) + 0.025(5) \\
&= 0.9875(4.075) + 0.125 \\
&= 4.1491 \\
n = 2 \Rightarrow H_A(3) &= 0.9875 H_A(2) + 0.025(5) \\
&= 0.9875(4.1491) + 0.125 \\
&= 4.2222
\end{aligned}
$$

(c) From Equation 1.33 with $n = 3$,

$$H_A(3) = (0.9875)^3(4) + 0.025 \sum_{k=0}^{2} (0.9875)^{3-k-1}(5)$$

$$= 3.8519 + 0.025[(0.9872)^2(5) + (0.9875)(5) + (5)]$$

$$= 4.2222$$

Due to the simple nature of the input, that is, $F_1(t) = \overline{F}$, $t \geq 0$, the analytical solution of the differential equation model

$$A\frac{dH}{dt} + \frac{1}{R}H = \overline{F} \tag{1.34}$$

is easily obtained. The solution is

$$H(t) = R\overline{F} + [H(0) - R\overline{F}]e^{-t/AR} \tag{1.35}$$

It is instructive to compare the approximate solution based on the difference equation approach with the exact solution shown in Equation 1.35. The results are shown in Table 1.1, which includes both solutions at equally spaced intervals for the first 2 min of the response.

Graphs of the continuous-time output $H(t)$ and discrete-time output $H_A(n)$, $n = 0, 8, 16, \ldots$ are shown in Figure 1.10.

By observation of Figure 1.10, it appears that the exact and approximate solutions for the tank level are in close agreement. The step size Δt is the determining factor in terms of how close the two solutions are at the discrete points in time where the approximate solution

TABLE 1.1
Comparison of Approximate and Exact Solutions

N	$t_n = n\Delta t$	$H_A(n)$	$H(t_n)$
0	0	4.0	4.0
1	0.25	4.0750	4.0745
2	0.5	4.1491	4.1481
3	0.75	4.2222	4.2208
4	1.0	4.2944	4.2926
5	1.25	4.3657	4.3635
6	1.5	4.4362	4.4335
7	1.75	4.5057	4.5027
8	2.0	4.5744	4.5710

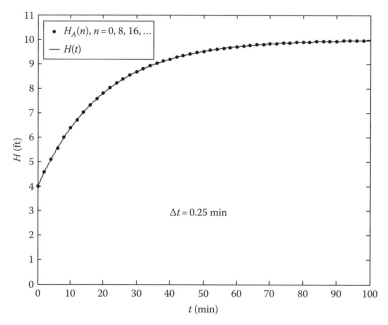

FIGURE 1.10 Exact and approximate solutions for tank level vs. time.

is defined. Choosing the step size Δt is generally a trade-off between the accuracy of the approximate solution and the computational effort required to obtain the approximate solution values.

Generally speaking, an assessment of whether the numerical value selected for Δt is reasonable cannot be made on the basis of comparing the approximate solution with the exact solution to the continuous-time model. Analytical solutions are rare due to the complexity of most real-world system models. A logical approach to finding an acceptable step size is to obtain approximate solutions with different step sizes (an order of magnitude apart) and comparing the results. If the approximate solutions are substantially identical, the smaller step size is eliminated from consideration. Conversely, if the approximate solutions are not close, the larger value of Δt is discarded. Eventually, a value of Δt will be found, which balances accuracy and computational requirements. This point will be revisited in greater detail after the subject of numerical integration is discussed.

1.4.1 INHERENTLY DISCRETE-TIME SYSTEMS

The dynamics of the liquid tank considered in Section 1.2 were classified as continuous because the variables associated with the tank's dynamic behavior were continuous time in nature. The continuous-time model of Equation 1.18 governs the relationship between physical quantities, that is, the flow in $F_1(t)$ and the liquid level $H(t)$.

Later on we learned that a discrete-time model (see Figure 1.9) could be obtained relating the approximate tank level and the sampled input flow. Both signals $F_1(n)$ and $H_A(n)$ were defined only at the discrete times $t_n = n\Delta t$, $n = 1, 2, 3, \ldots$.

Inherently discrete-time systems involve discrete-time signals, which are not the result of sampling a continuous-time signal. For example, consider the discrete-time system model given by

$$y(n) = \frac{1}{2}\left[y(n+1) + \frac{u(n)}{y(n-1)}\right], \quad n = 0, 1, 2, 3, \ldots \tag{1.36}$$

Equation 1.36 is simply a rule for transforming a discrete-time input signal $u(n)$ into an appropriate output signal $y(n)$. Is this discrete-time system useful? Let us investigate its behavior for the case where the input $u(n)$ is constant, for example, $u(n) = 25$, $n = 0, 1, 2, 3, \ldots$. First of all, we notice that the initial condition $y(-1)$ must be given before we can proceed to calculate subsequent output values $y(0)$, $y(1)$, $y(2)$, etc. Choosing $y(-1) = 1$ and solving for the first several outputs,

$$y(0) = \frac{1}{2}\left[y(-1) + \frac{u(0)}{y(n-1)}\right] = \frac{1}{2}\left[1 + \frac{25}{1}\right] = 13$$

$$y(1) = \frac{1}{2}\left[y(0) + \frac{u(1)}{y(0)}\right] = \frac{1}{2}\left[13 + \frac{25}{13}\right] = 7.4615$$

$$y(2) = \frac{1}{2}\left[y(1) + \frac{u(2)}{y(1)}\right] = \frac{1}{2}\left[7.4615 + \frac{25}{7.4615}\right] = 5.4060$$

$$y(3) = \frac{1}{2}\left[y(2) + \frac{u(3)}{y(2)}\right] = \frac{1}{2}\left[5.4060 + \frac{25}{5.4060}\right] = 5.0152$$

$$y(4) = \frac{1}{2}\left[y(3) + \frac{u(4)}{y(3)}\right] = \frac{1}{2}\left[5.0152 + \frac{25}{5.0152}\right] = 5.0000$$

Using different positive constants for $u(n)$ and other starting values for $y(-1)$ will reveal an interesting property of the system, namely, $\lim_{n\to\infty} y(n) = \sqrt{u}$. The discrete-time signals $u(n)$ and $y(n)$ are plotted in Figure 1.11. Hence, the primary purpose of the discrete-time system governed by Equation 1.36 is to compute the square root of its positive-valued constant input $u(n)$.

Another inherently discrete-time system is one we are all familiar with, namely, an interest-bearing account such as a bank account. The discrete-time signals of interest are $y(k)$, the account balance at the end of the kth interest period, and $u(k)$, the net deposit for the kth interest period (Figure 1.12). For this simple example, the net deposit during the kth interest period is assumed to have occurred at the end of the period.

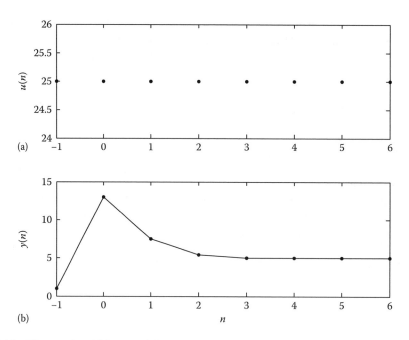

(a)

(b)

n

FIGURE 1.11 Discrete-time (a) input and (b) output of inherently discrete-time system for finding the square root of a positive number.

FIGURE 1.12 Example of an inherently discrete-time system.

Consider an account with an interest rate i (per interest period). The balance at the end of the kth interest period, $y(k)$, is the sum of

- The balance at the end of the $(k-1)$st period: $y(k-1)$
- The interest earned for the kth interest period: $i \cdot y(k-1)$
- The net deposit for the period: $u(k)$

Therefore, the model for this inherently discrete-time system is

$$y(k) = y(k-1) + iy(k-1) + u(k), \quad k = 1, 2, 3, \ldots \tag{1.37}$$

Example 1.2

A college trust fund is set up with $5000 on January 1, 2000. Starting on January 1, 2001, and every year thereafter, $1000 is added to the fund, which earns 7.5% interest annually.

(a) Track the end of year fund balance for the first several years.
(b) Find the account balance at the end of 18th year.

(a) The discrete-time model is

$$y(k) = y(k-1) + 0.075y(k-1) + u(k), \quad k = 1, 2, 3, \ldots$$

with input $u(k) = 1000$, $k = 1, 2, 3, \ldots$ and initial condition $y(0) = 5000$.

The account balance at the end of years 1, 2, and 3 are worked out as follows:

$$k = 1: \quad y(1) = y(0) + 0.075y(0) + u(1)$$
$$= 5000 + 0.075(5000) + 1000$$
$$= 6375$$
$$k = 2: \quad y(2) = y(1) + 0.075y(1) + u(2)$$
$$= 6375 + 0.075(6375) + 1000$$
$$= 7853.13$$
$$k = 3: \quad y(3) = y(2) + 0.075y(2) + u(3)$$
$$= 7853.13 + 0.075(7853.13) + 1000$$
$$= 9442.11$$

(b) The recursive solution could be continued for $k = 4, 5, 6, \ldots, 18$, resulting in the fund's balance at the end of the 18th year. However, a general solution of the discrete-time model is preferable since it can be evaluated for any value of the discrete-time variable k.

For the discrete-time model,

$$y(k) = y(k-1) + iy(k-1) + u(k), \quad k = 1, 2, 3, \ldots \tag{1.38}$$

$$= (1+i)y(k-1) + A \tag{1.39}$$

$$= \alpha y(k-1) + A \tag{1.40}$$

where $\alpha = 1 + i$ and A is the constant net deposit each interest period. The first several outputs are

$$y(1) = \alpha y(0) + A$$
$$y(2) = \alpha y(1) + A$$
$$= \alpha[\alpha y(0) + A] + A$$
$$= \alpha^2 y(0) + \alpha A + A$$
$$y(3) = \alpha y(2) + A$$
$$= \alpha[\alpha^2 y(0) + aA + A] + A$$
$$= \alpha[\alpha^3 y(0) + a^2 A + \alpha A] + A$$

suggesting the general expression for $y(k)$ is

$$y(k) = \begin{cases} y(0), & k = 0 \\ \alpha^k y(0) + (1 + \alpha + \alpha^2 + \alpha^3 + \cdots + \alpha^{k-1})A, & k = 1, 2, 3, \ldots \end{cases} \tag{1.41}$$

Further simplification is possible using the closed form of the finite geometric series in the previous equation. The general solution for $y(k)$ is

$$y(k) = \alpha^k y(0) + \frac{1 - \alpha^k}{1 - \alpha}A, \quad k = 1, 2, 3, \ldots \tag{1.42}$$

The account balance after 18 years is easily computed from the general solution above with $\alpha = 1.075$, $y(0) = 5,000$, and $A = 1,000$.

$$y(18) = (1.075)^{18}(5,000) + \left[\frac{1 - (1.075)^{18}}{1 - 1.075}\right]1,000 = 54,056.41$$

The results from part (a) can be verified using the general solution.

EXERCISES

1.7 Rework Example 1.1 using the trial-and-error method for determining a suitable value of Δt. Start with $\Delta t = 10$ min and calculate $H_A(n)$, $n = 0, 1, 2, \ldots, n_f$ where $n_f \Delta t = 100$ min. Repeat the steps with $\Delta t = 5$, 2.5, 1.25 min, and so forth until the approximations of $H(10)$, $H(20)$, $H(30), \ldots, H(100)$ are in agreement to at least one place after the decimal point. Use the following table for comparisons. Extend the table to smaller values of Δt if necessary.

$H_A(n)$		$H_A(n)$		$H_A(n)$		$H_A(n)$	
n	$\Delta t = 10$	n	$\Delta t = 5$	N	$\Delta t = 2.5$	N	$\Delta t = 1.25$
0		0		0		0	
1		2		4		8	
2		4		8		16	
3		6		12		24	
4		8		16		32	
5		10		20		40	
6		12		24		48	
7		14		28		56	
8		16		32		64	
9		18		36		72	
10		20		40		80	

1.8 Prove that the output of the discrete-time system in Equation 1.36 will approach the square root of the input, any positive constant "A." In other words, show that

$$\lim_{n \to \infty} y(n) = \sqrt{A}$$

where $u(n) = A$, $n = 0, 1, 2, 3, \ldots$.

1.9 An alternate model of the tank relates the outflow and liquid level according to

$$F_0(t) = \alpha[H(t)]^{1/2}$$

(a) Develop a new discrete-time model of the tank using the above relationship in conjunction with the differential equation $A(dH/dt) + F_0 = F_1$. The tank cross-sectional area is 10 ft^2 and the input flow is constant at 5 ft^3/min. The tank is initially filled to a level of 4 ft. Assume $\alpha = 2$ ft^3/min per ft$^{1/2}$.

(b) Calculate the approximate tank level for the first minute using a step size $\Delta t = 0.25$ min.

(c) Consider the same tank with zero in flow and an initial fluid level of 25 ft. Write a program to calculate the approximate level of the tank as it empties. Choose $\Delta t = 0.1$ min.

(d) The analytical solution for the level $H(t)$ when $F_1(t) = 0$, $t \geq 0$ is given by

$$H(t) = \left(H_0^{1/2} - \frac{\alpha t}{2A} \right)^2$$

where H_0 is the initial tank level. Compare the results from part (c) to the exact solution. Present the comparison of results in tabular and graphical form.

1.10 A holding tank serves as an effective way of smoothing variations in the flow of a liquid. For example, suppose the liquid flow rate from an upstream process is

$$F_1(t) = \overline{F} + f\sin\left(\frac{2\pi t}{T}\right), \quad t \geq 0$$

where
$\overline{F}$ is an average flow
f is the fluctuation about the average flow
T is the period of the fluctuations

Nominal parameter values for the input flow rate are $\overline{F} = 250$ ft^3/min, $f = 50$ ft^3/min, and $T = 15$ min.

A holding tank is placed between the source $F_1(t)$ and a downstream process that requires a more constant input flow rate, $F_0(t)$, as shown in Figure E1.10. The downstream process requires that the sustained fluctuations in the flow $F_0(t)$ be no larger than 10 ft^3/min. Assume the tank is linear and the fluid resistance $R = 0.25$ ft per ft^3/min.

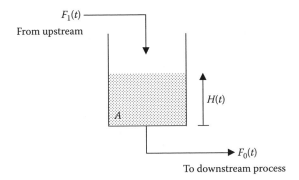

FIGURE E1.10

(a) Find the difference equation for $F_{0,A}(n)$, $n = 0, 1, 2, 3, \ldots$. Leave the tank cross-sectional area A as a parameter.
(b) Write a program to solve the difference equation with $\Delta t = 0.5$ min for a starting value of $A = 100$ ft^2. Graph both $F_{0,A}(n)$ and $H_A(n)$, $n = 0, 1, 2, \ldots$ for a period of time sufficient to determine if the design criterion is satisfied. Assume that the tank is initially empty.
(c) Repeat part (b) with a new value of A until the design criterion is satisfied.
(d) Graph the discrete-time signals $F_1(n)$ and $F_{0,A}(n)$, $n = 0, 1, 2, 3, \ldots$ for the tank whose area is the value determined in part (c).

1.5   CASE STUDY: POPULATION DYNAMICS (SINGLE SPECIES)

The population of a country is under investigation. Unlike the liquid tank example, there is no scientific principle to serve as a foundation for deriving a mathematical model that can be used to predict future populations. Instead, empirical observations of historical birth and death rates, immigration and emigration patterns, and a host of other pertinent data are utilized.

t (Years)	$P_{obs}(t)$, Millions
0	3.0000
10	3.2276
20	4.5759
30	6.9570
40	8.7618
50	9.1536
60	11.2669
70	14.5153
80	16.5059
90	17.9563
100	19.5078

One hundred years of observed population data, recorded at intervals of 10 years, are given in Table 1.2.

Based on the available data, researchers are convinced that the population is adequately modeled by the following differential equation, referred to in the literature as logistic growth (Haberman 1977).

$$\frac{dP}{dt} = cP(P_m - P) \tag{1.43}$$

$P = P(t)$ is the population "t" years after the initial population was recorded. The parameters c and P_m influence the specific growth pattern behavior. The model ignores immigration and emigration and all other external inputs, which influence dP/dt, the rate at which the population changes. The system model in Equation 1.43 is said to be autonomous, meaning there are no additional terms independent of P as might be the case if immigration or emigration inputs as a function of time were considered. The dynamics depend solely on initial conditions and the system parameters. It is also referred to as an unforced system since there are no external inputs.

Statistical analyses of the population data have resulted in estimated values for c and P_m to be 1.25×10^{-9} and 25 million, respectively. It is now 100 years since the initial population was measured. Government planners are interested in determining what the likely population will be over the next several decades. A method is needed to obtain an approximate solution of the model, that is, a difference equation for $P_A(n) \approx P(n\Delta t)$, $n = 0, 1, 2, \ldots$ is required.

When the continuous-time model is a first-order differential equation, a difference equation for approximating the dynamics at discrete points in time is easily obtained. Simply replace the first derivative term with an appropriate finite difference approximation, remembering to rename the dependent variable in some way since we are now dealing with an approximate solution. This is precisely the way a difference equation for approximating the liquid tank dynamics was obtained in Section 1.3.

Substituting a first-order finite difference approximation for dP/dt in Equation 1.43,

$$\frac{P_A(n+1) - P_A(n)}{\Delta t} = cP_A(n)[P_m - P_A(n)] \tag{1.44}$$

Note the appearance of $P_A(n)$ on the right-hand side of the equation in place of $P(t)$. Solving for $P_A(n+1)$ produces the following difference equation:

$$P_A(n+1) = P_A(n) + cP_A(n)[P_m - P_A(n)]\Delta t, \quad n = 0, 1, 2, 3, \ldots \tag{1.45}$$

Simplifying Equation 1.45 produces the following desired form:

$$P_A(n+1) = \{1 + c\Delta t[P_m - P_A(n)]\}P_A(n), \quad n = 0, 1, 2, 3, \ldots \tag{1.46}$$

Since our interest is in predicting populations for 101 years and beyond, we need to solve Equation 1.46 over a suitable range of values for the discrete-time variable "n." The appropriate integer values depend on the size of our time step Δt. For simplicity, we shall choose Δt equal to 1 year, necessitating the calculation of $P_A(101), P_A(102), \ldots, P_A(130)$ to obtain predictions for a 30 year time span.

TABLE 1.3
Comparison of Observed, Discrete-Time, and Continuous-Time
Populations

t (Years)	$P_{obs}(t)$, Millions	N (at = 1 Year)	$P_A(n)$, Millions	$P(t)$, Millions
0	3.0000	0	3.0000	3.0000
10	3.2276	10	3.9161	3.9276
20	4.5759	20	5.0493	5.0759
30	6.9570	30	6.4129	6.4570
40	8.7618	40	8.0003	8.0618
50	9.1536	50	9.7778	9.8536
60	11.2669	60	11.6834	11.7669
70	14.5153	70	13.6325	13.7153
80	16.5059	80	15.5321	15.6059
90	17.9563	90	17.2976	17.3563
100	19.5078	100	18.8671	18.9078
110	—	110	20.2076	20.2310
120	—	120	21.3139	21.3226
130	—	130	22.2012	22.1990

A recursive solution seems like our only alternative, since a general solution is not easily achievable. A computer program to generate the recursive solution is the way to proceed. We are starting from a known population $P_{obs}(0)$, so $P_A(0) = P_{obs}(0) = 3$ million. The results are computed in the MATLAB® script file "*Chap1_CaseStudy.m*" and shown in Table 1.3. A casual observation of this table indicates that the modelers were justified in choosing the logistic growth equation to model the country's population over the time period of one century. Naturally, this assumes that the approximate solution values $P_A(n)$ are reasonably close to the exact solution $P(t)$ for $t + n\Delta t$, $n = 0$, 10, 20, 30, ...

Ordinarily, models used to represent the dynamics of continuous-time systems are not amenable to exact solutions, even with the simplest types of input. However, an analytical solution to Equation 1.43 is as follows:

$$P(t) = \frac{P_m P(0)}{P(0) + [P_m - P(0)]e^{-cP_m t}}, \quad t \geq 0 \tag{1.47}$$

The solution can be verified by differentiation and substitution back into Equation 1.43. A quick glance at the solution shows the initial condition $P(0)$ results when t is equal to zero on the right-hand side of Equation 1.47. Knowing the exact solution to the continuous-time model, we can evaluate it at $t = 0, 10, 20, \ldots, 100$ years for comparison with the discrete-time model output to determine if our step size needs to be adjusted. The exact solution results are tabulated in the final column of Table 1.3.

Comparing the last two columns in the table should convince us that the step size Δt does not need to be reduced. While it is possible to reduce the discrepancy between the approximate and exact solutions by lowering Δt, it is hardly justified in view of the fact that the continuous-time model, Equation 1.43, is itself only an approximate representation of the true population dynamics.

The data in Table 1.3 are presented in graphical form in Figure 1.13. The discrete-time system model is used to predict future populations. The projected populations for Years 110, 120, and 130 are included in Table 1.3 and appear as data points in Figure 1.13.

The previous point relating to the accuracy of the approximate solution is worth reiterating. Extremely accurate solutions of nonlinear differential equation models are generally not warranted

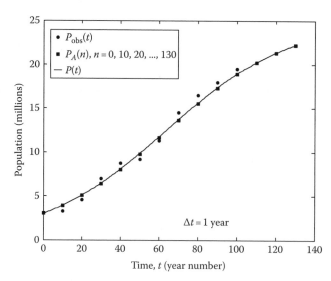

FIGURE 1.13 Observed, discrete-time (approximate), and continuous-time populations.

unless the continuous-time models were formulated to account for higher-order effects. Even then, one must limit the accuracy requirements in order to keep the computations manageable.

Exact solutions to continuous-time models are rare. How can we be certain if the solution of the discrete-time model is in agreement with the exact solution? There is no simple answer; however, there are some things we can do to check the validity of the approximate solution. We know the difference equations in the discrete-time model converge to the differential equations of the continuous-time model in the limiting case when the step size Δt approaches zero. Furthermore, the discrete-time solutions will approach the exact solutions of the continuous-time model as Δt is reduced to zero.

Systematically reducing the step size until the changes in the discrete-time outputs are within some tolerance demonstrates this convergence and is an effective way of selecting the step size Δt for future runs. We touched on this in the previous section as a way of choosing an appropriate value for the step size Δt. A word of caution—the step size may have to be readjusted as conditions of the discrete-time system model change.

Our intuition about the continuous-time system response may suggest we take a closer look at the discrete-time system model. For example, consider the tank model

$$A\frac{dH}{dt} = F_1(t) - F_0(t) \tag{1.48}$$

$$\Rightarrow \frac{d}{dt}H(t) = \frac{1}{A}[F_1(t) - F_0(t)] \tag{1.49}$$

Using the simple first-order difference approximation formula

$$\frac{d}{dt}H(t) \approx \frac{H_A(n+1) - H_A(n)}{\Delta t} \tag{1.50}$$

$$\Rightarrow H_A(n+1) = H_A(n) + \frac{\Delta t}{A}[F_1(n) - F_0(n)] \tag{1.51}$$

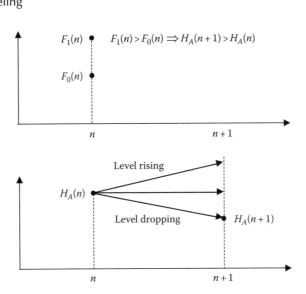

FIGURE 1.14 Change in discrete-time approximation of $H(t)$ in violation of Equation 1.52.

$$\Rightarrow H_A(n+1)\begin{cases} > H_A(n), & \text{when } F_1(n) > F_0(n) \\ = H_A(n), & \text{when } F_1(n) = F_0(n) \\ < H_A(n), & \text{when } F_1(n) > F_0(n) \end{cases} \tag{1.52}$$

Equation 1.52 is consistent with our expectation that the level in a tank is rising ($dH/dt > 0$) when the liquid is coming in faster than it is leaving and falling when the opposite is true. Consequently, a change in tank level like the one shown in Figure 1.14 is the reason to double check the calculations or the code that produced them.

Another check on the integrity of an approximate solution to a continuous-time model is to see whether the differential equation itself is satisfied within some tolerance. Since the logistic growth model in Equation 1.43 governs population growth at all times, it must apply at the discrete times $t_n = n\Delta t$, $n = 0, 1, 2, 3, \ldots$. Therefore,

$$\frac{d}{dt}P(t)\bigg|_{t=n\Delta t} = cP(n\Delta t)[P_m - P(n\Delta t)], \quad n = 0, 1, 2, 3, \ldots \tag{1.53}$$

We can approximate the first derivative term on the left-hand side of Equation 1.53 using a more accurate difference formula than the first-order difference quotient used to approximate dH/dt in the case of the liquid tank. Referring to Figure 1.15, dP/dt at $t = n\Delta t$ is approximated using an average of first-order difference approximations resulting in

$$\frac{d}{dt}P(t)\bigg|_{t=n\Delta t} \approx \frac{1}{2}\left[\frac{P_A(n+1) - P_A(n)}{\Delta t} + \frac{P_A(n) - P_A(n-1)}{\Delta t}\right] = \frac{P_A(n+1) - P_A(n-1)}{2\Delta t} \tag{1.54}$$

$$\Rightarrow \frac{d}{dt}P(t)\bigg|_{t=50\Delta t} \approx \frac{P_A(51) - P_A(49)}{2\Delta t}$$

$$\approx \frac{9.9638 \times 10^6 - 9.5930 \times 10^6}{2(1)}$$

$$\approx 0.1854 \times 10^6 \text{ people/year}$$

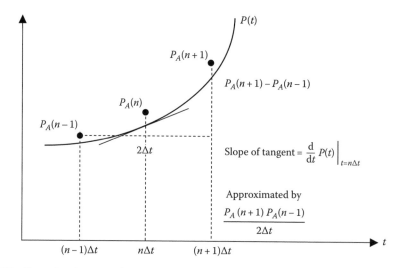

FIGURE 1.15 Second-order approximation of first derivative dP/dt.

The right-hand side of Equation 1.53 with $P(50\Delta t)$ replaced by $P_A(50)$ becomes

$$
\begin{aligned}
cP(n\Delta t)[P_m - P(n\Delta t)] &= cP_A(50)[P_m - P_A(50)] \\
&= 1.25 \times 10^{-9}(9.7778 \times 10^6)[25 \times 10^6 - 9.7778 \times 10^6] \\
&= 0.1860 \times 10^6 \text{ people/year}
\end{aligned}
\tag{1.55}
$$

in close agreement with the estimate of $(d/dt)P(t)|_{t=50\Delta t}$.

Further scrutiny of the logistic growth model, Equation 1.43, reveals several important and noteworthy characteristics of the underlying population dynamics. Expressing the model in a slightly different form

$$
g(P) = \frac{1}{p} \frac{dP}{dt} = c(P_m - P)
\tag{1.56}
$$

where $g(P)$, the rate of change in population dP/dt divided by the population P, is called the population growth rate. Different population models are normally characterized by the term(s) appearing on the right-hand side of Equation 1.56.

The growth rate function is plotted in Figure 1.16.

We expect the population to be increasing whenever the growth rate is positive, since a positive growth rate implies the instantaneous rate of change in the population, that is, the first derivative is also positive. The logistic population growth rate declines linearly with increasing population, eventually reaching zero when the population reaches P_m or 25 million in this case. In logistic growth models, P_m is called the carrying capacity.

Observe from Figure 1.13 that the discrete-time and continuous-time model outputs for 130 years ranged from the initial population of 3 million people to somewhere around 22 million people. Looking at the heavier line segment in Figure 1.16, corresponding to this range of populations, we notice that the growth rate is positive, and, hence, the population should be monotonically increasing, as indeed it was.

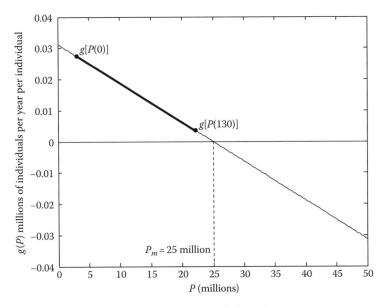

FIGURE 1.16 Plot of population growth rate $g(P)$ vs. population (P).

Is it possible for a population $P(t)$ governed by a logistic growth model to ever assume values on both sides of its carrying capacity? For example, is the population growth shown in Figure 1.13 capable of exceeding $P_m = 25$ million if we wait long enough? Figure 1.17 shows population time histories for the logistic model considered previously ($c = 1.25 \times 10^{-9}$, $P_m = 25 \times 10^6$) with different starting populations.

It is clear that the population approaches its carrying capacity from below or above in asymptotic fashion. We should not be surprised if we consider what happens to the population growth rate $g(P)$ as the population $P(t)$ approaches the carrying capacity from either direction (see Figure 1.16).

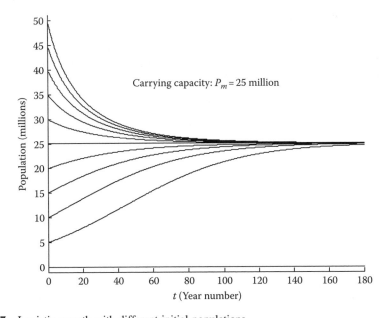

FIGURE 1.17 Logistic growth with different initial populations.

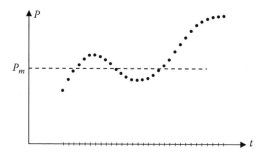

FIGURE 1.18   Discrete-time population model output inconsistent with logistic growth.

A discrete-time response $P_A(n)$ like the one appearing in Figure 1.18 is inconsistent with the properties of continuous-time logistic growth. However, crossing over the carrying capacity for one or two time increments is not inconsistent with the discrete-time nature of the approximate solution. Why not?

EXERCISES

1.11 Assume that the logistic growth population model accurately predicts future populations.
   (a) Some time in the future, the population will reach 98% of its carrying capacity. Find how many more years this will take by using the difference equation given in Equation 1.46. Does it make a difference whether you start from $P_A(0) = 3$ million or $P_A(130) = 22.2012$ million from Table 1.3.
   (b) Compare the answer obtained in part (a) with the analytical solution for $P(t)$.
   (c) The population growth rate $g(P)$ vs. P in Figure 1.16 does not explicitly involve time. Label the points on the growth rate curve corresponding to $\{t, P(t_i)\}$ where $t_0 = 0$, $t_1 = 25$, $t_2 = 50$, $t_3 = 75$, and $t_4 = 100$.
   (d) The carrying capacity P_m in a logistic growth model is an equilibrium population, meaning that if the population at some point in time were equal to P_m, it would remain there forever. Investigate whether it is stable or not by supposing the population is slightly less or slightly more than P_m, and determine whether the population returns to the carrying capacity. Obtain several approximate solutions corresponding to different initial populations reasonably close to P_m.
   (e) Find the other equilibrium population of the logistic growth model and determine if it is stable.

1.12 A simpler model for population growth of a species is one in which the growth rate is assumed constant, that is, independent of the population. Mathematically, this is represented by

$$\text{Growth rate} = g(P) = \frac{1}{P}\frac{dP}{dt} = k$$

1.13 Suppose a culture of bacteria is increasing in size according to the constant growth rate model above. The initial bacteria population is P_0.
   (a) Develop the difference equation for the discrete-time system approximation of the continuous-time model. Denote the discrete-time population as $P_A(n)$.
   (b) Find the general solution for $P_A(n)$, $n = 0, 1, 2, 3, \ldots$. Leave your answer in terms of k and P_0.
       The constant growth rate $k = 0.01$ bacteria/min per bacteria and the initial number of bacteria is 10,000.

(c) Solve the difference equation recursively using a step size $\Delta t = 1$ min for $P_A(n)$, $n = 1$, 2, 3, 4, 5. Compare the result for $P_A(5)$ to the value obtained from the general solution found in part (b).

(d) The analytical solution to the continuous-time model is $P(t) = P_0 e^{kt}$, $t \geq 0$. How long does it take for the population to reach 1 million?

(e) On the same graph, plot the continuous-time model output $P(t)$, $0 \leq t \leq 500$ and the discrete-time model output $P_A(n)$, $n = 0$, 50, 100, 150, $\ldots$, 1000 when $\Delta t = 0.5$ min.

(f) Explain what would happen to a population with constant growth rate k, if k were negative.

2 Continuous-Time Systems

2.1 INTRODUCTION

Before we start our exploration of simulation, it is important for us to have some basic knowledge of how linear time-invariant (LTI) dynamic systems behave. The analysis of linear systems and how they respond to elementary types of inputs is straightforward. Linear systems appear as building blocks in more complex systems. Our intuitive understanding of the entire system is enhanced by recognizing the fundamental behavior of its linear components. Control systems, for example, are oftentimes composed of linear continuous-time components interconnected to produce a desirable response to commanded as well as uncontrollable or disturbance inputs.

Speaking of control systems, the mathematical model of the process being controlled is often nonlinear; however, a properly designed regulatory control system will limit excursions of the process variables. In fact, the design of the controller may be based on a linearized model of the nonlinear process owing to the wealth of tools available in the field of linear control theory. Simulation can play a valuable role here by shedding light on the validity of using a linearized mathematical model to approximate a nonlinear system model.

Modern simulation software contains user interfaces employing graphical icons that serve as building blocks for representing the linear continuous- and discrete-time components within a system. In order to exploit this feature, the simulation builder must understand the meaning and differences between the assortment of linear system blocks (integrators, first-order lags, second-order systems, transfer functions, and state space models) at his or her disposal. The material on first- and second-order system response, and state variables covered in this chapter and Chapter 4, is intended as an introduction (or possibly a review) to the topic of linear continuous-time systems. There are literally dozens of excellent books on the subject of linear systems theory and linear control systems. Several are included in the references and the reader is encouraged to consult one or more as necessary.

In addition to the focus on linear systems in this chapter, one section includes several examples of nonlinear systems as well. A graphical illustration of how to linearize a nonlinear system model is presented as a preview of what is to come in Chapter 7 where the subject is revisited in more detail.

Simulation of continuous-time systems is not discussed in detail until Chapter 3 where the subject of numerical integration is introduced. However, a simulation model based on numerical differentiation, similar to what was done in Chapter 1, is presented. At the conclusion of this chapter, the reader will be capable of representing simple continuous-time systems in state variable form and generate discrete-time model approximations of them, which can be solved in a recursive fashion.

2.2 FIRST-ORDER SYSTEMS

Continuous-time dynamic systems are said to be first order if the highest derivative of the dependent variable appearing in the mathematical model is first order. Systems in which a quantity of material or energy changes at a rate dependent on the amount of material or energy present are typically first order in nature. The general representation of a scalar first-order system is

$$\frac{dy}{dt} = f(t, y, u) \tag{2.1}$$

where

t is the continuous-time variable

$u = u(t)$ is the system input

$y = y(t)$ is the system output

$f(t, y, u)$ is the derivative function, which relates the rate of change in y to all three arguments

Not all three arguments will be present in every first-order model. Furthermore, it is possible for multiple inputs $u_1(t), u_2(t), \ldots, u_r(t)$ to be present.

We begin our discussion of first-order systems with a special case, namely, where the derivative function is an explicit linear function of the input and output given by

$$f(t, y, u) = b_0 u(t) - a_0 y(t) \tag{2.2}$$

where a_0 and b_0 are constants. Combining Equations 2.1 and 2.2 gives

$$\frac{d}{dt} y(t) + a_0 y(t) = b_0 u(t) \tag{2.3}$$

Equation 2.3 is an LTI, ordinary differential equation. In the time-varying case, one or both of the linear system parameters a_0 and b_0 are functions of the independent variable t. Equation 2.3 is commonly expressed as

$$\tau \frac{d}{dt} y(t) + y(t) = Ku(t) \tag{2.4}$$

where τ and K are easily related to a_0 and b_0 by

$$\tau = \frac{1}{a_0} \quad \text{and} \quad K = \frac{b_0}{a_0} \tag{2.5}$$

Many simple real-world dynamic systems are modeled by the first-order differential equation (Equation 2.4). More complex systems often behave similarly to first-order systems under certain conditions. Furthermore, higher-order system models can be reduced to a system of coupled first-order models. Familiarity with first-order system response will prove useful later on when we undertake the task of simulating higher-order linear and nonlinear systems. For this reason, we explore some basic properties of first-order systems modeled by Equation 2.4.

2.2.1 Step Response of First-Order Systems

When the input $u(t)$ is constant, that is, $u(t) = A$, $t \geq 0$, the solution to Equation 2.4 for $y(t)$ is obtained using Laplace transform or classical time-domain methods. It is given below:

$$y(t) = y(0)e^{-t/\tau} + KA(1 - e^{-t/\tau}), \quad t \geq 0 \tag{2.6}$$

where $y(0)$ is the initial value of the output $y(t)$. Several graphs of $y(t)$ are shown in Figure 2.1 for the cases where $y(0) = 0$, $K = 5$, $A = 2$, and $\tau = 0.5$, 2, 5, and 10.

The graphs of $y(t)$ shown in Figure 2.1 are called the step response because the input resembles a step (changing from 0 to A at $t = 0$). Note that the initial condition is zero in all the step responses.

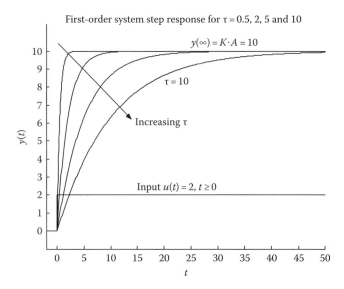

FIGURE 2.1   Step response of first-order system with different values of τ.

The constant A measures the amplitude of the input and is not an inherent system parameter. The system parameters are K and τ (or a_0 and b_0 from which they are computed). The first parameter K is called the system DC or steady-state gain. It is so named because the final value of the output, $y(\infty)$, is easily computed from

$$y(\infty) = K \cdot u(\infty) = K \cdot A \tag{2.7}$$

which in this case is $y(\infty) = 5 \cdot 2 = 10$ (see Figure 2.1). The final value $y(\infty)$ is unaffected by the initial condition $y(0)$. However, the graph of $y(t)$ in Equation 2.6 certainly depends on $y(0)$, since that is where it starts. A first-order system like the one in Equation 2.4 is called a first-order lag because of the way the step response in Figure 2.1 lags the step input.

There are situations when the input to a first-order system is not a step; however, the input remains constant for a period of time that is largely relative to the parameter τ. Equation 2.7 enables us to readily compute the final output value prior to a change in the input. In essence, we are tracking the first-order system from one steady-state level to another, and the transient response (portion of the overall step response that decays to zero) is ignored. Even without knowledge of the transient response, it is possible to predict the amount of time necessary for the new steady state to be established.

In the first-order system modeled by Equation 2.4, the first derivative vanishes when the system is at steady state, leaving $y_{ss} = K\bar{u}$, where y_{ss} is the output at steady state in response to the constant input $\bar{u}$. A similar result is obtained from Equation 2.6 with A replaced by and t approaching ∞.

The first-order system step responses shown in Figure 2.1 correspond to four distinct values for the parameter τ. It is apparent that while all approach the limiting value $y(\infty) = 10$, there is a noticeable difference in the amount of time required for each to get there. The individual step responses are correlated with the system parameter τ. This parameter is called the time constant of the first-order system. It is a measure of the speed of the step response as well as an indicator of the overall speed of the first-order system's dynamics. A "rule of thumb" for first-order systems is that the transient response vanishes after four or five time constants. The transient response component of the step response in Equation 2.6 with $y(0) = 0$ is

$$y_{tr}(t) = -KAe^{-t/\tau}, \quad t \geq 0 \tag{2.8}$$

when $t = 5\tau$,

$$y_{tr}(5\tau) = -KAe^{-5} = -KA(0.0067) \tag{2.9}$$

and the step response

$$y(5\tau) = KA(1 - e^{-5}) = 0.9933KA \tag{2.10}$$

is more than 99% complete. After four time constants have elapsed, the step response is slightly over 98% of its final value (see Figure 2.1).

First-order system models are commonplace in science, engineering, economics, business, etc. The liquid storage tank model in Section 1.2 and the population models considered in Section 1.5 are examples of first-order system models. Another example of a physical system described in terms of a first-order model is the simple electric circuit shown in Figure 2.2 along with the tank.

The circuit components are a capacitor C, a resistor R, and a voltage source $e_0(t)$. There is also a switch that connects the source to the rest of the circuit when it is in the closed position. Like the tank that stores its energy as a column of liquid, the circuit's capacitor stores energy in the form of electric charge. The potential energy of the fluid varies as the tank level changes and the electrical energy stored in the circuit varies with the amount of electrical charge stored in the capacitor. Both systems have a mechanism for dissipating energy. The tank does so whenever the level of fluid is dropping and the circuit dissipates energy in the resistor whenever there is current flowing.

The fluid resistance of the tank tells us the amount of effort, that is, height of liquid, required to produce a unit of flow from the tank. A typical unit for fluid resistance is ft per ft^3/min. The electrical counterpart is the electrical resistor that also measures the driving force, in this case, the voltage applied to the resistor, necessary to produce a unit of current flow, measured in amperes. The unit of electric resistance is volts/ampere, commonly called ohms.

Choosing the voltage across the capacitor $v_c(t)$ as the output, the circuit model is easily derived using basic principles of electrical circuits. The result is

$$RC\frac{\mathrm{d}}{\mathrm{d}t}v_c(t) + v_c(t) = e_0(t) \tag{2.11}$$

Comparison of Equation 2.11 with the standard form introduced in Equation 2.4 reveals the time constant of the circuit $\tau = RC$ and the steady-state gain $K = 1(\text{V/V})$. Hence, the transient response lasts for a period of time equal to approximately $5RC$. For a constant voltage applied to the circuit, that is, $e_0(t) = E_0$, $t \geq 0$, the steady-state voltage $v_c(\infty)$ is numerically equal to E_0 since $v_c(\infty) = KE_0 = 1 \cdot E_0$.

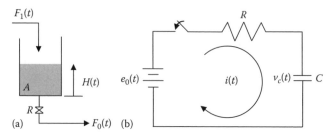

FIGURE 2.2 Examples of systems with first-order system models: (a) storage tank and (b) RC circuit.

The step response is obtained from Equation 2.6 with $y(0) = v_c(0) = 0$, $\tau = RC$, $K = 1$, and $A = E_0$. The result is

$$v_c(t) = E_0(1 - e^{-t/RC}), \quad t \geq 0 \tag{2.12}$$

The step response consists of the steady-state component

$$v_c(\infty) = E_0 \tag{2.13}$$

and the transient component

$$v_c(t)_{tr} = -E_0 e^{-t/RC}, \quad t \geq 0 \tag{2.14}$$

The transient response involves the exponential $e^{-t/RC}$, which is called the natural mode of the system. To understand this, consider the circuit response with zero applied voltage ($E_0 = 0$) and a nonzero initial voltage across the capacitor $v_c(0)$. From Equation 2.6, the solution for $v_c(t)$ is

$$v_c(t) = v_c(0)e^{-t/RC}, \quad t \geq 0 \tag{2.15}$$

a constant times the natural mode. Natural modes of linear systems are exponential functions of time involving the parameters of the system, in this case, R and C. The natural modes do not depend on the system inputs. The unforced response of higher-order system models is referred to as the natural response of the system. It contains a linear combination of the natural modes (only one for the first-order system model). In general, the natural modes of linear system models appear in the transient response independent of whether the system is being forced (excited by inputs) or simply responding to initial conditions as in the case of an autonomous system.

Example 2.1

A 12 V battery is used to charge the capacitor in the circuit shown in Figure 2.2. When the switch is closed at $t = 0$, the capacitor voltage is zero. Numerical values of the circuit parameters are $R = 5000 \ \Omega$ and $C = 0.125 \times 10^{-6}$ F (1 F = 1 A per V/s).

    (a) Find the time constant τ, steady-state gain K, and natural mode of the circuit.
    (b) Find the steady-state voltage $v_c(\infty)$ across the capacitor.
    (c) Determine how long it takes for the capacitor to charge up to 50% of $v_c(\infty)$.
    (d) Find and graph the transient component, steady-state component, and the complete response for the case where the capacitor is initially charged to 3 V.

(a) $\tau = RC = (5000 \ \Omega) \times 0.125 \times 10^{-6}$ F $= 0.000625$ s (625×10^{-6} s)

$$K = 1 \ \text{V/V}$$

Natural mode: $e^{-t/RC} = e^{-t/0.000625}$, $t \geq 0$

(b) $v_c(\infty) = KE_0 = (1 \ \text{V/V}) \times 12 \ \text{V} = 12 \ \text{V}$

(c) $v_c(t) = E_0(1 - e^{-t/RC}) \Rightarrow 6 = 12(1 - e^{-t/625 \times 10^{-6}})$, which can be solved using natural logarithms to give $t = 0004332$ s

(d) From Equation 2.6 with initial condition $v_c(0) = 3$ V, the complete response is

$$v_c(t) = v_c(0)e^{-t/RC} + KE_0(1 - e^{-t/RC}), \quad t \geq 0$$
$$= 3e^{-t/625 \times 10^{-6}} + (1)(12)(1 - e^{-t/625 \times 10^{-6}})$$

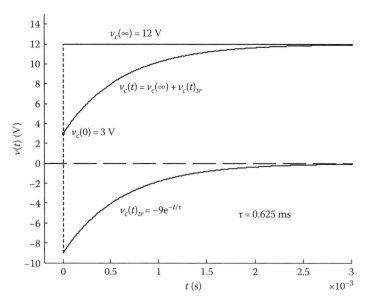

FIGURE 2.3   Steady-state, transient, and total response of an *RC* circuit.

The transient component is

$$v_c(t)_{tr} = [v_c(0) - KE_0]e^{-t/RC}, \quad t \geq 0$$
$$= [3 - (1)(12)]e^{-t/625 \times 10^{-6}}$$
$$= -9e^{-t/625 \times 10^{-6}}$$

and the steady-state component is

$$v_c(\infty) = KE = 1\frac{V}{V}(12\ V) = 12\ V$$

Graphs of the steady-state, transient, and complete responses are shown in Figure 2.3.

Note that the transient response has decayed to essentially zero after five time constants $(5 \times 625 \times 10^{-6} = 3.125 \times 10^{-3})$ have elapsed.

EXERCISES

2.1 The tank shown in Figure 2.2 has a constant cross-sectional area A and fluid resistance R.

    (a) Find expressions for the time constant τ and steady-state gain K of the tank in terms of the physical parameters A and R.

    (b) The empty tank is subject to a constant flow in of $\overline{F}$ ft³/min. Obtain an expression for the liquid level step response of the tank.

    (c) The cross-sectional area of the tank is 20 ft², and the fluid resistance is 0.5 ft per ft³/min. How high must the tank be if the inflow is constant at $\overline{F} = 15$ ft³/min for it not to overflow.

    (d) How long will it take for the tank level to reach 50% of its final height?

    (e) What size tank is needed if the time required to fill up is increased by 10%?

2.2 Consider the first-order system: $(d/dt)y(t) + a_0 y(t) = b_0 u(t)$

(a) Under what conditions does this system reduce to a pure integrator?

(b) For the continuous-time integrator in part (a), express the output $y(t)$ in terms of the input $u(t)$. Assume the initial condition is $y(0) = y_0$.

(c) When is a liquid storage tank a pure integrator?

2.3 The amount of salt Q in a well-stirred tank shown in Figure E2.3 depends on c_1, the concentration of salt in the brine solution entering the tank, as well as the flow rates F_1 and F_0 into and out of the tank. The continuous-time model is based on conservation of salt. It equates dQ/dt, the instantaneous rate of change in the amount of salt in the tank to the difference in the rate of salt entering the tank, $c_1 F_1$, and the rate of salt flowing out of the tank, cF_0.

The tank initially contains 100 lb of salt-free water. The concentration of salt in the brine solution flowing in is 0.25 lb/ft^3. Both the flow into and the flow out of the tank are both 1 ft^3/min. Note that 1 ft^3 of water weighs approximately 62.4 lb.

(a) Find $Q(t)$, the amount of salt in the tank as a function of time.

(b) Find the amount of salt in the tank at steady state.

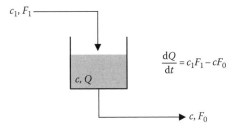

$$\frac{dQ}{dt} = c_1 F_1 - cF_0$$

FIGURE E2.3

2.4 A temperature-controlled chamber is shown in Figure E2.4:

The air temperature inside the chamber is assumed to be the same everywhere, namely, $T(t)$. The chamber walls are insulated to reduce heat loss or gain with its surroundings. Temperature control is achieved by circulating hot or cold water through pipes located inside the chamber. Heat exchange occurs between the air inside the chamber and the circulating water in the pipes. The heat flow from the circulating hot water is $Q_h(t)$, and $Q_c(t)$ is the heat flow to the cold water. Heat exchange $Q_0(t)$ also occurs between the air inside and outside the chamber. Ambient temperature outside the chamber is denoted $T_0(t)$.

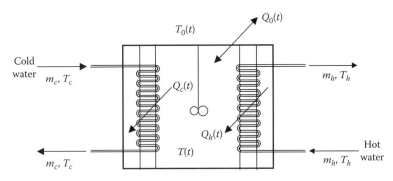

FIGURE E2.4

A suitable model for this thermal system is based on the conservation of energy.

$$c_A V \frac{dT}{dt} = Q_h - Q_c - Q_0$$

V is the volume (ft³) of air in the chamber, and c_A is the thermal capacitance of air (0.01375 Btu/°F/ft³). The heat flow terms on the right-hand side are given by

$$Q_h = \dot{m}_h c_p (T_h - T)$$

$$Q_c = \dot{m}_c c_p (T - T_c)$$

$$Q_0 = \frac{1}{R}(T - T_0)$$

where

$\dot{m}_h$ and $\dot{m}_c$ are the mass flow rates (lb/min) of the hot and cold water
c_p is the specific heat of water (1 Btu/lb/°F)
R is the thermal resistance (°F/Btu/min) of the chamber walls

The expressions for Q_h and Q_c assume that the flow rates of the circulating fluids are great enough that both fluids exit at the same temperature at which they entered the chamber.

(a) Express the mathematical model in the form of a differential equation relating the output T and its derivative to the inputs T_h, T_c, and T_0.
(b) Find the time constant and the three steady-state gains of the system. Check the units to verify that the time constant is in minutes and the steady-state gains are dimensionless (°F/°F).
(c) Show that the air temperatures inside and outside the chamber eventually equalize after both the hot and cold circulating water flows are turned off.
(d) Suppose the chamber air temperature is required to be higher than the outside ambient air temperature, which remains constant, that is, $T_0(t) = \overline{T}_0$, $t \geq 0$. The hot water temperature entering the chamber is three times greater than the ambient temperature. The initial air temperature inside the chamber is the same as the outside ambient temperature. Find the analytical solution for $T(t)$, $t \geq 0$, the air temperature inside the chamber.
(e) Graph the solution for $T(t)$, $t \geq 0$ in part (d) using the following values:

$$V = 5000 \text{ ft}^3, \quad R = 0.025°\text{F/Btu/min}, \quad \dot{m}_h = 50 \text{ lb/min}, \quad \text{and} \quad \overline{T}_0 = 60°\text{F}$$

2.3 SECOND-ORDER SYSTEMS

Input–output models of continuous-time dynamic systems where the highest derivative of the dependent variable is second order are classified as second-order systems. Second-order systems result when there are two energy storage elements present. Our interest for now is in linear second-order systems, which can be manipulated into the form shown in Equation 2.16 relating an output $y(t)$ to an input $u(t)$ involving generic system parameters ζ, ω_n, and K.

$$\frac{d^2}{dt^2} y(t) + 2\zeta\omega_n \frac{d}{dt} y(t) + \omega_n^2 y(t) = K\omega_n^2 u(t) \tag{2.16}$$

For an actual second-order system (mechanical, electrical, biological, etc.), the generic parameters can be expressed in terms of the system's physical parameters. The importance of each will be explained shortly.

The unit step response of the second-order system is the solution for $y(t)$ in Equation 2.16 when $y(0) = 0$ and the input $u(t) = 1$, $t \geq 0$, hereafter denoted by $\hat{u}(t)$. It can be found in any text related to linear systems or controls (Palm 1983; Franklin et al. 2002; Dorf and Bishop 2005). The unit step response assumes one of three forms depending on the location of the roots of the algebraic equation

$$s^2 + 2\zeta\omega_n s + \omega_n^2 = 0 \tag{2.17}$$

known as the characteristic equation of the system. The characteristic roots are the solution to Equation 2.17 and are given by

$$s_1, s_2 = -\zeta\omega_n \pm \sqrt{\zeta^2 - 1}\omega_n \tag{2.18}$$

The natural modes of the second-order system are $e^{s_1 t}$ and $e^{s_2 t}$. The step response depends on the value of the parameter ζ. There are three cases to consider.

Case 1: $\zeta > 1$
If we let $s_1 = -\zeta\omega_n - \sqrt{\zeta^2 - 1}\omega_n$ and $s_2 = -\zeta\omega_n - \sqrt{\zeta^2 - 1}\omega_n$, then both roots are negative (assuming $\omega_n > 0$) and $s_1 < s_2 < 0$. Introducing time constants τ_1 and τ_2 as the reciprocals of the characteristic roots s_1 and s_2, respectively,

$$\tau_1 = -\frac{1}{s_1}, \quad \tau_2 = -\frac{1}{s_2} \tag{2.19}$$

The unit step response is

$$y(t) = K\left[1 + \frac{\tau_2 e^{-t/\tau_2} - \tau_2 e^{-t/\tau_1}}{\tau_1 - \tau_2}\right], \quad t \geq 0 \tag{2.20}$$

Case 2: $0 < \zeta < 1$
The characteristic roots are complex conjugates and can be expressed as

$$s_1, s_2 = -\zeta\omega_n \pm j\sqrt{1 - \zeta^2}\omega_n \tag{2.21}$$

It is convenient to define a new quantity ω_d in terms of ζ and ω_n according to

$$\omega_d = \sqrt{1 - \zeta^2}\omega_n \tag{2.22}$$

The unit step response is

$$y(t) = K\left[1 - e^{-\zeta\omega_n t}\left(\cos\omega_d t + \frac{\zeta\omega_n}{\omega_d}\sin\omega_d t\right)\right], \quad t \geq 0 \tag{2.23}$$

An alternate form of Equation 2.23 is

$$y(t) = K\left[1 - \frac{\omega_n}{\omega_d}e^{-\zeta\omega_n t}\sin(\omega_d t + \varphi)\right], \quad t \geq 0 \tag{2.24}$$

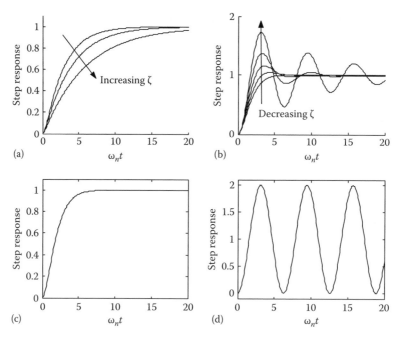

FIGURE 2.4 Unit step response of second-order system in Equation 2.16. (a) Overdamped, $\zeta = 1.5, 2, 3$. (b) Underdamped, $\zeta = 0.1, 0.3, \ldots, 0.9$. (c) Critically damped, $\zeta = 1$. (d) Zero damping, $\zeta = 0$.

where the phase angle term φ is given by

$$\varphi = \tan^{-1} \frac{\omega_d}{\zeta\omega_n} = \tan^{-1}\left(\frac{\sqrt{1-\zeta^2}}{\zeta}\right) \tag{2.25}$$

Case 3: $\zeta = 1$
From Equation 2.18, the characteristic roots are repeated, $s_1 = s_2 = -\omega_n$. The unit step response is

$$y(t) = K[1 - e^{-\omega_n t}(\omega_n t + 1)], \quad t \geq 0 \tag{2.26}$$

A graph of the unit step responses given in Equations 2.20, 2.23, and 2.26 with $K = 1$ is shown in Figure 2.4. The abscissa is $\omega_n t$, a dimensionless variable, which allows us to visualize the effect of the parameter ζ on the step response independent of ω_n. Note that all three step responses start from zero. Furthermore, the initial slope given by $dy(0)/dt$ is also zero for all three cases (see Exercise 2.6).

There are no oscillations in Case 1 ($\zeta > 1$), that is, the response is monotonically increasing without overshooting the final value $y(\infty) = K \cdot \bar{u} = 1$ for a unit step input. The transient period increases with increasing ζ. The system is said to be overdamped.

An oscillatory step response occurs in Case 2 ($0 < \zeta < 1$), and the system is referred to as underdamped. As the value of ζ decreases, the oscillations become more pronounced, and the settling time for the transient component to die out becomes larger.

The case when $\zeta = 1$ represents the transition from Case 1 to Case 2 (or vice versa). The second-order system is called critically damped in this situation.

The graph in Figure 2.4d is the unit step response for the case when $\zeta = 0$. From Equation 2.23 with $\zeta = 0$,

$$y(t) = K(1 - \cos \omega_n t), \quad t \geq 0 \tag{2.27}$$

resulting in sustained oscillations from 0 to 2 when $K = 1$. The differential equation of the unforced system is

$$\frac{d^2}{dt^2} y(t) + \omega_n^2 y(t) = 0 \tag{2.28}$$

and the natural response resulting from the presence of initial conditions is that of harmonic motion, that is, sustained oscillations about zero at a frequency of ω_n rad/s.

Except for the case when $\zeta = 0$, the unit step response approaches the limiting or steady-state value $y(\infty) = K$, which means that K is the DC or steady-state gain of the second-order system in Equation 2.16. The parameter ζ, which determines the existence and extent of the oscillations as well as the duration of the transient response, is called the damping ratio of the system. The last two parameters ω_n and ω_d are the natural frequency and damped natural frequency of the second-order system, respectively. The first, ω_n, is the frequency of the sustained oscillations ($\zeta = 0$) in Equation 2.27, and the second, ω_d, is the frequency of the decaying oscillations ($0 < \zeta < 1$) in Equation 2.24. It follows from Equation 2.22 that $\omega_d < \omega_n$. The natural frequency ω_n is an indication of the speed of the step response (and the system in general) since the oscillatory natural modes are damped by the exponential term with time constant $1/\zeta\omega_n$ in Equation 2.23.

Example 2.2

Figure 2.5 shows a delicate instrument placed on a table that moves as a result of a vertical force acting on it. Springs and dampers connect the table to the ground to limit the table's movement.

The combined mass of the table and instrument is m. The total stiffness of the springs is k and the total damping is c. The mechanical system is modeled by

$$m\frac{d^2}{dt^2} x(t) + c\frac{d}{dt} x(t) + kx(t) = f(t) \tag{2.29}$$

where
$x(t)$ is the displacement of the table (from its static equilibrium position)
$f(t)$ is the force acting on the platform resulting in the motion $x(t)$

(a) Find expressions for the steady-state gain K, the damping ratio ζ, and the natural frequency ω_n in terms of the physical parameters m, c, and k.
(b) Numerical values of the physical parameters are $m = 40$ lb$_m$, $k = 45$ lb$_f$/ft, and $c = 4$ lb$_f$ s/ft. Find the response of the table when the platform is subjected to a sudden deflection due to a force of 12 lb$_f$.
(c) Graph the solution and estimate the duration of the transient.
(d) The instrument is not usable if it is moving faster than 0.04 ft/s. How long a period of time must pass after the force is applied before the instrument will function properly?

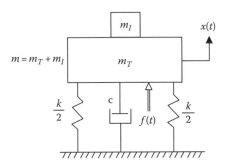

FIGURE 2.5   Mechanical system for Example 2.2.

(a) Dividing Equation 2.29 by m for comparison with the standard form of a second-order system in Equation 2.16 gives

$$\frac{d^2}{dt^2}x(t) + \frac{c}{m}\frac{d}{dt}x(t) + \frac{k}{m}x(t) = \frac{1}{m}f(t) \tag{2.30}$$

$$\Rightarrow 2\zeta\omega_n = \frac{c}{m}, \quad \omega_n^2 = \frac{k}{m}, \quad K\omega_n^2 = \frac{1}{m} \tag{2.31}$$

Solving for the parameters K, ω_n, and ζ yields

$$\omega_n = \sqrt{\frac{k}{m}}, \quad \zeta = \frac{c}{2\sqrt{km}}, \quad K = \frac{1}{k} \tag{2.32}$$

(b) Substituting the given values for m (in slugs), k, and c,

$$\omega_n = \sqrt{\frac{k}{m}} = \sqrt{\frac{45}{40/32.2}} = 6.0187 \text{ rad/s}$$

$$\zeta = \frac{c}{2\sqrt{km}} = \frac{4}{2\sqrt{45.40/32.2}} = 0.2675$$

$$K = \frac{1}{45} = 0.0222 \text{ in/lb}_f$$

The damping ratio $\zeta = 0.2675$ indicates the system is underdamped. From Equation 2.22, the damped natural frequency is

$$\omega_n = \sqrt{1-\zeta^2}\omega_n = \left(\sqrt{1-0.2675^2}\right)6.0187 = 5.7994 \text{ rad/s}$$

and the response to a step input $f(t) = \bar{F} = 12 \text{ lb}_f$, $t \geq 0$ is

$$x(t) = K\bar{F}\left[1 - e^{-\zeta\omega_n t}\left(\cos\omega_d t + \frac{\zeta\omega_n}{\omega_d}\sin\omega_d t\right)\right], \quad t \geq 0 \tag{2.33}$$

Substituting the numerical values for K, $\bar{F}$, ζ, ω_n, and ω_d results in

$$x(t) = 0.2667\left[1 - e^{-1.6100t}(\cos 5.7994t + 0.2776\sin 5.7994t)\right], \quad t \geq 0 \tag{2.34}$$

(c) A graph of the step response is generated in the script file "Chap2_Ex3_1.m" and shown in Figure 2.6.

The transient period can be approximated from the graph as roughly 3 s, or it can be computed from the time constant of the exponential envelope as

$$\text{Transient period} \approx 5 \times \frac{1}{\zeta\omega_n} = 5 \times \frac{1}{0.2675(6.0187)} = 3.1056 \text{ s}$$

(d) The first derivative is obtained by differentiation of the underdamped step response in Equation 2.24. The result is

$$\frac{d}{dt}y(t) = K\frac{\omega_n}{\sqrt{1-\zeta^2}}e^{-\zeta\omega_n t}\sin\omega_d t, \quad t \geq 0 \tag{2.35}$$

Substituting the numerical values for the system parameters K, ζ, ω_n, and ω_d gives

$$\frac{d}{dt}y(t) = 0.1388e^{-1.61t}\sin 5.7994t, \quad t \geq 0 \tag{2.36}$$

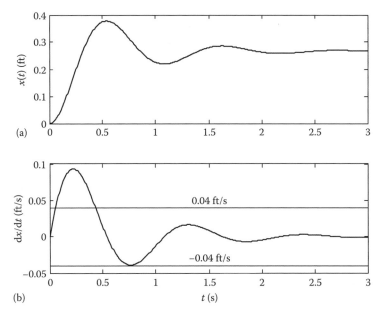

FIGURE 2.6 (a) Position and (b) velocity response of table and instrument ($\overline{F} = 12$ lb$_f$).

The first derivative is graphed in the lower half of Figure 2.6. From the graph, it appears that approximately 0.77 s must elapse for the instrument to be usable, that is, the instrument is moving at less than 0.04 ft/s in either direction after that period of time. (A closeup of the response in the neighborhood of dx/dt = −0.04 ft/s reveals that the instrument's velocity actually falls a bit short of −0.04 ft/s.)

2.3.1 Conversion of Two First-Order Equations to a Second-Order Model

A linear second-order system is sometimes represented as a system of two first-order differential equations like those in Equations 2.37 and 2.38:

$$\frac{dx}{dt} = ax + by + f(t) \tag{2.37}$$

$$\frac{dy}{dt} = cx + dy + g(t) \tag{2.38}$$

Suppose a single equation relating the dependent variable $x = x(t)$ and the inputs $f = f(t)$ and $g = g(t)$ is required. The first step is to solve for $y = y(t)$ in Equation 2.37,

$$y = \frac{1}{b}\left(\frac{dx}{dt} - ax - f\right) \tag{2.39}$$

Differentiating Equation 2.39,

$$\frac{dy}{dt} = \frac{1}{b}\left(\frac{d^2x}{dt^2} - a\frac{dx}{dt} - \frac{df}{dt}\right) = cx + dy + g \tag{2.40}$$

Replacing y in Equation 2.40 with Equation 2.39 gives

$$\frac{1}{b}\left(\frac{d^2x}{dt^2} - a\frac{dx}{dt} - \frac{df}{dt}\right) = cx + d\left[\frac{1}{b}\left(\frac{dx}{dt} - ax - f\right)\right] + g \tag{2.41}$$

and simplifying leads to the second-order differential equation,

$$\frac{d^2x}{dt^2} - (a + d)\frac{dx}{dt} + (ad - bc)x = \frac{df}{dt} - df + bg \tag{2.42}$$

A similar procedure is used to eliminate x from Equations 2.37 and 2.38 to give a second-order differential equation in y.

Example 2.3

The well-mixed tanks shown in Figure 2.7 contain uniform salt concentrations of $c_1 = c_1(t)$ and $c_2 = c_2(t)$, respectively. Concentration of salt in the input to the first tank is $c = c(t)$. The flow rates between the tanks are Q_1 and Q_2, where $Q_1 > Q_2 > 0$. The liquid volumes in both tanks remain constant at V_1 and V_2.

(a) Write the differential equations for the conservation of salt in each tank.
(b) Find the differential equation relating $c_2(t)$ and the input $c(t)$.
(c) Find expressions for the damping ratio, natural frequency, and steady-state gain.
(d) Find and plot the step response for c_2 under the following conditions:

$$Q_1 = 10 \text{ gal/min}, \quad Q_2 = 5 \text{ gal/min}, \quad V_1 = 15 \text{ gal}, \quad \text{and} \quad V_2 = 15 \text{ gal}$$

$$c_1(0) = c_2(0) = 0 \text{ lb of salt/gal}, \quad c(t) = \bar{c} = 0.25 \text{ lb salt/gal}, \quad t \geq 0$$

(a) Equating the accumulation of salt in each tank to the difference between the rates of salt in and out of the tanks,

$$\frac{d}{dt}(c_1 V_1) = Q_{in}c + Q_2 c_2 - Q_1 c_1 \tag{2.43}$$

$$\frac{d}{dt}(c_2 V_2) = Q_1 c_1 - Q_2 c_2 - Q_{out} c_2 \tag{2.44}$$

Since the holdup of liquid in both tanks is constant, the flows Q_{in} and Q_{out} are equal,

$$Q_{in} = Q_{out} = Q_1 - Q_2 \tag{2.45}$$

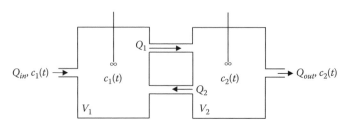

FIGURE 2.7 Two-tank mixing system.

And, therefore, Equations 2.43 and 2.44 become

$$V_1 \frac{dc_1}{dt} = (Q_1 - Q_2)c + Q_2 c_2 - Q_1 c_1 \tag{2.46}$$

$$V_2 \frac{dc_2}{dt} = Q_1 c_1 - Q_2 c_2 - (Q_1 - Q_2)c_2 \tag{2.47}$$

(b) Rearranging Equations 2.46 and 2.47 into the form of Equations 2.37 and 2.38,

$$\frac{dc_2}{dt} = -\frac{Q_1}{V_2}c_2 + \frac{Q_1}{V_2}c_1 \tag{2.48}$$

$$\frac{dc_1}{dt} = \frac{Q_2}{V_1}c_2 - \frac{Q_1}{V_1}c_1 + \frac{(Q_1 - Q_2)}{V_1}c \tag{2.49}$$

Comparing Equations 2.48, 2.49, and 2.37, Equation 2.38 implies

$$a = -\frac{Q_1}{V_2}, \quad b = \frac{Q_1}{V_2}, \quad c = \frac{Q_2}{V_1}, \quad d = -\frac{Q_1}{V_1}, \quad f(t) = 0, \quad g(t) = \frac{(Q_1 - Q_2)}{V_1}c(t) \tag{2.50}$$

From Equation 2.42, the second-order differential equation relating c_2 and c is

$$\frac{d^2 c_2}{dt^2} + Q_1 \left(\frac{1}{V_1} + \frac{1}{V_2} \right) \frac{dc_2}{dt} + \frac{Q_1(Q_1 - Q_2)}{V_1 V_2}c_2 = \frac{Q_1(Q_1 - Q_2)}{V_1 V_2}c \tag{2.51}$$

(c) Comparing the left-hand side of Equation 2.51 with the standard form in Equation 2.16 gives

$$2\zeta\omega_n = Q_1 \left(\frac{1}{V_1} + \frac{1}{V_2} \right), \quad \omega_n^2 = \frac{Q_1(Q_1 - Q_2)}{V_1 V_2} \tag{2.52}$$

$$\Rightarrow \omega_n = \left[\frac{Q_1(Q_1 - Q_2)}{V_1 V_2} \right]^{1/2}, \quad \zeta = \frac{(V_1 + V_2)}{2} \left[\frac{Q_1}{(Q_1 - Q_2)V_1 V_2} \right]^{1/2} \tag{2.53}$$

For $c(t) = \bar{c}$, the steady-state value of c_2 is obtained from Equation 2.51 by setting the derivatives equal to zero resulting in

$$\frac{Q_1(Q_1 - Q_2)}{V_1 V_2}(c_2)_{ss} = \frac{(Q_1 - Q_2)}{V_1} \left(\frac{Q_1}{V_2}\bar{c} \right) \tag{2.54}$$

$$\Rightarrow (c_2)_{ss} = \bar{c} \tag{2.55}$$

Hence, the steady-state gain $K = 1$ lb salt/lb salt as expected.

(d) For the given conditions, that is, $Q_2 = Q = 5$, $Q_1 = 2Q = 10$, and $V_1 = V_2 = V = 15$

$$\omega_n = \left[\frac{2Q(2Q - Q)}{VV} \right]^{1/2} = (2)^{1/2} \frac{Q}{V} = (2)^{1/2} \frac{5}{15} = 0.4714 \text{ rad/min} \tag{2.56}$$

$$\zeta = \frac{(V + V)}{2} \left[\frac{2Q}{(2Q - Q)VV} \right]^{1/2} = (2)^{1/2} = 1.4142 \tag{2.57}$$

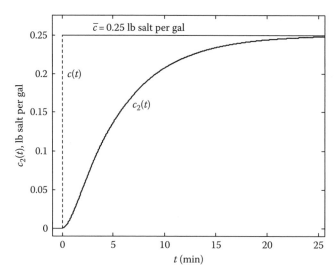

FIGURE 2.8 Response of salt concentration in second tank to step input $c(t) = 0.25$, $t \geq 0$.

From Equation 2.18, the characteristic roots of the overdamped system are

$$s_1, s_2 = -\zeta\omega_n \pm \sqrt{\zeta^2 - 1}\,\omega_n = -1.1381 \text{ rad/min}, \; -0.1953 \text{ rad/min} \qquad (2.58)$$

The time constants in Equation 2.19 are $\tau_1 = -1/s_1 = 0.8787$ min and $\tau_2 = -1/s_2 = 5.1213$ min, and from the unit step response in Equation 2.20, the response to a step of magnitude $\bar{c}$ is

$$c_2(t) = K\bar{c}\left[1 + \frac{\tau_2 e^{-t/\tau_2} - \tau_1 e^{-t/\tau_1}}{\tau_1 - \tau_2}\right] \qquad (2.59)$$

$$= 0.25\left[1 - \left(\frac{5.1213 e^{-t/5.1213} - 0.8787 e^{-t/0.8787}}{4.2426}\right)\right], \quad t \geq 0 \qquad (2.60)$$

The second-order differential equation in Equation 2.51 is in standard form; however, the second-order differential equation for $c_1(t)$ contains the first derivative dc/dt on the right-hand side of the equation (see Exercise 2.6). The implication of input derivatives in the system model will be discussed in a later section. A graph of the step response is shown in Figure 2.8.

EXERCISES

2.5 Starting with Equations 2.37 and 2.38, obtain the second-order differential equation relating the output $y = y(t)$ and its derivatives to the inputs $f = f(t)$ and $g = g(t)$.

2.6 In Example 2.3,

(a) Find the differential equation relating $c_1(t)$ and the input $c(t)$.

(b) Find the step response in $c_1(t)$ for the same initial conditions, system parameters, and input $c(t)$. Graph the step response for $c_1(t)$ and $c_2(t)$.

(c) Show that the first derivative dc_1/dt is discontinuous at $t = 0$ while the first derivative dc_2/dt is continuous at $t = 0$.

2.7 The two-tank system in Exercise 1.2 is second order.
   (a) Convert the model of the system from two first-order differential equations to one second-order differential equation with input $F_1(t)$ and output $H_2(t)$.
   (b) Find expressions for the damping ratio, natural frequency, and steady-state gain in terms of the physical parameters A_1, A_2, R_1, and R_2.
   (c) Use the results from part (b) to express the damping ratio in terms of the tank time constants $\tau_1 = A_1 R_1$ and $\tau_2 = A_2 R_2$.
   (d) Show that the system can never be underdamped.
      For parts (e) and (f), assume the following values for the system parameters:

$$A_1 = 100 \text{ ft}^2, \quad R_1 = 0.25 \text{ ft per ft}^3/\text{min}, \quad A_2 = 50 \text{ ft}^2, \quad \text{and} \quad R^2 = 0.1 \text{ ft per ft}^3/\text{min}$$

   (e) Find and graph the response $H_2(t)$ of the unforced system, that is, $F_1(t) = 0$, $t \geq 0$ starting from $H_1(0) = 40$ ft and $H_2(0) = 0$ ft.
   (f) Find and graph the step response of $H_2(t)$ when $F_1(t) = 75 \text{ ft}^3/\text{min}$. Both tanks are initially empty. Does the first tank achieve steady state in roughly $5\tau_1$? Does the second tank achieve steady state in roughly $5\tau_2$? Explain.

2.8 A fundamental difference between the step response of first- and second-order linear systems in standard form is the initial rate of change, that is, the first derivative at $t = 0$.
   (a) Show that the first-order system step response undergoes the maximum rate of change at $t = 0$.
   (b) Show that the initial derivative of the second-order system step response is zero regardless of whether the system is underdamped, critically damped, or overdamped.

2.4 SIMULATION DIAGRAMS

In many cases, dynamic systems are composed of individual components and subsystems. The relationship of a system's components to each other and the role they serve in the overall system design are oftentimes easier to comprehend when presented in visual form rather than by inspection of the mathematical models. Control systems for ground vehicles, aircraft, robotic devices, building environments, and so forth are typically presented in graphical form as block diagrams. The blocks are both static and dynamic depending on the component it represents. Modern continuous-time system simulation languages include extensive libraries of special purpose blocks to represent the dynamics of commonly occurring components.

It is useful to reduce the blocks in a block diagram of a continuous-time dynamic system to a level that exposes the pure integrators. The simulationist is then given the flexibility of approximating individual integrators using different numerical algorithms. This is especially useful in applications where simulation code is developed manually instead of relying on a general purpose simulation language. This point will be revisited in Chapter 3 following a discussion of numerical integration.

A block diagram of a continuous-time dynamic system comprising algebraic blocks and integrators is referred to as a simulation diagram. We begin with the first-order system of Equation 2.61:

$$\frac{\mathrm{d}}{\mathrm{d}t}y(t) + a_0 y(t) = b_1 \frac{\mathrm{d}}{\mathrm{d}t}u(t) + b_0 u(t) \tag{2.61}$$

Equation 2.61 is a more general form than the first-order models introduced in Section 2.2 due to the presence of the first derivative term on the right-hand side.

If we introduce a new variable $z = z(t)$ where

$$\frac{\mathrm{d}}{\mathrm{d}t}z(t) + a_0 z(t) = u(t) \tag{2.62}$$

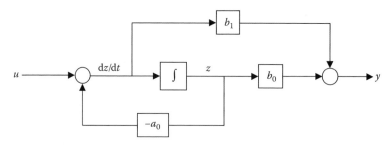

FIGURE 2.9   Simulation diagram of first-order system: $(d/dt)y(t) + a_0y(t) = b_1(d/dt)u(t) + b_0u(t)$.

the output y is related to z by

$$y(t) = b_0 z(t) + b_1 \frac{d}{dt} z(t) \tag{2.63}$$

It is left as an exercise to show that Equations 2.62 and 2.63 are equivalent to Equation 2.61. In addition to the blocks required to implement Equations 2.62 and 2.63, an integrator block is needed to integrate the first derivative dz/dt to generate $z(t)$, that is,

$$z(t) = \int \frac{dz}{dt} dt \tag{2.64}$$

The simulation diagram in Figure 2.9 is constructed by first drawing an integrator block and labeling the input dz/dt and output z corresponding to Equation 2.64. Next, we solve for the derivative term dz/dt in Equation 2.62 and draw a portion of the diagram to implement the result. Finally, the output y is generated from Equation 2.63 using the b_0 and b_1 gain blocks and a summing block.

The simulation diagram representation of the first-order system's dynamics involves a single dynamic block, namely, the integrator. The remaining blocks are sum blocks and gains that are algebraic in nature.

A block diagram for the same first-order system is shown in Figure 2.10. The block diagram is a direct implementation of Equation 2.61 after solving for the first derivative dy/dt. An additional variable z is not required in this case. The diagram in Figure 2.10 is not a simulation diagram because of the presence of the differentiator. In digital simulation, the differentiator (like the integrator) must be implemented using a numerical approximation. Numerical methods for approximating the derivative of a continuous-time function are available. However, they are rarely implemented in simulation applications due to their sensitivity to high-frequency noise components often present in continuous-time signals.

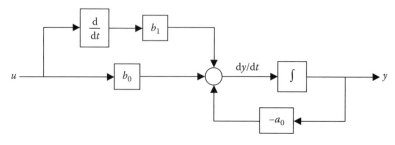

FIGURE 2.10   Block diagram of first-order system: $(d/dt)y(t) + a_0y(t) = b_1(d/dt)u(t) + b_0u(t)$.

A final observation relates to the special case when b_1 in Equation 2.61 is zero. The input derivative is absent, and the first-order system assumes the simpler form of Equation 2.3 or 2.4. Recall that this form was sufficient to model the dynamics of the linear tank in Chapter 1 and the simple RC circuit of Example 2.1.

Example 2.4

Draw a simulation diagram of the linear tank modeled by

$$A\frac{d}{dt}H(t) + \frac{1}{R}H(t) = F_1(t) \tag{2.65}$$

The diagram is shown in Figure 2.11.

Dividing Equation 2.65 by the parameter A and comparing the result to Equation 2.61 show

$$a_0 = \frac{1}{AR}, \quad b_0 = \frac{1}{A}, \quad b_1 = 0$$

leading to the simulation diagram shown in Figure 2.11.

The simulation diagram in Figure 2.11 is not unique. The "$1/A$" block can be moved from the location where z is its input to the left of the summer where F_1 becomes its input. In that case, z and H are identical. The alternate simulation diagram can be obtained directly by solving the differential equation of the tank for the first derivative,

$$\frac{d}{dt}H(t) = \frac{1}{A}\left[F_1(t) - \frac{1}{R}H(t)\right] \tag{2.66}$$

and implementing Equation 2.66 directly. Integrating the derivative dH/dt to get H completes the diagram.

Example 2.5

Suppose the current $i(t)$ in the RC circuit of Figure 2.2 is considered the output. The differential equation for the circuit becomes

$$\frac{d}{dt}i(t) + \frac{1}{RC}i(t) = \frac{1}{R}\frac{d}{dt}e_0(t) \tag{2.67}$$

Draw the simulation diagram for the circuit described by Equation 2.67.

From Equation 2.61, a_0, b_0, and b_1 are

$$a_0 = \frac{1}{RC}, \quad b_0 = 0, \quad b_1 = \frac{1}{R}$$

and the simulation diagram is drawn in Figure 2.12.

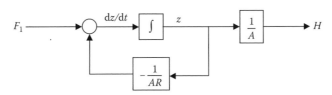

FIGURE 2.11   Simulation diagram of linear tank: $A(d/dt)H(t) + (1/R)H(t) = F_1(t)$.

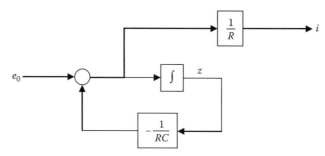

FIGURE 2.12   Simulation diagram for an RC circuit: $(d/dt)i(t) + (1/RC)i(t) = (1/R)(d/dt)e_0(t)$.

When the differential equation model of a first-order system contains a term involving the first derivative of the input, a direct link or coupling exists from the input directly to the output. In other words, when $b_1 \neq 0$ in Equation 2.61, sudden changes in the input are immediately reflected in the output. Notice the path of heavy solid lines in Figure 2.9 illustrating this point. A similar path in Figure 2.12 indicates the direct coupling from the applied voltage $e_0(t)$ to the output current $i(t)$.

In contrast, there is no direct connection from input to output in the simulation diagram shown in Figure 2.11 for the linear tank model. This is expected since changes in the inflow $F_1(t)$ must work their way through the tank dynamics, that is, the integrator, prior to affecting the output level $H(t)$. Hence, the tank prevents abrupt changes like a step or other inputs with high-frequency components from immediately causing any significant changes in the output $H(t)$. The tank behaves like a low-pass filter (see Exercise 1.10).

Obtaining a simulation diagram for a second-order system in the standard form

$$\frac{d^2}{dt^2}y(t) + 2\zeta\omega_n\frac{d}{dt}y(t) + \omega_n^2u(t) = K\omega_n^2u(t) \tag{2.68}$$

is straightforward. We begin by drawing two consecutive integrators, labeling the input and output of the first with d^2y/dt^2 and dy/dt, respectively. The second integrator integrates the first derivative dy/dt producing y and is labeled accordingly. The next step is to solve for the second derivative term in Equation 2.68 resulting in

$$\frac{d^2}{dt^2}y(t) = K\omega_n^2u(t) - 2\zeta\omega_n\frac{d}{dt}y(t) - \omega_n^2y(t) \tag{2.69}$$

Algebraic blocks (gains and summers) are used to implement Equation 2.69 leading to the simulation diagram shown in Figure 2.13.

The simulation diagram for a second-order system with first- or second-order derivatives of the input appearing in the differential equation model is not as straightforward. Starting with Equation 2.70

$$\frac{d^2}{dt^2}y(t) + a_1\frac{d}{dt}y(t) + a_0y(t) = b_0u(t) + b_1\frac{d}{dt}u(t) + b_2\frac{d^2}{dt^2}u(t) \tag{2.70}$$

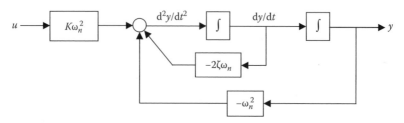

FIGURE 2.13   Simulation diagram of a second-order system in standard form.

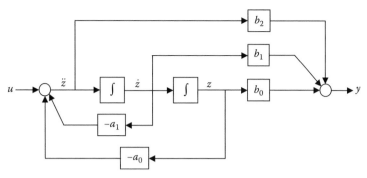

FIGURE 2.14 Simulation diagram for a second-order system with input derivatives present.

an approach similar to the method used for first-order systems with an input derivative term present is employed. An artificial variable $z(t)$ is introduced, and the output $y(t)$ is expressed as a linear combination of $z(t)$ and its two derivatives. The result is

$$\frac{d^2}{dt^2}z(t) + a_1\frac{d}{dt}z(t) + a_0z(t) = u(t) \tag{2.71}$$

$$y(t) = b_0z(t) + b_1\frac{d}{dt}z(t) + b_2\frac{d^2}{dt^2}z(t) \tag{2.72}$$

The simulation diagram of the second-order system in Equation 2.70 is shown in Figure 2.14. Note the use of the dot notation, short for differentiation with respect to time. It is clear that a direct connection from the input $u(t)$ to the output $y(t)$ exists only when b_2, the coefficient of the input second derivative in Equation 2.70, is nonzero.

Looking at the simulation diagrams in Figures 2.9 and 2.14 for the first- and second-order systems in Equations 2.61 and 2.70, a general pattern emerges for creating the simulation diagram of an nth-order system modeled by

$$\frac{dy^n}{dt^n} + a_{n-1}\frac{dy^{n-1}}{dt^{n-1}} + \cdots + a_1\frac{dy}{dt} + a_0y = b_0u + b_1\frac{du}{dt} + \cdots + b_{n-1}\frac{du^{n-1}}{dt^{n-1}} + b_n\frac{du^n}{dt^n} \tag{2.73}$$

The two equations equivalent to Equation 2.73 are

$$\frac{dz^n}{dt^n} + a_{n-1}\frac{dz^{n-1}}{dt^{n-1}} + \cdots + a_1\frac{dz}{dt} + a_0z = u \tag{2.74}$$

$$y = b_0z + b_1\frac{dz}{dt} + \cdots + b_{n-1}\frac{dz^{n-1}}{dt^{n-1}} + b_n\frac{dz^n}{dt^n} \tag{2.75}$$

The simulation diagram follows directly from Equations 2.74 and 2.75.

Example 2.6

A unicycle is traveling over an uneven road as shown in Figure 2.15.

The input is the road elevation $x_r(t)$ above some reference. The output is the vertical movement $x(t)$ of the rider and seat combination (with respect to its equilibrium position). Ignoring the compliance of the tire makes the wheel deflection $x_w(t) = x_r(t)$. Assume that the wheel remains

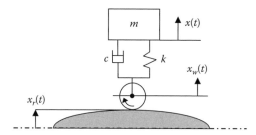

FIGURE 2.15 Unicycle traveling along an uneven road surface.

in contact with the road surface. The mass of the rider and seat is m, and c and k are suspension parameters.

(a) Find the differential equation relating the output $x(t)$ and input $x_r(t)$.
(b) Draw a simulation diagram of the system.
(c) Is there a direct coupling between the input and output? Explain.

(a) The differential equation is obtained by equating the sum of the suspension forces acting on the rider and seat to the product of its mass and acceleration.

$$m = \frac{d^2}{dt^2}x(t) = c\left[\frac{d}{dt}x_w(t) - \frac{d}{dt}x(t)\right] + k[x_w(t) - x(t)] \tag{2.76}$$

Replacing $x_w(t)$ with $x_r(t)$ gives

$$m\frac{d^2}{dt^2}x(t) = c\left[\frac{d}{dt}x_r(t) - \frac{d}{dt}x(t)\right] + k[x_r(t) - x(t)] \tag{2.77}$$

(b) Rearranging terms in Equation 2.77 gives

$$\frac{d^2}{dt^2}x(t) + \frac{c}{m}\frac{d}{dt}x(t) + \frac{k}{m}x(t) = \frac{k}{m}x_r(t) + \frac{c}{m}\frac{d}{dt}x_r(t) \tag{2.78}$$

Comparing Equations 2.78 and 2.70 leads to expressions for a_0, a_1, b_0, b_1, and b_2 in terms of the system parameters,

$$a_0 = \frac{k}{m}, \quad a_1 = \frac{c}{m}, \quad b_0 = \frac{k}{m}, \quad b_1 = \frac{c}{m}, \quad b_2 = 0 \tag{2.79}$$

and eventually the simulation diagram shown in Figure 2.16.

(c) Since both paths from x_r to x contain an integrator, there is no direct coupling between input and output. Consequently, an abrupt change in x_r such as a vertical jump in the road surface height does not result in a similar type of displacement of the rider and seat combination.

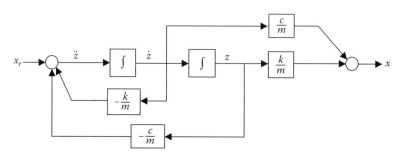

FIGURE 2.16 Simulation diagram for a unicycle suspension.

2.4.1 SYSTEMS OF EQUATIONS

System models can assume the form of coupled differential and algebraic equations. The simulation diagram representation is straightforward.

Example 2.7

A two-room building with temperatures $T_1(t)$ and $T_2(t)$ is shown in Figure 2.17.

The simplified model relating the uniform room temperatures $T_1(t)$ and $T_2(t)$ to the heat supplied from the furnace $Q_f(t)$ and outside temperature $T_0(t)$ is based on conservation of energy. It consists of the following differential and algebraic equations:

$$C_1 \frac{d}{dt} T_1(t) = Q_f(t) - Q_1(t) - Q_{12}(t) \tag{2.80}$$

$$C_2 \frac{d}{dt} T_2(t) = Q_{12}(t) - Q_2(t) \tag{2.81}$$

$$Q_{12}(t) = \frac{T_1(t) - T_2(t)}{R_{12}} \tag{2.82}$$

$$Q_1(t) = \frac{T_1(t) - T_0(t)}{R_1} \tag{2.83}$$

$$Q_2(t) = \frac{T_2(t) - T_0(t)}{R_2} \tag{2.84}$$

where C_1, C_2, R_1, R_2, and R_{12} are thermal parameters of the system. The simulation diagram shown in Figure 2.18 follows directly from Equations 2.80 through 2.84.

Combining Equations 2.80 through 2.84 and solving for the first derivatives give

$$\frac{d}{dt} T_1(t) = \frac{1}{C_1} \left[Q_f(t) - \frac{T_1(t) - T_0(t)}{R_1} - \frac{T_1(t) - T_2(t)}{R_{12}} \right] \tag{2.85}$$

$$= \frac{1}{C_1} \left[-\left(\frac{1}{R_1} + \frac{1}{R_{12}} \right) T_1(t) + \frac{1}{R_{12}} T_2(t) + \frac{1}{R_1} T_0(t) + Q_f(t) \right] \tag{2.86}$$

$$\frac{d}{dt} T_2(t) = \frac{1}{C_2} \left[\frac{T_1(t) - T_2(t)}{R_{12}} - \frac{T_2(t) - T_0(t)}{R_2} \right] \tag{2.87}$$

$$= \frac{1}{C_2} \left[\frac{1}{R_{12}} T_1(t) - \left(\frac{1}{R_2} + \frac{1}{R_{12}} \right) T_2(t) + \frac{1}{R_2} T_0(t) \right] \tag{2.88}$$

Equations 2.86 and 2.88 are of the form

$$\begin{aligned} \dot{x}_1 &= a_{11}x_1 + a_{12}x_2 + b_{11}u_1 + b_{12}u_2 \\ \dot{x}_2 &= a_{21}x_1 + a_{22}x_2 + b_{21}u_1 + b_{22}u_2 \end{aligned} \tag{2.89}$$

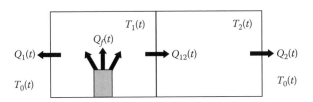

FIGURE 2.17 Heat flows and temperatures in a two-room building.

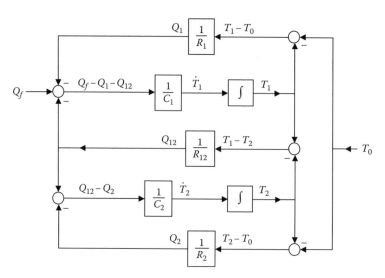

FIGURE 2.18 Simulation diagram for building room temperature model.

where $x_1 = T_1$, $x_2 = T_2$, $u_1 = T_0$, and $u_2 = Q_f$ and the coefficients a_{ij} and b_{ij} ($j = 1, 2$) depend on the system parameters according to

$$a_{11} = -\frac{1}{C_1}\left(\frac{1}{R_1} + \frac{1}{R_{12}}\right), \quad a_{12} = \frac{1}{R_{12}C_1}, \quad b_{11} = \frac{1}{R_1C_1}, \quad b_{12} = \frac{1}{C_1} \tag{2.90}$$

$$a_{21} = \frac{1}{R_{12}C_2}, \quad a_{22} = \frac{1}{C_2}\left(\frac{1}{R_2} + \frac{1}{R_{12}}\right), \quad b_{21} = \frac{1}{R_2C_2}, \quad b_{22} = 0 \tag{2.91}$$

Suppose we need to draw a simulation diagram for the system in Equation 2.89 with only x_1 or x_2 present. Using an approach similar to the one presented in Section 2.2 for converting two coupled first-order differential equations into a second-order differential equation, the second-order system in Equation 2.89 is equivalent to

$$\ddot{x}_1 + \alpha_1 \dot{x}_1 + \alpha_0 x_1 = \beta_{11}\dot{u}_1 + \beta_{10}u_1 + \beta_{21}\dot{u}_2 + \beta_{20}u_2 \tag{2.92}$$

where

$$\alpha_1 = -(a_{11} + a_{22}), \quad \alpha_0 = a_{11}a_{22} - a_{12}a_{21} \tag{2.93}$$

$$\beta_{11} = b_{11}, \quad \beta_{10} = a_{12}b_{21} - a_{22}b_{11}, \quad \beta_{21} = b_{12}, \quad \beta_{20} = a_{12}b_{22} - a_{22}b_{12} \tag{2.94}$$

The simulation diagram for Equation 2.92 is constructed in two steps. From superposition, the output x_1 can be viewed as the sum of x_{11} and x_{12} where

$$\ddot{x}_{11} + \alpha_1 \dot{x}_{11} + \alpha_0 x_{11} = \beta_{11}\dot{u}_1 + \beta_{10}u_1 \tag{2.95}$$

$$\ddot{x}_{12} + \alpha_1 \dot{x}_{12} + \alpha_0 x_{12} = \beta_{21}\dot{u}_2 + \beta_{20}u_2 \tag{2.96}$$

Simulation diagrams for Equations 2.95 and 2.96 are drawn separately, and outputs x_{11} and x_{12} are added to yield the complete output x_1. The result is shown in Figure 2.19.

Do not be misled into thinking that the simulation diagram shown in Figure 2.19 corresponds to a fourth-order system due to the presence of four integrators. There are two decoupled second-order systems, one with input u_1 and output x_{11} and the other with input u_2 and output x_{12}.

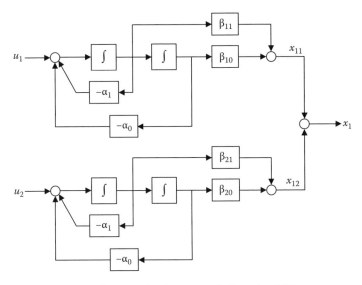

FIGURE 2.19 Simulation diagram for second-order system in Equation 2.92.

In reality, they are the same system, that is, the second-order system governed by the second-order model in Equation 2.92.

On the other hand, if the feedback coefficients in the two systems are not identical, that is, α_0 and α_1 in both cases, the result is indeed a fourth-order system (see Exercise 2.13).

EXERCISES

2.9 Show that the system of equations

$$\frac{d}{dt}z(t) + a_0 z(t) = u(t) \quad \text{and} \quad y(t) = b_0 z(t) + b_1 \frac{d}{dt}z(t)$$

used to construct the simulation diagram for the first-order system

$$\frac{d}{dt}y(t) + a_0 y(t) = b_1 \frac{d}{dt}u(t) + b_0 u(t)$$

is equivalent to the first-order differential equation above.

Hint: The variable $z(t)$ must be eliminated from the two equations.

2.10 An alternate simulation diagram for the second-order system

$$\frac{d}{dt^2}y(t) + 2\zeta\omega_n \frac{d}{dt}y(t) + \omega_n^2 y(t) = K\omega_n^2 u(t)$$

when it is critically damped or overdamped is shown in Figure E2.10:

Find expressions for K_1, α, and β in terms of the parameters ζ, ω_n, and K.

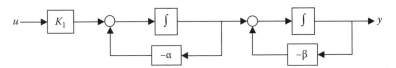

FIGURE E2.10

2.11 The circuit shown in Figure E2.11 is governed by the differential equation:

$$RC \frac{d^2}{dt^2} v_C + \frac{d}{dt} v_C + \frac{R}{L} v_C = \frac{d}{dt} e_S$$

Draw a simulation diagram for the circuit.

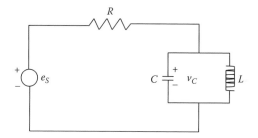

FIGURE E2.11

2.12 Consider the building temperature example with room temperatures described by Equations 2.86 and 2.88.
  (a) Find the second-order differential equation relating $T_2(t)$ and the system inputs $T_0(t)$ and $Q_f(t)$.
  (b) Draw a simulation diagram like the one shown in Figure 2.19.
2.13 Simulation diagrams are shown in Figure E2.13a through c.
  (a) Find the differential equation relating x and inputs u_1 and u_2 in Figure E2.13a.
  (b) Find the differential equation relating x and input u in Figure E2.13b.
  (c) Find the differential equation relating x and inputs u_1 and u_2 in Figure E2.13c.
  (d) Comment on the differences between the systems represented by each diagram.

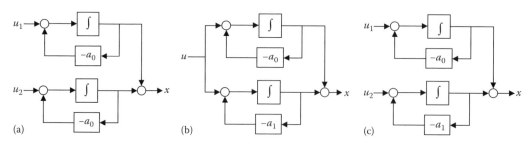

FIGURE E2.13

2.5 HIGHER-ORDER SYSTEMS

To this point, we have looked at linear continuous-time systems with first- and second-order dynamics only. Linear systems and linear controls texts include extensive coverage of lower-order system response. In particular, the response of first- and second-order systems to impulse, step, and sinusoidal inputs is fully developed.

The dynamics of complex systems with linear differential equation models are invariably higher than second order. One may question why so much attention is devoted to first- and second-order systems. The explanation is simple.

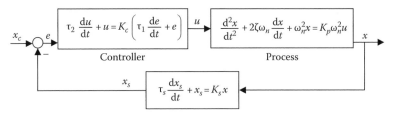

FIGURE 2.20 A control system consisting of first- and second-order components.

High-order linear systems are oftentimes a collection of components or subsystems that are intrinsically first or second order. An electrical circuit with several capacitors and inductors is a good example. The circuit dynamics will depend on the number of these energy storage elements and their location in the circuit. In general, its order will be equal to the number of energy storage elements since each element is itself modeled as a first-order component. With n nonredundant energy storage elements, an nth-order differential equation involving an output (a voltage or current in the circuit) and an input (if an independent source is present) governs the behavior of the circuit. The same principle applies to fluid, thermal, mechanical, chemical, and so forth, systems made up of components analogous to the resistor, capacitor, and inductor of the electrical circuit.

The block diagram of a simple feedback control system is shown in Figure 2.20. The controller, process, and sensor are the subsystem components, which are individually modeled as either first or second order.

The control system model comprises the three coupled differential equations

$$\tau_2 \frac{du}{dt} + u = K_c \left(\tau_1 \frac{de}{dt} + e \right) \tag{2.97}$$

$$\frac{d^2x}{dt^2} + 2\zeta\omega_n \frac{dx}{dt} + \omega_n^2 x = K_p \omega_n^2 u \tag{2.98}$$

$$\tau_s \frac{dx_s}{dt} + x_s = K_s x \tag{2.99}$$

and the summer equation

$$e = x_c - x_s \tag{2.100}$$

The command input $x_c = x_c(t)$ is the control system input, and the output of the process $x = x(t)$ is the control system output. Dependent variables $e(t)$, the error signal, $u(t)$, the output from the controller and input to the process, and $x_s(t)$, the sensor output are internal to the control system. Eliminating these variables produces a single fourth-order $(1 + 2 + 1)$ differential equation model of the control system in the form

$$\frac{d^4x}{dt^4} + a_3 \frac{d^3x}{dt^3} + a_2 \frac{d^2x}{dt^2} + a_1 \frac{dx}{dt} + a_0x = b_4 \frac{d^4x_c}{dt^4} + b_3 \frac{d^3x_c}{dt^3} + b_2 \frac{d^2x_c}{dt^2} + b_1 \frac{dx_c}{dt} + b_0x_c \tag{2.101}$$

where several of the coefficients a_i, $i = 0, 1, 2, 3$ and b_i, $i = 0, 1, 2, 3, 4$ may be zero.

A simulation diagram of the control system can be obtained from Equation 2.101 using the procedure from the previous section. Alternatively, simulation diagrams can be developed for the individual components in Figure 2.20 and properly connected to produce a simulation diagram for the control system. Simulation of the system based on a simulation diagram using the second approach is preferable since the internal variables are readily identifiable. We can check the simulation results to verify that inputs and outputs of the controller and sensor remain within proper operating ranges.

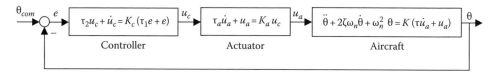

FIGURE 2.21   Control system for an aircraft pitch.

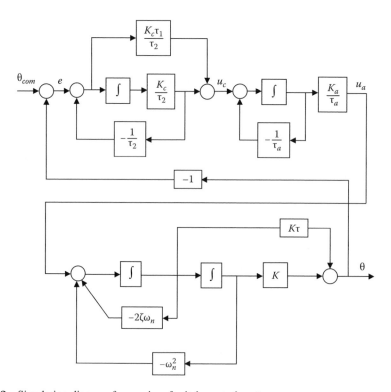

FIGURE 2.22   Simulation diagram for an aircraft pitch control system.

Example 2.8

The control system for the pitch of an aircraft is shown in Figure 2.21.

Draw a simulation diagram for the aircraft pitch control system block diagram.

Simulation diagrams of each component are connected to produce the simulation diagram of the entire control system shown in Figure 2.22.

EXERCISES

2.14 For the control system shown in Figure 2.20.

(a) Find the coefficients a_i, $i = 0, 1, 2, 3$ and b_i, $i = 0, 1, 2, 3, 4$ in Equation 2.101 in terms of the system parameters τ_1, τ_2, K_c, ζ, ω_n, K_p, τ_s and K_s.

Hint: The use of Laplace transforms (see Chapter 4) significantly reduces the amount of work necessary to eliminate the variables e, u, and x_s.

(b) Draw a simulation diagram based on the fourth-order differential equation model.

2.15 Find the differential equation for the control system in Figure 2.21 relating the output θ and its derivatives to the input θ_{com} and its derivatives. Draw the simulation diagram based on the resulting differential equation.

Hint: The use of Laplace transforms (see Chapter 4) significantly reduces the amount of work necessary to eliminate the variables e, u_c, and u_a.

2.16 For the railroad cars shown in Figure E2.16,
   (a) Write the differential equation expressing $\sum_k F_{i,k} = m_i \ddot{x}_i$, $i = 1, 2, 3$ for each car where $F_{i,k}$ is the kth force acting on the ith car.
   (b) Draw a simulation diagram of the system with integrators for x_i, $\dot{x}_i$, $i = 1, 2, 3$.
   (c) Find the differential equation relating the input $F(t)$ and output $x_1(t)$.

Hint: The use of Laplace transforms (see Chapter 4) significantly reduces the amount of work necessary to eliminate the variables x_2 and x_3.

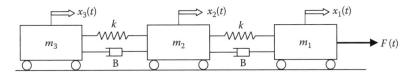

FIGURE E2.16

2.6 STATE VARIABLES

In everyday terms, one's state of mind on a given day is determined by the history of numerous psychological factors that influence our mental well-being. The state of the national economy (weak, moderate, strong) depends on numerous factors such as energy prices, inflation, trade balances, employment, productivity, housing, tax policies, corporate earnings, transportation, agriculture, and so forth. Imagine that all the economic factors (inputs) affecting the national economy were measurable and the complex interrelationships among those variables that determine the state of the economy were fully understood. If the state of the economy were known at some point in time and the complete set of aforementioned economic factors were observed from that time forward, knowledgeable economists would (in principle) be able to predict the state of the national economy at future times.

The essential point is that if we know the state of a system at some point in time and wish to predict its future, then knowledge of the system inputs only from that time onward is required. The current state of a system reflects the effect of prior inputs that are responsible for the system's transition from some previous state to the current state.

Consider a simple spring-mass-damper system subject to an applied force acting on the mass like the one shown in Figure 2.23. The spring and mass are both capable of storing energy. At any time,

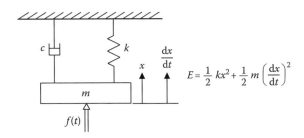

FIGURE 2.23   A spring-mass-damper system with applied force $f(t)$.

the instantaneous energy $E(t)$ stored in the system is given in Equation 2.102 where x is the position of the mass (relative to its equilibrium position) and dx/dt is the velocity of the mass.

$$E = \frac{1}{2}kx^2 + \frac{1}{2}m\left(\frac{dx}{dt}\right)^2 \tag{2.102}$$

A possible choice of state variables for the mechanical system is x and dx/dt. Given both state variables at time t_0 determines the energy $E(t_0)$. The applied force $f(t)$ for $t \geq t_0$ must be known to solve the initial value problem

$$m\frac{d^2x}{dt^2}(t) + c\frac{dx}{dt}(t) + kx(t) = f(t) \text{ given } x(t_0) \text{ and } \frac{dx}{dt}(t_0) \tag{2.103}$$

and determine both state variables x and dx/dt as well as $E(t)$ for $t \geq t_0$. The same cannot be said if only the position or the velocity of the mass were known at t_0. In that case, the initial energy in the system $E(t_0)$ would be unknown, and it would be impossible to predict future values of x and dx/dt even if the force $f(t)$ were known for $t \geq t_0$. Consequently, x or dx/dt alone is not a suitable choice for the state of the system.

The situation is illustrated for the general case of a system with two state variables $x_1(t)$ and $x_2(t)$ and single input $u(t)$ in Figure 2.24. Given $x_1(t_0)$, $x_2(t_0)$, and $u(t)$, $t \geq t_0$, both states can be determined from t_0 on.

The choice of state variables for a dynamic system model is not unique; however, the number of state variables is limited to the minimum number of variables, which satisfy the requirement of predicting future states given the current state and future inputs. This number of state variables is equal to the number of independent energy storage components present in the system. It is advantageous to choose physical (measurable) quantities as in the case of the mechanical system in Figure 2.23 whenever possible.

A simulation diagram is a valuable tool when it comes to choosing the state variables of a system. The outputs of each integrator in a simulation diagram representation of a system is a valid choice for the state variables. The choice of which integrator output is x_1, x_2, and so forth is arbitrary.

Consider a second-order system governed by

$$\frac{d^2}{dt^2}y(t) + a_1\frac{d}{dt}y(t) + a_0y(t) = b_0u(t) \tag{2.104}$$

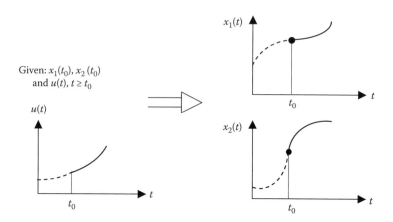

FIGURE 2.24   Dynamic system with state variables $x_1(t)$ and $x_2(t)$.

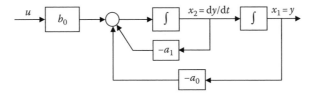

FIGURE 2.25 Simulation diagram of second-order system with state $x_1 = y$ and $x_2 = dy/dt$.

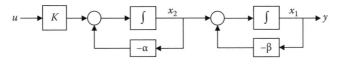

FIGURE 2.26 Simulation diagram for critically damped or overdamped second-order system using two first-order systems in a series.

A simulation diagram like the one shown in Figure 2.25 is easily constructed. State variables x_1 and x_2 are chosen as the output y and first derivative dy/dt, respectively.

The second-order system is critically damped or overdamped if $a_1^2 - 4a_0 \geq 0$. In this case, it is equivalent to two cascaded first-order systems as shown in Figure 2.26.

The parameters K, α, and β are related to a_0, a_1, and b_0 according to

$$K = b_0, \quad \alpha = \frac{a_1 \pm \sqrt{a_1^2 - 4a_0}}{2}, \quad \beta = \frac{a_1 \mp \sqrt{a_1^2 - 4a_0}}{2} \qquad (2.105)$$

State variable x_1 is again the system output y; however, the second state variable x_2 is no longer the output derivative dy/dt.

For an nth-order linear system model with constant coefficients, the state derivatives are expressible as a linear combination of the state variables and input(s). For example, from Figure 2.25, the state derivatives are equal to

$$\frac{dx_1}{dt} = x_2$$
$$\frac{dx_2}{dt} = b_0 u - a_0 x_1 - a_1 x_2 \qquad (2.106)$$

whereas in Figure 2.26, the appropriate expressions are

$$\frac{dx_1}{dt} = x_2 - \beta x_1$$
$$\frac{dx_2}{dt} = K u - \alpha x_2 \qquad (2.107)$$

In the general linear case with n states $x_1, x_2, \ldots, x_n$ and r inputs,

$$\frac{dx_1}{dt} = f_1(\underline{x}, \underline{u}) = a_{11} x_1 + a_{12} x_2 + \cdots + a_{1n} x_n + b_{11} u_1 + b_{12} u_2 + \cdots + b_{1r} u_r \qquad (2.108)$$

$$\frac{dx_2}{dt} = f_2(\underline{x}, \underline{u}) = a_{21} x_1 + a_{22} x_2 + \cdots + a_{2n} x_n + b_{21} u_1 + b_{22} u_2 + \cdots + b_{2r} u_r \qquad (2.109)$$

$$\frac{dx_n}{dt} = f_n(\underline{x}, \underline{u}) = a_{n1} x_1 + a_{n2} x_2 + \cdots + a_{nn} x_n + b_{n1} u_1 + b_{n2} u_2 + \cdots + b_{nr} u_r \qquad (2.110)$$

where

$$\underline{x} \text{ is the } n \times 1 \text{ state vector} \begin{bmatrix} x_1 \\ x_2 \\ \vdots \\ x_n \end{bmatrix}$$

$$\underline{u} \text{ is the } r \times 1 \text{ input vector} \begin{bmatrix} u_1 \\ u_2 \\ \vdots \\ u_r \end{bmatrix}$$

and $f_i(\underline{x}, \underline{u})$, $i = 1, 2, 3, \ldots, n$ is the state derivative function of the ith state variable.

Equations 2.108 through 2.110 can be written in the compact form

$$\dot{\underline{x}} = \underline{f}(\underline{x}, \underline{u}) = A\underline{x} + B\underline{u} \tag{2.111}$$

where

$$\dot{\underline{x}} = \begin{bmatrix} \dfrac{dx_1}{dt} \\ \dfrac{dx_2}{dt} \\ \vdots \\ \dfrac{dx_n}{dt} \end{bmatrix}, \quad A = \begin{bmatrix} a_{11} & a_{12} & \cdot & \cdot & a_{1n} \\ a_{21} & a_{22} & \cdot & \cdot & a_{2n} \\ \cdot & \cdot & \cdot & \cdot & \cdot \\ \cdot & \cdot & \cdot & \cdot & \cdot \\ \cdot & \cdot & \cdot & \cdot & \cdot \\ a_{n1} & a_{n2} & \cdot & \cdot & a_{nn} \end{bmatrix}, \quad B = \begin{bmatrix} b_{11} & b_{12} & \cdot & \cdot & b_{1r} \\ b_{21} & b_{22} & \cdot & \cdot & b_{2r} \\ \cdot & \cdot & \cdot & \cdot & \cdot \\ \cdot & \cdot & \cdot & \cdot & \cdot \\ \cdot & \cdot & \cdot & \cdot & \cdot \\ b_{n1} & b_{n2} & \cdot & \cdot & b_{nr} \end{bmatrix}$$

The $n \times n$ matrix A is called the system matrix, and the $n \times r$ matrix B is the input matrix.

Multivariable, LTI systems involve multiple inputs $u_1, u_2, \ldots, u_r$ and outputs $y_1, y_2, \ldots, y_p$. The outputs are linearly related to the states and the inputs according to

$$\underline{y} = C\underline{x} + D\underline{u} \tag{2.112}$$

where

$$\underline{y} = \begin{bmatrix} y_1 \\ y_2 \\ \vdots \\ y_p \end{bmatrix}, \quad C = \begin{bmatrix} c_{11} & c_{12} & \cdot & \cdot & c_{1n} \\ c_{21} & c_{22} & \cdot & \cdot & c_{2n} \\ \cdot & \cdot & \cdot & \cdot & \cdot \\ \cdot & \cdot & \cdot & \cdot & \cdot \\ \cdot & \cdot & \cdot & \cdot & \cdot \\ c_{p1} & c_{p2} & \cdot & \cdot & c_{pn} \end{bmatrix}, \quad D = \begin{bmatrix} d_{11} & d_{12} & \cdot & \cdot & d_{1r} \\ d_{21} & d_{22} & \cdot & \cdot & d_{2r} \\ \cdot & \cdot & \cdot & \cdot & \cdot \\ \cdot & \cdot & \cdot & \cdot & \cdot \\ \cdot & \cdot & \cdot & \cdot & \cdot \\ d_{p1} & d_{p2} & \cdot & \cdot & d_{pr} \end{bmatrix}$$

The $p \times n$ constant matrix C is called the output matrix, and the $p \times r$ matrix D is the direct transmission matrix.

Equations 2.111 and 2.112 taken together are the state equations of the system. Note that the states $x_1, x_2, \ldots, x_n$ are internal to the system as shown in Figure 2.27. Multivariable systems are easier to analyze in terms of state variables compared to the input–output model description of the system, that is, $dy_i/dt = f_i(\underline{y}, \underline{u})$, $i = 1, 2, \ldots, n$.

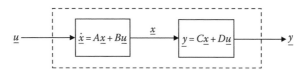

FIGURE 2.27 Dynamic system with input $\underline{u}$, output $\underline{y}$, and state $\underline{x}$.

Example 2.9

Interacting tanks with inflows into both tanks are shown in Figure 2.28. Choose the states to be the levels $H_1 = H_1(t)$ and $H_2 = H_2(t)$ and the single output as the volume of liquid in both tanks. Write the state equations for the system.

The continuous-time model of the linear tanks consists of the following equations:

$$A_1 \frac{dH_1}{dt} + F_{0,1} = F_1 \tag{2.113}$$

$$F_{0,1} = \frac{1}{R_{12}}(H_1 - H_2) \tag{2.114}$$

$$A_2 \frac{dH_2}{dt} + F_{0,2} = F_{0,1} + F_2 \tag{2.115}$$

$$F_{0,2} = \frac{1}{R_2}H_2 \tag{2.116}$$

Eliminating $F_{0,1}$ and $F_{0,2}$ from Equations 2.113 and 2.115 yields

$$A_1 \frac{dH_1}{dt} + \frac{1}{R_2}(H_2 - H_2) = F_1 \tag{2.117}$$

$$A_2 \frac{dH_2}{dt} + \frac{1}{R_2}H_2 = \frac{1}{R_{12}}(H_1 - H_2) + F_2 \tag{2.118}$$

Solving for the state derivatives in Equations 2.117 and 2.118

$$\frac{dH_1}{dt} = -\frac{1}{A_1 R_{12}}H_1 + \frac{1}{A_1 R_{12}}H_2 + \frac{1}{A_1}F_1 \tag{2.119}$$

$$\frac{dH_2}{dt} = -\frac{1}{A_2 R_{12}}H_1 - \left[\frac{1}{A_2 R_2} + \frac{1}{A_2 R_{12}}\right]H_2 + \frac{1}{A_2}F_2 \tag{2.120}$$

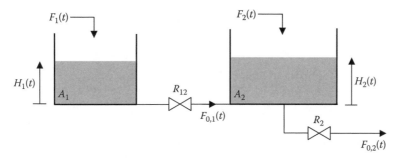

FIGURE 2.28 A system of interacting tanks.

Writing Equations 2.119 and 2.120 in matrix form gives the first part of the state equations,

$$
\begin{bmatrix} \dfrac{dH_1}{dt} \\ \dfrac{dH_2}{dt} \end{bmatrix} = \begin{bmatrix} -\dfrac{1}{A_1 R_{12}} & \dfrac{1}{A_1 R_{12}} \\ \dfrac{1}{A_2 R_{12}} & -\dfrac{1}{A_2 R_2} - \dfrac{1}{A_2 R_{12}} \end{bmatrix} \begin{bmatrix} H_1 \\ H_2 \end{bmatrix} + \begin{bmatrix} \dfrac{1}{A_1} & 0 \\ 0 & \dfrac{1}{A_2} \end{bmatrix} \begin{bmatrix} F_1 \\ F_2 \end{bmatrix}
\tag{2.121}
$$

The single output V_T, which represents the volume of liquid in both tanks, is

$$
V_T = A_1 H_1 + A_2 H_2 = [A_1 A_2] \begin{bmatrix} H_1 \\ H_2 \end{bmatrix}
\tag{2.122}
$$

The transmission matrix D is a 1×2 matrix of zeros due to the absence of a direct coupling from either input F_1 or F_2 to the output V_T.

2.6.1 Conversion from Linear State Variable Form to Single Input–Single Output Form

In Section 2.3, an example was presented illustrating the conversion of a second-order state variable model into a second-order differential equation by eliminating one of the state variables (see Equations 2.37, 2.38, and 2.42). The procedure involved manipulation and substitution of terms in the time domain, an approach that quickly becomes unwieldy as the number of state variables increases. Simpler methods are described in Chapter 4.

For a linear, third-order system with a single input, the starting point is the state variable model consisting of three coupled first-order differential equations expressing the state derivatives as a linear function of the states and input

$$
\begin{aligned}
\dot{x}_1 &= a_{11}x_1 + a_{12}x_2 + a_{13}x_3 + b_1 u \\
\dot{x}_2 &= a_{21}x_1 + a_{22}x_2 + a_{23}x_3 + b_2 u \\
\dot{x}_3 &= a_{31}x_1 + a_{32}x_2 + a_{33}x_3 + b_3 u
\end{aligned}
\tag{2.123}
$$

where the output y is x_1, x_2, or x_3.

A third order, input–output differential equation model equivalent to Equation 2.123 is

$$
\dddot{y} + \alpha_2 \ddot{y} + \alpha_1 \dot{y} + \alpha_0 y = \beta_2 \ddot{u} + \beta_1 \dot{u} + \beta_0 u
\tag{2.124}
$$

Expressions for the system coefficients α_2, α_1, and α_0 and input coefficients β_2, β_1, and β_0 are summarized in Equations 2.125 through 2.127 and Table 2.1.

TABLE 2.1
Input Coefficients on Right-Hand Side of Equation 2.125 for $y = x_1, x_2, x_3$

y	β_2	β_1	β_0
x_1	b_1	$-(a_{22}+a_{33})b_1 + (a_{12}b_2 + a_{13}b_3)$	$(a_{22}a_{33} - a_{23}a_{32})b_1 + (a_{13}a_{32} - a_{12}a_{33})b_2 + (a_{12}a_{23} - a_{13}a_{22})b_3$
x_2	b_2	$a_{21}b_1 - (a_{11}+a_{33})b_2 + a_{23}b_3$	$(a_{23}a_{31} - a_{21}a_{33})b_1 + (a_{11}a_{33} - a_{13}a_{31})b_2 + (a_{13}a_{21} - a_{11}a_{23})b_3$
x_3	b_3	$a_{31}b_1 + a_{32}b_2 - (a_{11}+a_{22})b_3$	$(a_{21}a_{32} - a_{22}a_{31})b_1 + (a_{12}a_{31} - a_{11}a_{32})b_2 + (a_{11}a_{22} - a_{12}a_{21})b_3$

$$\alpha_2 = -(a_{11} + a_{22} + a_{33}) \tag{2.125}$$

$$\alpha_1 = a_{11}(a_{22} + a_{33}) - a_{12}a_{21} - a_{13}a_{31} + a_{22}a_{33} - a_{23}a_{32} \tag{2.126}$$

$$\alpha_0 = a_{11}(a_{23}a_{32} - a_{22}a_{33}) + a_{12}(a_{21}a_{33} - a_{23}a_{31}) + a_{13}(a_{22}a_{31} - a_{21}a_{32}) \tag{2.127}$$

2.6.2 GENERAL SOLUTION OF THE STATE EQUATIONS

A solution to the state equation, Equation 2.111 can be found in any one of the texts on linear control theory listed in References. The solution is expressed in terms of an $n \times n$ matrix $\Phi(t)$, called the transition matrix of the system.

$$x(t) = \Phi(t)x(0) + \int_0^t \Phi(t - \tau)Bu(\tau) - d\tau \tag{2.128}$$

The transition matrix depends solely on the system matrix A. One method for finding $\Phi(t)$ uses a definition based on an infinite series,

$$\Phi(t) = I + (tA) + \frac{1}{2!}(tA)^2 + \frac{1}{3!}(tA)^3 + \cdots \tag{2.129}$$

As an illustration of how the transition matrix is used to solve the linear state equations, suppose the system matrix for an autonomous system ($u = 0$) is

$$A = \begin{bmatrix} 0 & 1 \\ -2 & -3 \end{bmatrix}$$

Using the infinite series expansion in Equation 2.129 or some other method (see Chapter 4) for finding $\Phi(t)$, the result is

$$\Phi(t) = \begin{bmatrix} 2e^{-t} - e^{-2t} & e^{-t} - e^{-2t} \\ -2e^{-t} + 2e^{-2t} & -e^{-t} + 2e^{-2t} \end{bmatrix} \tag{2.130}$$

and from Equation 2.128, the state $x(t)$, $t \geq 0$ is

$$\begin{bmatrix} x_1(t) \\ x_2(t) \end{bmatrix} = \begin{bmatrix} 2e^{-t} - e^{-2t} & e^{-t} - e^{-2t} \\ -2e^{-t} + 2e^{-2t} & -e^{-t} + 2e^{-2t} \end{bmatrix} \begin{bmatrix} x_1(0) \\ x_2(0) \end{bmatrix} \tag{2.131}$$

The state trajectory or state portrait is a plot showing the path of the state vector in state space. In the general case, there is a separate coordinate axis for each of the state variables. The time variable "t" does not appear explicitly; however, each point along the state trajectory corresponds to a specific point in time. Figure 2.29 shows four different state trajectories starting from different initial states. Note that the four state trajectories all terminate at the origin, the equilibrium point of the system.

EXERCISES

2.17 For the system of interacting tanks in Example 2.9.
   (a) Draw the simulation diagram of the system.
   (b) Choose a new set of state variables as

$$z_1 = H_1 + H_2, \quad z_2 = H_1 - H_2$$

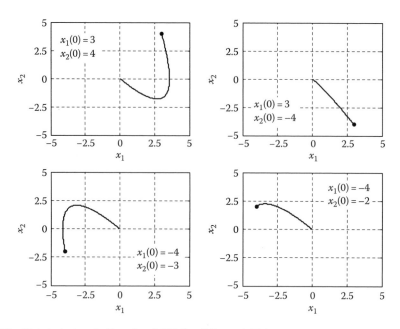

FIGURE 2.29 State trajectory in Equation 2.131 for different initial states.

and find the new system and input matrices A and B where

$$\begin{bmatrix} \dot{z}_1 \\ \dot{z}_2 \end{bmatrix} = A \begin{bmatrix} z_1 \\ z_2 \end{bmatrix} + B \begin{bmatrix} F_1 \\ F_2 \end{bmatrix}$$

Hint: Find H_1 and H_2 in terms of z_1 and z_2.

(c) Find the new output matrix C where

$$V_T = C \begin{bmatrix} z_1 \\ z_2 \end{bmatrix}$$

2.18 Write the state equations for the system of three railroad cars in Exercise 2.16. Choose the outputs to be the positions of each car.

2.19 An ecosystem consists of three species whose populations are denoted by F, S, and G. The growth rates of each specie are given by

$$\text{Growth rate of } F = \frac{1}{F}\frac{dF}{dt} = a - cS - u_F$$

$$\text{Growth rate of } S = \frac{1}{S}\frac{dS}{dt} = -k + \lambda F - m_G - u_S$$

$$\text{Growth rate of } G = \frac{1}{G}\frac{dG}{dt} = -e + \sigma S + u_G$$

Write the system in state variable form $\dot{x} = \underline{f}(\underline{x}, \underline{u})$ $y = g(\underline{x}, \underline{u})$ with the state $\underline{x} = [F\ S\ G]^T$, input $\underline{u} = [u_F\ u_S\ u_G]^T$, and output chosen as $y = F + S + G$.

2.20 Limestone is reduced to calcium oxide (CaO), magnesium oxide (MgO), and carbon dioxide (CO_2) by heating it in a reaction vessel maintained at a constant high temperature (McClamroch 1980). The limestone is made up of a fixed fraction β of calcium carbonate

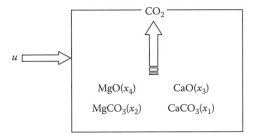

FIGURE E2.20

($CaCO_3$), and the rest is magnesium carbonate ($MgCO_3$). The process is described by the first-order irreversible chemical reactions

$$CaCO_3 \xrightarrow{k_1} CaO + CO_2$$
$$MgCO_3 \xrightarrow{k_2} MgO + CO_2$$

where k_1 and k_2 are the rate constants for the two reactions.

Limestone is added to the reaction vessel at a rate of u mol/h. The mass (in moles) of $CaCO_3$, $MgCO_3$, CaO, and MgO in the vessel are denoted by x_1, x_2, x_3, and x_4, respectively (see Figure E2.20).

Since each mole of reactant that decomposes yields one mole of product (plus one mole of carbon dioxide), the state equations are

$$\dot{x}_1 = -k_1 x_1 + \beta u$$
$$\dot{x}_2 = -k_2 x_2 + (1 - \beta)u$$
$$\dot{x}_3 = k_1 x_1$$
$$\dot{x}_4 = k_2 x_2$$

(a) Draw a simulation diagram of the system. What is the order of the system?
(b) Find the matrices A, B, C, and D in the state equation model if the outputs are $y_1 = x_3$ and $y_2 = x_4$.
(c) Find the differential equation relating y_1 and u. Comment on the result.
(d) Repeat part (c) for y_2 and u.
(e) The vessel is initially empty and $u(t) = A$, $t \geq 0$. Find analytic expressions for the state variables.

2.21 The populations of three species in a restricted area are governed by the differential equations

$$\dot{P}_1(t) = a_{11}P_1(t) + a_{12}P_2(t) + a_{13}P_3(t) + c_1 u(t)$$
$$\dot{P}_2(t) = a_{21}P_1(t) + a_{22}P_2(t) + a_{23}P_3(t) + c_2 u(t)$$
$$\dot{P}_3(t) = a_{31}P_1(t) + a_{32}P_2(t) + a_{33}P_3(t) + c_3 u(t)$$
$$0 \leq c_1 \leq 1, \quad 0 \leq c_2 \leq 1, \quad 0 \leq c_3 \leq 1, \quad \text{and} \quad c_1 + c_2 + c_3 = 1$$

where $u(t)$ is the total immigration rate for all species. The constants c_1, c_2, and c_3 represent the fraction of $u(t)$ immigrating to each of the species populations.

(a) Draw a simulation diagram of the system.
(b) Find the third-order differential equation relating $P_1(t)$ and $u(t)$.
(c) Draw a simulation diagram of the system containing three integrators in series where the input to the first integrator is $\dddot{P}_1(t)$

2.7 NONLINEAR SYSTEMS

Real-world dynamic systems exhibit nonlinear behavior. The continuous-time models that relate inputs and outputs of actual systems are (entirely or partially) composed of nonlinear algebraic and differential equations. We may well choose to employ a linear model as an approximation of a nonlinear system because it is far simpler to work with. A unified approach to solving nonlinear algebraic equations does not exist, to say nothing of nonlinear differential equations.

The principle of superposition states that if a system responds to inputs $u_1(t)$ and $u_2(t)$ with outputs $y_1(t)$ and $y_2(t)$, then the system's response to a linear combination of the inputs $u(t) = c_1 u_1(t) + c_2 u_2(t)$ is $y(t) = c_1 y_1(t) + c_2 y_2(t)$. Superposition is a property of linear system models. It is not applicable to models of nonlinear systems.

Unlike linear system models, a nonlinear system model exhibits dynamic response properties whose nature is dependent on the magnitude of its inputs and the initial state. Consider the two simple first-order systems, one linear and the other nonlinear, in Figure 2.30. Both systems are driven by the identical input.

Discrete-time system approximations for both continuous-time systems can be obtained by replacing the first derivative terms with divided differences, that is,

$$\frac{dy}{dt} \approx \frac{y_A[(n+1)T] - y_A[(nT)]}{(n+1)T - nT} = \frac{y_A(n+1) - y_A(n)}{T} \tag{2.132}$$

$$\frac{dz}{dt} \approx \frac{z_A[(n+1)T] - z_A[(nT)]}{(n+1)T - nT} = \frac{z_A(n+1) - z_A(n)}{T} \tag{2.133}$$

resulting in difference equations

$$y_A(n+1) = y_A(n) + T[u(n) - y_A(n)] \tag{2.134}$$

$$z_A(n+1) = z_A(n) + T[u(n) - z_A^2(n)] \tag{2.135}$$

Equations 2.134 and 2.135 can be solved recursively for $y_A(n)$ and $z_A(n)$, $n = 1, 2, 3, \ldots$ given initial values for $y_A(0)$ and $z_A(0)$. The results (every third point) are plotted in Figure 2.31 when the initial condition is zero for inputs $u(t) = 1$ and $u(t) = 10$.

Approximate responses $y_A(nT)$ for both inputs are typical linear first-order system step responses, namely, they each require roughly four to five time constants ($\tau = 1$ s) to reach steady state. Furthermore, the response $y_A(nT)$ in the lower left corner where $u(t) = 10$ is 10 times the response $y_A(nT)$ in the upper left corner where $u(t) = 1$. For a constant input $u(t) = \bar{u}$, the steady-state value is $y_A(\infty) = \bar{u}$ for the linear system.

In contrast to the linear system, the transient period of the nonlinear system is shorter when the input $u(t) = 10$ compared to when $u(t) = 1$. Furthermore, $z_A(\infty) = \bar{u}^{\frac{1}{2}}$ for the nonlinear system when the input is $u(t) = \bar{u}$, in violation of the principle of superposition.

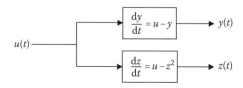

FIGURE 2.30   Linear and nonlinear system subject to identical input.

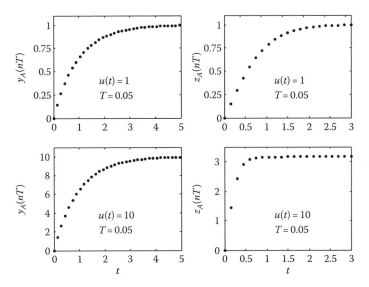

FIGURE 2.31   Approximation of linear and nonlinear system step responses.

A linear model approximation of a nonlinear system is often acceptable provided the system variables (inputs, states, outputs) are confined to a restricted operating region. A simple example serves to illustrate the point. Consider a system with input $u = u(t)$ and state $x = x(t)$ described by

$$\frac{dx}{dt} + 0.2x^{1/2} = u \tag{2.136}$$

The state derivative function is a nonlinear function of x, that is,

$$\frac{dx}{dt} = f(x, u) = -0.2x^{1/2} + u \tag{2.137}$$

For arbitrary input $u(t)$, the solution to Equation 2.137 can be approximated in a way similar to what we did in Chapter 1 using difference equations. However, suppose the input $u(t)$ is confined to a range that results in the state $x(t)$ varying between x_l and x_u as shown in Figure 2.32. It is reasonable to assume the term $0.2x^{1/2}$ in Equation 2.136 could be replaced by a linear function of x resulting in a simpler model. We will have more to say about linearization of nonlinear system models in Chapter 7.

Another distinguishing property of linear systems is the way they respond to sinusoidal inputs. At steady state, the output of a linear system forced by a sinusoidal input with radian frequency ω_0 is itself a sinusoid at the same frequency. In general, the output is shifted in time (out of phase) with respect to the input, and the amplitude is either attenuated or amplified compared to the amplitude of the input. This property is the foundation of linear AC steady-state analysis and the design of linear control systems by the method of frequency response. In the case of nonlinear systems, the output includes harmonics (sinusoidal terms at frequencies $n\omega_0$, $n = 1, 2, 3, \ldots$).

The type of nonlinearity portrayed in Figure 2.32 has been classified as "progressive" (Buckley 1964). The distinguishing characteristics of progressive nonlinearities are their monotonic continuous nature over the range of input and output values of interest. Furthermore, state derivative functions which are progressive nonlinearities can be approximated by linearization methods. "Essential" nonlinearities are those that cannot be represented by a simple continuous analytical function. Phenomena such as friction, dead zone and saturation in valves, and backlash in gears in mechanical systems; hysteresis in electrical components; and analog-to-digital quantization are examples of essential nonlinearities.

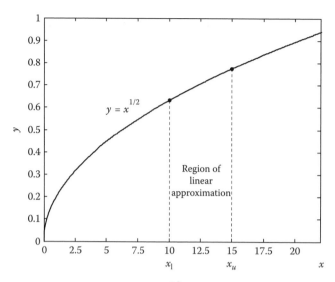

FIGURE 2.32 Linearizing the nonlinear function $0.2x^{1/2}$ in an interval $x_l \leq x \leq x_u$.

2.7.1 FRICTION

The first example illustrates a type of friction called coulomb friction. An object of mass m, resting on a flat surface, is subject to an external horizontal force $f(t)$ and a resisting frictional force f_μ as shown in Figure 2.33. The velocity of the mass obeys the relation in Equation 2.138

$$m\frac{dv}{dt} + f_\mu = f \tag{2.138}$$

The friction force f_μ is equal in magnitude to the force f until a breakaway force f_B is applied (see Figure 2.34), and the mass begins to slide along the surface. The breakaway force f_B depends on the coefficient of static friction μ_0 and the object's weight,

$$f_B = \mu_0 mg \tag{2.139}$$

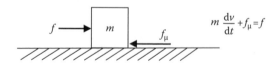

FIGURE 2.33 Nonlinear system example—coulomb friction.

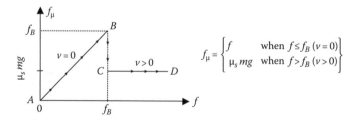

FIGURE 2.34 Friction force f_μ vs. increasing f applied to a mass initially at rest.

While in motion, the friction force f_μ is a constant dependent on the coefficient of sliding friction μ_s and the weight mg of the object as seen in Equation 2.140. Note that f_μ is also equal to $\mu_s mg$ when $f \leq f_B$ and $v > 0$.

$$f_\mu = \begin{cases} f & \text{when } f \leq f_B \ (v = 0) \\ \mu_s mg & \text{when } f > f_B \ (v > 0) \end{cases} \tag{2.140}$$

Example 2.10

The applied force $f(t)$ is shown in Figure 2.35. Find the velocity of the object.

$$f(t) = \begin{cases} 2f_B\left(\dfrac{t}{t_1}\right) & 0 \leq t < t_1 \\ 2f_B\left[2 - \left(\dfrac{t}{t_1}\right)\right] & t_1 \leq t < 2t_1 \\ 0 & t \geq 2t_1 \end{cases} \tag{2.141}$$

The difference equation resulting from the substitution of the divided difference $[v_A(n+1) - v_A(n)]/T$ for the first derivative dv/dt in Equation 2.138 is

$$v_A(n + 1) = v_A(n) + \frac{T}{m}[f(n) - f_\mu(n)] \tag{2.142}$$

A recursive solution for $v_A(n)$, $n = 1, 2, 3, \ldots$ given $v_A(0) = v(0) = 0$ is not as straightforward as it was in previous examples owing to the nature of the friction force. The MATLAB® M-file "*Chap2_Ex7_1.m*" includes the necessary conditional statements to handle the discontinuity in f_μ. Results are shown in Figure 2.36.

The analytical solution for the velocity $v(t)$ is plotted along with the approximate solution $v_A(n)$. It can be found by integrating the differential equation (Equation 2.138) over consecutive intervals using the appropriate value for the friction force (f or $\mu_s mg$) and the correct initial velocity for each interval. The details are left for an exercise; the results are as follows.

$$v(t) = \begin{cases} 0, & 0 \leq t \leq 0.5t_1 \\ gt_1\left[\mu_0\left(\dfrac{t}{t_1}\right)^2 - \mu_s\left(\dfrac{t}{t_1}\right) + \dfrac{2\mu_s - \mu_0}{4}\right], & 0.5t_1 \leq t < t_1 \\ gt_1\left[-\mu_0\left(\dfrac{t}{t_1}\right)^2 + (4\mu_0 - \mu_s)\left(\dfrac{t}{t_1}\right) - \dfrac{9\mu_0 - 2\mu_s}{4}\right], & t_1 \leq t < 2t_1 \\ gt_1\left[-\mu_s\left(\dfrac{t}{t_1}\right) + \dfrac{7\mu_0 + 2\mu_s}{4}\right], & 2t_1 \leq t < T_f \\ 0, & T_f < t \end{cases} \tag{2.143}$$

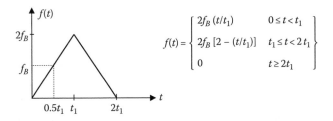

FIGURE 2.35 Applied force $f(t)$ vs. t.

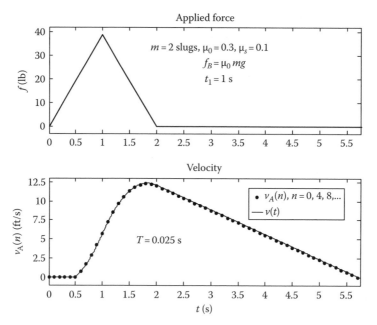

FIGURE 2.36 Approximate solution $v_A(n)$, $n = 0, 4, 8, \ldots$ and exact solution $v(t)$, $t \geq 0$.

The time T_f when the velocity returns to zero is obtained from

$$T_f = \frac{t_1}{4\mu_s}(7\mu_0 + 2\mu_s) \tag{2.144}$$

2.7.2 DEAD ZONE AND SATURATION

The next example of mechanical (pneumatic) nonlinearity is a valve that contains two nonlinear elements, dead zone and saturation. First, consider the nonlinear elements individually. An ideal dead zone nonlinearity is shown in Figure 2.37. The dead zone is the region between t_1 and t_2.

$$f(t) = \begin{cases} f_3\left(\dfrac{t - t_2}{t_3 - t_2}\right) & t_2 \leq t \\ 0 & t_1 < t < t_2 \\ f_0\left(\dfrac{t - t_1}{t_0 - t_1}\right) & t \leq t_1 \end{cases} \tag{2.145}$$

An ideal saturation nonlinearity is shown in Figure 2.38.

$$f(t) = \begin{cases} f_s\left(\dfrac{t}{t_s}\right) & |t| \leq t_s \\ \operatorname{sgn}(f_s) & |t| > t_s \end{cases} \tag{2.146}$$

The saturated regions are when $|t| > t_s$, that is, for $t < -t_s$, the value of $-f(t)$ does not change and for $t > t_s$, the value of $f(t)$ does not change.

Together, these nonlinearities (saturation and dead zone) form an approximation to the pneumatic behavior of a valve shown in Figure 2.39.

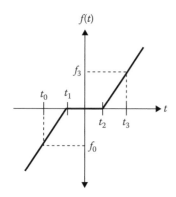

FIGURE 2.37 Dead zone nonlinearity.

$\pm I_o$ is the opening current, that is, the current needed to open the valve. $\pm I_s$ is the saturation current where any additional current (more than I_s or less than $-I_s$) does not open the valve any further. The region between I_o and $I_s(-I_o$ and $-I_s)$ is appropriately called the active region. The region between $-I_o$ and I_o is called the dead zone. However, in practice, leakage occurs below the opening current.

2.7.3 BACKLASH

Backlash nonlinearity often occurs in gears due to the spacing between individual teeth. The spacing is needed for the gears to mesh without binding. This spacing (d) is shown in Figure 2.40.

Figure 2.41 shows a plot of the backlash nonlinearity.

Assume the space d exists in the initial condition as in Figure 2.40. As the leading gear moves in one direction, the following gear does not move until contact is made after the leading gear is displaced by d. Then, the following gear tracks the leading gear as indicated by section 1 of Figure 2.41. When the leading gear reverses direction, it must be displaced by a distance $2d$ before contact is reestablished with the following gear, as indicated by section 2 of Figure 2.41. Similar to before, the following gear tracks the leading gear as indicated by section 3 of Figure 2.41. Another reversal of directions leads to section 4 in Figure 2.41.

2.7.4 HYSTERESIS

The graph of f_μ vs. f in Figure 2.34 is applicable so long as the applied force f and resulting velocity v are increasing along the path A-B-C-D. Once the block is in motion and the applied force f diminishes to zero, the return path does not follow D-C-B-A. That is, the sliding block does not abruptly stop when the applied force is reduced to f_B. Rather, the friction force remains at $\mu_s mg$ until the block decelerates to zero velocity. This type of nonlinear phenomenon is referred to as hysteresis.

An example of a real system with hysteresis, present by design, is a thermostatically controlled furnace supplying heat to a building. A simplified diagram of the system is depicted in Figure 2.42. An energy balance on the building interior space relates the accumulation of thermal energy to the heat flow from the furnace and heat loss to the outside.

The equation is

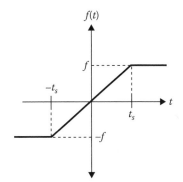

FIGURE 2.38 Saturation non-linearity.

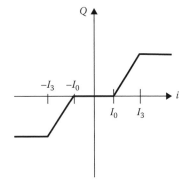

FIGURE 2.39 Valve flow vs. current.

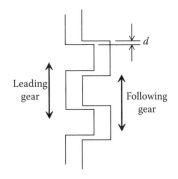

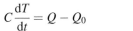

FIGURE 2.40 Backlash in gear teeth.

$$C\frac{dT}{dt} = Q - Q_0 \tag{2.147}$$

where C is the thermal capacitance of the air and contents inside the building, all of which are assumed to be at temperature T. The heat loss Q_0 is assumed proportional to the temperature difference $T - T_0$, that is,

$$Q_0 = \frac{T - T_0}{R} \tag{2.148}$$

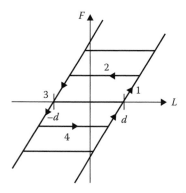

FIGURE 2.41 Backlash non-linearity.

with R the overall thermal resistance of the exterior walls and insulation. Combining Equations 2.147 and 2.148 and introducing the thermal system time constant $\tau = RC$ result in the first-order model

$$\tau \frac{dT}{dt} + T = RQ + T_0 \tag{2.149}$$

The furnace operates in one of two modes, on or off, depending on whether the building temperature T is below or above some tolerance Δ about a desired temperature T_d and whether the building temperature is increasing or decreasing. When it is on, a constant amount of heat $\overline{Q}$ is supplied; conversely, no heat is produced when the furnace is off. In mathematical terms,

$$Q = \begin{cases} \overline{Q}, T \leq T_d - \Delta & \text{or} \quad T_d - \Delta < T < T_d + \Delta \quad \text{and} \quad dT/dt > 0 \\ 0, T > T_d + \Delta & \text{or} \quad T_d - \Delta < T < T_d + \Delta \quad \text{and} \quad dT/dt < 0 \end{cases} \tag{2.150}$$

The hysteresis effect is evident from the graph in Figure 2.43 (McClamroch 1980).

From Equations 2.149 and 2.150, it follows that the state derivative dT/dt depends not only on the input T_0 and the state T but also on its own sign. Furthermore, since the furnace output Q in Figure 2.43 is multi-valued whenever the building temperature T falls within $T_d - \Delta$ to $T_d + \Delta$, the initial state $T(0)$ and the initial state of the furnace (on/off) must be specified to simulate or obtain analytical solutions for $T(t)$, $t \geq 0$.

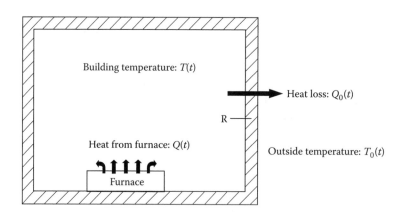

FIGURE 2.42 Temperature regulation in a building.

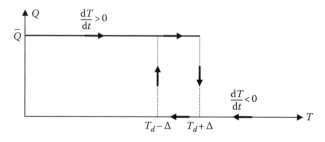

FIGURE 2.43 Hysteresis in furnace output vs. building temperature.

The example that follows illustrates a method for obtaining an approximate solution and the exact solution for the building temperature $T(t)$ when the outside temperature $T_0(t)$ is constant.

Example 2.11

A building's thermostat has been off for a period of time sufficient to allow the inside and outside temperatures to equalize. The thermostat is then set to 75°F. It is programmed to turn off when the interior temperature reaches 78°F and back on when it falls below 72°F. The furnace produces 36,000 Btu/h. Thermal capacitance of the occupied space and interior furnishings is 300 Btu/°F, and the thermal resistance of the walls is 8×10^{-4}°F per Btu/h. The outside temperature is a constant 50°F.

- (a) Find the time constant of the system.
- (b) Show that the furnace is capable of raising the building temperature to 78°F.
- (c) Find the temperature response and the time required for the building temperature to reach 78°F.
- (d) Find the temperature response and the time required for the building temperature to cool down to 72°F.
- (e) Find the temperature response and the time required for the building temperature to go back to 78°F.
- (f) Simulate the temperature responses in parts (c), (d), and (e) by solving a difference equation with appropriate step size and compare the approximate and exact solutions.

(a) The time constant, a measure of the speed of the system's dynamics is

$$\tau = RC = 8 \times 10^{-4} \, \frac{°F}{Btu/h} \cdot 300 \, \frac{Btu}{°F} = 0.24 \, h$$

(b) The steady-state temperature differential (inside minus outside) that the furnace is capable of maintaining is obtained from the first-order differential equation model in Equation 2.149 with the derivative set to zero and the furnace on, that is, $Q(t) = \overline{Q}$.

$$T_{ss} = R\overline{Q} + \overline{T}_0 \tag{2.151}$$

$$\Rightarrow T_{ss} - \overline{T}_0 = R\overline{Q} \tag{2.152}$$

where
$\overline{T}_0$ is the constant outside temperature
T_{ss} is the steady-state inside temperature

In this example,

$$T_{ss} - \overline{T}_0 = R\overline{Q} = 8 \times 10^{-4} \, \frac{°F}{Btu/h} \cdot 36{,}000 \, \frac{Btu}{h} = 28.8°F$$

Hence, the furnace is capable of raising the inside temperature from 50°F to 78.8°F, which is slightly higher than the 78°F shutoff setting of the thermostat.

(c) From Equation 2.6, the step response of the first-order system is

$$T(t) = T(0)e^{-t/\tau} + (\overline{T}_0 + R\overline{Q})(1 - e^{-t/\tau}) \tag{2.153}$$

which describes the building temperature from time $t = 0$ up to $t = t_1$ where

$$T(t_1) = T_d + \Delta = 75°F + 3°F = 78°F \tag{2.154}$$

Solving for t_1 gives

$$t_1 = \tau \ln \left[\frac{(\overline{T}_0 + R\overline{Q}) - T(0)}{(\overline{T}_0 + R\overline{Q}) - (T_d + \Delta)} \right]$$

$$= 0.24 \ln \left[\frac{(50 + 28.8) - 50}{(50 + 28.8) - (75 + 3)} \right] = 0.86 \, \text{h} \qquad (2.155)$$

From Equation 2.153 with $T(0) = 50°F$, the temperature response is

$$T(t) = 50e^{-t/0.24} + 78.8(1 - e^{-t/0.24}), \quad 0 \le t \le 0.86 \qquad (2.156)$$

(d) The furnace shuts off when the temperature reaches $T_d + \Delta = 78°F$ and the subsequent cooling from 78°F to $T_d - \Delta = 72°F$ follows the step response in Equation 2.153 with $\overline{Q} = 0$ and $T(0) = T_d + \Delta = 78°F$. Thus,

$$T(t) = (T_d + \Delta)e^{-(t-t_1)/\tau} + \overline{T}_0[1 - e^{-(t-t_1)/\tau}], \quad t_1 \le t \le t_2 \qquad (2.157)$$

$$= 78e^{-(t-0.86)/0.24} + 50[1 - e^{-(t-0.86)/0.24}], \quad 0.86 \le t \le t_2 \qquad (2.158)$$

where t_2 is the time when the building temperature is $T_d - \Delta - 72°F$. Note the $(t - t_1)$ in the exponent of Equation 2.157 since t_1 is the initial time of the step response. From Equation 2.157 with $t = t_2$, $T(t_2) = T_d - \Delta$, the time t_2 is given by

$$t_2 = t_1 + \tau \ln \left[\frac{(T_d + \Delta) - \overline{T}_0}{(T_d - \Delta) - \overline{T}_0} \right]$$

$$= 0.86 + 0.24 \ln \left[\frac{(75 + 3) - 50}{(75 - 3) - 50} \right] = 0.92 \, \text{h} \qquad (2.159)$$

(e) The cycle is completed when the building temperature returns to $T_d + \Delta = 78°F$. Using the same approach as before, the result is

$$T(t) = (T_d - \Delta)e^{-(t-t_2)/\tau} + (\overline{T}_0 + R\overline{Q})\left[1 - e^{-(t-t_2)/\tau}\right], \quad t_2 \le t \le t_3$$

$$= 72e^{-(t-0.92)/\tau} + 78.8[1 - e^{-(t-0.92)/\tau}], \quad 0.92 \le t \le t_3 \qquad (2.160)$$

Setting $T(t_3) = T_d + \Delta$ and solving for t_3,

$$t_3 = t_2 + \tau \ln \left[\frac{(\overline{T}_0 + R\overline{Q}) - (T_d - \Delta)}{(\overline{T}_0 + R\overline{Q}) - (T_d + \Delta)} \right]$$

$$= 0.92 + 0.24 \ln \left[\frac{(50 + 28.8) - (75 - 3)}{(50 + 28.8) - (75 + 3)} \right] = 1.43 \, \text{h} \qquad (2.161)$$

(f) The approximate solution for the building temperature is based on the difference equation obtained by replacing the first derivative dT/dt in Equation 2.149 with the finite difference $[T_A(n + 1) - T_A(n)]/T$. The result is

$$T_A(n + 1) = \left(1 - \frac{\Delta T}{\tau}\right) T_A(n) + \frac{\Delta T}{\tau}[RQ(n) + \overline{T}_0] \qquad (2.162)$$

where $Q(n)$ is based on the logic in Equation 2.150. The MATLAB M-file "Chap2_Ex7_2.m" evaluates the exact and approximate solutions and generates the graph shown in Figure 2.44.

The building temperature experiences periodic fluctuations between $T_d - \Delta = 72°F$ and $T_d + \Delta = 78°F$ as long as the outside temperature remains constant. The period is equal to $t_3 - t_1 = 1.43 - 0.86 = 0.57$ h.

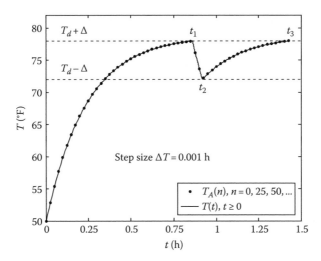

FIGURE 2.44 Exact and approximate solutions for building temperature.

2.7.5 QUANTIZATION

In digital control, it is often desired to discretize the continuous signal of a sensor for use by a computer or microprocessor. Conversion of this signal is achieved by an analog-to-digital converter (ADC) where the signal is quantized

The quantization nonlinearity is shown in Figure 2.45. In this example, a voltage range between V_0 and V_1 is designated as state zero, S_0; a voltage range between V_1 and V_2 is designated as state one, S_1; and so on. Each state is represented by a binary expression according to the number of bits used by the data type assigned to the state. For example, an 8-bit representation for state zero is 00000000, while state one is represented by 00000001. The more bits that are available for quantization yield a better resolution of the sensor's range, in this case, voltage. There are 2^n states where n is the number of bits. Therefore, an 8-bit ADC has 256 states, 0–255. The resolution is the sensor's range divided by the number of states. For example, a sensor with a voltage range from 0 to 10 V has a resolution of 0.04 V for an 8-bit ADC.

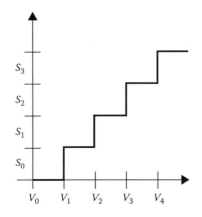

FIGURE 2.45 Quantization.

2.7.6 Sustained Oscillations and Limit Cycles

Both linear and nonlinear system differential equation models are capable of producing solutions involving sustained oscillations of the state variables. This comes as no surprise for linear systems. Indeed, we have already seen how the natural response of an undamped second-order system continues to oscillate forever (see Figure 2.4). Examples will be presented in Chapter 4 of forced linear systems with sustained sinusoidal oscillations in the output after the transient response has died out.

State trajectories of the autonomous system governed by the differential equation

$$\ddot{x} + \omega_n^2 x = 0 \quad \text{subject to } x(0) = x_0,\ \dot{x}(0) = \dot{x}_0 \tag{2.163}$$

are closed orbits in the $\dot{x}$ vs. x state space. Figure 2.46 shows state trajectories, also known as orbits, for the undamped system in Equation 2.163 with $\omega_n = 1$ rad/s starting from four different initial points in the state space.

The orbits are typically elliptical; however, those in Figure 2.46 are circular because the natural frequency $\omega_n = 1$ rad/s. Sustained oscillations of the state components x and dx/dt are shown in Figure 2.47.

Nonlinear systems can experience two types of sustained oscillations. The first class is similar to the case of linear systems. In the unforced case, the oscillations are sensitive to the initial conditions. That is, the particular points along the closed path of the state trajectory vary depending on the location of the initial point in state space. The initial point is always on the closed orbit. The amplitude and period of the oscillations depend on the system parameters and initial conditions.

The state trajectories of the nonlinear system described by the coupled first-order differential equations

$$\dot{x}_1 = x_1(a - bx_2) \tag{2.164}$$

$$\dot{x}_2 = x_2(cx_1 - d) \tag{2.165}$$

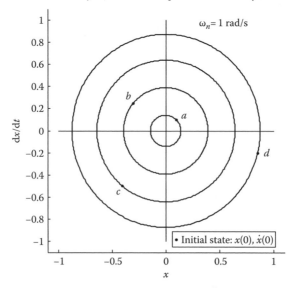

FIGURE 2.46 Closed orbits for the system $\ddot{x} + \omega_n^2 x = 0$ ($\omega_n = 1$ rad/s).

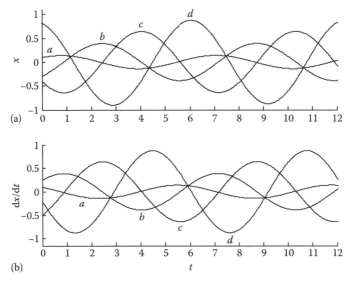

FIGURE 2.47 Sustained oscillations of x (a) and dx/dt (b) for undamped second-order systems.

are concentric closed curves "spun out" in a clockwise rotation from the initial point. The center of rotation is the equilibrium point located at $(d/c,\ a/b)$. The MATLAB M-file "*Chap2_Figs7_13and14.m*" uses a difference quotient with step size $T = 5 \times 10^{-5}$ to approximate the first derivatives in Equations 2.164 and 2.165. The approximate solutions in Figure 2.48 show four orbits starting from different initial states.

Time histories of the state variables are shown in Figure 2.49. In contrast to the sinusoidal oscillations of the LTI system governed by Equation 2.163, the oscillations of the nonlinear system in Equations 2.164 and 2.165 are not of a sinusoidal nature.

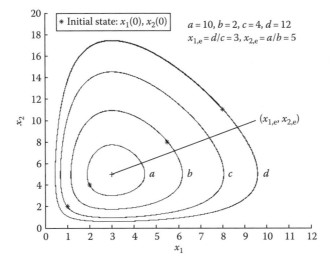

FIGURE 2.48 Closed orbits and sustained oscillations for the nonlinear system.

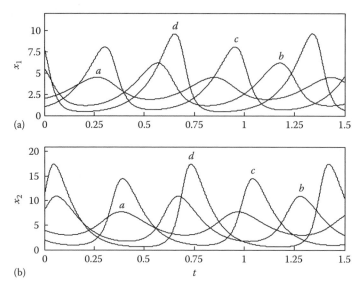

FIGURE 2.49   Sustained oscillations of x_1 (a) and x_2 (b) for nonlinear second-order system.

Another type of sustained oscillation is possible for an unforced nonlinear system. In this case, there is a single closed orbit in the state space independent of the initial conditions. If the initial state is located on this closed path, the state vector remains on it forever, periodically returning to the starting point. When the initial state is inside the closed curve, the state trajectory may be asymptotically attracted to the closed curve or repelled from it towards a stable equilibrium point in its interior. Should the initial state be located outside the closed curve, the state trajectory either converges to it in a finite time period or else spirals outward from it.

Sustained oscillations of this nature are called limit cycles. If the initial state is not on the limit cycle, the state trajectory is either attracted to or repelled from it. Limit cycles are either stable or unstable depending on which of the two situations applies.

An autonomous mechanical system with a stable limit cycle is given in Tse et al. (1963). Referring to Figure 2.50, the mass m is acted upon by a linear spring force F_k, a nonlinear damping force F_c, and a self-excitation force F, that is, a force with explicit dependence solely on the internal state of the system.

Note that there are no external forces present. The differential equation model is

$$m\ddot{x} = F - F_c - F_k = F_0\dot{x} - (cx^2)\dot{x} - kx \tag{2.166}$$

$$\Rightarrow \ m\ddot{x} + (cx^2 - F_0)\dot{x} + kx = 0 \tag{2.167}$$

The effective damping force is $(cx^2 - F_0)\dot{x}$. In the neighborhood of the equilibrium point $x = 0$, $\dot{x} = 0$, the term $(cx^2 - F_0) < 0$. The negative damping results in an increase of energy in the system

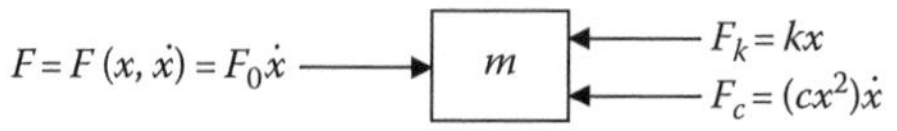

$$F = F(x, \dot{x}) = F_0\dot{x} \longrightarrow \boxed{\ m\ \} \begin{array}{l} \longleftarrow F_k = kx \\ \longleftarrow F_c = (cx^2)\dot{x} \end{array}$$

FIGURE 2.50   An autonomous nonlinear system with self-excitation force.

making the equilibrium point inherently unstable. Consequently, the state trajectory will move outwards from the origin in state space.

The reverse is true whenever $(cx^2 - F_0) > 0$. In this case, the damping term is positive and energy is dissipated from the system. The state trajectory spirals inward to points where the total energy in the system is lower. Clearly, a locus of points must exist in state space to function as a transition between the two phenomena. The locus must be a closed curve, namely, the limit cycle.

Example 2.12

For the mechanical system described by Equation 2.167

(a) Convert the system model to state variable form.
(b) Numerical values of the system parameters are $m = 1$, $k = 2$, $c = 0.5$, and $F_0 = 3$. Approximate the state derivatives numerically with appropriate step size to determine the state trajectories when the initial state is located at
  (i) $x(0) = -1$, $\dot{x}(0) = -5$
  (ii) $x(0) = 2$, $\dot{x}(0) = 5$
  (iii) $x(0) = -2$, $\dot{x}(0) = 15$
  (iv) $x(0) = 5$, $\dot{x}(0) = -20$
  Plot the trajectories in the state space.
(c) Estimate the period of the limit cycle.

(a) Choosing the state vector as $x_1 = x$, $x_2 = \dot{x}$ yields the state derivative functions

$$\dot{x}_1 = f_1(x_1, x_2) = x_2 \tag{2.168}$$

$$\dot{x}_2 = f_2(x_1, x_2) = -\frac{1}{m}[kx_1 + (cx_1^2 - F_0)x_2] \tag{2.169}$$

(b) Replacing $\dot{x}_1$ and $\dot{x}_2$ by difference quotients leads to the following difference equations for the discrete-time system

$$x_{1,A}(n+1) = x_{1,A}(n) + Tf_1[x_{1,A}(n), x_{2,A}(n)] \tag{2.170}$$

$$= x_{1,A}(n) + Tx_{2,A}(n) \tag{2.171}$$

$$x_{2,A}(n+1) = x_{2,A}(n) + Tf_2[x_{1,A}(n), x_{2,A}(n)] \tag{2.172}$$

$$= x_{2,A}(n) - \frac{T}{m}\left[kx_{1,A}(n) + \left\{cx_{1,A}^2(n) - F_0\right\}x_{2,A}(n)\right] \tag{2.173}$$

The difference equations are solved recursively in "Chap2_Ex7_3.m" for the given initial states. The limit cycle and the four state trajectories are shown in Figure 2.51. As expected, the state trajectories eventually converge to the limit cycle.

(c) Figure 2.52 shows the time responses for the state components starting from the initial state $x_1(0) = 5$, $x_2(0) = -20$. The period of sustained oscillations can be approximated from the graph by estimating the difference in successive zero crossings of either state component once the state "locks into" the limit cycle. By zooming in on Figure 2.46, the period is approximated as $11.94 - 6.43 = 5.51$. Can you determine the approximate time the state enters the limit cycle?

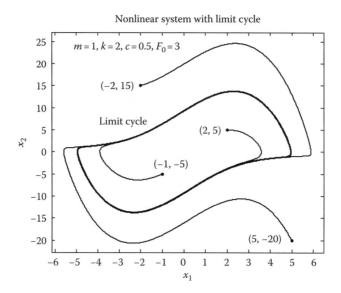

FIGURE 2.51 Approaches to limit cycle from several initial states.

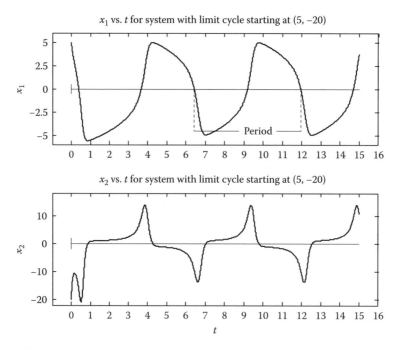

FIGURE 2.52 Time histories of state components [initial state: $x_1(0) = 5$, $x_2(0) = -20$].

EXERCISES

2.22 Examine the effect of changing the initial condition on the unit step response of the nonlinear system

$$\frac{\mathrm{d}x}{\mathrm{d}t} + x^2 = u, \quad u(t) = 1, \quad \geq 0$$

Plot $x_A(n)$, $n = 0, 1, 2, 3, \ldots$ when $x(0) = -2, -1, 0, 1, 5$ on the same graph. Use $T = 0.05$.

2.23 In Example 2.10, suppose instead of a constant friction force applied to the object as it slides, there is a variable friction force given by

$$f_\mu = \alpha v^\beta$$

Find and plot $v_A(n)$, $n = 1,2,3,\ldots$ in response to the force $f(t)$ in Example 2.10 when
  (i) $\alpha = 2$, $\beta = 0.5$
  (ii) $\alpha = 1$, $\beta = 1$
  (iii) $\alpha = 2$, $\beta = 2$

2.24 Nonlinear dynamic system is shown in the figure below. The input $u(t) = \sin 100\pi t$, $t \ge 0$.
  (a) Is the output $y(t)$ a sinusoidal function of the same frequency as the input like it would be in a linear system? Explain
  (b) Is the output $y(t)$ a periodic function? If so, what is the frequency?

$$u(t) \longrightarrow \boxed{y^{1/2} = u} \longrightarrow y(t)$$

2.25 In Example 2.10, find the displacement of the mass, $x(t)$, $t \ge 0$.

2.26 In Example 2.10, the applied force is

$$f(t) = \begin{cases} 2f_B \sin 0.25\pi t, & 0 \le t < 4 \\ 0, & t \ge 4 \end{cases}$$

  (a) Formulate a difference equation for $v_A(n)$ similar to Equation 2.142 and solve recursively.
  (b) Determine the analytical solution for $v(t)$.
  (c) Plot the approximate and analytical solutions on the same graph.

2.27 In Example 2.11, find the exact and approximate solutions for the building temperature $T(t)$ and furnace output $Q(t)$ if the desired setting T_d and tolerance Δ are
  (a) $T_d = 72°F$, $\Delta = 3°F$
  (b) $T_d = 78°F$, $\Delta = 1.5°F$

2.28 In Example 2.11, investigate the effect of lowering the desired temperature T_d on the thermostat and its effect on the furnace cycle time and the duty cycle, that is, percentage of time the furnace is on. Plot the results for $T_d = 68°F, 69°F, \ldots, 75°F$.

2.29 In Example 2.11, suppose the outside temperature $T_0(t)$ varies in a sinusoidal fashion with average value of $50°F$ and amplitude of $5°F$ as shown in Figure E2.29. The thermal capacitance of the room is $C = 500$ Btu/$°F$. The initial room temperature at 6 AM is $50°F$.
  (a) The thermostat is set at $T_d = 65°F$. Simulate the building temperature long enough to show several cycles of $T(t)$ after the initial transient response vanishes.
  (b) Find the energy cost per day if the cost of heating is $1.75¢$ per 1000 Btu's.
  (c) Repeat parts (a) and (b) for $T_d = 70°F$ and $75°F$.

2.30 The coulomb damping force acting on the mass shown in Figure E2.30 is given by $f_\mu = -\text{sgn}(\dot{x})\mu mg$. The initial condition is $x(0) = x_0 = -1$ m, $\dot{x}(0) = 0$ m/s. The equation of motion is

$$\ddot{x} + \omega_n^2 x = \frac{1}{m} f_\mu, \quad \omega_n^2 = \frac{k}{m} \quad (g = 9.81 \text{ m/s}^2)$$

The system parameters are $m = 6$ kg, $k = 300$ N/m, $\mu = 0.2$
  (a) Define the state as $x_1 = x$, $x_2 = \dot{x}$ and find difference equations for $x_{1,A}(n)$ and $x_{2,A}(n)$.
  (b) Solve the difference equation for four cycles of $x_{1,A}(n)$ and $x_{2,A}(n)$ using a step size of $T = 0.001$s. Plot the results.

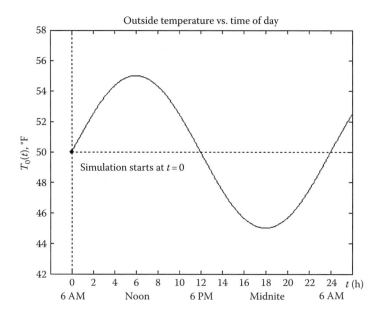

FIGURE E2.29

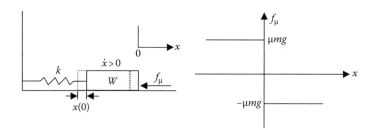

FIGURE E2.30

(c) The exact solution for $x(t)$ over the first cycle ($0 \leq t \leq t_2$) is

$$x(t) = \begin{cases} \left(x_0 + \dfrac{\mu mg}{k}\right) \cos \omega_n t - \dfrac{\mu mg}{k}, & 0 \leq t \leq t_1 \\ \left(x_1 - \dfrac{\mu mg}{k}\right) \cos \omega_n(t - t_1) + \dfrac{\mu mg}{k}, & t_1 \leq t \leq t_2 \end{cases}$$

where
$$t_1 = \frac{\pi}{\omega_n}$$
$$t_2 = t_1 + \frac{\pi}{\omega_n}$$
$$x_1 = -x_0 - 2\frac{\mu mg}{k}$$

Plot the exact solution for $x(t)$ over the first cycle and compare it to the approximate solution.

2.31 For the undamped second-order system modeled by

$$\ddot{x} + \omega_n^2 x = 0 \quad \text{subject to } x(0) = x_0, \dot{x}(0) = \dot{x}_0$$

show the equation of the closed trajectories are ellipses in the $x - \dot{x}$ plane that reduce to the circular orbits in Figure 2.40 when $\omega_n = 1$ rad/s.

2.32 Generate the state trajectory shown in Figure 2.45 starting at $(-2,15)$ by finding an approximate solution to the differential equation

$$\frac{dx_2}{dx_1} = -\frac{1}{m}\left[k\frac{x_1}{x_2} + cx_1^2 - F_0\right]$$

obtained as a result of dividing dx_2/dt in Equation 2.169 by dx_1/dt in Equation 2.168.

2.33 Generate 500 state trajectories starting from initial points randomly selected in the region $-10 \le x(0) \le 10$, $-10 \le \dot{x}(0) \le 10$ for the system governed by

$$m\ddot{x} + (F_0 - cx^2)\dot{x} + kx = 0$$

with the same parameter values as those in Example 2.12. Comment on the existence of a limit cycle and its effect on the trajectories.

2.34 Find the period of oscillations for the system modeled by

$$\dot{x}_1 = x_1(10 - 2x_2)$$

$$\dot{x}_2 = x_2(4x_1 - 12)$$

when the initial state is (i) $x_1(0) = 10$, $x_2(0) = 15$ and (ii) $x_1(0) = 4$, $x_2(0) = 6$.

2.8   CASE STUDY: SUBMARINE DEPTH CONTROL SYSTEM

Automatic depth control of a submarine is the focus of this section. Figure 2.53 illustrates a representative situation, where the actual depth of the submarine, denoted $c(t)$, is measured by a sensor and compared with the desired depth $r(t)$.

A simplified block diagram of the depth control system is shown in Figure 2.54. The error signal $e(t)$ is the difference between the commanded depth $r(t)$ and the actual depth $c(t)$. It is fed back to the controller that sends a signal to the stern plane actuator motor that controls the stern plane actuator angle $\theta(t)$. The submarine depth responds to changes in the stern plane angle.

The controller and stern plane actuator combination are modeled by

$$\theta = K_C e + K_I \int e\, dt \tag{2.174}$$

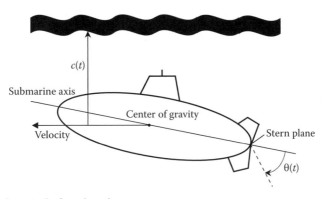

FIGURE 2.53   Depth control of a submarine.

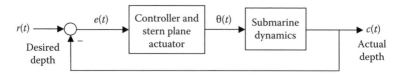

FIGURE 2.54   Block diagram of a submarine depth control system.

and the submarine dynamics are approximated by the simple first-order model

$$\tau \frac{dv}{dt} + v = K_{\dot\theta} \frac{d\theta}{dt} + K_\theta \theta \tag{2.175}$$

where $v = v(t)$ is the depth rate of the submarine. Integrating the depth rate yields the depth of the submarine

$$c = \int v \, dt \tag{2.176}$$

The error signal is output from the summer as

$$e = r - c \tag{2.177}$$

Equations 2.174 through 2.177 constitute the mathematical model of the simplified submarine depth control system. The goal is to choose the parameters K_C and K_I, so that the submarine responds to step changes in commanded depth in an acceptable manner.

A simulation diagram of the control system is a useful first step in helping choose a set of state variables. Employing the technique discussed in Section 2.4 for drawing a simulation diagram with input derivative terms present, the diagram is shown in Figure 2.55.

From the simulation diagram, the state equations are

$$\dot{x}_1 = v = K_\theta x_2 + K_{\dot\theta} \dot{x}_2 \tag{2.178}$$

$$= K_\theta x_2 + K_{\dot\theta} \left[\frac{1}{\tau}(\theta - x_2) \right] \tag{2.179}$$

The stern plane angle θ is expressible in terms of the states x_1, x_2, and x_3 and input r by

$$\theta = K_C e + K_I x_3 = K_C(r - x_1) + K_I x_3 \tag{2.180}$$

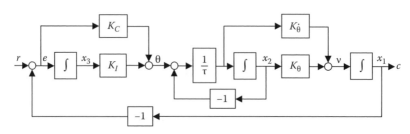

FIGURE 2.55   Simulation diagram of a submarine depth control system.

Combining Equations 2.179 and 2.180 gives

$$\dot{x}_1 = K_\theta x_2 + K_{\dot\theta} \left[\frac{1}{\tau} \{ K_C(r - x_1) + K_I x_3 - x_2 \} \right] \tag{2.181}$$

$$\Rightarrow \dot{x}_1 = \left(\frac{-K_{\dot\theta} K_C}{\tau} \right) x_1 + \left(K_\theta - \frac{K_{\dot\theta}}{\tau} \right) x_2 + \left(\frac{K_{\dot\theta} K_I}{\tau} \right) x_3 + \left(\frac{K_{\dot\theta} K_C}{\tau} \right) r \tag{2.182}$$

$$\dot{x}_2 = \left[\frac{1}{\tau} (\theta - x_2) \right] \tag{2.183}$$

$$= \left[\frac{1}{\tau} \{ K_C(r - x_1) + K_I x_3 - x_2 \} \right] \tag{2.184}$$

$$\Rightarrow \dot{x}_2 = \left(\frac{-K_C}{\tau} \right) x_1 - \left(\frac{1}{\tau} \right) x_2 + \left(\frac{K_I}{\tau} \right) x_3 + \left(\frac{K_C}{\tau} \right) r \tag{2.185}$$

$$\dot{x}_3 = r - x_1 \tag{2.186}$$

Equations 2.182, 2.185, and 2.186 represent the dynamic portion of the state variable model, that is, $\underline{\dot{x}} = A\underline{x} + Br$. Choosing the outputs as $y_1 = \theta$, $y_2 = v$, and $y_3 = c$ determines the matrices C and D in the output equation $\underline{y} = C\underline{x} + Dr$.

$$y_1 = \theta = K_C(r - x_1) + K_I x_3 \tag{2.187}$$

$$= -K_C x_1 + K_I x_3 + K_C r \tag{2.188}$$

$$y_2 = v = \dot{x}_1 = \left(\frac{-K_{\dot\theta} K_C}{\tau} \right) x_1 + \left(K_\theta - \frac{K_{\dot\theta}}{\tau} \right) x_2 + \left(\frac{K_{\dot\theta} K_I}{\tau} \right) x_3 + \left(\frac{K_{\dot\theta} K_C}{\tau} \right) r \tag{2.189}$$

$$y_3 = c = x_1 \tag{2.190}$$

In summary, the state equations are

$$\begin{bmatrix} \dot{x}_1 \\ \dot{x}_2 \\ \dot{x}_3 \end{bmatrix} = \begin{bmatrix} \dfrac{-K_{\dot\theta} K_C}{\tau} & K_\theta - \dfrac{K_{\dot\theta}}{\tau} & \dfrac{K_{\dot\theta} K_I}{\tau} \\ \dfrac{-K_C}{\tau} & \dfrac{-1}{\tau} & \dfrac{K_I}{\tau} \\ -1 & 0 & 0 \end{bmatrix} \begin{bmatrix} x_1 \\ x_2 \\ x_3 \end{bmatrix} + \begin{bmatrix} \dfrac{K_{\dot\theta} K_C}{\tau} \\ \dfrac{K_C}{\tau} \\ 1 \end{bmatrix} r \tag{2.191}$$

$$\begin{bmatrix} \theta \\ v \\ c \end{bmatrix} = \begin{bmatrix} -K_C & 0 & K_I \\ \dfrac{-K_{\dot\theta} K_C}{\tau} & K_\theta - \dfrac{K_{\dot\theta}}{\tau} & \dfrac{K_{\dot\theta} K_I}{\tau} \\ 1 & 0 & 0 \end{bmatrix} \begin{bmatrix} x_1 \\ x_2 \\ x_3 \end{bmatrix} + \begin{bmatrix} K_C \\ \dfrac{K_{\dot\theta} K_C}{\tau} \\ 0 \end{bmatrix} r \tag{2.192}$$

The exact solution for the outputs θ, v, and c in response to a given depth command r can be approximated by solving a system of difference equations obtained using the same approach we employed on previous occasions, that is, the first derivatives $\dot{x}_1$, $\dot{x}_2$, $\dot{x}_3$ in Equation 2.191 are replaced by first-order difference quotients, and the resulting difference equations are solved recursively for $x_{1,A}(n)$, $x_{2,A}(n)$, $x_{3,A}(n)$. The result is

$$
\begin{bmatrix}
\dfrac{1}{T}\{x_{1,A}(n+1) - x_{1,A}(n)\} \\[2ex]
\dfrac{1}{T}\{x_{2,A}(n+1) - x_{2,A}(n)\} \\[2ex]
\dfrac{1}{T}\{x_{3,A}(n+1) - x_{3,A}(n)\}
\end{bmatrix}
=
\begin{bmatrix}
\dfrac{-K_{\dot\theta}K_C}{\tau} & K_\theta - \dfrac{K_{\dot\theta}}{\tau} & \dfrac{K_{\dot\theta}K_I}{\tau} \\[2ex]
\dfrac{-K_C}{\tau} & \dfrac{-1}{\tau} & \dfrac{K_I}{\tau} \\[2ex]
-1 & 0 & 0
\end{bmatrix}
\begin{bmatrix}
x_{1,A}(n) \\[2ex]
x_{2,A}(n) \\[2ex]
x_{3,A}(n)
\end{bmatrix}
+
\begin{bmatrix}
\dfrac{K_{\dot\theta}K_C}{\tau} \\[2ex]
\dfrac{K_C}{\tau} \\[2ex]
1
\end{bmatrix}
r(n)
$$

$$(2.193)$$

The difference equations are updated according to

$$
x_{1,A}(n+1) = x_{1,A}(n) - \left(\frac{K_{\dot\theta}K_C T}{\tau}\right)x_{1,A}(n) + \left(K_{\dot\theta} - \frac{K_{\dot\theta}}{\tau}\right)Tx_{2,A}(n)
$$
$$
+ \left(\frac{K_{\dot\theta}K_I T}{\tau}\right)x_{3,A}(n) + \left(\frac{K_{\dot\theta}K_C T}{\tau}\right)r(n) \tag{2.194}
$$

$$
x_{2,A}(n+1) = x_{2,A}(n) - \left(\frac{K_C T}{\tau}\right)x_{1,A}(n) - \left(\frac{T}{\tau}\right)x_{2,A}(n) + \left(\frac{K_I T}{\tau}\right)x_{3,A}(n) + \left(\frac{K_C T}{\tau}\right)r(n) \tag{2.195}
$$

$$
x_{3,A}(n+1) = x_{3,A}(n) - Tx_{1,A}(n) + Tr(n) \tag{2.196}
$$

From Equation 2.192, the discrete-time outputs are

$$
\begin{bmatrix}
\theta_A(n) \\[2ex]
v_A(n) \\[2ex]
c_A(n)
\end{bmatrix}
=
\begin{bmatrix}
-K_C & 0 & K_1 \\[2ex]
\dfrac{-K_{\dot\theta}K_C}{\tau} & K_\theta - \dfrac{K_{\dot\theta}}{\tau} & \dfrac{K_{\dot\theta}K_1}{\tau} \\[2ex]
1 & 0 & 0
\end{bmatrix}
\begin{bmatrix}
x_{1,A}(n) \\[2ex]
x_{2,A}(n) \\[2ex]
x_{3,A}(n)
\end{bmatrix}
+
\begin{bmatrix}
K_C \\[2ex]
\dfrac{K_{\dot\theta}K_C}{\tau} \\[2ex]
0
\end{bmatrix}
r(n)
$$

$$(2.197)$$

Equations 2.194 through 2.197 are solved recursively in the M-file "*Chap2_Case_Study.m*" for the case where $r(t) = 100$, $t \geq 0$. The baseline parameter values are

Sub dynamics: $\tau = 10$ s, $K_{\dot\theta} = 20$ ft/s per deg/s, $K_\theta = 10$ ft/s per deg
Controller gains: $K_C = 0.6$ deg/ft, $K_I = 0.1$ deg/ft s
Step size: $T = 0.0025$ s

Graphs of the discrete-time outputs $\theta_A(n)$, $v_A(n)$, $c_A(n)$ are shown in Figure 2.56. For clarity, every 100th value of discrete-time output is plotted.

The discontinuity in stern plane angle θ at $t = 0$ is a consequence of lumping the controller and stern plane actuator dynamics into a single equation as we did in Equation 2.174. The first term $K_C e$ is responsible for the direct (strictly algebraic) connection from the error e to the stern plane angle θ and ultimately from r to θ in Figure 2.55. The discontinuity is calculated from

$$
\theta(0) = K_C e(0) = K_C[r(0) - c(0)] = 0.6 \, \text{deg/ft} \times (100 \, \text{ft} - 0) = 60 \, \text{deg} \tag{2.198}
$$

There is a direct connection from θ to v and, therefore, a direct path from r to v explaining the initial jump in depth rate as well. Figure 2.55 shows the term involving $K_{\dot\theta}$ in the sub dynamics is responsible for this. Exercise 2.36 presents an alternate representation of the stern plane actuator that eliminates the discontinuity in both θ and v.

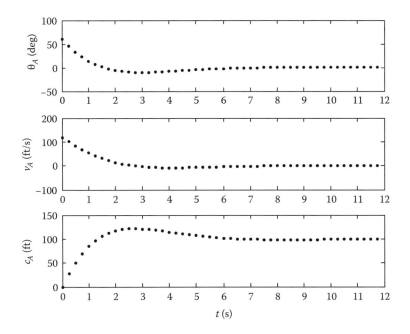

FIGURE 2.56 Discrete-time approximation of subdepth control system step response.

EXERCISES

2.35 Suppose the model of the controller and stern plane actuator in Equation 2.174 is replaced by the following equation:

$$\theta = K_C e + K_1 \int e(t)dt + K_D \frac{d}{dt} e(t)$$

The differential equation relating the control system output $c(t)$ and input $r(t)$ is

$$a_3 \dddot{c} + a_2 \ddot{c} + a_1 \dot{c} + a_0 c = b_3 \dddot{r} + b_2 \ddot{r} + b_1 \dot{r} + b_0 r$$

$$a_3 = \tau + K_D K_{\dot{\theta}} \quad b_3 = K_D K_{\dot{\theta}}$$
$$a_2 = 1 + K_C K_{\dot{\theta}} + K_D K_{\dot{\theta}} \quad b_2 = K_C K_{\dot{\theta}} + K_D K_{\theta}$$
$$a_1 = K_C K_{\theta} + K_I K_{\dot{\theta}} \quad b_1 = K_C K_{\theta} + K_I K_{\dot{\theta}}$$
$$a_0 = K_I K_{\theta} \quad b_0 = K_I K_{\theta}$$

(a) Draw a simulation diagram of the system with three integrators in series.
(b) Choose the state variables as $x_1 = z$, $x_2 = \dot{z}$, $x_3 = \ddot{z}$ where z, $\dot{z}$, $\ddot{z}$ are the outputs of the integrators. Define the output as $y = c$. Find the matrices A, B, C, and D in the state equations.
(c) Find the difference equations for the discrete-time states $x_{1,A}(n+1)$, $x_{2,A}(n+1)$, $x_{3,A}(n+1)$ and discrete-time output $c_A(n)$ similar to Equations 2.194 through 2.197.
(d) Choose the same values for K_C and K_I used to generate Figure 2.56 along with $K_D = 0$. Solve the difference equations recursively to obtain the sub response $y(n)$ for the same input $r(t)$ and compare your result with the graph for $c_A(n)$ in Figure 2.56.
(e) Experiment with new values for K_C, K_I, and K_D. Plot the results for $c_A(n)$.

2.36 The controller and stern plane actuator are modeled separately as shown Figure E2.36:

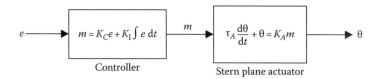

$$e \longrightarrow \boxed{m = K_C e + K_1 \int e \, dt} \xrightarrow{\;m\;} \boxed{\tau_A \frac{d\theta}{dt} + \theta = K_A m} \longrightarrow \theta$$

Controller   Stern plane actuator

FIGURE E2.36

(a) Redraw the simulation diagram of the subdepth control system. Comment on whether m, θ, or v is discontinuous at $t = 0$ when the commanded depth $r(t)$ is a step input.

(b) Determine the state variables and find the new matrices A, B, C, and D in the state equations assuming the output vector $\underline{y} = [m \; \theta \; v \; c]^T$.

3 Elementary Numerical Integration

3.1 INTRODUCTION

Dynamic systems with continuous-time models in the form of differential equations possess memory. For systems with memory, knowledge of the system's inputs at a given point in time is insufficient to determine the state of the system at the same time. For example, a circuit with capacitors and inductors possesses memory. The instantaneous energy stored in the circuit is a function of the current state (capacitor voltages and inductor currents), which depends on the history of its sources (inputs) from the time when the complete state was last known.

An nth-order dynamic system with state variables $x_1, x_2, \ldots, x_n$ and input $u_1, u_2, \ldots, u_m$ is modeled by expressions for the state derivatives, that is,

$$\underline{\dot{x}}(t) = \begin{bmatrix} \dfrac{\mathrm{d}x_1}{\mathrm{d}t} \\[6pt] \dfrac{\mathrm{d}x_2}{\mathrm{d}t} \\[6pt] \vdots \\[6pt] \dfrac{\mathrm{d}x_n}{\mathrm{d}t} \end{bmatrix} = \underline{f}(\underline{x}, \underline{u}) \tag{3.1}$$

where

$$\underline{x} = \begin{bmatrix} x_1 \\ x_2 \\ \vdots \\ x_n \end{bmatrix}, \quad \underline{u} = \begin{bmatrix} u_1 \\ u_2 \\ \vdots \\ u_m \end{bmatrix}, \quad \underline{f}(\underline{x}, \underline{u}) = \begin{bmatrix} f_1(x, u) \\ f_2(x, u) \\ \vdots \\ f_n(x, u) \end{bmatrix} \tag{3.2}$$

In a formal sense, n distinct integrations are required to obtain the state $\underline{x}$, namely

$$x_1(t) = x_1(t_0) + \int_{t_0}^{t} f_1(x, u)\mathrm{d}t' \tag{3.3}$$

$$x_2(t) = x_2(t_0) + \int_{t_0}^{t} f_2(x, u)\mathrm{d}t' \tag{3.4}$$

$$x_n(t) = x_n(t_0) + \int_{t_0}^{t} f_n(x, u)\mathrm{d}t' \tag{3.5}$$

For time-varying systems, a number of the system parameters are explicit functions of time. For example, the amount of fuel in a rocket or aircraft diminishes with time, thereby affecting its dynamic properties. The state derivative vector is generally denoted by $f(t, x, u)$, and Equations 3.3 through 3.5 are more appropriately expressed as

$$x_1(t) = x_1(t_0) + \int_{t_0}^{t} f_1(t', x, u) \mathrm{d}t' \tag{3.6}$$

$$x_2(t) = x_2(t_0) + \int_{t_0}^{t} f_2(t', x, u) \mathrm{d}t' \tag{3.7}$$

$$x_n(t) = x_n(t_0) + \int_{t_0}^{t} f_n(t', x, u) \mathrm{d}t' \tag{3.8}$$

Equations 3.3 through 3.8 remind us that if we know the complete state at some initial time t_0, then at some future time t, the state can be determined provided we know the inputs from t_0 to t. The n integrals to be evaluated in Equations 3.3 through 3.8 constitute the process of advancing or updating the state through time. This chapter looks at various alternatives for approximating these integrals.

3.2 DISCRETE-TIME SYSTEM APPROXIMATION OF A CONTINUOUS-TIME INTEGRATOR

The continuous-time integrator shown in Figure 3.1 is a special case of a first-order dynamic system in which the state derivative function is equal to the system input.

$$\frac{\mathrm{d}x}{\mathrm{d}t} = f(x, u) = f(u) = u(t) \tag{3.9}$$

Alternatively, it can be thought of as the simple linear first-order system

$$\frac{\mathrm{d}}{\mathrm{d}t} x(t) + a_0 x(t) = Ku(t) \tag{3.10}$$

discussed in Section 2.2 where $a_0 = 0$ and $K = 1$. The solution for $x(t)$ is given by

$$x(t) = x(t_0) + \int_{t_0}^{t} u(t') \mathrm{d}t' \tag{3.11}$$

where
   t_0 is some initial time
   $x(t_0)$ is the initial state

FIGURE 3.1 A continuous-time integrator.

Example 3.1

The input to an integrator is $u(t) = A \sin \omega t$, $t \geq 0$. Find the output if $x(0) = 0$.

$$x(t) = 0 + \int_0^t u(t')dt' \tag{3.12}$$

$$= \int_0^t A \sin \omega t' dt' \tag{3.13}$$

$$= A\left[-\frac{1}{\omega} \cos \omega t'\right]_0^t = \frac{A}{\omega}(1 - \cos \omega t) \tag{3.14}$$

The continuous-time input $u(t)$ and the integrator output are graphed in Figure 3.2 for the case where the amplitude A is unit and the radian frequency $\omega = 2\pi$ rad/s. The integrator output at any point in time t_1 is simply the area under the input from $t = 0$ to $t = t_1$. The output and corresponding area are shown in Figure 3.2 for $t_1 = 0.4$.

System simulation using analog computers was popular years ago. They were capable of implementing Equation 3.11 using electronic components (operational amplifiers, resistors, capacitors, and potentiometers). In fact, an analog computer simulation diagram is similar to the simulation diagram presented in the previous chapter. However, there are a number of hardware-related issues inherent in analog simulation, not present with digital simulation. The popularity and widespread use of digital computers has produced a shift from analog to digital as the primary means of continuous-time system simulation.

Digital computers, however, are sequential machines. Unlike analog computers, they are not capable of solving Equations 3.3 through 3.5 in a continuous fashion. Digital simulation of continuous-time systems relies on numerical algorithms to approximate the integral of the state derivative function using sampled values at discrete points in time. Figure 3.3 illustrates the process for the simple integrator in Equation 3.9.

The approximation block in Figure 3.3 is a discrete-time system with input $u(nT)$ and output $x_A(nT)$ designed to approximate the output of the continuous-time integral $x(t)$. A difference equation for the discrete-time system is needed.

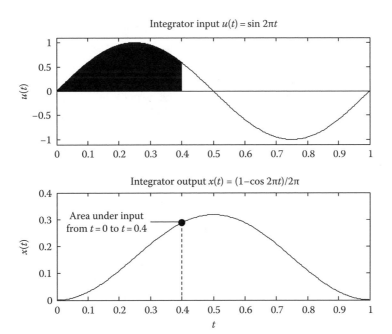

FIGURE 3.2 Continuous-time integrator $u(t) = \sin 2\pi t$, $x(t) = (1 - \cos 2\pi t)/2\pi$.

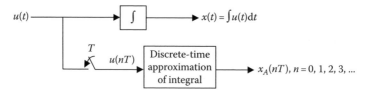

FIGURE 3.3 A continuous-time integrator and a discrete-time approximation.

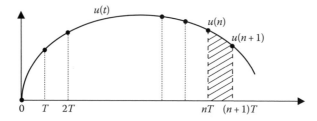

FIGURE 3.4 Interval of integration with subintervals of width T.

We begin by dividing the interval 0 to $(n+1)T$ into $n+1$ equal subintervals of width T as shown in Figure 3.4. Note that $u(n)$ and $u(n+1)$ are short for $u(nT)$ and $u[(n+1)]T$.

The integral of a continuous-time function $u(t)$ over the interval 0 to $(n+1)T$ is equal to the area bounded by the function and the t-axis. Dividing the area into two pieces gives

$$\int\limits_{0}^{(n+1)T} u(t)dt = \int\limits_{0}^{nT} u(t)dt + \int\limits_{nT}^{(n+1)T} u(t)dt \tag{3.15}$$

The integrals in Equation 3.15 with lower limits of zero represent the output of a continuous integrator with input $u(t)$ and both t_0 and $x(t_0)$ equal to zero (see Equation 3.11).

Consequently, Equation 3.15 is expressible as

$$x[(n + 1)T] = x(nT) + \int\limits_{nT}^{(n+1)T} u(t)dt \tag{3.16}$$

Several algorithms for approximating the integral in Equation 3.16 (shaded area in Figure 3.4) are presented in the following section. Each algorithm will result in a unique difference equation relating the discrete-time input $u(nT)$ and discrete-time output $x_A(nT)$ shown in Figure 3.3. The resulting discrete-time systems are the foundation of our venture into the field of numerical integration.

EXERCISES

3.1 Consider the first-order system

$$\frac{d}{dt}y(t) + a_0 y(t) = u(t)$$

(a) Find the response of the system to a step input $u(t) = 1$, $t \geq 0$.
(b) Find the response of the system to a ramp input $u(t) = t$, $t \geq 0$.
(c) In the limit as a_0 approaches zero, the first-order system reduces to a pure integrator. Show that the step and ramp responses in parts (a) and (b) approach $\int_0^t 1 \cdot dt'$ and $\int_0^t t'dt'$, respectively.

3.2 The signal $u(t) = c_0 + c_1(t - t_0)^2$, $t \geq 0$ in Figure E3.2 is input to a system governed by $dy/dt = u(t)$, that is, a continuous-time integrator:

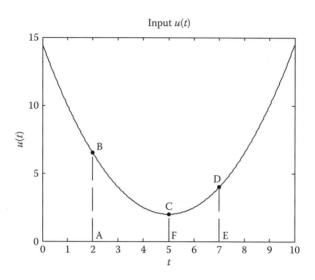

Input $u(t)$

FIGURE E3.2

The change in output $y(t)$ from $t = t_0$ to $t = t_0 + \Delta_1$ to $t = t_0 + \Delta_2$ is of interest. Using the values $c_0 = 2$, $c_1 = 1/2$ $t_0 = 5$, $\Delta_1 = 3$, and $\Delta_2 = 2$, approximate the difference $y(t_0 + \Delta_2) - y(t_0 - \Delta_1)$

(a) By replacing $u(t)$ with a piecewise linear function $u_1(t)$ through points B and C, and C and D and then integrating $u_1(t)$ between appropriate limits
(b) As the areas of trapezoids ABCF and CDEF
(c) By comparing your answers in parts (a) and (b) to the true value

$$y(t_0 + \Delta_2) - y(t_0 - \Delta_1) = \int_{t_0 - \Delta_1}^{t_0 + \Delta_2} u(t)\mathrm{d}t$$

3.3 A tank with cross-sectional area A_1 and resistance R_1 empties into a second tank with cross-sectional area A_2. The first tank has no inflow and is initially filled to a height $h_1(0)$. The second tank is initially empty and has no outflow. The flow between the tanks is denoted by $f_1(t)$, and the tank levels are $h_1(t)$ and $h_2(t)$.

(a) Find the first-order differential equations for $f_1(t)$ and $h_2(t)$.
(b) Show that the second tank is an integrator.
(c) Find expressions for the transient responses of $f_1(t)$ and $h_2(t)$.
(d) For system parameter values $A_1 = 100$ ft^2, $R_1 = 0.25$ ft per ft^3/min, $A_2 = 50$ ft^2, and $h_1(0) = 20$ ft, the responses $f_1(t)$ and $h_2(t)$ are plotted in Figure E3.3. Estimate the level in tank 2 after 50 min by approximating the area under $f_1(t)$, $0 \leq t \leq 50$ and dividing by A_2. Approximate the area using simple geometric shapes like rectangles and trapezoids.
(e) Compare your answer from part (d) with the true value $h_2(50)$.

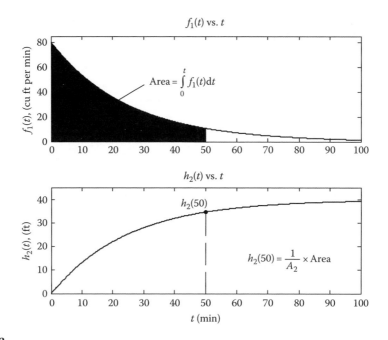

FIGURE E3.3

3.3 EULER INTEGRATION

The previous section presented a framework for finding a discrete-time system approximation of a continuous-time integrator. An approximation to the integral term in Equation 3.16 is needed. The simplest approach assumes the integrator input $u(t)$ is constant over the interval, that is, $u(t) \approx u(n)$, $nT \leq t \leq (n+1)T$ where $u(n)$ is short for $u(nT)$ as shown in Figure 3.5.

The exact area under the function $u(t)$, $nT \leq t \leq (n+1)T$ is being approximated by the area of the rectangle shown in Figure 3.5. Hence, Equation 3.16 becomes

$$x[(n+1)T] \approx x(nT) + Tu(n) \tag{3.17}$$

A difference equation results if we denote the approximation to $x(nT)$ by $x_A(n)$. By implication, $x[(n+1)T]$ is approximated by $x_A(n+1)$ and the difference equation reads

$$x_A(n+1) = x_A(n) + Tu(n) \tag{3.18}$$

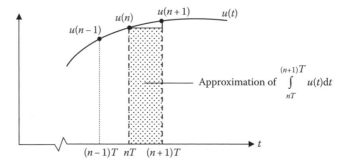

FIGURE 3.5 Approximation of area under $u(t)$ assuming $u(t) \approx u(nT)$.

Equation 3.18 is the difference equation of a numerical integrator that can be solved recursively to generate an approximation to the continuous-time integrator output $x(t)$ at discrete points in time, that is,

$$x_A(n) \approx x(nT), \quad n = 0, 1, 2, \ldots \tag{3.19}$$

The discrete-time system modeled in Equation 3.18 is commonly referred to as an Euler or rectangular integrator. The subinterval width T is termed the integration step size.

Euler integration can be derived by means other than approximating the integral in Figure 3.5 as the area of a rectangle. Alternatively, Euler integration is a consequence of assuming that the state derivative function is constant during each integration step. The starting point is the equation for the state derivative of a pure integrator

$$\frac{dx}{dt} = f(x, u) = u(t) \tag{3.20}$$

with initial condition $x(0)$ and input $u(t)$ known at the beginning of each integration step. Calculating the initial state derivative,

$$\frac{dx}{dt}(0) = f[x(0), u(0)] = u(0) \tag{3.21}$$

The approximation to the continuous-time state $x(t)$ is updated under the assumption the derivative $dx(0)/dt$ remains constant over the integration time step, that is,

$$x_A(T) = x(0) + T\frac{dx}{dt}(0) \tag{3.22}$$

$$= x(0) + Tu(0) \tag{3.23}$$

The situation is illustrated in Figure 3.6. The estimate $x_A(T)$ is the result of "riding" the tangent to $x(t)$ from the initial point $x(0)$ to the end of the interval.

The process is repeated to generate the updated states $x_A(2T)$, $x_A(3T)$, etc. A similar graphical interpretation applies with the exception that subsequent movements along the computed directions start from the approximate values $x_A(T)$, $x_A(2T)$, ... as opposed to the actual points $x(T)$, $x(2T)$, ... on the solution $x(t)$.

The result for $x_A(2T)$ is

$$x_A(2T) = x_A(T) + T\frac{dx}{dt}(T) \tag{3.24}$$

$$= x_A(T) + Tu(T) \tag{3.25}$$

Based on Equations 3.23 and 3.25, it follows that the $(n+1)$st state update is

$$x_A[(n+1)T] = x_A(nT) + Tu(nT) \tag{3.26}$$

Dropping T from the arguments in Equation 3.26 yields a result identical to Equation 3.18.

The two ways of deriving the difference equation for Euler integration are essentially the same. Approximating the shaded area in Figure 3.5 by a rectangle stems from assuming that the integrator input $u(t)$ is constant over each integration step. However, the derivative is equal to the input for a pure integrator. Hence, assuming that the input is constant is equivalent to making the same assumption about the derivative.

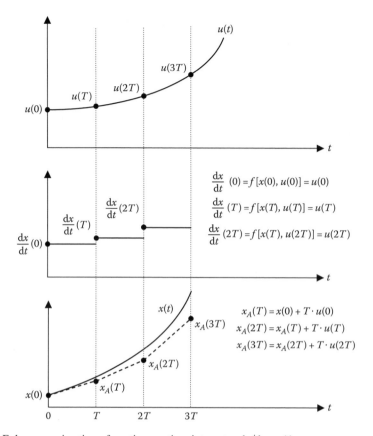

FIGURE 3.6 Euler approximation of continuous-time integrator $dx/dt = u(t)$.

According to Equation 3.18, the Euler integrator simply adds a rectangular area $Tu(n)$ to the current state $x_A(n)$ to produce the updated state $x_A(n+1)$. A general formula for $x_A(n+1)$ is easily obtained by observing

$$x_A(1) = x_A(0) + Tu(0) \tag{3.27}$$

$$x_A(2) = x_A(1) + Tu(1) \tag{3.28}$$

$$= [x_A(0) + Tu(0)] + Tu(1) \tag{3.29}$$

$$= x_A(0) + T[u(0) + u(1)] \tag{3.30}$$

leading to the general result

$$x_A(n+1) = x(0) + T[u(0) + u(2) + \cdots + u(n-1) + u(n)] \tag{3.31}$$

$$= x(0) + T\sum_{k=0}^{n} u(k) \tag{3.32}$$

The simplistic nature of Equation 3.32 results from the simple model describing the state derivative of a pure integrator, that is, $dx/dt = u(t)$.

Example 3.2

The input to a continuous-time integrator is a sinusoidal function $u(t) = \sin 2\pi t$, $0 \le t \le 0.5$. The initial condition is $x(0) = 0$.

(a) Use Euler integration with a step size $T = 0.05$ s to approximate the integrator output $x(t)$ at $t = 0.1, 0.2, \ldots, 0.5$ s.
(b) Compare your answers for $x_A(n)$ from part (a) with the continuous-time integrator output $x(t)$ at $t = 0.1, 0.2, \ldots, 0.5$ s.

(a)
$$x_A(n+1) = x_A(n) + Tu(n), \quad n = 0, 1, 2, 3, \ldots$$
$$= x_A(n) + T\sin(2\pi nT) \quad n = 0, 1, 2, 3, \ldots$$

$n = 0$: $\quad x_A(1) = x_A(0) + Tu(0)$
$$= 0 + 0.05\{[\sin(2\pi)(1)(0.05)]\}$$
$$= 0$$

$n = 1$: $\quad x_A(2) = x_A(1) + Tu(1)$
$$= 0 + 0.05\{[\sin(2\pi)(1)(0.05)]\}$$
$$= 0.0155$$

$n = 2$: $\quad x_A(3) = x_A(2) + Tu(2)$
$$= 0.0155 + 0.05\{[\sin(2\pi)(2)(0.05)]\}$$
$$= 0.0448$$

The process is continued for $n = 3, 4, \ldots, 9$ in order to obtain the required estimates of $x(t)$ at 0.1 s intervals. The results are tabulated in column 3 of Table 3.1.

(b) The exact values for $x(t)$ are calculated using Equation 3.14 with $A = 1$ and $\omega = 2\pi$ (see last column in Table 3.1).

3.3.1 BACKWARD (IMPLICIT) EULER INTEGRATION

If we can approximate the integrator input $u(t)$ by its numerical value at the beginning of an integration interval (see Figure 3.5), then we should be able to choose another value of the input at a different point in time within the interval. Two other points on the interval appear to be logical choices. One is the midpoint and the other is the endpoint of the interval. The latter choice will now be explored.

TABLE 3.1
Outputs of Euler Integrators ($T = 0.05$)
and Continuous-Time Integrator

N	Forward Euler $t_n = nT$	Backward Euler $x_A(n)$	Continuous-Time $x_A(n)$	$x(t_n)$
0	0	0	0	0
2	0.1	0.0155	0.0448	0.0304
4	0.2	0.0853	0.1328	0.1100
6	0.3	0.1828	0.2304	0.2083
8	0.4	0.2708	0.3002	0.2879
10	0.5	0.3157	0.3157	0.3183

Referring to Figure 3.5, suppose the input $u(t)$ is assumed equal to $u(n+1)$ instead of $u(n)$ in the interval $nT \leq t \leq (n+1)T$. The area of the rectangular strip intended to approximate the true area under the input is now $Tu(n+1)$, that is,

$$\int_{nT}^{(n+1)T} u(t)dt \approx Tu(n+1) \tag{3.33}$$

resulting in the numerical integrator

$$x_A(n+1) = x_A(n) + Tu(n+1) \tag{3.34}$$

Since the input is assumed constant over the integration interval, Equation 3.34 is also a difference equation for an Euler integrator. It differs from the previous Euler integrator in Equation 3.18 in that $u(n+1)$ replaces $u(n)$ in the calculation of the new state $x_A(n+1)$. The numerical integrator in Equation 3.18 is referred to as forward Euler whereas the difference equation in Equation 3.34 is that of a backward Euler integrator.

Unlike a pure continuous-time integrator, the derivative dx/dt of first and higher order systems is dependent on the state $x(t)$ and possibly one or more inputs. Difference equations for updating the discrete-time state using Euler integration depend on whether forward or backward integration is used. The two cases are

$$\text{Forward Euler: } x_A(n+1) = x_A(n) + Tf[(x_A(n), u(n)] \tag{3.35}$$

$$\text{Backward Euler: } x_A(n+1) = x_A(n) + Tf[(x_A(n+1), u(n+1)] \tag{3.36}$$

Equation 3.36 leads to implicit algebraic equations involving $x_A(n+1)$, which, depending on the state derivative function $f(x, u)$, may be difficult or impossible to solve analytically. For this reason, the backward Euler integrator in Equation 3.34 is also known as implicit Euler integration and the forward Euler integrator in Equation 3.18 is called explicit Euler integration.

Example 3.3

Rework Example 3.2 using the backward Euler integrator.

$$x_A(n+1) = x_A(n) + Tu(n+1), \quad n = 0, 1, 2, 3, \ldots$$
$$= x_A(n) + T\sin[2\pi(n+1)T], \quad n = 0, 1, 2, 3, \ldots$$

$$n = 0: \quad x_A(1) + x_A(0) + Tu(1)$$
$$= 0 + 0.05\{\sin[(2\pi)(1)(0.05)]\}$$
$$= 0.0155$$

$$n = 1: \quad x_A(2) = x_A(1) + Tu(2)$$
$$= 0.0155 + 0.05\{[\sin(2\pi)(2)(0.05)]$$
$$= 0.0448$$

The remaining values are presented in column 4 of Table 3.1.

Both numerical integrators produce significant errors in comparison to the analytical solution. Greater accuracy is possible by reducing the integration step size. The trade-off is, of course, the additional computations required.

EXERCISES

3.4 In Examples 3.2 and 3.3,

(a) Explain why the implicit Euler integrator produces higher estimates of the continuous-time integrator output than the explicit Euler integrator. Is this true in general?

(b) Find $x_A(5)$ for both numerical integrators and compare the results to $x(0.25)$. Explain why both integrators incur the maximum error $|x(nT) - x_A(n)|$ for $n = 5$.

(c) Repeat Examples 3.2 and 3.3 with a step size $T = 0.01$. Enter the numerical results in a table rounded to six places after the decimal point.

3.5 The RC circuit shown in Figure E3.5 is a first-order low-pass filter. The differential equation relating the output voltage $v_0(t)$ and input voltage $v_i(t)$ is

$$RC\frac{dv_0}{dt} + v_0 = v_i$$

A discrete-time integrator is used to approximate the continuous output $v_0(t)$ when the input $v_i(t)$ is an AC signal $\sin \omega t$.

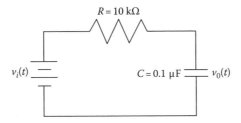

FIGURE E3.5

(a) Find the difference equation used to obtain $v_{0,A}(n)$ if forward Euler integration is used with a step size of T.

(b) For $v_i(t) = \sin \omega t$, find and plot $v_{0,A}(n)$ corresponding to $0 \le n \le 4\pi/\omega T$ when (i) $\omega = 100$ rad/s, $T = RC/10$ (ii) $\omega = 1000$ rad/s, $T = RC/100$.

3.6 The flow out of the tank shown in Figure E3.6 is given by $F_0 = cH^{1/2}$. The cross-sectional area of the tank $A = 50$ ft^2 and the constant $c = 2$ ft^3/min per ft$^{1/2}$. The tank is 25 ft in height and the initial level in the tank $H(0) = 16$ ft.

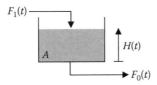

FIGURE E3.6

(a) The flow into the tank is $F_1(t) = \overline{F}_1 = 10$ ft^3/ min, $t \ge 0$. Find the steady-state height of liquid in the tank, $\underline{H}(\infty)$.

(b) Use forward Euler integration with a suitable step size and compare $\lim_{n \to \infty} H_A(n)$ with the result from part (a).

(c) The flow into the tank is $F_1(t) = 4 + (t/10)$, $t \ge 0$. Use forward Euler integration with a step size T and find the difference equation for updating the state $H_A(n)$. Leave your answer in terms of c, A, and T.

(d) For the input flow rate in part (c), using forward Euler integration with $T=0.1$ min, find n_f, where $n_f T$ is the time required to fill the tank, that is, $H_A(n_f-1) < 25$ and $H_A(n_f) \geq 25$. Plot the results.

3.7 The input to the integrator shown in Figure E3.7 is the continuous-time signal $u(t) = 1/(t+1)$, $t \geq 0$:

$u(t) \longrightarrow \boxed{\int} \longrightarrow x(t)$

FIGURE E3.7

(a) Find the difference equation for computing the state $x_A(n)$ recursively when implicit Euler integration with a step size T is used.
(b) Find $x_A(1)$, $x_A(2)$, and $x_A(3)$ if $T=0.1$.
(c) Compare your answer for $x_A(3)$ to the exact value $x(3T)$.

$$\text{Note:} \quad \int_0^t \frac{1}{(t'+1)\mathrm{d}t'} = 1n(1+t).$$

3.4 TRAPEZOIDAL INTEGRATION

Of the numerical integrators, the Euler integrators are the simplest to implement; however, for a given integration step size, they are also the least accurate. This is not necessarily a reason to choose another integrator since any desired level of accuracy is achievable with Euler integrators (in principle) simply by reducing the step size and performing additional calculations. Indeed, the simplicity of Euler integration is responsible for its widespread use in far-ranging applications.

There may be circumstances that dictate the integration step size in a simulation study and thus compel the developer to consider other methods for approximating the dynamics of a continuous-time system. Accordingly, we shall investigate other formulas and algorithms for numerical integration.

Starting with

$$x_A(n+1) = x_A(n) + \text{estimate of} \int_{nT}^{(n+1)T} u(t)\mathrm{d}t \tag{3.37}$$

a more accurate (compared with Euler integration) estimate of the integral in Equation 3.37 is attainable by approximating the input $u(t)$ by a piecewise linear function $u_1(t)$ where

$$u_1(t) = u(n) + \left[\frac{u(n+1) - u(n)}{T}\right](t - nT), \quad nT \leq t \leq (n+1)T \tag{3.38}$$

as shown in Figure 3.7.

It is left as an exercise to show that

$$\int_{nT}^{(n+1)T} u_1(t)\mathrm{d}t = \frac{T}{2}[u(n) + u(n+1)] \tag{3.39}$$

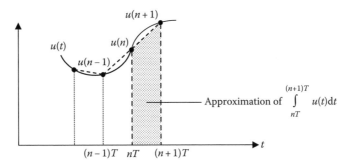

FIGURE 3.7   Trapezoidal approximation of area under $u(t)$, $nT \leq t \leq (n+1)T$.

The shaded area in Figure 3.7 used to approximate the true area under the input $u(t)$ is a trapezoid (rotated 90°) with bases $u(n)$ and $u(n+1)$ and height of T. The expression on the right-hand side of Equation 3.39 is simply the area of the corresponding trapezoid. Using the trapezoidal approximation, Equation 3.37 becomes

$$x_A(n+1) = x_A(n) + \frac{T}{2}[u(n) + u(n+1)] \tag{3.40}$$

Equation 3.40 is known as trapezoidal integration. Similar to backward Euler integration, the difference equation leads to an implicit algebraic equation in $x_A(n+1)$ for all continuous-time systems other than a pure integrator.

Example 3.4 demonstrates the use of trapezoidal integration to approximate a definite integral. The integrand can be thought of as the input $u(t)$ to a continuous-time integrator, while $x_A(n)$ $n = 0, 1, 2, 3, \ldots$ represents the discrete-time approximation to the output $x(t)$, at $t = nT$, $n = 0, 1, 2, 3, \ldots$

Example 3.4

Approximate the definite integral $x(t) = \int_0^t e^{-2t'} dt'$ at $t = 0, 1, 2, 3, \ldots, 1.0$ using trapezoidal integration with an integration step size $T = 0.1$.

$$u(n) = ut|_{t=nT} = e^{-2t}\big|_{t=nT} = e^{-2nT} \tag{3.41}$$

$$u(n+1) = u(t)|_{t=(n+1)T} = e^{2t}\big|_{t=(n+1)T} = e^{-2(n+1)T} \tag{3.42}$$

From Equation 3.40,

$$x_A(n+1) = x_A(n) + \frac{T}{2}\left[e^{-2nT} + e^{-2(n+1)T}\right] \tag{3.43}$$

Setting $x_A(0) = x(0) = 0$,

$$n = 0: \quad x_A(1) = 0 + \frac{0.1}{2}\left[e^{-2(0)(0.1)} + e^{-2(0+1)(0.1)}\right] = 0.09093654$$

$$n = 1: \quad x_A(2) = x_A(1) + \frac{0.1}{2}\left[e^{-2(1)(0.1)} + e^{-2(1+1)(0.1)}\right] = 0.16538908$$

TABLE 3.2

Approximations to a Definite Integral Using Three Numerical Integrators (Explicit and Implicit Euler, Trapezoidal) and the Exact Solution

n	$t_n = nT$	Euler (Explicit) $x_A(n)$	Euler (Implicit) $x_A(n)$	Trapezoidal $x_A(n)$	Exact $x(t_n)$
0	0.0	0.0	0.0	0.0	0.0
1	0.1	0.10000000	0.08187308	0.09093654	0.09063462
2	0.2	0.18187308	0.14890508	0.16538908	0.16483998
3	0.3	0.24890508	0.20378624	0.22634566	0.22559418
4	0.4	0.30378624	0.24871914	0.27625269	0.27533552
5	0.5	0.34871914	0.28550708	0.31711311	0.31606028
6	0.6	0.38550708	0.31562651	0.35056679	0.34940289
7	0.7	0.41562651	0.34028620	0.37795635	0.37670152
8	0.8	0.44028620	0.36047585	0.40038103	0.39905174
9	0.9	0.46047585	0.37700574	0.41874080	0.41735056
10	1.0	0.47700574	0.39053927	0.43377251	0.43233236

The remaining values $x_A(3)$, $x_A(4)$, ..., $x_A(10)$ are calculated in the same manner and shown in Table 3.2, which also includes the results obtained using both types of Euler integrators. The last column contains the exact values of the definite integral,

$$x(t) = \int_0^t e^{-2t'} dt' = \left[\frac{e^{-2t'}}{-2} \right]_0^t = \frac{1}{2}(1 - e^{-2t}) \tag{3.44}$$

For the same step size, the trapezoidal integrator is superior to the Euler integrators. An advantage of trapezoidal integration compared with Euler is the increased step size that can be used while maintaining comparable accuracy.

The following example illustrates the use of trapezoidal integration for a first-order system modeled by a differential equation with time-varying parameters.

Example 3.5

A nonlinear, time-varying dynamic system is modeled by the differential equation

$$t^2 \frac{dy}{dt} + y\frac{dy}{dt} + 2ty = u(t) \tag{3.45}$$

(a) Find the difference equation of the discrete-time system based on trapezoidal integration for approximating the response of the continuous-time system.

(b) Solve the difference equation for $y_A(n)$, $n = 0, 1, 2, \ldots$ when the continuous-time input is $u(t) = -3t^2/2$. The initial condition is $y(0) = 1$ and the step size is $T = 0.01$.

(c) Plot the discrete-time response $y_A(n)$, $n = 0, 1, 2, \ldots, 100$ and the exact solution $y(t) = -t^2 + (t^4 - t^3 + 1)^{1/2}$, $0 \le t \le 1$ on the same graph.

(a) Solving for the state derivative,

$$\frac{dy}{dt} = f(t, y, u) = \frac{1}{t^2 + y(t)} [u(t) - 2ty(t)] \tag{3.46}$$

From Equation 3.40 with u replaced by the derivative function $f(t, y, u)$, the difference equation for trapezoidal integration is

$$y_A(n + 1) = y_A(n) + \frac{T}{2} \{f[nT, y_A(nT), u(nT)] + f[nT + T, y_A(nT + T), u(nT + T)]\} \tag{3.47}$$

$$= y_A(n) + \frac{T}{2} \left\{ \frac{1}{(nT)^2 + y_A(n)} [u(n) - 2(nT)y_A(n)] \right.$$

$$\left. + \frac{1}{[(n+1)T]^2 + y_A(n+1)} [u(n+1) - 2[(n+1)T]y_A(n+1)] \right\} \tag{3.48}$$

(b) Equation 3.48 is an implicit equation for $y_A(n+1)$, which generally means some type of iterative, numerical root-solving algorithm is required to find $y_A(n+1)$ at each time step. This can increase the computational requirements dramatically, not to mention the additional programming required to implement the algorithm. In this example, however, Equation 3.48 can be manipulated to produce a quadratic function of the form

$$a[y_A(n+1)]^2 + by_A(n+1) + c = 0 \tag{3.49}$$

where a, b, and c are expressible in terms of $u(n)$, $y_A(n)$, and $u(n+1)$, all of which can be calculated at time $t_n = nT$. "Chap3_Ex4_2.m" includes the statements to determine a, b, and c and solve Equation 3.49 at each time step for the positive root.

(c) The approximate and exact solutions are shown in Figure 3.8. The exact continuous-time response $y(t)$ and the approximate discrete-time response $y_A(n)$ are indistinguishable from each other at times $t_n = nT$, $n = 0, 1, 2, \ldots, 100$. Let us not forget that the discrete-time signal $y_A(n)$ is defined solely at the discrete times $0, T, 2T, 3T, \ldots$, which explains why discrete-time signals should always be plotted as discrete data points. A dotted line should be used whenever the points are connected to emphasize this point.

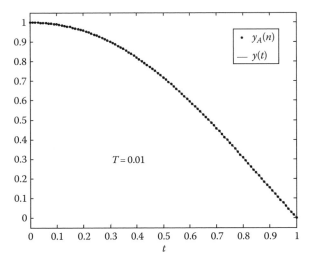

FIGURE 3.8   Graph of approximate (trapezoidal integration) and exact solutions.

EXERCISES

3.8 Referring to Figure 3.7,
  (a) Find the equation of the linear approximation $u_1(t)$ through the end points $[nT, u(n)]$ and $[(n+1)T, u(n+1)]$.
  (b) Verify Equation 3.39 by integrating $u_1(t)$ from nT to $(n+1)T$.

3.9 The first-order system $dx/dt = \lambda x$ with initial condition $x(0) = x_0$ is to be simulated using trapezoidal integration with step size T. The truncation error after n steps is $\varepsilon_n = x_A(n) - x(nT)$, where $x(t)$, $t \geq 0$ is the exact solution and $x_A(n)$, $n = 0, 1, 2, \ldots$ is the approximate (simulated) solution, that is, $x_A(n) \approx x(nT)$, $n = 0, 1, 2, 3, \ldots$. Suppose the truncation error after the first step is equal to a fraction of the initial condition, that is,

$$\varepsilon_1 = x_A(1) - x(T) = \alpha x_0 \quad (0 < \alpha \ll 1)$$

λT satisfies the condition

$$e^{\lambda T} = \frac{a\lambda T + b}{\lambda T + c}$$

Express the constants a, b, and c in terms of α and x_0.

3.10 The population of a city $P(t)$ is modeled by the differential equation $dP/dt = kP$.
  (a) Find the equation for updating $P_A(n)$, the approximate population at the end of year nT, using trapezoidal integration with step size T. Leave your answer in terms of k and T.
  (b) Suppose $k = 0.01$ people/year per person, the initial population is one million people and the step size $T = 1$ year. Find $P_A(1)$ and $P_A(2)$ to the nearest person.
  (c) Find the general solution for $P_A(n)$ and use it to find $P_A(100)$.
  (d) Compare the result from part (c) to the exact value $P(100)$.

3.11 The mass m in Figure E3.11 is subjected to a time-varying damping force $f_d(t)$. The differential equation describing the motion is $m(d/dt)v(t) = f_d(t)$ where $v(t)$ is the velocity of the mass and $f_d(t) = [-t/(1+t)]v(t)$.
  (a) Use trapezoidal integration with suitable step size T to approximate the velocity $v(t)$, $t \geq 0$. Note that $m = 1$ slug and the initial velocity $v(0) = 10$ ft/s.
  (b) Compare the simulated response $v_A(n)$, $n = 0, 1, 2, \ldots$ in part (a) to the exact solution $v(t) = 10(1+t)e^{-t}$, $t \geq 0$.

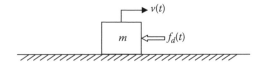

FIGURE E3.11

3.12 Find the largest step size T in Example 4.5 for which

$$|y(nT) - y_A(n)| < 0.005, \quad n = 0, 1, 2, \ldots, 1/T$$

3.5 NUMERICAL INTEGRATION OF FIRST-ORDER AND HIGHER CONTINUOUS-TIME SYSTEMS

The trapezoidal and two Euler integrators developed in the previous two sections were used to approximate the dynamics of the first-order system described by

$$\frac{dx}{dt} = f(x, u) = u \tag{3.50}$$

that is, a continuous-time integrator. We now consider the more general case when the state derivative function $f(x, u)$ is a function of the state x as well as the input u. For example,

$$\frac{dx}{dt} = f(x, u) = b_0 u - a_0 x \tag{3.51}$$

In the case of Euler integrators, the state derivative function $f(x, u)$ is assumed constant over the time interval corresponding to a single integration step. This assumption is responsible for Equations 3.35 and 3.36, which are repeated in Equations 3.52 and 3.55. The two equations are the starting points for deriving the difference equations for Euler integration to approximate the first-order system in Equation 3.51.

$$\text{Explicit Euler: } x_A(n+1) = x_A(n) + Tf[x_A(n), u(n)] \tag{3.52}$$

$$= x_A(n) + T[b_0 u(n) - a_0 x_A(n)] \tag{3.53}$$

$$\Rightarrow x_A(n+1) - (1 - a_0 T)x_A(n) = b_0 Tu(n) \tag{3.54}$$

$$\text{Implicit Euler: } x_A(n+1) = x_A(n) + Tf[x_A(n+1), u(n+1)] \tag{3.55}$$

$$= x_A(n) + T[b_0 u(n+1) - a_0 x_A(n+1)] \tag{3.56}$$

$$\Rightarrow (1 + a_0 T)x_A(n+1) - x_A(n) = b_0 Tu(n+1) \tag{3.57}$$

Note that $x_A(n+1)$ in Equation 3.52 is expressed explicitly in terms of $x_A(n)$ in contrast to Equation 3.55, which is an implicit equation with $x_A(n+1)$ appearing on both sides. In the case of nonlinear systems, $f(x, u)$ is a nonlinear function of x, and the implicit equation is more of a challenge to solve for $x_A(n+1)$ than is the explicit equation. For a linear first-order system, Equation 3.57 is easily solvable for $x_A(n+1)$ resulting in

$$x_A(n+1) = \frac{1}{1 + a_0 T}[x_A(n) + b_0 Tu(n+1)] \tag{3.58}$$

3.5.1 DISCRETE-TIME SYSTEM MODELS FROM SIMULATION DIAGRAMS

Recall that a simulation diagram represents the dynamics of a continuous-time system as a connection of algebraic blocks and integrators. Discrete-time systems for approximating the behavior of continuous-time systems can be obtained by replacing the continuous-time integrators with discrete-time (numerical) integrators. The continuous-time signals are converted to discrete-time signals.

To illustrate the process, consider the first-order system modeled by Equation 3.51. The simulation diagram is shown in Figure 3.9.

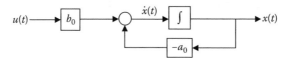

FIGURE 3.9   Simulation diagram of first-order system: $dx/dt = f(x, u) = b_0 u - a_0 x$.

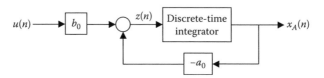

FIGURE 3.10   Discrete-time system approximation of first-order continuous-time system.

The discrete-time system approximation is shown in Figure 3.10. The continuous-time integrator is replaced by a discrete-time integrator, and all signals are discrete time. The input to the discrete-time integrator is labeled $z(n)$ for convenience.

The difference equation of the discrete-time integrator in Figure 3.10 depends on which numerical integrator is chosen to approximate the continuous-time integrator. For an explicit Euler integrator with input $z(n)$ and output $x_A(n)$,

$$x_A(n + 1) = x_A(n) + Tz(n) \tag{3.59}$$

where $z(n)$ is given by

$$z(n) = b_0 u(n) - a_0 x_A(n) \tag{3.60}$$

Substitution of Equation 3.60 into Equation 3.59 results in the explicit Euler integrator in Equation 3.53. For an implicit Euler integration, the continuous-time integrator is replaced by a discrete-time integrator block described by

$$x_A(n + 1) = x_A(n) + Tz(n + 1) \tag{3.61}$$

With $z(n)$ given by Equation 3.60, it follows that

$$z(n + 1) = b_0 u(n + 1) - a_0 x_A(n + 1) \tag{3.62}$$

Combining Equations 3.61 and 3.62 leads to the implicit Euler integrator Equation 3.56.

If trapezoidal integration is preferred, the discrete-time integrator in Figure 3.10 with input $z(n)$ and output $x_A(n)$ is governed by

$$x_A(n + 1) = x_A(n) \frac{T}{2}[z(n) + z(n + 1)] \tag{3.63}$$

Eliminating $z(n)$ and $z(n + 1)$ from Equations 3.60, 3.62, and 3.63 results in the implicit relation

$$x_A(n + 1) = \left(1 - \frac{a_0 T}{2}\right)x_A(n) - \frac{a_0 T}{2}x_A(n + 1) + \frac{b_0 T}{2}[u(n) + u(n + 1)] \tag{3.64}$$

Solving Equation 3.64 for $x_A(n+1)$ enables the state to be updated explicitly with trapezoidal integration according to

$$x_A(n+1) = \frac{(1 - a_0 T/2)}{(1 + a_0 T/2)} x_A(n) + \frac{b_0 T/2}{(1 + a_0 T/2)} [u(n) + u(n+1)] \tag{3.65}$$

Example 3.6

The velocity $v = v(t)$ of an object sinking in a body of water is described by

$$\frac{dv}{dt} + \frac{cg}{W} v = \frac{g}{W}(W - F_B) \tag{3.66}$$

where
    W is the weight of the object
    c is the drag coefficient
    F_B is the buoyant force
    g is the gravitational constant (32.2 ft/s^2)

The buoyant force is a constant that equals the weight of the volume of water displaced by the object. The object is a drum full of hazardous materials (Braun 1978) weighing 350 lb, and its volume is such that the buoyant force is 275 lb. The drag coefficient c was determined experimentally to be 0.8 lb/(ft/s). The drum is released at the surface with zero velocity.

(a) Find a difference equation based on trapezoidal integration to approximate the dynamics of the sinking drum. Choose a step size of $T = 0.5$ s.
(b) Find the approximate velocity, $v_A(nT)$, $n = 0, 10, 20, 30, \ldots, 150$.
(c) Find the true velocity $v(t)$. Use it to find the terminal velocity $v(\infty)$ and $v(nT)$, $n = 0, 10, 20, 30, \ldots, 150$.
(d) Graph the approximate and true velocity over a period of time sufficient for the drum to reach its terminal velocity.
(e) If the drum impacts the ocean floor, 1 mi below the surface, at greater than 60 mph, it will break apart. Comment on the possibility of this happening.

(a) Equation 3.66 can be expressed in the form

$$\frac{dv}{dt} = f(v, u) = b_0 u - a_0 v \tag{3.67}$$

where

$$a_0 = \frac{cg}{W} = \frac{0.8(32.2)}{350} = 0.0736, \quad b_0 = \frac{g}{W}(W - F_B) = \frac{32.2}{350}(350 - 275) = 6.9$$

and the input u treated as the function $u(t) = 1$, $t \geq 0$.
    Evaluating the coefficient terms in Equation 3.65,

$$1 - \frac{a_0 T}{2} = 1 - \frac{0.0736(0.5)}{2} = 0.9816, \quad 1 + \frac{a_0 T}{2} = 1 + \frac{0.0736(0.5)}{2} = 1.0184,$$

$$\frac{b_0 T}{2} = \frac{6.9(0.5)}{2} = 1.725$$

From Equation 3.65, the difference equation for approximating the dynamics of the sinking drum using trapezoidal integration is

$$v_A(n+1) = \frac{0.9816}{1.0184} v_A(n) + \frac{1.725}{1.0184} [1 + 1]$$

$$= 0.9639 v_A(n) + 3.3877, \quad n = 0, 1, 2, 3, \ldots$$

TABLE 3.3
Data Points from Trapezoidal Integration
($T=0.5$ s) and Exact Solution

n	$t_n = nT$	$v_A(n)$	$v(t_n)$
0	0	0.0	0.0
10	5	28.8667	28.8640
20	10	48.8450	48.8413
30	15	62.6718	62.6679
40	20	72.2411	72.2376
50	25	78.8640	78.8609
60	30	83.4475	83.4450
70	35	86.6198	86.6177
80	40	88.8153	88.8136
90	45	90.3347	90.3334
100	50	91.3863	91.3853
110	55	92.1141	92.1134
120	60	92.6178	92.6173
130	65	92.9664	92.9660
140	70	93.2077	93.2074
150	75	93.3747	93.3745

(b) Table 3.3 shows the results for $v_A(n)$ at discrete times $n = 0, 1, 2, 3, \ldots$ The numerical values were generated by running "Chap3_Ex5_1.m."

(c) The exact solution of Equation 3.67 is

$$v(t) = \frac{b_0}{a_0}(1 - e^{-a_0 t}) \tag{3.68}$$

$$= \frac{W - F_B}{c}[1 - e^{-(cg/W)t}] \tag{3.69}$$

The terminal velocity from Equation 3.69 is

$$v(\infty) = \frac{W - F_B}{c} = \frac{350 - 275}{0.8} = 93.75 \, \text{ft/s} \tag{3.70}$$

The analytical solution $v(t)$ is evaluated at $t = 0, 5, 10, \ldots, 75$ s and the values entered in Table 3.3.

(d) Graphs of $v(t)$ and the approximate solution (every fifth point) are shown in Figure 3.11.

(e) Since the terminal velocity of the drum exceeds 88 ft/s (60 mph), the possibility exists of it breaking when it reaches the ocean floor. It remains to be determined what the velocity of the drum is at the 1 mi depth of the ocean floor.

From Table 3.3, it is apparent that trapezoidal integration with a step size of $T = 0.5$ s results in a very accurate approximation of the true solution. However, in most simulation studies, an exact solution is not available. In that case, what can we do to assure accurate simulation results?

An iterative method to determine an acceptable integration step size requires that the simulation be executed with different values of T until changes in the output are deemed insignificant. For example, the step size can be continually reduced (say by one half, or a factor of 10) until graphs of consecutive outputs appear to coincide. The next to last step size is used in subsequent

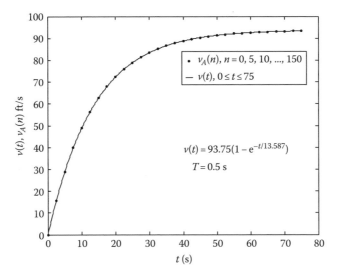

FIGURE 3.11 Approximate solution $v_A(n)$, $n = 0, 5, 10, \ldots, 150$ using trapezoidal integration ($T = 0.5$ s) and exact solution $v(t)$, $0 \le t \le 75$.

investigations. The method is not foolproof and should be repeated if the simulation conditions change as a result of significant changes in the system inputs or initial conditions. We will have more to say about how to select the integration step size in Chapters 6 and 8 when we investigate the subject of truncation errors and dynamic errors.

3.5.2 Nonlinear First-Order Systems

We now turn our attention to nonlinear first-order systems, that is, systems in which the state derivative $f(x, u)$ is a nonlinear function of the state x. The implicit numerical integrators produce implicit difference equations for updating the state.

Consider a first-order system governed by

$$\frac{\mathrm{d}x}{\mathrm{d}t} + N(x) = Ku \tag{3.71}$$

where $N(x)$ is a nonlinear function of the state x. The derivative function is

$$f(x, u) = Ku - N(x) \tag{3.72}$$

and the equation for updating the state using implicit Euler integration is from Equation 3.55

$$x_A(n + 1) = x_A(n) + T\{Ku(n + 1) - N[x_A(n + 1)]\} \tag{3.73}$$

Rearranging Equation 3.73 gives

$$x_A(n + 1) + TN[x_A(n + 1)] = x_A(n) + KTu(n + 1) \tag{3.74}$$

a nonlinear equation that may prove difficult or impossible to solve for $x_A(n + 1)$. To complicate matters further, multiple solutions may exist. The situation is illustrated in the following example.

Example 3.7

The continuous-time model for the sinking drum in Example 3.6 governed its motion $v(t)$ as a function of time t. A relationship between its velocity $v = v(t)$ and depth $y = y(t)$ is obtained by solving the differential equation (Braun 1978)

$$\frac{W}{g} v \frac{dv}{dy} + cv = W - F_B \tag{3.75}$$

(a) Find the difference equation to approximate the velocity of the drum as a function of depth using an implicit Euler integrator. Choose the integration step $T = 1$ ft.
(b) Find the approximate velocity $v_A(n)$ at depths of 0, 1000, 2000, 3000, 4000, 5000, and 6000 ft.
(c) Compare the results from part (b) to the true velocities $v(nT)$ at depths of 0, 1000, 2000, 3000, 4000, 5000, and 6000 ft.

(a) Dividing both sides of Equation 3.75 by Wv/g gives

$$\frac{dv}{dy} + \frac{g}{W}(F_B - W)\frac{1}{v} = -\frac{gc}{W}u \tag{3.76}$$

where the input $u = u(y) = 1$, $y \geq 0$. Comparing Equations 3.71 and 3.76, it follows that the nonlinear function $N(v)$ is

$$N(v) = \frac{g}{W}(F_B - W)\frac{1}{v} \tag{3.77}$$

and the constant K is expressible as

$$K = -\frac{gc}{W} \tag{3.78}$$

According to Equation 3.74, the implicit equation for $v_A(n+1)$ is

$$v_A(n+1) + T\frac{g}{W}(F_B - W)\frac{1}{v_A(n+1)} = v_A(n) - \frac{gc}{W}T(1) \tag{3.79}$$

Substituting the values $g = 32.2$, $c = 0.8$, $W = 350$, $F_B = 275$, and $T = 1$ ft yields

$$v_A(n+1) - 6.9\frac{1}{v_A(n+1)} = v_A(n) - 0.0736 \tag{3.80}$$

(b) Multiplying Equation 3.80 by $v_A(n+1)$ and collecting terms give

$$v_A^2(n+1) + [0.0736 - v_A(n)]v_A(n+1) - 6.9 = 0 \tag{3.81}$$

which can be solved using the quadratic formula. The result is

$$v_A(n+1) = \frac{[v_A(n) - 0.0736] \pm \sqrt{[v_A(n) - 0.0736]^2 + 27.6}}{2} \tag{3.82}$$

Hence, in this case, we are still able to update the state $v_A(n+1)$ explicitly in terms of the previous state $v_A(n)$. The first two iterations are illustrated in the following.

Starting from $v_A(0) = v(0) = 0$,

$$n = 0: \quad v_A(1) = \frac{[v_A(0) - 0.0736] \pm \sqrt{[v_A(0) - 0.0736]^2 + 27.6}}{2}$$

$$= \frac{0 - 0.0736 + \sqrt{[0 - 0.0736]^2 + 27.6}}{2}$$

$$= 2.5902$$

$$n = 1: \quad v_A(2) = \frac{[v_A(1) - 0.0736] \pm \sqrt{[v_A(1) - 0.0736]^2 + 27.6}}{2}$$

$$= \frac{2.5902 - 0.0736 + \sqrt{[2.5902 - 0.0736]^2 + 27.6}}{2}$$

$$= 4.1709$$

Note that since the velocity is increasing, the negative root of Equation 3.82 was discarded.

The M-file "Chap3__Ex5_2.m" generates the values of $v_A(n)$ for $n = 1$–6000. The approximate velocities at depths $y_n = nT$ ($n = 0$, 1000, 2000, 3000, 4000, 5000, and 6000) are listed in Table 3.4.

(c) An exact solution to Equation 3.75, $v = v(y)$, is not possible. However, it is possible to obtain an exact solution for depth y as a function of the velocity v, namely

$$y = -\frac{W}{g}\left[\frac{v}{c} + \frac{W - F_B}{c^2} \ln\left(\frac{W - F_B - cv}{W - F_B}\right)\right] \tag{3.83}$$

We are interested in the depths corresponding to velocities up to the terminal velocity of 93.75 ft/s. Equation 3.83 can be evaluated for $0 \leq v \leq 93.75$ and the results plotted with depth y along the abscissa and velocity v along the ordinate axis as in Figure 3.12.

From an observation of Figure 3.12, the true velocities at the required depths, 0, 1000, 2000, 3000, 4000, 5000, and 6000 ft, agree with the approximate values in Table 3.4.

The question in part (e) of Example 3.6 can now be answered. From Figure 3.12, the velocity of the drum at a depth of 1 mi (5280 ft) does exceed 60 mph (88 ft/s).

In the majority of cases, difference equations resulting from the use of implicit numerical integrators can only be solved by iterative schemes for finding the roots of nonlinear algebraic equations. For example, consider an object falling in a viscous medium where the drag force is a nonlinear function of velocity as shown in Figure 3.13. The continuous-time model describing the object's velocity $v(t)$ is given in Equation 3.87.

$$m\frac{dv}{dt} = W - f_D \tag{3.84}$$

$$\frac{dv}{dt} = \frac{W}{m} - \frac{1}{m}f(v) \tag{3.85}$$

$$\frac{dv}{dt} = g - \frac{1}{m}cv^p \tag{3.86}$$

$$\frac{dv}{dt} = g - \alpha v^p, \quad \alpha = \frac{c}{m} \tag{3.87}$$

A simulation diagram of the system is shown in Figure 3.14.

TABLE 3.4
Data Points from Implicit Euler Integration ($T = 1$ ft) of Continuous-Time Model in Equation 3.75

n	$y_n = nT$ (ft)	$v_A(n)$ (ft/s)
0	0	0
1000	1000	74.3629
2000	2000	85.9310
3000	3000	90.3467
4000	4000	92.2281
5000	5000	93.0618
6000	6000	93.4373

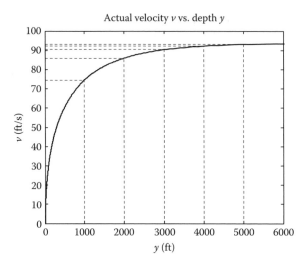

FIGURE 3.12   Graph of points obtained from exact solution, Equation 3.83.

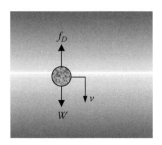

FIGURE 3.13   Object falling in a viscous medium with nonlinear drag force $f_d = cv^p$.

Replacing the continuous-time integrator with an implicit Euler integrator and making all the signals discrete time lead to a discrete-time system with difference equation

$$v_A(n + 1) = v_A(n) + T\{g - \alpha[v_A(n + 1)]^p\} \tag{3.88}$$

$$\Rightarrow v_A(n + 1) + \alpha T[v_A(n + 1)]^p = v_A(n) + Tg \tag{3.89}$$

Unless p is numerically equal to 1 or 2, a root-solving algorithm is required to solve Equation 3.89 for $v_A(n + 1)$ once $v_A(n)$ has been determined. This process can dramatically increase the amount of computational overhead in comparison to what would be required for an explicit numerical integrator.

3.5.3   Discrete-Time State Equations

Given the linear state equations

$$\dot{\underline{x}} = \underline{f}(\underline{x}, \underline{u}) = A\underline{x} + B\underline{u} \tag{3.90}$$

$$\underline{y} = \underline{g}(\underline{x}, \underline{u}) = C\underline{x} + D\underline{u} \tag{3.91}$$

for a continuous-time dynamic system, a discrete-time model approximation can be obtained in a straightforward manner. The approximation to the continuous-time state $\underline{x}(t)$ is $\underline{x}_A(nT)$ or simply $\underline{x}_A(n)$ for short. Difference equations for the discrete-time state $\underline{x}_A(n)$ using one of the numerical

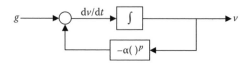

FIGURE 3.14   Simulation diagram for a falling object modeled by $dv/dt = g - \alpha v^p$.

integrators are obtained in exactly the same way as before. For example, using explicit Euler integration, the state derivative vector $\dot{x}$ is assumed constant over the integration interval. Thus,

$$x_A(n+1) = \underline{x}_A(n) + T\underline{f}[\underline{x}_A(n), \underline{u}(n)] \tag{3.92}$$

$$= \underline{x}_A(n) + T[A\underline{x}_A(n) + B\underline{u}(n)] \tag{3.93}$$

$$= (I + TA)\underline{x}_A(n) + TB\underline{u}(n) \tag{3.94}$$

The discrete-time output is determined from

$$\underline{y}_A(n) = C\underline{x}_A(n) + D\underline{u}(n) \tag{3.95}$$

An example involving the discrete-time state equations follows.

Example 3.8

A circuit used in control systems is the *RC* lead-lag network shown in Figure 3.15. The differential equation relating the output $v_0(t)$ and input $v_i(t)$ is

$$R_1 C_1 R_2 C_2 \ddot{v}_0 + (R_1 C_1 + R_1 C_2 + R_2 C_2)\dot{v}_0 + v_0 = R_1 C_1 R_2 C_2 \ddot{v}_i + (R_1 C_1 + R_2 C_2)\dot{v}_i + v_i \tag{3.96}$$

(a) Represent the circuit in state variable form.
(b) Find the discrete-time state equations for approximating the circuit dynamics based on the use of explicit Euler integration.
(c) The capacitor voltages are initially zero and the input is a step $v_i(t) = 1$ V, $t \geq 0$. Approximate the step response using explicit Euler integration with step size $T = 0.001$ s. The circuit parameter values are $R_1 = 10,000$ Ω, $R_2 = 5,000$ Ω, $C_1 = 7.5 \times 10^{-6}$ F, and $C_2 = 2.5 \times 10^{-6}$ F.
(d) An alternate form of the state equations is given by

$$\frac{dv_{C_1}}{dt} = -\frac{1}{C_1}\left(\frac{1}{R_1} + \frac{1}{R_2}\right)v_{C_1} - \frac{1}{R_2 C_1}v_{C_2} + \frac{1}{R_2 C_1}v_i \tag{3.97}$$

$$\frac{dv_{C_2}}{dt} = -\frac{1}{R_2 C_2}v_{C_1} - \frac{1}{R_2 C_2}v_{C_2} + \frac{1}{R_2 C_2}v_i \tag{3.98}$$

Find the matrices *A*, *B*, *C*, and *D* in the state variable model with the states equal to the capacitor voltages.
(e) Repeat part (c) using the new state equations. Compare the results in parts (c) and (e).

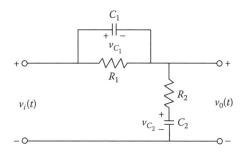

FIGURE 3.15   A lead-lag network.

(a) Dividing through by the lead coefficient term R_1, C_1, R_2, and C_2 and introducing new constants a_1, a_2, b_0, b_1, and b_2 give

$$\ddot{v}_0 + a_1\dot{v}_0 + a_0 v_0 = b_2\ddot{v}_i + b_1\dot{v}_i + b_0 v_i \tag{3.99}$$

where

$$a_0 = \frac{1}{R_1 C_1 R_2 C_2}, \quad a_1 = \frac{R_1 C_1 + R_1 C_2 + R_2 C_2}{R_1 C_1 R_2 C_2} \tag{3.100}$$

$$b_0 = \frac{1}{R_1 C_1 R_2 C_2}, \quad b_1 = \frac{R_1 C_1 + R_2 C_2}{R_1 C_1 R_2 C_2}, \quad b_2 = 1 \tag{3.101}$$

Constructing the simulation diagram for the system starts with the following two equations, which are equivalent to Equation 3.99 (see Section 2.4):

$$\ddot{z} + a_1\dot{z} + a_0 z = v_i \tag{3.102}$$

$$v_0 = b_0 z + b_1\dot{z} + b_2\ddot{z} \tag{3.103}$$

Solving for $\ddot{z}$ in Equation 3.102 and substituting the result in Equation 3.103 yield

$$v_0 = b_0 z + b_1\dot{z} + b_2[v_i - a_0 z - a_1\dot{z}] \tag{3.104}$$

$$= (b_0 - a_0 b_2)z + (b_1 - a_1 b_2)\dot{z} + b_2 v_i \tag{3.105}$$

The simulation diagram follows directly from Equations 3.102 and 3.105. It is presented in Figure 3.16. Choosing the outputs of the integrators in Figure 3.16 as the states results in

$$\dot{x}_1 = x_2 \tag{3.106}$$

$$\dot{x}_2 = a_0 x_1 - a_1 x_2 + v_i \tag{3.107}$$

$$v_0 = (b_0 - a_0 b_2)x_1 + (b_1 - a_1 b_2)x_2 + b_2 v_i \tag{3.108}$$

From Equations 3.106 through 3.108, the matrices A, B, C, and D in the linear state equations $\underline{\dot{x}} = A\underline{x} + B\underline{u}$, $\underline{y} = C\underline{x} + D\underline{u}$ are

$$A = \begin{bmatrix} 0 & 1 \\ -a_0 & -a_1 \end{bmatrix}, \quad B = \begin{bmatrix} 0 \\ 1 \end{bmatrix}, \quad C = [b_0 - a_0 b_2 \quad b_1 - a_1 b_2], \quad D = [b_2] \tag{3.109}$$

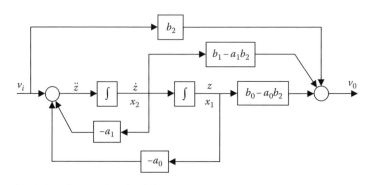

FIGURE 3.16   Simulation diagram for circuit in Figure 3.15.

In terms of the electrical parameters

$$A = \begin{bmatrix} 0 & 1 \\ \dfrac{-1}{R_1 C_1 R_2 C_2} & -\left(\dfrac{R_1 C_1 + R_1 C_2 + R_2 C_2}{R_1 C_1 R_2 C_2}\right) \end{bmatrix}, \quad B = \begin{bmatrix} 0 \\ 1 \end{bmatrix}, \quad C = \begin{bmatrix} 0 & \dfrac{-1}{R_2 C_1} \end{bmatrix}, \quad D = [1] \quad (3.110)$$

(b) From Equations 3.94 and 3.95, the discrete-time state equations are

$$\underline{x}_A(n+1) = \begin{bmatrix} 1 & T \\ \dfrac{-T}{R_1 C_1 R_2 C_2} & 1 - T\left(\dfrac{R_1 C_1 + R_1 C_2 + R_2 C_2}{R_1 C_1 R_2 C_2}\right) \end{bmatrix} \underline{x}_A(n) + \begin{bmatrix} 0 \\ T \end{bmatrix} v_i(n) \quad (3.111)$$

$$y_{A,1}(n) = v_0(n) \begin{bmatrix} 0 & \dfrac{-1}{R_2 C_1} \end{bmatrix} \underline{x}_A(n) + v_i(n) \quad (3.112)$$

(c) Equation 3.111 is solved recursively in "*Chap3_Ex5_3.m*" for the state $\underline{x}_A(n)$, which is used in Equation 3.112 to find the discrete-time step response $v_0(n)$, $n = 0, 1, 2, \ldots$. The first 25 discrete points and every 10th point after that until steady state are plotted in the top window in Figure 3.17.

(d) Solving Equations 3.97 and 3.98 for the state derivatives $\dot{v}_{C_1}$ and $\dot{v}_{C_2}$ leads to

$$\begin{bmatrix} \dot{v}_{C_1} \\ \dot{v}_{C_2} \end{bmatrix} = \begin{bmatrix} -\dfrac{1}{C_1}\left(\dfrac{1}{R_1} + \dfrac{1}{R_2}\right) & -\dfrac{1}{R_2 C_1} \\ -\dfrac{1}{R_2 C_2} & -\dfrac{1}{R_2 C_2} \end{bmatrix} \begin{bmatrix} v_{C_1} \\ v_{C_2} \end{bmatrix} + \begin{bmatrix} \dfrac{1}{R_2 C_1} \\ \dfrac{1}{R_2 C_2} \end{bmatrix} v_i \quad (3.113)$$

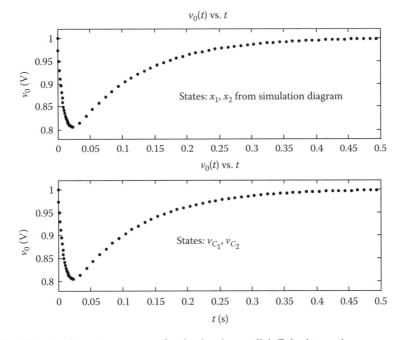

FIGURE 3.17   Discrete-time step response of a circuit using explicit Euler integration.

From the circuit, the output equation is

$$v_0 = v_i - v_{C_1} \tag{3.114}$$

$$= \begin{bmatrix} -1 & 0 \end{bmatrix} \begin{bmatrix} v_{C_1} \\ v_{C_2} \end{bmatrix} + [1]v_i \tag{3.115}$$

The matrices A, B, C, and D follow directly from Equations 3.113 and 3.115.

(e) The new state equations are discretized based on the use of explicit Euler integration and solved recursively in "Chap3_Ex5_3.m." The result is shown in the bottom window of Figure 3.17. The two step responses are identical.

The second choice of the state variables, namely, the capacitor voltages, is more intuitive than the state definition based on the simulation diagram in Figure 3.16. The output vector could be modified to include additional outputs $y_2 = v_{C_1}$ and $y_3 = v_{C_2}$ making $\underline{y} = [v_0 \, v_{C_1} \, v_{C_2}]^T$ to allow visualizing the capacitor voltages (see Exercise 3.21).

A recursive solution to Equation 3.111 requires the initial discrete-time state vector $\underline{x}_A(0) = [x_{1,A}(0) \, x_{2,A}(0)]^T = [x_1(0) \, x_2(0)]^T$. Since the states x_1 and x_2 are not physical quantities, their initial values must be calculated from knowledge of the initial capacitor voltages $v_{C_1}(0)$ and $v_{C_2}(0)$.

3.5.4 DISCRETE-TIME STATE SYSTEM MATRICES

If either of the two implicit numerical integrators is used instead of the explicit Euler integrator, Equation 3.92 is replaced with one of the following two equations:

$$\text{Implicit Euler: } \underline{x}_A(n+1) = \underline{x}_A(n) + T\underline{f}[\underline{x}_A(n+1), \underline{u}(n+1)] \tag{3.116}$$

$$\text{Trapezoidal: } \underline{x}_A(n+1) = \underline{x}_A(n) + \frac{T}{2}\{\underline{f}[\underline{x}_A(n), \underline{u}(n)] + \underline{f}[\underline{x}_A(n+1), \underline{u}(n+1)]\} \tag{3.117}$$

If the continuous-time system is linear, Equations 3.116 and 3.117 can be solved explicitly for $\underline{x}_A(n+1)$ in terms of $\underline{x}_A(n)$ and $\underline{u}(n+1)$. For the implicit Euler integrator,

$$\underline{x}_A(n+1) = \underline{x}_A(n) + T[A\underline{x}_A(n+1) + B\underline{u}(n+1)] \tag{3.118}$$

Solving *for* $\underline{x}_A(n+1)$ gives

$$\underline{x}_A(n+1) = (I + TA)^{-1}[\underline{x}_A(n) + TB\underline{u}(n+1)] \tag{3.119}$$

The state $\underline{x}_A(n)$ is updated recursively without the need to solve an implicit equation for $\underline{x}_A(n+1)$; however, the computations are more extensive than with explicit Euler integration because of the requirement to invert the matrix $I - TA$.

Using trapezoidal integration to update the state,

$$\underline{x}_A(n+1) = \underline{x}_A(n) + \frac{T}{2}[A\underline{x}_A(n) + B\underline{u}(n) + A\underline{x}_A(n+1) + B\underline{u}(n+1)] \tag{3.120}$$

Solving Equation 3.120 for $\underline{x}_A(n+1)$ gives

$$\underline{x}_A(n+1) = \left(I - \frac{1}{2}TA\right)^{-1}\left(I + \frac{1}{2}TA\right)\underline{x}_A(n) + \frac{1}{2}\left(I - \frac{1}{2}TA\right)^{-1}TB[\underline{u}(n) + \underline{u}(n+1)] \tag{3.121}$$

In summary, the use of the explicit Euler integrator to approximate the continuous-time system $\dot{\underline{x}} = A\underline{x} + B\underline{u}$ resulted in a discrete-time state variable model of the form

$$\underline{x}_A(n+1) = G\underline{x}_A(n) + H\underline{u}(n) \tag{3.122}$$

$$\underline{y}_A(n+1) = C\underline{x}_A(n) + D\underline{u}(n) \tag{3.123}$$

A similar result occurred for the two implicit numerical integrators, with the exception of $\underline{u}(n+1)$ appearing on the right-hand side of Equation 3.122 in place of $\underline{u}(n)$ with implicit Euler integration. Both $\underline{u}(n)$ and $\underline{u}(n+1)$ are present on the right-hand side in the case of trapezoidal integration. The matrices G and H are the discrete-time counterparts to A and B, the system and input matrices for the continuous-time case.

For a stable discrete-time system with state equations given by Equations 3.122 and 3.123, the steady-state response to a constant input $\underline{u}(n) = \underline{u}^0$, $n = 0, 1, 2, \ldots$ is obtained from Equation 3.122 by setting $\underline{x}_A(n) = \underline{x}_A(n+1) = \underline{x}_A(\infty)$ resulting in

$$\underline{x}_A(\infty) = G\underline{x}_A(\infty) + H\underline{u}^0 \tag{3.124}$$

$$= (I - G)^{-1} H\underline{u}^0 \tag{3.125}$$

The general solution of the scalar version of Equation 3.122 was given in Section 1.4. A similar approach using recursion works when the state and inputs are vectors and the coefficients of each are matrices. The result is (Ogata 1995)

$$\underline{x}_A(n) = \Phi(n)\underline{x}_A(0) + \sum_{k=0}^{n-1} \Phi(n-k-1)H\underline{u}(k) \tag{3.126}$$

$$\underline{y}_A(n) = C\Phi(n)\underline{x}_A(0) + C\sum_{k=0}^{n-1} \Phi(n-k-1)H\underline{u}(k) + D\underline{u}(n) \tag{3.127}$$

where the matrix $\Phi(n)$ is called the discrete-time state transition matrix. It is expressed in terms of the discrete-time system matrix G according to

$$\Phi(n) = G^n \tag{3.128}$$

From Equations 3.94, 3.119, and 3.121, the discrete-time state transition matrices for the three numerical integrators already considered are

$$\text{Explicit Euler: } \Phi(n) = (I + TA)^n \tag{3.129}$$

$$\text{Implicit Euler: } \Phi(n) = [(I - TA)^{-1}]^n \tag{3.130}$$

$$\text{Trapezoidal: } \Phi(n) = \left[\left(I - \frac{1}{2}TA\right)^{-1}\left(I + \frac{1}{2}TA\right)\right]^n \tag{3.131}$$

EXERCISES

3.13 Show that an approximate solution of the first-order continuous-time model

$$\frac{dx}{dt} = f(x, u)$$

based on replacing the derivative dx/dt with the finite difference $[x(n+1) - x(n)]/T$ is equivalent to using forward (explicit) Euler integration.

3.14 In Example 3.6, find the largest step size T for which

$$\text{Max}|v(nT) - v_A(nT)| \le 0.1$$

Start with $T = 0.025$ s and keep incrementing by 0.025 s until the condition is no longer satisfied.

3.15 Rework Example 3.6 using forward Euler integration. Choose the integration step size as $T = 0.5$ s, the same value used for trapezoidal integration. Prepare a similar table of results for the approximate and exact solutions.

3.16 The position of the sinking drum in Example 3.6 is related to its velocity by

$$y(t) = y(0) + \int_0^t v(t') dt'$$

Using trapezoidal integration and a step size $T = 2$ s, find the approximate solution $v_A(n)$ for 100 s and feed this discrete-time signal to another trapezoidal integrator to generate $y_A(n)$, the approximation to the actual position of the drum.

3.17 Consider the case of a liquid discharged from a tank at a rate proportional to the square root of the level in the tank. The continuous-time model is

$$A \frac{dH}{dt} + \alpha H^{1/2} = F_1$$

where $H = H(t)$ is the continuous-time tank level, $F_1 = F_1(t)$ is the flow in, and α is a constant dependent on the physical characteristics of the tank.

(a) Use implicit Euler integration to find a difference equation involving the discrete-time signals $H_A(n)$ and $F_1(n+1)$ where $H_A(n) \approx H(nT)$ and $F_1(n) = F_1(nT)$. Write the equation in implicit form with $H_A(n+1)$ on both sides.

(b) Show that the implicit equation can be solved explicitly for $H_A(n+1)$ in terms of $H_A(n)$ and $F_1(n+1)$ by making the substitution $x = [H_A(n+1)]^{1/2}$ and solving the resulting quadratic equation in x.

3.18 Suppose $\alpha = 0.5$ and $p = 1.2$ in the example of the object falling in a viscous medium. The object is initially at rest.

(a) Find the approximate velocity of the object after 5 s. Use an explicit Euler integrator with an appropriate step size.

(b) Repeat part (a) using an implicit Euler integrator.

Hint: Use a root-solving routine like the single point iteration or bisection method to solve the implicit equation.

3.19 Verify the solution for $x_A = (n+1)$ in Equation 3.121, which gives the updated state in the approximate solution of $\dot{x} = Ax + Bu$ by trapezoidal integration.

3.20 Find the discrete-time state equations for the circuit in Example 3.8 using

(a) Implicit Euler integration

(b) Trapezoidal integration

3.21 In the lead-lag circuit of Example 3.8, the outputs are $y_1 = v_0$, $y_2 = v_{C_1}$, and $y_3 = v_{C_2}$.

(a) Choose the states as the capacitor voltages v_{C_1} and v_{C_2}. Find expressions for the matrices A, B, C, and D in the state equations in terms of the electrical parameters R_1, R_2, C_1, and C_2.

(b) Find the difference equations based on trapezoidal integration with step size T for approximating the continuous-time system outputs to input $v_i(t)$.

(c) The capacitor voltages are both initially zero, and the input is a step voltage of 12 V applied at $t=0$. Solve the difference equations recursively, and plot the discrete-time outputs in the output vector $y_A(n) = [y_{1,A}(n) y_{2,A}(n) y_{3,A}(n)]^T$.

(d) The initial capacitor voltages are $v_{C_1}(0) = 1$ V $v_{C_2}(0) = 0$ V, and the input is $v_i(t) = 0$ V, $t \geq 0$. Solve the difference equations recursively and plot the discrete-time outputs in the output vector $\underset{\sim}{y}_A(n) = [y_{1,A}(n) y_{2,A}(n) y_{3,A}(n)]^T$.

3.22 For the circuit in Example 3.8 described by Equations 3.97 and 3.98

(a) Use the technique presented in Section 2.3 for converting two first-order differential equations into a single second-order differential equation to eliminate $v_{C_2}(t)$ from the two equations and obtain

$$\ddot{v}_{C_1} + \alpha_1 \dot{v}_{C_1} + \alpha_0 v_{C_1} = \beta_2 \ddot{v}_i + \beta_1 \dot{v}_i + \beta_0 v_i$$

Express the coefficients α_1, α_0, β_2, β_1, and β_0 in terms of the electrical parameters R_1, R_2, C_1, and C_2.

(b) The circuit output is $v_0(t)$. Find the matrices A, B, C, and D in the continuous-time state equation model. Express your answers in terms of the circuit parameters R_1, R_2, C_1, and C_2.

(c) Find the matrices G and H in the discrete-time state equations resulting from the use of explicit Euler integration to approximate the continuous-time response of the circuit.

(d) The input $v_i(t) = 1$ V, $t \geq 0$. Find and plot the discrete-time response $v_0(n)$, $n = 0, 1, 2, \ldots$ based on explicit Euler integration with step size $T = 0.001$ s and compare your answer to the results shown in Figure 3.17.

3.23 The dynamic interaction of rabbit and fox populations in a forest is under investigation. The predator–prey ecosystem is illustrated in Figure E3.23:

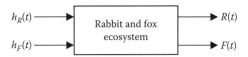

FIGURE E3.23

$R(t)$ is the population of rabbits after "t" weeks
$F(t)$ is the population of foxes after "t" weeks
$h_R(t)$ is the rate of rabbit hunting (rabbits/week)
$h_F(t)$ is the rate of fox hunting (foxes/week)

The mathematical model consists of the following coupled differential equations:

$$\frac{dR}{dt} = aR - bF - h_R$$

$$\frac{dF}{dt} = -cF + dR - h_F$$

a, b are constant parameters defining the growth rate of rabbits
c, d are constant parameters defining the growth rate of foxes

(a) Find the equilibrium point (R_e, F_e) when $h_R(t) = \bar{h}_R$, $t \geq 0$ and $h_F(t) = \bar{h}_F$, $t \geq 0$. Express your answers for R_e and F_e in terms of the system parameters a, b, c, and d and constant hunting rates $\bar{h}_R, \bar{h}_F$.

(b) Baseline values of the system parameters are given as follows:

$$a = 0.04 \; \frac{\text{rabbits/week}}{\text{rabbit}}, \quad b = 0.2 \; \frac{\text{rabbits/week}}{\text{fox}},$$

$$c = 0.1 \; \frac{\text{foxes/week}}{\text{fox}}, \quad d = 0.0075 \; \frac{\text{foxes/week}}{\text{rabbit}}$$

Foxes are endangered and hunting foxes is forbidden. Rabbits are hunted at a constant rate, and after a long period of time, the fox population stabilizes at 750. Find the constant rate of rabbit hunting. Find the rabbit population at the same time.

(c) Let the state be defined as $\underline{x}(t) = \begin{bmatrix} R(t) \\ F(t) \end{bmatrix}$ and the input vector $\underline{u}$ be defined as $\underline{u}(t) = \begin{bmatrix} h_R(t) \\ h_F(t) \end{bmatrix}$. Find the matrices A and B in the state equation $\underline{\dot{x}} = A\underline{x} + B\underline{u}$.

(d) Suppose neither rabbits nor foxes are hunted. Using explicit Euler integration with step size $T = 1$ week, find the 2×2 matrix G such that

$$\begin{bmatrix} R(n+1) \\ F(n+1) \end{bmatrix} = G \begin{bmatrix} R(n) \\ F(n) \end{bmatrix}, \quad n = 0, 1, 2, 3, \ldots$$

(e) Find the 2×2 transition matrix $\Phi(n)$ in the general solution

$$\begin{bmatrix} R(n) \\ F(n) \end{bmatrix} = \Phi(n) \begin{bmatrix} R(0) \\ F(0) \end{bmatrix}, \quad n = 0, 1, 2, 3, \ldots$$

(f) The initial populations of rabbits and foxes are $R(0) = 10,000$ and $F(0) = 1,000$. Use the general solution to find $R(10)$ and $F(10)$.

3.6 IMPROVEMENTS TO EULER INTEGRATION

Euler integration is popular in large measure due to its simplicity. A graphical interpretation of either explicit or implicit Euler integration is straightforward. A discussion of error characteristics for Euler integrators is deferred until a later chapter. However, it is apparent that serious errors can propagate as the discrete-time variable "n" increases with Euler integration as a result of the underlying assumption that the state derivative remains constant for an entire integration step. For systems in which one or more of the state variables experience frequent fluctuations (relative to the integration step size), this assumption is unjustified.

3.6.1 IMPROVED EULER METHOD

The inherent weakness of Euler integration can be overcome in ways other than by simply reducing the integration step size, which may not always be practical. An improved way of determining the new state $x_A(n+1)$ with explicit Euler integration is illustrated in Figure 3.18. Keep in mind that the current state $x_A(n)$ is generally not on the solution curve $x(t)$ as is shown in the figure.

With explicit Euler integration, advancing the state $x_A(n)$ is equivalent to projecting line segment L_1, whose slope is $f[x_A(n), u(n)]$, until it reaches the end of the interval at $(n+1)T$. The updated state is shown as $\hat{x}_A(n+1)$. From there, another forward Euler integration step would proceed along the line segment L_2 whose slope is $f[\hat{x}_A(n+1), u(n+1)]$.

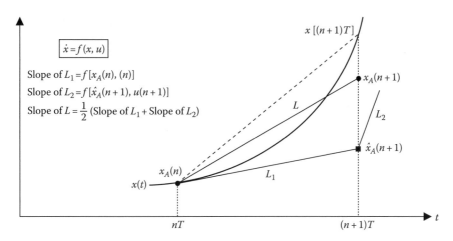

FIGURE 3.18 Illustration of improved Euler method.

Recognizing that L_1 may not be the most judicious direction to move along for approximating the continuous-time state $x[(n+1)T]$, the question to be asked is "Is there a better choice for determining the path from $x_A(n)$ to $x_A(n+1)$?" The line segment L starting from $x_A(n)$ with slope equal to the average of the slopes of L_1 and L_2 appears to be a more prudent choice.

The rationale for choosing the new direction along L is that the average of the slopes of L_1 and L_2 is more likely to reflect the direction of the chord from $x_A(n)$ to $x[(n+1)T]$ than the slope of line L_1 does. Alternatively, Euler integration is predicated on the assumption that the derivative function $f(x, u)$ is constant, which is true only when the solution $x(t)$ is a linear function of t. It makes sense to base the constant on evaluations of $f(x, u)$ at more than one point.

In summary, a new method for computing $x_A(n+1)$ consists of the following:

1. Prediction of the new state using forward Euler integration, that is, moving from $x_A(n)$ to $\hat{x}_A(n+1)$ along the line segment with slope L_1.

$$\hat{x}_A(n+1) = x_A(n) + Tf[x_A(n), u(n)] \tag{3.132}$$

2. Computing the derivative function $f[\hat{x}_A(n+1), u(n+1)]$ at $\hat{x}_A(n+1)$, that is, the slope of line segment L_2.
3. Improving the predicted value $\hat{x}_A(n+1)$, that is, moving from $x_A(n)$ along a line segment whose slope is the average of the slopes of line segments L_1 and L_2 to the new updated state $x_A(n+1)$.

$$x_A(n+1) = x_A(n) + \frac{T}{2}\{f[x_A(n), u(n)] + f[\hat{x}_A(n+1), u(n+1)]\} \tag{3.133}$$

The numerical integrator based on Equations 3.132 and 3.133 is called improved Euler integration, also known as Heun's method.

When the state is a vector and the system model is linear, that is,

$$\underline{\dot{x}} = \underline{f}(\underline{x}, \underline{u}) = A\underline{x} + B\underline{u} \tag{3.134}$$

the predicted state using forward Euler integration is given by Equation 3.94 of the previous section as

$$\hat{\underline{x}}_A(n + 1) = (I + TA)\underline{x}_A(n) + TB\underline{u}(n) \tag{3.135}$$

The improved state estimate is computed from

$$\underline{x}_A(n + 1) = \underline{x}_A(n) + \frac{T}{2} \{\underline{f}[\underline{x}_A(n), \underline{u}(n)] + \underline{f}[\hat{\underline{x}}_A(n + 1), \underline{u}(n + 1)]\} \tag{3.136}$$

Substituting Equation 3.135 into Equation 3.136 results in

$$\underline{x}_A(n + 1) = \left[I + TA + \frac{1}{2}(TA)^2\right]\underline{x}_A(n) + \frac{1}{2}T(I + TA)B\underline{u}(n) + \frac{1}{2}TB\underline{u}(n + 1) \tag{3.137}$$

The discrete-time system matrix using improved Euler integration is therefore

$$G = I + TA + \frac{1}{2}(TA)^2 \tag{3.138}$$

and the improved Euler discrete-time state transition matrix is from Equation 3.128

$$\phi(n) = G^n = \left[I + TA + \frac{1}{2}(TA)^2\right]^n \tag{3.139}$$

The difference in transition matrices between explicit Euler given in Equation 3.129 and improved Euler is the additional term $\frac{1}{2}(TA)^2$ in Equation 3.139.

The following example demonstrates the improved accuracy with improved Euler integration compared with ordinary Euler integration (explicit or implicit).

Example 3.9

Consider the autonomous second-order system

$$\ddot{x} + \omega^2 x = 0 \tag{3.140}$$

Choosing state variables $x_1(t) = x(t)$ and $x_2(t) = \dot{x}(t)$ leads to the state equations

$$\dot{x}_1 = f_1(x_1, x_2) = x_2 \tag{3.141}$$

$$\dot{x}_2 = f_2(x_1, x_2) = -\omega^2 x_1 \tag{3.142}$$

The initial conditions are $x_1(0) = x(0) = x_0$ and $x_2(0) = \dot{x}(0) = \dot{x}_0$.

(a) Find the system matrix A.
(b) Find the discrete-time state transition matrices for explicit and improved Euler integration.
(c) Find the general solution of the discrete-time state equations using both Euler integrators.
(d) Find the transient response using explicit and improved Euler integrators when $\omega = 1$ rad/s, $x_0 = 1$ ft, $\dot{x}_0 = 0$ ft/s, and $T = 0.25$ s. Plot the results.
(e) Find the exact solution for the transient response of the continuous-time system and compare it with the approximate solutions in part (d).

(a) From Equations 3.141 and 3.142, the system matrix is

$$A = \begin{bmatrix} 0 & 1 \\ -\omega^2 & 0 \end{bmatrix} \tag{3.143}$$

(b) The discrete-time state transition matrices are

Explicit Euler: $\Phi(n) = (I + TA)^n$ $\tag{3.144}$

$$= \begin{bmatrix} 1 & T \\ -\omega^2 T & 1 \end{bmatrix}^n \tag{3.145}$$

Improved Euler: $\Phi(n) = \left[I + TA + \frac{1}{2}(TA)^2 \right]^n$ $\tag{3.146}$

$$= \begin{bmatrix} 1 - \frac{1}{2}(\omega T)^2 & T \\ -\omega^2 T & 1 - \frac{1}{2}(\omega T)^2 \end{bmatrix}^n \tag{3.147}$$

(c) General solutions for the discrete-time states for each integrator are

Explicit Euler: $\underline{x}_A(n) = \Phi(n)\underline{x}(0) = \begin{bmatrix} 1 & T \\ -\omega^2 T & 1 \end{bmatrix}^n \begin{bmatrix} x_0 \\ \dot{x}_0 \end{bmatrix}$ $\tag{3.148}$

Improved Euler: $\underline{x}_A(n) = \Phi(n)\underline{x}(0) = \begin{bmatrix} 1 - \frac{1}{2}(\omega T)^2 & T \\ -\omega^2 T & 1 - \frac{1}{2}(\omega T)^2 \end{bmatrix}^n \begin{bmatrix} x_0 \\ \dot{x}_0 \end{bmatrix}$ $\tag{3.149}$

(d) The transient responses of the discrete-time states $x_{1,A}(n)$ and $x_{2,A}(n)$ when $\omega = 1$ rad/s, $x_0 = 1$ ft, $\dot{x}_0 = 0$ ft/s, and $T = 0.25$ s are plotted in Figures 3.19 and 3.20 for the explicit and improved Euler integrators.

(e) The exact solution for the continuous-time states of the undamped second-order system in Equation 3.140 is given in the following and plotted in Figures 3.19 and 3.20.

$$x_1(t) = x_0 \cos \omega t, \quad x_2(t) = -\omega x_0 \sin \omega t \tag{3.150}$$

Note the considerable improvement in accuracy obtained with the improved Euler integrator. The discrete-time state $\underline{x}_A(n) = [x_{1,A}(n)\, x_{2,A}(n)]^T$ based on explicit Euler integration is a poor approximation to the continuous-time state $\underline{x}(t)$, to say the least. This is not surprising in light of the fact that the state derivatives $\dot{x}_1$ and $\dot{x}_2$ vary significantly over the interval T in violation of the basic assumption underlying explicit Euler integration. (See graph of $x_2 = \dot{x}_1$ in Figure 3.19.)

In Chapter 8, we will learn that explicit Euler integration of an undamped second-order system is never stable and should not be used. However, lightly damped second-order systems, that is, those with high natural frequencies, require smaller integration steps for accurate results. The controlling parameter for dynamic accuracy is ωT, the product of natural frequency and the integration time step.

3.6.2 MODIFIED EULER INTEGRATION

In general, forward Euler integration does not result in the "best" direction for advancing the state from $\underline{x}_A(n)$ to $\underline{x}_A(n+1)$ (see Figure 3.18). As the name suggests, improved Euler integration

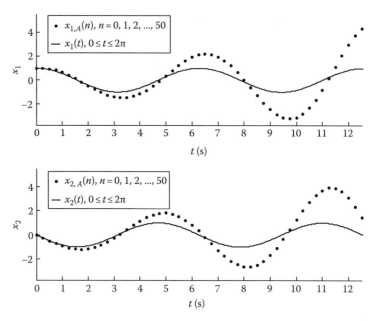

FIGURE 3.19 Continuous-time solution and approximate solution using explicit Euler integration ($T = 0.25$ s) to second-order system $\ddot{x} + \omega^2 x = 0$, $\omega = 1$ rad/s.

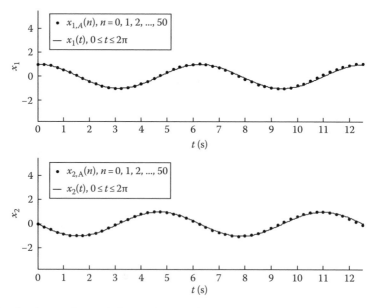

FIGURE 3.20 Continuous-time solution and approximate solution using improved Euler integration ($T = 0.25$ s) to second-order system $\ddot{x} + \omega^2 x = 0$, $\omega = 1$ rad/s.

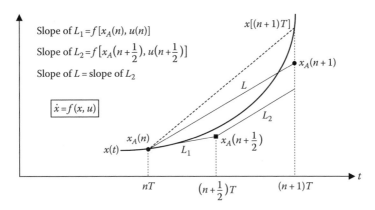

FIGURE 3.21 Illustration of the modified Euler method.

represents an improvement although it comes with a penalty of requiring twice as many state derivative function evaluations compared with explicit Euler integration for the identical step size.

Another method for finding a better direction (compared with explicit Euler integration) to proceed from the current state $\underline{x}_A(n)$ is portrayed in Figure 3.21. It is called the midpoint or modified Euler method because the line segment L, which determines the new approximate state, is based on a state derivative calculation at the midpoint of the interval.

A forward Euler step is taken along line segment L_1 ending up at the point $[(n+1/2)T, \underline{x}_A(n+1/2)]$. A new direction is calculated, namely, $\underline{f}[\underline{x}_A(n+1/2), \underline{u}(n+1/2)]$, which represents the slope of line L_2. Finally, the updated state $\underline{x}_A(n+1)$ is obtained by starting from the current state $\underline{x}_A(n)$ and moving in the direction of line segment L, which is parallel to line segment L_2, until the end of the interval.

A discrete-time state equation can be obtained for the modified Euler integration solution of $\dot{\underline{x}} = A\underline{x} + B\underline{u}$ in the same way it was obtained for the improved Euler integrator. First, the state $\underline{x}_A(n+1/2)$ is calculated from

$$\underline{x}_A\left(n + \frac{1}{2}\right) = \underline{x}_A(n) + \frac{T}{2}\underline{f}[\underline{x}_A(n), \underline{u}(n)] \tag{3.151}$$

The updated state $\underline{x}_A(n+1)$ is based on the derivative function $\underline{f}(\underline{x}, \underline{u})$ evaluated at the point $[(n+1/2)T, \underline{x}_A(n+1/2)]$. The updated state is therefore

$$\underline{x}_A(n + 1) = \underline{x}_A(n) + T\underline{f}\left[\underline{x}_A\left(n + \frac{1}{2}\right), \underline{u}\left(n + \frac{1}{2}\right)\right] \tag{3.152}$$

From Equations 3.151, 3.152, and $\underline{f}(\underline{x}, \underline{u}) = A\underline{x} + B\underline{u}$, the discrete-time state equation is

$$\underline{x}_A(n + 1) = \left[I + (TA) + \frac{1}{2}(TA)^2\right]\underline{x}_A(n) + \frac{1}{2}T^2AB\underline{u}(n) + TB\underline{u}\left(n + \frac{1}{2}\right) \tag{3.153}$$

and the discrete-time state transition matrix using the modified Euler method is

$$\Phi(n) = \left[I + (TA) + \frac{1}{2}(TA)^2\right]^n \tag{3.154}$$

Hence, the state transition matrices with improved Euler integration, Equation 3.139, and modified Euler integration, Equation 3.154, are identical. The modified Euler integrator requires input sampling at twice the normal frequency of $1/T$ due to the presence of the term $\underline{u}(n + 1/2)$ in Equation 3.153.

Trapezoidal, improved, and modified Euler integrators are roughly comparable in accuracy and are superior to the explicit and implicit Euler integrators. The state transition matrices for improved and modified Euler integration are only slightly more involved than the state transition matrix for forward Euler integration, the difference being the additional term $1/2(TA)^2$. The added computations necessary to include the squared term depend on the dimension of the square matrix A, which in turn is related to the size of the state vector.

The discrete-time transition matrices for the explicit integrators (forward Euler, improved Euler, and modified Euler) bear a striking similarity to the transition matrix for the continuous-time state equations. The state transition matrix $\Phi(t)$ for the system $\underline{\dot{x}} = A\underline{x} + B\underline{u}$ was introduced in Equation 2.129. It was expressed in terms of an infinite series of matrices, that is,

$$\Phi(t) = I + (tA) + \frac{1}{2!}(tA)^2 + \frac{1}{3!}(tA)^3 + \cdots \tag{3.155}$$

At the end of a single integration step, the exact solution to $\underline{\dot{x}} = A\underline{x}$ is

$$\underline{x}(T) = \Phi(T)\underline{x}(0) = \left[I + (TA) + \frac{1}{2!}(TA)^2 + \frac{1}{3!}(TA)^3 + \cdots\right]\underline{x}(0) \tag{3.156}$$

With improved or modified Euler integration, the discrete-time state vector approximation at the same time is

$$\underline{x}_A(1) = \Phi(1)\underline{x}_A(0) = \left[I + TA + \frac{1}{2}(TA)^2\right]\underline{x}(0) \tag{3.157}$$

The difference or error in the discrete-time state approximation is therefore

$$\underline{x}_A(1) - x(T) = -\left[\frac{1}{3!}(TA)^3 + \frac{1}{4!}(TA)^4 + \cdots\right]\underline{x}(0) \tag{3.158}$$

The importance of choosing T small is evident from Equation 3.158.

Example 3.10

In Section 2.6, a second-order system with system matrix A and transition matrix $\Phi(t)$ was given. They are repeated as follows:

$$A = \begin{bmatrix} 0 & 1 \\ -2 & -2 \end{bmatrix}, \quad \Phi(t) = \begin{bmatrix} 2e^{-t} - e^{-2t} & e^{-t} - e^{-2t} \\ -2e^{-t} + 2e^{-2t} & -e^{-t} + 2e^{-2t} \end{bmatrix} \tag{3.159}$$

The initial state is $\underline{x}(0) = [x_1(0) \ x_2(0)]^T$. Modified Euler integration is to be used to compute $\underline{x}_A(1) = [x_{1,A}(1) \ x_{2,A}(1)]^T$.

(a) Find the continuous-time state transition matrix at $t = T$, that is, $\Phi(t)|_{t=T}$.
(b) Find the discrete-time state transition matrix at $n = 1$, that is, $\Phi(n)|_{n=1}$.
   For parts (c) and (d), the initial state $\underline{x}(0) = [1 \ 1]^T$ and the step size $T = 0.25$ s.
(c) Find $\Phi(t)|_{t=1}$ and $\underline{x}(T)$
(d) Find $\Phi(n)|_{n=1}$ and $\underline{x}_A(1)$.

(a) From Equation 3.159,

$$\Phi(t)|_{t=T} = \begin{bmatrix} 2e^{-T}e^{-2T} & e^{-T} - e^{-2T} \\ -2e^{-T} + 2e^{-2T} & -e^{-T} + 2e^{-2T} \end{bmatrix} \tag{3.160}$$

(b) From Equation 3.154,

$$\Phi(n)|_{n=1} = I + TA + \frac{1}{2}(TA)^2 \tag{3.161}$$

$$= I + T\begin{bmatrix} 0 & 1 \\ -2 & -3 \end{bmatrix} + \frac{1}{2}\left(T\begin{bmatrix} 0 & 1 \\ -2 & -3 \end{bmatrix}\right)^2 \tag{3.162}$$

$$= \begin{bmatrix} 1 - T^2 & T\left(1 - \frac{3}{2}T\right) \\ T(-2 + 3T) & 1 - 3T + \frac{7}{2}T^2 \end{bmatrix} \tag{3.163}$$

(c) $\Phi(t)|_{t=0.25} = \begin{bmatrix} 2e^{-0.25} - e^{-2(0.25)} & e^{-(0.25)} - e^{-2(0.25)} \\ -2e^{-0.25} + 2e^{-2(0.25)} & -e^{-0.25} + 2e^{-2(0.25)} \end{bmatrix} = \begin{bmatrix} 0.9511 & 0.1723 \\ -0.3445 & 0.4343 \end{bmatrix}$

$$\begin{bmatrix} x_1(0.25) \\ x_2(0.25) \end{bmatrix} = \Phi(t)|_{t=0.25}\underline{x}(0)$$

$$= \begin{bmatrix} 0.9511 & 0.1723 \\ -0.3445 & 0.4343 \end{bmatrix}\begin{bmatrix} 1 \\ 1 \end{bmatrix} = \begin{bmatrix} 1.1234 \\ 0.0898 \end{bmatrix} \tag{3.164}$$

(d) $\Phi(n)|_{n=1} = \begin{bmatrix} 1 - (0.25)^2 & 0.25\left\{1 - \frac{3}{2}(0.25)\right\} \\ 0.25[-2 + 3(0.25)] & 1 - 3(0.25) + \frac{7}{2}(0.25)^2 \end{bmatrix} = \begin{bmatrix} 0.9375 & 0.1563 \\ -0.3125 & 0.4688 \end{bmatrix}$

$$\begin{bmatrix} x_{1,A}(1) \\ x_{2,A}(1) \end{bmatrix} = \Phi(n)|_{n=1}\underline{x}_A(0)$$

$$= \begin{bmatrix} 0.9375 & 0.1563 \\ -0.3125 & 0.4688 \end{bmatrix}\begin{bmatrix} 1 \\ 1 \end{bmatrix} = \begin{bmatrix} 1.0938 \\ 0.1563 \end{bmatrix} \tag{3.165}$$

The discrete-time state transition matrix $\Phi(n)$ for $n=1$ and the continuous-time state transition matrix $\Phi(t)$ at $t=T$ differ significantly. The discrepancy is attributable to the integration step size T, which must be reduced to make $[x_{1,A}(1)\ x_{2,A}(1)]^T$ closer to the exact solution $[x_1(0.25)\ x_1(0.25)]^T$ and assure substantial agreement of subsequent vectors $[x_{1,A}(n)\ x_{2,A}(n)]^T$ and $[x_1(nT)\ x_2(nT)]^T$, $n=2, 3, 4,\dots$.

Table 3.5 shows the effect of reducing the step size T on the discrete-time state transition matrix and state vector. As expected, the difference between the discrete- and continuous-time quantities diminishes as the step size is reduced.

The next example looks at the transient and steady-state responses of a second-order system using modified Euler integration.

Example 3.11

The input to the second-order system in Figure 3.22 is a unit step $u(t) = 1$, $t \geq 0$.
    System parameters are $\zeta = 0.5$, $\omega_n = 0.4$ rad/s, and $K = 2$. Both initial conditions are zero.

(a) Write state equations for the system if $x_1 = y$ and $x_2 = dy/dt$ and the output $y_1 = y$.
(b) Find the discrete-time system matrix G based on the use of modified Euler integration to approximate the solution of the continuous-time state equations. Leave your answers in terms of the system parameters ζ, ω_n, and K and integration step size T.

TABLE 3.5
Effect of Reduced Step Size on the Continuous-Time and Discrete-Time State Transition Matrices and State Vectors

T	$\Phi(t)\|_{t=T}$	$x(T)$	$\Phi(n)\|_{n=1}$	$\underline{x}_A(n)$
0.5	$\begin{bmatrix} 0.9511 & 0.1723 \\ -0.3445 & 0.4343 \end{bmatrix}$	$\begin{bmatrix} 1.1233 \\ 0.0898 \end{bmatrix}$	$\begin{bmatrix} 0.9375 & 0.1563 \\ -0.3125 & 0.4688 \end{bmatrix}$	$\begin{bmatrix} 1.0938 \\ 0.1563 \end{bmatrix}$
0.05	$\begin{bmatrix} 0.9976 & 0.4664 \\ -0.0928 & 0.8584 \end{bmatrix}$	$\begin{bmatrix} 1.0440 \\ 0.7657 \end{bmatrix}$	$\begin{bmatrix} 0.9975 & 0.0462 \\ -0.0925 & 0.8588 \end{bmatrix}$	$\begin{bmatrix} 1.0438 \\ 0.7662 \end{bmatrix}$
0.01	$\begin{bmatrix} 0.9999 & 0.0099 \\ -0.0197 & 0.9703 \end{bmatrix}$	$\begin{bmatrix} 1.0098 \\ 0.9506 \end{bmatrix}$	$\begin{bmatrix} 0.9999 & 0.0098 \\ -0.0197 & 0.9703 \end{bmatrix}$	$\begin{bmatrix} 1.0097 \\ 0.9506 \end{bmatrix}$

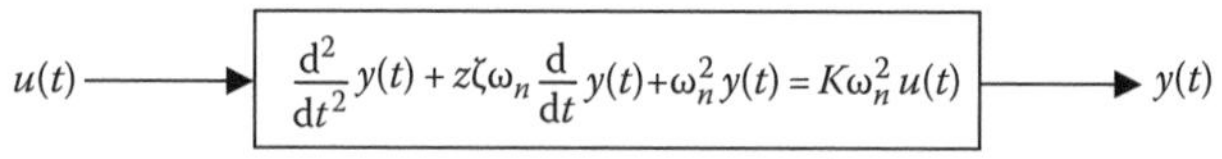

$$u(t) \longrightarrow \boxed{\dfrac{d^2}{dt^2}y(t) + z\zeta\omega_n\dfrac{d}{dt}y(t) + \omega_n^2 y(t) = K\omega_n^2 u(t)} \longrightarrow y(t)$$

FIGURE 3.22   A second-order system with a unit step input.

(c) Find the continuous-time response for $x_1(t)$.
(d) Find the steady state $\underline{x}(\infty)$.
(e) Choose the integration step $T = 0.5$ s and find the discrete-time system matrix G.
(f) Find the steady-state vector $\underline{x}_A(\infty)$. Compare the results from parts (d) and (f).
(g) Find the discrete-time signal $x_{1,A}(n)$ and compare it with $x_1(t)$.

(a) The state equations for this second-order system are easily found from a simulation diagram using cascaded integrators with outputs dy/dt and y. The result is

$$\frac{dx_1}{dt} = x_2 \tag{3.166}$$

$$\frac{dx_2}{dt} = K\omega_n^2 u - \omega_n^2 x_1 - 2\zeta\omega_n x_2 \tag{3.167}$$

Since $y_1 = y = x_1$, the matrices A, B, C, and D in the state equations are

$$A = \begin{bmatrix} 0 & 1 \\ -\omega_n^2 & -2\zeta\omega_n \end{bmatrix}, \quad B = \begin{bmatrix} 0 \\ K\omega_n^2 \end{bmatrix}, \quad C = \begin{bmatrix} 1 & 0 \end{bmatrix}, \quad D = [0] \tag{3.168}$$

(b) From Equation 3.153, the discrete-time system matrix is

$$G = I + (TA) + \frac{1}{2}(TA)^2 \tag{3.169}$$

$$= I + T\begin{bmatrix} 0 & 1 \\ -\omega_n^2 & -2\zeta\omega_n \end{bmatrix} + \frac{1}{2}T^2\begin{bmatrix} 0 & 1 \\ -\omega_n^2 & -2\zeta\omega_n \end{bmatrix}^2 \tag{3.170}$$

$$= \begin{bmatrix} 1 - \frac{1}{2}(\omega_n T)^2 & T(1 - \zeta\omega_n T) \\ -\omega_n^2 T(1 - \zeta\omega_n T) & 1 - 2\zeta\omega_n T + \frac{1}{2}(\omega_n T)^2(4\zeta^2 - 1) \end{bmatrix} \tag{3.171}$$

(c) The unit step response is (see Equation 2.23)

$$x_1(t) = K\left[1 - e^{-\zeta\omega_n t}\left(\cos \omega_d t + \frac{\zeta\omega_n}{\omega_d} \sin \omega_d t\right)\right], \quad t \geq 0 \tag{3.172}$$

The damped natural frequency ω_d is computed from its definition

$$\omega_d = \left(\sqrt{1 - \zeta^2}\right)\omega_n = \left(\sqrt{1 - 0.5^2}\right)0.4 = \frac{\sqrt{3}}{5} \text{ rad/s}$$

Substituting the system parameter values into Equation 2.23 and simplifying lead to

$$x_1(t) = 2\left[1 - e^{-t/5}\left\{\cos\left(\frac{\sqrt{3}}{5}\right)t + \left(\frac{\sqrt{3}}{3}\right)\sin\left(\frac{\sqrt{3}}{5}\right)t\right\}\right], \quad t \geq 0 \tag{3.173}$$

(d) The continuous-time state vector at steady state $\underline{x}(\infty)$ is obtained from

$$\underline{\dot{x}}(\infty) = A\underline{x}(\infty) + B\underline{u}(\infty) = \underline{0} \tag{3.174}$$

$$\underline{x}(\infty) = -A^{-1}B\underline{u}(\infty) \tag{3.175}$$

where

$$A = \begin{bmatrix} 0 & 1 \\ -\omega_n^2 & -2\zeta\omega_n \end{bmatrix} = \begin{bmatrix} 0 & 1 \\ -(0.4)^2 & -2(0.5)(0.4) \end{bmatrix} = \begin{bmatrix} 0 & 1 \\ -0.16 & -0.4 \end{bmatrix}$$

$$B = \begin{bmatrix} 0 \\ K\omega_n^2 \end{bmatrix} = \begin{bmatrix} 0 \\ 2(0.4)^2 \end{bmatrix} = \begin{bmatrix} 0 \\ 0.32 \end{bmatrix}$$

$$\Rightarrow \underline{x}(\infty) = -\begin{bmatrix} 0 & 1 \\ -0.16 & -0.4 \end{bmatrix}^{-1}\begin{bmatrix} 0 \\ 0.32 \end{bmatrix}[1] = \begin{bmatrix} 2 \\ 0 \end{bmatrix}$$

(e) The discrete-time system matrix is computed from Equation 3.171 as

$$G = \begin{bmatrix} 1 - \dfrac{1}{2}[0.4(0.5)]^2 & 0.5[1 - 0.5(0.4)(0.5)] \\ -(0.4)^2 0.5[1 - 0.5(0.4)(0.5)] & 1 - 2(0.5)(0.4)(0.5) + \dfrac{1}{2}[0.4(0.5)]^2[4(0.5)^2 - 1] \end{bmatrix}$$

$$= \begin{bmatrix} 0.980 & 0.45 \\ -0.072 & 0.80 \end{bmatrix}$$

(f) The discrete-time state is updated using Equation 3.153.

$$\underline{x}_A(n+1) = G\underline{x}_A(n) + \frac{1}{2}T^2 AB\underline{u}(n) + TB\underline{u}\left(n + \frac{1}{2}\right) \tag{3.176}$$

$$= \begin{bmatrix} 0.980 & 0.45 \\ -0.072 & 0.80 \end{bmatrix}\underline{x}_A(n) + \frac{1}{2}(0.5)^2\begin{bmatrix} 0 & 1 \\ -0.16 & -0.4 \end{bmatrix}\begin{bmatrix} 0 \\ 0.32 \end{bmatrix}[1]$$

$$+ (0.5)\begin{bmatrix} 0 \\ 0.32 \end{bmatrix}[1] \tag{3.177}$$

$$\underline{x}_A(n+1) = \begin{bmatrix} 0.980 & 0.45 \\ -0.072 & 0.80 \end{bmatrix}\underline{x}_A(n) + \begin{bmatrix} 0.040 \\ 0.144 \end{bmatrix}[1] \tag{3.178}$$

Note that $\underline{u}(n)$ and $\underline{u}(n + 1/2)$ in Equation 3.176 are both equal to the 1×1 vector [1].

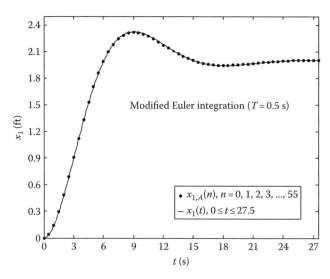

FIGURE 3.23 Continuous- and discrete-time step responses.

The discrete-time state vector at steady state $\underline{x}_A(\infty)$ is obtained by substituting $\underline{x}_A(n+1) = \underline{x}_A(n) = \underline{x}_A(\infty)$ in the previous equation.

$$\Rightarrow \underline{x}_A(\infty) = \left(I - \begin{bmatrix} 0.980 & 0.45 \\ -0.072 & 0.80 \end{bmatrix} \right)^{-1} \begin{bmatrix} 0.040 \\ 0.144 \end{bmatrix} [1] = \begin{bmatrix} 2 \\ 0 \end{bmatrix}$$

in agreement with the continuous-time state vector at steady state $\underline{x}(\infty)$.

(g) The difference equation for $\underline{x}_A(n)$ in Equation 3.178 is solved recursively in the MATLAB® script file "Chap3_Ex6_3.m." The continuous-time state variable $x_1(t)$ and the discrete-time state variable $x_{1,A}(n)$ are plotted in Figure 3.23.

A sample of the results for $x_{1,A}(n)$ along with the exact solution for $x_1(t)$ are compiled in Table 3.6.

Our last example is that of a nonlinear second-order system. The equations developed in this and previous sections for linear systems are not applicable; however, the implementation of numerical integration is nonetheless straightforward. A state variable model of the nonlinear system is required. The discrete-time state is updated using the state derivative functions in accordance with the desired numerical integration routine.

TABLE 3.6
Summary of Results for $x_1(t)$ and $x_{1,A}(n)$

n	$x_{1,A}(n)$	t_n	$x_1(t_1)$	N	$x_{1,A}(n)$	t_n	$x_1(t_n)$
0	0	0	0	30	2.0038	15	2.0046
5	0.6904	2.5	0.6806	35	1.9509	17.5	1.9487
10	1.7046	5	1.6989	40	1.9604	20	1.9580
15	2.2447	7.5	2.2487	45	1.9869	22.5	1.9859
20	2.2979	10	2.3062	50	2.0042	25	2.0043
25	0.1434	12.5	2.1492	55	2.0080	27.5	2.0086

Example 3.12

A simple pendulum is shown in Figure 3.24. The mass of the rod is negligible compared to the mass m of the sphere. Linear damping at the fixed end is assumed.

The angular position of the rod $\theta(t)$ satisfies the nonlinear differential equation

$$J\ddot{\theta} + c\dot{\theta} + mgr \sin\theta = 0 \qquad (3.179)$$

(a) Find the nonlinear state equations when $x_1 = \theta$ and $x_1 = \dot{\theta}$.
(b) Find the difference equations for updating the discrete-time state components $x_{1,A}(n)$ and $x_{2,A}(n)$ when explicit Euler integration is used.

Suppose the system parameters are $m = 0.25$ slugs, $r = 0.75$ ft, and $c = 0.1$ ft lb per rad/s. The moment of inertia $J = mr^2 = 0.1406$ ft lb s^2. Find a suitable value for T and solve the discrete-time state equations recursively under the following conditions:

(c) $\theta(0) = \pi/6$ rad, $\dot{\theta}(0) = 0$ rad/s. Graph $x_{1,A}(n)$ and $x_{2,A}(n)$.
(d) $\theta(0) = 0$ rad, $\dot{\theta}(0) = 0.5$ rad/s. Graph $x_{1,A}(n)$ and $x_{2,A}(n)$.

(a)

$$\dot{x}_1 = \dot{\theta} = x_2 \qquad (3.180)$$

$$\dot{x}_2 = \ddot{\theta} = \frac{1}{J}[-mgr \sin\theta - c\dot{\theta}] \qquad (3.181)$$

$$= \frac{1}{J}(-mgr \sin x_1 - cx_2) \qquad (3.182)$$

The continuous-time state equations are

$$\dot{x}_1 = f_1(x_1, x_2) = x_2 \qquad (3.183)$$

$$\dot{x}_2 = f_2(x_1, x_2) = \frac{1}{J}(-mgr \sin x_1 - cx_2) \qquad (3.184)$$

(b) Using explicit Euler integration, the difference equations for updating the discrete-time state are

$$x_{1,A}(n+1) = x_{1,A}(n) + Tf_1[x_{1,A}(n), x_{2,A}(n)] \qquad (3.185)$$

$$\Rightarrow x_{1,A}(n+1) = x_{1,A}(n) + Tx_{2,A}(n) \qquad (3.186)$$

$$x_{2,A}(n+1) = x_{2,A}(n) + Tf_2[x_{1,A}(n), x_{2,A}(n)] \qquad (3.187)$$

$$\Rightarrow x_{2,A}(n+1) = x_{2,A}(n) - \frac{T}{J}[mgr \sin x_{1,A}(n) + cx_{2,A}(n)] \qquad (3.188)$$

(c) Choosing $T = 0.0025$ s, a recursive solution of Equations 3.186 and 3.188 is easily obtained. The initial state is $x_{1,A}(0) = x_1(0) = \pi/6$ rad and $x_{2,A}(0) = x_2(0) = 0$ rad/s. To make it easier to visualize the discrete nature of the response, graphs of $x_{1,A}(n)$ and $x_{2,A}(n)$ are shown for $n = 0, 20, 40, \ldots, 4000$ in Figure 3.25 corresponding to approximations of each state at times $t_n = 0, 0.05, 0.1, \ldots, 10$ s. As expected, the pendulum returns to its equilibrium position.

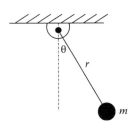

FIGURE 3.24 A simple nonlinear pendulum with damping.

(d) Simulation results for $x_{1,A}(0) = 0$ rad and $x_{2,A}(0) = 0.5$ rad/s are shown in Figure 3.26.

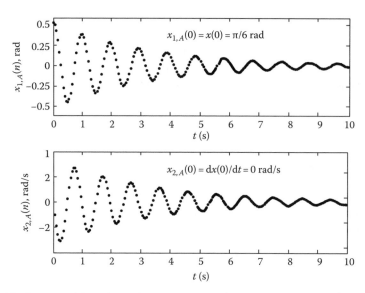

FIGURE 3.25 Nonlinear pendulum simulation with $x_1(0) = \pi/6$ rad and $x_2(0) = 0$ rad/s.

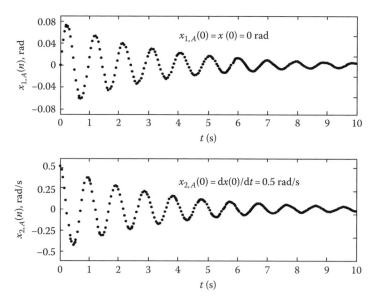

FIGURE 3.26 Nonlinear pendulum simulation with $x_1(0) = 0$ rad and $x_2(0) = 0.5$ rad/s.

Exact solutions for the state components are not easily obtained owing to the nonlinearity in Equation 3.179. A "quasi exact" solution could be found by choosing an exceedingly small value of T and plotting the results on the same graph for comparison with the discrete-time approximations shown in Figures 3.25 and 3.26. It is left as an exercise to show that the discrete-time and "quasi exact" responses are in basic agreement.

Looking at the graphs in Figures 3.25 and 3.26, we might be inclined to believe that the integration step size $T = 0.0025$ s is a "one size fits all" value for simulating the pendulum dynamics. However, Figure 3.27 will quickly dispel this thinking. The results shown

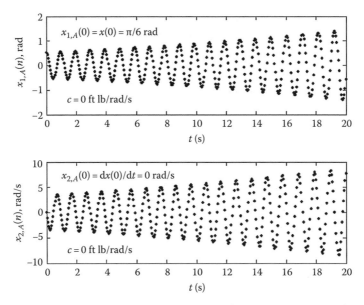

FIGURE 3.27 Undamped pendulum simulation with $x_1(0) = \pi/6$ rad and $x_2(0) = 0$ rad/s.

in Figure 3.27 correspond to an undamped pendulum ($c = 0$) with the same initial conditions as in part (c) and the same step size of 0.0025 s. Every 20th point of the discrete-time state responses is plotted.

Clearly, explicit Euler integration using a step size of $T = 0.0025$ s is not advisable since the discrete-time state responses bear no resemblance whatsoever to the real (continuous-time) system responses. A valuable lesson of this example is the need to exercise caution when choosing the integration step size for numerical integration. If we are not careful, the integrators may be "unstable" under certain conditions. This point is revisited in detail in Chapter 8.

EXERCISES

3.24 A mass is suspended from a stationary support by a spring as shown in Figure E3.24. The mass is displaced from its equilibrium position 1 ft and released with zero velocity. The continuous-time model of the system is $m\ddot{x} + kx = 0$.

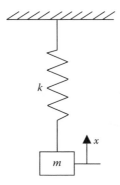

FIGURE E3.24

(a) Find the matrix A in the state equations $\dot{\underline{x}} = A\underline{x}$ for the continuous-time model.

(b) Find the matrix G in the discrete-time state equations $\underline{x}_A(n+1) = G\underline{x}_A(n)$ resulting from the use of improved Euler integration to approximate the response of the continuous-time system.

(c) The system parameters are $k = 4$ lb/ft and $m = 1$ slug. Fill in the following table:

N	$x_A(n)$ $T = 0.05$ s	$x(nT)$	n	$x_A(n)$ $T = 001$ s	$x(nT)$
0			0		
1			5		
2			10		
3			15		
4			20		
5			25		
6			30		
7			35		
8			40		
9			45		
10			50		

3.25 By trial and error, determine an acceptable value for the step size T in simulating the nonlinear pendulum response in Example 3.12 using implicit Euler integration. The initial conditions are $x_1(0) = \pi/6$ rad and $x_2(0) = 0.5$ rad/s. Plot the discrete-time state $x_{1,A}(n)$, $n = 0, 1, 2, \ldots, n_f$ where $n_f T = 10$ s for each value of T.

3.26 Repeat Exercise 3.25 using trapezoidal integration instead of implicit Euler.

3.27 Choose a very small time step, for example, $T = 0.0001$ s, in Example 3.12 to obtain the "quasi exact" solution and plot the results on the same graph with the discrete-time responses in Figures 3.25 and 3.26. Comment on the results.

3.28 The nonlinear pendulum model in Example 3.12 is often approximated by

$$J\ddot{\theta} + c\dot{\theta} + mgr\theta = 0$$

when the angular displacement θ is small, that is, the small angle approximation $\theta = \sin\theta$ is used resulting in the linear differential equation model above. Compare the results of simulating the linear and nonlinear models using modified Euler integration. The initial angle $\theta(0) = 5°$ and the initial angular velocity $\dot{\theta}(0) = 0°/s$.

3.29 A logistic population growth model

$$\frac{dP}{dt} = cP(P_m - P)$$

is to be simulated in order to approximate the population $P(t)$ for a period of time.

(a) Find the difference equation for $P_A(n)$ intended to approximate $P(t)$ based on the use of the following numerical integrators:

(i) Explicit Euler ($T = 0.25$ year)

(ii) Trapezoidal ($T = 0.5$ year)

(iii) Improved Euler ($T = 0.5$ year)

(b) Fill in the following table with the simulated populations based on the three numerical integrators and the exact solution. Note that $c = 1.25 \times 10^{-9}$, $P_m = 25$ million, and $P(0) = 5$ million. The exact solution is given by

$$P(t) = \frac{P_m P(0)}{P(0) + [P_m - P(0)]e^{-cP_m t}}, \quad t \geq 0$$

t (Years)	0	50	100	150	200	250
Explicit Euler	5.0000					
Trapezoidal	5.0000					
Improved Euler	5.0000					
Exact	5.0000					

3.30 The tank in Figure E3.30 has a brine solution flowing into it. The solution is stirred well to ensure that the concentration of salt in the tank is uniform.

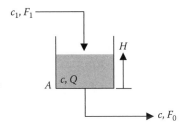

FIGURE E3.30

c_1 is the brine concentration (lb/gal)
F_1 is the brine flow (gal/min)
c is the salt concentration in tank (lb/gal)
Q is the quantity of salt in tank (lb)
H is the liquid level in tank (ft)
V is the volume of liquid in tank (gal)
F_0 is the flow rate from tank (gal/min)

The mathematical model consists of the following equations:

$$\frac{dQ}{dt} = c_1 F_1 - c F_0$$

$$c \frac{Q}{V}, \quad V = AH$$

$$A\frac{dH}{dt} + F_0 = F_1, \quad K F_0 = \alpha H^{1/2}$$

The system baseline parameter values are $A = 25$ ft^2 and $\alpha = 0.75$ gal/min per ft$^{1/2}$.

Note: 1 ft^3 of water is roughly 8.3 gal.

(a) Draw a simulation diagram of the system.
(b) Choose the state variables as $x_1 = Q$ and $x_2 = H$ and the outputs $y_1 = c$, $y_2 = Q$, and $y_3 = V$. Write the state equations in the form

$$\dot{x}_1 = f_1(x_1, x_2, c_1, F_1), \quad y_1 = g_1(x_1, x_2, c_1, F_1)$$
$$\dot{x}_2 = f_2(x_1, x_2, c_1, F_1) \quad y_2 = g_2(x_1, x_2, c_1, F_1)$$
$$y_3 = g_3(x_1, x_2, c_1, F_1)$$

(c) Find expressions for the steady-state values of the states $x_1(\infty)$ and $x_2(\infty)$ and the outputs $y_1(\infty)$, $y_2(\infty)$, and $y_3(\infty)$ assuming c_1 and F_1 are constant.
(d) The tank is initially filled with 100 gal of water (no salt). Brine starts flowing in to the tank at a rate of 2 gal/min. The salt concentration of the brine is 0.25 lb/gal. Both the flow rate and salt concentration of the brine flow remain constant. Using explicit Euler and improved Euler integration, find the discrete-time state equations.

$$\underline{x}_A(n + 1) = \underline{f}[(\underline{x}_A(n), \underline{u}(n)]$$

$$\underline{y}_A(n) = \underline{g}[\underline{x}_A(n), \underline{u}(n)]$$

which are used to obtain an approximate solution for the continuous-time states and outputs.
(e) Solve the discrete-time state equations recursively for the discrete-time states and $x_{1,A}(n)$ and $x_{2,A}(n)$ and outputs $y_{1,A}(n)$, $y_{2,A}(n)$, and $y_{3,A}(n)$. Graph the transient responses. Comment on the values of T used for each type of numerical integrator.
(f) Compare the steady-state results obtained in part (e) with the predicted values from part (c). Comment on the results.

3.7 CASE STUDY: VERTICAL ASCENT OF A DIVER

As a diver submerges, pressure increases in direct proportion to the depth. This pressure is caused by the combined weight of the surrounding water and the atmosphere above and is called ambient pressure. At a depth of 70 ft, ambient pressure is equal to more than three atmospheres (three times the atmospheric pressure at sea level). In order to overcome this pressure and fill his lungs with vital air, the diver must breathe air supplied to him at the ambient pressure.

The air is a mixture of approximately 20% vital oxygen and 80% inert nitrogen. The oxygen component of the air is used by the body, and waste carbon dioxide is exhaled. Under normal atmospheric conditions, the nitrogen component of the mixture has no effect. But under pressure, it dissolves in the bloodstream and in tissues and remains there after the diver begins to ascend. If the diver ascends too quickly, the nitrogen expands and equalizes with the decreasing ambient pressure. Nitrogen bubbles form in the bloodstream and the tissues, leading to an extremely painful condition known as decompression sickness (DCS), more commonly known as the "bends," which can cause paralysis and even death.

The focus of this study is an investigation of the types of cable forces that can be used to bring a deep-sea diver safely to the surface. The mathematical model governing the diver's ascent consists of differential equations relating the forces acting on the diver and the dynamics of the diver's internal body pressure (McClamroch 1980). The following notation is used:

$h = h(t)$ is the depth of diver below sea level, ft
$\dot{h} = (d/dt)h(t)$ is the velocity of diver, ft/s
$\ddot{h} = (d^2/dt^2)h(t)$ is the acceleration of diver, ft/s^2

$p = p(t)$ is the internal body pressure of diver, relative to atmospheric pressure at sea level, lb/ft^2

$\dot{p} = (d/dt)p(t)$ is the rate of change of diver's internal body pressure, lb/ft^2 per s

$\Delta p = \Delta p(t)$ is the difference between body pressure and local underwater pressure, lb/ft^2

$f_c = f_c(t)$ is the external cable force on diver, lb

$f_d = f_d(t)$ is the drag force on diver, lb

f_B is the buoyant force on diver, lb

m is the mass of diver, slugs

W is the weight of diver and gear at sea level, lb

V is the volume of diver and gear, ft^3

K is the body tissue constant of diver, s^{-1}

μ is the drag coefficient of diver under water, lb s/ft

γ is the weight density of water ($62.4\ \text{lb}/\text{ft}^3$)

g is the gravitational constant ($32.2\ \text{ft}/\text{s}^2$)

The forces acting on the diver are a cable force f_c, a drag force f_d, a buoyant force f_B, and the diver's weight W. From Newton's second law with h and all forces measured positive in the downward direction,

$$m\ddot{h} = W - f_B + f_d - f_c \tag{3.189}$$

The drag force is modeled by

$$f_d = -\mu\dot{h} \tag{3.190}$$

The buoyant force is equal to the weight of water displaced by the diver and gear

$$f_B = \gamma V \tag{3.191}$$

Combining Equations 3.189, 3.190, and 3.191 gives

$$\frac{W}{g}\ddot{h} + \mu\dot{h} = (W - \gamma V) - f_c \tag{3.192}$$

The right-hand side of Equation 3.192 is the difference between the diver's effective weight in the water $(W - \gamma V)$ and the cable force f_c. Denoting the net cable force by

$$f_n = (W - \gamma V) - f_c \tag{3.193}$$

leads to the second-order differential equation

$$\frac{W}{g}\ddot{h} + \mu\dot{h} = f_n \tag{3.194}$$

The rate of change of the diver's internal body pressure is assumed proportional to the difference between the local underwater (ambient) pressure and the diver's internal body pressure. That is,

$$\dot{p} = K(\gamma h - p) \tag{3.195}$$

We are interested in h, the diver's depth below the surface, and Δp, the difference between the internal body pressure of the diver and the ambient underwater pressure. The dynamic system under investigation is portrayed in Figure 3.28.

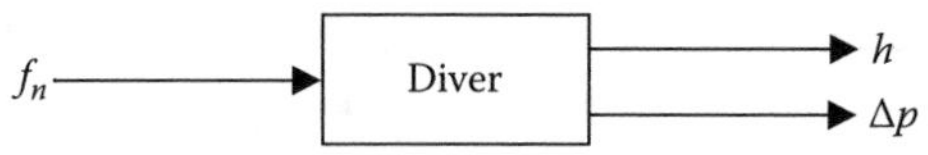

FIGURE 3.28 Dynamic system with input f_n and outputs h and Δp.

The third-order linear dynamic system can be modeled in state variable form. The state variables are chosen as

$$x_1 = h, \quad x_2 = \dot{h}, \quad x_3 = p \tag{3.196}$$

Solving for the state derivatives

$$\dot{x}_1 = \dot{h} = x_2 \tag{3.197}$$

$$\dot{x}_2 = \ddot{h} = -\frac{\mu g}{W}\dot{h} + \frac{g}{W}f_n \tag{3.198}$$

$$= -\frac{\mu g}{W}x_2 + \frac{g}{W}f_n \tag{3.199}$$

$$\dot{x}_3 = \dot{p} = K\gamma x_1 - Kx_3 \tag{3.200}$$

$$\Rightarrow \begin{bmatrix} \dot{x}_1 \\ \dot{x}_2 \\ \dot{x}_3 \end{bmatrix} = \begin{bmatrix} 0 & 1 & 0 \\ 0 & \dfrac{-\mu g}{W} & 0 \\ K\gamma & 0 & -K \end{bmatrix} \begin{bmatrix} x_1 \\ x_2 \\ x_3 \end{bmatrix} + \begin{bmatrix} 0 \\ \dfrac{g}{W} \\ 0 \end{bmatrix} [f_n] \tag{3.201}$$

The outputs are expressed in terms of the states as

$$y_1 = h = x_1 \tag{3.202}$$

$$y_2 = P - \gamma h = x_3 - \gamma x_1 \tag{3.203}$$

$$\Rightarrow \begin{bmatrix} y_1 \\ y_2 \end{bmatrix} = \begin{bmatrix} 1 & 0 & 0 \\ -\gamma & 0 & 1 \end{bmatrix} \begin{bmatrix} x_1 \\ x_2 \\ x_3 \end{bmatrix} \tag{3.204}$$

The state equation matrices A, B, C, and D are given by

$$A = \begin{bmatrix} 0 & 1 & 0 \\ 0 & \dfrac{-\mu g}{W} & 0 \\ K\gamma & 0 & -K \end{bmatrix}, \quad B = \begin{bmatrix} 0 \\ \dfrac{g}{W} \\ 0 \end{bmatrix}, \quad C = \begin{bmatrix} 1 & 0 & 0 \\ -\gamma & 0 & 1 \end{bmatrix}, \quad D = \begin{bmatrix} 0 \\ 0 \end{bmatrix} \tag{3.205}$$

In order to obtain a numerical solution to the state equations, the initial conditions, that is, the initial state $\underline{x}(0)$, must be known. Assuming the diver is initially in equilibrium with his or her surroundings leads to

$$\dot{h}(0) = x_2(0) = 0 \tag{3.206}$$

$$\dot{p}(0) = K[\gamma h(0) - p(0)] \tag{3.207}$$

$$= K[\gamma x_1(0) - x_3(0)] \tag{3.208}$$

Setting $\dot{p}$ equal to zero in Equation 3.208 gives

$$x_3(0) = \gamma x_1(0) \tag{3.209}$$

Initial depth $x_1(0)$ is arbitrary; however, to be in equilibrium, the diver's effective weight in the water $W - \gamma V$ must be counterbalanced by the initial cable force $f_c(0)$.

$$f_c(0) = W - \gamma V \tag{3.210}$$

Note that the initial net force to maintain the diver in equilibrium is

$$f_n(0) = (W - \gamma V) - f_c(0) = 0 \tag{3.211}$$

A simulation of the diver's ascent subject to a constant cable force in excess of $f_c(0)$ in Equation (3.210) is needed. The discrete-time state equation is

$$\underline{x}_A(n+1) = G\underline{x}_A(n) + H\underline{u}(n) \tag{3.212}$$

where G and H depend on the choice of numerical integrator. Using trapezoidal integration for now and leaving the other discrete-time integrators for the exercise problems, the discrete-time state is updated according to Equation 3.121

$$\underline{x}_A(n+1) = \left(I + \frac{1}{2}TA\right)^{-1}\left(I + \frac{1}{2}TA\right)\underline{x}_A(n) + \frac{1}{2}\left(I - \frac{1}{2}TA\right)^{-1}TB[\underline{u}(n) + \underline{u}(n+1)] \tag{3.213}$$

With a constant cable force $f_c = \bar{f}_c$, $t \geq 0$, the input f_n is likewise constant, that is,

$$f_n = \bar{f}_n = (W - \gamma V)\bar{f}_c, \quad t \geq 0 \tag{3.214}$$

The second term in Equation 3.213 can be simplified, that is,

$$\frac{1}{2}\left(I - \frac{1}{2}TA\right)^{-1}TB[\underline{u}(n) + \underline{u}(n+1)] = \frac{1}{2}\left(I - \frac{1}{2}TA\right)^{-1}TB[\bar{f}_n + \bar{f}_n] \tag{3.215}$$

$$= T\left(I - \frac{1}{2}TA\right)^{-1}B\bar{f}_n \tag{3.216}$$

Hence, for the special case where the input $u(n)$ is constant for all n, the discrete-time matrices G and H in Equation 3.212 using trapezoidal integration are

$$G = \left(I - \frac{1}{2}TA\right)^{-1}\left(I + \frac{1}{2}TA\right) \tag{3.217}$$

$$H = T\left(I = \frac{1}{2}TA\right)^{-1}B \tag{3.218}$$

Baseline numerical values for the system parameters are $K = 0.2$, $\mu = 6.5$, $W = 300$, $V = 3$, and the step size $T = 0.25$ s. Evaluating matrices A and B,

$$A = \begin{bmatrix} 0 & 1 & 0 \\ 0 & -0.6977 & 0 \\ 12.48 & 0 & -0.2 \end{bmatrix}, \quad B = \begin{bmatrix} 0 \\ 0.1073 \\ 0 \end{bmatrix}$$

The discrete-time matrices G and H are obtained from Equations 3.217 and 3.218.

$$G = \begin{bmatrix} 1 & 0.2299 & 0 \\ 0 & 0.8396 & 0 \\ 3.0439 & 0.3500 & 0.9512 \end{bmatrix}, \quad H = \begin{bmatrix} 0.0031 \\ 0.0247 \\ 0.0047 \end{bmatrix}$$

The discrete-time state equation, Equation 3.212, is

$$\underline{x}_A(n+1) = \begin{bmatrix} 1 & 0.2299 & 0 \\ 0 & 0.8396 & 0 \\ 3.0439 & 0.3500 & 0.9512 \end{bmatrix} \underline{x}_A(n) + \begin{bmatrix} 0.0031 \\ 0.0247 \\ 0.0047 \end{bmatrix} \bar{f}_n \qquad (3.219)$$

Before simulating the diver's ascent to the surface, we can make the cable force equal to its equilibrium value in Equation 3.210 and observe whether the system remains in equilibrium. Setting $f_c = W - \gamma V = 112.8$ lb makes the net force $\bar{f}_n = 0$. Additionally, we must remember to make $x_3(0) = \gamma x_1(0)$ where $x_1(0)$ is the arbitrary initial depth.

Figure 3.29 shows the results of solving Equation 3.219 under these conditions with the diver starting at 500 ft below the surface. As expected, the system remains in an equilibrium state.

Suppose the cable force is increased by 10% above its equilibrium value to $1.1(W - \gamma V) = 1.1$ $(112.8) = 124.08$ lb. The MATLAB script file "*Chap3_CaseStudy.m*" generates a recursive solution to the discrete-time system difference equations in Equation 3.219. The results are plotted in Figure 3.30 for a duration of time sufficient to bring the diver to the surface. The integration step size T could be varied an order of magnitude in either direction and the results compared with those in Figure 3.30 to determine if the current value $T = 0.25$ s needs to be adjusted.

Since the system dynamics are linear, analytical solutions for the continuous-time state variables in Equation 3.201 are easily determined and given in Equations 3.220 through 3.222.

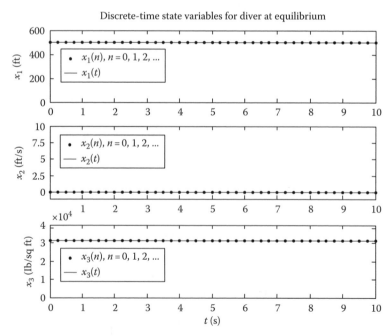

FIGURE 3.29 System at equilibrium: $x_1(0) = 500$ ft, $x_2(0) = 0$ ft/s, $x_3(0) = \gamma x_1(0) = 3120$ lb/ft², $f_c = W - \gamma V = 112.8$ lb.

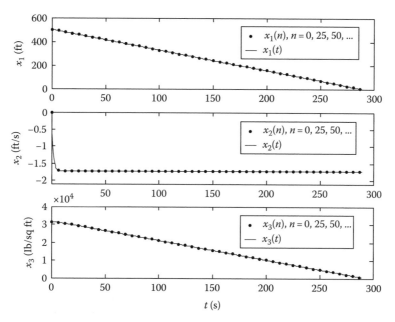

FIGURE 3.30 State variables for diver's ascent: $f_c = 1.1\,(W - \gamma V)$.

The derivation is left as an exercise problem at the end of this section.

$$x_1(t) = h(0) + \frac{gf_n}{\alpha W}\left[t - \frac{1 - e^{-at}}{\alpha}\right]$$ (3.220)

$$x_2(t) = \frac{gf_n}{\alpha W}(1 - e^{-\alpha t})$$ (3.221)

$$x_3(t) = \gamma\left[h(0) + \frac{gf_n}{\alpha W}\left\{t + \frac{K(1 - e^{-at})}{\alpha(\alpha - K)} - \frac{\alpha(1 - e^{-Kt})}{K(\alpha - K)}\right\}\right]$$ (3.222)

where the constant $\alpha = \mu g/W$. The analytical solutions for the states are plotted in Figure 3.30 along with the discrete-time states. There is close agreement between the numerical (discrete time) and analytical solutions for each of the state variables. Notice that after about 6 s, the diver is surfacing at a constant velocity, and both depth and internal body pressure are decreasing linearly with time.

Equation 3.220 can be used to estimate the time required for the diver to surface. If the initial depth is great enough, at the time the diver surfaces, the exponential term will have died out. Consequently, the time to surface t_s can be estimated from

$$0 = h(0) + \frac{gf_n}{\alpha W}\left[t_s - \frac{1}{\alpha}\right]$$ (3.223)

$$\Rightarrow t_s = \frac{W}{\mu g} - \frac{\mu h(0)}{f_n} = \frac{W}{\mu g} - \frac{\mu h(0)}{(W - \gamma V) - f_c}$$

$$= \frac{300}{6.5(32.2)} - \frac{6.5(500)}{-0.1(112.8)} = 289.6$$ (3.224)

in agreement with the graphs of $x_{1,A}(n)$ and $x_1(t)$ shown in Figure 3.30.

We have yet to look at the differential pressure $\Delta p = p - \gamma h$, the second component of the output vector y_2 in Equation 3.203. The discrete-time output $\underline{y}_A(n)$ is a linear combination of the discrete-time state $\underline{x}_A(n)$ and input $\underline{u}(n)$, that is,

$$\underline{y}_A(n) = C\underline{x}_A(n) + D\underline{u}(n) \tag{3.225}$$

The second component of $y_A(n)$ reduces to

$$y_{2,A}(n) = C_{2,1}x_{1,A}(n) + C_{2,2}x_{2,A}(n) + C_{2,3}x_{3,A}(n) \tag{3.226}$$

since the direct transmission matrix D is zero. Substituting the components of C in Equation 3.205 into Equation 3.226 gives

$$y_{2,A}(n) = -\gamma x_{1,A}(n) + x_{3,A}(n) \tag{3.227}$$

3.7.1 Maximum Cable Force for Safe Ascent

Suppose a safe ascent implies that the differential pressure Δp is never to exceed a value denoted by Δp_{max}. The maximum cable force for a safe ascent $(f_c)_{max}$ can be obtained in one of two ways.

3.7.1.1 Trial and Error

The constant cable force can be initialized to a value slightly more than the equilibrium force $W_{eff} = W - \gamma V$ and the diver's ascent simulated. If the maximum differential pressure during the ascent is less than Δp_{max}, the ascent is simulated again with a larger cable force. The reverse is true if the maximum differential pressure exceeds Δp_{max}. The process is repeated until the cable force producing a maximum differential pressure of Δp_{max} (within some tolerance) is obtained. Figure 3.31 shows the result of simulating several ascents to find $(f_c)_{max}$ for the case when $\Delta p_{max} = 4$ psi.

The discrete-time differential pressure responses are labeled and graphed as if they were continuous time in nature; however, the points along each plot were obtained by recursive solution of difference equations. Note the dramatic increase in ascent time as the cable force approaches the equilibrium value of 112.8 lb.

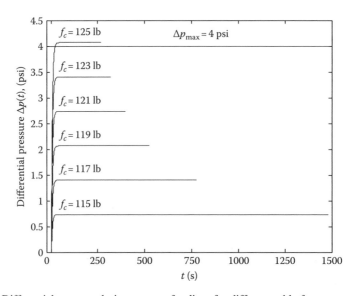

FIGURE 3.31 Differential pressure during ascent of a diver for different cable forces.

3.7.1.2 Analytical Solution

A second approach to finding the maximum cable force for a safe ascent is based on analytical solutions for the state variables $x_1(t)$ and $x_3(t)$. From Equations 3.220 and 3.222, the steady-state responses are

$$x_1(t)_{ss} = h(0) + \frac{g\bar{f}_n}{\alpha W}\left[t - \frac{1}{\alpha}\right] \tag{3.228}$$

$$x_3(t)_{ss} = \gamma\left[h(0) + \frac{g\bar{f}_n}{\alpha W}\left\{t + \frac{K}{\alpha(\alpha - K)} - \frac{\alpha}{K(\alpha - K)}\right\}\right] \tag{3.229}$$

where $\bar{f}_n$ is the constant net force applied during ascent of the diver. The differential pressure at steady state is

$$\Delta p_{ss} = x_3(t)_{ss} - \gamma x_1(t)_{ss} \tag{3.230}$$

Substituting Equations 3.228 and 3.229 into Equation 3.230 and simplifying the expression result in

$$\Delta p_{ss} = -\frac{\gamma\bar{f}_n}{\mu K} \tag{3.231}$$

The cable force $(\overleftarrow{f_c})_{max}$ responsible for $\Delta p_{ss} = \Delta p_{max}$ is obtained from

$$\Delta p_{max} = -\frac{\gamma\bar{f}_n}{\mu K} = -\frac{\gamma[(W - \gamma V) - (\bar{f}_c)_{max}]}{\mu K} \tag{3.232}$$

$$\Rightarrow (\bar{f}_c)_{max} = (W - \gamma V) + \frac{\mu K \Delta p_{max}}{\gamma}$$

$$= [300 - 62.4(3)] + \frac{6.5(0.2)(4 \times 144)}{62.4}$$

$$= 124.8\,\text{lb} \tag{3.233}$$

in agreement with the response shown in Figure 3.31.

3.7.2 DIVER ASCENT WITH DECOMPRESSION STOPS

Ordinarily, a deep-sea diver ascending from several 100 ft or more down makes several decompression stops to allow the nitrogen gas to be released in a slow and controlled manner. His internal pressure is given time to equalize with the ambient pressure at different depths. A typical cable force for accomplishing this is shown in Figure 3.32.

Alternating the cable force between a value larger than the diver's effective weight, that is, $(1 + \beta)W_{eff}$, $\beta > 0$, and the diver's effective weight W_{eff} results in the diver remaining at certain depths for a fixed period of time before continuing the ascent to the surface.

To illustrate, suppose the diver is initially at a depth of 500 ft and is to be brought to the ocean surface in stages allowing for decompression. The difference in ambient water pressure and his internal pressure is not to exceed 4 psi. Figure 3.33 shows the results of a simulation for the same diver as before brought up by a periodic cable force of 124 lb for $L = 100$ s followed by a value of 112.8 lb for 200 s and then repeated. Once again, the discrete-time signals are plotted as if they were continuous time.

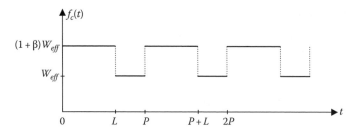

FIGURE 3.32 Cable force profile for raising a diver with decompression stops.

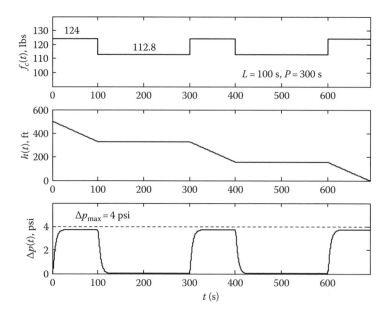

FIGURE 3.33 Diver ascent with stops for decompression.

Referring to Figure 3.31 or Equation 3.233, it follows that a constant cable force of 124 lb for the entire ascent would bring the diver to the surface with a differential pressure never exceeding the safe limit of 4 psi. The ascent would be considerably faster than the 691.75 s shown in Figure 3.33. However, the profile of $\Delta p(t)$ shown in Figure 3.33 is a safer alternative than the exponential rise shown in Figure 3.31.

It may appear from Figure 3.33 that the diver's depth becomes constant when the cable force is switched from $f_c = 124$ lb to $f_c = W_{eff} = 112.8$ lb. Realistically, the diver cannot come to an abrupt stop when the cable force changes in step fashion. A close-up look at the net cable force $f_n(t)$ in Equation 3.193 and the diver's velocity during a portion of the ascent is shown in Figure 3.34.

It is clear from observing the velocity that the diver continues moving toward the surface for a short period of time immediately following the change in cable force from 124 to 112.8 lb (or equivalently the net force changing from -11.2 to 0 lb).

The explanation of decompression and the plots shown in Figure 3.33 oversimplify the problem. Onset of the "bends" is caused by an excess of dissolved nitrogen in the blood to the point where it cannot be disposed of in a normal manner. The amount of nitrogen that dissolves in the blood is related to the time a diver remains at a given depth. That is, it takes longer to absorb a dangerous amount of nitrogen at a shallow depth compared to a depth further down from the surface. Reference tables provide empirical data relating required decompression times with duration of time spent at a given depth (Reseck 1990).

A final observation about the diver model is relevant. The coupling between the second-order differential equation for h in Equation 3.192 and the first-order differential equation for p in

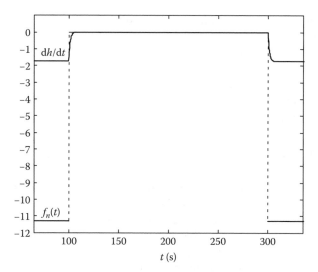

FIGURE 3.34   Net cable force $f_n(t)$ in lbs and diver velocity dh/dt in ft/s with decompression stops.

Equation 3.195 is one way, that is, the diver's internal pressure p does not affect the depth h, and a second-order state model is suitable if the pressure is not of interest. On the other hand, the depth h influences the diver's internal pressure p, and hence the first-order differential Equation (3.195) cannot be solved independently of the second-order differential Equation (3.192).

EXERCISES

3.31 A simple study can be conducted to find the "best" value for T, the integration step size. Since a graph of the analytical solutions for the continuous-time state variables and the discrete-time state approximation are in close agreement (Figure 3.30) when $T = 0.25$ s, we would like to know if a larger value of T can be used without sacrificing significant accuracy. With this in mind, for $T = 0.5, 1, 2, 4, \ldots$.
  (a) Find the system and input matrices G and H of the discrete-time system using trapezoidal integration.
  (b) Solve the resulting discrete-time state equations for the discrete-time state vector $\underline{x}_A(n) = [x_{1,A}(n), x_{2,A}(n), x_{3,A}(n)]^T$ and plot the results on the same graph as the continuous-time solution, similar to Figure 3.30. Stop when a noticeable difference between $\underline{x}_A(n)$ and $\underline{x}(nT)$ occurs.
3.32 Using the baseline conditions for the system parameters unless stated otherwise,
  (a) Find the cable force $(f_c)_{max}$ to bring up divers (plus gear) weighing 200, 250, 300, 350, and 400 lb while not exceeding a maximum differential pressure $\Delta P_{max} = 4$ psi. Enter the results in the following table. Prepare a graph of $(f_c)_{max}$ vs. W. Comment on the results.
  (b) In part (a), record the time required for the diver to surface t_s and enter in the following table. Plot a graph of t_s vs. W.

W (lb)	200	250	300	350	400
$(f_c)_{max}$					
t_s (s)					

  (c) Suppose the volume V of the diver and gear vary with the diver's weight according to $V = 1 + (W/150)$. Repeat parts (a) and (b).

3.33 For the velocity during diver with baseline conditions, find the "best" step size T for simulating the diver's ascent from 250 ft using
  (i) Explicit Euler integration
  (ii) Improved Euler integration
  (iii) Modified Euler integration
  Specify your criterion for determining the "best" step size.

3.34 Derive the analytical solutions for the continuous-time states given in Equation 3.220 through 3.222.

  Hint: It may be necessary to defer this problem until after reading Section 4.2 on the Laplace transform.

3.35 Using the baseline conditions given for the diver, simulate the response using explicit Euler integration when the constant cable force is 15% below the equilibrium value. Prepare plots of the continuous-time and discrete-time states for a duration of 100 s.

3.36 It is suggested that a sinusoidal cable force $f_c(t) = \overline{F} + A\sin(2\pi t/p)$ be more effective in bringing the diver to the surface safely, that is, $\Delta p(t) \leq \Delta p_{max}$, $t \geq 0$, and in less time compared to a constant force. Using the baseline system parameters, choose a numerical integration method to approximate the system dynamics with the suggested type of cable force, that is, experiment with different values of $\overline{F}$, A, and P, and comment on the validity of the claim about using the sinusoidal cable force.

3.37 Suppose the diver with baseline condition parameter values is initially at a depth of 750 ft at equilibrium conditions.
  (a) The cable force bringing the diver to the surface is as shown in Figure 3.32 with $\beta = 0.1$ and $P = 300$ s. Vary the duty cycle $100 \times (L/P)$ from 20% to 100% in increments of 20%, and simulate the diver's ascent using a numerical integrator with appropriate step size. Fill in the following table and plot the results.

Duty Cycle (%)	Time to Surface (s)	Maximum Differential Pressure (psi)
20		
40		
60		
80		
100		

  (b) Repeat part (a) with the duty cycle fixed at 50% and vary the parameter β as shown in the table and fill in the table.

β	Time to Surface (s)	Maximum Differential Pressure (psi)
0.03		
0.06		
0.09		
0.12		
0.15		

3.38 For a diver with system parameters $W = 300$, $K = 0.2$, $\mu = 6.5$, and $V = 3$,
  (a) Plot the inverse relationship in Equation 3.233, that is, Δp_{max} vs. $(f_c)_{max}$.
  (b) Simulate several diver ascents from different initial depths using constant cable forces and compare the simulated maximum differential pressure with the values from the graph.

3.39 A 250 lb diver with gear weighing another 100 lb is 400 ft below the surface in equilibrium with his surroundings. A winch cable begins bringing him to the surface using a constant force.

(a) Using the analytical solution for the state variables, find the required force needed for the diver to be ascending at a constant rate of 1.5 ft/s ($\dot{h} = -1.5$ ft/s) when he reaches the surface. The remaining parameter values are $K = 0.25$, $\mu = 5$, and $V = 3.25$.

(b) Simulate the diver ascent using the force determined in part (a) to verify the result. Use Euler integration with step size $T = 0.1$ s.

(c) Plot the state variables for the simulation in part (b).

3.40 A diver initially in equilibrium at a depth of 150 ft is ascending to the surface under the influence of a constant cable force equal to 10% greater than the equilibrium force. The cable snaps when the diver is 50 ft from the surface. Simulate the diver's depth, velocity, and differential pressure for 120 s. Use any of the numerical integrators presented.

System parameters are $W = 325$, $K = 0.23$, $\mu = 4.8$, and $V = 3.15$.

4 Linear Systems Analysis

4.1 INTRODUCTION

Chapter 2 introduced first- and second-order linear time-invariant (LTI) systems in a very superficial way. A general form for the family of step responses, in the absence of input derivative terms, was presented for both types of systems. Alternate representations of LTI systems, namely, simulation diagrams and state-space models, were also discussed.

Chapters 1 and 3 outlined methods for transforming continuous-time differential equation models into discrete-time system models comprising difference equations. In doing so, the grounds were laid for the foundation of continuous-time system simulation.

A natural question that arises is "How accurate is the simulation?" In the case of continuous-time systems with LTI models, it helps to have a solid grasp of how LTI systems respond to elementary inputs such as a step, polynomials, exponentials, and periodic functions. The analytical solutions serve as a benchmark in comparing different simulation (discrete-time system) models.

This chapter begins with a review of the Laplace transformation and its use in finding the free and forced response of continuous-time LTI system models. The counterpart of the Laplace transform for discrete-time systems is the z-transform, and it is covered in the later sections along with examples of how it facilitates the process of finding the response of discrete-time LTI systems. Time and frequency domain characteristics of continuous- and discrete-time LTI system models are discussed. Mappings from the s-plane to the z-plane corresponding to specific numerical integrators are introduced as a quick way of obtaining discrete-time model approximations of continuous-time systems.

4.2 LAPLACE TRANSFORM

The Laplace transform, as the name implies, is a transformation of functions between two domains. The independent variables in the two domains are commonly denoted "t" and "s" as shown in Figure 4.1, and the domains are referred to as the time domain (or t-domain) and s-domain, respectively.

A class of functions $f(t)$ defined for $t \geq 0$ in the time domain are transformed into functions $F(s)$ in the s-domain according to

$$F(s) = \int_{0}^{\infty} f(t) e^{-st} dt \qquad (4.1)$$

Equation 4.1 is the definition of the one-sided Laplace transform of a function $f(t)$. The definition of $f(t)$ for $t < 0$ is irrelevant since the interval of integration in Equation 4.1 is 0 to ∞. It is valid for functions $f(t)$, which are said to be of exponential order, that is, functions that are bounded by increasing exponentials as $t \to \infty$, assuring the convergence of the integral in Equation 4.1. This includes all real-world signals as well as certain functions for which $\lim_{t \to \infty} f(t) = \infty$.

The notation $\mathcal{L}\{f(t)\}$ is interpreted as the Laplace transform of $f(t)$, that is, the function of "s" resulting from evaluating the integral in Equation 4.1. The function $f(t)$ and its Laplace transform

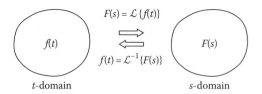

FIGURE 4.1 The Laplace transform $\mathcal{L}\{f(t)\} = F(s)$ and its inverse $f(t) = \mathcal{L}^{-1}\{F(s)\}$.

$F(s)$ are referred to as a Laplace transform pair using the symbol $\Leftrightarrow$ with the function $f(t)$ on one side and its transform $F(s)$ on the other side. To illustrate, consider the unit step function $\hat{u}(t)$ that equals 1 for $t \geq 0$ and zero for $t < 0$.

$$\hat{U}(s) = \mathcal{L}\{\hat{u}(t)\} = \int_0^\infty \hat{u}(t)e^{-st}dt = \int_0^\infty 1e^{-st}dt = \left.\frac{e^{-st}}{-s}\right|_0^\infty = 0 - \left(\frac{1}{-s}\right) = \frac{1}{s} \tag{4.2}$$

The contribution from the upper limit, $e^{-s(\infty)}$ in Equation 4.2, is zero provided $s > 0$. More specifically, $\text{Re}(s) > 0$ because s is a complex variable $s = \sigma + j\omega$. Therefore,

$$\mathcal{L}\{\hat{u}(t)\} = \frac{1}{s}, \quad \text{Re}(s) > 0 \tag{4.3}$$

indicating the integral in Equation 4.2 converges so long as the complex variable s is located in the right half of the complex plane. Note that the constant function $u(t) = 1$, $-\infty < t < \infty$ is identical to $\hat{u}(t)$ for $t \geq 0$ and consequently has the same Laplace transform.

Henceforth, we shall omit reference to the region of convergence for the integral in Equation 4.1 and simply be concerned with the result. The region of convergence is only of interest when we perform the inverse Laplace transformation using an integration formula to transform $F(s)$ into $f(t)$. Returning to the example of the unit step function, the Laplace transform pair is

$$\hat{u}(t) \Leftrightarrow \frac{1}{s} \tag{4.4}$$

TABLE 4.1
Table of Laplace Transform Pairs for Elementary Continuous-Time Signals

$f(t)$	$F(s) = \mathcal{L}\{f(t)\}$
$\hat{u}(t) = \begin{cases} 0, & t < 0 \\ 1, & t \geq 0 \end{cases}$	$\frac{1}{s}$
$e^{\pm at}$	$\frac{1}{s \mp a}$
t^n	$\frac{n}{s^{(n+1)}}$
$\cos \omega t$	$\frac{s}{s^2 + \omega^2}$
$\sin \omega t$	$\frac{\omega}{s^2 + \omega^2}$

The Laplace transform of other continuous-time functions $f(t)$, $t \geq 0$ is handled in the same manner. For example, the exponential function $f(t) = e^{at}$ has a Laplace transform

$$F(s) = \mathcal{L}\{f(t)\} = \mathcal{L}\{e^{at}\} = \int_0^\infty e^{at}e^{-st}dt = \int_0^\infty e^{-(s-a)t}dt = \frac{1}{s - a} \tag{4.5}$$

Additional time signals of importance are $f(t) = t^n$ $(n = 0, 1, 2, \ldots)$ along with the trigonometric functions $f(t) = \cos \omega t$ and $f(t) = \sin \omega t$. Applying the definition for the Laplace transform of $f(t)$ in Equation 4.1 produces the results shown in Table 4.1.

4.2.1 PROPERTIES OF THE LAPLACE TRANSFORM

Certain properties of the Laplace transform enable $F(s)$ to be determined without resorting to the definition in Equation 4.1. Several of these properties are presented without proof. The first is the linearity property, which states that the Laplace transform of a linear combination of continuous-time functions is equal to the same linear combination of respective transforms.

P1:

$$\text{Given } \mathcal{L}\{f_1(t)\} = F_1(s) \quad \text{and} \quad \mathcal{L}\{f_2(t)\} = F_2(s)$$

$$\Rightarrow \mathcal{L}\{a_1 f_1(t) + a_2 f_2(t)\} = a_1 \mathcal{L}\{f_1(t)\} + a_2 \mathcal{L}\{f_2(t)\} = a_1 F_1(s) + a_2 F_2(s) \tag{4.6}$$

In properties P2 to P6 that follow, we start with $\mathcal{L}\{f(t)\} = F(s)$.

P2:

$$\mathcal{L}\{e^{\pm at}f(t)\} = F(s \mp a) \tag{4.7}$$

P3:

$$\mathcal{L}\{f(t - t_0)\hat{u}(t - t_0)\} = e^{-t_0 s}F(s) \tag{4.8}$$

P2 and P3 are shifting theorems with P2 applying to a function $f(t)$ multiplied by an exponential time function $e^{\pm at}$. Its Laplace transform $F(s)$ is shifted by an amount "a" in the s-domain. P3 applies to functions $f(t)$ delayed t_0 units, that is, shifted an amount t_0 to the right. Note the presence of the delayed unit step function $\hat{u}(t - t_0)$ that zeros out the portion of the signal $f(t)$, $-t_0 \leq t < 0$ shifted from the negative to the positive t-axis. The $\hat{u}(t - t_0)$ can be omitted in P3 if $f(t) = 0$, $t < 0$ (see Figure 4.2).

P4 applies to continuous-time functions in the t-domain expressible in product form when one of the factors is t^n.

P4:

$$\mathcal{L}\{t^n f(t)\} = (-1)^n \frac{d^n}{ds^n}F(s) \tag{4.9}$$

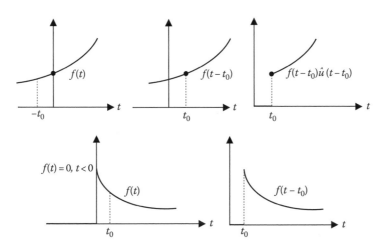

FIGURE 4.2 Illustration of the shifting property P3.

P5 shows that integration of functions in the t-domain is equivalent to division by the Laplace variable "s" in the s-domain.

P5:

$$\mathcal{L}\left\{\int_0^t f(t')dt'\right\} = \frac{F(s)}{s} \tag{4.10}$$

P6 expresses the Laplace transform of derivatives of $f(t)$ in terms of $F(s)$ and initial conditions. This property is central to solving linear differential equations using algebraic techniques in contrast to the classical time-domain approach.

P6:

$$\mathcal{L}\left\{\frac{d^n}{dt^n}f(t)\right\} = s^n F(s) - s^{n-1}f(0) - s^{n-2}\frac{d}{dt}f(0) - \cdots - s\frac{d^{n-2}}{dt^{n-2}}f(0) - \frac{d^{n-1}}{dt^{n-1}}f(0) \tag{4.11}$$

Periodic signals occur frequently as inputs to dynamic systems. The following property applies to functions that are periodic for $t \geq 0$.

P7:
If $f(t)$ is periodic with period T, that is, $f(t-T)=f(t)$, $t \geq 0$

$$F(s) = \frac{1}{1-e^{-Ts}}\int_0^T e^{-st}f(t)dt \tag{4.12}$$

The convolution of two functions $f(t)$ and $g(t)$ is defined in terms of an integral

$$f(t) * g(t) = \int_0^t f(t-\tau)g(\tau)d\tau \tag{4.13}$$

where $f(t) * g(t)$ denotes the operation of convolving the two continuous-time functions $f(t)$ and $g(t)$. The convolution $g(t) * f(t)$ is equivalent to $f(t) * g(t)$ because a change of variable $\lambda = t - \tau$ in Equation 4.13 leads directly to

$$\int_0^t f(t-\tau)g(\tau)d\tau = \int_0^t g(t-\lambda)f(\lambda)d\lambda = g(t) * f(t) \tag{4.14}$$

It will be shown in the following section that convolution can be used to represent the response of an LTI system to an arbitrary input, and the following property is useful in determining the response of LTI systems.

P8:

$$\mathcal{L}\{f(t) * g(t)\} = \mathcal{L}\left\{\int_0^t f(t-\tau)g(\tau)d\tau\right\} = F(s)G(s) \tag{4.15}$$

The convolution property P8 is a useful reminder that

$$\mathcal{L}\{f(t)g(t)\} \neq F(s)G(s) \tag{4.16}$$

that is, the Laplace transform of the product of two functions is not equal to the product of the individual Laplace transforms. To illustrate, suppose $g(t)$ is the unit step function.

$$\mathcal{L}\{f(t)g(t)\} = \mathcal{L}\{f(t)\hat{u}(t)\} = \mathcal{L}\{f(t)\} = F(s) \tag{4.17}$$

If Equation 4.16 were an equality, Equation 4.17 would lead to

$$F(s)G(s) = F(s) \tag{4.18}$$

$$\Rightarrow G(s) = 1 \tag{4.19}$$

Equation 4.19 is false and the inequality in Equation 4.16 is correct. Several examples are presented to illustrate the use of these properties.

Example 4.1

Find $\mathcal{L}\{4e^{-2t} \sin 3t - 5t \cos 3t\}$.

$$\mathcal{L}\{4e^{-2t} \sin 3t - 5t \cos 3t\} = 4\mathcal{L}\{e^{-2t} \sin 3t\} - 5\mathcal{L}\{t \cos 3t\} \quad \text{(P1)} \tag{4.20}$$

$$\mathcal{L}\{\sin 3t\} = \frac{3}{s^2 + 9} \tag{4.21}$$

$$\Rightarrow \mathcal{L}\{e^{-2t} \sin 3t\} = \frac{3}{s^2 + 9}\bigg|_{s \leftarrow s+2} = \frac{3}{(s+2)^2 + 9} \quad \text{(P2)} \tag{4.22}$$

$$\mathcal{L}\{\cos 3t\} = \frac{s}{s^2 + 9} \tag{4.23}$$

$$\Rightarrow \mathcal{L}\{t \cos 3t\} = -\frac{d}{ds}\left(\frac{s}{s^2 + 9}\right) = \frac{s^2 - 9}{(s^2 + 9)^2} \quad \text{(P3)} \tag{4.24}$$

Hence,

$$\mathcal{L}\{4e^{-2t} \sin 3t - 5t \cos 3t\} = 4\left[\frac{3}{(s+2)^2 + 9}\right] - 5\left[\frac{s^2 - 9}{(s^2 + 9)^2}\right]$$

Example 4.2

$f(t) = (t - 2)\hat{u}(t - 2)$. Graph $f(t)$ and find $F(s)$.

First, the ramp function t, $-\infty < t < \infty$ is delayed (right shifted) two units of time to the right to produce the function $t - 2$, $-\infty < t < \infty$. Second, the shifted function is zeroed out for $t \leq 2$ as a result of the multiplicative term $\hat{u}(t - 2)$. The result is shown in Figure 4.3.

Shifting property P3 is used to find the Laplace transform of $f(t)$.

$$t \Leftrightarrow \frac{1}{s^2} \Rightarrow (t - 2)\hat{u}(t - 2) \Leftrightarrow e^{-2s}\left(\frac{1}{s^2}\right) \tag{4.25}$$

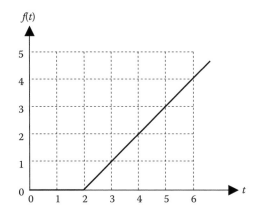

FIGURE 4.3   Graph of $f(t) = (t - 2)\hat{u}(t - 2)$.

Example 4.3

Find the Laplace transform of the signal $f(t)$ shown in Figure 4.4.

The piecewise continuous function $f(t)$ is defined in different intervals by

$$f(t) = \begin{cases} 0, & t < 1 \\ 2(t - 1), & 1 \leq t < 2 \\ 3, & 2 \leq t < 4 \\ -t + 7, & 4 \leq t < 7 \\ 0, & 7 \leq t \end{cases} \qquad (4.26)$$

Next, we write the function $f(t)$ in a single expression using unit step functions. The procedure is straightforward. The first nonzero term in Equation 4.26 is multiplied by the appropriate step function, so that it "turns on" at the correct time. This gives

$$f(t) = 2(t - 1)\hat{u}(t - 1) \qquad (4.27)$$

that is valid for the first two intervals $t < 1$ and $1 \leq t < 2$. The description of $f(t)$ changes for $t \geq 2$ necessitating a new term that is activated (goes from 0 to 1) for $t \geq 2$. Suppose we write

$$f(t) = 2(t - 1)\hat{u}(t - 1) + [\quad]\hat{u}(t - 2) \qquad (4.28)$$

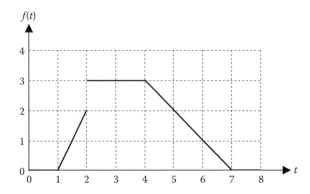

FIGURE 4.4   A piecewise continuous-time function.

The missing term in brackets must subtract out the previous expression for $f(t)$, that is, $2(t-1)$ and add the expression that holds for $2 \leq t < 4$—in this case the constant 3.

$$f(t) = 2(t-1)\hat{u}(t-1) + [-2(t-1) + 3]\hat{u}(t-2) \tag{4.29}$$

that is correct for $t < 1$, $1 \leq t < 2$, and $2 \leq t < 4$. You should check this yourself by choosing values of t from $-\infty < t < 4$. The same procedure is repeated until the function $f(t)$ is defined as follows:

$$f(t) = 2(t-1)\hat{u}(t-1) + [-2(t-1)+3]\hat{u}(t-2) + [-3+(-t+7)]\hat{u}(t-4)$$
$$+ [-(-t+7)+0]\hat{u}(t-7) \tag{4.30}$$

Simplifying Equation 4.30 yields

$$f(t) = 2(t-1)\hat{u}(t-1) + (-2t+5)\hat{u}(t-2) - (t-4)\hat{u}(t-4) + (t-7)\hat{u}(t-7) \tag{4.31}$$

The first, third, and fourth terms in Equation 4.31 have the form $f(t-t_0)\hat{u}(t-t_0)$ and can be Laplace transformed using property P3. The second term in Equation 4.31 can be manipulated into a similar form by doing the following:

$$(-2t+5)\hat{u}(t-2) = [-2(t-2)+1]\hat{u}(t-2) \tag{4.32}$$
$$= -2(t-2)\hat{u}(t-2) + \hat{u}(t-2) \tag{4.33}$$

Consequently, $f(t)$ is expressible as

$$f(t) = 2(t-1)\hat{u}(t-1) - 2(t-2)\hat{u}(t-2) + \hat{u}(t-2) - (t-4)\hat{u}(t-4) + (t-7)\hat{u}(t-7) \tag{4.34}$$

Using property P3, the Laplace transform of $f(t)$ in Equation 4.34 is

$$F(s) = 2\frac{e^{-s}}{s^2} - 2\frac{e^{-2s}}{s^2} + \frac{e^{-2s}}{s} - \frac{e^{-4s}}{s^2} + \frac{e^{-7s}}{s^2} \tag{4.35}$$

Note that the second term in Equation 4.31, the only one not of the form $f(t-t_0)\hat{u}(t-t_0)$, is present due to the discontinuity in $f(t)$ at $t=2$.

Example 4.4

Find the Laplace transform of the periodic signal $u(t)$ shown in Figure 4.5.

The signal $u(t)$ is periodic for $t \geq 0$ with period $T=3$. Let $u_1(t)$ represent the first cycle of $u(t)$, that is,

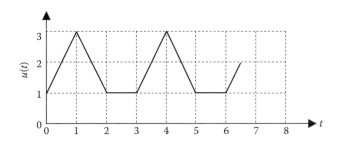

FIGURE 4.5 Graph of signal $u(t)$, periodic for $t \geq 0$.

$$u_1(t) = \begin{cases} 2t + 1, & 0 \leq t < 1 \\ -2t + 5, & 1 \leq t < 2 \\ 1, & 2 \leq t < 3 \\ 0, & 3 \leq t \end{cases} \tag{4.36}$$

From property P7,

$$U(s) = \frac{1}{1 - e^{-3s}} \int_0^3 e^{-st} u(t) dt \tag{4.37}$$

Since $u(t) = u_1(t)$, $0 \leq t < 3$, Equation 4.37 can be written with $u(t)$ replaced by $u_1(t)$,

$$U(s) = \frac{1}{1 - e^{-3s}} \int_0^\infty e^{-st} u_1(t) dt \tag{4.38}$$

$$= \frac{1}{1 - e^{-3s}} U_1(s) \tag{4.39}$$

Using the same approach as in Example 4.3, the piecewise continuous function $u_1(t)$ is decomposed into a sum of terms involving step functions.

$$u_1(t) = (2t + 1)\hat{u}(t) + [-(2t + 1) + (-2t + 5)]\hat{u}(t - 1) + [-(-2t + 5) + 1]\hat{u}(t - 2) - \hat{u}(t - 3) \tag{4.40}$$

$$= 2t\hat{u}(t) + \hat{u}(t) - 4(t - 1)\hat{u}(t - 1) + 2(t - 2)\hat{u}(t - 2) - \hat{u}(t - 3) \tag{4.41}$$

The shifting property P3 is used repeatedly to obtain the Laplace transform of $u_1(t)$ in Equation 4.41.

$$U_1(s) = \frac{2}{s^2} + \frac{1}{s} - 4\frac{e^{-s}}{s^2} + 2\frac{e^{-2s}}{s^2} - \frac{e^{-3s}}{s} \quad (P3) \tag{4.42}$$

From Equation 4.39, $U(s)$ is therefore

$$U(s) = \frac{1}{1 - e^{-3s}} \left[\frac{1 - e^{-3s}}{s} + \frac{2 - 4e^{-s} + 2e^{-2s}}{s^2} \right] \tag{4.43}$$

An alternate approach to finding $U(s)$ is to evaluate the integral in Equation 4.37 in pieces, namely

$$U(s) = \frac{1}{1 - e^{-3s}} \left[\int_0^1 e^{-st}(2t + 1) dt + \int_1^2 e^{-st}(-2t + 5) dt + \int_2^3 e^{-st}(1) dt \right] \tag{4.44}$$

In general, the second approach is more time-consuming because of the need to evaluate definite integrals.

Example 4.5

Find the function $y(t)$ whose Laplace transform is $Y(s) = 1/(s^2(s + 2))$.

$Y(s)$ is expressed as a product of terms,

$$Y(s) = \frac{1}{s^2} \cdot \frac{1}{(s + 2)} = F(s) \cdot G(s) \tag{4.45}$$

where

$$F(s) = \frac{1}{s^2}, \quad G(s) = \frac{1}{(s+2)} \tag{4.46}$$

From Table 4.1, the functions $f(t)$ and $g(t)$ are

$$f(t) = t, \quad t \geq 0, \quad \text{and} \quad g(t) = e^{-2t}, \quad t \geq 0 \tag{4.47}$$

From the convolution property P8, $y(t)$ is the convolution of $f(t)$ and $g(t)$.

$$y(t) = f * g = \int_0^t f(t-\tau)g(\tau)d\tau = \int_0^t (t-\tau)e^{-2\tau}d\tau \tag{4.48}$$

Evaluating the definite integral in Equation 4.48 gives

$$y(t) = t \int_0^t e^{-2\tau}d\tau - \int_0^t \tau e^{-2\tau}d\tau = \frac{1}{4}(2t - 1 + e^{-2t}), \quad t \geq 0 \tag{4.49}$$

The same result is obtained using the alternate form of the convolution integral,

$$y(t) = g * f = \int_0^t g(t-\tau)f(\tau)d\tau = \int_0^t e^{-2(t-\tau)}\tau d\tau = e^{-2t}\int_0^t \tau e^{2\tau}d\tau \tag{4.50}$$

4.2.2 INVERSE LAPLACE TRANSFORM

The last example required us to find the function $y(t)$ satisfying $\mathcal{L}\{y(t)\} = y(s)$. In this context, $y(t)$ is referred to as the inverse Laplace transform of $Y(s)$. In general, given a Laplace transform $F(s) = \mathcal{L}\{f(t)\}$, $f(t)$ can be determined using the inverse function $\mathcal{L}^{-1}\{F(s)\}$ (Ogata 1998),

$$f(t) = \mathcal{L}^{-1}\{F(s)\} = \frac{1}{2\pi j} \int_{z_c - j\infty}^{z_c + j\infty} F(s)e^{st}ds, \quad t > 0 \tag{4.51}$$

where z_c is a real constant chosen to assure convergence of the integral. Finding $f(t)$ in Equation 4.51 requires evaluating an integral of a function of a complex variable over a contour in the s-plane. Fortunately, there is a simpler approach to performing the inverse Laplace transformation when $F(s)$ involves a ratio of polynomials in s. To illustrate, suppose we are asked to find the inverse Laplace transform of $F(s)$ where

$$F(s) = \frac{1/2}{s^2} - \frac{1/4}{s} + \frac{1/4}{s+2} \tag{4.52}$$

Inverse Laplace transforming both sides of Equation 4.52 and using the linearity property P1, we can write

$$f(t) = \mathcal{L}^{-1}\{F(s)\} = \mathcal{L}^{-1}\left\{\frac{1/2}{s^2} - \frac{1/4}{s} + \frac{1/4}{s+2}\right\} \tag{4.53}$$

$$= \mathcal{L}^{-1}\left\{\frac{1/2}{s^2}\right\} - \mathcal{L}^{-1}\left\{\frac{1/4}{s}\right\} + \mathcal{L}^{-1}\left\{\frac{1/4}{s+2}\right\} \tag{4.54}$$

$$= \frac{1}{2}\mathcal{L}^{-1}\left\{\frac{1}{s^2}\right\} - \frac{1}{4}\mathcal{L}^{-1}\left\{\frac{1}{s}\right\} + \frac{1}{4}\mathcal{L}^{-1}\left\{\frac{1}{s+2}\right\} \tag{4.55}$$

From Table 4.1,

$$\mathcal{L}^{-1}\left\{\frac{1}{s^2}\right\} = t, \quad t \geq 0, \quad \mathcal{L}^{-1}\left\{\frac{1}{s}\right\} = 1, \quad t \geq 0, \quad \mathcal{L}^{-1}\left\{\frac{1}{s+2}\right\} = e^{-2t}, \quad t \geq 0 \tag{4.56}$$

$$\Rightarrow f(t) = \frac{1}{2}t - \frac{1}{4} + \frac{1}{4}e^{-2t}, \quad t \geq 0 \tag{4.57}$$

Note $y(t)$ in Equation 4.49 and $f(t)$ in Equation 4.57 are the same functions. Therefore, $Y(s)$ in Equation 4.45 and $F(s)$ in Equation 4.52 are equal, that is,

$$\frac{1}{s^2(s+2)} = \frac{1/2}{s^2} - \frac{1/4}{s} + \frac{1/4}{s+2} \tag{4.58}$$

This suggests that the inverse Laplace transform of a quotient like the one on the left-hand side of Equation 4.58 can be found by first expressing it as a summation of terms and then resorting to a table of Laplace transform pairs. This method is termed partial fraction expansion and will be discussed shortly. First, we establish the need for inverting Laplace transforms expressed in terms of proper fractions with polynomials in "s" in the numerator and denominator.

4.2.3 Laplace Transform of the System Response

The significance of Laplace transforms in the analysis of LTI dynamic systems is in large part a consequence of property P6, which relates the Laplace transform of $df(t)/dt$ and higher derivatives to $F(s)$, the Laplace transform of $f(t)$. For example, consider a linear second-order system with input $u(t)$ and output $y(t)$ with initial conditions $y(0)$, $\dot{y}(0)$ modeled by

$$\frac{d^2}{dt^2}y(t) + a_1\frac{d}{dt}y(t) + a_0y(t) = b_2\frac{d^2}{dt^2}u(t) + b_1\frac{d}{dt}u(t) + b_0u(t) \tag{4.59}$$

Laplace transforming both sides of Equation 4.59, with $\mathcal{L}\{u(t)\} = U(s)$ and $\mathcal{L}\{y(t)\} = Y(s)$ results in

$$s^2Y(s) - sy(0) - \dot{y}(0) + a_1[sY(s) - y(0)] + a_0Y(s)$$
$$= b_2[s^2U(s) - su(0) - \dot{u}(0)] + b_1[sU(s) - u(0)] + b_0U(s) \tag{4.60}$$

where $u(0)$ and $\dot{u}(0)$ are the initial values of the input and its first derivative.

Collecting terms and solving for $Y(s)$ give

$$Y(s) = \underbrace{\left[\left(\frac{b_2 s^2 + b_1 s + b_0}{s^2 + a_1 s + a_0}\right)U(s) - \frac{b_2 u(0)s + b_2 \dot{u}(0) + b_1 u(0)}{s^2 + a_1 s + a_0}\right]}_{\text{terms involving input } u(t)} + \underbrace{\frac{y(0)s + \dot{y}(0) + a_1 y(0)}{s^2 + a_1 s + a_0}}_{\text{term involving initial state, } [y(0), \dot{y}(0)]^T}$$

(4.61)

The complete response $y(t)$ consists of two components. The first,

$$y_{zs}(t) = \mathcal{L}^{-1}\left\{\left(\frac{b_2 s^2 + b_1 s + b_0}{s^2 + a_1 s + a_0}\right)U(s) - \frac{b_2 u(0)s + b_2 \dot{u}(0) + b_1 u(0)}{s^2 + a_1 s + a_0}\right\}$$

(4.62)

is called the zero-state response, so named because it represents the system's response when the initial state is zero, that is, $y(0) = \dot{y}(0) = 0$. The terms in Equation 4.62 result from the presence of a forcing function $u(t)$, which explains why the zero-state response is also referred to as the forced response.

The second component is the zero-input response, which is the response of the unforced system, that is, $u(t) = 0$, $t \geq 0$. It is also known as the free response.

$$y_{zi}(t) = \mathcal{L}^{-1}\left\{\frac{y(0)s + \dot{y}(0) + a_1 y(0)}{s^2 + a_1 s + a_0}\right\}$$

(4.63)

For an elementary type of input $u(t)$, its Laplace transform $U(s)$ will be a ratio of polynomials in s (see Table 4.1) with the order of the denominator higher than the order of the numerator by at least one. Thus, the terms inside the brackets in Equations 4.62 and 4.63 are also proper fractions with numerator and denominator polynomials in s.

For example, suppose $y(0) = y_0$, $\dot{y}(0) = \dot{y}_0$ and $u(t) = \sin \omega t$, $t \geq 0$. Since $u(0) = 0$ and $\dot{u}(0) = \omega$, $Y(s)$ becomes

$$Y(s) = \left[\left(\frac{b_2 s^2 + b_1 s + b_0}{s^2 + a_1 s + a_0}\right)\frac{\omega}{s^2 + \omega^2} - \frac{b_2 \omega}{s^2 + a_1 s + a_0}\right] + \frac{y_0 s + \dot{y}_0 + a_1 y_0}{s^2 + a_1 s + a_0}$$

(4.64)

$$= \frac{y_0 s^3 + (a_1 y_0 + \dot{y}_0)s^2 + \omega(b_1 + y_0\omega)s + \omega(a_1 y_0\omega + \dot{y}_0\omega + b_0 - b_2\omega^2)}{s^4 + a_1 s^3 + (a_0 + \omega^2)s^2 + a_1\omega^2 s + a_0\omega^2}$$

(4.65)

Inverting $Y(s)$ is facilitated by decomposing the right-hand side of Equation 4.65 into a sum of terms for which the inverse Laplace transform is readily determined from tables such as Table 4.1. The same applies for higher order systems with arbitrary inputs.

4.2.4 Partial Fraction Expansion

The second-order system example demonstrates that $Y(s)$ will ordinarily be a proper fraction with polynomials in "s" in the numerator and denominator, that is,

$$Y(s) = \frac{N(s)}{D(s)} = \frac{a_m s^m + a_{m-1}s^{m-1} + \cdots + a_1 s + a_0}{s^n + b_{n-1}s^{n-1} + \cdots + b_1 s + b_0}, \quad (n > m)$$

(4.66)

We begin the process of expanding $Y(s)$ into a sum of terms by determining the roots of $D(s) = 0$. The nature of the n roots will dictate the form of the expansion. A number of cases will be considered.

Case I: All roots of $D(s) = 0$ are real and distinct

Let the n distinct roots be the real numbers $p_1, p_2, \ldots, p_n$ obtained by factoring the denominator $D(s)$ into

$$D(s) = (s - p_1)(s - p_2) \cdots (s - p_n) \tag{4.67}$$

$$\Rightarrow Y(s) = \frac{N(s)}{D(s)} = \frac{a_m s^m + a_{m-1} s^{m-1} + \cdots + a_1 s + a_0}{(s - p_1)(s - p_2) \cdot s(s - p_n)} \tag{4.68}$$

$p_1, p_2, \ldots, p_n$ are called the poles of $Y(s)$. The partial fraction expansion of $Y(s)$ is

$$Y(s) = \frac{a_m s^m + a_{m-1} s^{m-1} + \cdots + a_1 s + a_0}{(s - p_1)(s - p_2) \cdots (s - p_n)} = \frac{c_1}{s - p_1} + \frac{c_2}{s - p_2} + \cdots + \frac{c_n}{s - p_n} \tag{4.69}$$

where the constants c_i, $i = 1, 2, \ldots, n$, referred to as the residues of $Y(s)$ at the respective poles p_i, $i = 1, 2, \ldots, n$, are obtained from

$$c_i = \left[(s - p_i) \frac{N(s)}{D(s)} \right]_{s = p_i} \tag{4.70}$$

$$= \left[\frac{a_m s^m + a_{m-1} s^{m-1} + \ldots + a_1 s + a_0}{(s - p_1)(s - p_2) \ldots (s - p_{i-1})(s - p_{i+1}) \ldots (s - p_n)} \right]_{s = p_i}, \quad i = 1, 2, \ldots, n \tag{4.71}$$

Example 4.6

Find the inverse Laplace transform of

$$Y(s) = \frac{s^2 + 1}{s^3 + 10.5 s^2 + 14s + 4.5} \tag{4.72}$$

Factoring the denominator leads to the partial fraction expansion as follows:

$$Y(s) = \frac{s^2 + 1}{(s + 0.5)(s + 1)(s + 9)} = \frac{c_1}{s + 0.5} + \frac{c_2}{s + 1} + \frac{c_3}{s + 9} \tag{4.73}$$

The constants c_1, c_2, and c_3 are obtained from Equation 4.71 as follows:

$$c_1 = \left[(s = 0.5) \frac{s^2 + 1}{(s + 0.5)(s + 1)(s + 9)} \right]_{s = -0.5} = \frac{(-0.5)^2 + 1}{(-0.5 + 1)(-0.5 + 9)} = \frac{5}{17} \tag{4.74}$$

$$c_2 = \left[(s + 1) \frac{s^2 + 1}{(s + 0.5)(s + 1)(s + 9)} \right]_{s = -1} = \frac{(-1)^2 + 1}{(-1 + 0.5)(-1 + 9)} = -\frac{1}{2} \tag{4.75}$$

$$c_3 = \left[(s + 9) \frac{s^2 + 1}{(s + 0.5)(s + 1)(s + 9)} \right]_{s = -9} = \frac{(-9)^2 + 1}{(-9 + 0.5)(-9 + 1)} = \frac{41}{34} \tag{4.76}$$

$$\Rightarrow Y(s) = \frac{5/17}{s + 0.5} - \frac{1/2}{s + 1} + \frac{41/34}{s + 9} \tag{4.77}$$

$y(t)$ is obtained by inverse Laplace transforming the terms in Equation 4.77,

$$y(t) = \mathcal{L}^{-1} \left\{ \frac{5/17}{s + 0.5} - \frac{1/2}{s + 1} + \frac{41/34}{s + 9} \right\} \tag{4.78}$$

$$= \frac{5}{17} e^{-0.5t} - \frac{1}{2} e^{-t} + \frac{41}{34} e^{-9t}, \quad t \geq 0 \tag{4.79}$$

Case II: All roots of $D(s) = 0$ are real and at least one is a multiple root
Suppose p_1 has multiplicity m_1 and p_2 multiplicity m_2. There are a total of $n - m_1 - m_2 + 2$ distinct pole values, that is, $p_1, p_2, p_3, \ldots, p_{n-m_1-m_2+2}$. In factored form, $D(s)$ is

$$D(s) = (s - p_1)^{m_1}(s - p_2)^{m_2}(s - p_3)\cdots(s - p_{n-m_1-m_2+2}) \tag{4.80}$$

The partial fraction expansion of $Y(s)$ is

$$Y(s) = \frac{a_{m_1}}{(s - p_1)^{m_1}} + \frac{a_{m_1-1}}{(s - p_1)^{m_1-1}} + \cdots + \frac{a_1}{(s - p_1)} + \frac{b_{m_2}}{(s - p_2)^{m_2}} + \frac{b_{m_2-1}}{(s - p_2)^{m_2-1}} + \cdots + \frac{b_1}{(s - p_2)}$$

$$+ \frac{c_1}{s - p_3} + \frac{c_2}{s - p_4} + \cdots + \frac{c_{n-m_1-m_2}}{s - p_{n-m_1-m_2+2}} \tag{4.81}$$

Note the number of terms in the expansion corresponding to a particular pole is identical to the order (multiplicity) of the pole. The constants c_i, $i = 1, 2, \ldots, n - m_1 - m_2$ are evaluated in the same way as in Case I. For example, c_1 is obtained from

$$c_1 = \left[(s - p_3)\frac{N(s)}{D(s)} \right]_{s=p_3} = \left[\frac{a_m s^m + a_{m-1} s^{m-1} + \cdots + a_1 s + a_0}{(s - p_1)^{m_1}(s - p_2)^{m_2}\cdots(s - p_4)(s - p_5)\cdots(s - p_{n-m_1-m_2+2})} \right]_{s=p_3} \tag{4.82}$$

The constants $a_{m_1}, a_{m_1-1}, \ldots, a_2, a_1$ are evaluated using

$$a_k = \frac{1}{(m_1 - k)}\frac{d^{m_1-k}}{ds^{m_1-k}}\left[(s - p_1)^{m_1}\frac{N(s)}{D(s)} \right]_{s=p_1}, \quad k = m_1, m_1 - 1, \ldots, 2, 1 \tag{4.83}$$

A similar formula applies for the constants $b_{m_2}, b_{m_2-1}, \ldots, b_2, b_1$.

Example 4.7

$$Y(s) = \frac{1}{s^5 + 14s^4 + 75s^3 + 194s^2 + 244s + 120} \tag{4.84}$$

(a) Find the partial fraction expansion of $Y(s)$.
(b) Find $y(t)$.

(a) The poles of $Y(s)$ are found by using a root-solving program such as the MATLAB® function "roots" that returns the roots of a polynomial. The call is "roots(a)" where "a" is the array of coefficients in descending order of the polynomial. With a = [1 14 75 194 244 120], "roots (a)" returns $-5, -3, -2, -2, -2$. $Y(s)$ is written with its denominator in factored form and then expanded as follows:

$$Y(s) = \frac{1}{(s + 2)^3(s + 3)(s + 5)} = \frac{a_3}{(s + 2)^3} + \frac{a_2}{(s + 2)^2} + \frac{a_1}{s + 2} + \frac{c_1}{s + 3} + \frac{c_2}{s + 5} \tag{4.85}$$

Evaluating c_1 and c_2 first,

$$c_1 = \left[(s+3)\frac{1}{(s+2)^3(s+3)(s+5)}\right]_{s=-3} = \left[\frac{1}{(-3+2)^3(-3+5)}\right] = -\frac{1}{2} \tag{4.86}$$

$$c_2 = \left[(s+5)\frac{1}{(s+2)^3(s+3)(s+5)}\right]_{s=-5} = \left[\frac{1}{(-5+2)^3(-5+3)}\right] = \frac{1}{54} \tag{4.87}$$

Next, the coefficients a_3, a_2, and a_1 are computed from Equation 4.83

$$a_3 = \frac{1}{(3-3)}\frac{d^{3-3}}{ds^{3-3}}\left[(s+2)^3\frac{1}{(s+2)^3(s+3)(s+5)}\right]_{s=-2}$$

$$= \left[\frac{1}{(-2+3)(-2+5)}\right] = \frac{1}{3} \tag{4.88}$$

$$a_2 = \frac{1}{(3-2)}\frac{d^{3-2}}{ds^{3-2}}\left[(s+2)^3\frac{1}{(s+2)^3(s+3)(s+5)}\right]_{s=-2} \tag{4.89}$$

$$= \frac{d}{ds}\left[\frac{1}{(s+3)(s+5)}\right]_{s=-2} = \left[\frac{-1(2s+8)}{(s^2+8s+15)^2}\right]_{s=-2} = -\frac{4}{9} \tag{4.90}$$

$$a_1 = \frac{1}{(3-1)!}\frac{d^{3-1}}{ds^{3-1}}\left[(s+2)^3\frac{1}{(s+2)^3(s+3)(s+5)}\right]_{s=-2} \tag{4.91}$$

$$= \frac{1}{2}\frac{d^2}{ds^2}\left[\frac{1}{(s+3)(s+5)}\right]_{s=-2} = \left[\frac{3s^2+24s+49}{(s^2+8s+15)^3}\right]_{s=-2} = \frac{13}{27} \tag{4.92}$$

An alternative approach to finding the constants c_1, c_2, a_3, a_2, and a_1 is to use the "residue" function in MATLAB that finds the poles of $Y(s)$ and the residues as well. $Y(s)$ is defined by arrays $n = [1]$, $d = [1\ 14\ 75\ 194\ 244\ 120]$, and the statement "$[R, P] = residue\ (n, d)$" returns the poles -5, -3, -2, -2, -2 in array "P" and the residues $1/54$, $-1/2$, $13/27$, $-4/9$, $1/3$ in array "R."

(b) From Table 4.1 and property P2, the inverse transform of $Y(s)$ in Equation 4.85 is

$$y(t) = a_3 e^{-2t}\frac{t^2}{2} + a_2 e^{-2t}t + a_1 e^{-2t} + c_1 e^{-3t} + c_2 e^{-5t} \tag{4.93}$$

Substituting the values for c_1 and c_2 from Equations 4.86 and 4.87 as well as a_3, a_2, and a_1 from Equations 4.88, 4.90, and 4.92 into Equation 4.93 gives

$$y(t) = \frac{1}{6}e^{-2t}t^2 - \frac{4}{9}e^{-2t}t + \frac{13}{27}e^{-2t} - \frac{1}{2}e^{-3t} + \frac{1}{54}e^{-5t} \tag{4.94}$$

Case III: Complex roots of $D(s) = 0$

When the polynomial $D(s)$ possesses nonrepeated complex roots, it is possible to apply Case I or II and obtain the partial fraction expansion. However, the coefficients will be complex numbers, and the partial fraction expansion will include complex exponentials, which have to be combined in a way to produce real-valued functions. There are two alternatives that eliminate the need for complex number arithmetic. Both are presented followed by an illustrative example.

Suppose the denominator of $Y(s)$ in Equation 4.66 has a single pair of complex roots. Factoring the denominator into a product of linear factors and a quadratic factor,

$$Y(s) = \frac{N(s)}{D(s)} = \frac{a_m s^m + a_{m-1} s^{m-1} + \cdots + a_1 s + a_0}{(s - p_1)(s - p_2)\cdots(s - p_{n-2})(as^2 + bs + c)} \tag{4.95}$$

where $p_1, p_2, \ldots, p_{n-2}$ are real and $as^2 + bs + c = 0$ has complex roots $\alpha \pm j\beta$ ($\beta > 0$). For simplicity, assume the poles $p_1, p_2, \ldots, p_{n-2}$ are distinct. The partial fraction expansion is

$$Y(s) = \frac{N(s)}{D(s)} = \frac{c_1}{s - p_1} + \frac{c_2}{s - p_2} + \cdots + \frac{c_{n-2}}{s - p_{n-2}} + \frac{d_1 s + d_2}{as^2 + bs + c} \tag{4.96}$$

The constants $c_1, c_2, c_3, \ldots, c_{n-2}$ are obtained as before (Case I). The constants d_1 and d_2 are obtained by recombining the terms on the right-hand side of Equation 4.96 and then equating the coefficients of powers of s in the numerator with the coefficients of like powers of s in the original form of the numerator $N(s)$. The inverse Laplace transform of the last term in Equation 4.96 is

$$\mathcal{L}^{-1}\left\{\frac{d_1 s + d_2}{s^2 + as + b}\right\} = e^{\alpha t}\left[d_1 \cos \beta t \left(\frac{d1\alpha + d2}{\beta}\right) \sin \beta t\right] \tag{4.97}$$

To illustrate, consider

$$Y(s) = \frac{s + 1}{s^4 + 5s^3 + 11s^2 + 15s} = \frac{s + 1}{s(s + 3)(s^2 + 2s + 5)} \tag{4.98}$$

$$= \frac{c_1}{s} + \frac{c_2}{s + 3} + \frac{d_1 s + d_2}{s^2 + 2s + 5} \tag{4.99}$$

From the quadratic formula, the roots of $s^2 + 2s + 5$ are $-1 \pm j2$. Thus, $\alpha = -1$ and $\beta = 2$. The constants c_1 and c_2 are calculated from

$$c_1 = \left[s \frac{s + 1}{s(s + 3)(s^2 + 2s + 5)}\right]_{s=0} = \frac{s + 1}{(s + 3)(s^2 + 2s + 5)}\bigg|_{s=0} = \frac{1}{15} \tag{4.100}$$

$$c_2 = \left[(s + 3)\frac{s + 1}{s(s + 3)(s^2 + 2s + 5)}\right]_{s=-3} = \frac{s + 1}{s(s^2 + 2s + 5)}\bigg|_{s=-3} = \frac{1}{12} \tag{4.101}$$

Combining terms in Equation 4.99 over a common denominator and equating the numerator to $s + 1$, the numerator in Equation 4.98 gives

$$s + 1 = \frac{1}{15}(s + 3)(s^2 + 2s + 5) + \frac{1}{12}s(s^2 + 2s + 5) + (d_1 s + d_2)s(s + 3) \tag{4.102}$$

$$\Rightarrow s + 1 = \left(\frac{1}{15} + \frac{1}{12} + d_1\right)s^3 + \left(\frac{1}{3} + \frac{1}{6} + 3d_1 + d_2\right)s^2 + \left(\frac{11}{15} + \frac{5}{12} + 3d_2\right)s + 1 \tag{4.103}$$

Equating coefficients of like powers of s on both sides of Equation 4.103,

$$s^3: \ 0 = \frac{1}{15} + \frac{1}{12} + d_1 \Rightarrow d_1 = -\frac{1}{15} - \frac{1}{12} = -\frac{3}{20}$$

$$s^2: \ 0 = \frac{1}{3} + \frac{1}{6} + 3d_1 + d_2 \Rightarrow d_2 = -\frac{1}{3} - \frac{1}{6} - 3d_1 = -\frac{1}{2} - 3\left(-\frac{3}{20}\right) = -\frac{1}{20}$$

$$s^1: \ 1 = \frac{11}{15} + \frac{5}{12} + 3d_2 \Rightarrow d_2 = \frac{1}{3}\left(1 - \frac{11}{15} - \frac{5}{12}\right) = -\frac{1}{20}$$

$$s^0: \ 1 = 1$$

Note, only two of the first three equations are needed to solve for d_1 and d_2, and the remaining two equations serve as a check. Solving for c_1 and c_2 directly in Equations 4.100 and 4.101 eliminates the need to solve four simultaneous equations for the unknown constants c_1, c_2, d_1, and d_2. Substituting the values for c_1, c_2, d_1, and d_2 into Equation 4.99 yields

$$Y(s) = \frac{1/15}{s} + \frac{1/12}{s+3} + \frac{(-3/20)s - 1/20}{s^2 + 2s + 5} \tag{4.104}$$

$$= \frac{1}{15}\left(\frac{1}{s}\right) + \frac{1}{12}\left(\frac{1}{s+3}\right) - \frac{1}{20}\left(\frac{3s+1}{s^2+2s+5}\right) \tag{4.105}$$

The last term is inverted using Equation 4.97 with $d_1 = 3$, $d_2 = 1$, $\alpha = -1$, and $\beta = 2$.

$$Y(t) = \frac{1}{15} + \frac{1}{12}e^{-3t} - \frac{1}{20}e^{-t}\left[3\cos 2t + \left(\frac{3(-1)+1}{2}\right)\sin 2t\right] \tag{4.106}$$

$$= \frac{1}{15} + \frac{1}{12}e^{-3t} - \frac{1}{20}e^{-t}(3\cos 2t - \sin 2t) \tag{4.107}$$

The second method for inverse Laplace transforming terms like the one in Equation 4.97 is based on decomposing it into two terms that can be readily inverted. Starting with an expression containing a quadratic in the denominator with complex roots, the first step is to complete the square as illustrated in the following equations:

$$F(s) = \frac{d_1 s + d_2}{s^2 + as + b} \tag{4.108}$$

$$= \frac{d_1 s + d_2}{(s^2 + as + a^2/4) + (b - a^2/4)} \tag{4.109}$$

$$= \frac{d_1 s + d_2}{(s + a/2)^2 + \omega^2} \quad \left(\omega^2 = b - \frac{a^2}{4}\right) \tag{4.110}$$

After completing the square in the denominator, Equation 4.110 is expressed as the sum of two terms that are the Laplace transforms of shifted trigonometric functions.

$$F(s) = \frac{d_1[(s + a/2) - a/2] + d_2}{(s + a/2)^2 + \omega^2} \tag{4.111}$$

$$= d_1 \frac{(s + a/2)}{(s + a/2)^2 + \omega^2} + \left[\frac{d_2 - (a/2)d_1}{\omega}\right]\frac{\omega}{(s + a/2)^2 + \omega^2} \tag{4.112}$$

From Table 4.1 and the shifting property P2, $f(t) = \mathcal{L}^{-1}\{F(s)\}$ is

$$f(t) = d_1 e^{-(a/2)t} \cos \omega t + \left[\frac{d_2 - (a/2)d_1}{\omega}\right]e^{-(a/2)t}\sin \omega t \tag{4.113}$$

Returning to the previous example,

$$\frac{1}{20}\left(\frac{3s+1}{s^2+2s+5}\right) = \frac{1}{20}\left(\frac{d_1 s + d_2}{s^2 + as + b}\right) \tag{4.114}$$

making $d_1 = 3$, $d_2 = 1$, $a = 2$, $b = 5$, and $\omega^2 = b - a^2/4 = 4$. Substituting the values for d_1, d_2, a, b, and ω into Equation 4.113 leads to the inverse Laplace transform,

$$\frac{1}{20}\left\{d_1 e^{-(a/2)t}\cos \omega t + \left[\frac{d_2 - (a/2)d_1}{\omega}\right]e^{-(a/2)t}\sin \omega t\right\} = \frac{1}{20}(3e^{-t}\cos 2t - e^{-t}\sin 2t) \tag{4.115}$$

in agreement with the result shown in Equation 4.107.

Rather than having to remember

$$\mathcal{L}^{-1}\left\{\frac{d_1 s + d_2}{s^2 + as + b}\right\} = d_1 e^{-(a/2)t} \cos \omega t + \left[\frac{d_2 - (a/2)d_1}{\omega}\right] e^{-(a/2)t} \sin \omega t \qquad (4.116)$$

the inverse Laplace transform in the last example can be obtained directly by completing the square, that is,

$$F(s) = \frac{1}{20}\left(\frac{3s + 1}{s^2 + 2s + 5}\right) \qquad (4.117)$$

$$= \frac{1}{20}\left\{\frac{3[(s+1) - 1] + 1}{(s+1)^2 + 2^2}\right\} \qquad (4.118)$$

$$= \frac{1}{20}\left[\frac{3(s+1) - 2}{(s+1)^2 + 2^2}\right] \qquad (4.119)$$

$$= \frac{1}{20}\left[3\frac{(s+1)}{(s+1)^2 + 2^2} - \frac{2}{(s+1)^2 + 2^2}\right] \qquad (4.120)$$

Inverse Laplace transformation of $F(s)$ gives the same $f(t)$ in Equation 4.115.

EXERCISES

4.1 Find the Laplace transforms of the functions $f(t)$ given below. Note that $\hat{u}(t - t_0)$ is the unit step function delayed t_0 units of time.
(a) $t^2 \sin 2t$ (b) $t\hat{u}(t - 1)$ (c) $(t - 1)\hat{u}(t)$ (d) $2[\hat{u}(t - 1) - \hat{u}(t - 4)]$
(e) $(d/dt)(te^{-t})$ (f) $\sin(2t + \pi/4)$ (g) $(1 - 3t)e^{-3t}$ (h) $e^{-3t}\int_0^t \sin 2\tau \cos 2\tau d\tau$
(i) $\sin^2 t$ (j) $te^{-2t}\sin 3t$ (k) $\int_0^t \tau e^{-2\tau} \cos 3(t - \tau)d\tau$

4.2 Find the inverse Laplace transforms of the functions $F(s)$ given in the following:
(a) $\dfrac{1}{(s^2 - 1)}$ (b) $\dfrac{1}{(s + 2)^2 + 9}$ (c) $\dfrac{s + 1}{(s + 2)(s + 3)}$ (d) $\dfrac{s + 1}{s(s^2 - 4)^3}$

(e) $\dfrac{e^{-s} - e^{-3s}}{s(s + 1)}$ (f) $\dfrac{s + 1}{s^2 + 1}$ (g) $\dfrac{2s + 1}{(s^2 + s + 1)^2}$ (h) $\dfrac{s}{(s^2 + 2s + 5)(s^2 + 5s + 6)}$

4.3 Find the Laplace transform of the functions $f(t)$ in Figure E4.3a and b. In Figure E4.3b, the function is parabolic over the intervals $0 \le t < 2$ and $4 \le t < 6$ and passes through the points $(0, 0)$, $(1, 3)$, $(2, 4)$ and $(4, 4)$, $(5, 3)$, $(6, 0)$.

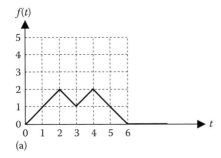

(a)

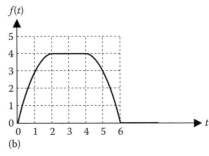

(b)

FIGURE E4.3

4.4 Graph the function $f(t)$ defined by

$$f(t) = t\hat{u}(t) + (t-1)\hat{u}(t-1) - 2t\hat{u}(t-2) + \hat{u}(t-3)$$

and find its Laplace transform.

4.5 Find the Laplace transform of the periodic function $f(t)$ shown in Figure E4.5:

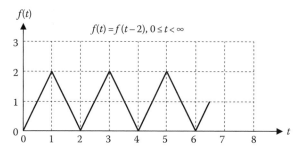

FIGURE E4.5

4.3 TRANSFER FUNCTION

Before we introduce the transfer function, the concept of an impulse function is presented because of its relevance to the response of LTI systems.

4.3.1 IMPULSE FUNCTION

An impulse function at $t = t_0$, denoted $\delta(t - t_0)$, is defined by its property of sifting the value of a function $f(t)$ at t_0 inside an integral, that is,

$$\int_{-\infty}^{\infty} \delta(t - t_0)f(t)dt = f(t_0), \quad -\infty < t_0 < \infty \tag{4.121}$$

$$f * g$$

The impulse function $\delta(t - t_0)$ is equal to zero wherever $t \neq t_0$ and is not finite at $t = t_0$. No such function exists in a physical sense; however, it can be used to approximate real signals $x(t)$, which occur over a very short duration Δ and satisfy the condition

$$\int_{t_0}^{t_0+\Delta} x(t)dt = 1 \tag{4.122}$$

as illustrated in Figure 4.6.

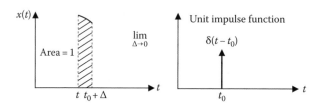

FIGURE 4.6 The unit impulse function $\delta(t - t_0)$ as a limit of a real function $x(t)$.

From Equation 4.121, the Laplace transform of $\delta(t - t_0)$ is given by

$$\int\limits_0^\infty \delta(t - t_0)e^{-st}dt = e^{-st0}, \quad t_0 > 0 \tag{4.123}$$

When $t_0 = 0$, Equation 4.123 with lower limit 0^- reduces to

$$\mathcal{L}\{\delta(t)\} = 1 \tag{4.124}$$

4.3.2 Relationship between Unit Step Function and Unit Impulse Function

The unit step function $\hat{u}(t)$ that equals 1 for $t > 0$ and 0 for $t > 0$ is discontinuous at $t = 0$. Although it cannot be implemented in a physical sense, it serves as an approximation to actual signals, which switch from one level to another in a very short period of time. The first derivative of a unit step function is zero everywhere except at the origin where it fails to exist as a result of the discontinuity. The unit impulse function $\delta(t)$ is likewise zero for all values of t except $t = 0$ where it is infinite.

The unit impulse function $\delta(t)$ can be thought of as the derivative of the unit step function $\hat{u}(t)$. This provides a framework for analyzing systems with discontinuous inputs that result when input derivatives are present in the mathematical model (Ogata 1998). To illustrate, consider the first-order system differential equation model

$$\frac{dy}{dt} + 2y = \frac{du}{dt} + u \tag{4.125}$$

where the input $u = \hat{u}(t)$ and the system is initially at rest, that is, $y(0^-) = 0$. Note that the initial time is taken as 0^- to indicate the initial state value prior to application of the step input at $t = 0$. Substituting $\hat{u}(t)$ for u in Equation 4.125 and replacing $(d/dt)\hat{u}(t)$ with $\delta(t)$,

$$\frac{dy}{dt} + 2y = \frac{d}{dt}\hat{u}(t) + \hat{u}(t) = \delta(t) + \hat{u}(t) \tag{4.126}$$

We learned in Chapter 2 that differential equations, where the highest order derivatives of the input and output are identical, possess a direct path between the input and output. We should therefore expect the output $y(t)$ in Equation 4.125 to be discontinuous at $t = 0$, that is, $y(0^+) \neq y(0^-)$ when the input is a unit step $\hat{u}(t)$. The impulse function on the right-hand side of Equation 4.126 is infinite at $t = 0$ accounting for the jump in $y(t)$ over the infinitesimal time period from $t = 0^-$ to $t = 0^+$.

It is possible to demonstrate this behavior without actually solving Equation 4.126 for $y(t)$. Solving for dy/dt in Equation 4.126,

$$\frac{dy}{dt} = \delta(t)\hat{u}(t) - 2y \tag{4.127}$$

Integrating both sides of Equation 4.127 from 0^- to t,

$$y(t) = \int\limits_{0^-}^t [\delta(\lambda) + \hat{u}(\lambda) - 2y(\lambda)]d\lambda \tag{4.128}$$

Decomposing the integral in Equation 4.128 into two separate integrals,

$$y(t) = \int_{0^-}^{0^+} [\delta(\lambda) + \hat{u}(\lambda) - 2y(\lambda)]d\lambda + \int_{0^+}^{t} [\delta(\lambda) + \hat{u}(\lambda) - 2y(\lambda)]d\lambda \tag{4.129}$$

The first integral simplifies because $\hat{u}(t)$ and $y(t)$ are both finite at $t=0$. The second integral simplifies by virtue of $\delta(t) = 0$ and $\hat{u}(t) = 1$ for $t \geq 0^+$. Equation 4.129 becomes

$$y(t) = \int_{0^-}^{0^+} \delta(\lambda)d\lambda + \int_{0_+}^{t} [1 - 2y(\lambda)]d\lambda \tag{4.130}$$

From the sifting property of the impulse function, Equation 4.121, the first term on the right-hand side of Equation 4.130 is 1. Evaluating $y(t)$ at $t = 0^+$,

$$y(0^+) = 1 + \int_{0^+}^{0^+} [1 - 2y(\lambda)]d\lambda = 1 \tag{4.131}$$

proving that $y(t)$ is discontinuous at $t = 0$ since $y(0^-) = 0$.

For functions that are discontinuous at the origin, the initial conditions in the differentiation property of Laplace transforms (P6) apply at $t = 0^-$. Hence, for $n = 1$

$$\mathcal{L}\left\{\frac{dy}{dt}\right\} = sY(s) - y(0^-) \tag{4.132}$$

Returning to Equation 4.126, Laplace transformation of both sides yields

$$sY(s) - y(0^-) + 2Y(s) = 1 + \frac{1}{s} \tag{4.133}$$

$$\Rightarrow Y(s) = \frac{s+1}{s(s+2)} + \frac{y(0^-)}{s+2} = \frac{s+1}{s(s+2)} = \frac{1}{2}\left[\frac{1}{s} + \frac{1}{s+2}\right] \tag{4.134}$$

$$\Rightarrow Y(t) = \frac{1}{2}(1 + e^{-2t}), \quad t \geq 0^+ \tag{4.135}$$

Substituting $t = 0^+$ in Equation 4.135 gives $y(0^+) = 1$ in agreement with Equation 4.131.

The initial condition $y(0^+)$ can be obtained by applying the initial value property of Laplace transforms that states

P9:

$$y(0^+) = \lim_{t \to 0^+} y(t) = \lim_{s \to \infty} sY(s) \tag{4.136}$$

In this example,

$$y(0^+) = \lim_{s \to \infty} sY(s) = \lim_{s \to \infty} s\left[\frac{s+1}{s(s+2)}\right] = 1 \tag{4.137}$$

4.3.3 IMPULSE RESPONSE

The response of LTI systems to an impulse forcing function is of great interest. We shall see why momentarily, but first an example is presented illustrating the process of finding the response of a simple system to an "impulse-like" input and comparing it with the true impulse response of the system.

Example 4.8

A spring-mass-damper system is struck by a hammer resulting in a force $f(t)$ like the one shown in Figure 4.7.

 (a) Find and graph the response $x(t)$ for $T = 1$, 0.5, 0.1, 0.01 s.
 (b) Find and graph the impulse response.

(a) The differential equation model of the system is

$$m\ddot{x} + c\dot{x} + kx = f \Rightarrow \ddot{x} + 2\dot{x} + 4x = \frac{1}{T}[\hat{u}(t) - \hat{u}(t - T)] \tag{4.138}$$

Laplace transforming Equation 4.138 with zero initial conditions,

$$(s^2 + 2s + 4)X(s) = \frac{1}{T}\left(\frac{1}{s} - \frac{e^{-Ts}}{s}\right) \tag{4.139}$$

$$X(s) = \frac{1}{T}\left[\frac{1 - e^{-Ts}}{s(s^2 + 2s + 4)}\right] \tag{4.140}$$

Inverse Laplace transformation of Equation 4.140 eventually results in

$$x(t) = \frac{1}{4T}\left[1 - e^{-t}\left(\cos\sqrt{3}t + \frac{1}{\sqrt{3}}\sin\sqrt{3}t\right)\right]$$
$$- \frac{1}{4T}\left[1 - e^{-(t-T)}\left\{\cos\sqrt{3}(t - T) + \frac{1}{\sqrt{3}}\sin\sqrt{3}(t - T)\right\}\right]\hat{u}(t - T) \tag{4.141}$$

which is simply a linear combination of the step response and a delayed version of the step response. Graphs of Equation 4.141 for $T = 1$, 0.5, 0.1, 0.01 s are generated in the M-file "Chap4_Ex3_1.m" and shown in Figure 4.8.

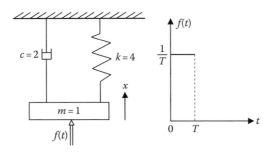

FIGURE 4.7 Mechanical system with pulse input $f(t)$.

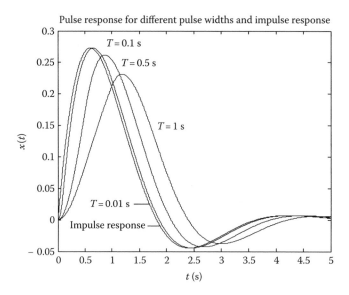

FIGURE 4.8 Pulse response of mechanical system for $T = 1, 0.5, 0.1, 0.01$ s and the impulse response.

(b) The true impulse response is obtained by Laplace transforming

$$\ddot{x} + 2\dot{x} + 4x = \delta(t) \tag{4.142}$$

$$\Rightarrow (s^2 + 2s + 4)X(s) = 1 \tag{4.143}$$

Solving for $X(s)$ followed by inverse Laplace transformation results in

$$x_{\text{impulse response}}(t) = \frac{1}{\sqrt{3}} e^{-t} \sin \sqrt{3}t, \quad t \geq 0 \tag{4.144}$$

It is graphed in Figure 4.8 and appears identical to the response $x(t)$ when the pulse width $T = 0.01$ s. Hence, the impulse response provides an accurate approximation of how the mechanical system responds to inputs of short (relative to the time constants of the system's natural modes) duration.

In Section 4.4.2, a mathematical model of a second-order system with input $u(t)$ and output $y(t)$ was introduced as an example of how Laplace transforms can be used to solve for the system response. The second-order LTI system shown in Figure 4.9 is referred to as a single input–single output (SISO) system. The mathematical model is given by

$$\frac{d^2}{dt^2} y(t) + a_1 \frac{d}{dt} y(t) a_0 y(t) = b_2 \frac{d^2}{dt^2} u(t) + b_1 \frac{d}{dt} u(t) + b_0 u(t) \tag{4.145}$$

Laplace transforming Equation 4.145, with zero initial conditions for the input, output, and their derivatives, leads to

$$Y(s) = \left(\frac{b_2 s^2 + b_1 s + b_0}{s^2 + a_1 s + a_0} \right) U(s) \tag{4.146}$$

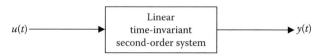

FIGURE 4.9 Second-order system with input $u(t)$ and output $y(t)$.

The ratio of $Y(s)$ to $U(s)$, when all initial conditions are identically zero, is called the transfer function of the system. Denoting it by $H(s)$,

$$H(s) = \frac{Y(s)}{U(s)} = \left(\frac{b_2 s^2 + b_1 s + b_0}{s^2 + a_1 s + a_0} \right) \tag{4.147}$$

Consider an nth-order LTI system with transfer function expressible as the ratio of two polynomials in proper fraction form, that is, the denominator polynomial is higher order than the numerator polynomial as in

$$H(s) = \frac{b_m s^m + b_{m-1} s^{m-1} + \cdots + b_1 s + b_0}{s^n + a_{n-1} s^{n-1} + \cdots + a_1 s + a_0} \quad (n > m) \tag{4.148}$$

The transfer function in Equation 4.148 is an alternative to the differential equation model representation of the system dynamics. It offers a convenient way of determining the forced response of an LTI system. From Equation 4.147, $Y(s)$ is equal to the product of the system transfer function $H(s)$ and the Laplace transform of the input,

$$Y(s) = H(s)U(s) \tag{4.149}$$

and the response is obtained by inverse Laplace transformation $y(t) = \mathcal{L}^{-1}\{Y(s)\}$.

The following examples illustrate the use of the transfer function to obtain the forced response of an LTI system.

Example 4.9

A first-order system is governed by the differential equation

$$\frac{dy}{dt} + 2y = u, \quad y(0) = 0 \tag{4.150}$$

(a) Find $H(s)$, the transfer function of the system.
(b) Find $y(t)$, the response when $u(t)$ is (i) $\sin 3t$, $t \geq 0$ and (ii) $\hat{u}(t)$.

(a) From Equation 4.148 with $n = 1$, $m = 0$, $b_0 = 1$, and $a_0 = 2$

$$H(s) = \frac{b_0}{s + a_0} = \frac{1}{s + 2} \tag{4.151}$$

(b) For $u(t) = \sin 3t$, $U(s) = 3/(s^2 + 9)$ and Equation 4.149 becomes

$$Y(s) = H(s)U(s) = \frac{1}{s + 2} \cdot \frac{3}{s^2 + 9} \tag{4.152}$$

$$= \frac{3}{13} \left[\frac{1}{s + 2} - \frac{s - 2}{s^2 + 9} \right] \tag{4.153}$$

$$= \frac{3}{13} \left[\frac{1}{s + 2} - \frac{s}{s^2 + 9} + \frac{2}{3} \frac{3}{s^2 + 9} \right] \tag{4.154}$$

$$y(t) = \mathcal{L}^{-1}\{Y(s)\} = \frac{3}{13} \left(e^{-2t} - \cos 3t + \frac{2}{3} \sin 3t \right), \quad t \geq 0 \tag{4.155}$$

For $u(t) = \hat{u}(t)$, $U(s) = 1/s$ and Equation 4.149 reduces to

$$Y(s) = H(s)U(s) = \frac{1}{s+2} \cdot \frac{1}{s} = \frac{1}{2}\left[\frac{1}{s} - \frac{1}{s+2}\right] \tag{4.156}$$

$$y(t) = \mathcal{L}^{-1}\{Y(s)\} = \frac{1}{2}(1 - e^{-2t}), \quad t \geq 0 \tag{4.157}$$

Example 4.10

For the system with transfer function,

$$H(s) = \frac{s^2 + 3s + 1}{(s+1)(s+3)(s+5)} \tag{4.158}$$

(a) Find the differential equation model of the system.
(b) Find the forced response to the input $u(t) = t$, $t \geq 0$.

(a) The differential equation of the system is obtained from $H(s)$ as follows:

$$H(s) = \frac{Y(s)}{U(s)} = \frac{s^2 + 3s + 1}{(s+1)(s+3)(s+5)} = \frac{s^2 + 3s + 1}{s^3 + 9s^2 + 23s + 15} \tag{4.159}$$

$$\Rightarrow (s^3 + 9s^2 + 23s + 15)Y(s) = (s^2 + 3s + 1)U(s) \tag{4.160}$$

$$\Rightarrow s^3 Y(s) + 9s^2 Y(s) + 23sY(s) + 15Y(s) = s^2 U(s) + 3sU(s) + U(s) \tag{4.161}$$

Performing the inverse Laplace transformation of the individual terms with all initial conditions zero results in the differential equation

$$\frac{d^3}{dt^3}y(t) + 9\frac{d^2}{dt^2}y(t) + 23\frac{d}{dt}y(t) + 15y(t) = \frac{d^2}{dt^2}u(t) + 3\frac{d}{dt}u(t) + u(t) \tag{4.162}$$

(b) Substituting $U(s) = 1/s^2$ in Equation 4.149 gives

$$Y(s) = \frac{s^2 + 3s + 1}{s^3 + 9s^2 + 23s + 15} \cdot \frac{1}{s^2} \tag{4.163}$$

The MATLAB statements

```
n = [1 3 1]; d = [1 9 23 15 0 0];
[R,P] = residue(n,d)
```

result in the residues and poles of the partial fraction expansion leading to the following expansion for $Y(s)$,

$$Y(s) = \frac{1}{15}\left(\frac{1}{s^2}\right) + \frac{22}{225}\left(\frac{1}{s}\right) - \frac{1}{8}\left(\frac{1}{s+1}\right) - \frac{1}{36}\left(\frac{1}{s+3}\right) + \frac{11}{200}\left(\frac{1}{s+5}\right) \tag{4.164}$$

and the forced response is obtained by inverse Laplace transformation of $Y(s)$,

$$y(t) = \frac{1}{15}t + \frac{22}{225} - \frac{1}{8}e^{-t} - \frac{1}{36}e^{-3t} + \frac{11}{200}e^{-5t}, \quad t \geq 0 \tag{4.165}$$

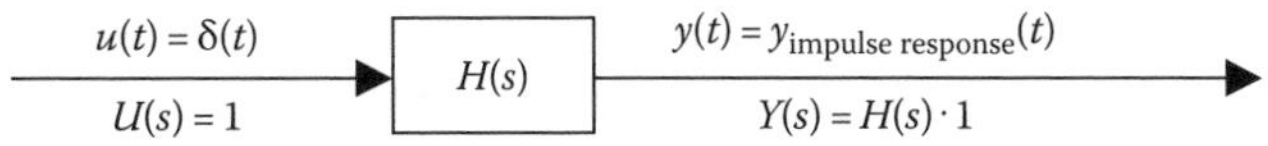

FIGURE 4.10 Linear time-invariant system with unit impulse input.

4.3.4 RELATIONSHIP BETWEEN IMPULSE RESPONSE AND TRANSFER FUNCTION

The impulse response function and the transfer function of an LTI system are related. Suppose the input to an LTI system is a unit impulse function as illustrated in Figure 4.10.

Since $Y(s) = H(s)U(s) = H(s) \cdot 1 = H(s)$, it follows that

$$y_{\text{impulse response}}(t) = \mathcal{L}^{-1}\{H(s)\} = h(t) \tag{4.166}$$

In other words, the impulse response of an LTI system is simply the inverse Laplace transform of the system transfer function. It is denoted $h(t)$ and referred to as the impulse response function. The impulse response function serves as alternative way of describing the dynamics of an LTI system. It can be used to find the forced response to an arbitrary input by first finding the transfer function $H(s) = \mathcal{L}\{h(t)\}$ and then proceeding in a similar manner to Example 4.10.

Alternatively, the forced response of an LTI system can be obtained directly from

$$y(t) = \mathcal{L}^{-1}\{H(s)U(s)\} = \int_0^t h(t-\tau)u(\tau)d\tau \tag{4.167}$$

that is, by convolution of the impulse response function $h(t)$ and the input $u(t)$. To illustrate, the unit step response of the third-order system in Example 4.10 is obtained using the convolution integral in Equation 4.167.

$$H(s) = \frac{s^2 + 3s + 1}{(s+1)(s+3)(s+5)} = \frac{1}{8}\left[-\frac{1}{s+1} - \frac{2}{s+3} + \frac{11}{s+5}\right] \tag{4.168}$$

$$h(t) = \mathcal{L}^{-1}\{H(s)\} = \frac{1}{8}(-e^{-t} - 2e^{-3t} + 11e^{-5t}) \tag{4.169}$$

$$y(t) = \int_0^t h(\tau)u(t-\tau)d\tau = \int_0^t \frac{1}{8}(-e^{-\tau} - 2e^{-3\tau} + 11e^{-5\tau}) \cdot 1 d\tau \tag{4.170}$$

$$= \frac{1}{8}\left[\left(e^{-\tau} + \frac{2}{3}e^{-3\tau} - \frac{11}{5}e^{-5\tau}\right)\right]_0^t \tag{4.171}$$

$$= \frac{1}{15} + \frac{1}{8}e^{-t} + \frac{1}{12}e^{-3t} - \frac{11}{40}e^{-5t}, \quad t \geq 0 \tag{4.172}$$

The initial condition $y(0^-) = 0$ and from Equation 4.172, $y(0^+) = 0$ as well. The response $y(t)$ is therefore continuous at $t = 0$ despite the discontinuity in the step input. In other words, a direct coupling from the input to the output does not exist. We should expect this result by observing that the third-order differential equation in Equation 4.162 does not contain a term on the right-hand side involving the third derivative of the input. If we express the system model in state variable form, the 1×1 direct coupling matrix D would be zero.

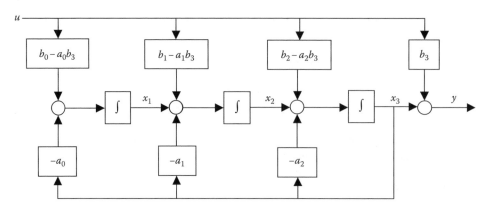

FIGURE 4.11 Simulation diagram of third-order system in observer canonical form.

A simulation diagram for an LTI system offers a convenient way of defining the states and revealing whether a direct path (no integrators) exists from the input to the output. Figure 4.11 shows a simulation diagram for the third-order system

$$\frac{d^3}{dt^3}y(t) + a_2\frac{d^2}{dt^2}y(t) + a_1\frac{d}{dt}y(t) + a_0y(t) = b_3\frac{d^3}{dt^3}u(t) + b_2\frac{d_2}{dt^2}u(t) + b_1\frac{d}{dt}u(t) + b_0u(t) \quad (4.173)$$

in what is known as observer canonical form (Ogata 1998). This form clearly shows the direct path from the input u to the output y when $b_3 \neq 0$. For the case when $b_3 = 0$, the state x_3 is equal to the output y and a direct path exists from u to $\dot{x}_3$. For a unit step input, the following is true if $b_3 = 0$:

$$y(0^+) = y(0^-), \quad \dot{y}(0^+) = \dot{y}(0^-) + (b_2 - a_2b_3)u(0^+) = \dot{y}(0^-) + b_2 \quad (4.174)$$

Consider the third-order system with transfer function given in Equation 4.159 and modeled by the differential equation in Equation 4.162. Comparing Equations 4.162 and 4.173 implies $a_2 = 9$, $a_1 = 23$, and $a_0 = 15$ and $b_3 = 0$, $b_2 = 1$, $b_1 = 3$, and $b_0 = 1$. Assuming zero initial conditions, the first derivative jumps from $\dot{y}(0^-) = 0$ to $\dot{y}(0^+) = \dot{y}(0^-) + b_2 = 1$ at $t = 0$.

Differentiating the solution for the unit step response $y(t)$ in Equation 4.172 gives

$$\frac{dy}{dt} = \frac{1}{40}(-5e^{-t} - 10e^{-3t} + 55e^{-5t}) \quad (4.175)$$

At $t = 0^+$,

$$\frac{dy}{dt}(0^+) = \frac{1}{40}(-5 - 10 + 55) = 1 \quad (4.176)$$

The system transfer function provides a convenient way of finding the forced response of an SISO LTI system. However, finding the transfer function can be a challenge when the mathematical model of the system consists of coupled algebraic and differential equations as opposed to a single nth-order differential equation relating the system input and output. Fortunately, the Laplace transform can be used to reduce the problem of finding the transfer function into one of an algebraic nature. The alternative, namely, eliminating dependent signals and their derivatives in the time domain, is far more cumbersome.

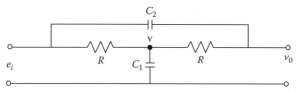

FIGURE 4.12 Circuit with input e_i and output v_0.

For example, consider the bridged-T network shown in Figure 4.12.
The node voltage method for analyzing the circuit results in the following equations:

$$\frac{e_i - v}{R} = C_1 \frac{dv}{dt} + \frac{v - v_0}{R} \tag{4.177}$$

$$C_2 \frac{d}{dt}(e_i - v_0) + \frac{v - v_0}{R} = 0 \tag{4.178}$$

Rearranging Equations 4.177 and 4.178 with node voltage terms on one side and the input terms on the other gives

$$RC_1 \frac{dv}{dt} + 2v - v_0 = e_i \tag{4.179}$$

$$RC_2 \frac{dv_0}{dt} + v_0 - v = RC_2 \frac{de_i}{dt} \tag{4.180}$$

The node voltage v must be eliminated from Equations 4.179 and 4.180 to arrive at a second-order differential equation involving e_i and v_0. Laplace transforming both equations with initial conditions set to zero and collecting terms produces the algebraic system of equations

$$\begin{aligned}(RC_1s + 2)V(s) - V_0(s) &= E_i(s) \\ -V(s) + (RC_2s + 1)V_0(s) &= RC_2sE_i(s)\end{aligned} \tag{4.181}$$

Using Cramer's rule, the solution for $V_0(s)$ is

$$V_0(s) = \frac{\begin{vmatrix} RC_1s + 2 & E_i(s) \\ -1 & RC_2sE_i(s) \end{vmatrix}}{\begin{vmatrix} RC_1s + 2 & -1 \\ -1 & RC_2s + 1 \end{vmatrix}} = \frac{R^2C_1C_2s^2 + 2RC_2s + 1}{R^2C_1C_2s^2 + R(C_1 + 2C_2)s + 1}E_i(s) \tag{4.182}$$

The transfer function is

$$\frac{V_0(s)}{E_i(s)} = \frac{R^2C_1C_2s^2 + 2RC_2s + 1}{R^2C_1C_2s^2 + R(C_1 + 2C_2)s + 1} \tag{4.183}$$

Inverse Laplace transformation of Equation 4.183 leads to the differential equation

$$R^2C_1C_2 \frac{d^2v_0}{dt^2} + R(C_1 + 2C_2) \frac{dv_0}{dt} + v_0 = R^2C_1C_2 \frac{d^2e_i}{dt^2} + 2RC_2 \frac{de_i}{dt} + e_i \tag{4.184}$$

4.3.5 Systems with Multiple Inputs and Outputs

In general, linear systems (and nonlinear systems) have more than a single input and output. Those systems and their models are designated multiple input–multiple output, abbreviated as MIMO. The transfer function concept still applies.

Suppose, for example, an LTI system such as an electric circuit is driven by independent voltage sources $e_1(t)$ and $e_2(t)$, and signals $i_R(t)$, $v_C(t)$, and $v_{\text{load}}(t)$ appearing at various points in the circuit are defined as outputs. A total of six transfer functions exist, one from each of two inputs to each of three outputs. We can write

$$I_R(s) = G_{1,1}(s)E_1(s) + G_{1,2}(s)E_2(s) \tag{4.185}$$

$$V_C(s) = G_{2,1}(s)E_1(s) + G_{2,2}(s)E_2(s) \tag{4.186}$$

$$V_{\text{load}}(s) = G_{3,1}(s)E_1(s) + G_{3,2}(s)E_2(s) \tag{4.187}$$

where

$$G_{1,1}(s) = \left.\frac{I_R(s)}{E_1(s)}\right|_{E_2(s)=0}, \quad G_{1,2}(s) = \left.\frac{I_R(s)}{E_2(s)}\right|_{E_1(s)=0} \tag{4.188}$$

$$G_{2,1}(s) = \left.\frac{V_C(s)}{E_1(s)}\right|_{E_2(s)=0}, \quad G_{2,2}(s) = \left.\frac{V_C(s)}{E_2(s)}\right|_{E_1(s)=0} \tag{4.189}$$

$$G_{3,1}(s) = \left.\frac{V_{\text{load}}(s)}{E_1(s)}\right|_{E_2(s)=0}, \quad G_{3,2}(s) = \left.\frac{V_{\text{load}}(s)}{E_2(s)}\right|_{E_1(s)=0} \tag{4.190}$$

The notation $G_{ij}(s)$ denotes the transfer function from the jth input to the ith output.

Equations 4.185 through 4.187 are a consequence of the principle of superposition that applies to linear systems. Superposition implies that the response of a system to multiple inputs applied simultaneously is equivalent to the sum of the system responses to the individual inputs applied one at a time.

An MIMO system and a method for finding its transfer functions are the focus of the following example.

Example 4.11

The amount of solute (drug or metabolite) introduced to or produced in the human body is often assumed to be stored in different compartments of the body. A separate equation for each compartment relates the rate of solute removal to the amount or concentration of the solute in the compartment. The solute can either be transported to another compartment or eliminated from the body by metabolism or excretion. Consider the linear compartment model described in Riggs (1970) for describing the quantities of iodine in humans. The state variables are

    x_1: Amount of inorganic iodine in the thyroid gland
    x_2: Amount of organic iodine in the thyroid gland
    x_3: Amount of hormonal iodine in the extrathyroidal tissue
    x_4: Amount of iodine in the inorganic iodide compartment

and the inputs are

    q_3: Rate of entry of exogenous iodide
    q_4: Rate of entry of exogenous hormonal iodine

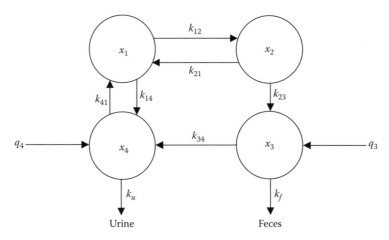

FIGURE 4.13 Compartmental model for iodine distribution in a human.

The model equations are summarized by the diagram illustrated in Figure 4.13, where k_{12}, k_{13}, k_{21}, k_{24}, k_{31}, k_{43}, k_u, and k_f are the rate constants governing the transfer of iodine between the compartments and its excretion from the body.

The outputs are

$y_1 = x_1 + x_2 + x_3 + x_4$, total iodine in the body
$y_2 = k_f x_3 + k_u x_4$, rate of iodine excretion from the body

(a) Write the state equations for the system and find the matrices A, B, C, and D.
(b) Draw a block diagram of the system, and label the Laplace transforms of the states x_1, x_2, x_3, and x_4 and outputs y_1 and y_2.
(c) Find the transfer function $Y_2(s)/Q_4(s)$.
(d) Baseline values of the system parameters are

$$k_{12} = 0.8/\text{day}, \quad k_{21} = 0.005/\text{day}, \quad k_{23} = 0.01/\text{day}, \quad \text{and} \quad k_{34} = 0.3/\text{day}$$

$$k_{14} = 0.15/\text{day}, \quad k_{41} = 0.5/\text{day}, \quad k_f = 0.02/\text{day}, \quad \text{and} \quad k_u = 1.2/\text{day}$$

Find the steady-state iodine levels in each compartment in response to a daily intake of iodine, $q_4 = 150$ μg/day. Assume $q_3 = 0$ μg/day.
(e) Find and graph the step response of $x_2(t)$ if the daily intake of iodine drops from 150 (where it has been for a long time) to 50 μg/day.

(a) From Figure 4.13, the state equations are

$$\left.\begin{aligned}
\dot{x}_1 &= -(k_{12} + k_{14})x_1 + k_{21}x_2 + k_{41}x_4 \\
\dot{x}_2 &= k_{12}x_1 - (k_{21} + k_{23})x_2 \\
\dot{x}_3 &= k_{23}x_2 - (k_{34} + k_f)x_3 + q_3 \\
\dot{x}_4 &= k_{14}x_1 + k_{34}x_3 - (k_{14} + k_u)x_4 + q_4
\end{aligned}\right\} \tag{4.191}$$

$$\left.\begin{aligned}
y_1 &= x_1 + x_2 + x_3 + x_4 \\
y_2 &= k_f x_3 + k_u x_4
\end{aligned}\right\} \tag{4.192}$$

The matrices A, B, C, and D in $\underline{\dot{x}} = A\underline{x} + B\underline{u}$ and $\underline{y} = C\underline{x} + D\underline{u}$ where $\underline{u} = [q_3 \; q_4]^T$ are

$$A = \begin{bmatrix} -(k_{12} + k_{14}) & k_{21} & 0 & k_{41} \\ k_{12} & -(k_{21} + k_{23}) & 0 & 0 \\ 0 & k_{23} & -(k_{34} + k_f) & 0 \\ k_{14} & 0 & k_{34} & -(k_{41} + k_u) \end{bmatrix}, \quad B = \begin{bmatrix} 0 & 0 \\ 0 & 0 \\ 1 & 0 \\ 0 & 1 \end{bmatrix} \qquad (4.193)$$

$$C = \begin{bmatrix} 1 & 1 & 1 & 1 \\ 0 & 0 & k_f & k_u \end{bmatrix}, \quad D = \begin{bmatrix} 0 & 0 \\ 0 & 0 \end{bmatrix} \qquad (4.194)$$

(b) The block diagram is obtained by Laplace transforming the state Equations 4.193 and 4.194, then solving for $X_1(s)$, $X_2(s)$, $X_3(s)$, and $X_4(s)$ in the respective equations. Introducing the notation $k_1 = k_{12} + k_{14}$, $k_2 = k_{21} + k_{23}$, $k_3 = k_{34} + k_f$, and $k_4 = k_{41} + k_u$ yields

$$X_1(s) = \frac{1}{s + k_1} [k_{21} X_2(s) + k_{41} X_4(s)] \qquad (4.195)$$

$$X_2(s) = \left(\frac{k_{12}}{s + k_2} \right) X_1(s) \qquad (4.196)$$

$$X_3(s) = \frac{1}{s + k_3} [k_{23} X_2(s) + Q_3(s)] \qquad (4.197)$$

$$X_4(s) = \frac{1}{s + k_4} [k_{14} X_1(s) + k_{34} X_3(s) + Q_4(s)] \qquad (4.198)$$

The block diagram follows immediately from Equations 4.195 through 4.198 and Equation 4.192. It is shown in Figure 4.14.

(c) The transfer function $Y_2(s)/Q_4(s)$ can be obtained by graphical methods from the block diagram or directly from the model equations. The latter approach is illustrated. Laplace transforming the second output equation in Equation 4.192 followed by division of each term by $Q_4(s)$,

$$Y_2(s) = k_f X_3(s) + k_u X_4(s) \qquad (4.199)$$

$$\Rightarrow \frac{Y_2(s)}{Q_4(s)} = k_f \frac{X_3(s)}{Q_4(s)} + k_u \frac{X_4(s)}{Q_4(s)} \qquad (4.200)$$

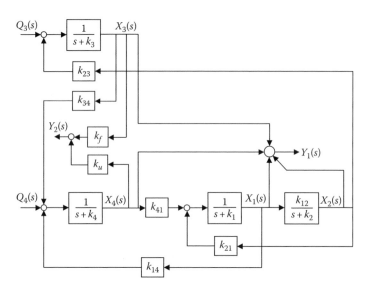

FIGURE 4.14   Block diagram of system modeled by Equations 4.195 through 4.198 and 4.192.

Setting $Q_3(s) = 0$ in Equation 4.197 and solving Equations 4.195 through 4.198 for $X_3(s)$ and $X_4(s)$,

$$X_3(s) = \frac{\begin{vmatrix} s+k_1 & -k_{21} & 0 & -k_{41} \\ -k_{12} & s+k_2 & 0 & 0 \\ 0 & -k_{23} & 0 & 0 \\ -k_{14} & 0 & Q_4(s) & s+k_4 \end{vmatrix}}{\begin{vmatrix} s+k_1 & -k_{21} & 0 & -k_{41} \\ -k_{12} & s+k_2 & 0 & 0 \\ 0 & -k_{23} & s+k_3 & 0 \\ -k_{14} & 0 & -k_{34} & s+k_4 \end{vmatrix}} \tag{4.201}$$

$$X_4(s) = \frac{\begin{vmatrix} s+k_1 & -k_{21} & 0 & 0 \\ -k_{12} & s+k_2 & 0 & 0 \\ 0 & -k_{23} & s+k_3 & 0 \\ -k_{14} & 0 & -k_{34} & Q_4(s) \end{vmatrix}}{\begin{vmatrix} s+k_1 & -k_{21} & 0 & -k_{41} \\ -k_{12} & s+k_2 & 0 & 0 \\ 0 & -k_{23} & s+k_3 & 0 \\ -k_{14} & 0 & -k_{34} & s+k_4 \end{vmatrix}} \tag{4.202}$$

Evaluation of the determinants in Equations 4.201 and 4.202 is a tedious process left as an exercise problem. The results are as follows:

$$X_3(s) = \left[\frac{\alpha_0}{s^4 + a_3 s^3 + a_2 s^2 + a_1 s + a_0}\right] Q_4(s) \tag{4.203}$$

$$X_4(s) = \left[\frac{s^3 + \beta_2 s^2 + \beta_1 s + \beta_0}{s^4 + a_3 s^3 + a_2 s^2 + a_1 s + a_0}\right] Q_4(s) \tag{4.204}$$

$$\left.\begin{array}{l} \alpha_0 = k_{12} k_{23} k_{41}, \quad \beta_0 = k_1 k_2 k_3 - k_{12} k_{21} k_3 \\ \beta_1 = k_1 k_2 + k_1 k_3 + k_2 k_3 - k_{12} k_{21}, \quad \beta_2 = k_1 + k_2 + k_3 \end{array}\right\} \tag{4.205}$$

$$\left.\begin{array}{l} a_0 = k_1 k_2 k_3 k_4 - k_{14} k_{41} k_2 k_3 - k_{12} k_{21} k_3 k_4 - k_{12} k_{23} k_{34} k_{41} \\ a_1 = k_1 k_2 k_3 + k_1 k_2 k_4 + k_1 k_3 k_4 + k_2 k_3 k_4 - k_{12} k_{21}(k_3 + k_4) - k_{14} k_{41}(k_2 + k_3) \\ a_2 = k_1 k_2 + k_1 k_3 + k_1 k_4 + k_2 k_3 + k_2 k_4 + k_3 k_4 - k_{12} k_{21} - k_{14} k_{41} \\ a_3 = k_1 + k_2 + k_3 + k_4 \end{array}\right\} \tag{4.206}$$

Combining Equations 4.200, 4.203, and 4.204 produces the desired transfer function,

$$\frac{Y_2(s)}{Q_4(s)} = \frac{k_f \alpha_0 + k_u(s^3 + \beta_2 s^2 + \beta_1 s + \beta_0)}{s^4 + a_3 s^3 + a_2 s^2 + a_1 s + a_0} \tag{4.207}$$

(d) The steady-state iodine levels in each compartment are obtained from the state equations $\dot{\underline{x}} = A\underline{x} + B\underline{u}$ with $\dot{\underline{x}} = 0$.

$$\underline{x}_{ss} = -A^{-1} B \underline{u}_{ss} \quad \text{where } \underline{u}_{ss} = \begin{bmatrix} (q_3)_{ss} \\ (q_4)_{ss} \end{bmatrix} = \begin{bmatrix} 0 \\ 150 \ \mu g/day \end{bmatrix} \tag{4.208}$$

For the given values of the rate constants,

$$
\underline{X}_{ss} = - \begin{bmatrix} -0.95 & 0.005 & 0 & 0.5 \\ 0.8 & -0.015 & 0 & 0 \\ 0 & 0.01 & -0.32 & 0 \\ 0.15 & 0 & 0.3 & -1.7 \end{bmatrix}^{-1} \begin{bmatrix} 0 & 0 \\ 0 & 0 \\ 1 & 0 \\ 0 & 1 \end{bmatrix} \begin{bmatrix} 0 \\ 150 \end{bmatrix} = \begin{bmatrix} 89.6 \\ 4780.9 \\ 149.4 \\ 122.5 \end{bmatrix} \tag{4.209}
$$

(e) Using the same method we employed to find $X_3(s)/Q_4(s)$ and $X_4(s)/Q_4(s)$, the transfer function $X_2(s)/Q_4(s)$ is

$$
\frac{X_2(s)}{Q_4(s)} = \frac{\gamma_1 s + \gamma_0}{s^4 + a_3 s^3 + a_2 s^2 + a_1 s + a_0} \tag{4.210}
$$

$$
\gamma_0 = k_{12} k_{41} k_3, \quad \gamma_1 = k_{12} k_{41} \tag{4.211}
$$

Working backward from the transfer function $X_2(s)/Q_4(s)$, the differential equation relating $x_2(t)$ and $q_4(t)$ is

$$
\dddot{x}_2 + a_3 \dddot{x}_2 + a_2 \ddot{x}_2 + a_1 \dot{x}_2 + a_0 x_2 = \gamma_1 \dot{q}_4 + \gamma_0 q_4 \tag{4.212}
$$

Once the initial conditions are established, Equation 4.212 can be solved to find the complete step response.

Let us assume the input $q_4(t)$ has been constant at 150 μg/day long enough for the system to reach the steady-state levels given in Equation 4.209. It is possible to redefine $t = 0$ as the instant when $q_4(t)$ switches from 150 to 50 μg/day. Figure 4.15 shows the input dropping from $q_4(0^-) = 150$ μg/day to $q_4(0^+) = 50$ μg/day.

With the system at steady-state at $t = 0^-$, the initial conditions are $x_2(0^-) = 4780.9$ μg, $\dot{x}_2(0^-) = \ddot{x}_2(0^-) = \dddot{x}_2(0^-) = 0$. Laplace transforming Equation 4.212,

$$
s^4 X_2(s) - s^3 x_2(0^-) + a_3 [s^3 X_2(s) - s^2 x_2(0^-)] + a_2 [s^2 X_2(s) - s x_2(0^-)]
$$
$$
+ a_1 [s X_2(s) - x_2(0^-)] + a_0 X_2(s) = \gamma_1 [s Q_4(s) - q_4(0^-)] + \gamma_0 Q_4(s) \tag{4.213}
$$

Solving for $X_2(s)$,

$$
X_2(s) = \frac{\gamma_1 s + \gamma_0}{s^4 + a_3 s^3 + a_2 s^2 + a_1 s + a_0} Q_4(s) + \frac{x_2(0^-)(s^3 + a_3 s^2 + a_2 s + a_1) - \gamma_1 q_4(0^-)}{s^4 + a_3 s^3 + a_2 s^2 + a_1 s + a_0} \tag{4.214}
$$

where

$$
Q_4(s) = \mathcal{L}\{q_4(t)\} = \frac{q_4(0^+)}{s} \tag{4.215}
$$

M-file "Chap3_Ex3_4.m" uses the "residue" function to evaluate the partial fraction expansion of each term on the right-hand side of Equation 4.214. The final expression for $X_2(s)$ is of the form

$$
X_2(s) = \sum_{i=1}^{5} \frac{c_i}{s - p_i} \tag{4.216}
$$

FIGURE 4.15　Step change in input $q_4(t)$.

where the system poles are $p_1 = -1.7901$, $p_2 = -0.8621$, $p_3 = -0.3248$, and $p_4 = -0.0080$ and the input pole $p_5 = 0$. The residues are $c_1 = 13.6$, $c_2 = -59.1$, $c_3 = 2.4$, $c_4 = 3230.4$, and $c_5 = 1593.6$. The partial fraction expansion of $X_2(s)$ is

$$X_2(s) = \frac{13.6}{s + 1.7901} - \frac{59.1}{s + 0.8621} + \frac{2.4}{s + 0.3248} + \frac{3230.4}{s + 0.0080} + \frac{1593.6}{s} \tag{4.217}$$

Inverting $X_2(s)$ gives

$$x_2(t) = 13.6e^{-1.7901t} - 59.1e^{-0.8621t} + 2.4e^{-0.3248t} + 3230.4e^{-0.0080t} + 1593.6 \tag{4.218}$$

Note that a convenient check of $x_2(t)$ in Equation 4.218 is

$$x_2(0^-) = 13.6 - 59.1 + 2.4 + 3230.4 + 1593.6 = 4780.9$$

which agrees with the initial condition. The step response is shown in Figure 4.16.

The natural modes of the system are $e^{-1.7901t}$, $e^{-0.8621t}$, $e^{-0.3248t}$, and $e^{-0.0080t}$ and the dominant time constant $\tau_{dominant} = 1/0.008 = 125$ days. It takes approximately $5 \times \tau_{dominant} = 625$ days for x_2 to attain the new steady-state value of 1593.6 μg.

There is another property of Laplace transforms that is particularly useful when it comes to finding the steady-state response of a system. Known as the Final Value Theorem, it relates the steady state or final value of a signal to its Laplace transform, that is,

P10:
Given $Y(s) = \mathcal{L}\{y(t)\}$, if a final value $y(\infty)$ exists, it is given by

$$y(\infty) = \lim_{t \to \infty} y(t) = \lim_{s \to 0} sY(s) \tag{4.219}$$

For a system with transfer function $G(s)$, the steady-state response to a step input of magnitude U_0 is

$$y(\infty) = \lim_{s \to 0} sY(s) = \lim_{s \to 0} sG(s)U(s) = \lim_{s \to 0} sG(s)\frac{U_0}{s} = G(0)U_0 \tag{4.220}$$

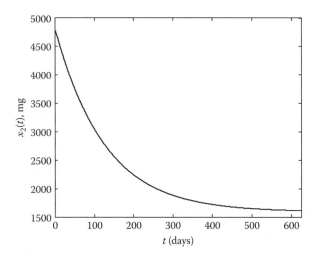

FIGURE 4.16 Step response for $x_2(t)$ following step change in $q_4(t)$ from 150 to 50 mg/day.

$G(0)$ is referred to as the steady-state gain of the system. The final value property makes it possible to determine the final value $y(\infty)$ from $Y(s)$ without having to find $y(t)$. This is particularly useful when trying to find the steady-state response of a system to a constant input. The input must be constant long enough to allow the transient response to vanish. Practically speaking, this is roughly four to five times the largest effective time constant of the system.

In Example 4.11, the transfer function $Y_1(s)/Q_4(s)$ can be expressed as

$$\frac{Y_1(s)}{Q_4(s)} = \frac{X_1(s) + X_2(s) + X_3(s) + X_4(s)}{Q_4(s)} \tag{4.221}$$

$$= \frac{X_1(s)}{Q_4(s)} + \frac{X_2(s)}{Q_4(s)} + \frac{X_3(s)}{Q_4(s)} + \frac{X_4(s)}{Q_4(s)} \tag{4.222}$$

where the last three terms on the right-hand side are obtained from Equations 4.210, 4.203, and 4.204. The remaining term is left as an exercise. The result is

$$\frac{X_1(s)}{Q_4(s)} = \frac{\delta_2 s^2 + \delta_1 s + \delta_0}{s^4 + a_3 s^3 + a_2 s^2 + a_1 s + a_0} \tag{4.223}$$

$$\delta_0 = k_{41} k_2 k_3, \quad \delta_1 = k_{41}(k_2 + k_3), \quad \delta_2 = k_{41} \tag{4.224}$$

making the transfer function

$$G(s) = \frac{Y_1(s)}{Q_4(s)} = \frac{s^3 + (\beta_2 + \delta_2)s^2 + (\beta_1 + \gamma_1 + \delta_1)s + \alpha_0 + \beta_0 + \gamma_0 + \delta_0}{s^4 + a_3 s^3 + a_2 s^2 + a_1 s + a_0} \tag{4.225}$$

The final value $y_1(\infty) = x_1(\infty) + x_2(\infty) + x_3(\infty) + x_4(\infty)$ when $q_4(t) = 150$, $t \geq 0$ (same initial input in Example 4.11) is

$$y_1(\infty) = G(0) \cdot 150 = \left(\frac{\alpha_0 + \beta_0 + \gamma_0 + \delta_0}{a_0} \right) \cdot 150$$

$$= \left(\frac{0.004 + 0.00328 + 0.128 + 0.0024}{0.004016} \right) \cdot 150 = 5142.4 \text{ μg} \tag{4.226}$$

in agreement with the sum of the components of x_{ss} in Equation 4.209.

A word of caution when applying the final value property. A function $y(t)$ could theoretically grow without bound, that is, $\lim_{t \to \infty} y(t) = \infty$ or have an undamped oscillatory component, and the final value property will nevertheless produce a finite value. Clearly, the result does not represent a final or steady-state value. We shall investigate the conditions that produce theoretical unbounded outputs of a linear system in a future section.

4.3.6 TRANSFORMATION FROM STATE VARIABLE MODEL TO TRANSFER FUNCTION

The state-space representation offers several advantages over the input–output transfer function method of describing the dynamics of a linear system. For one, it is a more complete representation since the states provide useful information about the internal behavior of the system. Properties of linear systems such as observability and controllability as well as system identification and state feedback are topics normally covered in modern control theory, which rely on state-space models.

However, there are times when the transfer function of an SISO system (or transfer functions if the system is MIMO) is required for a system modeled in state variable form.

Consider an MIMO system with inputs $u_1, u_2, \ldots, u_r$ and outputs $y_1, y_2, \ldots, y_m$ modeled in state space by

$$\dot{\underline{x}} = A\underline{x} + B\underline{u} \qquad (4.227)$$

$$\underline{y} = C\underline{x} + D\underline{u} \qquad (4.228)$$

where $\underline{x}$ is the n-dimensional state vector $[x_1 \quad x_2 \quad \ldots \quad x_n]^T$ and the matrices A, B, C, and D are appropriately dimensioned. Laplace transformation of Equation 4.227 with $\underline{x}(0) = \underline{0}$ gives

$$s\underline{X}(s) = A\underline{X}(s) + B\underline{U}(s) \qquad (4.229)$$

$$\Rightarrow \underline{X}(s) = (sI - A)^{-1} B\underline{U}(s) \qquad (4.230)$$

Laplace transforming $\underline{y} = C\underline{x} + D\underline{u}$ and substituting $\underline{X}(s)$ from Equation 4.230 gives

$$\underline{Y}(s) = [C(sI - A)^{-1} B + D]\underline{U}(s) \qquad (4.231)$$

$$= G(s)\underline{U}(s) \qquad (4.232)$$

where $G(s)$, known as the transfer matrix, is a matrix of transfer functions from each of the r inputs to each of the m outputs, that is,

$$G_{ij}(s) = \frac{Y_i(s)}{U_j(s)}, \quad i = 1, 2, \ldots, m, \quad j = 1, 2, \ldots, r \qquad (4.233)$$

To illustrate, let us revisit the state variable model for iodine storage in Example 4.11 where the matrices A, B, C, and D are given in Equations 4.193 and 4.194. There are two inputs $u_1(t) = q_3(t)$ and $u_2(t) = q_4(t)$, and outputs $y_1(t)$ and $y_2(t)$ are defined in Equation 4.192. One of the four transfer functions, namely, $Y_1(s)/Q_4(s)$, is given in Equation 4.225. Using the baseline parameter values in Example 4.11 results in

$$\frac{Y_1(s)}{Q_4(s)} = \frac{s^3 + 1.785s^2 + 0.88655s + 0.13768}{s^4 + 2.985s^3 + 2.42855s^2 + 0.52054s + 0.004016} \qquad (4.234)$$

The matrix $\Phi(s) = (sI - A)^{-1}$ in Equation 4.231 is computed according to

$$\Phi(s) = (sI - A)^{-1} = \left(sI - \begin{bmatrix} -0.95 & 0.005 & 0 & 0.5 \\ 0.8 & -0.015 & 0 & 0 \\ 0 & 0.01 & -0.32 & 0 \\ 0.15 & 0 & 0.3 & -1.7 \end{bmatrix} \right)^{-1} \qquad (4.235)$$

$$= \begin{bmatrix} s+0.95 & -0.005 & 0 & -0.5 \\ -0.8 & s+0.015 & 0 & 0 \\ 0 & -0.01 & s+0.32 & 0 \\ -0.15 & 0 & -0.3 & s+1.7 \end{bmatrix}^{-1} \qquad (4.236)$$

$\Phi(s)$ is the Laplace transform of the continuous-time system transition matrix $\Phi(t)$, used to obtain the state response in the time domain. Inverting $(sI - A)$ results in

$$\Phi(s) = \begin{bmatrix} \phi_{11}(s) & \phi_{12}(s) & \phi_{13}(s) & \phi_{14}(s) \\ \phi_{21}(s) & \phi_{22}(s) & \phi_{23}(s) & \phi_{24}(s) \\ \phi_{31}(s) & \phi_{32}(s) & \phi_{33}(s) & \phi_{34}(s) \\ \phi_{41}(s) & \phi_{42}(s) & \phi_{43}(s) & \phi_{44}(s) \end{bmatrix} \tag{4.237}$$

where

$$\phi_{11}(s) = \frac{1}{\Delta(s)}[(s + 0.015)(s + 0.32)(s + 1.7)] \tag{4.238}$$

$$\phi_{12}(s) = \frac{1}{\Delta(s)}[-0.003(s + 1.2)] \tag{4.239}$$

$$\phi_{13}(s) = \frac{1}{\Delta(s)}[0.15(s + 0.015)] \tag{4.240}$$

$$\phi_{14}(s) = \frac{1}{\Delta(s)}[0.5(s + 0.015)(s + 0.32)] \tag{4.241}$$

$$\phi_{21}(s) = \frac{1}{\Delta(s)}[0.8(s + 0.32)(s + 1.7)] \tag{4.242}$$

$$\phi_{22}(s) = \frac{1}{\Delta(s)}[s^3 + 2.97s^2 + 2.388s + 0.4928] \tag{4.243}$$

$$\phi_{23}(s) = \frac{1}{\Delta(s)}[-0.12] \tag{4.244}$$

$$\phi_{24}(s) = \frac{1}{\Delta(s)}[0.4(s + 0.32)] \tag{4.245}$$

$$\phi_{31}(s) = \frac{1}{\Delta(s)}[0.008(s + 1.7)] \tag{4.246}$$

$$\phi_{32}(s) = \frac{1}{\Delta(s)}[0.01(s^2 + 2.65s + 0.865)] \tag{4.247}$$

$$\phi_{33}(s) = \frac{1}{\Delta(s)}[s^3 + 2.665s^2 + 1.57575s + 0.0163] \tag{4.248}$$

$$\phi_{34}(s) = \frac{1}{\Delta(s)}[0.004] \tag{4.249}$$

$$= A^k \underline{x}_0 + \sum_{i=0}^{k-1} A^{k-i-1} B \underline{u}_i, \quad k = 0, 1, 2, 3, \ldots \tag{4.250}$$

$$\phi_{42}(s) = \frac{1}{\Delta(s)} = [0.00375(s + 0.824)] \tag{4.251}$$

$$\phi_{43}(s) = \frac{1}{\Delta(s)}[0.3(s^2 + 0.965s + 0.01025)] \tag{4.252}$$

$$\phi_{44}(s) = \frac{1}{\Delta(s)}[s^3 + 1.285s^2 + 0.31905s + 0.00328] \tag{4.253}$$

$$\Delta(s) = |sI - A| = s^4 + 2.985s^3 + 2.42855s^2 + 0.52054s + 0.004016 \tag{4.254}$$

Finally, the transfer function matrix $G(s)$ in Equation 4.232 is given by

$$G(s) = C\Phi(s)B + D \tag{4.255}$$

$$= \begin{bmatrix} 1 & 1 & 1 & 1 \\ 0 & 0 & k_f & k_u \end{bmatrix} \begin{bmatrix} \phi_{11}(s) & \phi_{12}(s) & \phi_{13}(s) & \phi_{14}(s) \\ \phi_{21}(s) & \phi_{22}(s) & \phi_{23}(s) & \phi_{24}(s) \\ \phi_{31}(s) & \phi_{32}(s) & \phi_{33}(s) & \phi_{34}(s) \\ \phi_{41}(s) & \phi_{42}(s) & \phi_{43}(s) & \phi_{44}(s) \end{bmatrix} \begin{bmatrix} 0 & 0 \\ 0 & 0 \\ 1 & 0 \\ 0 & 1 \end{bmatrix} + \begin{bmatrix} 0 & 0 \\ 0 & 0 \end{bmatrix} \tag{4.256}$$

$$= \begin{bmatrix} \phi_{13}(s) + \phi_{23}(s) + \phi_{33}(s) + \phi_{43}(s) & \phi_{14}(s) + \phi_{24}(s) + \phi_{34}(s) + \phi_{44}(s) \\ k_f\phi_{33}(s) + k_u\phi_{43}(s) & k_f\phi_{34}(s) + k_u\phi_{44}(s) \end{bmatrix} \tag{4.257}$$

The component $G_{12}(s)$ in Equation 4.257 is the transfer function $Y_1(s)/Q_4(s)$ previously obtained in Equation 4.234. The reader can verify that the two are identical.

EXERCISES

4.6 Show that the step response of a system whose impulse response function $h(t) = 3e^{-2t} + 5\delta(t)$ is discontinuous at $t = 0$.

4.7 The differential equation of an LTI system is

$$\frac{d^3y}{dt^3} + 5\frac{d^2y}{dt^2} + 11\frac{dy}{dt} + 15y = 2\frac{d^3u}{dt^3} + u$$

(a) Find the transfer function $H(s) = Y(s)/U(s)$ of the system.
(b) Find the impulse response function $h(t)$ for the system.
(c) Find the step response when the initial conditions at $t = 0^-$ are identically zero.
(d) Find $y(\infty)$ using the final value property, and check your answer with the result obtained in part (c) as $t \to \infty$.
(e) Find $y(0^+)$ using the initial value property and check your answer with the result obtained in part (c) as $t \to 0^+$.
(f) Find the step response by convolution and compare your answer to the step response found in part (c).
(g) Draw a simulation diagram for the system in observer canonical form.
(h) Represent the system in state variable form $\dot{x} = Ax + Bu$, $y = Cx + Du$.
(i) Find the 1×1 transfer function $G(s) = Y(s)/U(s)$ using Equation 4.255.

4.8 Repeat Exercise 4.7 when the system differential equation is

(a) $\dfrac{dy}{dt} + 5y = 10u$

(b) $\dfrac{d^2y}{dt^2} + 5\dfrac{dy}{dt} + 6y = u$

(c) $\dfrac{d^3y}{dt^3} + 5\dfrac{d^2y}{dt^2} + 11\dfrac{dy}{dt} + 15y = u$

4.9 Use convolution to find the response of the systems with transfer functions

(a) $H(s) = \dfrac{s+3}{s^2 + 2s + 1}$,   (b) $H(s) = \dfrac{1}{s^2 + 3s + 2}$,   (c) $H(s) = \dfrac{s+1}{s^2 + 2s + 2}$

to the following inputs: (i) $u(t) = \hat{u}(t)$, (ii) $u(t) = \hat{u}(t) - \hat{u}(t-2)$, and (iii) $u(t) = t\hat{u}(t)$.

4.10 The circuit in Figure E4.10 is governed by the differential equation

$$\frac{d^2v_0}{dt^2} + \frac{1}{RC}\frac{dv_0}{dt} + \frac{1}{LC}v_0 = \frac{1}{C}\frac{di_g}{dt}$$

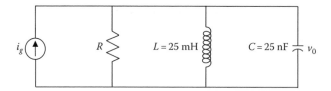

FIGURE E4.10

Find the impulse response function and plot the results when (a) $R = 400\ \Omega$, (b) $R = 500\ \Omega$, and (c) $R = 625\ \Omega$.

4.11 Repeat Example 4.10 with $H(s) = 1/[(s+1)(s+3)(s+5)]$.

4.12 Find the transfer function of the bridged-T circuit in Figure 4.12 using equations in the time domain only to find the differential equation of the circuit.

4.13 For the system of interacting tanks shown in Figure E4.13:

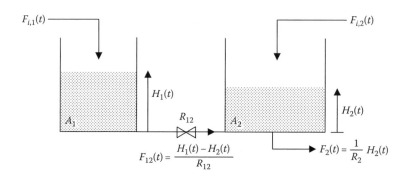

FIGURE E4.13

(a) Find the transfer functions $\dfrac{H_2(s)}{F_{i,1}(s)}$, $\dfrac{H_1(s)}{F_{i,2}(s)}$, $\dfrac{H_1(s)}{F_{i,1}(s)}$, $\dfrac{F_2(s)}{F_{i,2}(s)}$

(b) Find the differential equation relating $H_1(t)$ and $F_{i,2}(t)$.

4.14 The unit step response of a system is

$$y(t) = 1 + e^{-2t}(\cos 3t + 4\sin 3t)$$

(a) Find the transfer function of the system.
(b) Find the impulse response of the system.
(c) Find the differential equation of the system.

4.15 In Example 4.11, find $x_3(\infty)$ and $x_4(\infty)$ when $q_4(t) = 150\ \mu g/day$, $t \geq 0$ and $q_3(t) = 0$, $t \geq 0$ using the final value property and the expressions for $X_3(s)$ and $X_4(s)$ in Equations 4.203 and 4.204. Compare your answer with the results in Equation 4.209.

4.16 In Example 4.11, find $X_1(s)/Q_4(s)$ and compare your answer with the expression in Equation 4.223.

4.17 In Example 4.11,
(a) Find the transfer functions $Y_1(s)/Q_3(s)$, $Y_2(s)/Q_3(s)$ in a form similar to Equation 4.225.
(b) Find the step responses for $y_1(t)$ and $y_2(t)$ to inputs $q_4(t) = 50\ \mu g/day$, $t \geq 0$ and $q_3(t) = 0$, $t \geq 0$. Assume the initial state is x_{ss} in Equation 4.209.

4.18 In Example 4.11, verify that the transfer function $Y_2(s)/Q_4(s)$ in Equation 4.207 is the same as $G_{22}(s)$ in Equation 4.257.

4.4   STABILITY OF LINEAR TIME INVARIANT CONTINUOUS-TIME SYSTEMS

In order for a physical system to operate as intended, it must be capable of generating output(s) in a stable fashion. Regulation of a process temperature is unsatisfactory if the heat source cycles continuously between extremes, that is, off or operating at maximum output, unless it is designed to operate that way like a room thermostat. A control system for maintaining a fixed amount of material in a storage tank in the presence of a fluctuating input may not be performing as intended if the regulating valve in the input line continually cycles between its limits. Each is a real-world example of a control system operating in an unstable manner.

The starting point of an investigation concerning the stability of a system is its mathematical model. The discussion is confined to LTI systems. Excluding nonlinear systems may appear to significantly limit the range of systems considered. However, nonlinear systems can be linearized about specific operating points and stability analyses performed with respect to each operating point. The subject of linearization is treated in Chapter 7.

Consider the second-order system model from the previous section,

$$\frac{d^2}{dt^2}y(t) + a_1\frac{d}{dt}y(t) + a_0y(t) = b_2\frac{d^2}{dt^2}u(t) + b_1\frac{d}{dt}u(t) + b_0u(t) \tag{4.258}$$

Applying the differentiation property of the Laplace transform and collecting terms, the Laplace transform of the system output is

$$Y(s) = \left[H(s)U(s) - \frac{b_2u(0^-)s + b_2\dot{u}(0^-) + b_1u(0^-)}{s^2 + a_1s + a_0}\right] + \frac{y(0^-)s + \dot{y}(0^-) + a_1y(0^-)}{s^2 + a_1s + a_0} \tag{4.259}$$

where $H(s)$ is the transfer function

$$H(s) = \frac{Y(s)}{U(s)} = \frac{b_2s^2 + b_1s + b_0}{s^2 + a_1s + a_0} \tag{4.260}$$

For zero input, $Y(s)$ reduces to the Laplace transform of the free response, that is,

$$Y_{\text{free}}(s) = \frac{y(0)s + \dot{y}(0) + a_1y(0)}{s^2 + a_1s + a_0} \tag{4.261}$$

Note that in the absence of an input, the "$-$" superscript on the initial conditions is no longer necessary. The free response $y_{\text{free}}(t) = \mathcal{L}^{-1}\{Y_{\text{free}}(s)\}$ depends on the roots of the equation $s^2 + a_1s + a_0 = 0$. Denoting the roots as p_1 and p_2, $y_{\text{free}}(t)$ assumes one of the forms in

$$y_{\text{free}}(t) = \begin{cases} c_1e^{p_1t} + c_2e^{p_2t}, & p_1, p_2 \text{ real and distinct} \\ e^{\sigma t}[c_1\cos\omega t + c_2\sin\omega t], & p_1, p_2 \text{ complex} \\ (c_1 + c_2t)e^{pt}, & p_1 = p_2 = p \end{cases} \tag{4.262}$$

Constants c_1 and c_2 depend on the initial conditions $y(0)$ and $\dot{y}(0)$. The constants σ, ω, p_1, p_2, and p depend on the values of a_0 and a_1, which are related to the physical parameters of the system. For example, a_0 and a_1 depend on M, B, and K in a mechanical system or R, L, and C for an electrical circuit. The free response in Equation 4.262 is also referred to as the natural response of the system. It consists of a linear combination of the system's natural modes.

4.4.1 CHARACTERISTIC POLYNOMIAL

The denominator of the transfer function $H(s)$ in Equation 4.260 is

$$\Delta(s) = s^2 + a_1 s + a_0 = (s - p_1)(s - p_2) \tag{4.263}$$

It is called the characteristic polynomial of the system and $\Delta(s) = 0$ is the characteristic equation. The roots of the characteristic polynomial are referred to as the poles of the system transfer function, and from Equations 4.260 and 4.263, $H(p_1) = H(p_2) = \infty$.

The stability of the system is related to the free response, specifically the limit $L = \lim_{t \to \infty} y_{\text{free}}(t)$ when one or both initial conditions are nonzero. The following possibilities exist:

1. $L = 0$.
2. $L = \text{constant} \neq 0$.
3. L fails to exist because the free response oscillates with constant amplitude.
4. L fails to exist because the magnitude of the free response approaches infinity.

The system is said to be asymptotically stable in the first case, marginally stable in the second and third cases, and unstable in the last case.

Since the poles p_1 and p_2 dictate the behavior of the free response, they also determine the nature of the system's stability. As a result, we can infer that the stability of the second-order linear system in Equation 4.258 is an inherent system property, that is, it depends on the values of the system parameters and not on the system inputs. The previous statement is entirely general and not restricted to the second-order system under consideration. The different possibilities for the poles of $H(s)$ in Equation 4.260 are illustrated in Figure 4.17.

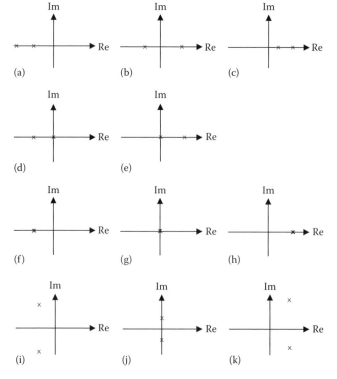

FIGURE 4.17 Possible locations for transfer function poles of a second-order system.

In (a), (b), (c), (d), and (e), the poles p_1 and p_2 are real and distinct. From Equation 4.262, the free response is the linear combination of natural modes $e^{p_1 t}$ and $e^{p_2 t}$. Since

$$\lim_{t \to \infty} e^{pt} = \begin{cases} 0, & p < 0 \\ 1, & p = 0 \\ \infty, & p > 0 \end{cases} \tag{4.264}$$

the two natural modes decay to zero in (a), and the limit $L = 0$. Therefore, (a) corresponds to an asymptotically stable system. In (b), one of the natural modes grows monotonically over time and L fails to exist. Hence, (b) represents an unstable system. A similar analysis of the remaining cases (c) through (k) leads to the results shown in Table 4.2.

In summary, the second-order system with transfer function in Equation 4.260 is asymptotically stable provided the two poles are located entirely in the left half of the complex plane. The system is unstable if one or both of its poles lie in the right half of the complex plane or if it has a double pole at the origin. Lastly, it is marginally stable if there is a single pole at the origin and the other pole is negative or there exists a pair of purely imaginary poles located on the imaginary axis. The Routh–Hurwitz stability condition is a simple test for the presence of right-half-plane poles of the transfer function for an nth order LTI system (Dorf and Bishop 2005).

An alternate definition of asymptotic stability is based on the system's forced response. It states that for a system to be asymptotically stable, its response to any bounded input must remain bounded over time. The same conclusions with respect to the pole locations of an asymptotically stable system shown in Table 4.2 apply to this alternate definition as well.

Systems that are not asymptotically stable according to this definition, that is, bounded input–bounded output (BIBO), are classified as marginally stable or unstable. In the case of a marginally stable system, the forced response to a bounded input may or may not be bounded depending on the input. Consider case (d) in Figure 4.17 where one of the poles is $s = 0$ and the other is located along the negative real axis. In particular, suppose the second pole is $s = -2$ and the second-order system transfer function is

$$H(s) = \frac{s + 3}{s(s + 2)} \tag{4.265}$$

TABLE 4.2
Poles, Natural Modes, and Stability
for a Second-Order System

Poles	Natural Modes	System Stability
(a) $p_1 < 0$, $p_2 < 0$	$e^{p_1 t}, e^{p_2 t}$	Asymptotically stable
(b) $p_1 < 0$, $p_2 > 0$	$e^{p_1 t}, e^{p_2 t}$	Unstable
(c) $p_1 > 0$, $p_2 > 0$	$e^{p_1 t}, e^{p_2 t}$	Unstable
(d) $p_1 < 0$, $p_2 = 0$	$e^{p_1 t}, 1$	Marginally stable
(e) $p_1 = 0$, $p_2 > 0$	$1, e^{p_2 t}$	Unstable
(f) $p_1 = p_2 = p < 0$	e^{pt}, te^{pt}	Asymptotically stable
(g) $p_1 = p_2 = p = 0$	$1, t$	Unstable
(h) $p_1 = p_2 = p > 0$	e^{pt}, te^{pt}	Unstable
(i) $p_1, p_2 = \sigma \pm j\omega$ $(\sigma < 0)$	$e^{\sigma t} \cos \omega t, e^{\sigma t} \sin \omega t$	Asymptotically stable
(j) $p_1, p_2 = \pm j\omega$	$\cos \omega t, \sin \omega t$	Marginally stable
(k) $p_1, p_2 = \sigma \pm j\omega$ $(\sigma > 0)$	$e^{\sigma t} \cos \omega t, e^{\sigma t} \sin \omega t$	Unstable

---

The forced response to input $u_1(t) = \sin t$, $t \geq 0$ is obtained as follows:

$$Y_1(s) = H(s)U_1(s) = \frac{s+3}{s(s+2)} \frac{1}{s^2+1} = \frac{1.5}{s} - \frac{0.1}{s+2} - \frac{1.4s}{s^2+1} - \frac{0.2}{s^2+1} \tag{4.266}$$

$$y_1(t) = 1.5 - 0.1e^{-2t} - 1.4\cos t - 0.2\sin t, \quad t \geq 0 \tag{4.267}$$

The forced response to input $u_2(t) = 1$, $t \geq 0$ is obtained in similar fashion.

$$Y_2(s) = H(s)U_2(s) = \frac{s+3}{s(s+2)s} = \frac{1.5}{s^2} - \frac{0.25}{s} + \frac{0.25}{s+2} \tag{4.268}$$

$$y_2(t) = 1.5t - 0.25 + 0.25e^{-2t}, \quad t \geq 0 \tag{4.269}$$

In both instances, the input is a bounded function of time. The output $y_1(t)$ remains bounded while the system response $y_2(t)$ is unbounded as a result of the first term. Careful examination of the system transfer function in Equation 4.265 reveals that the only bounded inputs capable of producing an unbounded output are those whose Laplace transform contains a pure "s" term in the denominator. In other words, the input must either be a constant or a sum of bounded time functions containing a constant.

The forced response of an unstable system to a bounded input is always unbounded due to the presence of an unstable natural mode (see Table 4.2) which appears in the response. For example, the forced response of a second-order system with a double pole at $s = 0$ (case [g] in Figure 4.17) to any bounded input contains the unstable mode "t" and is always unbounded.

A higher order LTI system is unstable if the transfer function contains one or more right-half-plane poles, the same as for a second-order system. It is not surprising since the characteristic polynomial of an nth-order system can always be factored into a number of linear and quadratic factors with real coefficients. Using partial fraction expansion, the transfer function with factored denominator can be decomposed into a sum of first- and second-order systems. For example, consider the fifth-order system with transfer function given by

$$H(s) = \frac{Y(s)}{U(s)} = \frac{7s^4 + 19s^3 + 45s^2 + 62s + 52}{s^5 + 5s^4 + 12s^3 + 26s^2 + 32s + 24} \tag{4.270}$$

With the help of the MATLAB "residue" function,

$$H(s) = \frac{s}{s^2+4} + \frac{s+1}{s^2+2s+2} + \frac{5}{s+3} \tag{4.271}$$

and the output $Y(s) = H(s)U(s)$ of the fifth-order system can be expressed as

$$Y(s) = \frac{s}{s^2+4}U(s) + \frac{s+1}{s^2+2s+2}U(s) + \frac{5}{s+3}U(s) \tag{4.272}$$

The system is marginally stable as a result of the complex poles at $s = \pm j2$ located on the imaginary axis. The remaining poles at $s = -1 \pm j$ and $s = -3$ are associated with stable natural modes. The step response of the system with transfer function in Equation 4.270 remains bounded. However, the bounded inputs $u(t) = \sin 2t$ or $u(t) = \cos 2t$ result in an $(s^2+4)^2$ term in the denominator of $Y(s)$ and $t\sin 2t$ or $t\cos 2t$ terms in the output $y(t)$. Hence, a bounded step response is necessary but not a sufficient condition for asymptotic stability of LTI systems.

For MIMO systems, the number of transfer functions can grow quickly. However, since stability is an intrinsic property of the system, that is, independent of the system inputs, it is not necessary to investigate each and every transfer function to determine if the system is stable. We shall soon see that the denominator polynomial of each transfer function is identical and, therefore, must be the characteristic polynomial of the system, $\Delta(s)$.

The transfer function matrix $G(s)$ of an MIMO system is the matrix whose ijth element is the transfer function $Y_i(s)/U_j(s)$. From the previous section,

$$G(s) = C(sI - A)^{-1}B + D = C\Phi(s)B + D \tag{4.273}$$

where
A is the $n \times n$ coefficient matrix
B, C, and D are the other matrices in the state variable model description

The inverse of $sI - A$ is $\Phi(s)$, which can be expressed in terms of the adjoint of matrix $sI - A$ and its determinant according to

$$\Phi(s) = (sI - A)^{-1} = \frac{1}{|sI - A|} \text{Adj}(sI - A) \tag{4.274}$$

It follows from Equations 4.273 and 4.274 that every component transfer function of $G(s)$ has the same denominator, that is, the nth-order polynomial

$$|sI - A| = s^n + a_{n-1}s^{n-1} + a_{n-2}s^{n-2} + \cdots + a_1s + a_0 \tag{4.275}$$

Hence, the stability of a linear system described by the state variable model $\dot{\underline{x}} = A\underline{x} + B\underline{u}$, $\underline{y} = C\underline{x} + D\underline{u}$ depends solely on the coefficient matrix A. Furthermore, it is immaterial whether the system is SISO with one transfer function or MIMO with several transfer functions; the coefficient matrix A is all we need to determine whether the system is asymptotically stable, marginally stable, or unstable.

This is consistent with the earlier statement that the stability of the second-order system modeled by the differential equation in Equation 4.258 depends strictly on the constants a_0 and a_1. After all, the 2×2 coefficient matrix A, while not unique, is determined entirely by a_0 and a_1. One choice for the states is $x_1 = y$ and $x_2 = \dot{y}$ that leads to

$$A = \begin{bmatrix} 0 & 1 \\ -a_0 & -a_1 \end{bmatrix} \tag{4.276}$$

The characteristic polynomial in Equation 4.263 and the nth-order polynomial in Equation 4.275 with $n = 2$ are identical, that is,

$$\Delta(s) = s^2 + a_1s + a_0 = |sI - A| \tag{4.277}$$

A compartment model for iodine storage in humans was presented in Example 4.11. The M-file "*Chap4_iodine.m*" computes the coefficient matrix

$$A = \begin{bmatrix} -0.95 & 0.005 & 0 & 0.5 \\ 0.8 & -0.015 & 0 & 0 \\ 0 & 0.01 & -0.32 & 0 \\ 0.15 & 0 & 0.3 & -1.7 \end{bmatrix}$$

The characteristic polynomial was given as

$$\Delta(s) = s^4 + 2.985s^3 + 2.42855s^2 + 0.52054s + 0.004016 \tag{4.278}$$

It is left as an exercise (Exercise 4.21) to show that expansion of the determinant $|sI - A|$ produces the characteristic polynomial given in Equation 4.278. The characteristic roots (poles of the system transfer functions) can be obtained by finding the roots of $\Delta(s) = 0$ in Equation 4.278 or equivalently the roots of

$$\Delta(s) = |sI - A| = 0 \tag{4.279}$$

that are also referred to as the eigenvalues of matrix A. The MATLAB functions "`roots(1 2.985 2.43855 0.52054 0.004016)`" and "`eig(A)`" both return the characteristic roots -1.7901, -0.8621, -0.3248, and -0.0080. Since all the characteristic roots are in the left half of the complex plane, the system is asymptotically stable.

4.4.2 FEEDBACK CONTROL SYSTEM

Real-world processes are nonlinear and may possess one or more equilibrium states. Linear models used to approximate the dynamics in the neighborhood of the equilibrium points are for the most part stable. However, control systems designed to improve some aspect of the system's performance may in fact produce the opposite effect. An example is presented of a stable open-loop system under closed-loop control, which can produce unstable modes in the natural response if the control system parameters are chosen incorrectly.

Figure 4.18 shows a simplified block diagram of a feedback control system for controlling the heading or yaw angle of a small ship. The open-loop system consists of the power converter (motor and gears that control the ship's rudder) modeled by a first-order lag with gain $K_P = 10°$ (rudder)/V and time constant $\tau_p = 0.2$ s. The ship's yaw dynamics include a gain $K_S = 0.5°$ (heading)/s/° (rudder) and time constant $\tau_S = 7.5$ s resulting in a sluggish response to changes in rudder position. A feedback closed-loop control system is implemented to improve the response. $\theta_{com}(s)$ and $\theta(s)$ are Laplace transforms of the commanded and actual ship headings, respectively. $E(s)$ is the Laplace transform of the error signal input to the controller.

The closed-loop system transfer function $\theta(s)/\theta_{com}(s)$ is obtained by eliminating $E(s)$ and $U(s)$ from the following three equations:

$$E(s) = \theta_{com}(s) - \theta(s) \tag{4.280}$$

$$U(s) = K_C \left(\frac{s+1}{s+10} \right) E(s) \tag{4.281}$$

$$\theta(s) = \left[\frac{0.5}{s(7.5s+1)} \right] \left[\frac{10}{(0.2s+1)} \right] U(s) \tag{4.282}$$

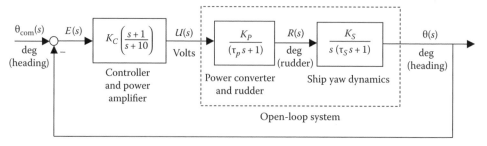

FIGURE 4.18 Block diagram of control system for ship heading.

The result is

$$\frac{\theta(s)}{\theta_{com}(s)} = \frac{5K_C(s+1)}{1.5s^4 + 22.7s^3 + 78s^2 + 5(K_C+2)s + 5K_C}$$
(4.283)

The characteristic polynomial is

$$\Delta(s) = 1.5s^4 + 22.7s^3 + 78s^2 + 5(K_C+2)s + 5K_C$$
(4.284)

For every value of controller gain K_C, there are four closed-loop system poles, which are the solutions to the characteristic equation, $\Delta(s) = 0$. Root-locus (Dorf and Bishop 2005) is a graphical design method used by control system engineers to plot the poles as the gain parameter K_C varies from 0 to ∞. There are four branches or loci, each containing one of the poles.

The M-file "*Chap4_feedback_yaw.m*" produces a root-locus plot shown in Figure 4.19a. When the gain $K_C = 10$, $\Delta(s)$ has two linear factors with real poles at $s = -3.922$ and $s = -10.525$ and a quadratic factor with a pair of complex poles located at $-0.343 \pm j0.831$ (see Figure 4.19b).

The quadratic factor damping ratio, natural frequency, damped natural frequency, and effective time constant are shown in Table 4.3.

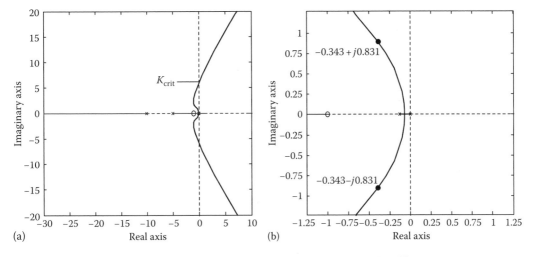

FIGURE 4.19 (a) Root-locus plot. (b) Zoom in near complex poles where $K_C = 10$.

TABLE 4.3
Closed-Loop System Properties ($K_C = 10$)

Characteristic polynomial	$\Delta(s) = 1.5s^4 + 22.7s^3 + 78s^2 + 60s + 50$
Poles	$p_1 = -10.525, p_2 = -3.922, p_3, p_4 = -0.343 \pm j0.831$
Factory	$s^2 + 0.686s + 0.808, s + 10.53, s + 3.92$
Damping ratio	$\zeta = 0.382$
Natural frequency	$\omega_n = 0.899$ rad/s
Damped natural frequency	$\omega_d = 0.831$ rad/s
Time constants	$\tau_1 = = \dfrac{1}{-p_1} = 0.095$ s, $\tau_2 = \dfrac{1}{-p_2} = 0.255$ s, $\tau = \dfrac{1}{\zeta\omega_n} = 2.914$ s

The natural response of the closed-loop system ($K_C = 10$) is given by

$$\theta_{nat}(t) = c_1 e^{-t/0.095} + c_2 e^{-t/0.255} + e^{-t/2.914}[c_3 \cos(0.831t) + c_4 \sin(0.831t)] \qquad (4.285)$$

The closed-loop system response, when $K_C = 10$, is faster than the open-loop system as evidenced by the reduction in dominant time constant from 7.5 to 2.914 s.

Suppose the ship is maintaining a heading of 0° (with the rudder angle at 0°) when it becomes necessary to increase the heading by 5°. In the open-loop system, a pulse input to the power converter and rudder subsystem is selected to produce the new desired heading. A pulse is specified rather than a step input because the rudder angle must return to zero once the new heading is achieved. What would happen if a step input were applied? For a pulse input of magnitude A and duration T,

$$u(t) = A - Au(t - T), \quad t \geq 0 \qquad (4.286)$$

the ship's heading is from Equation 4.282

$$\theta_{open\text{-}loop}(s) = \left[\frac{0.5}{s(7.5s + 1)}\right]\left[\frac{10}{(0.2s + 1)}\right]\frac{A(1 - e^{-Ts})}{s} \qquad (4.287)$$

The inverse Laplace transform, $\theta_{open\text{-}loop}(t) = \mathcal{L}^{-1}\{\theta_{open\text{-}loop}(s)\}$, is obtained by partial fraction expansion of Equation 4.287 without the $1 - e^{-Ts}$ followed by the shifting property P3 introduced in Section 4.4.2. It is left as an exercise to find $\theta_{open\text{-}loop}(t)$ and show that the final value, that is, new heading, is

$$\theta_{open\text{-}loop}(\infty) = K_P K_S A T = 5AT \qquad (4.288)$$

The closed-loop system response with $K_C = 10$ to a command heading of 5° is obtained from Equation 4.283 as

$$\theta_{closed\text{-}loop}(s) = \frac{50(s + 1)}{1.5s^4 + 22.7s^3 + 78s^2 + 60s + 50} \cdot \frac{5}{s} \qquad (4.289)$$

Using the MATLAB "residue" function to find the residues (partial fraction expansion coefficients) and poles of $\theta_{closed\text{-}loop}(s)$ in Equation 4.289 results in

$$R_1 = -0.2188, \quad R_2 = 1.3934, \quad R_3, R_4 = -3.0873 \mp j0.6270, \quad R_5 = 5$$

$$p_1 = -10.5254, \quad p_2 = -3.9215, \quad p_3, p_4 = -0.3432 \mp j0.8305, \quad p_5 = 0$$

enabling $\theta_{closed\text{-}loop}(s)$ to be expressed as the sum

$$\theta_{closed\text{-}loop}(s) = \sum_{i=1}^{5}\left(\frac{R_i}{s - p_i}\right) \qquad (4.290)$$

Invert Laplace transforming Equation 4.290 gives the time domain response

$$\theta_{closed\text{-}loop}(t) = \sum_{i=1}^{5} R_i e^{p_i t}, \quad t \geq 0 \qquad (4.291)$$

The third and fourth terms involve complex coefficients and complex exponentials,

$$R_3 e^{p_3 t} + R_4 e^{p_4 t} = (-3.087 - j0.627)e^{(-0.343+j0.831)t} + (-3.087 + j0.627)e^{(-0.343-j0.831)t} \quad (4.292)$$

It is inadvisable to express the real-valued closed-loop response $\theta_{\text{closed-loop}}(t)$ in terms of complex exponentials with complex coefficients. However, computing and plotting the response using MATLAB to evaluate the terms in Equation 4.292 produce real numbers because $R_3 e^{p_3 t} + R_4 e^{p_4 t}$ is real-valued for all values of t. In fact, it is easily shown that $\theta_{\text{closed-loop}}(t)$ reduces to the real expression

$$\begin{aligned}\theta_{\text{closed-loop}}(t) = \ & -0.2188\,e^{-10.5254t} + 1.3934\,e^{-3.9215t} \\ & -e^{-0.3432t}[6.175\cos(0.8305t) - 1.254\sin(0.8305t)] + 5, \quad t \geq 0 \end{aligned} \quad (4.293)$$

The open-loop response with $A = 0.1$, $T = 10$ s and closed-loop response with $K_C = 10$ are plotted in Figure 4.20.

Figure 4.19a shows that the quadratic factor poles migrate to the right-half plane producing a pair of unstable modes when the gain K_C is larger than the critical gain K_{crit}. An approximation of K_{crit} is possible by varying K_C in Equation 4.284 until the MATLAB "roots" function indicates the presence of a pair of imaginary poles located on the imaginary axis. After several attempts at locating the critical gain, the approximate result is $K_C = 166.19$, and the poles of the marginally stable closed-loop system are located at approximately -14.0705, $-0.000011 \pm j6.086566$, 1.0627.

Increasing K_C further produces an unstable system. Figure 4.21 shows the heading response for the closed-loop system with $K_C = 166.19$. Note the sustained oscillations in the marginally stable system. An unstable response corresponding to $K_C = 175$ is also shown in Figure 4.21. The increasing magnitude of oscillations in the unstable system results from a pair of complex poles in the right-half plane at $0.0601 \pm j6.2285$.

Applying the final value property to the closed-loop transfer function in Equation 4.283 gives

$$\theta_{ss} = \lim_{s \to 0} s \left\{ \frac{5K_C(s+1)}{1.5s^4 + 22.7s^3 + 78s^2 + 5(K_C + 2)s + 5K_C} \right\} \frac{\theta_{\text{com}}}{s} = \theta_{\text{com}} \quad (4.294)$$

Equation 4.294 holds as long as the control system is asymptotically stable, that is, $K_C < K_{\text{crit}}$.

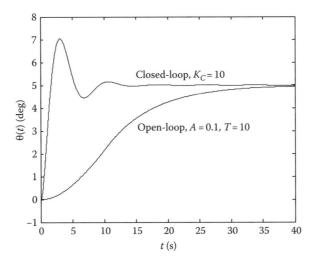

FIGURE 4.20 Ship heading response with open- and closed-loop control.

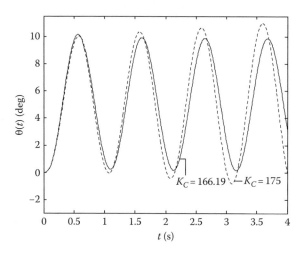

FIGURE 4.21 Heading response for marginally stable and unstable closed-loop system.

The previous example illustrates the concept of stability for an LTI system. The results are predicated on the system response being confined to a range of values for which the linear model is an accurate representation of the actual system's dynamics. Furthermore, limitations on power consumption, component displacements, velocities, etc., must also be satisfied. For example, the design of the ship heading control system using the proportional controller with gain $K_C = 10$ could result in an unrealizable rudder response. A strong argument for simulation is that it allows us to check and monitor such assumptions.

EXERCISES

4.19 For the systems governed by the following differential equations:

(a) $\dot{y} = u$ (an integrator) (b) $\ddot{y} = u$ (a double integrator)

(c) $\dot{y} + 2y = u$ (d) $\dot{y} - 2y = u$

(e) $\ddot{y} + 1.5\dot{y} + 0.5y = u$ (f) $\ddot{y} + 4y = u$

(g) $\ddot{y} - 9y = u$ (h) $\dddot{y} + 4\ddot{y} + 6\ddot{y} + 5\dot{y} + 2y = u$

(i) $\dddot{y} + 2.5\dddot{y} + 2\ddot{y} + 2.5\dot{y} + y = u$

determine whether the system is asymptotically stable, marginally stable, or unstable, and find the natural response, that is, a linear combination of the natural modes.

4.20 Find the characteristic polynomial and characteristic roots of the system with state equations

(a) $\dot{\underline{x}} = \begin{bmatrix} 0 & 1 \\ 2 & -3 \end{bmatrix} \underline{x} + \begin{bmatrix} 0 \\ 1 \end{bmatrix} u, \quad y = \begin{bmatrix} 1 & 0 \end{bmatrix} \underline{x}$

(b) $\dot{\underline{x}} = \begin{bmatrix} 0 & 0 & 1 \\ 0 & 1 & 0 \\ -2 & -1 & -2 \end{bmatrix} \underline{x} + \begin{bmatrix} 0 & 0 \\ 1 & 0 \\ 0 & 1 \end{bmatrix} \begin{bmatrix} u_1 \\ u_2 \end{bmatrix}, \quad \underline{y} = \begin{bmatrix} 1 & 0 \\ 0 & 1 \end{bmatrix} \begin{bmatrix} x_1 \\ x_2 \end{bmatrix}$

(c) $\dot{\underline{x}} = \begin{bmatrix} 20 & -4 & 8 \\ -40 & 8 & -20 \\ -60 & 12 & -26 \end{bmatrix} \underline{x}, \quad y = \begin{bmatrix} 1 & 0 & 1 \end{bmatrix} \underline{x}$

4.21 Show that $|sI - A| = s^4 + 2.985s^3 + 2.42855s^2 + 0.52054s + 0.004016$ when A is the coefficient matrix given by

$$A = \begin{bmatrix} -0.95 & 0.005 & 0 & 0.5 \\ 0.8 & -0.015 & 0 & 0 \\ 0 & 0.01 & -0.32 & 0 \\ 0.15 & 0 & 0.3 & -1.7 \end{bmatrix}$$

4.22 Derive the expression for the closed-loop transfer function $\theta(s)/\theta_{com}(s)$ in Equation 4.283.

4.23 Starting with the Laplace transform $\theta_{open-loop}(s)$ of the open-loop system

$$\theta_{open-loop}(s) = \left[\frac{K_P}{s(\tau_P s + 1)} \right] \left[\frac{K_S}{(\tau_S s + 1)} \right] U(s)$$

(a) Find $\theta_{open-loop}(t)$ in response to the pulse input given in Equation 4.286. Leave your answer in terms of the K_P, K_S, τ_P, τ_S and the pulse parameters A and T.
(b) Verify Equation 4.288 for the final value $\theta_{open-loop}(\infty)$.
(c) Verify the open-loop pulse response shown in Figure 4.20.
(d) Find and plot the open-loop step response
 (i) As the limit as $T \to \infty$ of the open-loop pulse response.
 (ii) By inverse Laplace transformation of $\theta_{open-loop}(s)$ when $U(s) = A/s$.

4.24 In the ship heading example, the input to the ship yaw dynamics in Figure 4.18 is $R(s)$, the rudder angle in degree.
(a) Find the transfer function $R(s)/\theta_{com}(s)$.
(b) Find and plot a graph of $r(t)$ for the case where $\theta_{com}(t) = 5°$, $t \geq 0$ and $K_C = 10$. Comment on the results.
(c) For the same command input $\theta_{com}(t) = 5°$, $t \geq 0$ as in part (b), find the maximum controller gain K_C for which the rudder deflection never exceeds $30°$. Plot $r(t)$ and $\theta(t)$ for a time sufficient for the system to reach steady state.

4.25 For the closed-loop system to control the ship's heading
(a) Find the fourth-order differential equation relating the output $\theta(t)$ and input $\theta_{com}(t)$.
(b) Find a suitable choice for matrices A, B, C, and D in the state variable form $\dot{x} = Ax + Bu$, $y = Cx$ where $u = \theta_{com}$ and $y = \theta$. Leave your answers in terms of the system parameters K_C, K_P, K_S, τ_P, and τ_S.

 Hint: Draw a simulation diagram.

(c) Choose the same values for K_P, K_S, τ_P, and τ_S as in the example. Find the characteristic polynomial $\Delta(s)$ as a function of K_C by evaluating $|sI - A|$.
(d) Prepare a table with two columns. The first column contains values of $K_C = 1, 5, 10, 25, 50, 75, \ldots, 200$ V/deg heading, and the second column lists the four closed-loop system poles.
(e) Use the MATLAB M-file "*Chap4_ feedback_yaw.m*" or write your own to find the value(s) of K_C that results in an underdamped quadratic factor of $\Delta(s)$ with damping ratio equal to 0.5.

4.26 The water current speed $v_W(t)$ influences the angle of the ship's rudder and is considered a load variable or disturbance. The open-loop system is redrawn to reflect the disturbance input in Figure E4.26:

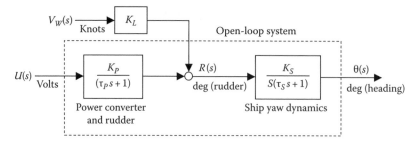

FIGURE E4.26

The load gain K_L can be assumed constant if the angle between the ship's rudder and the water current direction is relatively constant.

(a) Find the closed-loop transfer functions $\dfrac{\theta(s)}{\theta_{\text{com}}(s)}\bigg|_{V_W(s)=0}$ and $\dfrac{\theta(s)}{V_W(s)}\bigg|_{\theta_{\text{com}(s)}=0}$

where

$$\theta(s) = \left[\frac{\theta(s)}{\theta_{\text{com}}(s)}\bigg|_{V_W(s)=0}\right]\theta_{\text{com}}(s) + \left[\frac{\theta(s)}{V_W(s)}\bigg|_{\theta_{\text{com}(s)}=0}\right]V_W(s)$$

(b) Find $\theta(t)$ when $\theta_{\text{com}}(t)=0$, $t \geq 0$ and $v_W(t)=2$ kn, $t \geq 0$. Assume the parameter values K_P, K_S, τ_P, and τ_S are the same as in the example. The controller gain $K_C=7.5$ V/deg heading and the load gain $K_L=0.5°$ rudder/kn.

4.27 A ship with parameters K_P, K_S, τ_P, and τ_S given in the text is traveling in its intended direction, due North as shown in Figure E4.27. The ship cruising speed is 20 kn. The ocean current suddenly switches from zero to five knots in an east-to-west direction. Find the ship's heading $\theta(t)$ with the control system gain $K_C=5$ V/deg heading.

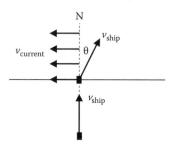

FIGURE E4.27

Hint: Find the new command heading to keep the ship traveling due north.

4.5 FREQUENCY RESPONSE OF LTI CONTINUOUS-TIME SYSTEMS

The response of LTI continuous-time systems to sinusoidal inputs is of interest because it provides an alternative to time domain methods based on the impulse response function to characterize the system's dynamics. A nonperiodic signal $f(t)$ can be resolved into sinusoidal functions over a continuum of frequencies according to Jackson (1991)

$$f(t) = \frac{1}{2\pi}\int_{-\infty}^{\infty} F(j\omega)e^{j\omega t}d\omega \tag{4.295}$$

where the sinusoidal functions are the complex exponentials

$$e^{j\omega t} = \cos \omega t + j \sin \omega t \quad (-\infty < \omega < \infty) \tag{4.296}$$

and the function $F(j\omega)$ is given by

$$F(j\omega) = \int_{-\infty}^{\infty} f(t)e^{-j\omega t}\,dt \tag{4.297}$$

The complex-valued function $F(j\omega)$ is called the Fourier integral or Fourier transform of the signal $f(t)$. Entire books have been written on the Fourier transform and its applications (Papoulis 1962; Bracewell 1986) while other books in the area of signals and systems (Kailath 1980; Jackson 1991; Kraniauskas 1992) include considerable coverage of the topic. $F(j\omega)$ is a function that assumes complex values over the frequency range $(-\infty, \infty)$. In polar form, $F(j\omega)$ is written as

$$F(j\omega) = A(j\omega)e^{j\phi(j\omega)}, \quad A(j\omega) = |F(j\omega)| \quad \text{and} \quad \phi(j\omega) = \text{Arg}[F(j\omega)] \tag{4.298}$$

where the magnitude $A(j\omega)$ is called the Fourier spectrum of $f(t)$.

In rectangular form,

$$F(j\omega) = R(j\omega) + jX(j\omega), \quad R(j\omega) = \text{Re}\{F(j\omega)\}, \quad X(j\omega) = \text{Im}\{F(j\omega)\} \tag{4.299}$$

If $f(t)$ is causal, that is, $f(t) = 0$, $t < 0$, it can be expressed as a continuum of the real sinusoidal functions $\cos \omega t$ or $\sin \omega t$ (Papoulis 1962)

$$f(t) = \frac{2}{\pi}\int_{0}^{\infty} R(j\omega)\cos \omega t\,d\omega = -\frac{2}{\pi}\int_{0}^{\infty} X(j\omega)\sin \omega t\,d\omega, \quad t > 0 \tag{4.300}$$

implying that $R(j\omega)$ and $X(j\omega)$ are not independent.

Suppose an LTI system with transfer function $H(s)$ is subjected to an input $u(t)$ with Fourier transform $U(j\omega)$. By a convolution property similar to the one for Laplace transforms, the Fourier transform of the output $y(t)$ is given by

$$Y(j\omega) = H(j\omega)U(j\omega) \tag{4.301}$$

where $H(j\omega)$ is the system transfer function with s replaced by $j\omega$. $H(j\omega)$ is called the frequency response function of the system. It follows from Equation 4.295

$$y(t) = \frac{1}{2\pi}\int_{-\infty}^{\infty} H(j\omega)U(j\omega)e^{j\omega t}\,d\omega \tag{4.302}$$

and, therefore, each input component $(1/2\pi)U(j\omega)e^{j\omega t}$ in the continuum of frequencies from $-\infty$ to ∞ is scaled by $H(j\omega)$ and integrated over $(-\infty, \infty)$ to form the output $y(t)$. If the input $u(t) = U_0 \cos \omega_0 t$, its Fourier transform is (Jackson 1991)

$$U(j\omega) = U_0\pi[\delta(\omega + \omega_0) + \delta(\omega + \omega_0)] \tag{4.303}$$

and Equation 4.302 reduces to (see Exercise 4.28)

$$y(t) = U_0 \cdot |H(j\omega_0)| \cos \{\omega_0 t + \text{Arg}[H(j\omega_0)]\} \tag{4.304}$$

The amplitude of the output is equal to the amplitude of the input multiplied by the magnitude of the frequency response function evaluated at ω_0. The phase angle (with respect to the input) equals the argument of the frequency response function at ω_0. Equation 4.304 is an essential property of linear systems and the foundation of AC steady-state analysis of electric circuits. Equation 4.304, valid for stable LTI systems, applies only in the steady state, that is, after the system's natural response has vanished.

In the case of nonlinear systems, the steady-state output in response to a sinusoidal input with frequency ω_0 contains sinusoids at harmonic frequencies $2\omega_0, 3\omega_0, 4\omega_0, \ldots$ along with a sinusoidal component at the fundamental frequency ω_0. Example 4.12 illustrates the property in Equation 4.304 for a simple first-order system.

Example 4.12

For the first-order system in Figure 4.22,

    (a) Find the transient and steady-state responses to the input $u(t) = A \sin \omega_0 t$. Leave your answer in terms of the system parameters K and τ and input parameters A and ω_0.
    (b) Find the frequency response function of the system.
    (c) $A = 1$, $\omega_0 = 2$ rad/s, $K = 3$, and $\tau = 0.5$ s. Plot $u(t)$ and $y(t)$ on the same graph.
    (d) Find the time lag between the input and output at steady state, and verify the result from the graphs of $u(t)$ and $y(t)$.

(a) For input $u(t) = A \sin \omega_0 t$, $Y(s)$ is given by

$$Y(s) = \frac{K}{\tau s + 1} U(s) = \frac{K}{\tau s + 1} \left(\frac{A\omega_0}{s^2 + \omega_0^2} \right) = \frac{KA\omega_0}{\tau} \left[\frac{1}{(s + 1/\tau)(s^2 + \omega_0^2)} \right] \tag{4.305}$$

Performing a partial fraction expansion of the last term in Equation 4.305 and simplifying,

$$Y(s) = \frac{KA\omega_0}{1 + (\omega_0\tau)^2} \left[\frac{\tau}{s + 1/\tau} + \frac{1}{s^2 + \omega_0^2} - \frac{\tau s}{s^2 + \omega_0^2} \right] \tag{4.306}$$

The inverse Laplace transform of $Y(s)$ is

$$y(t) = \frac{KA\omega_0}{1 + (\omega_0\tau)^2} [\tau e^{-t/\tau} + \frac{1}{\omega_0} \sin \omega_0 t - \tau \cos \omega_0 t] \tag{4.307}$$

Using the trigonometric relationship

$$A \cos \omega_0 t + B \sin \omega_0 t = C \sin (\omega_0 t + \varphi) \tag{4.308}$$

where

$$C = (A^2 + B^2)^{1/2}, \quad \varphi = \tan^{-1} (A/B) \tag{4.309}$$

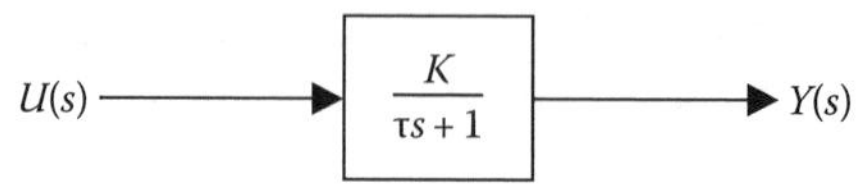

U(s) ──────▶ $\boxed{\dfrac{K}{\tau s + 1}}$ ──────▶ Y(s)

FIGURE 4.22 First-order system ($K > 0$).

the $\sin \omega_0 t$ and $\cos \omega_0 t$ terms in Equation 4.307 may be combined into a single term, that is,

$$y(t) = KA\left\{\frac{\omega_0\tau}{1+(\omega_0\tau)^2}e^{-t/\tau} + \frac{1}{[1+(\omega_0 t)]^{1/2}}\sin(\omega_0 t + \varphi)\right\} \tag{4.310}$$

where

$$\varphi = -\tan^{-1}(\omega_0\tau) \tag{4.311}$$

From Equation 4.310, the transient and steady-state responses are

$$y_{tr}(t) = \frac{KA\omega_0\tau}{1+(\omega_0\tau)^2}e^{-t/\tau} \tag{4.312}$$

$$y_{ss}(t) = \frac{KA}{[1+(\omega_0\tau)^2]^{1/2}}\sin(\omega_0 t + \varphi) \tag{4.313}$$

(b) The frequency response function is

$$H(j\omega) = H(s)\big|_{s=j\omega} = \frac{K}{\tau s + 1}\bigg|_{s=j\omega} \tag{4.314}$$

$$= \frac{K}{1+j\omega\tau} \tag{4.315}$$

From Equation 4.314, the magnitude and phase angle of $H(j\omega)$ are

$$|H(j\omega)| = \left|\frac{K}{1+j\omega\tau}\right| \tag{4.316}$$

$$= \frac{K}{[1+(\omega\tau)^2]^{1/2}} \quad (K > 0) \tag{4.317}$$

$$\mathrm{Arg}H(j\omega) = -\tan^{-1}(\omega\tau) \tag{4.318}$$

(c) Substituting the given values for A, K, τ, and $\omega = \omega_0$ gives

$$y_{tr}(t) = \frac{(3)(1)(2)(0.5)}{1+[(2)(0.5)]^2}e^{-t/\tau} = 1.5e^{-2t} \tag{4.319}$$

$$y_{ss}(t) = \frac{(3)(1)}{\{1+[(2)(0.5)]^2\}^{1/2}}\sin\{2t - \tan^{-1}[(2)(0.5)]\} \tag{4.320}$$

$$= 1.5\sqrt{2}\sin\left(2t - \frac{\pi}{4}\right) \tag{4.321}$$

The input $u(t) = \sin 2t$ and output $y(t) = 1.5e^{-2t} + 1.5\sqrt{2}\sin(2t - \pi/4)$ are shown in Figure 4.23. The transient response dies out in approximately $5\tau = 5(0.5) = 2.5$ s.

(d) Figure 4.24 is a close-up of Figure 4.23 near the peaks of $u(t)$ and $y(t)$. The lag time T is estimated as $T \approx 4.31 - 3.92 = 0.39$ s in agreement with the exact value

$$\omega_0 T = \varphi \Rightarrow T = \frac{\varphi}{\omega_0} = \frac{\pi/4}{2} = \frac{\pi}{8} = 0.393 \text{ s} \tag{4.322}$$

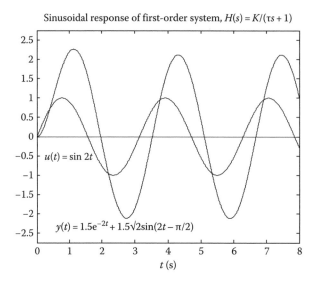

FIGURE 4.23   Graph of input $u(t)$ and output $y(t)$.

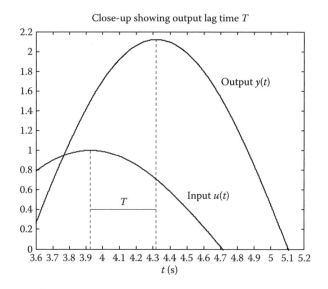

FIGURE 4.24   Close-up of input and response near peaks.

This example illustrates how the steady-state sinusoidal response of an LTI system can be obtained considerably faster using the frequency response function compared to methods that determine the complete response.

Graphical tools exist for conveying the magnitude and phase properties of an LTI continuous-time system with transfer function $H(s)$. The simplest one consists of graphs of $|H(j\omega)|$ and Arg $H(j\omega)$ vs. ω. The graphs are typically plotted over a frequency range of interest. Control systems engineers and analog filter designers prefer a variation of the frequency response plots in which $20 \log|H(j\omega)|$, the magnitude measured in decibels (db), is plotted vs. ω on a logarithmic scale. The result (along with the phase plot) is called a Bode diagram or Bode plot.

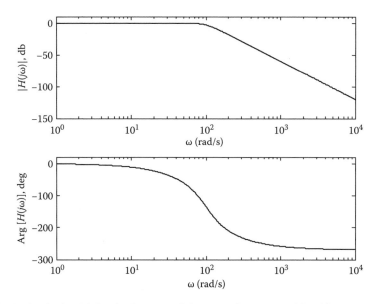

FIGURE 4.25   Bode plot for third-order Butterworth low-pass filter ($\omega_b = 100$ rad/s).

To illustrate, consider a system with transfer function

$$H(s) = \frac{\omega_b^3}{(s + \omega_b)(s^2 + \omega_b s + \omega_b^2)} \tag{4.323}$$

which describes a third-order low-pass Butterworth filter designed to pass frequencies in the band 0 (DC) to ω_b and reject all others. The M-file "*Chap4_Fig5_4.m*" includes statements to evaluate the magnitude and phase of $H(s)$ when $\omega_b = 100$ rad/s for frequencies between 10^0 and 10^4 rad/s. The Bode plot is shown in Figure 4.25.

The control system toolbox, a complementary suite of utilities designed for use with the MATLAB environment, includes a function "bode" for drawing the Bode plot of an LTI system. The control system toolbox is covered later in Section 4.4.10.

The magnitude measured in db (sometimes referred to as the gain) is close to zero, and, hence, the magnitude is close to 1 over a considerable portion of the interval $0 \leq \omega \leq \omega_b$. At $\omega = \omega_b$,

$$|H(j\omega_b)| = \left| \frac{\omega_b^3}{(s + \omega_b)(s^2 + \omega_b s + \omega_b^2)} \right|_{s=j\omega_b} = \frac{1}{|-1 + j|} = \frac{1}{\sqrt{2}} \tag{4.324}$$

$$\Rightarrow 20 \log |H(j\omega_b)| = 20 \log \frac{1}{\sqrt{2}} \approx -3 \text{ db} \tag{4.325}$$

The gain is -3 db at $\omega = \omega_b$ and starts falling off from ω_b at approximately 60 db for every 10-fold increase in frequency (decade) (see Figure 4.25).

The frequency response function of a system dictates the extent to which sinusoidal inputs at specific frequencies are passed or rejected by the system, and coupled with the fact that input time signals can be resolved into sinusoids over a continuum of frequencies, explains why linear systems are often called linear filters.

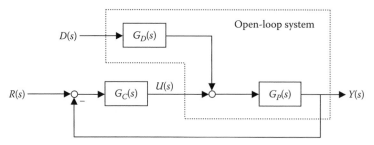

FIGURE 4.26 A feedback control system with command and disturbance inputs.

The individual components in a linear feedback control system such as sensors, controllers, and power converters are examples of continuous-time filters, which transmit the range of frequencies in the input according to their frequency response function. Control system design based on frequency response relies on assumptions related to the frequency content of the command inputs and the uncontrollable inputs, referred to as load variables or disturbances.

A simple unity feedback control system is shown in Figure 4.26. $R(s)$ and $D(s)$ are the reference (command) and disturbance inputs.

The open-loop system model is

$$Y(s) = G_P(s)[U(s) + G_D(s)D(s)] \tag{4.326}$$

The control system output $Y(s)$ can be written as

$$Y(s) = T_R(s)R(s) + T_D(s)D(s) \tag{4.327}$$

where

$$T_R(s) = \frac{G_C(s)G_P(s)}{1 + G_C(s)G_P(s)}, \quad T_D(s) = \frac{G_D(s)G_P(s)}{1 + G_C(s)G_P(s)} \tag{4.328}$$

It frequently happens that the command input $r(t)$ is a slow varying signal compared to the disturbance input $d(t)$. Assuming $G_P(s)$ and $G_D(s)$ are fixed, proper design entails selecting a controller transfer function $G_C(s)$ to simultaneously make $|T_R(j\omega)|$ close to 1 at the lower frequencies contained in $r(t)$ and $|T_D(j\omega)|$ close to zero for the frequencies present in $d(t)$. Suppose the command input is band-limited from 0 to 0.25 Hz (1.57 rad/s) and the disturbance frequencies start at roughly 10 Hz (62.8 rad/s) and the open-loop system transfer functions are

$$G_P(s) = \frac{K}{s^2 + 2\zeta\omega_n s + \omega_n^2} = \frac{1}{s^2 + 2.25s + 0.5625} \tag{4.329}$$

$$G_D(s) = K_D = 40 \tag{4.330}$$

The controller is of the proportional plus integral (P-I) type,

$$G_C(s) = K_C + \frac{K_1}{s} = 5 + \frac{2}{s} \tag{4.331}$$

Bode plots of $T_R(j\omega)$ and $T_D(j\omega)$ are generated in "Chap4_Fig5_6.m" and shown in Figure 4.27. The frequency content of the command input $r(t)$ is confined primarily to frequencies below 1.57 rad/s. The output will track the input closely since the gain $20\log|T_R(j\omega)|$ is roughly 0 db,

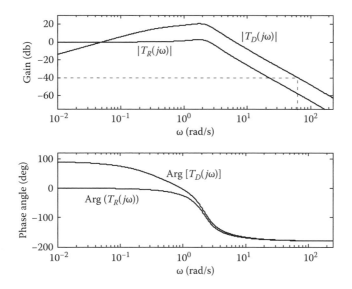

FIGURE 4.27   Bode plot for closed-loop frequency response functions $T_R(j\omega)$ and $T_D(j\omega)$.

corresponding to a magnitude of 1 from DC ($\omega = 0$) to approximately 1 rad/s. The phase angle Arg $[T_R(j\omega)]$ is close to $0°$ from $\omega = 0$ to $\omega \approx 0.5$ rad/s and is $-36.1°$ at $\omega = 1.57$ rad/s.

Conversely, the gain $20\log|T_D(j\omega)| = -40$ db, which is equivalent to a magnitude of 0.01 at approximately 62 rad/s. The control system effectively filters out the disturbances by attenuating all frequencies above 62.8 rad/s.

The steady-state error, $e_{ss} = y(\infty) - r(\infty)$, is zero when $r(t)$ or $d(t)$ is constant. This can be demonstrated by showing that the DC gains $T_R(j0) = 1$ and $T_D(j0) = 0$, a direct consequence of the open-loop gain $G_C(0)G_P(0) = \infty$. The infinite open-loop gain results from the presence of the integrator in $G_C(s)$. While zero steady-state error is a desirable condition, we must still be mindful of the location of the control system's characteristic roots since it determines the transient response.

The transfer functions of real-world components and complete systems possess Bode plots in which the gain "rolls off" at high frequencies. Properly designed closed-loop control systems track low-frequency command inputs reasonably well. Further increases in frequency require excessive power be delivered to control system components, thus limiting the system's ability to track higher frequency command inputs.

Any component or system with transfer function $G(s)$ given by the ratio of polynomials in proper fraction form, that is, numerator polynomial, is lower order than denominator will satisfy

$$\lim_{\omega \to \text{currency}} |G(j\omega)| = 0 \Rightarrow \lim_{\omega \to \infty} 20\log|G(j\omega)| = -\infty \tag{4.332}$$

A common measure of the frequency where "roll off" begins is ω_b and the interval $(0, \omega_b)$ is called the bandwidth of the system. The frequency ω_b satisfies

$$|G(j\omega_b)| = \frac{1}{2^{1/2}}|G(j0)| \Rightarrow 20\log|G(j\omega_b)| = 20\log|G(j0)| - 3\,\text{db} \tag{4.333}$$

Consequently, ω_b is the (lowest) frequency at which the gain (magnitude function measured in db) is 3 db below the DC gain of the system.

Consider the first-order system in Figure 4.22 with magnitude function $|H(j\omega)|$ given in Equation 4.316. The frequency ω_b is obtained from

$$|H(j\omega_b)| = \frac{K}{[1 + (\omega_b\tau^2)^{1/2}]} = \frac{1}{2^{1/2}} \cdot |H(j0)| = \frac{1}{2^{1/2}} \cdot K \tag{4.334}$$

$$\Rightarrow 1 + (\omega_b\tau)^2 = 2 \tag{4.335}$$

$$\Rightarrow \omega_b = \frac{1}{\tau} \tag{4.336}$$

Equation 4.336 is important because it relates ω_b, a frequency domain parameter to the time constant τ, which characterizes the system's transient response in the time domain. Furthermore, being inversely proportional to the system, time constant tells us that the bandwidth frequency ω_b is a measure of the speed of response of the first-order system. Hence, first-order systems like the one in Figure 4.22 with a fast natural mode (τ small) exhibit larger bandwidths.

For a second-order system with transfer function

$$G(s) = \frac{K\omega_n^2}{s^2 + 2\zeta\omega_n s + \omega_n^2} \tag{4.337}$$

increasing the natural frequency ω_n (with ζ constant) decreases the transient response time regardless of whether the system is underdamped, overdamped, or critically damped (see expressions for step response in Section 2.3). It is left as an exercise to show that the bandwidth frequency ω_b for the system with transfer function in Equation 4.337 is proportional to ω_n. Specifically,

$$\omega_b = \left[1 - 2\zeta^2 + (2 - 4\zeta^2 + 4\zeta^4)^{1/2}\right]^{1/2} \omega_n, \quad (K = 1) \tag{4.338}$$

and, therefore, ω_b is a measure of the speed of response for a second-order system as well.

A Bode plot for three second-order systems, all with $\omega_n = 1$ rad/s and damping ratios of $\zeta = 0.25$, 1, 2, is shown in Figure 4.28. Also shown is an enlargement of the plots for the purpose of

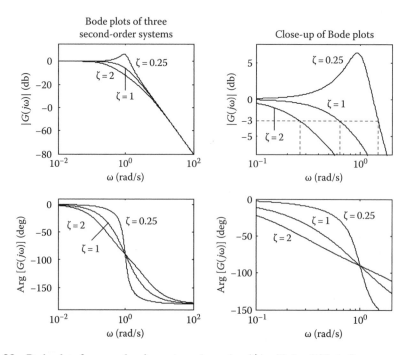

FIGURE 4.28   Bode plots for second-order systems ($\omega_n = 1$ rad/s) with $\zeta = 0.25, 1, 2$.

estimating the corresponding bandwidths. The calculated values of ω_b from Equation 4.338 are 1.4845 rad/s ($\zeta = 0.25$), 0.6436 rad/s ($\zeta = 1$), and 0.2666 rad/s ($\zeta = 2$) in agreement with the values estimated from Figure 4.28.

Figure 4.28 shows a peak in the gain (and magnitude function) for the underdamped system indicating the presence of a resonant frequency. The resonant frequency is $\omega_r = 0.935$ rad/s with $|G(j\omega_r)| = 2.0656$ (6.3 db). Not all underdamped second-order systems exhibit resonance (see Exercise 4.32).

The Bode plots and bandwidth calculations are handled in the MATLAB script file "*Chap4_Fig5_7.m*."

The step responses of the three second-order systems are shown in Figure 4.29. The rise time is defined as $t_r = t_{0.9} - t_{0.1}$, where $t_{0.1}$ and $t_{0.9}$ are the times required for the step response to reach 10% and 90% of its final value, respectively. The rise time is another measure of the system's speed of response. The times $t_{0.1}$ and $t_{0.9}$ and the approximate rise times are shown on the zoomed-in plots of the step responses. As expected, the lightly damped system ($\zeta = 0.25$) with the greatest bandwidth responds the quickest (shortest rise time) while the overdamped system ($\zeta = 2$) with the smallest bandwidth is the most sluggish and least responsive.

The step responses are generated in the M-file "*Chap4_Fig5_8.m*."

LTI systems modeled by transfer functions where the order of the numerator and that of the denominator polynomials are equal, that is, a direct connection exists from the input to the output, exhibit finite gain at frequencies approaching infinity. That is,

$$\lim_{\omega \to \infty} |H(j\omega)| = \lim_{\omega \to \infty} \left| \frac{\alpha_n s^n + \alpha_{n-1} s^{n-1} + \cdots + \alpha_1 s + \alpha_0}{\beta_n s^n + \beta_{n-1} s^{n-1} + \cdots + \beta_1 s + \beta_0} \right|_{s=j\omega} = \frac{\alpha_n}{\beta_n} \tag{4.339}$$

Since a real system cannot respond in a way suggested by Equation 4.339, the transfer function $H(s)$ with equal order polynomials in the numerator and denominator, or equivalently the same number of finite zeros and poles, is an ideal approximation that breaks down above a certain frequency. Nonetheless, it is a useful approximation to the transfer function of a system that readily passes

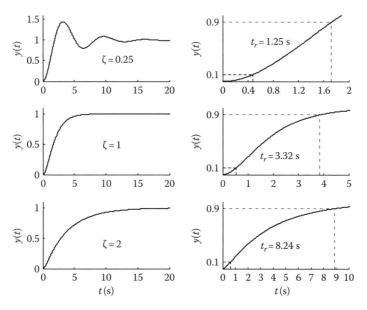

FIGURE 4.29 Step responses and rise times for three second-order systems.

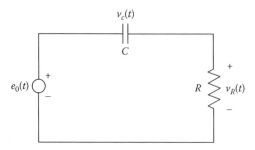

FIGURE 4.30 Circuit with high-pass filter transfer function.

high-frequency components present in its input(s), as in the case of a high-pass filter. Of course, when the high-frequency signals represent unwanted noise, which is invariably present in control systems, the closed-loop transfer function should be designed to attenuate the noise (see Exercise 4.34).

The simple *RC* circuit in Figure 4.30 with the voltage $v_R(t)$ as output is an example of a high-pass filter. The transfer function is

$$H(s) = \frac{V_R(s)}{E_0(s)} = \frac{RC_s}{RC_s + 1} \tag{4.340}$$

At high frequencies ($\omega \gg 1/RC$), the magnitude $|H(j\infty)| \approx 1$ (0 db). Note that the capacitor behaves like a short circuit at high frequencies.

4.5.1 STABILITY OF LINEAR FEEDBACK CONTROL SYSTEMS BASED ON FREQUENCY RESPONSE

Linear control systems are a class of LTI systems, and the basic premises of stability presented in the previous section are applicable. The following is a brief introduction to stability, as it applies to simple feedback control systems from the viewpoint of frequency response. For a more detailed discussion of the subject, the reader is encouraged to refer to any of the texts in linear feedback control systems listed in the References.

Figure 4.31 is a block diagram of a servo control system with transfer functions for the controller, actuator, plant, and sensor/transmitter.

Insight into the stability of the system can be ascertained by tracking the response to the error signal $e(t) = \mathcal{L}^{-1}\{E(s)\}$ as it propagates around the loop. Suppose the loop is broken immediately following the transmitter and a test signal $e(t) = \sin \omega t$ is inserted at the controller input. Each component along the open-loop path processes a sinusoidal input and delivers a sinusoidal output (both at radian frequency ω) to the next component. Magnitude and phase shift of the individual sinusoids are determined by the frequency response functions of each component at radian frequency ω.

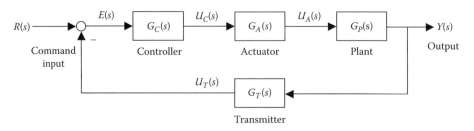

FIGURE 4.31 Block diagram of representative linear feedback control system.

The closed-loop control system is unstable if $-u_T(t) = -\mathcal{L}^{-1}\{U_T(s)\}$ is ever in phase with $e(t)$ and its amplitude is greater than one. When this occurs, the error signal propagates around the loop and increases in magnitude while doing so. Conversely, when $e(t)$ and $-u_T(t)$ are in phase and $|-u_T(t)| = |u_T(t)| < 1$, a stable system results. Finally, a marginally stable system exists when $e(t)$ and $-u_T(t)$ are in phase and $|-u_T(t)| = |u_T(t)| = 1$.

Since the negative sign in $-u_T(t)$ is equivalent to $-180°$ phase shift, $-u_T(t)$ will be in phase with $e(t)$ whenever $u_T(t)$ lags $e(t)$ by $-180°$, that is, there is a combined total of $-180°$ phase lag in the open-loop system. The frequency at which this occurs is called the phase crossover frequency ω_{cp}. Hence, for a closed-loop, negative feedback control system to be marginally stable (or unstable), there must exist at least one frequency where the open-loop phase lag is equal to $-180°$.

The open-loop transfer function is

$$G_{OL}(s) = G_C(s)G_A(s)G_P(s)G_T(s) \tag{4.341}$$

For this example, assume the dynamics of each component are described by

$$G_C(s) = K_C, \quad G_A(s) = \frac{K_A}{\tau_A s + 1}, \quad G_P(s) = \frac{K_P}{s(\tau_P s + 1)}, \quad G_T(s) = \frac{K_T}{\tau_T s + 1} \tag{4.342}$$

where $K_C = 0.25$, $K_A = 2$, $\tau_A = 0.25$, $K_P = 8$, $\tau_P = 4$, $K_T = 0.1$, and $\tau_T = 0.003$.

The open-loop transfer function becomes

$$G_{OL}(s) = K_C \frac{K_A}{\tau_A s + 1} \cdot \frac{K_P}{s(\tau_P s + 1)} \cdot \frac{K_T}{\tau_T s + 1} \tag{4.343}$$

$$= (0.25)\frac{2}{0.25s + 1} \cdot \frac{8}{s(4s + 1)} \cdot \frac{0}{0.003s + 1} \tag{4.344}$$

$$= \frac{0.4}{s(0.25s + 1)(4s + 1)(0.003s + 1)} \tag{4.345}$$

A Bode plot of the open-loop transfer function is shown in Figure 4.32.

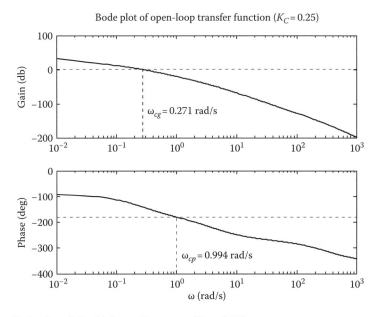

Bode plot of open-loop transfer function ($K_C = 0.25$)

FIGURE 4.32 Bode plot of $G_{OL}(s)$ for stable system ($K_C = 0.25$).

Inspection of Equation 4.345 reveals the open-loop phase varies from $-90°$ at $\omega = 0$ to $-360°$ at $\omega \to \infty$ indicating the possibility of a marginally stable or unstable system.

The phase crossover frequency ω_{cp} was determined by trial and error to be approximately 0.9936 rad/s. As a check,

$$\text{Arg}[G_{OL}(j0.9936)] \approx -180° \tag{4.346}$$

The magnitude function evaluated at $\omega_{cp} \approx 0.9936$ rad/s is

$$|G_{OL}(j\omega_{cp})| = G_{OL}(j0.9936)| = 0.0953(-20.4 \text{ db}) \tag{4.347}$$

The system is stable since the magnitude function is less than one, or equivalently the gain is less than 0 db, at the phase crossover frequency. The gain of -20.4 db is a measure of stability. Control engineers would say the "gain margin" is 20.4 db.

Another indicator of stability, the "phase margin," is the difference between the open-loop phase lag and $-180°$ at the frequency where the gain is 0 db. This frequency, called the gain crossover frequency ω_{cg}, is approximately 0.271 rad/s for the stable system in Figure 4.32. Since Arg $[G_{OL}(j\omega_{cg})] = \text{Arg}[G_{OL}(j0.271)] = -141.2°$, the phase margin is equal to $-142.1 - (-180) = 37.9°$. Higher phase margins imply a greater measure of relative stability.

Increasing the controller gain K_C generally makes the system more responsive. Consider raising the gain K_C by an amount sufficient to make the system marginally stable, that is, $|G_{OL}(j\omega_{cp})| = 1 \Rightarrow 20 \log |G_{OL}(j\omega_{cp})| = 0$ db. From Equation 4.347, it follows that if we multiply the current gain $K_C = 0.25$ by $1/|G_{OL}(j\omega_{cp})| = 1/0.0953$, the new open-loop gain will be equal to 0 db at ω_{cp} (which remains unchanged at 0.9936 rad/s). The Bode plot of the open-loop system transfer function when $K_C = 0.25(1/0.0953) = 2.62$ is shown in Figure 4.33.

The gain crossover frequency is identical to the phase crossover frequency, and the two stability margins have been reduced to zero. The control system is marginally stable, and there will be persistent oscillations at the crossover frequency 0.9936 rad/s in the natural response of the system.

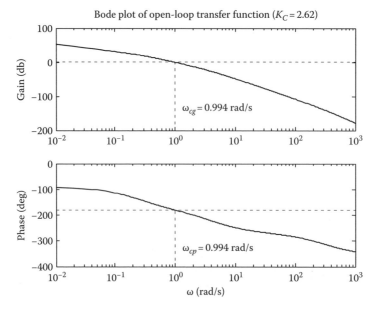

FIGURE 4.33 Bode plot of $G_{OL}(s)$ for marginally stable system.

The closed-loop transfer function is

$$G_{CL}(s) = \frac{G_C(s)G_A(s)G_P(s)}{1 + G_C(s)G_A(s)G_P(s)G_T(s)} \tag{4.348}$$

and the closed-loop system poles are the roots of

$$1 + G_C(s)G_A(s)G_P(s)G_T(s) = 1 + (2.6224)\frac{2}{0.25s+1} \cdot \frac{8}{s(4s+1)} \cdot \frac{0.1}{0.003s+1} = 0 \tag{4.349}$$

$$\Rightarrow (0.25s+1)s(4s+1)(0.003s+1) + (2.6224)(2)(8)(0.1) = 0 \tag{4.350}$$

$$\Rightarrow 0.003s^4 + 1.01275s^3 + 4.253s^2 + s + 4.1958 = 0 \tag{4.351}$$

Solving the characteristic equation above produces the four closed-loop system poles,

$$s_1 = -333.3, \quad s_2 = -4.25, \quad s_3 = j0.9936, \quad s_4 = -j0.9936$$

demonstrating the marginal stability (poles on the imaginary axis) of the system as well as the frequency of sustained oscillations, namely, $\omega_{cp} = 0.9936$ rad/s.

Further increase in controller gain K_C produces an unstable system resulting in negative stability margins (gain and phase) as well as closed-loop system poles in the right-half plane. Superior performance requires a different type of controller, that is, one which provides sufficient phase lead in the vicinity of the gain crossover frequency for adequate stability and possibly phase lag at lower frequencies to improve steady-state response. Indeed, this is the essence of synthesizing controllers for feedback control systems using frequency response methods. Simulation is an indispensable tool for verifying control system design.

EXERCISES

4.28 Use Equations 4.302 and 4.303 to derive Equation 4.304.

4.29 The Fourier spectrum $|F(j\omega)|$ of a signal $f(t)$ can be used to find the energy in the signal in the frequency spectrum (ω_1, ω_2) according to

$$E_f(\omega_1, \omega_2) = \int_{\omega_1}^{\omega_2} |F(j\omega)|^2 d\omega$$

(a) Find the Fourier transform of the exponential $f(t) = \begin{cases} 0, & t < 0 \\ e^{-\alpha t}, & t \geq 0 \end{cases}$.

(b) Find and graph $|F(j\omega)|$.

(c) Find ω_0 such that $E_f(0, \omega_0) = 1/2 E_f(0, \infty)$.

4.30 For the third-order Butterworth filter in Equation 4.323 with $\omega_b = 2\pi$ rad/s, find

(a) The poles of $H(s)$.

(b) The impulse response function $h(t)$.

(c) The filter output at steady state when the input is $u(t) = \sin(0.5\omega_b t) + \sin(2\omega_b t)$.

4.31 Derive Equation 4.338 relating the bandwidth and natural frequency of a second-order system in standard form.

4.32 For a second-order system with natural frequency $\omega_n = 1$ rad/s, find

(a) The maximum value of ζ for which the system has a resonant frequency.

(b) The resonant frequency if $\zeta = 0$.

(c) The response when $\zeta = 0$ to a sinusoidal input at the resonant frequency.

4.33 The circuit shown in Figure E4.33 is designed to block 60 Hz noise in the input $v_i(t)$ from appearing in the output $v_o(t)$.

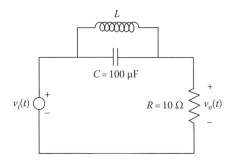

FIGURE E4.33

(a) Show that the transfer function $H(s) = V_o(s)/V_i(s) = (R(LCs^2 + 1))/(RLCs^2 + Ls + R)$.
(b) Find the frequency response function $H(j\omega)$.
(c) Find the inductance L for which $|H(j2\pi \cdot 60)| = 0$.
(d) Write an M-file to draw a Bode plot for $10^2 \leq \omega \leq 10^4$ rad/s.
(e) Find and graph $v_o(t)$ when $v_i(t) = \sin(2\pi \cdot 55)t + \sin(2\pi \cdot 60)t$.
(f) Find and graph $v_o(t)$ when $v_i(t) = \sin(2\pi \cdot 100)t + \sin(2\pi \cdot 60)t$.

4.34 A system for controlling the attitude of a rigid satellite is shown Figure E4.34:

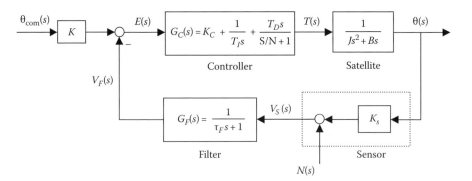

FIGURE E4.34

The controller determines the torque $T(t)$ developed by a pair of thrusters to control the satellite's attitude $\theta(t)$. The controller input is an error voltage signal $e(t)$, which is the difference between the commanded attitude $\theta_{com}(t)$ converted to a voltage and the filtered sensor output $v_F(t)$. The sensor output voltage $v_S(t)$ contains an additive noise component $n(t)$.

A low-pass filter is inserted between the comparator and the sensor output to attenuate the noise in the feedback signal. The gain K converts the commanded angle (deg) to a voltage for comparison to the output voltage from the filter $v_F(t)$. The numerical value of K is the same as the sensor gain K_S.

The command and noise inputs are

$$\theta_{com}(t) = \begin{cases} 0, & t < 0 \\ At, & 0 \leq t < t_0 \\ At_0, & t \geq t_0 \end{cases}$$
$$n(t) = N_0 \sin \omega_0 t, \quad t \geq 0$$

Baseline parameter values are

$$K = K_S = 0.1\,\text{V/deg},$$

$$J = 150\,\text{ft lb/deg/s}^2, \quad B = 15\,\text{ft lb/deg/s}$$

$$K_C = 10\,\text{ft lb/V}, \quad T_1 = 60\,\text{s}, \quad T_D = 30\,\text{s}, \quad N = 10$$

$$A = 1.2°/\text{s}, \quad t_0 = 2.5\,\text{s}$$

$$N_0 = 1\,\text{V}, \quad \omega_0 = 50\,\text{Hz}$$

In parts (a) through (d), assume the filter is not present, that is, $V_S(s)$ is input to the summer.

(a) Find the transfer functions $H_{com}(s) = \theta(s)/\theta_{com}(s)$ and $H_N(s) = \theta(s)/N(s)$. Leave your answers in terms of the parameters $K, K_S, J, K_C, T_I, T_D, N$.

(b) Obtain Bode Plots for $H_{com}(j\omega)$ and $H_N(j\omega)$.

(c) Find and graph $\theta(t)$, $t \geq 0$.

(d) Find and graph the torque $T(t)$, $t \geq 0$.

In parts (e) through (i) the filter is present.

(e) Find the filter time constant τ_F if the filter gain is -40 db at the noise frequency.

(f) Find the transfer functions $H_{com}(s) = \theta(s)/\theta_{com}(s)$ and $H_N(s) = \theta(s)/N(s)$. Leave your answers in terms of $K, K_S, J, K_C, T_I, T_D, N, \tau_F$.

(g) Obtain Bode Plots for $H_{com}(j\omega)$ and $H_N(j\omega)$ using the value for τ_F.

(h) Find and graph $\theta(t)$, $t \geq 0$.

(i) Find the gain and phase margins of the closed-loop system.

4.35 For the control system shown in Figure 4.31,

(a) Use the given baseline parameter values (except K_C), and fill in the missing values in the following table:

	$K_C = 0.1$	$K_C = 0.25$	$K_C = 1$	$K_C = 2.5$
Phase margin				
Gain margin				
Band margin				

(b) Compare the step responses for each of the cases in the table.

4.6 z-TRANSFORM

Difference equations result from approximation of continuous-time differential equation models. Inputs to the difference equations are commonly discrete-time signals resulting from sampling a continuous-time signal (sample data systems). Inherently discrete-time systems are modeled by difference equations relating inputs and outputs that change only at discrete points in time, as in the case of a numeric processor with a fixed cycle time or a loan balance with monthly payments to reduce the outstanding balance.

In the same way, we characterized continuous-time signals and continuous-time systems; discrete-time counterparts (signals and systems) can be analyzed with the help of a mathematical transformation. Instead of an integral transformation from a continuous-time signal $f(t)$, $t \geq 0$ to its Laplace transform $F(s)$, a different type of mapping is applied to a discrete-time function $f(k)$ or f_k, $k = \cdots -3, -2, -1, 0, 1, 2, 3, \ldots$. Similar to $F(s)$, the z-transform $F(z)$ is a complex-valued function, that is, s and z are both complex variables. Only causal signals, those that satisfy $f_k = 0$, $k = \cdots -3, -2, -1$, will be considered.

The z-transform of a causal discrete-time signal f_k, $k = 0, 1, 2, 3, \ldots$ denoted $F(z)$ or $z\{f_k\}$ is defined by the infinite series

$$F(z) = \mathfrak{z}\{f_k\} = \sum_{k=0}^{\infty} f_k z^{-k} \tag{4.352}$$

The region of convergence of $F(z)$ in the z-plane is all complex numbers greater than a certain distance from the origin, that is, $|z| > R$ where R depends on the particular sequence of numbers (discrete-time signal) f_k (Kuo 1980). As in the case of the Laplace transformation, the region of convergence of the z-transform for a particular discrete-time signal is of passing interest. The main consideration is that the sum in Equation 4.352 converges to a complex number somewhere in the z-plane. Several simple discrete-time signals and their z-transforms follow. The derivations follow directly from the definition in Equation 4.352.

Example 4.13

Find the z-transform of the unit step $\hat{u}_k = 1$, $k = 0, 1, 2, 3, \ldots$ shown in Figure 4.34.

$$U(z) = \mathfrak{z}\{\hat{u}_k\} = \sum_{k=0}^{\infty} 1 \cdot z^{-k} = 1 + z^{-1} + (z^{-1})^2 + (z^{-1})^3 + \cdots \tag{4.353}$$

The infinite series converges to a sum, that is,

$$U(z) = \sum_{k=0}^{\infty} (z^{-1})^k = \frac{1}{1 - z^{-1}} = \frac{z}{z - 1} \tag{4.354}$$

provided $|z^{-1}| < 1$ or equivalently $|z| > 1$. Hence, the region of convergence is outside the Unit Circle, $|z| = 1$. A closed form for $U(z)$ is preferable to the infinite series and often easy to recognize when u_k is a simple expression.

Example 4.14

(a) Find the z-transform of the discrete-time signal u_k resulting from sampling the continuous-time function $u(t) = e^{-at}$, $t \geq 0$ every T s.
(b) Suppose $u(t)$ and u_k are as shown in Figure 4.35. Find $U(z)$.

(a) Sampling a continuous-time signal $u(t)$ every T s results in a discrete-time signal u_k where $u_k = u(t)|_{t=kT} = u(kT)$, $k = 0, 1, 2, 3, \ldots$. Hence, from the definition of the z-transform and $u_k = e^{-akT}$, $k = 0, 1, 2, \ldots$,

$$U(z) = \sum_{k=0}^{\infty} e^{-akT} z^{-k} = \sum_{k=0}^{\infty} (e^{-aT} z^{-1})^k = \frac{1}{1 - e^{-akT} z^{-1}} = \frac{z}{z - e^{-aT}}, \quad |z| > e^{-aT} \tag{4.355}$$

Note the dependence of $U(z)$ on the sampling interval T.

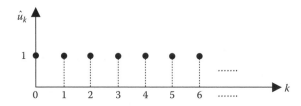

FIGURE 4.34   The discrete-time unit step.

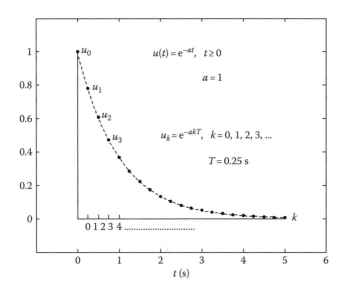

FIGURE 4.35   Uniform sampling of a continuous-time exponential function.

(b) For $a = 1$, $T = 0.25$,

$$U(z) = \frac{z}{z - e^{-0.25}}, \quad |z| > e^{-0.25} \tag{4.356}$$

The next example looks at a discrete-time signal, which occurs frequently in the analysis of linear discrete-time systems, namely, the geometric sequence.

Example 4.15

Find the z-transform of the discrete-time signal

$$u_k = a^k, \quad k = 0, 1, 2, 3, \ldots \tag{4.357}$$

Once again, our starting point is the definition of the z-transform in Equation 4.352.

$$U(z) = \sum_{k=0}^{\infty} a^k z^{-k} = \sum_{k=0}^{\infty} (az^{-1})^k = \frac{1}{1 - az^{-1}} = \frac{z}{z - a}, \quad |z| > |a| \tag{4.358}$$

The result is easily checked by long division, that is, if the denominator in Equation 4.358 is divided into the numerator, the result is

$$\frac{z}{z - a} = 1 + az^{-1} + a^2 z^{-2} + a^3 z^{-3} + \cdots + a^k z^{-k} + \cdots \tag{4.359}$$

From the definition of $U(z)$ as an infinite series,

$$U(z) = \sum_{k=0}^{\infty} u_k z^{-k} = u_0 + u_1 z^{-1} + u_2 z^{-2} + u_3 z^{-3} + \cdots + u_k z^{-k} + \cdots \tag{4.360}$$

Comparing Equation 4.359 and Equation 4.360, it follows that $u_0 = 1$, $u_1 = a$, $u_2 = a^2$, $u_3 = a^3, \ldots$, and, therefore, $u_k = a^k$, $k = 0, 1, 2, 3, \ldots$. The long division method provides a quick check on $U(z)$ for a discrete-time signal u_k, $k = 0, 1, 2, 3, \ldots$. Typically, the first several coefficients in the infinite series expression for $U(z)$ are compared to the corresponding values of the discrete-time signal u_k with an equivalence necessary (but not sufficient) for $U(z) = \mathfrak{z}\{u_k\}$.

Depending on the numerical value of the constant "a," the discrete-time signal u_k in Equation 4.357 can asymptotically approach zero in magnitude ($|a| < 1$), remain constant in magnitude ($|a| = 1$), or increase in magnitude without bound ($|a| > 1$). All six cases are shown in Figure 4.36. Note that when $a = 1$, the discrete-time unit step (Figure 4.34) results and Equation 4.358 reduces to Equation 4.354.

The exponential sequence in Example 4.14 is also a geometric sequence. This is evident by expressing it in a slightly different way, that is,

$$u_k = e^{-akT} = (e^{-aT})^k = (b)^k, \quad k = 0, 1, 2, 3, \ldots \quad \text{where } b = e^{-aT} \tag{4.361}$$

The sequences resulting from uniform sampling of continuous-time sine and cosine functions are fundamental discrete-time signals with z-transforms that follow directly from the basic definition. The results are

$$\sin k\omega T \Leftrightarrow \frac{(\sin \omega T)z}{z^2 - (2 \cos \omega T)z + 1} \tag{4.362}$$

$$\cos k\omega T \Leftrightarrow \frac{z(z - \cos \omega T)}{z^2 - (2 \cos \omega T)z + 1} \tag{4.363}$$

where the symbol $\Leftrightarrow$ denotes a z-transform pair, that is, a discrete-time signal and its z-transform.

The discrete-time signals in Equations 4.362 and 4.363 produce interesting results when the sampling occurs at certain frequencies as shown in Example 4.16.

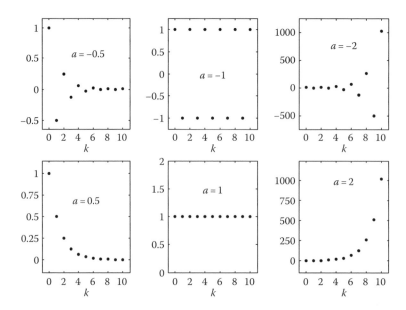

FIGURE 4.36 Discrete-time signal $u_k = a^k$, $k = 0, 1, 2, 3, \ldots$ for $a = -0.5, -1, -2, 0.5, 1, 2$.

Example 4.16

Find the z-transform of the discrete-time signal obtained from sampling

     (a) $x(t) = \sin 3t,\ t \geq 0$ when $T = \pi/6$ s
     (b) $x(t) = \sin 3t,\ t \geq 0$ when $T = \pi/3$ s
     (c) $x(t) = \cos \omega t,\ t \geq 0$ when $T = 2\pi/\omega$ s

From Equations 4.362 and 4.363,

(a) $$x_k = \sin 3kT \Leftrightarrow \frac{(\sin 3 \cdot \pi/6)z}{z^2 - (2 \cos 3 \cdot \pi/6)z + 1} = \frac{(\sin \pi/2)z}{z^2 - (2 \cos \pi/2)z + 1} = \frac{z}{z^2 + 1} \tag{4.364}$$

(b) $$x_k = \sin 3kT \Leftrightarrow \frac{(\sin 3 \cdot \pi/3)z}{z^2 - (2 \cos 3 \cdot \pi/3)z + 1} = \frac{(\sin \pi)z}{z^2 - (2 \cos \pi)z + 1} = 0 \tag{4.365}$$

(c) $$\cos k\omega T \Leftrightarrow \frac{z(z - \cos \omega \cdot 2\pi/\omega)}{z^2 - (2 \cos \omega \cdot 2\pi/\omega)z + 1} = \frac{z(z - \cos 2\pi)}{z^2 - (2 \cos 2\pi)z + 1}$$
$$= \frac{z(z - 1)}{z^2 - 2z + 1} = \frac{z}{z - 1} \tag{4.366}$$

Figure 4.37 shows the continuous-time signal $x(t) = \sin 3t,\ t \geq 0$ and the discrete-time signals $x_k = \sin 3kT,\ k = 0, 1, 2, 3, \ldots$ resulting from sampling in parts (a) and (b).

    Note, in part (a), the frequency of sampling $\omega_s = 2\pi/T = 12$ rad/s is four times the frequency of the signal $x(t)$. The result given in Equation 4.364 is easily verified by long division of $z^2 + 1$ into z giving the infinite series

$$U(z) = \frac{z}{z^2 + 1} z^{-1} - z^{-3} + z^{-5} - z^{-7} + z^{-9} - z^{11} + \cdots \tag{4.367}$$

$$\Rightarrow u_k = \begin{cases} 0, & k = 0, 2, 4, 6, \ldots \\ 1, & k = 1, 5, 9, \ldots \\ -1, & k = 3, 7, 11, \ldots \end{cases} = \begin{cases} 0, & k = 0, 2, 4, 6, \ldots \\ (-1)^{(k+3)/2}, & k = 1, 3, 5, 7, \ldots \end{cases} \tag{4.368}$$

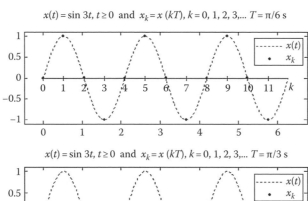

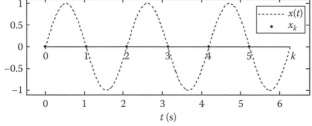

FIGURE 4.37   Uniform sampling of $x(t) = \sin 3t$ ($T = \pi/6$ s and $T = \pi/3$ s).

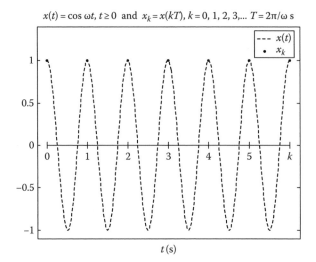

FIGURE 4.38 Uniform sampling of cos ωt ($T = 2\pi/\omega$ s).

At the slower sampling frequency of 6 rad/s in part (b), the discrete-time signal is identically zero for all k. In part (c), the cosine function is sampled once per cycle resulting in the discrete-time unit step function shown in Figure 4.38.

Table 4.4 is a brief listing of elementary continuous-time functions and their Laplace transforms along with the discrete-time signals resulting from uniform sampling of the continuous-time signals and the corresponding z-transforms (Jacquot 1981).

TABLE 4.4
Table of Laplace and z-Transforms

$f(t), t \geq 0$	$F(s) = L\{f(t)\}$	$f_k = f(kT), k = 0, 1, 2, \ldots$	$F(z) = z\{f_k\}$
1	$\dfrac{1}{s}$	1	$\dfrac{z}{z-1}$
T	$\dfrac{1}{s^2}$	KT	$\dfrac{Tz}{(z-1)^2}$
e^{-at}	$\dfrac{1}{s+a}$	e^{-akT}	$\dfrac{z}{z-e^{-aT}}$
te^{-at}	$\dfrac{1}{(s+a)^2}$	kTe^{-akT}	$\dfrac{Te^{-aT}z}{(z-e^{-aT})^2}$
$\sin \omega t$	$\dfrac{\omega}{s^2+\omega^2}$	$\sin k\omega T$	$\dfrac{(\sin \omega T)z}{z^2 - 2(\cos \omega T)z + 1}$
$\cos \omega t$	$\dfrac{s}{s^2+\omega^2}$	$\cos k\omega T$	$\dfrac{z^2 - (\cos \omega T)z}{z^2 - 2(\cos \omega T)z + 1}$
$e^{-at}\sin \omega t$	$\dfrac{\omega}{(s+a)^2+\omega^2}$	$e^{-akT}\sin k\omega T$	$\dfrac{(e^{-aT}\sin \omega T)z}{z^2 - 2(e^{-aT}\cos \omega T)z + e^{-2at}}$
$e^{-at}\cos \omega t$	$\dfrac{s+a}{(s+a)^2+\omega^2}$	$e^{-akT}\cos k\omega T$	$\dfrac{z^2 - (e^{-aT}\cos \omega T)z}{z^2 - 2(e^{-aT}\cos \omega T)z + e^{-2aT}}$

4.6.1 DISCRETE-TIME IMPULSE FUNCTION

We now introduce a discrete-time function, which plays a prominent role in analyzing the behavior of linear discrete-time systems. The unit strength discrete-time impulse occurring at discrete-time $k = 0$ is defined by

$$\delta_k = \begin{cases} 1, & k = 0 \\ 0, & k = 1, 2, 3, \ldots \end{cases} \tag{4.369}$$

Delaying the discrete-time impulse by n units of discrete-time produces

$$\delta_{k-n} = \begin{cases} 1, & k = n \\ 0, & k = 0, 1, 2, \ldots, n-1, n+1, \ldots \end{cases} \tag{4.370}$$

It follows directly from the definition of the z-transform that

$$\mathfrak{z}\{\delta_k\} = 1 \quad \text{and} \quad \mathfrak{z}\{\delta_{k-n}\} = z^{-n} \tag{4.371}$$

An arbitrary discrete-time signal f_k, $k = 0$, 1, 2,$\ldots$ can be expressed as a weighted sum of unit discrete-time impulses, that is,

$$f_k = \sum_{i=0}^{\infty} f_i \delta_{k-i} = f_0 \delta_k + f_1 \delta_{k-1} + f_2 \delta_{k-2} + f_3 \delta_{k-3} + \quad k = 0, 1, 2, 3, \ldots \tag{4.372}$$

The output of a linear discrete-time system subject to a unit discrete-time impulse is termed the unit impulse response. Just like in the case of continuous-time systems, the discrete-time impulse response reflects the natural dynamics of the system. This will be demonstrated after the z-domain transfer function is introduced.

Example 4.17

Represent the discrete-time signal u_k, $k = 0$, 1, 2, 3,$\ldots$ shown in Figure 4.39 in terms of discrete-time impulses and find $U(z)$.

$$u_k = \begin{cases} 0, & k = 0, 1, 2, 6, 7, \ldots \\ 1, & k = 3, 5 \\ 2, & k = 4 \end{cases} \tag{4.373}$$

From Equation 4.372,

$$u_k = 1 \cdot \delta_{k-3} + 2 \cdot \delta_{k-4} + 1 \cdot \delta_{k-5} \tag{4.374}$$

$$U(z) = \mathfrak{z}\{u_k\} = \mathfrak{z}\{\delta_{k-3} + 2\delta_{k-4} + \delta_{k-5}\} = z^{-3} + 2z^{-4} + z^{-5} = \frac{z^2 + 2z + 1}{z^5} \tag{4.375}$$

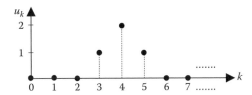

FIGURE 4.39   Graph of discrete-time signal u_k, $k = 0$, 1, 2, 3,$\ldots$.

Note in Equation 4.375 we employed the linearity property of z-transforms, that is,

$$\mathfrak{z}\{\delta_{k-3} + 2\delta_{k-4} + \delta_{k-5}\} = \mathfrak{z}\{\delta_{k-3}\} + 2\mathfrak{z}\{\delta_{k-4}\} + \mathfrak{z}\{\delta_{k-5}\} \qquad (4.376)$$

In the general case,

$$\mathfrak{z}\{au_k + by_k\} = \sum_{k=0}^{\infty}(au_k + by_k)z^{-k} = a\sum_{k=0}^{\infty}u_k z^{-k} + b\sum_{k=0}^{\infty}yk z^{-k} = aU(z) + bY(z) \qquad (4.377)$$

Other useful properties (analogous to those of the Laplace transform) of the z-transform are included in Table 4.5.

The "delay" property is especially important. Suppose a discrete-time signal u_k for which $u_k = 0$ when $k < 0$ is delayed n units of discrete-time. The delayed signal, denoted u_{k-n}, is expressed in terms of u_k in Table 4.5. The case where $n = 1$ and 2 along with the general case is illustrated in Figure 4.40a through d.

The unit-delay operator, as the name suggests, delays its input by one unit of discrete-time. The symbol for a unit-delay operator is a block with z^{-1} inside. If the input to a unit-delay operator is the discrete-time signal u_k shown in Figure 4.40a, the output would be u_{k-1} in Figure 4.40b. A pair of unit-delay operators in series is shown in Figure 4.41.

The outputs x_k and y_k are related to the input u_k by

$$x_k = u_{k-1} = \begin{cases} 0, & k = 0 \\ u_0, & k = 1 \\ u_1, & k = 2 \\ \cdots & \cdots \end{cases} \qquad (4.378)$$

TABLE 4.5
Useful Properties of the z-Transform

Description	Discrete-Time Signal	Property
Linearity	$u_k = ax_k + by_k$	$U(z) = aX(z) + bY(z)$
Delay (right shifting)	Given u_k, $k = 0, 1, 2, \ldots$, where $u_k = 0$ for $k < 0$	$\mathfrak{z}\{u_{k-n}\} = z^{-n}U(z)$
	$u_{k-n} = \begin{cases} 0, & k = 0, 1, 2, \ldots, n-1 \\ u_0, & k = n \\ u_1, & k = n+1 \\ u_2, & k = n+2 \\ \text{etc.} \end{cases}$	
Summation	$y_k = \sum_{i=0}^{k} u_i$	$Y(z) = \dfrac{z}{z-1}U(z)$
Multiplication by geometric sequence	$y_k = a^k u_k$	$Y(z) = U\left(\dfrac{z}{a}\right)$
Multiplication by k property	$y_k = ku_k$	$Y(z) = -z\dfrac{d}{dz}U(z)$
Initial value property	$f_k = \sum_{i=0}^{\infty} f_i \delta_{k-i}, \quad k = 0, 1, 2, 3, \ldots$	$f_0 = \lim_{|z| \to \infty} F(z)$
Final value property	$f_k = \sum_{i=0}^{\infty} f_i \delta_{k-i}, \quad k = 0, 1, 2, 3, \ldots$	$f_\infty = \lim_{|z| \to 1}(z-1)F(z)$
Periodic signal	$f_k = f_{k+n}, \quad k = 0, 1, 2, 3, \ldots$	$F(z) = \dfrac{z^n}{z^n - 1}\hat{F}(z)$ where $\hat{F}(z) = \sum_{k=0}^{n-1} f_k z^{-k}$

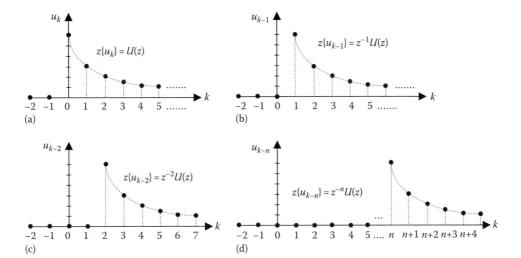

FIGURE 4.40 Illustration of the delay property in Table 4.5.

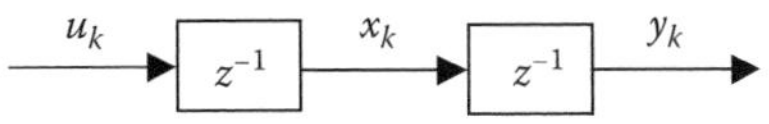

$$\xrightarrow{u_k}\ \boxed{z^{-1}}\ \xrightarrow{x_k}\ \boxed{z^{-1}}\ \xrightarrow{y_k}$$

FIGURE 4.41 Unit-delay operators in series.

$$y_k = x_{k-1} = u_{k-2} = \begin{cases} 0, & k = 0, 1 \\ u_0, & k = 2 \\ u_1, & k = 3 \\ \ldots & \ldots \end{cases} \tag{4.379}$$

In a later section when we introduce simulation diagrams for discrete-time systems, it will be apparent that the unit delay is the counterpart to a continuous-time integrator in the simulation diagram of continuous-time systems.

Several examples illustrating the properties in Table 4.5.

Example 4.18

A unit alternating sequence ($a = -1$ in Figure 4.36) is the input to a summer as shown in Figure 4.42.

(a) Find the output y_k, $k = 0, 1, 2, 3, \ldots$.
(b) Find $Y(z)$.

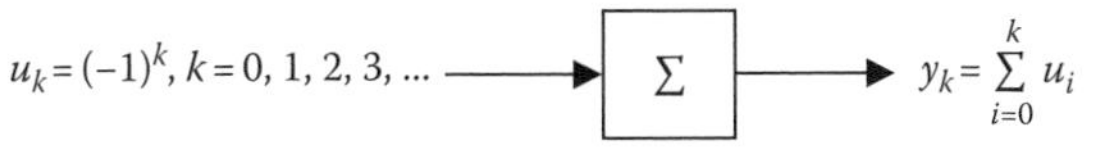

$$u_k = (-1)^k,\ k = 0, 1, 2, 3, \ldots\ \longrightarrow\ \boxed{\Sigma}\ \longrightarrow\ y_k = \sum_{i=0}^{k} u_i$$

FIGURE 4.42 A summer with a unit alternating sequence input.

(a) Referring to the graphs of the geometric sequence in Figure 4.36 for the case when $a = -1$, it is apparent that the output of the summer is

$$
y_k = \begin{cases} 1, & k = 0, 2, 4, \ldots \\ 0, & k = 1, 3, 5, \ldots \end{cases} \tag{4.380}
$$

(b) From the definition of the z-transform as an infinite series in z^{-1},

$$
Y(z) = 1 + 1 \cdot z^{-2} + 1 \cdot z^{-4} + 1 \cdot z^{-6} + \cdots \tag{4.381}
$$

$$
= 1 + (z^{-2}) + (z^{-2})^2 + (z^{-2})^3 + \cdots \tag{4.382}
$$

$$
= \frac{1}{1 - (z^{-2})} \tag{4.383}
$$

$$
= \frac{z^2}{z^2 - 1} \tag{4.384}
$$

Alternatively, from the summation property in Table 4.5 and knowing $\mathfrak{z}\{a^k\} = z/(z - a)$,

$$
Y(z) = \frac{z}{z - 1} U(z) = \frac{z}{z - 1} \left(\frac{z}{z + 1} \right) = \frac{z^2}{z^2 - 1} \tag{4.385}
$$

Example 4.19

Find the z-transform of the discrete-time signal resulting from sampling the output of a half-wave rectifier whose input is the continuous-time function $\sin \omega_0 t$. Sampling starts at $t = 0$ at a frequency of $8\omega_0$, where $\omega_0 = 2\pi$ rad/s.

The output of the half-wave rectifier is

$$
v(t) = \begin{cases} \sin \omega_0 t, & k\pi/\omega_0 \leq t \leq (k+1)\pi/\omega_0 & \text{for } k = 0, 2, 4, \ldots \\ 0, & (k+1)\pi/\omega_0 \leq t \leq (k+2)\pi/\omega_0 & \text{for } k = 1, 3, 5, \ldots \end{cases} \tag{4.386}
$$

Both $v(t)$ and $v_k = v(kT)$, $k = 0, 1, 2, 3, \ldots$ are shown in Figure 4.43. The discrete-time signal v_k is periodic, and the period is $n = 8$, that is, $v_{k+8} = v_k$, $k = 0, 1, 2, 3, \ldots$.

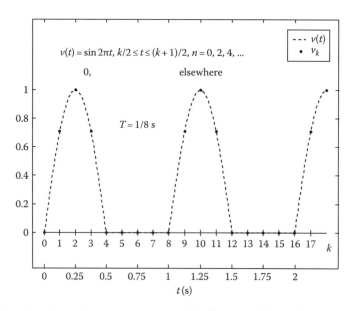

FIGURE 4.43 Sampling the continuous-time output of a half-wave rectifier with input $\sin 2\pi t$.

The z-transform of the first cycle of v_k is

$$\hat{V}(z) = \sum_{k=0}^{7} v_k z^{-k} = \sum_{k=0}^{3} \sin\,(2\pi kT)z^{-k} + \sum_{k=4}^{7} 0 \cdot z^{-k} = \sum_{K=0}^{3} \sin\left(\frac{k\pi}{4}\right)z^{-k} \tag{4.387}$$

$$= 0 + \sin\left(\frac{\pi}{4}\right)z^{-1} + \sin\left(\frac{\pi}{2}\right)z^{-2} + \sin\left(\frac{3\pi}{4}\right)z^{-3} \tag{4.388}$$

$$= 0 + \frac{\sqrt{2}}{2}z^{-1} + z^{-2} + \frac{\sqrt{2}}{2}z^{-3} \tag{4.389}$$

Applying the property in Table 4.5 for periodic signals gives

$$V(z) = \frac{z^n}{z^n - 1}\hat{V}(z) = \frac{z^8}{z^8 - 1}\left(\frac{\sqrt{2}}{2}z^{-1} + z^{-2} + \frac{\sqrt{2}}{2}z^{-3}\right) \tag{4.390}$$

$$= \frac{z^5}{z^8 - 1}\left(\frac{\sqrt{2}}{2}z^2 + z + \frac{\sqrt{2}}{2}\right) \tag{4.391}$$

Long division of $z^8 - 1$ into $(\sqrt{2}/2)z^7 + z^6 + (\sqrt{2}/2)z^5$ will generate a power series in z^{-1} with coefficients corresponding to the sampled values shown in Figure 4.43.

4.6.2 Inverse z-Transform

The analysis of discrete-time system dynamics requires the capability of inverting a z-transform $F(z)$ to find the discrete-time signal f_k, $k = 0, 1, 2, 3, \ldots$. It is similar to the way in which the inverse Laplace transform was obtained, that is, by exploiting the basic properties of the z-transform, referring to tables of z-transform pairs, partial fraction expansion, and using one additional method not applicable to continuous-time systems, namely, long division. A simple example of finding the inverse z-transform based on some of the methods outlined above follows.

Example 4.20

Find the inverse z-transform of

$$F(z) = \frac{z + 1}{z(z + 2)} \tag{4.392}$$

(a) Using properties of the z-transform along with the lookup table of z-transform pairs.
(b) By the method of long division.

(a)
$$F(z) = \frac{z + 1}{z(z + 2)} = z^{-1}\left(\frac{z + 1}{z + 2}\right) \tag{4.393}$$

$$= z^{-1}\left(\frac{z}{z + 2}\right) + z^{-2}\left(\frac{z}{z + 2}\right) \tag{4.394}$$

From Table 4.4, the term $(z/(z + 2))$ is the z-transform of the discrete-time signal $g_k = (-2)^k$, $k = 0$, $1, 2, 3, \ldots$. From the delay property in Table 4.5, f_k is the sum of g_k delayed one unit of time and g_k delayed two units of discrete-time. Denoting the delayed signals by $\tilde{g}_{k,1}$ and $\tilde{g}_{k,2}$, we can write

$$f_k = \tilde{g}_{k,1} + \tilde{g}_{k,2} \quad k = 0, 1, 2, 3, \ldots \tag{4.395}$$

where

$$\tilde{g}_{k,1} = \begin{cases} 0, & k = 0 \\ (-2)^{k-1}, & k = 1, 2, 3, \ldots \end{cases} \tag{4.396}$$

$$\tilde{g}_{k,2} = \begin{cases} 0, & k = 0, 1 \\ (-2)^{k-2}, & k = 2, 3, 4, \ldots \end{cases} \tag{4.397}$$

Combining Equations 4.395 and 4.396, the inverse z-transform is

$$f_k = \begin{cases} 0, & k = 0 \\ 1, & k = 1 \\ (-2)^{k-1} + (-2)^{k-2}, & k = 2, 3, 4, \ldots \end{cases} \tag{4.398}$$

Simplifying the expression in Equation 4.398 when $k = 2, 3, 4, \ldots$ gives

$$f_k = \begin{cases} 0, & k = 0 \\ 1, & k = 1 \\ -(-2)^{k-2}, & k = 2, 3, 4, \ldots \end{cases} \tag{4.399}$$

(b) Long division of the denominator in Equation 4.392 into the numerator results in an infinite series. The first few terms are

$$\frac{z+1}{z^2+2z} = z^{-1} - z^{-2} + 2z^{-3} - 4z^{-4} + 8z^{-5} - \cdots \tag{4.400}$$

Looking at Equation 4.400, it is possible to recognize a pattern in the coefficients starting with the z^{-2} term. This pattern results in the expression in the third line of the general solution in Equation 4.399. The reader should verify that Equations 4.399 and 4.400 generate identical values for f_k, $k = 0, 1, 2, 3, \ldots$ as they must.

4.6.3 PARTIAL FRACTION EXPANSION

Causal signals, that is, discrete-time signals f_k that are identically zero for negative values of discrete-time k, possess z-transforms of the form

$$F(z) = \frac{N(z)}{D(z)} = \frac{b_0 z^n + b_1 z^{n-1} + \cdots + b_m z^{n-m}}{z^n + a_1 z^{n-1} + \cdots + a_{n-1} z + a_n} \quad (n \geq m) \tag{4.401}$$

The partial fraction expansion of $F(z)$ depends on the nature of the roots of $D(z)$. Equation 4.401 is rewritten with the denominator $D(z)$ in factored form,

$$F(z) = \frac{b_0 z^n + b_1 z^{n-1} + \cdots + b_m z^{n-m}}{(z - p_1)(z - p_2) \cdots (z - p_n)} \quad (n \geq m) \tag{4.402}$$

where $p_1, p_2, \ldots, p_n$ are the poles of $F(z)$. Three cases are considered for finding the inverse z-transform of $F(z)$ by partial fractions.

Case I: Poles of $F(z)$ are real and distinct
When the poles $p_1, p_2, \ldots, p_n$ are real and unequal, $F(z)$ in partial fraction form is

$$F(z) = c_0 + c_1 \left(\frac{z}{z - p_1}\right) + c_2 \left(\frac{z}{z - p_2}\right) + \cdots + c_n \left(\frac{z}{z - p_n}\right) \tag{4.403}$$

The constant c_0 is easily determined by substituting $z=0$ in Equations 4.402 and 4.403.

$$c_0 = F(z)|_{z=0} = F(0) = \begin{cases} 0, & n > m \\ \dfrac{b_n}{(-p_1)(-p_2)\cdots(-p_n)}, & n = m \end{cases} \tag{4.404}$$

The remaining coefficients $c_1, c_2, \ldots, c_n$ are obtained from (Cadzow 1973)

$$c_i = \left(\frac{z - p_i}{z}\right) F(z) \bigg|_{z=p_i}, \quad i = 1, 2, 3, \ldots, n \tag{4.405}$$

From the z-transform pairs $\delta_k \Leftrightarrow 1$, $a^k \Leftrightarrow z/(z-a)$ and the linearity property of the z-transform, the inverse z-transform of $F(z)$ in Equation 4.403 is

$$f_k = c_0 \delta_k + c_1 p_1^k + c_2 p_2^k + \cdots + c_n p_n^k, \quad k = 0, 1, 2, 3, \ldots \tag{4.406}$$

Example 4.21

Find the discrete-time signal with z-transform given by

$$F(z) = \frac{z^2 + z + 1}{z^2 - 4} \tag{4.407}$$

Factoring the denominator leads to the partial fraction expansion

$$\frac{z^2 + z + 1}{z^2 - 4} = \frac{z^2 + z + 1}{(z - 2)(z + 2)} = c_0 + c_1 \left(\frac{z}{z - 2}\right) + c_2 \left(\frac{z}{z + 2}\right) \tag{4.408}$$

where

$$c_0 = F(0) = \frac{z^2 + z + 1}{(z - 2)(z + 2)} \bigg|_{z=0} = -\frac{1}{4} \tag{4.409}$$

$$c_1 = \left(\frac{z - 2}{z}\right) F(z) \bigg|_{z=2} = \frac{z^2 + z + 1}{z(z + 2)} \bigg|_{z=2} = \frac{7}{8} \tag{4.410}$$

$$c_2 = \left(\frac{z + 2}{z}\right) F(z) \bigg|_{z=-2} = \frac{z^2 + z + 1}{z(z - 2)} \bigg|_{z=-2} = \frac{3}{8} \tag{4.411}$$

$$\Rightarrow F(z) = -\frac{1}{4} + \frac{7}{8} \left(\frac{z}{z - 2}\right) + \frac{3}{8} \left(\frac{z}{z + 2}\right) \tag{4.412}$$

$$\Rightarrow f_k = -\frac{1}{4} \delta_k + \frac{7}{8} (2)^k + \frac{3}{8} (-2)^k, \quad k = 0, 1, 2, 3, \ldots \tag{4.413}$$

It is left as an exercise to show that the first several values of f_k, $k = 0$, 1, 2, 3, ... are in agreement with the values obtained by long division of $z^2 - 4$ into $z^2 + z + 1$.

If the denominator $D(z)$ in Equation 4.401 has a factor z^p, the inverse z-transform of $z^p F(z)$ should be determined first, followed by use of the delay property to obtain the final result. To illustrate, suppose $F(z)$ is given by

$$F(z) = \frac{z^2 + z + 1}{z^3 (z^2 - 4)} \tag{4.414}$$

We start by inverting $z^3 F(z)$,

$$z^3 F(z) = \frac{z^2 + z + 1}{z^2 - 4} \tag{4.415}$$

From Example 4.21, we know

$$\mathfrak{z}^{-1}\left\{\frac{z^2 + z + 1}{z^2 - 4}\right\} = -\frac{1}{4}\delta k + \frac{7}{8}(2)^k + \frac{3}{8}(-2)^k \tag{4.416}$$

Hence, the inverse z-transform of $F(z)$ in Equation 4.414 is the discrete-time signal in Equation 4.416 delayed three units of discrete-time, that is,

$$f_k = \begin{cases} 0, & k = 0, 1, 2, \\ -\dfrac{1}{4}\delta_{k-3} + \dfrac{7}{8}(2)^{k-3} + \dfrac{3}{8}(-2)^{k-3}, & k = 3, 4, 5, \ldots \end{cases} \tag{4.417}$$

Case II: Repeated real poles of $F(z)$

Suppose the pole p_1 has multiplicity m_1. The partial fraction expansion contains the m_1 terms

$$c_1\left(\frac{z}{z - p_1}\right) + \cdots + c_{m_1-1}\left(\frac{z}{z - p_1}\right)^{m_1-1} + c_{m_1}\left(\frac{z}{z - p_1}\right)^{m_1} \tag{4.418}$$

associated with the factor $(z - p_1)^{m1}$ in the denominator of $F(z)$. Simultaneous equations are developed for the constants $c_1, c_2, \ldots, c_{m_1-1}$. An illustrative example follows.

Example 4.22

Find f_k, $k = 0, 1, 2, 3, \ldots$ when

$$F(z) = \frac{2z^2 + z}{(z - 1)^3(z + 1)} \tag{4.419}$$

The partial fraction expansion of $F(z)$ is

$$F(z) = \frac{2z^2 + z}{(z - 1)^3(z + 1)} = c_0 + c_1\left(\frac{z}{z - 1}\right)c_2\left(\frac{z}{z - 1}\right)^2 + c_3\left(\frac{z}{z - 1}\right)^3 + c_4\left(\frac{z}{z + 1}\right) \tag{4.420}$$

The constants c_0 and c_4 are obtained as they would in Case I, that is,

$$c_0 = F(0) = 0 \tag{4.421}$$

$$c_4 = \left(\frac{z + 1}{z}\right)F(z)\bigg|_{z=-1} = \frac{2z + 1}{(z - 1)^3}\bigg|_{z=-1} = \frac{1}{8} \tag{4.422}$$

The coefficient of the highest order term is evaluated directly from

$$c_3 = \left(\frac{z - 1}{z}\right)^3 F(z)\bigg|_{z=1} = \frac{2z + 1}{z^2(z + 1)}\bigg|_{z=1} = \frac{3}{2} \tag{4.423}$$

Substituting the values for c_0, c_3, and c_4 into Equation 4.420 yields

$$F(z) = \frac{2z^2 + z}{(z-1)^3(z+1)} = c_1\left(\frac{z}{z-1}\right) + c_2\left(\frac{z}{z-1}\right)^2 + \frac{3}{2}\left(\frac{z}{z-1}\right)^3 + \frac{1}{8}\left(\frac{z}{z+1}\right) \tag{4.424}$$

Combining the terms on the right-hand side of Equation 4.424 into a single term with common denominator $(z-1)^3(z+1)$ and then equating the numerators give

$$2z^2 + z = c_1 z(z-1)^2(z+1) + c_2 z^2(z-1)(z+1) + \frac{3}{2}z^3(z+1) + \frac{1}{8}z(z-1)^3 \tag{4.425}$$

Expanding the right-hand side of Equation 4.425 and equating coefficients of like powers of z on both sides lead to

$$\begin{cases} z^4 : 0 = c_1 + c_2 + \dfrac{3}{2} + \dfrac{1}{8} \\[2mm] z^3 : 0 = -c_1 + \dfrac{3}{2} - \dfrac{3}{8} \\[2mm] z^2 : 2 = -c_1 - c_2 + \dfrac{3}{8} \\[2mm] z : 1 = c_1 - \dfrac{1}{8} \end{cases} \tag{4.426}$$

Selecting two of the above equations for simultaneous solution results in $c_1 = 9/8$ and $c_2 = -11/4$. Substituting the known values for c_1 and c_2 in Equation 4.424 gives

$$F(z) = \frac{9}{8}\left(\frac{z}{z-1}\right) - \frac{11}{4}\left(\frac{z}{z-1}\right)^2 + \frac{3}{2}\left(\frac{z}{z-1}\right)^3 + \frac{1}{8}\left(\frac{z}{z+1}\right) \tag{4.427}$$

Inverting $F(z)$ is accomplished using Table 4.6 (Cadzow 1973)

$$f_k = \frac{9}{8} - \frac{11}{4}(k+1) + \frac{3}{2}\left[\frac{(k+1)(k+2)}{2}\right] + \frac{1}{8}(-1)^k, \quad k = 0, 1, 2, 3, \ldots \tag{4.428}$$

TABLE 4.6
Table for Inverting z-Transforms of the
Form $[z/(z-a)]^n$, $n = 1, 2, 3, \ldots$

$F(z)$	f_k, $k = 0, 1, 2, 3, \ldots$
$\dfrac{z}{(z-a)}$	a^k
$\left[\dfrac{z}{(z-a)}\right]^2$	$(k+1)a^k$
$\left[\dfrac{z}{(z-a)}\right]^3$	$\dfrac{(k+1)(k+2)}{2}a^k$
$\left[\dfrac{z}{(z-a)}\right]^4$	$\dfrac{(k+1)(k+2)(k+3)}{3}a^k$
$\left[\dfrac{z}{(z-a)}\right]^5$	$\dfrac{(k+1)(k+2)(k+3)(k+4)}{4}a^k$

Evaluating the first several values of f_k gives

$$f_0 = \frac{9}{8} - \frac{11}{4} + \frac{3}{2} + \frac{1}{8} = 0, \quad f_1 = \frac{9}{8} - \frac{22}{4} + \frac{18}{4} - \frac{1}{8} = 0, \quad f_2 = \frac{9}{8} - \frac{33}{4} + \frac{36}{4} + \frac{1}{8} = 2$$

$$f_3 = \frac{9}{8} - \frac{44}{4} + \frac{60}{4} - \frac{1}{8} = 5, \quad f_4 = \frac{9}{8} - \frac{55}{4} + \frac{90}{4} + \frac{1}{8} = 10, \quad f_5 = \frac{9}{8} - \frac{66}{4} + \frac{126}{4} - \frac{1}{8} = 16$$

Checking the above by long division confirms the numerical values above.

$$
\begin{array}{r}
2z^{-2} + 5z^{-3} + 10z^{-4} + 16z^{-5} \\
\hline
z^4 - 2z^3 + 0z^2 + 2z - 1\,\big|\,2z^2 + z \\
\end{array}
$$

$$
\begin{aligned}
&2z^2 - 4z + 0 \quad + 4z^{-1} \quad - 2z^{-2} \\
&\underline{\ 5z + 0 \quad\ \ - 4z^{-1} \quad + 2z^{-2}} \\
&\ 5z - 10 \quad\ + 0z^{-1} \quad + 10z^{-2} \quad - 5z^{-3} \\
&\underline{\ 10 \quad - 4z^{-1} \quad\ - 8z^{-2} \quad + 5z^{-3}} \\
&\ 10 \quad - 20z^{-1} \quad - 0z^{-1} \quad + 20z^{-3} - 10z^{-4} \\
&\underline{\ 16z^{-1} \quad - 8z^{-2} \quad\ - 15z^{-3} + 10z^{-4}}
\end{aligned}
$$

Case III: Complex poles of $F(z)$

When $F(z)$ possesses complex poles, the partial fraction expansion is dictated by the last two z-transform pairs in Table 4.4. An example serves to illustrate the procedure.

$$F(z) = \frac{z^2 + z}{(z - 1)(z^2 - 3z + 9)} \tag{4.429}$$

The first step is to decompose $F(z)$ in two parts,

$$F(z) = \frac{Az^2 + Bz}{(z^2 - 3z + 9)} + C\left(\frac{z}{z - 1}\right) \tag{4.430}$$

The constant C is evaluated from

$$C = \left(\frac{z - 1}{z}\right)\frac{z^2 + z}{(z - 1)(z^2 - 3z + 9)}\bigg|_{z=1} = \frac{z + 1}{z^2 - 3z + 9}\bigg|_{z=1} = \frac{2}{7} \tag{4.431}$$

Constants A and B are obtained by combining the terms in Equation 4.430 into a single term with common denominator $(z - 1)(z^2 - 3z + 9)$ and then equating the numerator to $z^2 + z$ the numerator in Equation 4.429. The resulting expression for $F(z)$ is

$$F(z) = \frac{-(2/7)z^2 + (11/7)z}{z^2 - 3z + 9} + \frac{2}{7}\left(\frac{z}{z - 1}\right) \tag{4.432}$$

$$= -\frac{1}{7}\left(\frac{2z^2 - 11z}{z^2 - 3z + 9}\right) + \frac{2}{7}\left(\frac{z}{z - 1}\right) \tag{4.433}$$

The quadratic factor in the denominator of Equation 4.433 implies that inverting $F(z)$ will require a linear combination of $e^{-akT}\sin k\omega T$ and $e^{-akT}\cos k\omega T$ (see Table 4.4). Comparing the standard form of the denominator in the last row of Table 4.4 and the quadratic denominator in Equation 4.433,

$$z^2 - 2(e^{-aT}\cos\omega T)z + e^{-2aT} = z^{-2} - 3z + 9 \tag{4.434}$$

Equating like powers of z and solving for $e^{-\alpha T}$ and ωT,

$$e^{-2aT} = 9 \Rightarrow e^{-aT} = 3 \tag{4.435}$$

$$-2(e^{-aT} \cos \omega T) = -3 \Rightarrow \cos(\omega T) = \frac{1}{2} \Rightarrow \omega T = \frac{\pi}{3} \tag{4.436}$$

The quadratic numerator in $F(z)$ in Equation 4.433 must be expressed as a linear combination of the standard numerator forms in the last two rows of Table 4.4, that is,

$$2z^2 - 11z = c_1(e^{-\alpha T} \sin \omega T)z + c_2[z^2 - (e^{-\alpha T} \cos \omega T)z] \tag{4.437}$$

Solving for c_1 and c_2 in Equation 4.437 leads to $c_1 = -16\sqrt{3}/9$, $c_2 = 2$. $F(z)$ is now written in a form where Table 4.4 can be used to find f_k.

$$F(z) = -\frac{1}{7}\left\{\frac{c_1(e^{-\alpha T} \sin \omega T)z}{z^2 - 2(e^{-\alpha T} \cos \omega T)z + e^{-2\alpha T}} + \frac{c_2[z^2 - (e^{-\alpha T} \cos \omega T)z]}{z^2 - 2(e^{-\alpha T} \cos \omega T)z + e^{-2\alpha T}}\right\} + \frac{2}{7}\left(\frac{z}{z-1}\right) \tag{4.438}$$

$$f(k) - \frac{1}{7}[c_1 e^{-\alpha kT} \sin k\omega T + c_2 e^{-\alpha T} \cos k\omega T] + \frac{2}{7}, \quad k = 0, 1, 2, 3, \ldots \tag{4.439}$$

$$= -\frac{1}{7}\left[\frac{-16\sqrt{3}}{9}(3)^k \sin\left(\frac{k\pi}{3}\right) + 2(3)^k \cos\left(\frac{k\pi}{3}\right)\right] + \frac{2}{7}, \quad k = 0, 1, 2, 3, \ldots \tag{4.440}$$

By observation of $F(z)$ in Equation 4.429, it follows that $f_0 = 0$ and $f_1 = 1$. The reader can readily verify these values from Equation 4.440 with $k = 0, 1$.

An alternative approach when $F(z)$ contains complex poles is to proceed the same way as in Case 1 where all the poles were real and distinct. The key is appropriate conversion between rectangular and polar representations of the complex roots of $F(z)$ and the complex coefficients arising from partial fraction expansion.

Suppose $F(z)$ is of the form

$$F(z) = \frac{N(z)}{D(z)} = \frac{N(z)}{(z - p_1)(z - p_2)} \tag{4.441}$$

where the complex poles expressed in polar form are $p_1 = Re^{j\theta}$, $p_2 = Re^{-j\theta}$.

Expanding $F(z)$ as we did in Case 1 (real and distinct poles),

$$F(z) = A_1\left(\frac{z}{z - p_1}\right) + A_2\left(\frac{z}{z - p_2}\right) \tag{4.442}$$

where

$$A_1 = \frac{(z - p_1)}{z}\left[\frac{N(z)}{(z - p_1)(z - p_2)}\right]_{z=p_1} = \frac{N(p_1)}{p_1(p_1 - p_2)} \tag{4.443}$$

and A_2 is the conjugate of A_1. In polar form, $A_1 = Ce^{j\phi}$, $A_2 = Ce^{-j\phi}$ Equation 4.442 becomes

$$F(z) = Ce^{j\phi}\left(\frac{z}{z - Re^{j\theta}}\right) + Ce^{-j\phi}\left(\frac{z}{z - Re^{-j\theta}}\right) \tag{4.444}$$

The inverse z-transform of $F(z)$ in Equation 4.444 is

$$f_k = Ce^{j\phi}(R\,e^{j\theta})^k + Ce^{-j\phi}(Re^{-j\theta})^k \tag{4.445}$$

$$= CR^k[e^{j\phi}(e^{j\theta})^k + e^{-j\phi}(e^{-j\theta})^k] \tag{4.446}$$

$$= 2CR^k\left[\frac{e^{j(k\theta+\phi)} + e^{-j(k\theta+\phi)}}{2}\right] \tag{4.447}$$

$$= 2CR^k \cos(k\theta + \phi), \quad k = 0, 1, 2, \ldots \tag{4.448}$$

Thus, $F(z)$ in Equation 4.441 can be inverted simply by finding polar coordinates of the poles p_1, p_2, and complex coefficients A_1, A_2.

Example 4.23

Find the inverse z-transform of $F(z) = z^2 - z/z^2 - 0.6z + 0.25$.

Factoring the denominator to find the poles p_1 and p_2,

$$F(z) = \frac{z^2 - z}{z^2 - 0.6z + 0.25} = \frac{z^2 - z}{(z - p_1)(z - p_2)}, \quad p_{1,2} = 0.3 \pm j0.4 \tag{4.449}$$

Converting the complex poles to polar form gives

$$p_{1,2} = 0.3 \pm j0.4 = Re^{\pm j\theta} \tag{4.450}$$

where

$$R = [(0.3)^2 + (0.4)^2]^{1/2} = 0.5$$

$$\theta = \tan^{-1}\left(\frac{4}{3}\right) = 0.9273 \text{ rad}$$

From Equation 4.443, the constant A_1 in the partial fraction expansion of $F(z)$ is

$$A_1 = \frac{N(p_1)}{p_1(p_1 - p_2)} = \frac{p_1^2 - p_1}{p_1(p_1 - p_2)} = \frac{p_1 - 1}{p_1 - p_2}$$

$$= \frac{0.3 + j0.4 - 1}{0.3 + j0.4 - (0.3 - j0.4)}$$

$$= \frac{1}{2} + j\frac{7}{8} \tag{4.451}$$

Converting A_1 polar form,

$$C = |A_1| = \left|\frac{1}{2} + j\frac{7}{8}\right| = \frac{\sqrt{65}}{8} \tag{4.452}$$

$$\phi = \text{Arg}(A_1) = \text{Arg}\left(\frac{1}{2} + j\frac{7}{8}\right) = \tan^{-1}\left(\frac{7}{4}\right) = 1.0517 \text{ rad} \tag{4.453}$$

From Equation 4.448, the discrete-time signal f_k is

$$f_k = 2\frac{\sqrt{65}}{8}(0.5)^k \cos(0.9273k + 1.0517), \quad k = 0, 1, 2, \ldots \tag{4.454}$$

$$= 2.0156(0.5)^k \cos(0.9273k + 1.0517), \quad k = 0, 1, 2, \ldots \tag{4.455}$$

The reader should check that the first several values of f_k obtained from Equation 4.455 agree with the numerical values obtained by long division of the denominator $z^2 - 0.6z + 0.25$ of $F(z)$ into the numerator $z^2 - z$.

EXERCISES

4.36 Find the z-transforms of the following causal sequences f_k, $k = 0$, 1, 2, 3,.... Use long division to check the first two nonzero values of f_k.

(a) ka^k (b) $k^2(-1)^k$ (c) $\delta_k + (0.5)^k$ (d) $\sin k\pi$ (e) $(-1)^k \cos(2k\pi/3)$

(f) $(k+1)\delta_{k-1}$ (g) $f_k = \begin{cases} 0, & k = 0, 2, 4, 6, \ldots \\ 1, & k = 0, 1, 3, 5, \ldots \end{cases}$ (h) $f_k = \begin{cases} 0, & k = 0, 1, 2, 3, \ldots \\ k, & k = 4, 5, 6, \ldots \end{cases}$

(i) $f_k = \begin{cases} 1, & k = 0 \\ 0, & k = 0, 1, 3, 5, 7, \ldots \\ \dfrac{1}{2}(-2)^{k/2}, & k = 2, 4, 6, \ldots \end{cases}$

4.37 Find the inverse z-transform of the expressions below. Use long division to check the first two nonzero values of f_k.

(a) $\dfrac{z+a}{z+b}$ (b) $\dfrac{z^2+1}{z^2(z^2-1)}$ (c) $\dfrac{z^2}{(z-3)^3}$ (d) $\dfrac{z^2+1}{(z+1)^2}$

(e) $\dfrac{z^3+z}{(z^2-1)^2}$ (f) $\dfrac{z^2+1}{z^3+z^2}$ (g) $\dfrac{z+2}{z^2-z+4}$ (h) $\dfrac{z(z-2)}{z^2-z+(3/4)}$

(i) $\dfrac{z^4}{z^4-1}$ (j) $\dfrac{z^2+1}{z^2+2}$ (k) $\dfrac{z+1}{z(z^2+z+2)}$

4.38 Find the z-transforms of the discrete-time signals resulting from uniform sampling of the continuous-time functions below. All functions are zero for $t < 0$.

(a) $1 + 2t$ (b) te^{-2t} (c) $e^{-at} - e^{-bt}$ (d) $t^2 \sin 2t$ (e) $e^{-2t} \cos t$ (f) $1/2^t$

4.39 (a) Find the z-transforms $U(z)$ and $F(z)$ of the discrete-time signals pictured in Figure E4.39.

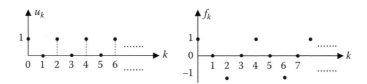

FIGURE E4.39

(b) Express the signal u_k in Figure E4.39 as $u_k = \alpha + \beta(-1)^k$, $k = 0$, 1, 2, 3, ... and determine α and β. Use the linearity property and the z-transforms of the unit step and unit alternating sequence, that is, $\mathfrak{z}\{\hat{u}k\} = z/(z-1)$ and $\mathfrak{z}\{(-1)^k\} = z/(z+1)$, to find $U(z)$.

4.40 Graph the discrete-time signals f_k, $k = 0, 1, 2, 3, \ldots$ below and find $F(z)$.

(a) $f_k = \sum_{i=0}^{\infty} \delta_{k-i}$ (b) $f_k = \sum_{i=0}^{k} 1$ (c) $f_k = \sum_{i=0}^{k} i$ (d) $f_k = \sum_{i=0}^{k} (-1)^i$

4.41 Use the two methods discussed for inverting z-transforms with complex poles to find the discrete-time signal f_k, $k = 0, 1, 2, \ldots$ with z-transform

$$F(z) = \frac{3z^2 + z}{(z-1)(z^2 + 2z + 2)}$$

4.42 Write a MATLAB function to invert

$$F(z) = \frac{b_3 z^3 + b_2 z^2 + b_1 z + b_0}{(z - p_1)(z^2 + a_1 z + a_0)}$$

where the quadratic term $z^2 + a_1 z + a_0 = (z - p_2)(z - p_3)$, $p_2, p_3 = \alpha \pm j\beta$.
The inverse z-transform is given by

$$f_k = F_0 \delta_k + A_1 (p_1)^k + 2CR^k \cos(k\theta + \phi), \quad k = 0, 1, 2, 3, \ldots$$

The function input parameters are p_1, a_1, a_0, b_3, b_2, b_1, and b_0 and the outputs are A_1, C, R, θ, ϕ, and F_0. The function declaration line is

```
[A1, C, R, theta, phi, F0] = invert(p1,a1,a0,b3,b2,b1,b0)
```

Check the function by running it for

$$(i) \quad F(z) = \frac{3z^2 + z}{(z-1)(z^2 + 2z + 2)}$$

$$(ii) \quad F(z) = \frac{2z^3 + z^2 + 4z + 5}{(z-3)(z^2 + 2z + 4)}$$

and comparing the first several values of f_k, $k = 0, 1, 2, \ldots$ with the values obtained by long division of the cubic denominator into the quadratic numerator.

4.7 *z*-DOMAIN TRANSFER FUNCTION

We have seen how the transfer function of a linear continuous-time system is used to find the system's response to elementary inputs. Stability and frequency response characteristics of the system can be inferred from the transfer function as well. A discrete-time system transfer function does the same for linear discrete-time systems. We begin with the nth-order, linear, constant coefficient difference equation

$$y_k + a_1 y_{k-1} + \cdots + a_n y_{k-n} = b_0 u_{k-1} + \cdots + b_m u_{k-m}, \quad n \geq m \tag{4.456}$$

z-Transforming both sides and applying the linearity property gives

$$\mathfrak{z}\{y_k\} + a_1 \mathfrak{z}\{y_{k-1}\} + \cdots + a_n \mathfrak{z}\{y_{k-n}\} = b_0 \mathfrak{z}\{u_k\} + b_1 \mathfrak{z}\{u_{k-1}\} + \cdots + b_m \mathfrak{z}\{u_{k-m}\} \tag{4.457}$$

Assuming the input is applied at $k = 0$ and the initial values $y_{-1}, y_{-2}, \ldots, y_{-n}$ are zero, we can use the delay property in Table 4.5 in the previous section to arrive at

$$Y(z) + a_1 z^{-1} Y(z) + \cdots + a_n z^{-n} Y(z) = b_0 U(z) + b_1 z^{-1} U(z) + \cdots + b_m z^{-m} U(z) \tag{4.458}$$

The z-domain transfer function is defined as the ratio of $Y(z)$ to $U(z)$. Thus,

$$H(z) = \frac{Y(z)}{U(z)} = \frac{b_0 + b_1 z^{-1} + \cdots + b_m z^{-m}}{1 + a_1 z^{-1} + \cdots + a_{n-1} z^{-n+1} + a_n z^{-n}}, \quad n \geq m \tag{4.459}$$

$$= \frac{b_0 z^n + b_1 z^{n-1} + \cdots + b_m z^{n-m}}{z^n + a_1 z^{n-1} + \cdots + a_{n-1} z + a_n}, \quad n \geq m \tag{4.460}$$

Depending on the application, one of the two forms given in Equations 4.459 and 4.460 for the transfer function, also called the pulse transfer function, is usually preferable. A good example to illustrate how to find the z-domain transfer function of a discrete-time system is an Euler integrator. Recall from Section 3.3 that the difference equation for approximating a continuous-time integrator using explicit Euler integration is

$$x_A(n+1) = x_A(n) + Tu(n), \quad n = 0, 1, 2, \ldots \tag{4.461}$$

where $u(n)$ and $x_A(n)$ are the discrete-time input and outputs and $x_A(0) = 0$. Employing the notation of this chapter, the difference equation is written

$$x_k - x_{k-1} = Tu_{k-1}, \quad k = 1, 2, 3, \ldots \tag{4.462}$$

where $u_k = u(kT)$, $k = 0, 1, 2, \ldots$ are sampled values of the input signal and x_k, $k = 0, 1, 2, \ldots$ is the discrete-time output intended to approximate the continuous-time integrator output $x(t)$ at the end of each integration step. The initial condition is $x_0 = x(0) = 0$ and the first computed value is x_1.

z-transforming Equation 4.462,

$$\mathfrak{z}\{x_k\} - \mathfrak{z}\{x_{k-1}\} = T\mathfrak{z}\{u_{k-1}\} \tag{4.463}$$

Since x_k and u_k are both zero for $k < 0$, the delay property in Table 4.5 applies.

$$X(z) - z^{-1}X(z) = Tz^{-1}U(z) \tag{4.464}$$

$$H(z) = \frac{X(z)}{U(z)} = \frac{Tz^{-1}}{1 - z^{-1}} = \frac{T}{z - 1} \tag{4.465}$$

Example 4.24

The input to a continuous-time integrator is $u(t) = \sin \pi t$.

(a) Approximate the output $x(t)$ using Euler integration with step size $T = 0.1$ s.
(b) Find the exact solution $x(t)$ and plot on the same graph with x_k.

(a) Solving for $X(z)$ in Equation 4.465 and looking up $U(z) = z\{u_k\} = z\{\sin k\omega T\}$ from Table 4.4 give

$$X(z) = H(z)U(z) = \frac{T}{z - 1} \left[\frac{(\sin \omega T)z}{z^2 - 2(\cos \omega T)z + 1} \right] \Bigg|_{T=0.1, \omega = \pi} \tag{4.466}$$

$$= \frac{0.1}{z - 1} \left[\frac{(\sin 0.1\pi)z}{z^2 - 2(\cos 0.1\pi)z + 1} \right] \tag{4.467}$$

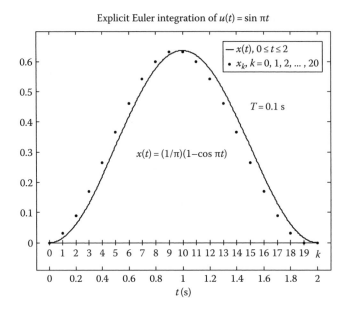

FIGURE 4.44 Continuous- and discrete-time (explicit Euler) integrator outputs.

Using the method of partial fraction expansion presented in Section 4.4.6, the inverse z-transform of $X(z)$ is (details are left as an exercise)

$$x_k = 0.05 \left[\frac{\sin 0.1\pi}{1 - \cos 0.1\pi} (1 - \cos 0.1k\pi) - \sin 0.1k\pi \right], \quad k = 0, 1, 2, 3, \ldots \qquad (4.468)$$

(b) The continuous-time integrator output is obtained by integration of the input $u(t)$,

$$x(t) = \int_0^t u(\lambda)d\lambda = \int_0^t \sin \pi\lambda \, d\lambda = \frac{1}{\pi}(1 - \cos \pi t) \qquad (4.469)$$

The discrete-time signal x_k and the continuous-time integrator output $x(t)$ are plotted in Figure 4.44 for one cycle of the input.

4.7.1 Nonzero Initial Conditions

Using the z-transform to solve a difference equation with nonzero initial conditions requires additional terms to account for the nonzero values. Suppose y_k is a discrete-time signal for which $y_{-1} \neq 0$ like the one shown in Figure 4.45. Also shown is y_{k-1}.

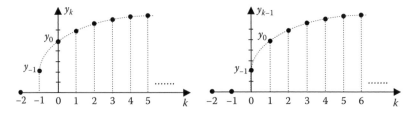

FIGURE 4.45 Discrete-time signals y_k and $y_{k-1}(y_{-1} \neq 0)$.

$3\{y_{k-1}\}$ is obtained from the basic definition of the z-transform, that is,

$$3\{y_{k-1}\} = \sum_{k=0}^{\infty} y_{k-1} z^{-k} = y_{-1} + y_0 z^{-1} + y_1 z^{-2} + y_2 z^{-3} + \cdots \tag{4.470}$$

$$= y_{-1} + z^{-1}[y_0 + y_1 z^{-1} + y_2 z^{-2} + \cdots] \tag{4.471}$$

$$= y_{-1} + z^{-1} Y(z) \tag{4.472}$$

Consider the first-order difference equation

$$y_k = \beta u_k + \alpha y_{k-1}, \quad k = 0, 1, 2, 3, \ldots \tag{4.473}$$

with input u_k applied at $k=0$ and nonzero initial condition y_{-1}. It will be shown later that Equation 4.473 is the difference equation of a low-pass digital filter. z-transforming Equation 4.473 and using the result in Equation 4.472 give

$$Y(z) = \beta U(z) + \alpha[y_{-1} + z^{-1} Y(z)] \tag{4.474}$$

Multiplying Equation 4.474 by z and solving for $Y(z)$ give

$$Y(z) = \frac{\beta z}{z - \alpha} U(z) + \alpha y_{-1}\left(\frac{z}{z - \alpha}\right) \tag{4.475}$$

The first term on the right-hand side of Equation 4.475 is $H(z)\,U(z)$. The additional term results from the nonzero initial condition. A similar procedure is employed for higher order difference equations with several nonzero initial conditions.

Example 4.25

For the discrete-time system described by Equation 4.473,

(a) Find the response to a unit step when the initial condition $y_{-1} \neq 0$.
(b) Find the response to a unit alternating input when $y_{-1} \neq 0$.
(c) For $\alpha = 0.9$ and $\beta = 0.1$, graph the responses in parts (a) and (b) when $y_{-1} = 2$.

(a) $u_k = 1$, $k = 0, 1, 2, 3, \ldots$, and $U(z) = z/(z-1)$.

$$Y(z) = \frac{\beta z}{z - \alpha}\left(\frac{z}{z - 1}\right) + \alpha y_{-1}\left(\frac{z}{z - \alpha}\right) \tag{4.476}$$

$$= \frac{\beta}{1 - \alpha}\left[\frac{z}{z - 1} - \alpha \frac{z}{z - \alpha}\right] + \alpha y_{-1}\left(\frac{z}{z - \alpha}\right) \tag{4.477}$$

$$y_k = \frac{\beta}{1 - \alpha}(1 - \alpha^{k+1}) + y_{-1}\alpha^{k+1}, \quad k = 0, 1, 2, \ldots \tag{4.478}$$

(b) $u_k = (-1)^k$, $k = 0, 1, 2, 3, \ldots$, and $U(z) = z/(z+1)$.

$$Y(z) = \frac{\beta z}{z - \alpha}\left(\frac{z}{z + 1}\right) + \alpha y_{-1}\left(\frac{z}{z - \alpha}\right) \tag{4.479}$$

$$= \frac{\beta}{1 + \alpha}\left[\frac{z}{z + 1} + \alpha \frac{z}{z - \alpha}\right] + \alpha y_{-1}\left(\frac{z}{z - \alpha}\right) \tag{4.480}$$

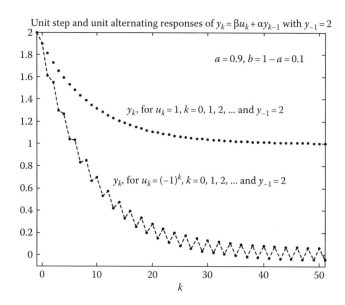

Unit step and unit alternating responses of $y_k = \beta u_k + \alpha y_{k-1}$ with $y_{-1} = 2$

$a = 0.9, b = 1 - a = 0.1$

y_k, for $u_k = 1, k = 0, 1, 2, \ldots$ and $y_{-1} = 2$

y_k, for $u_k = (-1)^k, k = 0, 1, 2, \ldots$ and $y_{-1} = 2$

FIGURE 4.46 Responses of discrete-time system with nonzero initial condition.

$$yk = \frac{\beta}{1 + \alpha}[(-1)^k + \alpha^{k+1}] + y_{-1}\alpha^{k+1}, \quad k = 0, 1, 2, \ldots \qquad (4.481)$$

Note that the solutions in Equations 4.478 and 4.481 reduce to the given initial condition y_{-1} for $k = -1$.

(c) Graphs of y_k, $k = -1, 0, 1, 2, \ldots$ in Equations 4.478 and 4.481 are shown in Figure 4.46.

Note how the system passes the low-frequency unit step and effectively blocks the higher frequency unit alternating sequence once the transient component α^{k+1} dies out. Setting $\beta = 1 - \alpha$, the normalized unit step and unit alternating steady-state responses are

$$(y_k)_{ss} = 1 \quad \text{for } u_k = 1, \ k = 0, 1, 2, 3, \ldots \qquad (4.482)$$

$$(y_k)_{ss} = \frac{1 - \alpha}{1 + \alpha}(-1)^k = \frac{1 - 0.9}{1 - 0.9}(-1)^k = \frac{1}{19}(-1)^k \quad \text{for } u_k = (-1)^k, \ k = 0, 1, 2, 3, \ldots \qquad (4.483)$$

4.7.2 Approximating Continuous-Time System Transfer Functions

It is common practice to start with a block diagram representation of a continuous-time system and transform it to a block diagram of a discrete-time system with comparable dynamics. The discrete-time signals are intended to approximate the corresponding signals in the continuous-time system at discrete points in time. To illustrate, suppose we have a need to approximate the behavior of a second-order system

$$H(s) = \frac{Y(s)}{U(s)} = \frac{K\omega_n^2}{s^2 + 2\zeta\omega_n s + \omega_n^2} \qquad (4.484)$$

A simulation diagram is shown in Figure 4.47.

The continuous-time integrator blocks with transfer function $H_1(s) = 1/s$ are replaced by discrete-time (numerical) integrators with z-domain transfer functions $H_1(z)$, and the signals become discrete-time in nature (see Figure 4.48).

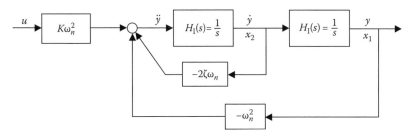

FIGURE 4.47 Simulation diagram for second-order system in Equation 4.484.

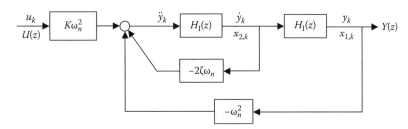

FIGURE 4.48 Discrete-time system with numerical integrator z-domain transfer functions.

Block diagram reduction or any other suitable method, for example, Mason's Gain Formula (Dorf and Bishop 2005), using signal flow graphs or solution of simultaneous equations results in the pulse transfer function of the discrete-time system given in Equation 4.485.

$$H(z) = \frac{Y(z)}{U(z)} = \frac{K\omega_n^2 H_I^2(z)}{\omega_n^2 H_I^2(z) + 2\zeta\omega_n H_1(z) + 1} \qquad (4.485)$$

Choosing $H_I(z)$ as the z-domain transfer function for an explicit Euler integrator (see Equation 4.465) gives

$$H(z) = \frac{Y(z)}{U(z)} = \frac{K\omega_n^2 (T/(z-1))^2}{\omega_n^2 (T/(z-1))^2 + 2\zeta\omega_n (T/(z-1)) + 1} \qquad (4.486)$$

Simplifying the above expression yields

$$H(z) = \frac{Y(z)}{U(z)} = \frac{K(\omega_n T)^2}{z^2 - 2(1 - \zeta\omega_n T)z + 1 - 2\zeta\omega_n T + (\omega_n T)^2} \qquad (4.487)$$

The difference equation for the discrete-time system is obtained directly from the z-domain transfer function expressed in terms of negative power of z.

$$\frac{Y(z)}{U(z)} = \frac{K(\omega_n T)^2 z^{-2}}{1 - 2(1 - \zeta\omega_n T)z^{-1} + [1 - 2\zeta\omega_n T + (\omega_n T)^2]z^{-2}} \qquad (4.488)$$

$$\Rightarrow Y(z) - 2(1 - \zeta\omega_n T)z^{-1}Y(z) + [1 - 2\zeta\omega_n T + (\omega_n T)^2]z^{-2}Y(z) = K(\omega_n T)^2 z^{-2}U(z) \qquad (4.489)$$

$$\Rightarrow Y_{k-2}(1 - \zeta\omega_n T)yk - 1 + [1 - 2\zeta\omega_n T + (\omega_n T)^2]y_{k-2} = K(\omega_n T)^2 u_{k-2} \qquad (4.490)$$

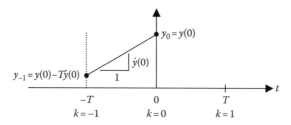

FIGURE 4.49   Initial conditions y_0, y_{-1} obtained from $y(0)$ and $\dot{y}(0)$.

Assuming the initial conditions are y_{-1} and y_0, the discrete-time variable k in Equation 4.490 assumes the values $k = 1, 2, 3, \ldots$. The first computed value is y_1.

Initial conditions in the discrete-time system model are based on the initial conditions for the continuous-time system, $y(0)$ and $\dot{y}(0)$. Figure 4.49 illustrates a derivation for y_{-1} using backward extrapolation from the point $y(0)$ along the line with slope $\dot{y}(0)$. Note the dependence on T in the result for y_{-1}. A similar approach is used to extrapolate y_1 when the initial conditions are y_0 and y_1. The first computed value is y_2. What is the starting value for k in Equation 4.490?

Example 4.26

Consider a second-order system with parameters $K = 1$, $\omega_n = 2$ rad/s, and $\zeta = 0.5$.

(a) Using explicit Euler integration with step size $T = 0.025$ s, find a difference equation that can be solved recursively to approximate the unit step response of the continuous-time system.
(b) Find the analytical solution for the step response of the continuous-time system.
(c) Plot the continuous- and discrete-time responses on the same graph.

(a) A recursive solution for y_k, $k = 1, 2, 3, \ldots$ is obtained from Equation 4.490 as follows.

$$y_k = 2(1 - \zeta\omega_n T)y_{k-1} - [1 - 2\zeta\omega_n T + (\omega_n T)^2]y_{k-2} + K(\omega_n T)^2 u_{k-2}, \quad k = 1,2,3,\ldots \quad (4.491)$$

$$\Rightarrow y_k = 1.95y_{k-1} - 0.9525y_{k-2} + 0.0025u_{k-2}, \quad k = 1,2,3,\ldots \quad (4.492)$$

(b) The continuous-time step response can be obtained from the transfer function of the second-order system by inverse Laplace transformation of $Y(s) = H(s)U(s)$. Alternatively, we can use Equation 2.23 or 2.24 for the step response of an underdamped second-order system. Adopting the latter approach,

$$y(t) = K\left[1 - e^{-\zeta\omega_n t}\left(\cos \omega_d t + \frac{\zeta\omega_n}{\omega_d}\sin \omega_d t\right)\right], \quad t \geq 0 \quad \left(\omega_d = \sqrt{1 - \zeta^2}\,\omega_n\right) \quad (4.493)$$

$$\Rightarrow y(t) = 1 - e^{-t}\left(\cos \sqrt{3}t + \frac{1}{\sqrt{3}}\sin \sqrt{3}t\right) \quad (4.494)$$

(c) Graphs of the solution to Equations 4.492 (every other point) and 4.494 are plotted in Figure 4.50, and selected values are presented in Table 4.7 for comparison (see MATLAB M-file "Chap4_Ex7_3.m").

From the numerical values in Table 4.7, it appears that the discrete- and continuous-time transient responses are in agreement to one place after the decimal point. Greater accuracy requires we reduce the step size or consider a more accurate numerical integrator like the ones discussed in Chapter 3.

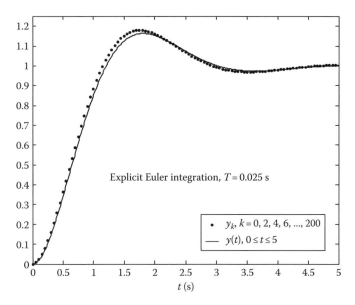

FIGURE 4.50 Continuous- and discrete-time (Euler integration) second-order system step responses ($K = 1$, $\omega_n = 2$ rad/s, $\zeta = 0.5$).

TABLE 4.7
Continuous- and Discrete-Time
(Euler Integration) Responses

K	y_k	t_k	$y(t_k)$
0	0	0	0
10	01170	0.25	0.1044
20	0.3643	0.5	0.3403
30	0.6425	0.75	0.6105
40	0.8845	1.0	0.8494
50	1.0562	1.25	1.0234
60	1.1506	1.5	1.1244
70	1.1787	1.75	1.1616
80	1.1605	2.0	1.1531
90	1.1174	2.25	1.1184
100	1.0677	2.5	1.0746

A general approach to deriving the z-domain transfer function $H(z)$ for a discrete-time system intended to approximate a linear continuous-time system with transfer function $H(s)$ is now given. Starting with a simulation diagram of the continuous-time system, each integrator block with transfer function $H_I(s) = 1/s$ is replaced by a discrete-time transfer function block $H_I(z)$ corresponding to a specific numerical integrator. For example, replacing $H_I(s)$ by $H_I(z)$ for explicit Euler integration,

$$\frac{1}{s} \leftarrow H_I(z) = \frac{T}{z-1} \Rightarrow s \leftarrow \frac{1}{H_I(z)} = \frac{z-1}{T} \tag{4.495}$$

Hence, when explicit Euler integration is used to approximate the continuous-time integrators in an LTI system with transfer function $H(s)$, the z-domain transfer function of the discrete-time system is obtained by replacing s in $H(s)$ with $(z-1)/T$. That is,

$$H(z) = H(s)|_{s \leftarrow (z-1)/T} \tag{4.496}$$

For the continuous-time second-order system of Equation 4.484,

$$H(z) = \frac{K\omega_n^2}{s^2 + 2\zeta\omega_n s + \omega_n^2}\bigg|_{s \leftarrow (z-1)/T} = \frac{K\omega_n^2}{((z-1)/T)^2 + 2\zeta\omega_n((z-1)/T) + \omega_n^2} \tag{4.497}$$

Simplifying Equation 4.497 results in Equation 4.487.

Example 4.27

Use trapezoidal integration in place of explicit Euler to approximate the unit step response of the second-order system in Example 4.26.

Approximating a continuous-time integrator with input $u(t)$ and output $y(t)$ using trapezoidal integration results in (see Equation 3.40)

$$y_k = y_{k-1} + \frac{T}{2}[u_{k-1} + u_k] \tag{4.498}$$

z-transforming Equation 4.498,

$$Y(z) - z^{-1}Y(z) = \frac{T}{2}[z^{-1}U(z) + U(z)] \tag{4.499}$$

and solving for $H_I(z)$ give

$$H_I(z) = \frac{Y(z)}{U(z)} = \frac{T}{2}\left[\frac{1 + z^{-1}}{1 - z^{-1}}\right] \tag{4.500}$$

$$= \frac{T}{2}\left[\frac{z + 1}{z - 1}\right] \tag{4.501}$$

The z-domain transfer function of the discrete-time system is therefore

$$H(z) = H(s)|_{s \leftarrow \frac{1}{H_I(z)}} = \frac{K\omega_n^2}{s^2 + 2\zeta\omega_n s + \omega_n^2}\bigg|_{s \leftarrow (2/T)((z-1)/(z+1))} \tag{4.502}$$

Replacing s by $(2/T)((z-1)/(z+1))$ in Equation 4.502 and simplifying result in

$$H(z) = \frac{K(\omega_n T)^2(z^2 + 2z + 1)}{[4(1 + \zeta\omega_n T) + (\omega_n T)^2]z^2 + 2[(\omega_n T)^2 - 4]z + 4(1 - \zeta\omega_n T) + (\omega_n T)^2} \tag{4.503}$$

Multiplying the numerator and denominator of $H(z)$ in Equation 4.503 by z^{-2} leads to the difference equation of the discrete-time system,

$$[4(1 + \zeta\omega_n T) + (\omega_n T)^2]y_k + 2[(\omega_n T)^2 - 4]y_{k-1} + [4(1 - \zeta\omega_n T) + (\omega_n T)^2]y_{k-2}$$
$$= K(\omega_n T)^2(u_k + 2u_{k-1} + u_{k-2}) \tag{4.504}$$

Substituting the given values for K, ζ, and ω_n gives

$$y_k = \frac{1}{4.1025}[7.995y_{k-1} - 3.9025y_{k-2} + 0.0025(u_k + 2u_{k-1} + u_{k-2})], \quad k = 1, 2, 3, \dots \quad (4.505)$$

where $u_k = 1$, $k = 0$, 1, 2, 3, … (zero otherwise) and $y_{-1} = y_0 = 0$.

The unit step responses of the continuous- and discrete-time system approximation in Equation 4.505 are calculated in the M-file "*Chap4_Ex7_4.m*." The results are graphed in Figure 4.51 and tabulated in Table 4.8. As expected, the trapezoidal integrator is more accurate than the explicit Euler.

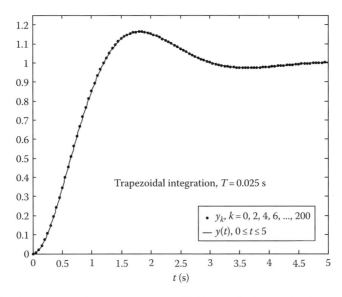

FIGURE 4.51 Continuous- and discrete-time (trapezoidal integration) second-order system step responses ($K = 1$, $\omega_n = 2$ rad/s, $\zeta = 0.5$).

TABLE 4.8
Continuous- and Discrete-Time
(Trapezoidal Integration) Responses

K	y_k	t_k	$y(t_k)$
0	0	0	0
10	0.1090	0.25	0.1044
20	0.3468	0.5	0.3403
30	0.6169	0.75	0.6105
40	0.8546	1.0	0.8494
50	1.0268	1.25	1.0234
60	1.1261	1.5	1.1244
70	1.1620	1.75	1.1616
80	1.1526	2.0	1.1531
90	1.11175	2.25	1.1184
100	1.0736	2.5	1.0746

4.7.3 SIMULATION DIAGRAMS AND STATE VARIABLES

When a discrete-time system is modeled by one or more difference equations, a simulation diagram represents a more visual description of the system's dynamics. Furthermore, a simulation diagram leads directly to an equivalent discrete-time state-space model, in much the same way a continuous-time state variable model was developed from a simulation diagram of the continuous-time system. As in the continuous-time case, the simulation diagram and state-space models of a discrete-time system are not unique.

The dynamic block in a simulation diagram representation of a continuous-time system is the integrator or $1/s$ block. In a discrete-time system, delaying y_k for one time step results in y_{k-1}. If $y_{-1} = 0$, the delay property states $\mathfrak{z}\{y_{k-1}\} = z^{-1}\mathfrak{z}\{y_k\}$. For a discrete-time system, the unit-delay block is the counterpart to the integrator block. Figure 4.52 shows several common ways of representing a unit-delay block.

A block diagram implementation of the nth-order difference Equation 4.456 is shown in Figure 4.53.

The block diagram shown in Figure 4.53 contains $n + m$ unit delays to implement the nth-order discrete-time system governed by the difference equation in Equation 4.456. Only block diagrams with the minimum number of n delays are classified as simulation diagrams. A simulation diagram serves as a convenient way of identifying the discrete-time states in much the same way continuous-time simulation diagrams were used to define the continuous-time states. The discrete-time states $x_{1,k}, x_{2,k}, \ldots, x_{n,k}$ are chosen as the outputs of the n unit delays. As in the case of continuous-time systems, the simulation diagram and, hence, the states are not unique.

When the past input terms $u_{k-1}, u_{k-2}, \ldots, u_{k-m}$ are not present in Equation 4.456, the constants $b_1 = b_2 = \cdots = b_m = 0$ and the block diagram in Figure 4.53 reduces to a simulation diagram.

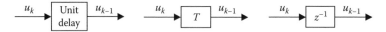

FIGURE 4.52 Graphical representation of the delay property.

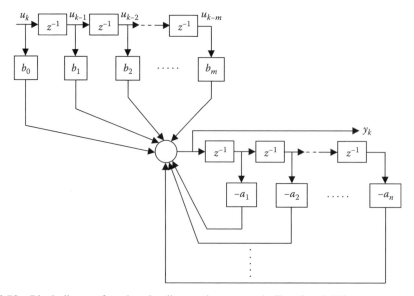

FIGURE 4.53 Block diagram for nth-order discrete-time system in Equation 4.456.

When one or more past input terms appear in the difference equation, a simulation diagram can be constructed by starting with the z-domain transfer function in Equation 4.459 expressed as

$$\frac{Y(z)}{U(z)} = \frac{Y(z)}{W(z)} \frac{W(z)}{U(z)} \tag{4.506}$$

where

$$\frac{W(z)}{U(z)} = \frac{1}{1 + a_1 z^{-1} + a_2 z^{-2} + \cdots + a_n z^{-n}} \tag{4.507}$$

$$\frac{Y(z)}{W(z)} = b_0 + b_1 z^{-1} + \cdots + b_m z^{-m} \tag{4.508}$$

Difference equations corresponding to Equations 4.507 and 4.508 are

$$w_k = u_k - a_1 w_{k-1} - a_2 w_{k-2} - \cdots - a_n w_{k-n} \tag{4.509}$$

$$y_k = b_0 w_k + b_1 w_{k-1} + \cdots + b_m w_{k-m} \tag{4.510}$$

Implementation of Equations 4.509 and 4.510 results in the simulation diagram shown in Figure 4.54. Discrete-time state equations relate the state vector at time $k+1$ to the discrete-time state vector and input vector at time k. In the single input case with the states as shown in Figure 4.54, the result is

$$\begin{cases} x_{1,k+1} = x_{2,k} \\ x_{2,k+1} = x_{3,k} \\ \vdots \\ x_{n-1,k+1} = x_{n,k} \\ x_{n,k+1} = w_k = -a_n x_{1,k} - \cdots - a_2 x_{n-1,k} - a_1 x_{n,k} + u_k \end{cases} \tag{4.511}$$

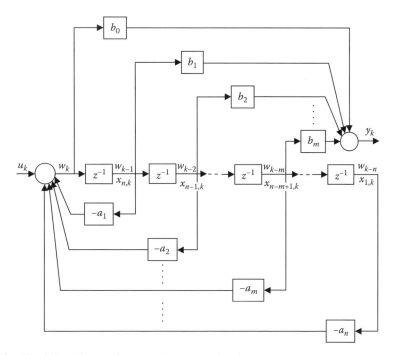

FIGURE 4.54 Simulation diagram for nth-order system showing states $x_{1,k}, x_{2,k}, \ldots, x_{n,k}$.

The output y_k is expressed in terms of the state and input according to

$$y_k = b_m x_{n-m+1,k} + \cdots + b_2 x_{n-1,k} + b_1 x_{n,k} + b_0 w_k \tag{4.512}$$

$$= b_m x_{n-m+1,k} + \cdots + b_2 x_{n-1,k} + b_1 x_{n,k} + b_0 [u_k - a_n x_{1,k} - \cdots a_2 x_{n-1,k} - a_1 x_{n,k}] \tag{4.513}$$

$$y_k = \begin{cases} -a_n b_0 x_{1,k} - a_{n-1} b_0 x_{2,k} - \cdots - a_1 b_0 x_{n,k} + b_0 u_k, & m = 0 \\ -a_n b_0 x_{1,k} - a_{n-1} b_0 x_{2,k} \cdots - a_{m+1} b_0 x_{n-m,k} \\ \quad + (b_m - a_m b_0) x_{n-m+1,k} + \cdots + (b_1 - a_1 b_0) x_{n,k} + b_0 u_k, & m = 1, \ldots, n-1 \\ (b_n - a_n b_0) x_{1,k} + (b_{n-1} - a_{n-1} b_0) x_{n-1,k} + \cdots + (b_1 - a_1 b_0) x_{n,k} + b_0 u_k, & m = n \end{cases} \tag{4.514}$$

In the general case of a linear discrete-time system with r inputs and p outputs, the discrete-time state equations are of the form

$$\underline{x}_{k+1} = A \underline{x}_k + B \underline{u}_k, \quad \underline{y}_k = C \underline{x}_k + D \underline{u}_k \tag{4.515}$$

where the system matrix A is $n \times n$, the input matrix B is $n \times r$, the output matrix C is $p \times n$, and the direct coupling matrix D is $p \times r$. For the discrete-time system described by Equations 4.511 and 4.514, the system matrix A and input matrix B are

$$A = \begin{bmatrix} 0 & 1 & 0 & \cdots & 0 & 0 \\ 0 & 0 & 1 & \cdots & 0 & 0 \\ \vdots & \vdots & \vdots & \ddots & \vdots & \vdots \\ 0 & 0 & 0 & \cdots & 0 & 1 \\ -a_n & -a_{n-1} & -a_{n-2} & \cdots & -a_2 & -a_1 \end{bmatrix}, \quad B = \begin{bmatrix} 0 \\ 0 \\ \vdots \\ 0 \\ 1 \end{bmatrix} \tag{4.516}$$

and the output matrix C and direct transmission matrix D are

$$C = \begin{cases} [-a_n b_0 & -a_{n-1} b_0 & \cdots & -a_1 b_0], & m = 0 \\ [-a_n b_0 & -a_{n-1} b_0 & \cdots & -a_{m+1} b_0 & b_m - a_m b_0 & \cdots & b_1 - a_1 b_0], & m = 1, \ldots, n-1 \\ [b_n - a_n b_0 & b_{n-1} - a_{n-1} b_0 & \cdots & b_1 - a_1 b_0], & m = n \end{cases} \tag{4.517}$$

$$D = [b_0] \tag{4.518}$$

The simulation diagram is redrawn for the case where $m = n$ in Figure 4.55.

To illustrate the use of the state equations, consider the discrete-time approximation to a second-order continuous-time system using trapezoidal integration. From Equation 4.504, the difference equation is

$$y_k + a_1 y_{k-1} + a_2 y_{k-2} = b_0 u_k + b_1 u_{k-1} + b_2 u_{k-2} \quad (m = n = 2) \tag{4.519}$$

$$a_1 = \frac{2[(\omega_n T)^2 - 4]}{4(1 + \zeta \omega_n T) + (\omega_n T)^2}, \quad a_2 = \frac{4(1 - \zeta \omega_n T) + (\omega_n T)^2}{4(1 + \zeta \omega_n T) + (\omega_n T)^2} \tag{4.520}$$

$$b_0 = \frac{K(\omega_n T)^2}{4(1 + \zeta \omega_n T) + (\omega_n T)^2}, \quad b_1 = \frac{2K(\omega_n T)^2}{4(1 + \zeta \omega_n T) + (\omega_n T)^2}, \quad b_2 = \frac{K(\omega_n T)^2}{4(1 + \zeta \omega_n T) + (\omega_n T)^2} \tag{4.521}$$

The simulation diagram is shown in Figure 4.56.

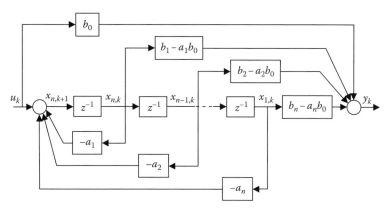

FIGURE 4.55 Simulation diagram for nth-order discrete-time system in Equation 4.456 ($m = n$).

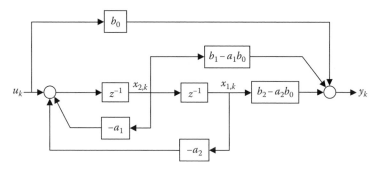

FIGURE 4.56 Simulation diagram for trapezoidal integration of second-order system.

The state equations follow directly from Figure 4.56.

$$\begin{cases} x_1, k+1 = x_2, k \\ x_2, k+1 = -a_2 x_{1,k} - a_1 x_{2,k} + u_k \end{cases} \tag{4.522}$$

$$y_k = (b_2 - a_2 b_0) x_{1,k} + (b_1 - a_1 b_0) x_{2,k} + b_0 u_k \tag{4.523}$$

and the matrices A, B, C, and D in Equation 4.515 are

$$A = \begin{bmatrix} 0 & 1 \\ -a_2 & -a_1 \end{bmatrix} = \begin{bmatrix} 0 & 1 \\ -\dfrac{4(1 - \zeta\omega_n T) + (\omega_n T)^2}{4(1 + \zeta\omega_n T) + (\omega_n T)^2} & -\dfrac{2[(\omega_n T)^2 - 4]}{4(1 + \zeta\omega_n T) + (\omega_n T)^2} \end{bmatrix}, \quad B = \begin{bmatrix} 0 \\ 1 \end{bmatrix} \tag{4.524}$$

$$C = [\,b_2 - a_2 b_0 \quad b_1 - a_1 b_0\,] = 8K \left(\frac{\omega_n T}{4(1 + \zeta\omega_n T) + (\omega_n T)^2} \right)^2 [\,\zeta\omega_n T \quad 2 + \zeta\omega_n T\,] \tag{4.525}$$

$$D = [b_0] = \left[\frac{K(\omega_n T)^2}{4(1 + \zeta\omega_n T) + (\omega_n T)^2} \right] \tag{4.526}$$

Using the same second-order system parameter values as in Examples 4.26 and 4.27, recursive solution of Equations 4.522 and 4.523 produces identical results to those shown in Figure 4.51 and Table 4.8 (see M-file "*Chap4_trapezoidal_state.m*").

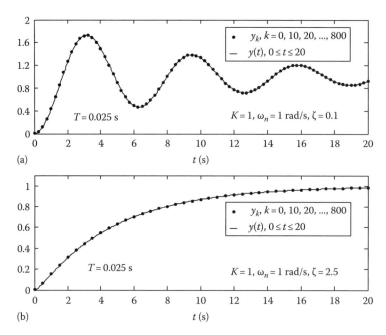

FIGURE 4.57 Trapezoidal integration of (a) light and (b) heavily damped second-order systems.

Discrete-time approximations to the step responses of two additional continuous-time second-order systems, one with light damping ($K = 1$, $\zeta = 0.1$, $\omega_n = 1$ rad/s) and the other heavily damped ($K = 1$, $\zeta = 2.5$, $\omega_n = 1$ rad/s) are shown in Figure 4.57. Results are based on recursive solution of the state equations for trapezoidal integration (see M-file "*Chap4_Fig7_14.m*"). Agreement between the exact and approximate solutions for both systems appears to be acceptable. More detailed comparisons require numerical outputs from the continuous- and discrete-time systems.

4.7.4 SOLUTION OF LINEAR DISCRETE-TIME STATE EQUATIONS

A general solution to the discrete-time state equations gives the state $\underline{x}_k$ for any value of discrete-time k without resorting to a recursive (sequential) solution. Solving for the first several values of $\underline{x}_k$ in Equation 4.515 leads to the observation

$$\underline{x}_k = A^k \underline{x}_0 + A^{k-1} B \underline{u}_0 + A^{k-2} B \underline{u}_1 + \cdots + AB \underline{u}_{k-2} + B \underline{u}_{k-1}, \quad k = 0, 1, 2, 3, \ldots \qquad (4.527)$$

$$= A^k \underline{x}_0 + \sum_{i=0}^{k-1} A^{k-i-1} B \underline{u}_i, \quad k = 0, 1, 2, 3, \ldots \qquad (4.528)$$

Equation 4.528 for the state $\underline{x}_k$ is substituted in Equation 4.515 to obtain the general solution for the output $\underline{y}_k$, $\underline{k} = 0, 1, 2, 3, \ldots$. The result is

$$\underline{y}_k = CA^k \underline{x}_0 + C \left(\sum_{i=0}^{k-1} A^{k-i-1} B \underline{u}_i \right) + D \underline{u}_k, \quad k = 0, 1, 2, 3, \ldots \qquad (4.529)$$

The discrete-time state transition matrix Φ_k is defined as

$$\Phi_k = A^k, \quad k = 0, 1, 2, 3, \ldots \tag{4.530}$$

Solutions for $\underline{x}_k$ and $\underline{y}_k$, in terms of Φ_k, are

$$\underline{x}_k = \Phi_k \underline{x}_0 + \sum_{i=0}^{k-1} \Phi_{k-i-1} B \underline{u}_i, \quad k = 0, 1, 2, 3, \ldots \tag{4.531}$$

$$\underline{y}_k = C\Phi_k \underline{x}_0 + C \sum_{i=0}^{k-1} (\Phi_{k-i-1} B \underline{u}_i) + D \underline{u}_k, \quad k = 0, 1, 2, 3, \ldots \tag{4.532}$$

Observe that an unforced system ($\underline{u}_k = 0$, $k = 0, 1, 2, 3, \ldots$) transitions from its initial state x_0 to a new state $\underline{x}_k$ at time k according to $\underline{x}_k = \Phi_k \underline{x}_0$. The discrete-time state equations and solutions are analogous to the results for continuous-time systems.

An expression for evaluating the discrete-time transition matrix can be obtained by z-transforming the first equation in Equation 4.515 resulting in

$$\mathfrak{z}\{\underline{x}_{k+1}\} = \mathfrak{z}\{A\underline{x}_k + B\underline{u}_k\} = A\underline{X}(z) + B\underline{U}(z) \tag{4.533}$$

It is left as an exercise to show that

$$\mathfrak{z}\{x_{k+1}\} = z[X(z) - x_0] \tag{4.534}$$

Combining Equations 4.533 and 4.534 gives

$$z[\underline{X}(z) - \underline{x}_0] = A\underline{X}(z) + B\underline{U}(z) \tag{4.535}$$

$$(zI - A)\underline{X}(z) = z\underline{x}_0 + B\underline{U}(z) \tag{4.536}$$

$$\underline{X}(z) = (zI - A)^{-1}[z\underline{x}_0 + B\underline{U}(z)] \tag{4.537}$$

$$\underline{x}_k = \mathfrak{z}^{-1}\{(zI - A)^{-1}(z\underline{x}_0)\} + \mathfrak{z}^{-1}\{(zI - A)^{-1} B\underline{U}(z)\} \tag{4.538}$$

Comparison of Equations 4.531 and 4.538 with $\underline{u}_k = 0$, $k = 0, 1, 2, \ldots$ implies

$$\Phi_k = \mathfrak{z}^{-1}\{\Phi(z)\} = \mathfrak{z}^{-1}\{z(zI - A)^{-1}\} \tag{4.539}$$

An example using the discrete-time state equations follows.

Example 4.28

The yearly increase in a monetary fund is a weighted sum of the increases over the prior 2 years plus an end-of-year (EOY) deposit. The fund starts with an initial amount P_0.

(a) Write the difference equation for y_k, $k = 0, 1, 2, 3, \ldots$ the fund balance at the end of the kth year. Let u_k, $k = 0, 1, 2, 3, \ldots$ be the EOY deposit in the fund. The weights are α (previous year increase) and β (increase 2 years ago).

(b) Draw a simulation diagram and convert the difference equation to state variable form.

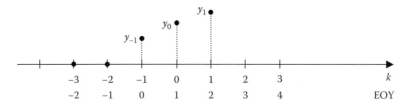

FIGURE 4.58   Relationship between discrete-time variable k and end of year.

(c) Given $\alpha = 0.5$, $\beta = 0.25$, and $P_0 = \$100$, and all EOY deposits are zero, find the components of the discrete-time state transition matrix needed to solve for y_k, $k = 0, 1, 2, 3, \ldots$.

(d) Find and plot the fund balance y_k, $k = 0, 1, 2, 3, \ldots$.

(a) The time line in Figure 4.58 shows the relationship between the discrete-time variable k and the EOY marker. Note that the initial fund amount is y_{-1}.

The difference equation for y_k, $k = 0, 1, 2, 3, \ldots$ is

$$y_k - y_{k-1} = \alpha(y_{k-1} - y_{k-2}) + \beta(y_{k-2} y_{k-3}) + u_k, \quad k = 0, 1, 2, 3, \ldots \tag{4.540}$$

The initial conditions are $y_{-1} = P_0$, $y_{-2} = y_{-3} = 0$.

Rewriting Equation 4.540 in the standard from introduced in Equation 4.456

$$y_k + a_1 y_{k-1} + a_2 y_{k-2} + a_3 y_{k-3} = b_0 u_k, \quad k = 0, 1, 2, 3, \ldots \tag{4.541}$$

where $a_1 = -(1 + \alpha)$, $a_2 = \alpha - \beta$, $a_3 = \beta$, and $b_0 = 1$.

(b) Referring to Figure 4.53 or 4.54 with $n = 3$, $m = 0$, and $b_0 = 1$, the simulation diagram reduces to Figure 4.59.

The state equations follow from the simulation diagram.

$$\begin{cases} x_{1,k+1} = x_{2,k} \\ x_{2,k+1} = x_{3,k} \\ x_{3,k+1} = -a_3 x_{1,k} - a_2 x_{2,k} - a_1 x_{3,k} + u_k \end{cases} \tag{4.542}$$

$$y_k = -a_3 x_{1,k} - a_2 x_{2,k} - a_1 x_{3,k} + u_k \tag{4.543}$$

$$A = \begin{bmatrix} 0 & 1 & 0 \\ 0 & 0 & 1 \\ -a_3 & -a_2 & -a_1 \end{bmatrix} = \begin{bmatrix} 0 & 1 & 0 \\ 0 & 0 & 1 \\ -\beta & \beta - \alpha & 1 + \alpha \end{bmatrix}, \quad B = \begin{bmatrix} 0 \\ 0 \\ 1 \end{bmatrix} \tag{4.544}$$

$$C = [-a_3 \quad -a_2 \quad -a_1] = [-\beta \quad \beta - \alpha \quad 1 + \alpha], \quad D = [1] \tag{4.545}$$

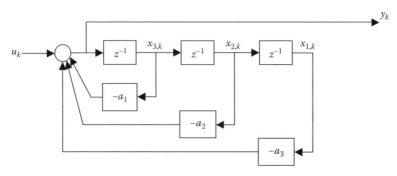

FIGURE 4.59   Simulation diagram for monetary fund example.

(c) $a_1 = -(1 + \alpha) = -(1 + 0.5) = -1.5$, $a_2 = \alpha - \beta = 0.5 - 0.25 = 0.25$, $a_3 = \beta = 0.25$

$$zI - A = z \begin{bmatrix} 1 & 0 & 0 \\ 0 & 1 & 0 \\ 0 & 0 & 1 \end{bmatrix} - \begin{bmatrix} 0 & 1 & 0 \\ 0 & 0 & 1 \\ -a_3 & -a_2 & -a_1 \end{bmatrix} \tag{4.546}$$

$$= \begin{bmatrix} z & -1 & 0 \\ 0 & z & -1 \\ a_3 & a_2 & z + a_1 \end{bmatrix} = \begin{bmatrix} z & -1 & 0 \\ 0 & z & -1 \\ 0.25 & 0.25 & z - 1.5 \end{bmatrix} \tag{4.547}$$

$$\Phi(z) = z(zI - A)^{-1} \tag{4.548}$$

Inverting $(zI - A)$ followed by multiplication by z results in

$$\Phi(z) = \frac{z}{z^3 - 1.5z^2 + 0.25z + 0.25} \begin{bmatrix} z^2 - 1.5z + 0.25 & z - 1.5 & 1 \\ -0.25 & z(z - 1.5) & z \\ -0.25z & -0.25(z + 1) & z^2 \end{bmatrix} \tag{4.549}$$

From Equation 4.532, with $\underline{u}_k = \underline{0}$, $k = 0, 1, 2, \ldots$ the solution for y_k is

$$y_k = C\Phi_k \underline{x}_0 \tag{4.550}$$

where the initial state

$$\underline{x}_0 = \begin{bmatrix} y_{-3} \\ y_{-2} \\ y_{-1} \end{bmatrix} = \begin{bmatrix} 0 \\ 0 \\ P_0 \end{bmatrix}$$

The transition matrix Φ_k is obtained by inverse z-transforming $\Phi(z)$ in Equation 4.549. The last column of Φ_k is all that is necessary to determine y_k as a result of the zeros in the first and second rows of $\underline{x}_0$. The last column of Φ_k comprises

$$(\Phi_k)_{1,3} = \mathfrak{z}^{-1}\{\Phi_{1,1}(z)\} = \mathfrak{z}^{-1}\left\{\frac{z}{z^3 - 1.5z^2 + 0.25z + 0.25}\right\} \tag{4.551}$$

$$(\Phi_k)_{2,3} = \mathfrak{z}^{-1}\{\Phi_{2,1}(z)\} = \mathfrak{z}^{-1}\left\{\frac{z^2}{z^3 - 1.5z^2 + 0.25z + 0.25}\right\} \tag{4.552}$$

$$(\Phi_k)_{3,3} = \mathfrak{z}^{-1}\{\Phi_{3,1}(z)\} = \mathfrak{z}^{-1}\left\{\frac{z^3}{z^3 - 1.5z^2 + 0.25z + 0.25}\right\} \tag{4.553}$$

The roots of $z^3 - 1.5z^2 + 0.25z + 0.25 = 0$ are $p_1 = 1$, $p_2 = 0.8090$, $p_3 = -0.3090$. $(\Phi_k)_{1,3}$, $(\Phi_k)_{2,3}$, $(\Phi_k)_{3,3}$ are linear combinations of the geometric sequences $(p_1)^k$, $p_2{}^k$, $(p_3)^k$, that is,

$$(\Phi_k)_{1,3} = A_1(p_1)^k + A_2(p_2)^k + A_3(p_3)^k \tag{4.554}$$

$$(\Phi_k)_{2,3} = B_1(p_1)^k + B_2(p_2)^k + B_3(p_3)^k \tag{4.555}$$

$$(\Phi_k)_{3,3} = C_1(p_1)^k + C_2(p_2)^k + C_3(p_3)^k \tag{4.556}$$

The partial fraction expansion coefficients are evaluated in M-file "*Chap4_Ex7_5.m*." The results are

$$A_1 = 4, \quad A_2 = -4.6833, \quad A_3 = 0.6833$$

$$B_1 = 4, \quad B_2 = -3.7889, \quad B_3 = -0.2111$$

$$C_1 = 4, \quad C_2 = -3.065, \quad C_3 = 0.0652$$

(d) From Equations 4.545 and 4.550, the fund balance is

$$y_k = [-a_3 \quad -a_2 \quad -a_1] \begin{bmatrix} (\Phi_k)_{1,3} \\ (\Phi_k)_{2,3} \\ (\Phi_k)_{3,3} \end{bmatrix} P_0 \tag{4.557}$$

$$= [-\beta \quad \beta - \alpha \quad 1 + \alpha] \begin{bmatrix} A_1(p_1)^k + A_2(p_2)^k + A_3(p_3)^k \\ B_1(p_1)^k + B_2(p_2)^k + B_3(p_3)^k \\ C_1(p_1)^k + C_2(p_2)^k + C_3(p_3)^k \end{bmatrix} P_0 \tag{4.558}$$

$$= \{-\beta[A_1(p_1)^k + A_2(p_2)^k + A_3(p_3)^k] + (\beta - \alpha)[B_1(p_1)^k + B_2(p_2)^k + B_3(p_3)^k]$$

$$+ (1 + \alpha)[C_1(p_1)^k + C_2(p_2)^k + C_3(p_3)^k]\} P_0 \tag{4.559}$$

A graph of y_k, $k = -3, -2, -1, 0, 1, 2, \ldots$ is show in Figure 4.60.

The limiting value, $y_\infty = \$400$ from the root $p_1 = 1$. Since the magnitude of roots p_2 and p_3 are less than 1, it follows from Equation 4.559 at steady state that the output y_∞ is given by

$$y_\infty = \lim_{k \to \infty} y_k = [-\beta A_1 + (\beta - \alpha)B_1 + (1 + \alpha)C_1] P_0$$

$$= [-(0.25)(4) + (0.25 - 0.5)(4) + (1 + 0.5)(4)]100 = 400 \tag{4.560}$$

We will have a lot more to say about the location of these roots in the next section on stability.

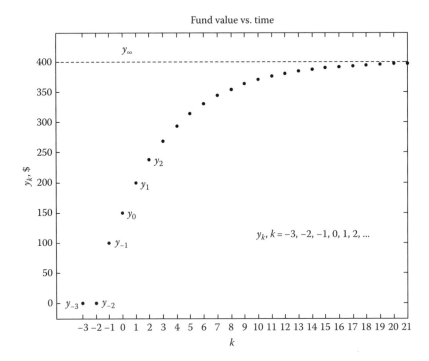

FIGURE 4.60 Discrete-time system output in Example 4.28.

The complete transition matrix is left as an exercise problem. However, a suitable check on the correctness of Φ_k is that it satisfies Φ_0, where I is the $n \times n$ identity matrix. This follows from Equation 4.530 with $k=0$ as well as Equation 4.531 with zero input and $k=0$. A quick glance at $\Phi(z)$ in Equation 4.549 should be enough to convince you that $\Phi(z)$ (*Hint:* only the diagonal terms of $\Phi(z)$ contain cubic polynomials in z in the numerator). Keep in mind that $\Phi_0 = I$ is necessary but not sufficient for Φ_k to be correct.

4.7.5 WEIGHTING SEQUENCE (IMPULSE RESPONSE FUNCTION)

A difference equation and a z-domain transfer function are but two of several different ways of characterizing a discrete-time system. A third approach is based on the system's impulse response function, similar to the case of continuous-time systems. Recall from our discussion of linear continuous-time systems that the response to an arbitrary input $u(t)$, $t \geq 0$ is expressible in the form of a convolution integral, that is,

$$y(t) = \int_0^t h(\tau)u(t-\tau)d\tau \tag{4.561}$$

where $h(t)$, $t \geq 0$ is the impulse response function. It is related to the continuous-time system transfer function $H(s)$ according to $h(t) = \mathcal{L}^{-1}\{H(s)\}$.

We now demonstrate the existence of a sequence, h_k, $k=0, 1, 2, 3, \ldots$ which allows us to find the forced response of a linear discrete-time system to an arbitrary input u_k, $k=0, 1, 2, 3, \ldots$ similar to the convolution integral in Equation 4.561 for linear continuous-time systems. The only restriction is that the initial conditions prior to application of the input, namely, $y_{-1}, y_{-2}, \ldots, y_{-n}$, are zero for an nth-order linear discrete-time system.

Consider the first-order system

$$y_k + a_1 y_{k-1} = b_0 u_k + b_1 u_{k-1} \tag{4.562}$$

where $y_{-1} = 0$ and the input $u_k = 0$, $k < 0$. Evaluating the first several values of y_k,

$$k = 0: \quad y_0 = b_0 u_0 \tag{4.563}$$

$$k = 1: \quad y_1 + a_1 y_0 = b_0 u_1 + b_1 u_0 \tag{4.564}$$

$$y_1 = b_0 u_1 + (b_1 - a_1 b_0) u_0 \tag{4.565}$$

$$k = 2: \quad y_2 + = a_1 y_1 = b_0 u_2 + b_1 u_1 \tag{4.566}$$

$$y_2 = b_0 u_2 + (b_1 - a_1 b_0) u_1 - a_1(b_1 - a_1 b_0) u_0 \tag{4.567}$$

$$k = 3: \quad y_3 + a_0 y_2 = b_1 u_3 + b_0 u_2 \tag{4.568}$$

$$y_3 = b_0 u_3 + (b_1 - a_1 b_0) u_2 - a_1(b_1 - a_1 b_0) u_1 + a_1^2(b_1 - a_1 b_0) u_0 \tag{4.569}$$

By induction, a general solution for y_k, $k=0, 1, 2, 3, \ldots$ is

$$y_k = \sum_{i=0}^{k} h_i u_{k-i}, \quad k = 0, 1, 2, 3, \ldots \tag{4.570}$$

where

$$h_i = \begin{cases} b_0, & i = 0 \\ (b_1 - a_1 b_0)(-a_1)^{i-1}, & i = 1, 2, 3, \ldots \end{cases} \tag{4.571}$$

The discrete-time variable in Equation 4.571 is written as "i" instead of "k" to avoid confusion; however, it is helpful to think of the sequence as h_k, $k = 0, 1, 2, 3, \ldots$. Equation 4.570 reveals that the current output y_k is a linear combination of the current and past inputs, that is, writing out the terms in the sum

$$y_k = h_0 u_k + h_1 u_{k-1} + h_2 u_{k-2} + \cdots + h_k u_0, \quad k = 0, 1, 2, 3, \ldots \tag{4.572}$$

The weights are in fact the numerical values of the sequence h_k, $k = 0, 1, 2, 3, \ldots$ with the current input u_k weighted by h_0, the previous input u_{k-1} weighted by h_1 up to the oldest input u_0 with a weight of h_k. The sequence h_k, $k = 0, 1, 2, 3, \ldots$ in Equations 4.570 and 4.572 is called the weighting sequence of the discrete-time system.

The sum in Equation 4.570 is called the convolution sum, the counterpart to the convolution integral for continuous-time systems in Equation 4.561. The weighting sequence and convolution sum representation are not restricted to the simple first-order discrete-time system in Equation 4.562. They are applicable to nth-order LTI discrete-time systems. Fortunately, a more efficient technique for determining the weighting sequence than was previously illustrated exists. The method is deferred until after the following example.

Example 4.29

The low-pass filter in Equation 4.473 is a first-order discrete-time system.

  (a) Find the weighting sequence h_k, $k = k = 0, 1, 2, 3, \ldots$.
  (b) Graph the weighting sequence for $\alpha = 0.9$ and $\beta = 0.1$.
  (c) Find the unit step response by convolution, and compare the result with the response in Equation 4.478 with $y_{-1} = 0$.

(a) For the discrete-time system in Equation 4.473, $a_1 = -\alpha$, $b_0 = \beta$, and $b_1 = 0$. The weighting sequence given in Equation 4.571 reduces to

$$h_k = \begin{cases} \beta, & k = 0 \\ (\alpha\beta)(\alpha)^{k-1}, & k = 1, 2, 3, \ldots \end{cases} \tag{4.573}$$

$$= \beta(\alpha)^k, \quad k = 0, 1, 2, 3, \ldots \tag{4.574}$$

(b) The weighting sequence with $\alpha = 0.9$ and $\beta = 0.1$ is graphed in Figure 4.61.

(c) From Equation 4.570 with $u_k = 1$, $k = 0, 1, 2, 3, \ldots$, the unit step response is

$$y_k = \sum_{i=0}^{k} h_i u_{k-i} = \sum_{i=0}^{k} h_i = \sum_{i=0}^{k} \beta\alpha^i \tag{4.575}$$

$$= \beta\left(\frac{1 - \alpha^{k+1}}{1 - \alpha}\right) \tag{4.576}$$

$$= 1 - (0.9)^{k+1}, \quad k = 0, 1, 2, 3, \ldots \tag{4.577}$$

in agreement with the unit step response obtained from Equation 4.478 with $y_{-1} = 0$.

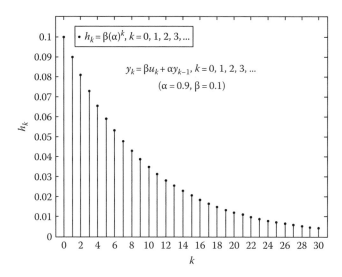

FIGURE 4.61 Weighting sequence h_k, $k = 0, 1, 2, 3, \ldots$ for first-order system in Equation 4.473.

The memory and transient response of a stable linear discrete-time system are reflected in its weighting sequence. Loosely speaking, the memory in a discrete-time system depends on how far back past inputs affect the current output in a significant way, that is, if the current output is predominantly influenced by only the last several inputs, then the system is said to exhibit a relatively short memory. Conversely, if distant inputs are influential in determining the current output, the system possesses a longer memory.

From the convolution sum representation for the current output y_k in Equation 4.572, it is readily apparent that the amount of memory in the system is directly related to how fast the weighting sequence approaches zero. (Discrete-time systems with weighting sequences that do not approach zero as k approaches infinity are considered in the next section dealing with stability.) Transient and steady-state response will also be considered at the same time; however, it should be clear even now that a fast responding system, that is, one with a short transient response must have a weighting sequence that approaches zero quickly and is, therefore, characterized as a system with a short memory.

For the first-order system considered in Example 4.29, the rate of decay to zero in the weighting sequence depends solely on the parameter α. Figure 4.62 shows the unit step responses of three first-order systems with different values of α and $\beta = 1 - \alpha$.

One is a fast responding system ($\alpha = 0.3$), one with moderate speed ($\alpha = 0.7$), and the last one is seen to have a sluggish response ($\alpha = 0.9$).

The response of an LTI discrete-time system to an impulse δ_k is quite significant. From the convolution sum in Equation 4.570, the unit impulse response is

$$(y_k)_{\text{impulse response}} = \sum_{i-0}^{k} h_i \delta_{k-i} = h_k, \quad k = 0, 1, 2, 3, \ldots \tag{4.578}$$

In other words, the impulse response is identical to the weighting sequence. Furthermore, for a system with z-domain transfer function $H(z)$, the z-transform of the impulse response is given by

$$Y_{\text{impulse response}}(z) = H(z)\mathfrak{z}\{\delta_k\} = H(z) \cdot 1 = H(z) \tag{4.579}$$

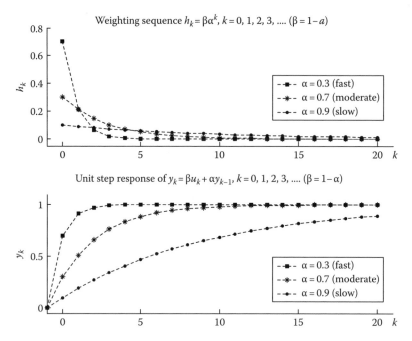

FIGURE 4.62 Weighting sequences and unit step responses of three first-order discrete-time systems governed by $y_k = (1 - \alpha)u_k + \alpha y_{k-1}$, $k = 0, 1, 2,\ldots$

Invert z-transforming Equation 4.579,

$$(y_k)_{\text{impulse response}} = h_k = \mathfrak{z}^{-1}\{H(z)\} \tag{4.580}$$

Equation 4.580 tells us the impulse response of an LTI discrete-time system is equal to the inverse z-transform of the z-domain transfer function of the system. Henceforth, the impulse response sequence will be denoted h_k, $k = 0, 1, 2,\ldots$ This most important property of discrete-time systems is illustrated in Figure 4.63.

The z-domain transfer function of the first-order system in Example 4.29 is

$$H(z) = \frac{Y(z)}{U(z)} = \frac{\beta z}{z - \alpha} \tag{4.581}$$

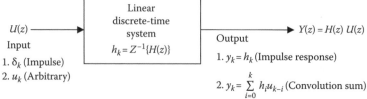

FIGURE 4.63 Relationship of impulse response to z-domain transfer function.

$$h_k = \mathfrak{z}^{-1}\{H(z)\} \tag{4.582}$$

$$= \mathfrak{z}^{-1}\left\{\frac{\beta z}{z - \alpha}\right\} \tag{4.583}$$

$$= \beta\alpha^k, \quad k = 0, 1, 2, 3, \ldots \tag{4.584}$$

The impulse response (weighting sequence) is therefore the same as in Equation 4.574.

The impulse response is fundamental to the design of digital filters implemented by linear difference equations. The two major categories of such filters are FIR and IIR, which stand for "finite impulse response" and "infinite impulse response," respectively (Orfanidis 1996).

EXERCISES

4.43 Find the z-domain transfer function of the discrete-time system, which results from an approximation to a continuous-time integrator using
   (a) Implicit Euler integration
   (b) Improved Euler integration

4.44 Find the z-domain transfer function $H(z)$ of the discrete-time system resulting from approximation of the first-order system $\tau\dot{y}(t) + y(t) = ku(t)$ using the following numerical integrators:
   (a) Explicit Euler
   (b) Implicit Euler
   (c) Trapezoidal

4.45 Let u_k, $k = 0$, 1, 2, 3, ... be uniformly spaced samples of an input $u(t)$ and y_k, $k = 0$, 1, 2, 3, ... be an approximation to $y(t) = \int_0^t u(t)\mathrm{d}t$ based on trapezoidal integration.
   (a) Find a difference equation relating u_k and y_k.
   (b) Solve the difference equation recursively using an appropriate step size to approximate the area under
      (i) $u(t) = te^{-t/2}$, $1 \le t \le 2$
      (ii) $u(t) = (1/\sqrt{2\pi})e^{-t^2/2}$, $0 \le t \le 5$

4.46 Prove Equation 4.534 for the scalar case, that is, show that $\mathfrak{z}\{x_{k+1}\} = z[X(z) - x_0]$, where x_0 is the value of x_k at $k = 0$.

4.47 In Example 4.28, find the complete transition matrix and verify that $\Phi_0 = I$.

4.48 In Example 4.28,
   (a) Find $Y(z)$, the z-transform of the response, by z-transforming the difference equation of the system with appropriate initial conditions.
   (b) Find y_∞ by applying the final value property (see Table 4.5).
   (c) Find y_0 by applying the initial value property (see Table 4.5).
   (d) Find $y_k = \mathfrak{z} - 1\{Y(z)\}$.

4.49 In Example 4.28, assume the initial conditions $y_{-3} = y_{-2} = y_{-1} = 0$.
   (a) Find $H(z) = Y(z)/U(z)$, the z-domain transfer function of the system.
   (b) The input is $u_k = A_0\delta_k + A_1\delta_{k-1} + A_2\delta_{k-2}$ and $y_{-1} = y_{-2} = y_{-3} = 0$. Find A_0, A_1, A_2 if the response is identical to the case when $u_k = 0$, $k = 0$, 1, 2, ... and $y_{-1} = P_0$, $y_{-2} = y_{-3} = 0$.

4.50 The unit step response of a discrete-time system is $y_k = -1 + 3^{k+1}$, $k = 0$, 1, 2, 3, ...
   (a) Find the difference equation relating u_k and y_k.
   (b) Find the impulse response, h_k, $k = 0$, 1, 2, 3, ...

4.51 The discrete-time signal $u_k = 1 + k$, $k = 0$, 1, 2, 3, ... is delayed one unit of discrete-time and then input to a discrete-time system with z-domain transfer function $H(z) = Y(z)/U(z) = z^2/(z+1)^2$. Find the output y_k at $k = 3$ and $k = 6$.

4.52 A discrete-time system with input u_k and output y_k is governed by the difference equation
$y_k = \alpha_1 y_{k-1} + \beta_1 u_{k-1} + \beta_0 u_k$, $k = 0, 1, 2, 3, \ldots$
(a) Find the z-domain transfer function of the system
(b) Find the impulse response sequence h_k, $k = 0, 1, 2, 3, \ldots$
　　(i) By inverse z-transformation of $H(z)$
　　(ii) By recursive solution of the difference equation with $u_k = \delta_k$
(c) Find the final value of the unit step response in terms of α_1, β_0, and β_1.
　　(i) By letting $k \to \infty$ in the unit step response
　　(ii) By applying the final value property
　　(iii) By setting $u_k = 1$, $k = 0, 1, 2, 3, \ldots$ and solving for $y_\infty = \lim_{k \to \infty} y_k = \lim_{k \to \infty} y_{k-1}$
　　　in the difference equation

4.53 Use the same approach for finding $\mathfrak{z}\{y_{k-1}\}$ when $y_{-1} \neq 0$ resulting in Equation 4.472 to find
(a) $\mathfrak{z}\{y_{k-2}\}$, $y_{-1}, y_{-2} \neq 0$
(b) $\mathfrak{z}\{y_{k-n}\}$, $y_{-1}, y_{-2}, \ldots, y_{-n} \neq 0$

4.54 A discrete-time system is described by $y_k + a_1 y_{k-1} + a_2 y_{k-2} = 0$, $k = 0, 1, 2, 3, \ldots$
(a) Find $Y(z)$ for the case when $y_{-1} = 0$ and $y_{-2} = 0$.
(b) Find $Y(z)$ for the case when the right-hand side is $b_0 \delta_k + b_1 \delta_{k-1}$ and $y_{-1} = y_{-2} = 0$.
(c) Find expressions for the weights b_0 and b_1 in terms of a_1, a_2, y_{-1}, and y_{-2}, so that the response y_k, $k = 0, 1, 2, 3, \ldots$ is the same in parts (a) and (b). Comment on the implication of replacing initial conditions with impulse forcing functions.

4.55 A simulation diagram for an M-B-K mechanical system governed by the second-order differential equation $M\ddot{y}(t) + B\dot{y}(t) + Ky(t) = f(t)$ is shown in Figure E4.55:

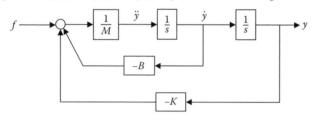

FIGURE E4.55

(a) Find a difference equation relating y_k and f_k based on the use of explicit Euler integration. Convert the difference equation to state variable form.
(b) Find a difference equation relating y_k and f_k based on the use of implicit Euler integration. Convert the difference equation to state variable form.
(c) Find a difference equation relating y_k and f_k based on the use of trapezoidal Euler integration. Convert the difference equation to state variable form.
(d) Find a difference equation relating y_k and f_k based on the use of explicit Euler integration for the first integrator ($\dot{y}$) and implicit Euler integration for the second integrator (y). Convert the difference equation to state variable form.
(e) Approximate the unit step response of the system for parts (a) through (d) when $M = 1$, $B = 2$, and $K = 1$, and compare each with the continuous-time response.

4.56 Consider the double integrator shown in Figure E4.56:

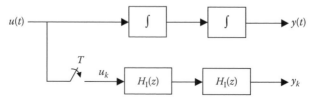

FIGURE E4.56

(a) Write the differential equation relating $y(t)$ and $u(t)$.

(b) Find the difference equation relating y_k and u_k if both numerical integrators are based on explicit Euler integration.

(c) Find dy/dt and $y(t)$ when the initial conditions are $y(0) = 0$, $\dot{y}(0) = 1$ and the input $u(t) = 10 - e^{-t/2}$, $t \geq 0$.

(d) Find the z-domain transfer function and impulse response of the discrete-time system.

(e) Find the output y_k, $k = 1, 2, 3, \ldots$ when the integration step size $T = 0.1$ s.

(f) Plot the continuous- and discrete-time outputs on the same graph, and comment on the results.

4.57 A discrete-time system is described by $y_k + a_1 y_{k-1} + a_2 y_{k-2} = 0$, $k = 0, 1, 2, 3, \ldots$.

(a) Find $Y(z)$ for the case when $y_{-1} \neq 0$, $y_{-2} \neq 0$.

(b) Find $Y(z)$ for the case when the right-hand side is $b_0 \delta_k + b_1 \delta_{k-1}$ and $y_{-1} = y_{-2} = 0$.

(c) Find expressions for the weights b_0 and b_1 in terms of a_1, a_2, y_{-1}, y_{-2} so that the response y_k, $k = 0, 1, 2, 3, \ldots$ is the same in Parts (a) and (b). Comment on the implication of replacing initial conditions with impulse forcing functions.

4.58 Show that the unit step response of a discrete-time system with z-domain transfer function $H(z)$ is given by

$$y_k = \mathfrak{z}^{-1}\left\{ \frac{z}{z-1} H(z) \right\}, \quad k = 0, 1, 2, \ldots$$

4.8 STABILITY OF LTI DISCRETE-TIME SYSTEMS

One way of characterizing the stability of a discrete-time system is by the way it responds to a bounded input. When the response remains bounded, the system is said to exhibit BIBO stability. The implications of BIBO stability on the system's z-domain transfer function, impulse response (weighting sequence), and natural response will be explored.

Consider an nth-order LTI discrete-time system described by Equation 4.456 in the previous section. The z-domain transfer function is

$$H(z) = \frac{Y(z)}{U(z)} = \frac{b_0 z^n + b_1 z^{n-1} + \cdots + b_m z^{n-m}}{z^n + a_1 z^{n-1} + \cdots + a_{n-1} z + a_n}, \quad n \geq m \tag{4.585}$$

Suppose the poles of $H(z)$ are real and distinct. Then

$$Y(z) = H(z)U(z) = \frac{b_0 z^n + b_1 z^{n-1} + \cdots + b_m z^{n-m}}{(z - p_1)(z - p_2) \cdots (z - p_n)} U(z) \tag{4.586}$$

In the case where the poles of $U(z)$ are different from $p_1, p_2, \ldots p_n$,

$$Y(z) = A_0 \left\{ A_1 \frac{z}{z - p_1} + A_2 \frac{z}{z - p_2} + \cdots + A_n \frac{z}{z - p_n} \right\}$$
$$+ \text{ terms due poles of } U(z) \mathfrak{z}^{-1}\{U(z)\} \tag{4.587}$$

The response y_k $k = 0, 1, 2, 3, \ldots$ is therefore

$$y_k = A_0 \delta_k + \left\{ A_1 p_1^k + A_2 p_2^k + \cdots + A_n p_n^k \right\}$$
$$+ \text{ terms generated from } \mathfrak{z}^{-1}\{U(z)\} \text{ terms generated from } \mathfrak{z}^{-1}\{U(z)\} \tag{4.588}$$

The bracketed expression is the natural response, that is, a linear combination of the natural modes $p_1^k, p_2^k, \cdots, p_n^k$, while the terms arising from the inverse z-transformation of $U(z)$ are similar in nature to the input and comprise the forced component of the overall response. Since the natural response is excited by the presence of an input, it must obviously be a bounded sequence for a BIBO stable system.

The impulse response $h_k = \mathfrak{z}^{-1}\{H(z)\}$ is also a linear combination of the system's natural modes $p_1^k, p_2^k, \ldots, p_n^k$, (plus in some cases, a weighted impulse at the origin). Imagine a discrete-time system with impulse response h_k, $k = 0, 1, 2, \ldots$ subject to a unit step input $u_k = 1$, $k = 0, 1, 2, \ldots$. Using the convolution sum form of the output,

$$|y_k| = \left| \sum_{i=0}^{k} h_i u_{k-i} \right| = \left| \sum_{i=0}^{k} h_i \right| < \sum_{i=0}^{k} |h_i|, \quad k = 0, 1, 2, \ldots \tag{4.589}$$

From Equation 4.589, the step response at discrete-time k remains finite provided the sum of the first $k+1$ values of the magnitude of the impulse response satisfies

$$\sum_{i=0}^{k} |h_i| < \infty, \quad k = 0, 1, 2, 3, \ldots \tag{4.590}$$

It follows that the entire response y_k, $k = 0, 1, 2, 3, \ldots$ is bounded whenever the impulse response sequence satisfies

$$\sum_{k=0}^{\infty} |h_k| < \infty \tag{4.591}$$

While Equation 4.591 was derived for the case where the input is a unit step, it applies to any bounded input. Equation 4.591 is a necessary and sufficient condition for the output of an LTI discrete-time system to remain bounded in response to any bounded input. A consequence of Equation 4.591 is that the weighting sequence of a BIBO stable system must decay to zero as $k \to \infty$.

From Equation 4.588, an nth-order LTI discrete-time system with z-domain transfer function having real and distinct poles is BIBO stable when the poles satisfy

$$-1 < p_i < 1, \quad i = 1, 2, 3, \ldots, n \tag{4.592}$$

The expression for the output y_k in Equation 4.588 assumed that the poles of $H(z)$ were real and distinct. A real pole p with multiplicity m generates a weighted sum of the natural modes $p^k, kp^k, \ldots, k^{m-1}p^k$ in the output; however, Equation 4.592 still applies for BIBO stability.

When a pair of complex poles of $H(z)$ is present, y_k contains trigonometric terms like $R^k \cos(k\theta + \varphi)$ where R is the magnitude of the complex poles. In order to include the possibility of complex poles of $H(z)$, Equation 4.592 is appropriately expressed as

$$|p_i| < 1, \quad i = 1, 2, \ldots, n \tag{4.593}$$

Consequently, a sufficient condition for BIBO stability of LTI discrete-time systems is that all of its z-domain transfer function poles have a magnitude less than 1, that is, all poles are located inside the Unit Circle in the complex plane.

In Example 4.30, we look at a second-order system with real and distinct poles subject to a bounded input. The effect of moving one of the poles is investigated. Following that, we consider the ramifications of various locations of the z-domain transfer function's poles in the complex plane.

Example 4.30

A discrete-time system is described by the difference equation

$$y_k + a_1 y_{k-1} + a_2 y_{k-2} = b_0 u_k, \quad k = 0, 1, 2, 3, \ldots \tag{4.594}$$

Initial conditions $y_{-1} = y_{-2} = 0$. The input sequence is given by

$$u_k = 1 + (0.1)^k, \quad k = 0, 1, 2, 3, \ldots \tag{4.595}$$

Find the z-domain transfer function $H(z)$ and its poles, the impulse response h_k, $k = 0, 1, 2, 3, \ldots$, the total response y_k, $k = 0, 1, 2, 3, \ldots$, and the natural and forced components of the total response, and comment on stability for the following cases:

$$\text{(a) } a_1 = 0, \quad a_2 = -0.25, \quad b_0 = 1$$
$$\text{(b) } a_1 = -0.5, \quad a_2 = -0.5, \quad b_0 = 1$$
$$\text{(c) } a_1 = -1.5, \quad a_2 = -1, \quad b_0 = 1$$

(a) z-transforming the difference equation $y_k - 0.25 y_{k-2} = u_k$, $k = 0, 1, 2, 3, \ldots$ yields

$$H(z) = \frac{Y(z)}{U(z)} = \frac{z^2}{z^2 - 0.25} = \frac{z^2}{(z - 0.5)(z + 0.5)} \tag{4.596}$$

with poles $p_1 = -0.5$, $p_2 = 0.5$. The impulse response is obtained from

$$h_k = \mathfrak{z}^{-1}\{H(z)\} = \mathfrak{z}^{-1}\left\{\frac{z^2}{(z + 0.5)(z - 0.5)}\right\} \tag{4.597}$$

$$= \mathfrak{z}^{-1}\left\{\frac{0.5z}{z + 0.5} + \frac{0.5z}{z - 0.5}\right\} \tag{4.598}$$

$$= 0.5[(-0.5)^k + (0.5)^k], \quad k = 0, 1, 2, 3, \ldots \tag{4.599}$$

$$= (0.5)^{k+1}[(-1)^k + 1], \quad k = 0, 1, 2, 3, \ldots \tag{4.600}$$

$$= (0.5)^k, \quad k = 0, 2, 4, 6, \ldots \tag{4.601}$$

The complete response y_k, $k = 0, 1, 2, \ldots$ is determined by inverse z-transformation of

$$Y(z) = \frac{z^2}{(z^2 - 0.25)}\left[\frac{z}{z - 1} + \frac{z}{z - 0.1}\right] \tag{4.602}$$

$$= \frac{z^3(2z - 1.1)}{(z + 0.5)(z - 0.5)(z - 1)(z - 0.1)} \tag{4.603}$$

$$= \frac{7}{12}\left(\frac{z}{z + 0.5}\right) + \frac{1}{8}\left(\frac{z}{z - 0.5}\right) + \frac{4}{3}\left(\frac{z}{z - 1}\right) - \frac{1}{24}\left(\frac{z}{z - 0.1}\right) \tag{4.604}$$

$$\Rightarrow y_k = \frac{7}{12}(-0.5)^k + \frac{1}{8}(0.5)^k + \frac{4}{3} - \frac{1}{24}(0.1)^k, \quad k = 0, 1, 2, 3, \ldots \tag{4.605}$$

From Equation 4.605, the natural (free) response and forced response are

$$(y_k)_{\text{natural}} = \frac{7}{12}(-0.5)^k + \frac{1}{8}(0.5)^k, \quad k = 0, 1, 2, \ldots \tag{4.606}$$

$$(y_k)_{\text{forced}} = \frac{4}{3} - \frac{1}{24}(0.1)^k, \quad k = 0, 1, 2, \ldots \tag{4.607}$$

The system is stable as evidenced by the natural response decaying to zero as $k \to \infty$. This was expected since the two poles of $H(z)$ are located between -1 and $+1$. Can you show that Equation 4.591 is satisfied as well? Note the similarity between the natural response in Equation 4.606 and the impulse response in Equation 4.599.

(b) The difference equation becomes $y_k - 0.5y_{k-1} - 0.5y_{k-2} = u_k$, $k = 0, 1, 2, 3, \ldots$. The results for this system are obtained in an analogous fashion to part (a).

$$H(z) = \frac{z^2}{z^2 - 0.5z - 0.5} = \frac{z^2}{(z + 0.5)(z - 1)} \quad (p_1 = -0.5, p_2 = 1) \tag{4.608}$$

$$h_k = \frac{1}{3}[(-0.5)^k + 2], \quad k = 0, 1, 2, 3, \ldots \tag{4.609}$$

$$y_k = \frac{7}{18}(-0.5)^k + \frac{44}{27} + \frac{2}{3}k - \frac{1}{54}(0.1)^k, \quad k = 0, 1, 2, 3, \ldots \tag{4.610}$$

$$(y_k)_{\text{nat}} = \frac{7}{18}(-0.5)^k + \frac{44}{27}, \quad k = 0, 1, 2, 3, \ldots \tag{4.611}$$

$$(y_k)_{\text{forced}} = \frac{2}{3}k - \frac{1}{54}(0.1)^k, \quad k = 0, 1, 2, 3, \ldots \tag{4.612}$$

The forced response also contains a constant component resulting from the pole of $U(z)$ at $z = 1$. This constant is combined with the constant in the natural response, and the sum of $44/27$ is shown entirely in the natural response in Equation 4.611.

The second pole of $H(z)$, namely, $p_2 = 1$, does not satisfy Equation 4.592, and the system is not BIBO stable. In this case, a bounded input produced an unbounded output. The impulse response in Equation 4.609 does not asymptotically decay to zero.

(c) The difference equation is $y_k - 1.5y_{k-1} - y_{k-2} = u_k$, $k = 0, 1, 2, 3, \ldots$.

$$H(z) = \frac{z^2}{z^2 - 1.5z - 1} = \frac{z^2}{(z + 0.5)(z - 2)}, \quad (p_1 = -0.5, p_2 = 2) \tag{4.613}$$

$$h_k = \frac{1}{5}(-0.5)^k + \frac{4}{5}(2)^k, \quad k = 0, 1, 2, 3, \ldots \tag{4.614}$$

$$y_k = \frac{7}{30}(-0.5)^k + \frac{232}{95}(2)^k - \frac{2}{3} - \frac{1}{114}(0.1)^k, \quad k = 0, 1, 2, 3, \ldots \tag{4.615}$$

$$(y_k)_{\text{natural}} = \frac{7}{30}(-0.5)^k + \frac{232}{95}(2)^k, \quad k = 0, 1, 2, 3, \ldots \tag{4.616}$$

$$(y_k)_{\text{forced}} = -\frac{2}{3} - \frac{1}{114}(0.1)^k, \quad k = 0, 1, 2, 3, \ldots \tag{4.617}$$

Once again, the system is unstable. The natural response and, by implication, the impulse response are unbounded as $k \to \infty$.

The real poles of an nth-order LTI discrete-time system transfer function are located on the real axis in the complex plane. Figure 4.64 shows real poles located at (from right to left) 1.25, 1, 0.75, -0.5, -1, and -1.5 along the real axis.

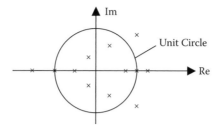

FIGURE 4.64   The Unit Circle and various locations of real and complex poles.

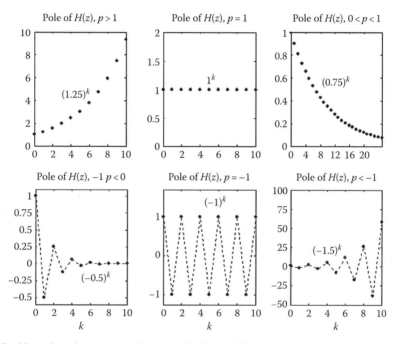

FIGURE 4.65   Natural modes corresponding to real poles of $H(z)$.

There are six distinct regions for location of real poles along the real axis, each with a different type of natural mode. According to Equation 4.592, only the poles at 0.75 and -0.5 located inside the Unit Circle correspond to stable natural modes. The impulse response h_k, $k = 0$, 1, 2,... approaches zero as $k \to \infty$ in both cases. When the poles are located on the Unit Circle at $+1$ and 1, the impulse response sequence remains finite as $k \to \infty$; however, a linear discrete-time system with a pole at either location is not BIBO stable.

The remaining two cases correspond to real poles located outside the Unit Circle, either in the region $p > 1$ or $p < -1$. The natural response of an LTI discrete-time system with poles located in either region is unbounded, and, hence, the system is not BIBO stable.

Figure 4.65 illustrates the natural modes corresponding to each of the real poles.

4.8.1   COMPLEX POLES OF $H(z)$

Three pairs of complex poles are also shown in Figure 4.64. The z-domain transfer function $H(z)$ possesses a pair of complex poles if there is a quadratic factor in its denominator with complex roots. Figure 4.66 illustrates the case where $H(z)$ has complex poles at $z = a \pm jb$. The transformation to polar form $z = re^{\pm j\theta}$ is shown as well.

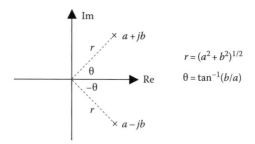

FIGURE 4.66 Complex poles of discrete-time system transfer function $H(z)$.

In terms of polar coordinates, the quadratic factor is

$$Q(z) = (z - re^{j\theta})(z - re^{-j\theta}) = z^2 - (2r \cos \theta)z + r^2 \tag{4.618}$$

Consider a second-order discrete-time system with z-domain transfer function

$$H(z) = \frac{Az^2 + Bz}{z^2 - (2r \cos \theta)z + r^2} \tag{4.619}$$

For reasons that will soon become apparent, $H(z)$ is expressed as

$$H(z) = \frac{c_1(r \sin \theta)z + c_2[z^2 - (r \cos \theta)z]}{z^2 - (2r \cos \theta)z + r^2} \tag{4.620}$$

where c_1 and c_2 are obtained by equating like powers of z in the numerators of Equations 4.619 and 4.620. The result is

$$c_1 = \frac{Ar \cos \theta + B}{r \sin \theta}, \quad c_2 = A \tag{4.621}$$

$H(z)$ in Equation 4.620 is expressed as

$$H(z) = c_1 \left[\frac{(r \sin \theta)z}{z^2 - (2r \cos \theta)z + r^2} \right] + c_2 \left[\frac{z^2 - (r \cos \theta)z}{z^2 - (2r \cos \theta)z + r^2} \right] \tag{4.622}$$

Referring to Table 4.4 with $e^{-aT} = r$ and $\omega T = \theta$ suggests the impulse response $h_k = \mathfrak{z}^{-1}\{H(z)\}$ is

$$h_k = c_1 r^k \sin k\theta + c_2 r^k \cos k\theta = r^k(c_1 \sin k\theta + c_2 \cos k\theta), \quad k = 0, 1, 2, 3, \ldots \tag{4.623}$$

There are three cases to consider, which are illustrated in Figure 4.64. The three cases correspond to the region inside the Unit Circle ($r < 1$), all points on the Unit Circle ($r = 1$), and the exterior of the Unit Circle ($r > 1$). It follows from Equation 4.623 that the impulse response satisfies the necessary condition for BIBO stability in Equation 4.591 only in the first case, $r < 1$, that is, when the poles are located inside the Unit Circle. The natural response, being of similar form to the impulse response, decays to zero as $k \to \infty$. Hence, the system is BIBO stable.

When the poles are either on the Unit Circle or outside, Equation 4.591 is not satisfied, and the system is therefore not BIBO stable. The natural response consists of sustained oscillations when $r = 1$ and oscillations of increasing magnitude when $r > 1$.

Example 4.31

A second-order discrete-time system has a z-domain transfer function given by

$$H(z) = \frac{z^2 + 3z}{Q(z)} \tag{4.624}$$

where $Q(z)$ is a quadratic with complex roots located in the three different regions like the ones shown in Figure 4.64. Suppose the roots are

(a) $-0.25 \pm j0.5$ (b) $0.5(1 \pm j\sqrt{3})$ (c) $1 \pm j$

(a) Find the z-domain transfer function $H(z)$ for each case.
(b) Find the impulse response h_k, $k = 0, 1, 2, 3, \ldots$ for each case.
(c) Graph the impulse response for each case.

(a)

(i) ($a = -0.25$, $b = 0.5$). The polar coordinates of the transfer function poles are

$$r = [(-0.25)^2 + (0.5)^2]^{1/2} = 0.5990, \quad \theta = \tan^{-1}\left(\frac{0.5}{-0.25}\right) = 2.0344 \text{ rad}$$

$$Q(z) = z^2 - (2r \cos \theta)z + r^2$$
$$= z^2 - [2(0.5990) \cos(2.0344)]z + (0.5990)^2$$
$$= z^2 + 0.5z + 0.3125$$

$$\Rightarrow H(z) = \frac{z^2 + 3z}{Q(z)} = \frac{z^2 + 3z}{z^2 + 0.5z + 0.3125} \tag{4.625}$$

(ii) ($a = 0.5$, $b = 0.5\sqrt{3}$) $\Rightarrow r = 1$, $\theta = 1.0472$ rad, $H(z) = \dfrac{z^2 + 3z}{z^2 - z + 1}$ \hfill (4.626)

(iii) ($a = 1, b = 1$) $\Rightarrow r = \sqrt{2}$, $\theta = \dfrac{\pi}{4}$ rad, $H(z) = \dfrac{z^2 + 3z}{z^2 - 2z + 2}$ \hfill (4.627)

(b)

(i) $c_1 = \dfrac{Ar \cos \theta + B}{r \sin \theta} = \dfrac{1(0.5990) \cos(2.0344) + 3}{0.5990 \sin(2.0344)} = 5.5, \quad c_2 = A = 1$

The constants c_1 and c_2 for (ii) and (iii) are determined in similar fashion. From Equation 4.623, the impulse responses are

(i) $h_k = (0.5990)^k[5.5 \sin(2.0344k) + \cos(2.0344k)], \quad k = 0, 1, 2, 3, \cdots$ \hfill (4.628)

(ii) $h_k = 4.0415 \sin(1.0472k) + \cos(1.0472k), \quad k = 0, 1, 2, 3, \cdots$ \hfill (4.629)

(iii) $h_k = (\sqrt{2}^k)\left[4 \sin\left(\dfrac{k\pi}{4}\right) + \cos\left(\dfrac{k\pi}{4}\right)\right], \quad k = 0, 1, 2, 3, \cdots$ \hfill (4.630)

(c) Graphs of the impulse responses in Equations 4.628 through 4.630 are shown in Figure 4.67.
The discrete-time system with poles located inside the Unit Circle is BIBO stable. The impulse response given in Equation 4.628 satisfies the necessary and sufficient condition for BIBO stability in Equation 4.591. Poles of the transfer functions in Equations 4.626 and 4.627 are situated on the Unit Circle and outside it, respectively. Neither system is BIBO stable.

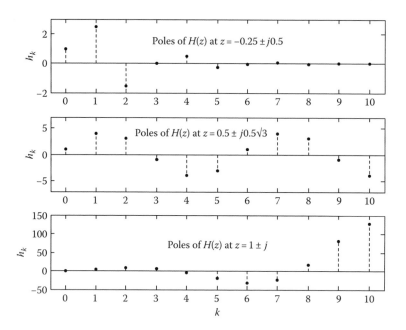

FIGURE 4.67 Impulse responses for discrete-time systems with different complex poles of $H(z)$.

Consider a system with a pair of complex poles of $H(z)$ on the Unit Circle at $e^{\pm j\theta}$. Its response to the bounded input $u_k = \sin k\theta$, $k = 0, 1, 2, 3, \ldots$ is obtained from

$$Y(z) = H(z)U(z) = \frac{N(z)}{(z - e^{j\theta})(z - e^{-j\theta})} \cdot \frac{\sin \theta \cdot z}{z^2 - (2 \cos \theta)z + 1} \tag{4.631}$$

$$= \frac{N(z)}{(z - e^{j\theta})(z - e^{-j\theta})} \cdot \frac{\sin \theta \cdot z}{(z - e^{j\theta})(z - e^{-j\theta})} \tag{4.632}$$

$$= \frac{\sin \theta \cdot zN(z)}{(z - e^{j\theta})^2(z - e^{-j\theta})^2} \tag{4.633}$$

It is left as an exercise to show that y_k contains a linear combination of the terms, $\cos k\theta$, $\sin k\theta$, $k \cos k\theta$, and $k \sin k\theta$. Consequently, the response is unbounded, and the system is not BIBO stable.

When a real pole of $H(z)$ is located on the Unit Circle at $z = -1$ or $z = +1$, and the input is $u_k = (-1)^k$, $k = 0, 1, 2, 3$, or the unit step $u_k = 1$, $k = 0, 1, 2, 3, \ldots$, respectively, the response is unbounded due to the presence of $(z + 1)^2$ or $(z - 1)^2$ in the denominator of the output $Y(z)$. The first case results in the term $k(-1)^k$ (multiplied by a constant) appearing in the output. In the second case, y_k contains a term proportional to $k(1)^k = k$ (see Example 4.30, part [b]).

We conclude this section with a simulation of the continuous-time control system in Figure 4.68.

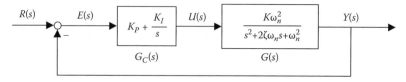

FIGURE 4.68 P–I control of a second-order continuous-time process.

The analog P–I (Proportional–Integral) controller $G_C(s)$ is approximated by a discrete-time controller with transfer function $G_C(z)$ based on the use of trapezoidal integration. It was shown in Section 4.4.7 that the z-domain transfer function of a trapezoidal integrator is

$$H_I(z) = \frac{T}{2}\left[\frac{z+1}{z-1}\right] \tag{4.634}$$

And, therefore, $G_C(z)$ is obtained by replacing s with $1/H_I(z)$, that is,

$$G_C(z) = K_P + \left.\frac{K_I}{s}\right|_{s \leftarrow (2/T)((z-1)/(z+1))} = \frac{(2K_P + K_I T)z - 2K_P + K_I T}{2(z-1)} \tag{4.635}$$

Several discrete-time approximations to the second-order system in Figure 4.68 were developed in Section 4.4.7. Explicit Euler approximation resulted in

$$G(z) = \frac{Y(z)}{U(z)} = \frac{K(\omega_n T)^2}{z^2 - 2(1 - \zeta\omega_n T)z + 1 - 2\zeta\omega_n T + (\omega_n T)^2} \tag{4.636}$$

The block diagram of the discrete-time system intended to simulate the continuous-time control system is shown in Figure 4.69.

The closed-loop transfer function is

$$H(z) = \frac{Y(z)}{R(z)} = \frac{G_C(z)G(z)}{1 + G_C(z)G(z)} \tag{4.637}$$

and the poles of $H(z)$ are the roots of the characteristic equation

$$\Delta(z) = 1 + G_C(z)G(z) = 0 \tag{4.638}$$

Substituting Equations 4.635 and 4.636 into Equation 4.638 yields

$$\Delta(z) = z^3 + \alpha_1 z^2 + \alpha_2 z + \alpha_3 = 0 \tag{4.639}$$

where

$$\left.\begin{aligned}
\alpha_1 &= -3 + 2\zeta\omega_n T \\
\alpha_2 &= 3 - 4\zeta\omega_n T + (\omega_n T)^2[1 + K(K_p + 0.5K_1 T)] \\
\alpha_3 &= -12\zeta\omega_n T + (\omega_n T)^2[-1 + K(-K_p + 0.5K_1 T)]
\end{aligned}\right\} \tag{4.640}$$

Table 4.9 summarizes the results for different combinations of continuous-time second-order systems, controllers, and integration step size.

The continuous-time system and discrete-time poles are shown in Figure 4.70. All three continuous-time control systems are stable since the poles are located in the left-half plane. The discrete-time systems for simulating them, however, are not all BIBO stable. In fact, the discrete-time systems in Rows 2 and 3 in Table 4.9 possess a pair of complex poles located on the Unit Circle and outside it, respectively.

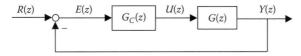

FIGURE 4.69 Block diagram of discrete-time system.

TABLE 4.9
Continuous- and Discrete-Time Control System Poles

System Parameters and Integration Step Size	Poles of $H(s)$	Poles of $H(z)$	Mag of $H(z)$ Complex Poles
$K = 1, \omega_n = 10, \zeta = 1.0$			
$K_P = 0.5, K_I = 2$	$-9.1616 \pm j5.9448$	$0.5398 \pm j0.3201$	0.628
$T = 0.05$	-1.6768	0.9205	
$K = 1, \omega_n = 5, \zeta = 0.5124$			
$K_P = 1, K_I = 3$	$-1.7133 \pm j6.4225$	$0.8674 \pm j0.4979$	1.000
$T = 0.075$	-1.6975	0.8808	
$K = 1, \omega_n = 20, \zeta = 0.15$			
$K_P = 1, K_I = 3$	$-2.2436 \pm j28.0745$	$0.9436 \pm j0.7085$	1.180
$T = 0.025$	-1.5128	0.9629	

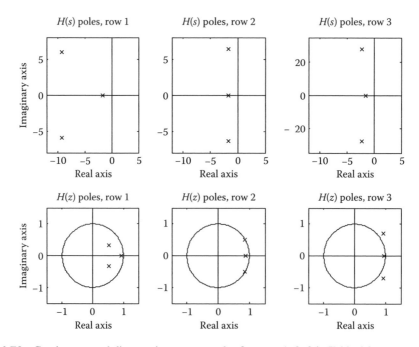

FIGURE 4.70 Continuous- and discrete-time system poles for rows 1, 2, 3 in Table 4.9.

It is important to keep in mind that while the discrete-time system approximation in Row 1 of Table 4.9 is stable, its accuracy in approximating the continuous-time system response to various inputs is another matter. Suppose the input to the control system in Figure 4.68 is $r(t)$, $t \geq 0$ and the output is $y(t)$, $t \geq 0$. If the discrete-time system response to $r_k = r(kT)$, $k = 0, 1, 2, \ldots$ is y_k, $k = 0, 1, 2, \ldots$, an accurate simulation requires that $y_k \approx y(kT)$, $k = 0, 1, 2, \ldots$.

The locations of the continuous- and discrete-time poles corresponding to Row 1 in Figure 4.70 imply that the natural responses consist of a monotonically decaying and damped oscillatory modes. The question still remains whether the time constants and damped natural frequencies are comparable. A thorough examination of this point is deferred to a later chapter; however, we can gain insight by looking at the step responses of each system.

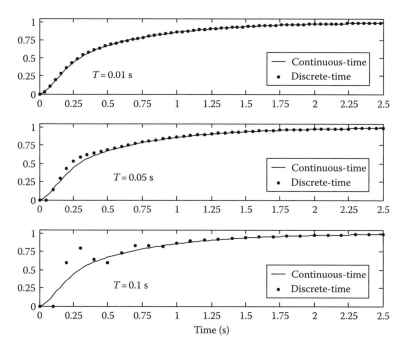

FIGURE 4.71 Unit step response of continuous-time system (Figure 4.68) and discrete-time system (Figure 4.69) with $T = 0.01, 0.05, 0.1$ s.

Example 4.32

Find and graph the unit step response of the continuous-time system in Figure 4.68 ($K = 1$, $\omega_n = 10$, $\zeta = 1.0$, $K_P = 0.5$, $K_I = 2$) and the discrete-time system shown in Figure 4.69 with integration step size $T = 0.01, 0.05, 0.1$ s.

The step responses are computed in M-file "Chap8_Ex8_3.m" and shown in Figure 4.71. The top graph is a plot of every fourth point of the discrete-time system step response.

The discrete-time system is stable for all three values of integration step size T and the steady-state values are identical to the continuous-time steady-state value. However, the transient response of the discrete-time system when $T = 0.1$ s varies considerably from the continuous-time system transient response.

EXERCISES

4.59 Find the poles of the z-domain transfer functions $H(z)$ below, and comment on the stability of the corresponding discrete-time systems.

(a) $\dfrac{z^2 + 2z}{32z^3 - 16z^2 - 22z + 1}$ (b) $\dfrac{4z^2}{z^3 - (3/2)z^2 + (3/4)z - (1/8)}$ (c) $\dfrac{3z + 1}{4z^3 - 3z + 1}$

(d) $\dfrac{z^4 - z}{16z^4 - 28z^3 + 22z^2 - 8z + 1}$ (e) $\dfrac{1}{4z^4 + 3z^2 - 1}$ (f) $\dfrac{z^3 + 2z^2 + z}{2z^3 - 5z^2 + 6z - 2}$

4.60 Prove $\sum_{k=0}^{\infty} |h_k| < \infty$ is a sufficient condition for BIBO stability of an LTI discrete-time system.

4.61 A discrete-time system is described by the difference equation

$$4y_k - 3y_{k-2} + y_{k-3} = u_k, \quad k = 0, 1, 2, 3, \ldots \quad (y_{-1} = y_{-2} = y_{-3} = 0)$$

(a) Find the weighting sequence h_k, $k = 0, 1, 2, 3, \ldots$ of the system.

(b) Check whether the condition $\sum_{k=0}^{\infty} |h_k| < \infty$ is satisfied. Is the system stable?

(c) Find and graph the system response to the input $u_k = 1 + 2(-1)^k$, $k = 0, 1, 2, 3, \ldots$.

4.62 Show the work required to establish Equations 4.608 through 4.610 in part (b) and Equations 4.613 through 4.615 in part (c) of Example 4.30.

4.63 Find the inverse z-transform of $Y(z) = \dfrac{(\sin \theta \cdot zN(z))}{((z - e^{j\theta})^2(z - e^{-j\theta})^2)}$ in Equation 4.633.

4.64 For a discrete-time system with z-domain transfer function given by

$$H(z) = \frac{Y(z)}{U(z)} = \frac{z^2 + z + 1}{z^3 - 0.5z^2 - z + 0.5}$$

(a) Find the zeros and poles of $H(z)$.

(b) Find the impulse response sequence h_k, $k = 0, 1, 2, 3, \ldots$.

(c) Find the unit step response.

(d) Find the forced response to $u_k = (-1)^k$, $k = 0, 1, 2, 3, \ldots$.

4.65 For the control system in Figure 4.68,

(a) Find the transfer function $H_E(s) = E(s)/R(s)$.

(b) Use explicit Euler integration with integration step T to obtain $H_E(z)$, an approximation to the continuous-time transfer function $H_E(s)$.

(c) Assume $K = 1$, $\omega_n = 10$, $\zeta = 1.0$, $K_P = 0.5$, $K_I = 2$, and $T = 0.05$, and find the poles of $H_E(z)$. Compare your answer with the results shown in Table 4.9 for the same parameter values.

(d) Find the difference equation relating e_k and r_k, $k = 0, 1, 2, 3, \ldots$.

(e) Solve the difference equation when $r_k = 1$, $k = 0, 1, 2, 3, \ldots$.

4.66 For the control system in Figure 4.69 with baseline parameters specified in the last row of Table 4.9, find the poles of $H(z)$ and plot the magnitude of the most distant pole(s) from the origin when

(a) ζ varies from 0 to 2

(b) T varies from 0.01 to 0.5 s

(c) K_P varies from 0.5 to 5

(d) ω_n varies from 5 to 50 rad/s

4.67 End-of-month deposits d_k, $k = 0, 1, 2, 3, \ldots$ are placed in an investment account paying interest at a rate of i per month. The initial account balance is P_0. The difference equation for P_k, the account balance after k months, is

$$P_{k+1} = (1 + i)P_k + d_{k+1}, \quad k = 0, 1, 2, 3, \ldots$$

(a) Find the z-domain transfer function $H(z) = P(z)/D(z)$ and its pole.

Hint: Use the left shifting property, $_3\{P_{k+1}\} = zP(z) - zP_0$. Comment on the stability of the discrete-time system.

(b) Sketch P_k, $k = 0, 1, 2, 3, \ldots$ when no deposits are made and $i > 0$. Repeat for $i < 0$.

(c) Find the general solution for P_k, $k = 1, 2, 3, \ldots$ when

(i) $d_k = \begin{cases} 0, & k = 0 \\ D, & k = 1, 2, 3, \ldots \end{cases}$

(ii) $d_k = \begin{cases} 0, & k = 0, 2, 4, \ldots \\ D, & k = 1, 3, 5, \ldots \end{cases}$

(iii) $d_k = \begin{cases} 0, & k = 0 \\ 2D, & k = 1, 3, 5, \ldots \\ -D, & k = 2, 4, 6, \ldots \end{cases}$

4.68 Figure E4.68 shows the relationship between acceleration, velocity, and position of a particle moving along a straight line.

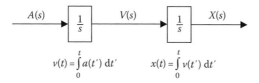

FIGURE E4.68

(a) Write the differential equations relating $v(t)$ and $a(t)$, $x(t)$ and $v(t)$, and $x(t)$ and $a(t)$.
(b) Use trapezoidal integration to approximate the three differential equations. That is, find the difference equations relating v_k and a_k, x_k and v_k, and x_k and a_k.
(c) Find the poles of the three transfer functions $V(z)/A(z)$, $X(z)/V(z)$, and $X(z)/A(z)$. Comment on the stability of each.
(d) Suppose the acceleration is given by

$$a(t) = \begin{cases} t, & 0 \le t < 1 \\ 1, & 1 \le t < 2 \\ 3 - t, & 2 \le t < 3 \\ 0, & 3 < t \end{cases}$$

Find analytical expressions for $v(t)$ and $x(t)$.
(e) Solve the difference equations recursively for a suitable value of T. Plot $v(t)$, $t \ge 0$ along with v_k, $k = 0, 1, 2, 3, \ldots$ on the same graph, and do the same for $x(t)$, $t \ge 0$ and x_k, $k = 0$, 1, 2, 3,....

4.69 The motion of a mass suspended from a spring without friction is governed by $md^2x/dt^2 + kx = f$, where $f = f(t)$ is the applied force acting on the mass.

(a) Find the transfer function $H(s) = X(s)/F(s)$ in terms of the natural frequency $\omega_n = \sqrt{k/m}$ and the constant $c = 1/m$. Where are the poles located?
(b) Use explicit Euler, implicit Euler, and trapezoidal integration to obtain discrete-time approximations, that is, find $H(z) = X(z)/F(z)$. Leave your answers in terms of c, ω_n, and the integration step size T.
(c) Find the poles for each z-domain transfer function $H(z)$ in part (b). Comment on the stability in each case.
(d) Let $m = 1$ slug, $k = 0.5$ lb/in., $x(0) = 2$ in., $\dot{x}(0) = 0$ in./s, and $f(t) = 0$, $t \ge 0$. Find and graph the continuous-time response $x(t)$.
(e) Choose the integration step T, so that $\omega_n T = 0.01$. Find the poles of each transfer function $H(z)$ and the discrete-time responses x_k, $k = 0, 1, 2, 3, \ldots$ for the same conditions in part (d). Plot the discrete-time responses on the same graph as $x(t)$.

4.9 FREQUENCY RESPONSE OF DISCRETE-TIME SYSTEMS

By now, it should be apparent that the methods for describing and analyzing the behavior of LTI continuous-time and discrete-time systems are similar. Indeed, both types of systems possess a natural response, independent of the system's input (or inputs), and similar in form to the impulse response of the system. The impulse response and the system transfer function form a Laplace transform pair for continuous-time systems and a z-transform pair in the case of discrete-time systems. The forced response of each is expressible by convolution, an integral for continuous-time systems and a sum for discrete-time systems.

BIBO stable systems are characterized by the location of transfer function poles in the s and z complex planes. Alternatively, the impulse response function of the continuous-time system and the impulse response (weighting sequence) of the discrete-time system satisfy equivalent types of constraints (in integral or summation form) when the systems are stable. In both instances, stability is an inherent property requiring the asymptotic decay of the natural modes.

The analogy continues into the realm of frequency response. The response of discrete-time systems to sinusoidal inputs characterizes the system dynamics in the same way as it does for continuous-time systems. Moreover, methods based on frequency response often play a critical role in the overall design of a discrete-time system.

4.9.1 STEADY-STATE SINUSOIDAL RESPONSE

We begin by considering the response of a stable LTI discrete-time system with z-domain transfer function $H(z)$ to the sinusoidal input

$$u_k = \sin k\omega T, \quad k = 0, 1, 2, \ldots \tag{4.641}$$

The z-transform of the output y_k, $k = 0, 1, 2, \ldots$, is expressed as

$$Y(z) = H(z)U(z) = H(z)\left[\frac{(\sin \omega T)z}{z^2 - (2 \cos \omega T)z + 1}\right] \tag{4.642}$$

$$= H(z)\left[\frac{(\sin \omega T)z}{(z - e^{j\omega T})(z - e^{-j\omega T})}\right] \tag{4.643}$$

A partial fraction expansion of $Y(z)$ in Equation 4.643 includes the first two terms shown in Equation 4.644 along with additional terms resulting from the poles of $H(z)$.

$$Y(z) = c_1 \frac{z}{z - e^{j\omega T}} + c_2 \frac{z}{z - e^{-j\omega T}} + \{\text{terms due to poles of } H(z)\} \tag{4.644}$$

where c_1 and c_2 are a complex conjugate pair. The constant c_1 is obtained from

$$c_1 = \frac{z - e^{j\omega T}}{z}\left[H(z)\frac{(\sin \omega T)z}{(z - e^{j\omega T})(z - e^{-j\omega T})}\right]_{z=e^{j\omega T}} = H(e^{j\omega T})\frac{\sin \omega T}{e^{j\omega T} - e^{-j\omega T}} \tag{4.645}$$

From Euler's identity, $e^{j\theta} = \cos \theta + j \sin \theta$, the denominator reduces to $2j \sin \omega T$, and the constants c_1 and c_2 become

$$c_1 = \frac{H(e^{j\omega T})}{2j}, \quad c_2 = \bar{c}_1 = \frac{H(e^{-j\omega T})}{-2j} \tag{4.646}$$

Combining Equations 4.644 and 4.646 yields

$$Y(z) = \frac{H(e^{j\omega T})}{2j} \left[\frac{z}{z - e^{j\omega T}} \right] - \frac{H(e^{-j\omega T})}{2j} \left[\frac{z}{z - e^{-j\omega T}} \right] + \{\text{terms due to poles of } H(z)\} \quad (4.647)$$

$H(e^{j\omega T})$ and $H(e^{-j\omega T})$ are complex conjugates. In polar form,

$$H(e^{j\omega T}) = Me^{j\theta}, \quad H(e^{-j\omega T}) = Me^{-j\theta} \quad (4.648)$$

where

$$M = |H(e^{j\omega T})|, \quad \theta = \text{Arg}\{H(e^{j\omega T})\} \quad (4.649)$$

Substituting both parts of Equation 4.648 into Equation 4.647 gives

$$Y(z) = \frac{M}{2j} \left[\frac{e^{j\theta} z}{z - e^{j\omega T}} - \frac{e^{-j\theta} z}{z - e^{-j\omega T}} \right] + \{\text{terms to poles of } H(z)\} \quad (4.650)$$

Inverting $Y(z)$ produces the discrete-time system response

$$y_k = \frac{M}{2j} [e^{j\theta} (e^{j\omega T})^k - e^{-j\theta} (e^{-j\omega T})^k] + \mathfrak{z}^{-1} \{\text{terms due to poles of } H(z)\} \quad (4.651)$$

$$= M \left[\frac{e^{j(k\omega T + \theta)} - e^{-j(k\omega T + \theta)}}{2j} \right] + \mathfrak{z}^{-1} \{\text{terms due to poles of } H(z)\} \quad (4.652)$$

The first term is equal to $M \sin(k\omega T + \theta)$, and $\mathfrak{z}^{-1} \{\text{terms due to poles of } H(z)\}$ is the transient response, which decays to zero at steady state for a stable system. Hence, at steady state, the response of a stable LTI discrete-time system with transfer function $H(z)$ to the sinusoidal input in Equation 4.641 is

$$(y_k)_{ss} = M \sin(k\omega T + \theta) \quad (4.653)$$

$$= |H(e^{j\omega T})| \sin[k\omega T + \text{Arg}\{H(e^{j\omega T})\}] \quad (4.654)$$

$H(e^{j\omega T})$ is the discrete-time frequency response function of the system. Note the dependence of M and θ in Equation 4.649 on the period T as well as on the frequency ω.

An expression similar to Equation 4.654 applies to LTI continuous-time systems (see Equation 4.304). Once the transient response has vanished, the steady-state response consists of a sinusoid at the same frequency as the input. The frequency response function, $H(j\omega)$, in the case of continuous-time systems, and $H(e^{j\omega T})$ for discrete-time systems, establishes the frequency dependent amplitude and phase shift in the steady-state response.

Linear systems are effectively filters that pass certain frequency components in their inputs more readily than others. Analog and digital filters are examples of how a system designer can exploit frequency response to produce a system with desirable frequency discrimination characteristics.

Frequency response is important in the study of continuous-time system simulation. It allows us to characterize the dynamic errors of a discrete-time system model intended to approximate (simulate) the dynamics of a continuous-time system. In particular, the frequency response function $H_1(e^{j\omega T})$ of a numerical integrator can be compared with the frequency response function of a continuous-time integrator $H_I(j\omega) = (1/s)|_{s=j\omega} = 1/j\omega$. For the most part, this is deferred until Chapter 8; however, we will lay some of the groundwork for what is to come later in this section.

4.9.2 PROPERTIES OF THE DISCRETE-TIME FREQUENCY RESPONSE FUNCTION

There are several important properties of $H(e^{j\omega T})$ worthy of discussion. First and foremost is its periodic nature. The argument $e^{j\omega T}$ is a unit vector beginning at $(1,0)$ in the complex plane when $\omega = 0$. As the frequency ω increases, the vector rotates counterclockwise around the Unit Circle as shown in Figure 4.72. At the sampling frequency $\omega_s = 2\pi/T$, the unit vector has completed one revolution, that is,

$$H(e^{j\omega_s T}) = H(e^{j(2\pi/T)T}) = H(e^{j2\pi}) = H(e^{j0}) = H(1) \qquad (4.655)$$

The complex values of $H(e^{j\omega T})$ generated by the first revolution around the Unit Circle are repeated during subsequent revolutions, that is, as ω increases from $k\omega_s$ to $(k+1)\omega_s$, $k = 1, 2, 3, \ldots$. In mathematical terms, the periodicity property is

$$H(e^{j(\omega+k\omega_s)T}) = H(e^{j\omega T}), \quad k = 1, 2, 3, \ldots \qquad (4.656)$$

Another property of the frequency response function is the symmetry of $|H(e^{j\omega T})|$ about the angles $\omega T = \pi, 3\pi, 5\pi, \ldots$, that is, the magnitude of $H(e^{j\omega T})$ is symmetric or folded about the frequencies $\omega = \pi/T, 3\pi/T, 5\pi/T, \ldots$. $|H(e^{j\omega T})|$ is referred to as "even" function about the radian frequencies $\omega = 0.5\omega_s, 1.5\omega_s, 2.5\omega_s, \ldots$. In mathematical terms,

$$\left| H(e^{j[n\omega_s+\Delta]}) \right| = \left| H(e^{j[n\omega_s-\Delta]}) \right|, \quad n = 0.5, 1.5, 2.5, \ldots \qquad (4.657)$$

The phase of $H(e^{j\omega T})$ is an "odd" function about the same frequencies, that is,

$$\text{Arg } H(e^{j[n\omega_s+\Delta]}) = -\text{Arg } H(e^{j[n\omega_s-\Delta]}), \quad n = 0.5, 1.5, 2.5, \ldots \qquad (4.658)$$

Due to the periodic behavior of $H(e^{j\omega T})$, it is unnecessary to plot $|H(e^{j\omega T})|$ and Arg $[H(e^{j\omega T})]$ outside $(0 \le \omega < \omega_s)$. In fact, it is customary to draw a Bode plot of $H(e^{j\omega T})$ from a lower frequency (greater than or equal to zero) up to the so-called Nyquist frequency $\omega_N = 0.5\omega_s$, from the theory of sampling. We shall have more to say about the sampling theorem after the following example, which illustrates the frequency response function of a discrete-time system and its aforementioned properties.

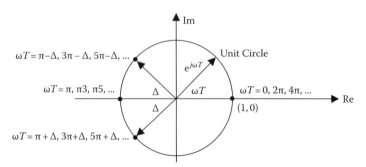

FIGURE 4.72 Rotation of unit vector $e^{j\omega T}$ around the Unit Circle.

Example 4.33

A first-order system described by $\tau dy/dt + y = u$ is to be simulated using explicit Euler integration with a step size T.

    (a) Find the z-domain transfer function $H(z)$ of the resulting discrete-time system.

    (b) Plot the magnitude (in db) and phase components of the frequency response function $H(e^{j\omega T})$ over the interval $0.1 \leq \omega \leq 0.5\omega_s$. The system time constant τ is 0.25 s, and the step size T is chosen according to $T/\tau = 0.1$.

    (c) Find the transient and steady-state response of the continuous-time system when the input is given by $u(t) = 2\sin 3t$, $t \geq 0$.

    (d) Use the discrete-time frequency response function $H(e^{j\omega T})$ to determine the steady-state output of the simulated (discrete-time) system. Verify the results graphically.

    (e) Compare the steady-state sinusoidal responses of the continuous- and discrete-time systems.

(a) The transfer function of the continuous-time system is

$$H(s) = \frac{1}{\tau s + 1} \tag{4.659}$$

The z-domain transfer function of the discrete-time system approximation obtained by the use of explicit Euler integration is

$$H(z) = \left.\frac{1}{\tau s + 1}\right|_{s \leftarrow (z-1)/T} = \frac{T/\tau}{z - 1 + (T/\tau)} \tag{4.660}$$

(b) The frequency response function of the discrete-time system is given by

$$H(e^{j\omega T}) = \left.\frac{T/\tau}{z - 1 + (T/\tau)}\right|_{z \leftarrow e^{j\omega T}} = \frac{0.1}{e^{j\omega T} - 0.9} \tag{4.661}$$

$$= \frac{0.1}{(\cos \omega T - 0.9) + j \sin \omega T} \tag{4.662}$$

The magnitude function is

$$\left|H(e^{j\omega T})\right| = \frac{0.1}{[(\cos \omega T - 0.9)^2 + \sin^2 \omega T]^{1/2}} \tag{4.663}$$

Using the trigonometric identity $\sin^2 \theta + \cos^2 \theta = 1$, the magnitude function becomes

$$\left|H(e^{j\omega T})\right| = \frac{0.1}{[1.81 - 1.8 \cos \omega T]^{1/2}} \tag{4.664}$$

From Equation 4.662, the phase angle of $H(e^{j\omega T})$ is

$$\text{Arg}[H(e^{j\omega T})] = -\tan^{-1}\left(\frac{\sin \omega T}{\cos \omega T - 0.9}\right) \tag{4.665}$$

The sampling frequency $\omega_s = 2\pi/T = 80\pi$ rad/s. The Nyquist frequency $\omega_N = 0.5\omega_s = 40\pi = 125.67$ rad/s. "Chap4_Ex9_1.m" contains the MATLAB code to generate the Bode plot shown in Figure 4.73.

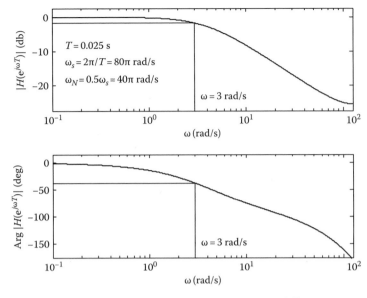

FIGURE 4.73 Bode plot for system with frequency response function $H(e^{j\omega T})$ in Equation 4.661.

(c) The continuous-time response $y(t)$ to the input $u(t) = 2 \sin 3t$, $t \geq 0$ is obtained from the transfer function of the continuous-time system

$$H(s) = \frac{Y(s)}{U(s)} = \frac{1}{\tau s + 1} = \frac{1}{0.25s + 1} \tag{4.666}$$

$$Y(s) = H(s)U(s) = \frac{1}{0.25s + 1}\left[2\left(\frac{3}{s^2 + 9}\right)\right] \tag{4.667}$$

$$= \frac{24}{(s + 4)(s^2 + 9)} \tag{4.668}$$

Inverting $Y(s)$ by partial fractions leads to

$$y(t) = \frac{24}{25}\left(e^{-t/0.25} + \frac{4}{3}\sin 3t - \cos 3t\right) \tag{4.669}$$

The transient and steady-state components of $y(t)$ are

$$y_{tr} = \frac{24}{25}e^{-t/0.25}, \quad y_{ss} = \frac{24}{25}\left(\frac{4}{3}\sin 3t - \cos 3t\right) \tag{4.670}$$

(d) The magnitude and phase of the discrete-time frequency response function at $\omega = 3$ rad/s and $T = 0.025$ s are obtained from Equations 4.664 and 4.665, respectively.

$$|H(e^{j3(0.025)})| = \frac{0.1}{[1.81 - 1.8 \cos 3(0.025)]^{1/2}} = 0.815 \tag{4.671}$$

$$\text{Arg}[H(e^{j3(0.025)})] = -\tan^{-1}\left[\frac{\sin 3(0.025)}{\cos 3(0.025) - 0.9}\right] = -0.657 \text{ rad} \tag{4.672}$$

The dashed lines in Figure 4.73 show the gain $20 \log(0.815) = -1.78$ db and phase angle -0.657 rad $= -37.63°$ at $\omega = 3$ rad/s. For a sinusoidal input with magnitude $|u(t)| = 2$, the discrete-time system output at steady state is from Equation 4.654

$$(y_k)_{ss} = 2[0.815 \sin(0.075k - 0.657)] \tag{4.673}$$

A graph of u_k and $(y_k)_{ss}$ is shown in Figure 4.74.

The steady-state output component lags the input by $37.63°$, and its amplitude is

$$2|H(e^{j3(0.025)})| = 2(0.815) = 1.63.$$

In order to verify the frequency response values in Equations 4.671 and 4.672, a blown-up portion of Figure 4.74 near consecutive peaks is shown in Figure 4.75.

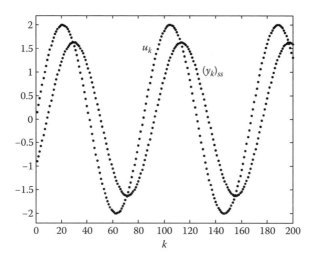

FIGURE 4.74 Sinusoidal input and steady-state sinusoidal output of discrete-time system.

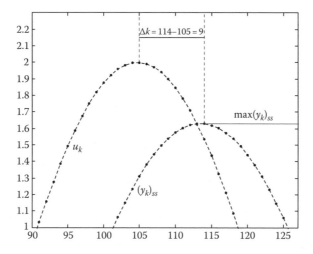

FIGURE 4.75 Graphical enlargement of Figure 4.74 for verifying frequency response characteristics.

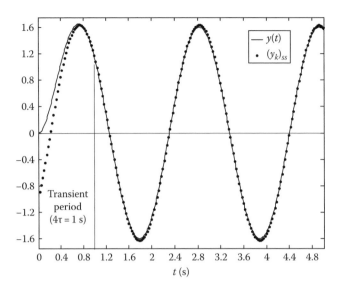

FIGURE 4.76 Continuous-time system response and sinusoidal steady-state response of discrete-time system obtained using frequency response function.

The amplitude of $(y_k)_{ss}$ is in agreement with the predicted value of 1.63. The peak amplitudes occur at approximately $k = 105$ for u_k and $k = 114$ for $(y_k)_{ss}$. The measured phase lag is $\Delta k\omega T = (114 - 105)(3 \text{ rad/s})(0.025 \text{ s}) = 0.675$ rad, which compares favorably with the analytical value of 0.657 rad.

(e) The continuous-time system response $y(t)$ in Equation 4.669 for $0 \leq t \leq 5$ and $(y_k)_{ss}$, the steady-state component of the discrete-time system output, are plotted in Figure 4.76.

Note the agreement between the two outputs once the continuous-time transient response has vanished.

The discrete-time system transient component cannot be obtained solely from the discrete-time frequency response function $H(e^{j\omega T})$. However, from Equation 4.660, we know it has the form cp^k where $p = 1 - T/\tau = 0.9$. Finding the constant c requires partial fraction expansion of $Y(z) = H(z)U(z)$ where $U(z)$ is the z-transform of the discrete-time input $u_k = 2 \sin k\omega T$, $k = 0, 1, 2, \ldots$.

4.9.3 SAMPLING THEOREM

The Bode plot in Figure 4.73 displays the frequency response characteristics of the discrete-time system used to simulate a first-order continuous-time system. $H(e^{j\omega T})$ is a periodic function with period $\omega_s = 2\pi/T$. The maximum frequency on Bode plots of discrete-time systems is generally limited to $\omega_N = \omega_s/2 = \pi/T$, where ω_N is called the Nyquist frequency. The limitation is based on the sampling theorem, which we shall explore in a very rudimentary fashion.

First, let us verify the periodic nature of $H(e^{j\omega T})$ as well as its symmetry about the frequencies $0.5\omega_s$, $1.5\omega_s$, $2.5\omega_s, \ldots$ or equivalently ω_N, $3\omega_N$, $5\omega_N, \ldots$. Figure 4.77 shows the magnitude $|H(e^{j\omega T})|$ and phase $\text{Arg}[H(e^{j\omega T})]$ corresponding to the frequency response function $H(e^{j\omega T})$ in Equation 4.662. The sampling frequency $\omega_s = 2\pi/T = 251.3$ rad/s and the plots in Figure 4.77 extend for three periods, that is $(0 \geq \omega \leq 3\omega_s)$. The symmetry of $H(e^{j\omega T})$ in both magnitude and phase about the Nyquist frequency, $\omega_N = \omega_s/2 = 125.7$ rad/s, $3\omega_N$, $5\omega_N$, $7\omega_N, \ldots$ is evident.

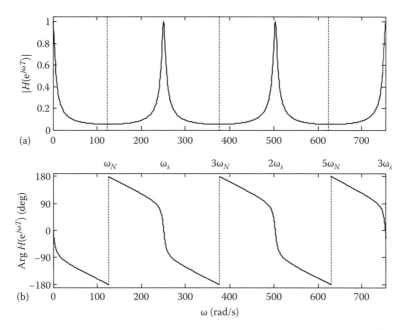

FIGURE 4.77   (a) Periodic nature and (b) symmetry of frequency response function $H(e^{j\omega T})$.

The sampling theorem, as the name suggests, applies to sample data systems where a continuous-time signal is periodically sampled. The theorem, however, has important ramifications for discrete-time systems and continuous system simulation.

Continuous-time sinusoids $\sin \omega t$ are sampled every T s as shown in Figure 4.78. In the top frame, the frequency of the sinusoid is $\omega = 0.2\pi$ rad/s, and sampling occurs at a rate of one sample per second ($\omega_s = 2\pi$ rad/s). The period of the sinusoid is $2\pi/\omega = 10$ s, and the sampling rate of 10 samples per period is sufficient to reconstruct the original sinusoid.

In the middle plot, the frequency of the continuous-time sinusoid is 1.4π rad/s while the sampling remains fixed at 2π rad/s. The sampled points appear to come from the lower frequency ($\omega = 0.4\pi$ rad/s) continuous-time sinusoid shown in dashed form. In the bottom graph, the continuous-time sinusoid has a frequency of 2.2π rad/s. Sampling it at the rate of 2π rad/s produces the identical set of data points obtained in the top graph making it appear as if the continuous-time sinusoid being sampled is the one at $\omega = 0.2\pi$ rad/s.

There is no ambiguity in identifying the correct sinusoid being sampled provided sampling occurs more than twice as fast as the frequency of the sinusoid, that is, $\omega_s > 2\omega$. In other words, the sampled points uniquely determine any continuous-time sinusoid whose frequency is less than one half the sampling frequency, that is, $\omega < \omega_s/2$.

By definition, the Nyquist frequency is $\omega_N = \omega_s/2$. Hence, for a given sampling frequency ω_s, only sinusoids with frequency less than the Nyquist frequency can be distinguished from lower frequency sinusoids. A sinusoid with frequency greater then the Nyquist frequency will be "aliased" into a lower frequency sinusoid as in the last two cases shown in Figure 4.78. This explains why Bode plots of discrete-time frequency response functions range from a lower frequency up to the Nyquist frequency ω_N.

The sampling theorem applies to sampling of continuous-time signals in general. Aliasing occurs when $\omega_s \leq 2\omega_0$, where ω_0 represents the highest frequency present in the band-limited signal.

Effect of sampling sin ωt (ω = 0.2π, 1.4π, 2.2π rad/s) at ω_s = 2π rad/s

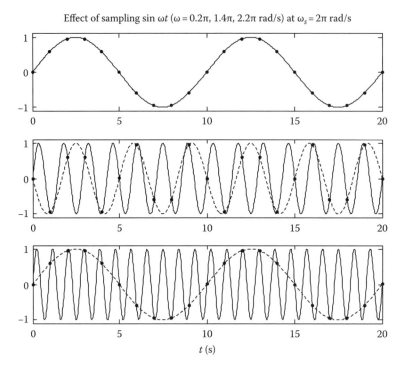

FIGURE 4.78   Illustration of aliasing of a sampled sinusoid.

In terms of the sampling period, $T < \pi/\omega_0$ to prevent aliasing. The sampling theorem presents a formula, albeit difficult to implement, for reconstructing the band-limited continuous-time signal from the numerical values of the samples (Cadzow 1973).

The sampling theorem extends to simulation of continuous-time systems. The sampling interval T becomes the integration step size. Continuous-time inputs to the differential equations are sampled in the process of generating the discrete-time inputs to the difference equations. Consequently, the frequency content of the continuous-time input signals influences the choice of appropriate step size in the simulation.

Example 4.34

The continuous-time first-order system in Example 4.33 is to be simulated using trapezoidal integration

   (a) Find the z-domain transfer function of the discrete-time system and the difference equation. Leave your answers in terms of the continuous-time system time constant τ and the integration step size T.
   (b) Find the sampling frequency and Nyquist frequency when $\tau = 5$ s and $T = 0.25$ s.
   (c) Find the continuous-time output $y(t)$ when the input $u(t) = \sin \omega t$, $t \geq 0$.
   (d) Plot the continuous-time and discrete-time outputs on the same graph when

       (i) $\omega = \pi$ rad/s $\omega = 7\pi$ rad/s $\omega = 8\pi$ rad/s.

   (e) Compare $H(j\omega)$ and $H(e^{j\omega T})$ at $\omega = \pi$, 7π, and 8π rad/s.

(a) The z-domain transfer function of the discrete-time system is

$$H(z) = H(s)\big|_{s \leftarrow (2/T)((z-1)/(z+1))} = \frac{1}{\tau s + 1}\bigg|_{s \leftarrow (2/T)((z-1)/(z+1))} \tag{4.674}$$

$$= \frac{1}{\tau[(2/T)((z-1)/(z+1))] + 1} \tag{4.675}$$

$$= \frac{T(z+1)}{(2\tau + T)z - (2\tau - T)} \tag{4.676}$$

$$\Rightarrow H(z) = \frac{Y(z)}{U(z)} = \frac{T(1 + z^{-1})}{(2\tau + T) - (2\tau - T)z^{-1}} \tag{4.677}$$

Inverting Equation 4.677 produces the difference equation

$$(2\tau + T)y_k - (2\tau - T)y_{k-1} = T(u_k + u_{k-1}), \quad k = 1, 2, 3, \ldots \tag{4.678}$$

(b) $\omega_s = 2\pi/T = 2\pi/0.25 = 8\pi$ rad/s, $\omega_N = \omega_s/2 = 8\pi/2 = 4\pi$ rad/s.

(c) The continuous-time output $y(t)$ is obtained by inverse Laplace transformation of

$$Y(s) = H(s)U(s) = \frac{1}{\tau s + 1}\left[\frac{\omega}{s^2 + \omega^2}\right] \tag{4.679}$$

Following partial fraction expansion and inverse Laplace transformation, the result is

$$y(t) = \frac{\tau \omega}{1 + (\tau\omega)^2}\left[e^{-t/\tau} - \cos\omega t + \frac{1}{\tau\omega}\sin\omega t\right] \tag{4.680}$$

(d) Substituting $\tau = 5$ s, $\omega = \pi$, 7π, and 8π rad/s gives the continuous-time output for the three cases enumerated. The simulated output is obtained by recursive solution of the difference equation after solving explicitly for y_k in Equation 4.678.

$$y_k = \frac{1}{2\tau + T}[(2\tau - T)y_{k-1} + T(u_k + u_{k-1})], \quad k = 1, 2, 3, \ldots \tag{4.681}$$

The continuous- and discrete-time outputs are evaluated in "Chap4_Ex9_2.m." Plots of $y(t)$, $t \geq 0$ and y_k, $k = 0, 1, 2, \ldots$ for the three input sinusoids are presented in Figures 4.79 through 4.81 along with the continuous- and discrete-time inputs.

In Figure 4.79, the input frequency $\omega = \pi$ rad/s is well below the Nyquist frequency $\omega_N = 4\pi$ rad/s. Sampled values of the discrete-time input $u_k = \sin(k\omega T)$ are an accurate reflection of the continuous-time sinusoidal input $u(t) = \sin\omega t$. As a result, the continuous-time response $y(t)$ and simulated response y_k are in close agreement at the sample times $t_k = kT$, $k = 0, 1, 2, \ldots$.

In Figure 4.80, $\omega = 7\pi$ rad/s exceeds the Nyquist frequency $\omega_N = 4\pi$ rad/s. The simulated output is the response to the alias term whose frequency is π rad/s (shown dotted in Figure 4.80). Understandably, the simulated response y_k bears no resemblance to the continuous-time response $y(t)$.

In Figure 4.81, the sampling frequency is the same as the frequency of the sinusoid, that is, $\omega_s = \omega = 8\pi$ rad/s. The same value of zero is sampled once per cycle making the effective input to the discrete-time system $u_k = 0$, $k = 0, 1, 2, \ldots$. The simulated (discrete-time) output is identically zero as well. The continuous-time response is also shown.

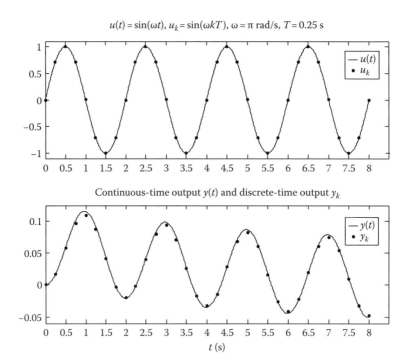

FIGURE 4.79   Continuous- and discrete-time inputs and outputs ($\omega = \pi$ rad/s).

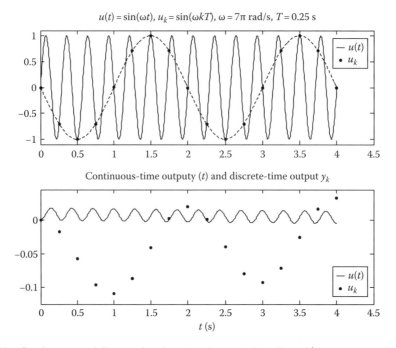

FIGURE 4.80   Continuous- and discrete-time inputs and outputs ($\omega = 7\pi$ rad/s).

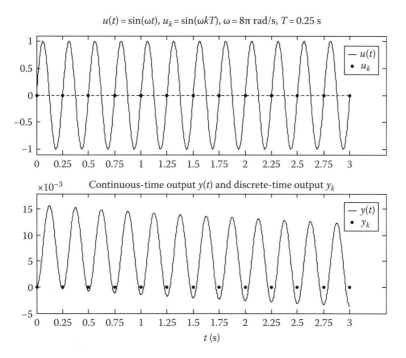

FIGURE 4.81   Continuous- and discrete-time system inputs and outputs ($\omega = 8\pi$ rad/s).

(e) There are several ways to determine the magnitude and phase of the continuous-time and discrete-time frequency response functions. One is to start with $H(j\omega)$ and $H(e^{j\omega T})$ and express the magnitude and phase components in terms of ω with τ and T (in the discrete-time case) as parameters.

$$H(j\omega) = H(s)|_{s=j\omega} = \left.\frac{1}{\tau s + 1}\right|_{s=j\omega} = \frac{1}{\tau j\omega + 1} \tag{4.682}$$

$$= \frac{1}{[(\tau\omega)^2 + 1]^{1/2}} \quad \angle -\tan^{-1}(\tau\omega) \tag{4.683}$$

$$H(e^{j\omega T}) = H(z)|_{z=e^{j\omega T}} = \left.\frac{T(z + 1)}{(2\tau + T)z - (2\tau - T)}\right|_{z=e^{j\omega T}} \tag{4.684}$$

$$= \frac{T(e^{j\omega T} + 1)}{(2\tau + T)e^{j\omega T} - (2\tau - T)} \tag{4.685}$$

$$= \frac{T(\cos\omega T + j\sin\omega T + 1)}{(2\tau + T)(\cos\omega T + j\sin\omega T) - (2\tau - T)} \tag{4.686}$$

$$|H(e^{j\omega T})| = \frac{T[(1 + \cos\omega T)^2 + \sin^2\omega T]^{1/2}}{\{[(2\tau + T)\cos\omega T - (2\tau - T)]^2 + [(2\tau + T)\sin\omega T]^2\}^{1/2}} \tag{4.687}$$

$$\text{Arg}[H(e^{j\omega T})] = \tan^{-1}\left(\frac{\sin\omega T}{1 + \cos\omega T}\right) - \tan^{-1}\left(\frac{(2\tau + T)\sin\omega T}{(2\tau + T)\cos\omega T - (2\tau - T)}\right) \tag{4.688}$$

A simpler alternative for computing either frequency response function for a given frequency ω is by direct substitution of $s = j\omega$ and $z = e^{j\omega T}$. The resulting complex numbers in rectangular form are then expressed in polar form. To illustrate, consider the continuous-time frequency response function $H(j\omega)$ when $\omega = \pi$ rad/s.

$$H(j\pi) = \frac{1}{5j\pi + 1} = \frac{1 - j5\pi}{1 - j5\pi} \cdot \frac{1}{1 + j5\pi} \tag{4.689}$$

$$= \frac{1}{1 + (5\pi)^2} - j\frac{5\pi}{1 + (5\pi)^2} \tag{4.690}$$

$$= 0.0040 - j0.0634$$

$$|H(j\pi)| = [(0.0040)^2 + (-0.0634)^2]^{1/2} = 0.0635$$

$$\text{Arg}[H(j\pi)] = \tan^{-1}\left(\frac{-0.0634}{0.0040}\right) = -1.5072 \text{ rad} (-86.4°)$$

The discrete-time frequency response function is evaluated in the same fashion.

$$H(e^{j\pi 0.25}) = \frac{0.25(e^{j\pi 0.25} + 1)}{[(2)(5) + 0.25]e^{j\pi 0.25} - [(2)(5) - 0.25)]}\bigg| \tag{4.691}$$

$$|H(e^{j\pi 0.25})| = 0.0036 - j0.0601 = [(0.0036)^2 + (-0.0601)^2]^{1/2} = 0.0602$$

$$\text{Arg}[H(e^{j\pi 0.25})] = \tan^{-1}\left(\frac{-0.0601}{0.0036}\right) = -1.5105 \text{ rad} (-86.6°)$$

Table 4.10 lists the magnitude and phase of the continuous-time and discrete-time frequency response functions evaluated at $\omega = \pi$, 7π, and 8π rad/s. The DC ($\omega = 0$) values are also included. It follows from Equation 4.682 that the DC gain and phase of the continuous-time system are 1° and 0°, respectively. For the discrete-time system, the DC gain and phase are obtained from $H(e^{j0T}) = H(1) = 1$. The period of $H(e^{j\omega T})$ is $\omega_s = 8\pi$ rad/s and its symmetric about the Nyquist frequency $\omega_N = 4\pi$ rad/s.

In order to verify the magnitudes shown in Table 4.10, it is necessary to extend the time scale in Figures 4.79 through 4.81. For example, in Figure 4.79, when the input is $u(t) = \sin \pi t$, $t \geq 0$, the continuous- and discrete-time steady-state responses are sinusoids with amplitudes in the neighborhood of 0.06 (see Table 4.10). Looking at Figure 4.79, the continuous- and discrete-time responses have yet to reach steady state. Figure 4.82 shows both responses for a period of time equal to five time constants ($5\tau = 25$ s). The steady-state amplitudes are in agreement with the values in the table.

4.9.4 DIGITAL FILTERS

Linear discrete-time systems process signals with known frequency content in a predictable fashion. Digital filters are designed to block or pass selected frequencies present in the discrete-time inputs.

TABLE 4.10
Continuous- and Discrete-Time Frequency Response
($\omega = 0$, π, 7π, 8π)

ω, rad/s	$|H(j\omega)|$	Arg[$H(j\omega)$] (°)	$|H(e^{j\omega 0.25})|$	Arg[$H(e^{j\omega 0.25})$] (°)
0	1	0	1	0
π	0.0635	−86.36	0.0602	−86.54
7π	0.0091	−89.48	0.0602	−86.54
8π	0.0080	−89.54	1	0

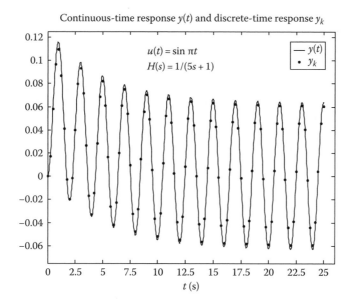

FIGURE 4.82 Continuous- and discrete-time responses showing steady state.

Numerous references in the area of digital signal processing are available for further reading about digital filters (Oppenheim 1999; Parks 1987).

Low-pass filters, as the name suggests, are intended to readily pass low frequencies and attenuate all others. Figure 4.73 showed the discrete-time frequency response function of a low-pass filter. The cutoff frequency (bandwidth) of a low-pass filter with DC gain of 0 db is the frequency at which the gain equals -3 db. A low-pass filter can be thought of as passing those frequencies within its bandwidth.

Two ways of implementing a low-pass digital filter are illustrated next.

Example 4.35

Twenty-five years of end-of-month lake water temperature readings T_k, $k = 0, 1, 2, 3, \ldots, 300$ months, are stored in MATLAB data file "*Chap4_LakeTemp.mat.*" Researchers would like to determine if the lake temperature, adjusted for monthly variations, has changed over that time.

(a) A moving average of the past 12 readings is used to smooth the seasonal temperature variations, that is,

$$\hat{T}_k = \frac{1}{12}[T_{k-1} + T_{k-2} + \cdots + T_{k-12}], \quad k = 12, 13, \ldots, 300 \qquad (4.692)$$

where $\hat{T}_k$ is the seasonally adjusted end-of-month lake temperature starting with end-of-month 12. (Note that T_0 is the lake temperature on December 31 of a given year and $\hat{T}_{12}$ represents the seasonally adjusted lake temperature on December 31 of the following year.) Find the z-domain transfer function $H(z) = \hat{T}(z)/T(z)$, and plot the magnitude of the discrete-time frequency response function.

(b) The period of the seasonal variation is $P = 12$ months. The frequency is $\omega_0 = 2\pi/P = \pi/6$ rad/month. Find $|H(e^{j\omega_0 T})|$ where $T = 1$ month (sampling period).

(c) Find $\hat{T}_k$, $k = 12, 13, \ldots, 300$ and plot the values on the same graph with T_k, $k = 0, 1, 2, \ldots, 300$. Estimate the yearly increase in lake temperature.

(d) A first-order low-pass digital filter with z-domain transfer function

$$H(z) = \frac{\hat{T}(z)}{T(z)} = \frac{(1-\alpha)z}{z-\alpha} = \frac{(1-\alpha)}{1-\alpha z^{-1}} \tag{4.693}$$

is used to filter the monthly temperature variations and pass any low-frequency lake temperature variation with time. Plot the discrete-time frequency response magnitude and phase for $\alpha = 0.99$. Find the cutoff frequency and determine whether it is less or greater than $\omega_0 = \pi/6$ rad/month. In addition, find $|H(e^{j\omega_0 T})|$.

(e) Find the difference equation relating $\hat{T}_k$ and T_k. Solve it recursively for $\hat{T}_k$, $k = 1, 2, \ldots, 300$ and plot both input and output on the same graph.

(a) Taking the z-transform of Equation 4.692 gives

$$\hat{T}(z) = \frac{1}{12}[z^{-1}T(z) + z^{-2}T(z) + \cdots + z^{-12}T(z)] \tag{4.694}$$

$$\Rightarrow H(z) = \frac{\hat{T}(z)}{T(z)} = \frac{1}{12}(z^{-1} + z^{-2} + \cdots + z^{-12}) \tag{4.695}$$

The magnitude function is shown in Figure 4.83. The data points for generating the graph in Figure 4.83 are computed in M-file "*Chap4_Ex9_3.m.*"

(b) From Figure 4.83, we see that the zeros of $|H(e^{j\omega T})|$ are located at $\omega_0 = \pi/6$ and multiples of ω_0, namely, $2\pi/6$, $3\pi/6$, $4\pi/6$, $5\pi/6$, $\pi, \ldots$.

(c) The smoothing algorithm, Equation 4.692, is applied to the monthly lake temperature data, and the results $\hat{T}_k$, $k = 12, 13, \ldots, 300$ are plotted along with the discrete-time input T_k, $k = 0, 1, 2, \ldots, 300$ in Figure 4.84.

The estimated annual increase in lake temperature is

$$m_1 = \frac{\hat{T}_{300} - \hat{T}_{12}}{24} = \frac{71.016 - 65.353}{24} = 0.236 \ \frac{°F}{year} \tag{4.696}$$

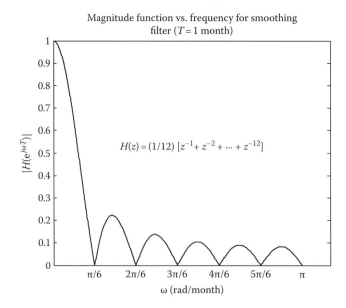

FIGURE 4.83 Magnitude function for smoothing filter.

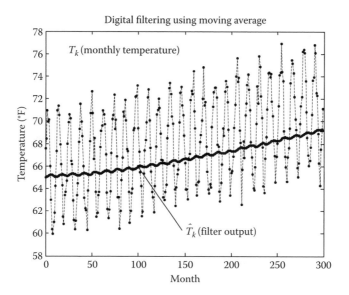

FIGURE 4.84 Input T_k, $k = 0, 1, 2, \ldots, 300$ and smoothing filter output $\hat{T}_k$, $k = 12, 13, \ldots, 300$.

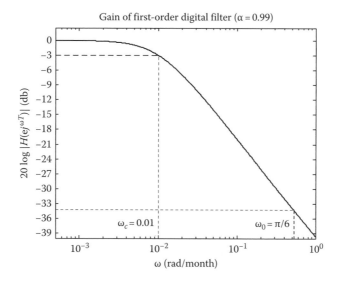

FIGURE 4.85 Gain of first-order low-pass filter with $H(z)$ in Equation 4.693.

(d) The gain of the filter with z-domain transfer function in Equation 4.693 and $\alpha = 0.99$ is shown in Figure 4.85.

The cutoff frequency ω_c is obtained from

$$20 \log |H(e^{j\omega_c T})| = -3 \tag{4.697}$$

It is left as an exercise problem to show that

$$\omega_c = \frac{1}{T} \cos^{-1} \left(\frac{1 + \alpha^2 - 10^{0.3 + 2\log(1-\alpha)}}{2\alpha} \right) \tag{4.698}$$

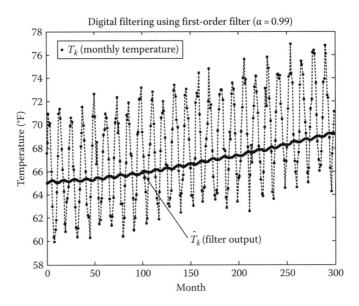

FIGURE 4.86　Input T_k, $k = 0, 1, 2, \ldots, 300$ and first-order filter output $\hat{T}_k$, $k = 1, 2, \ldots, 300$.

Substituting $\alpha = 0.99$, $T = 1$ into Equation 4.698 yields $\omega_c = 0.01$ rad/month. Referring to Figure 4.85, $\omega_0 = \pi/6$ rad/month is well beyond the cutoff frequency, and we should expect the seasonal fluctuations to be removed by the digital filter. The magnitude of $H(e^{j\omega_0 T})$ is 0.0194 (-34.2 db).

(e) The difference equation is obtained from Equation 4.693 by inverse z-transformation.

$$\hat{T}_k - \alpha \hat{T}_{k-1} = (1 - \alpha)T_k, \quad k = 1, 2, 3, \ldots, 300 \tag{4.699}$$

where $\hat{T}_0 = T_0$. Equation 4.699 is solved recursively for $\hat{T}_k$, $k = 1, 2, 3, \ldots, 300$ in "*Chap4_Ex9_3.m.*" The discrete-time input and output are shown in Figure 4.86.

Example 4.35 illustrates the use of FIR and IIR filters. From Equation 4.695, the FIR smoothing filter impulse response is

$$h_k = \frac{1}{12}(\delta_{k-1} + \delta_{k-2} + \cdots + \delta_{k-12}) \tag{4.700}$$

$$= \begin{cases} 0, & k = 0 \\ \dfrac{1}{12}, & k = 1, 2, \ldots, 12 \\ 0, & k = 13, 14, \ldots \end{cases} \tag{4.701}$$

From Equation 4.693, the first-order IIR low-pass digital filter impulse response is

$$h_k = (1 - \alpha)\alpha^k, \quad k = 0, 1, 2, 3, \ldots \tag{4.702}$$

Based on the convolution sum for the output of a discrete-time system, the FIR filter output depends solely on the past 12 inputs (not surprising) while the infinite memory IIR filter output relies on the entire set of past inputs.

Choosing $\alpha = 0.99$ places the pole of $H(z)$ precariously close to the Unit Circle, the stability boundary in the z-plane. As a consequence, discrete-time input signals with poles near $z = 0.99$, for example, a step input with pole at $z = 1$, are readily passed.

The transient response period is considerable since the natural mode $\alpha^k = 0.99^k$ takes a long while to decay to zero. In Figure 4.86, if we arbitrarily assume the transient period to be 150 months $[0.99^{150} = 0.22]$, the estimated slope of the linear rise in lake temperature is computed as

$$m_2 = \frac{(\hat{T}_{300} - \hat{T}_{150})°\text{F}}{(300 - 150)\,\text{months} \times 1/12\,\text{year/month}} = \frac{69.179 - 66.649}{12.5} = 0.202\,\frac{°\text{F}}{\text{year}} \quad (4.703)$$

which is close to the value obtained using the FIR smoothing filter.

EXERCISES

4.70 Repeat Example 4.33 using implicit Euler instead of explicit Euler integration for approximating the continuous-time system.

4.71 A second-order system with damping ratio ζ and natural frequency ω_n is simulated using trapezoidal integration. The DC gain of the system is unity.
   (a) Find the discrete-time frequency response function $H(e^{j\omega T})$. Leave your answer in terms of ζ, ω_n, T, and ω.
   (b) Draw a Bode plot of $H(e^{j\omega T})$ when the continuous-time system poles are as shown in Figure E4.71. Assume $\omega_n T = 0.1$.

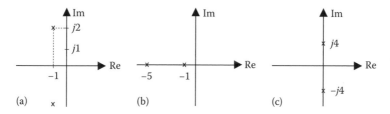

FIGURE E4.71

4.72 The electrical circuit shown in Figure E4.72 is that of a biquad filter, so named because the transfer function from the input to the output contains quadratic factors in the numerator and denominator. The differential equation of the circuit is

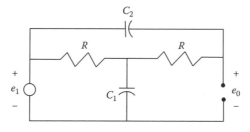

FIGURE E4.72

$$a_2 \ddot{e}_0 + a_1 \dot{e}_0 + a_0 e_0 = b_2 \ddot{e}_1 + b_1 \dot{e}_1 + b_0 e_1$$

where the constants a_0, a_1, a_2 and b_0, b_1, b_2 are related to R, C_1, C_2 by

$$a_0 = 1, \quad a_1 = RC_1 + 2RC_2, \quad a_2 = RC_1 RC_2, \quad b_0 = 1, \quad b_1 = 2RC_2, \quad b_2 = RC_1 RC_2$$

(a) Find the transfer function $G(S) = E_0(S)/E_1(S)$.

(b) A discrete-time system approximation based on trapezoidal integration has a z-domain transfer function $G(z)$ given by

$$G(z) = \frac{\beta_2 z^2 + \beta_1 z + \beta_0}{\alpha_2 z^2 + \alpha_1 z + \alpha_0}$$

Show that

$$\beta_0 = 4\tau_1\tau_2 - 4\tau_2 T + T^2, \quad \beta_1 = -8\tau_1\tau_2 + 2T^2, \quad \beta_2 = 4\tau_1\tau_2 + 4\tau_2 T + T^2$$
$$\alpha_0 = 4\tau_1\tau_2 - 2(\tau_1 + 2\tau_2)T + T^2, \quad \alpha_1 = -8\tau_1\tau_2 + 2T^2$$
$$\alpha_2 = 4\tau_1\tau_2 + 2(\tau_1 + 2\tau_2)T + T^2$$

where $\tau_1 = RC_1$ and $\tau_2 = RC_2$ and T is the integration step size.

(c) Draw a Bode plot for the discrete-time frequency response $G(e^{j\omega T})$ when $\tau_1 = 0.1$ s, $\tau_2 = 0.001$ s, and $T = 2 \times 10^{-4}$ s.

(d) Fill in the following table.

ω, rad/s	$|G(j\omega)|$	$\text{Arg}[G(j\omega)]$	$|G(e^{j\omega T})|$	$\text{Arg}[G(e^{j\omega T})]$
0				
5				
100				
5000				

4.73 An analog signal $r(t)$ is the command input to a digital control system, part of which is shown in Figure E4.73. The signal $r(t)$ must be sampled and converted to a discrete-time signal for use by the digital controller. The command input consists of a signal component $s(t)$ and a high-frequency (compared to the sampling rate $1/T_s$) noise component $n(t)$. An antialiasing filter is inserted before sampling to eliminate aliasing in $\hat{r}_k$ the input to the controller.

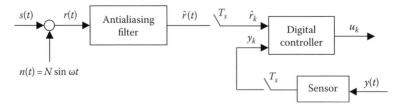

FIGURE E4.73

A fourth-order Butterworth low-pass filter is chosen. The transfer function is

$$G(s) = \frac{\hat{R}(s)}{R(s)} = \left(\frac{\omega_n^2}{s^2 + 2\cos(\pi\omega_n/8)s + \omega_n^2}\right)\left(\frac{\omega_n^2}{s^2 + 2\cos(3\pi\omega_n/8)s + \omega_n^2}\right)$$

(a) The control system sampling rate is 1000 Hz. Find the Nyquist frequency ω_N.

(b) Find ω_n, so that the magnitude of $G(j\omega)$ is -60 db at the Nyquist frequency.

Hint: Use trial and error guesses for ω_n along with Bode plots until the condition $|G(j\omega_N)| = -60$ db is approximately satisfied.

(c) The signal and noise components of the command input $r(t)$ are $s(t) = 1$, $t \geq 0$ and $n(t) = 5 \times 10^{-3} \sin(2 \times 10^6 t)$, $t \geq 0$. Find the filter output $\hat{r}(t)$ at steady state.

(d) Find $G(z)$, the z-domain transfer function of the discrete-time system approximation to $G(s)$ using explicit Euler integration. Leave your answer in terms of the integration step size T.

(e) Comment on the choice of T necessary to simulate the filter response by recursive solution of the difference equation corresponding to $G(z)$.

4.74 A method for approximating a continuous-time system with transfer function $G(s)$ is illustrated in Figure E4.74. A continuous-time input $u(t)$ is sampled every T s to produce the discrete-time input u_k. A zero-order hold (ZOH) reconstructs a piecewise continuous approximation to $u(t)$ denoted $\hat{u}(t)$, which is the input to the continuous-time system. The continuous-time output $y(t)$ is sampled every T s resulting in the discrete-time output y_k. The discrete-time system with input u_k and output y_k serves as an approximation to the continuous-time system with input $u(t)$ and output $y(t)$. The z-domain transfer function of the discrete-time system is (Jacquot)

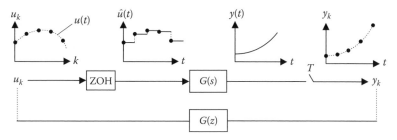

FIGURE E4.74

$$G(z) = \frac{Y(z)}{U(z)} = \left(\frac{z-1}{z}\right) \mathfrak{z} \left\{ \mathcal{L}^{-1} \left[\frac{G(s)}{s} \right] \right\}$$

where $\mathfrak{z} \{ \mathcal{L}^{-1}[G(s)/s] \}$ stands for the z-transform of the discrete-time signal resulting from sampling the continuous-time signal $\mathcal{L}^{-1}[G(s)/s]$.

(a) Find the z-domain transfer function using the ZOH approximation method when the continuous-time system is first order with transfer function $G(s) = 1/(\tau s + 1)$. Leave your answer in terms of the time constant τ and sampling period T.

(b) Find the discrete-time frequency response function $G(e^{j\omega T})$, and obtain expressions for the magnitude $|G(e^{j\omega T})|$ and phase $\text{Arg}[G(e^{j\omega T})]$.

(c) Plot the magnitude and phase of $G(e^{j\omega T})$ when $\tau = 1$ s and $T = 0.1$ s.

(d) Compare the continuous- and discrete-time unit step responses and comment on the results.

(e) Find $|G(e^{j\omega T})|$ and $\text{Arg}[G(e^{j\omega T})]$ and compare with the values given in Table 4.10 where $\tau = 5$ s and $T = 0.25$ s.

4.75 Derive Equation 4.698 for the cutoff frequency of the first-order low-pass digital filter with z-domain transfer function $H(z) = (1 - \alpha)z/(z - \alpha)$.

4.76 A notch filter is designed to attenuate input signals at one specific frequency called the notch frequency. The transfer function of a notch filter is

$$G(s) = \frac{s^2 + \omega_n^2}{s^2 + 2\varsigma \omega_n s + \omega_n^2} \quad (\omega_n \text{ is notch frequency})$$

(a) Find $G(z)$, the z-domain transfer function of a digital filter obtained by approximation of $G(s)$ using trapezoidal integration. Leave your answer in terms of ζ, ω_n and the integration step size T.

(b) The digital filter is to be used to filter out the monthly lake temperature fluctuations in Example 4.35. The notch frequency is $\omega_n = \pi/6$ rad/month and the sampling period is $T = 1$ month. On the same graph, plot $|G(e^{j\omega T})|$ vs. ω from zero to the Nyquist frequency for $\zeta = 0.25, 0.5, 0.75$.

(c) Choose the value of ζ, which produces the largest attenuation at the notch frequency, and use the digital notch filter to filter out the monthly lake temperature fluctuations in the dataset "*Chap4_LakeTemp.mat.*" Prepare a graph similar to the ones in Figures 4.84 and 4.86.

4.77 The design of a digital filter calls for the placement of a pair of poles and zeros as shown in Figure E4.77.

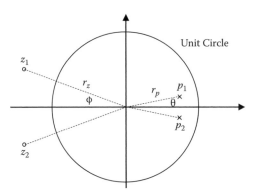

FIGURE E4.77

(a) Find the difference equation relating the filter's input $u(n)$ and output $y(n)$. The filter coefficients should be expressed in terms of r_p, θ, r_z, and ϕ.

(b) Express the magnitude function $|H(e^{j\omega T})|$ in terms of the parameters r_p, θ, r_z, ϕ and the sampling time T.

(c) Plot the magnitude function for the case when $T = 1$ s, $r_p = 0.9, r_z = 2, \phi = \pi/4$ and $\theta = 0.2, 0.4, 0.6, 0.8, 1$ rad. Comment on the results.

4.10 CONTROL SYSTEM TOOLBOX

This chapter has emphasized analytical methods for obtaining continuous- and discrete-time system response to elementary types of inputs. In this section, we explore the use of MATLAB functions in the control system toolbox designed to facilitate the process of modeling and simulation of LTI dynamic systems. The control system toolbox is a supplement to MATLAB. The reader is encouraged to check out the entire suite of available functions either online or in the control system toolbox lab manual (from The Mathworks, Inc.). Many of the functions are discussed and illustrated in recent linear controls texts and companion lab manuals (D'Azzo and Houpis, 1995; Ogata 1998; Nekoogar 1999; Dorf and Bishop 2005).

Continuous- and discrete-time transfer functions are defined by specifying numerator and denominator polynomials in vector form. SISO and MIMO dynamic systems portrayed in block diagram form can be reduced to obtain specific transfer functions, which can be analyzed (by other control system toolbox functions) in the time and frequency domain. Impulse and step responses as well as responses to arbitrary inputs of both types of systems are easily obtained. The z-domain transfer functions for simulating continuous-time systems based on various methods of approximation are available. Conversion between state-space and transfer function descriptions of a system is accomplished using specific toolbox commands.

This section contains some relatively simple examples of the control system toolbox functions. Exposition is kept to a minimum. For more information, the reader should check out the robust set of online interactive demos, tutorials, and case studies illustrating how the toolbox can be used to support modeling and simulation functions.

4.10.1 TRANSFER FUNCTION MODELS

Continuous- and discrete-time transfer functions are constructed using "tf" with proper arguments and stored as a named MATLAB object such as "sys." For example, the transfer function

$$G_1(s) = 25 \left[\frac{(10s + 1)(s + 2)}{2s^4 + 5s^3 + 4s + 1} \right] \tag{4.704}$$

is implemented by the following statements:

```
num = 25*conv([10 1],[1 2])
den = [2 5 0 4 1]
sys_G1 = tf (num, den)
```

Note conv([10 1],[1 2]) produces the numerator vector [10 21 2]. A more intuitive way of creating the same transfer function is

```
s = tf('s')
sys_G1 = 25*(10*s^2+21*s+2)/(2*s^4+5*s^3+4*s+1)
```

A discrete-time system with sampling period $T = 0.01$ s and pulse (z-domain) transfer function

$$G_2(z) = \frac{5z^2 + 3z + 2}{z^2 + 10z + 4} \tag{4.705}$$

is created from either of the two sets of statements below:

```
num = [5 3 2]; den = [1 10 4]
sys_G2 = tf (num, den, 0.01)
z = tf('z',0.01)
sys_G2 = (5*z^2+3*z+2)/(z^2+10*z+2)
```

The poles and zeros of a continuous- or discrete-time system transfer function are obtained using the "pzmap (sys)" command where "sys" refers to the MATLAB description of the transfer function. A pole-zero map of the transfer function $G_1(s)$ in Equation 4.704 is obtained from the command "pzmap (sys_G1)" and shown in Figure 4.87.

The numerical values of the poles and zeros shown in Figure 4.87 are returned in "P" and "Z" after issuing the command " [P,Z] = pzmap (sys_G1)." The result is

$$P = -2.7418, 0.2385 + 0.8475i\ 0.2385 - 0.8475i, -0.2353$$
$$Z = -2.0000, -0.1000$$

4.10.2 STATE-SPACE MODELS

State-space models of continuous-time systems are described by matrices A, B, C, and D appearing in the state equations. The same holds for a discrete-time system, which also requires a sampling

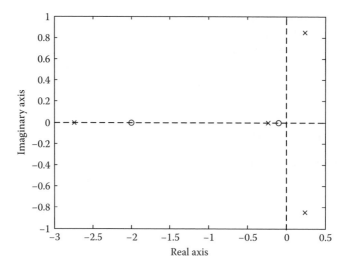

FIGURE 4.87   Pole-zero map for $G_1(s)$ in Equation 4.704.

time T for a complete representation. State-space models for continuous-time systems are created using "$sys = ss(A,B,C,D)$," while discrete-time models in state space are generated by

$$``sys = ss(A,B,C,D,T).''$$

A continuous-time second-order system with damping ratio $\zeta = 0.5$ and natural frequency $\omega_n = 2$ rad/s was approximated using trapezoidal integration with step size $T = 0.025$ s in Section 4.7 resulting in discrete-time system state equations

$$\underline{x}_{k+1} = A\underline{x}_k + B\underline{u}_k \tag{4.706}$$

$$\underline{y}_k = C\underline{x}_k + D\underline{u}_k \tag{4.707}$$

with A, B, C, and D given in Equations 4.516 through 4.518.

The resulting matrices A, B, C, and D and sampling time T appear in the M-file "*Chap4_Tustin.m*" statement "$sys = ss(A,B,C,D,T)$" to create a discrete-time system state-space model with numerical values

```
a =
                         x1              x2
           x1          1.949         -0.9512
           x2            1              0
b =
                         u1
           x1            1
           x2            0
c =
                         x1              x2
           y1        0.002406       2.971e-005
d =
                         u1
           y1        0.0006094
```

Sampling time: 0.025 discrete-time model.

The object "sys" can be referenced by other control system toolbox commands to investigate frequency response characteristics of the discrete-time system as well as dynamic response to specific types of forcing functions. It is also instrumental in the process of converting a state-space model to a transfer function representation, the next subject of discussion.

4.10.3   State-Space/Transfer Function Conversion

The state equations for a submarine depth control system were developed in Section 2.8. The closed-loop control system is third order with three outputs, θ (stern plane angle), v (depth rate), c (depth), and a single input r (commanded depth). The MATLAB file "*Chap4_sub.m*" below illustrates several commands for converting between state-space models of the system and the transfer function form.

```
% Chap4_sub.m
KC = 0.6; KI = 0.1;
tau = 10; Kthd = 20; Kth = 10;
a11 = -Kthd*KC/tau; a12 = (Kth- (Kthd/tau)); a13 = Kthd*KI/tau;
a21 = -KC/tau; a22 = -1/tau; a23 = KI/tau;
a31 = -1; a32 = 0; a33 = 0;
b1 = Kthd*KC/tau; b2 = KC/tau; b3 = 1;
c11 = -KC; c12 = 0; c13 = KI;
c21 = -Kthd*KC/tau; c22 = Kth- (Kthd/tau); c23 = Kthd*KI/tau;
c31 = 1; c32 = 0; c33 = 0;
d1 = KC; d2 = Kthd*KC/tau; d3 = 0;
A1 = [a11 a12 a13; a21 a22 a23; a31 a32 a33];
B1 = [b1; b2; b3];
C1 = [c11 c12 c13; c21 c22 c23; c31 c32 c33];
D1 = [d1; d2; d3];

sys_ss_1 = ss(A1,B1,C1,D1)%          creates state-space system
                                       object for (A1,B1,C1,D1)
sys_tf = tf(sys_ss_1)%               converts state-space
                                       system object to transfer
                                       function system object

[num1,den1] = ss2tf(A1,B1,C1,D1)%    alternate method for
                                       converting state space
                                       (A1,B1,C1,D1) to transfer
                                       function

sys_ss_2 = ss(sys_tf)%               converts transfer function
                                       object to state-space
                                       object

[A3,B3,C3,D3] = tf2ss(num1,den1)%    converts transfer function
                                       to state-space control
                                       canonical form with
                                       matrices (A3,B3,C3,D3)

[num2,den2] = ss2tf(A3,B3,C3,D3)%    converts state-space (A3,
                                       B3,C3,D3) to transfer
                                       functions
```

Numerical values are assigned to matrices A_1, B_1, C_1, and D_1 using the baseline system parameter values from Section 2.8. The system matrices are

$$
\begin{array}{llll}
\text{A1} = & -1.2000 & 8.0000 & 0.2000 \\
& -0.0600 & -0.1000 & 0.0100 \\
& -1.0000 & 0 & 0 \\
\text{B1} = & 1.2000 & & \\
& 0.0600 & & \\
& 1.0000 & & \\
\text{C1} = & -0.6000 & 0 & 0.1000 \\
& -1.2000 & 8.0000 & 0.2000 \\
& 1.0000 & 0 & 0 \\
\text{D1} = & 0.6000 & & \\
& 1.2000 & & \\
& 0 & &
\end{array}
$$

The statement "sys_ss_1 = ss(A1,B1,C1,D1)" creates the object "sys_ss_1" associated with the continuous-time system matrices A_1, B_1, C_1, and D_1. The next statement "sys_tf''
(sys_ss_1)''" creates the transfer function object "sys_tf" with embedded information about the three system transfer functions, one each from the command input to the three outputs. The transfer functions are displayed as

```
Transfer function from input to output . . .
```

$$
\#1: \frac{0.6\ s^3 + 0.16\ s^2 + 0.01\ s - 1.506\text{e-}018}{s^3 + 1.3\ s^2 + 0.8\ s + 0.1}
$$

$$
\#2: \frac{1.2\ s^3 + 0.8\ s^2 + 0.1\ s - 3.474\text{e-}017}{s^3 + 1.3\ s^2 + 0.8\ s + 0.1}
$$

$$
\#3: \frac{1.2\ s^2 + 0.8\ s + 0.1}{s^3 + 1.3\ s^2 + 0.8\ s + 0.1}
$$

Note that the first two transfer functions are consistent with the control system simulation diagram (Figure 2.55), which shows direct paths from the input r to outputs θ and v. The numerator of transfer function #3 is second order due to the presence of the integrator in the path from r to c.

An alternative approach to finding the same three transfer functions uses " [num1, den1] = ss2tf(A1,B1,C1,D1)." Output matrix "num1" (with three rows, one for each output) stores the coefficients of the three numerator polynomials, and row vector "den1" contains the coefficients of the denominator, that is, characteristic polynomial. The result is

$$
\begin{array}{lllll}
\text{num1} = & 0.6000 & 0.1600 & 0.0100 & 0.0000 \\
& 1.2000 & 0.8000 & 0.1000 & 0.0000 \\
& 0 & 1.2000 & 0.8000 & 0.1000 \\
\text{den1} = & 1.0000 & 1.3000 & 0.8000 & 0.1000
\end{array}
$$

Converting the transfer function of an SISO system to a state-space model is achieved using either "ss" or "tf2ss." The command "sys_ss_2 = ss(sys_tf)" computes a state-space realization of the transfer function object "sys_tf" displayed as

```
a =
                    x1              x2              x3
        x1         -1.3           -0.4           -0.1
        x2          2              0              0
        x3          0              0.5            0
b =
                    u1
        x1          1
        x2          0
        x3          0
c =
                    x1              x2              x3
        y1         -0.62          -0.235         -0.06
        y2         -0.76          -0.43          -0.12
        y3          1.2            0.4            0.1
d =
                    u1
        y1          0.6
        y2          1.2
        y3          0
```

Referring to the above matrices as A_2, B_2, C_2, and D_2, it is not surprising that they differ from A_1, B_1, C_1, and D_1 since the state-space model representation of a continuous-time system is not unique.

An alternative method for creating a state-space model from a transfer function is to use "[A3,B3,C3,D3] = tf2ss(num1,den1)" where "num1" and "den1" are the numerator and denominator arrays, respectively, created previously by the command "ss2tf." This results in creation of output matrices A_3, B_3, C_3, and D_3 given below:

```
A3 =       -1.3000       -0.8000       -0.1000
            1.0000        0             0
            0             1.0000        0
B3 =        1
            0
            0
C3 =       -0.6200       -0.4700       -0.0600
           -0.7600       -0.8600       -0.1200
            1.2000        0.8000        0.1000
D3 =        0.6000
            1.2000
            0
```

State-space models created by "tf2ss" are in controller canonical form (Ogata 1998).

The last statement [num2,den2] = ss2tf(A3,B3,C3,D3) in "*Chap4_sub.m*" converts the state-space model in controller canonical form back to the three transfer functions.

TABLE 4.11
Three Different State-Space Models of Submarine Depth Control System

i	A_i	B_i	C_i	D_i
1	$\begin{bmatrix} -1.2 & 8 & 0.2 \\ -0.06 & -0.1 & 0.01 \\ -1 & 0 & 0 \end{bmatrix}$	$\begin{bmatrix} 1.2 \\ 0.06 \\ 1 \end{bmatrix}$	$\begin{bmatrix} -0.6 & 0 & 0.1 \\ -1.2 & 8 & 0.2 \\ 1 & 0 & 0 \end{bmatrix}$	$\begin{bmatrix} 0.6 \\ 1.2 \\ 0 \end{bmatrix}$
2	$\begin{bmatrix} -1.3 & -0.4 & -0.1 \\ 2 & 0 & 0 \\ 0 & 0.5 & 0 \end{bmatrix}$	$\begin{bmatrix} 1 \\ 0 \\ 0 \end{bmatrix}$	$\begin{bmatrix} -0.62 & -0.235 & -0.06 \\ -0.76 & -0.43 & -0.12 \\ 1.2 & 0.4 & 0.1 \end{bmatrix}$	$\begin{bmatrix} 0.6 \\ 1.2 \\ 0 \end{bmatrix}$
3	$\begin{bmatrix} -1.3 & -0.8 & -0.1 \\ 1 & 0 & 0 \\ 0 & 1 & 0 \end{bmatrix}$	$\begin{bmatrix} 1 \\ 0 \\ 0 \end{bmatrix}$	$\begin{bmatrix} -0.62 & -0.47 & -0.06 \\ -0.76 & -0.86 & -0.12 \\ 1.2 & 0.80 & 0.1 \end{bmatrix}$	$\begin{bmatrix} 0.6 \\ 1.2 \\ 0 \end{bmatrix}$

The state-space models for the submarine control system are summarized in Table 4.11. A good way of checking the results is to compute the eigenvalues of the coefficient matrices A_1, A_2, and A_3 in the table. The MATLAB command "eig(A)" returns the same characteristic roots, namely, $-0.5687 \pm j0.5400$ and -0.1626, for all three matrices.

4.10.4 SYSTEM INTERCONNECTIONS

Block diagrams can be systematically reduced in complexity using control system toolbox functions such as "parallel," "series," and "feedback." Consider the block diagram shown in Figure 4.88.

$$G_c(s) = 5\left(\frac{10s+1}{2s+1}\right), \quad H(s) = \frac{1}{50s+1} \tag{4.708}$$

$$G_1(s) = \frac{8}{3s+1}, \quad G_2(s) = \frac{s+5}{s^2+12s+25}, \quad G_3(s) = \frac{1}{0.2s+1}, \quad G_4(s) = \frac{1}{s} \tag{4.709}$$

Using block diagram algebra, the transfer function $Y(s)/R(s)$ can be found by executing the statements below found in M-file "*Chap4_block_diagram.m*."

```
1. s=tf('s');
2. Gc=5*(10*s+1)/(2*s+1);
3. G1=8/(3*s+1);
4. G2=(s+5)/(s^2+12*s+25);
5. G3=1/(0.2*s+1);
```

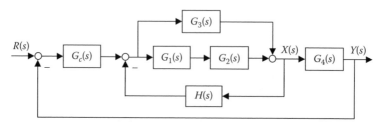

FIGURE 4.88 Block diagram of a continuous-time system.

```
 6. G4 = 1/s;
 7. H = 1/(50*s+1);
 8. G1G2 = series(G1,G2);
 9. G1G2_plus_G3 = parallel(G1G2,G3);
10. TF_inner_loop = feedback(G1G2_plus_G3,H)
11. G = series(Gc,TF_inner_loop);
12. G_forward_path_1 = series(G,G4);
13. TF_outer_loop_1 = feedback(G_forward_path_1,1)
```

The inner loop transfer function "TF_inner_loop" and outer loop transfer function TF_outer_loop_1 are

```
Transfer function:
```
$$150 s^4 + 1933 s^3 + 5189 s^2 + 3353 s + 65$$

$$30 s^5 + 520.6 s^4 + 2733 s^3 + 4693 s^2 + 1445 s + 90$$

Transfer function:
$$7500 s^5 + 97400 s^4 + 269095 s^3 + 193593 s^2 + 20015 s + 325$$

$$60 \quad s^7 + 1071 \quad s^6 + 1.349e004 \quad s^5 + 1.095e005 \quad s^4 + 276678 \quad s^3 + 195218 \quad s^2 + 20105 s + 325$$

Other transfer functions may be obtained by proper use of the three system interconnection commands. For example, $X(s)/R(s)$ in Figure 4.88 can be found by deleting statement 11 and changing statements 12 and 13 to read

```
14. G_forward_path_2 = series(Gc,TF_inner_loop);
15. TF_outer_loop_2 = feedback(G_forward_path_2,G4)
```

An alternate implementation of the transfer function $X(s)/R(s)$ is possible by expressing it in terms of $Y(s)/R(s)$. Starting with

$$Y(s) = G_4(s)X(s) \tag{4.710}$$

$$\Rightarrow \frac{Y(s)}{R(s)} = G_4(s)\frac{X(s)}{R(s)} \tag{4.711}$$

$$\Rightarrow \frac{X(s)}{R(s)} = \frac{1}{G_4(s)}\frac{Y(s)}{R(s)} \tag{4.712}$$

The transfer function $X(s)/R(s)$ can now be obtained by statement 14 below:

```
16. TF_outer_loop_2 = series(1/G4,TF_outer_loop_1)
```

The functions "parallel," "series," and "feedback" to reduce a system with forward and feedback connections apply to discrete-time system block diagrams as well.

4.10.5 SYSTEM RESPONSE

The impulse and step response of continuous- and discrete-time LTI systems can be generated in either graphical form or stored in an array of data points. To illustrate, suppose we are interested in the step response of the submarine depth control system considered earlier. Unit step responses of the stern plane angle θ, depth rate v, and depth c are obtained by appending "step (sys_ss_1)" or "step(sys_tf)" at the end of M-file "*Chap4_sub.m*." The graphs are shown in Figure 4.89.

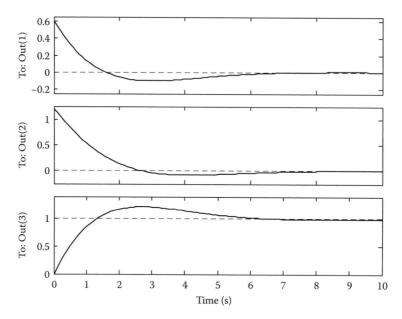

FIGURE 4.89 Unit-step response in θ, *v*, and *c*.

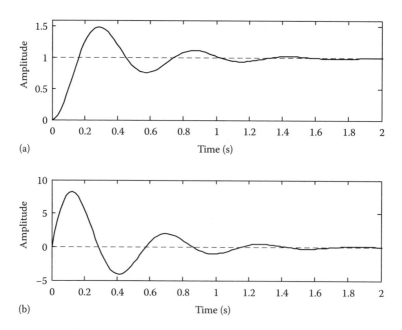

FIGURE 4.90 (a) Step and (b) impulse response of continuous-time system in Figure 4.88.

Step and impulse responses of the system in Figure 4.88 with *y(t)* as output are obtained by issuing the control system toolbox commands "step(TF_outer_loop_1)" and "impulse (TF_outer_loop_1)" in M-file "*Chap4_block_diagram.m*." The step and impulse responses are shown in Figure 4.90.

The response of an LTI system to an arbitrary input is obtained using "LSIM(SYS,U,T)" where "SYS" represents a MATLAB system object. "U" and "T" are arrays used to define the input (s) values and corresponding regularly spaced values of time, respectively.

The case study in Section 3.7 involved the ascent of a diver subject to a vertical cable force f_c. A state-space model was formulated and repeated as follows:

$$\begin{bmatrix} \dot{x}_1 \\ \dot{x}_2 \\ \dot{x}_3 \end{bmatrix} = \begin{bmatrix} 0 & 1 & 0 \\ 0 & -\dfrac{\mu g}{W} & 0 \\ K\gamma & 0 & -K \end{bmatrix} \begin{bmatrix} x_1 \\ x_2 \\ x_3 \end{bmatrix} + \begin{bmatrix} 0 \\ \dfrac{g}{W} \\ 0 \end{bmatrix} [f_n] \tag{4.713}$$

$$\begin{bmatrix} y_1 \\ y_2 \end{bmatrix} = \begin{bmatrix} 1 & 0 & 0 \\ -\gamma & 0 & 1 \end{bmatrix} \begin{bmatrix} x_1 \\ x_2 \\ x_3 \end{bmatrix} \tag{4.714}$$

The input $f_n = W - \gamma V - f_c$ is the net force (weight – buoyant force – cable force) acting on the diver. The output y_1 is depth below the surface, and y_2 is the difference between the internal body pressure of the diver and the local (same depth as diver) underwater pressure. The states x_1, x_2, and x_3 are depth, velocity, and internal pressure of the diver, respectively. The system parameters are μ, W, and K; and g and γ are physical constants.

Suppose the diver's ascent from an initial equilibrium state $x_{1,e} = 500$ ft, $x_{2,e} = 0$ ft/s, and $x_{3,e} = \gamma x_{1,e} = 62.4 \text{ lb/ft}^3 \times 500 \text{ ft} = 31{,}200 \text{ lb/ft}^2$ (216.7 psi) is required. A cable force

$$f_c(t) = (W - \gamma V) + \overline{F}(1 - e^{-t/\tau}), \quad t \geq 0 \tag{4.715}$$

where $\overline{F}$ and τ are design parameters is under investigation. The cable force $f_c(t)$ and the resulting net force $f_n(t)$ are plotted in Figure 4.91 for the case where $\overline{F} = 25$ lb and $\tau = 40$ s (see M-file "*Chap4_diver.m*.")

The M-file "*Chap4_diver.m*" includes a statement to create the state-space object "sys" from matrices A, B, C, and D in Equations 4.713 and 4.714. The time vector "t" is defined and input

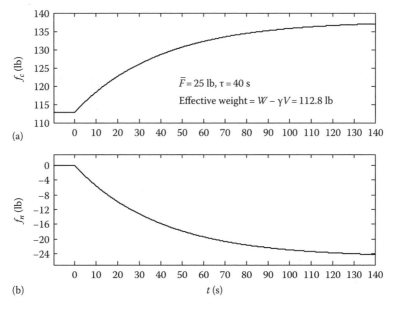

(a)

(b)

FIGURE 4.91 (a) Cable force and (b) net force on diver vs. time.

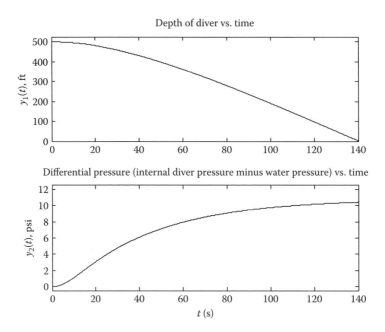

FIGURE 4.92 Outputs y_1 and y_2 from diver state-space model with input f_n.

vector "fn" is calculated from the equation $f_n = (W - \gamma V) - f_c$. The statement "y = LSIM(sys, fn, t, x0)," where "x0" is the initial state vector, returns data points for outputs y_1 and y_2 in the array "y." Graphs of $y_1(t)$ and $y_2(t)$ are shown in Figure 4.92.

4.10.6 CONTINUOUS-/DISCRETE-TIME SYSTEM CONVERSION

We are well aware of the need to approximate the dynamics of continuous-time systems using discrete-time systems. Replacing the differential equations of LTI continuous-time system models with difference equations is an important aspect of continuous system simulation. Section 4.4.7 introduced a technique for accomplishing the task based on substitution of a suitable function of z for the Laplace variable s in the continuous-time system transfer function. Examples were presented illustrating how to obtain the z-domain transfer function of the discrete-time system based on the use of explicit Euler integration and trapezoidal integration, also known as Tustin's method.

Additional transformations $s = f(z)$ for other methods are discussed in a later chapter. For all but the simplest continuous-time systems, the algebraic manipulation required to obtain the z-domain or pulse transfer function in a suitable form is unwieldy at best. The MATLAB control system toolbox "c2d" function expedites the process of converting continuous-time models to discrete-time approximations. The required arguments are a MATLAB system object for the continuous-time system, the sample time (integration step size), and an optional string to select one of the five available approximation methods listed below:

```
'zoh' Zero-order hold on the inputs.
'foh' Linear interpolation of inputs (triangle appx.).
'imp' Impulse-invariant discretization.
'tustin' Bilinear (Tustin) approximation.
'prewarp' Tustin approximation with frequency prewarping. The crit-
    ical frequency Wc (rad/sec) is specified as 4th input by SYSD = C2D
    (SYSC,Ts,'prewarp',Wc)
'matched' Matched pole-zero method (for SISO systems only).
```

To illustrate, consider the problem of approximating a second-order system with natural frequency $\omega_n = 2$ rad/s, $\zeta = 0.5$, and DC gain of unity. Example 4.27 presented solutions based on the use of explicit Euler integration and trapezoidal integration (Tustin's method), also known as the bilinear transform method. The following statements are from the M-file "*Chap4_Tustin.m*," which creates the continuous-time transfer function "H_s" and generates the discrete-time system transfer function "H_z" using Tustin's method.

```
T=0.025; wn=2; zeta=0.5; K=1;
H_s=tf(K*wn^2, [1 2*zeta*wn wn^2])
H_z=c2d(H_s,T,'tustin')
```

The continuous- and discrete-time transfer functions appear in the MATLAB Command Window as

```
Transfer function:
       4
-------------------
s^2+2 s+4
Transfer function
0.0006094 z^2+0.001219 z+0.0006094
------------------------------------------
        z^2 - 1.949 z+0.9512
Sampling time: 0.025
```

The pulse transfer function approximation of the continuous-time second-order system using Tustin's method is (see Equation 4.503)

$$H(z) = \frac{K(\omega_n T)^2 (z^2 + 2z + 1)}{\left[4(1 + \zeta\omega_n T) + (\omega_n T)^2\right]z^2 + 2\left[(\omega_n T)^2 - 4\right]z + 4(1 - \zeta\omega_n T) + (\omega_n T)^2} \tag{4.716}$$

Substituting the numerical values $\omega_n = 2$, $\zeta = 0.5$, $K = 1$, and $T = 0.025$ for the system parameters gives

$$H(z) = \frac{0.0025(z^2 + 2z + 1)}{4.1025z^2 - 7.9950z + 3.9025} \tag{4.717}$$

$$= \frac{0.00060938z^2 + 0.0012187z + 0.00060938}{z^2 - 1.9488z + 0.9512} \tag{4.718}$$

in agreement with the result from using the "c2d" function.

There is also a function called "d2c" for converting a discrete-time transfer function previously created as an object "sysd" to an equivalent continuous-time transfer function object "sysc." The syntax is "SYSC=D2C(SYSD,METHOD)" where the second argument is a string signifying the method of approximation.

4.10.7 Frequency Response

The magnitude and gain of a system transfer function at a particular frequency ω were evaluated in earlier sections by substituting $j\omega$ for s in continuous-time transfer functions and $e^{j\omega T}$ for z in discrete-time transfer functions. Choosing a range of values for ω led to plots of magnitude, gain $= 20(\log[\text{magnitude}])$ and phase vs. frequency.

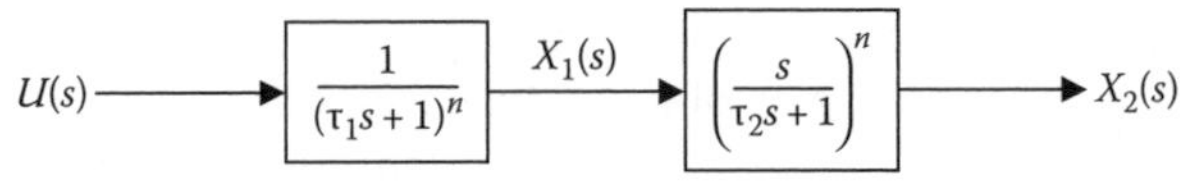

$$U(s) \longrightarrow \boxed{\dfrac{1}{(\tau_1 s + 1)^n}} \xrightarrow{X_1(s)} \boxed{\left(\dfrac{s}{\tau_2 s + 1}\right)^n} \longrightarrow X_2(s)$$

FIGURE 4.93 Low- and high-pass filters in series.

The control system toolbox provides an easier way of obtaining the frequency response characteristics of both continuous- and discrete-time system models. Assuming an LTI model object called "sys" has been created using "tf" or possibly "ss," a Bode plot is drawn by execution of the command "BODE(sys)." If "sys" represents a discrete-time system, the call is modified to include an additional argument for the sampling time T, namely, "BODE(sys,T)."

Optional arguments permit specifying multiple systems with different line plot characteristics and a user selectable range of frequencies. To illustrate, consider the two blocks in series shown in Figure 4.93.

The first component is a low-pass filter with transfer function

$$G_1(s) = \frac{X_1(s)}{U(s)} = \frac{1}{(\tau_1 s + 1)^n} \tag{4.719}$$

and break frequency $\omega_1 = 1/\tau_1$. The second component transfer function

$$G_2(s) = \frac{X_2(s)}{X_1(s)} = \left(\frac{s}{\tau_2 s + 1}\right)^n \tag{4.720}$$

represents a high-pass filter with break frequency $\omega_2 = 1/\tau_2$. The frequency response characteristics of the series combination with transfer function

$$G_{12}(s) = \frac{1}{(\tau_1 s + 1)^n} \left(\frac{s}{\tau_2 s + 1}\right)^n \tag{4.721}$$

$$= \left[\frac{s}{(\tau_1 s + 1)(\tau_2 s + 1)}\right]^n \tag{4.722}$$

are obtained using the "BODE" function for a model object "sys" corresponding to Equation 4.722. The following M-file statements generate plots of the gain (magnitude in db) for the low-pass filter ($\tau_1 = 1$ s), high-pass filter ($\tau_2 = 0.01$ s), and the band-pass filter with pass band ($\omega_1 \leq \omega \leq \omega_2$) resulting from the combination of the two filters in series. The plots are shown in Figure 4.94. The exponent n was chosen to be three.

```
tau1=1; tau2=0.01; n=3;
sys1=tf(1,[tau1 1])
sys2=tf([1 0],[tau2 1])
for i=1:n-1
sysG1=SERIES(sys1,sys1)
sysG2=SERIES(sys2,sys2)
end
sysG12=SERIES(sysG1,sysG2)
BODEMAG(sysG1,'b',sysG2,'r',sysG12,'k')
```

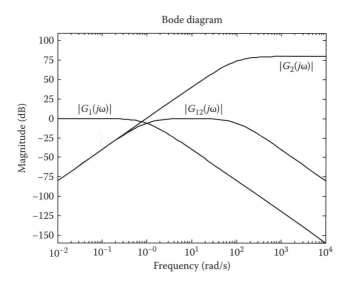

FIGURE 4.94 Gain of individual and combined blocks in Figure 4.93.

A discrete-time approximation of the continuous-time band-pass filter using Tustin's method is obtained by adding the statements

```
T=pi/1e4; % sample time to make wN=10^4 rad/sec
sysG12_d=C2D(sysG12,T, 'tustin');% converts continuous-time filter
    % to discrete-time filter using Tustin's method
BODEMAG(sysG12_d, 'r') % plot gain of discrete-time filter
BODEMAG(sysG12, 'b') % plot gain of continuous-time filter
```

The sample time should be at least an order of magnitude less than $\tau_2 = 0.01$ s and possibly smaller depending on the frequency content of the continuous-time input. A value of $T = \pi/10^4$ s was chosen to make the Nyquist frequency $\omega_N = \pi/T = 10^4$ s, the same as the upper limit in Figure 4.94. Selecting appropriate values of T for discrete-time models is deferred until Chapter 8.

A comparison of the continuous-time and discrete-time band-pass filter gains for ($10^{-2} \leq \omega \leq \omega_N = 10^4$) is shown in Figure 4.95. The two gains are nearly identical for ω up to 10^3 rad/s. Frequency response includes phase characteristics as well as gain. The phase properties of the two filters are left for an exercise problem.

4.10.8 ROOT LOCUS

For simple feedback control systems with a controller gain K_C, the closed-loop system poles depend on the value of K_C. A root-locus plot displays the location of all the poles as the design parameter K_C varies from zero to infinity. The starting point is creation of the open-loop system model "sys" followed by a call to the control system toolbox function "rlocus(sys)." The following example illustrates the use of "BODE" and "rlocus" to determine the limits of stability for a simple control system.

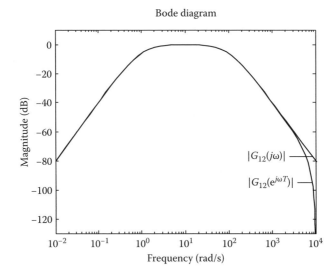

FIGURE 4.95 Gain of continuous- and discrete-time band-pass filters.

Example 4.36

An overdamped second-order system is subject to proportional control as shown in Figure 4.96. A sensor is present in the feedback loop.

Baseline values of the system and sensor parameters are

$$K_p = 15, \quad \tau_1 = 3 \text{ s}, \quad \tau_2 = 15 \text{ s}, \quad K_T = 0.1, \quad \tau_T = 0.25 \text{ s}$$

(a) Create a model "sys" for the open-loop system with $K_C = 1$.
(b) Use the control system toolbox to draw a Bode plot of the open-loop system.
(c) Determine the stability margins of the control system and the critical gain K_{cr}.
(d) Find ω_0, the frequency of oscillations for the marginally stable system.
(e) Check the results for K_{cr} using a root-locus plot and the characteristic equation.
(f) Plot step responses of the closed-loop system for $K_C = 0.25K_{cr}, 0.5K_{cr}, 0.75K_{cr}, K_{cr}$.

(a) The model object "sys" is created in "*Chap4_Ex10_1.m*" with the statements

```
KP = 15; tau1 = 3; tau2 = 15; KT = 0.5; tauT = 0.25; KC = 1;
denG = conv([tau1 1],[tau2 1])
G = tf(KP,denG); % process transfer function
denH = [tauT 1];
H = tf(KT,denH); %sensor transfer function
sys = KC*SERIES(G,H)
```

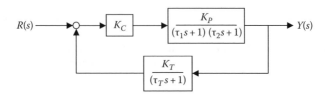

FIGURE 4.96 Feedback control system with proportional control.

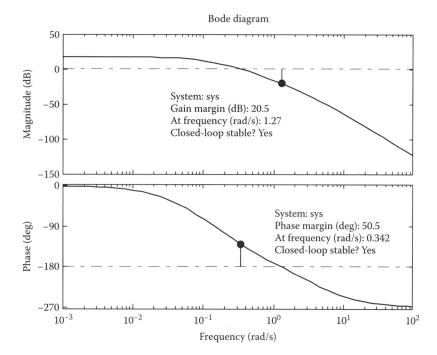

FIGURE 4.97   Bode plot for control system in Figure 4.96.

(b) The command "BODE(sys)" results in the Bode plot in Figure 4.97.

(c) The stability margins were defined in Section 4.4.5. The gain margin is the open-loop system gain at the frequency where the phase of the open-loop system equals −180°. The phase margin is the difference between the open-loop phase and −180° at the frequency where the gain is 0 db. Figure 4.97 shows the gain margin is 20.5 db and the phase margin is 50.5°. Increasing the controller gain K_C by the equivalent of 20.5 db moves the gain plot in a vertical direction to a point where the system is marginally stable, that is, the new gain margin is 0 db. Solving for K_{cr} in magnitude,

$$20 \log K_{cr} = 20.5 \ \Rightarrow \ K_{cr} = 10^{20.5/20} = 10.5925$$

(d) The 0 db gain margin would occur at the same frequency as the 20.5 db gain margin in Figure 4.97, that is, 1.27 rad/s, which is also ω_0, the frequency of oscillations of the marginally stable system.

(e) The root-locus plot is shown in Figure 4.98. The approximate value of K_{cr} is 10.6, that is, the value of K_C where the locus intersects the imaginary axis. Note that the imaginary part of the complex pole is $\omega_0 = 1.27$ rad/s, in agreement with the crossover frequency shown in Figure 4.97.

As a check on the value of K_{cr} from part (c), the statement

[R,K] =rlocus(sys,Kcr)

returns the three closed-loop poles in array $R = [-4.4006, 0.003 \pm j1.2747]$. The real part of the complex poles should be zero when $K_c = K_{cr}$; however, 0.003 results because of the round-off in the gain margin value of 20.5 shown in Figure 4.97.

The exact values of K_{cr} and ω_0 can be obtained from the characteristic equation

$$K_C K_P K_T + (\tau_1 s + 1)(\tau_2 s + 1)(\tau_T + 1) = 0 \tag{4.723}$$

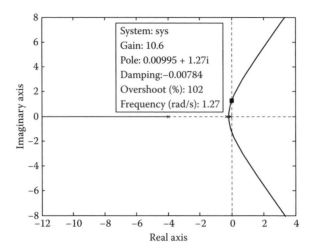

FIGURE 4.98 Root-locus plot for control system in Figure 4.96.

with $K_C = K_{cr}$ and $s = j\omega_0$. Setting the real and imaginary components of the resulting equation to zero leads to the following two equations:

$$\omega_0^2 = \frac{\tau_1 + \tau_2 + \tau_T}{\tau_1 \tau_2 \tau_T} \tag{4.724}$$

$$K_{cr} = \frac{[\tau_1 \tau_1 + \tau_T(\tau_1 + \tau_2)]\omega_0^2 - 1}{K_P K_T} \tag{4.725}$$

The solution is (see "*Chap4_Ex10_1.m*") $K_r = 10.5733$, $\omega_0 = 1.273665$ rad/s.

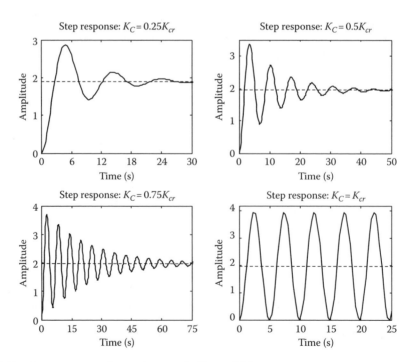

FIGURE 4.99 Step responses of control system in Figure 4.96.

(f) Step responses of the closed-loop system with $K_C = 0.25K_{cr}$, $0.5K_{cr}$, $0.75K_{cr}$, K_{cr} are generated by the statements

```
for i = 1:4
subplot(2,2,i)
sys_cl = FEEDBACK(0.25*i*KCR*G,H); % closed-loop system
step(sys_cl) % step response
end
```

where "KCR" is the exact value for K_{cr}. The step responses, shown in Figure 4.99, exhibit less damping as the controller gain increases. The step response of the marginally stable system ($K_C = K_{cr}$) contains an oscillatory component at the frequency $\omega_0 = 1.27$ rad/s.

EXERCISES

Use the control system toolbox whenever possible to do the following problems:

4.78 The block diagram of a typical feedback control system was presented in Figure 4.31 and redrawn below (Figure E4.78):

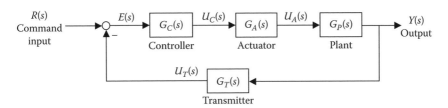

FIGURE E4.78

Use the transfer functions given in Section 4.4.5 and the baseline parameter values unless stated otherwise.
(a) Find the magnitude and phase of each component $G_C(s)$, $G_A(s)$, $G_P(s)$, and $G_T(s)$ at the open-loop system phase crossover frequency $\omega_0 = 0.9936$ rad/s. Compare the results to the magnitude and phase of the open-loop transfer function $G_{OL}(s) = G_C(s)G_A(s)G_P(s)G_T(s)$ at the same frequency.
(b) Input to the open-loop system (feedback path broken at summer) is $r(t) = \sin \omega_0 t$. Generate graphs of $e(t) = r(t)$, along with $u_C(t)$, $u_A(t)$, $y(t)$, and $u_T(t)$. Comment on the stability of the closed-loop system.

Hint: Recall the closed-loop system is unstable if the magnitude of $u_T(t)$ is greater than or equal to 1 at the phase crossover frequency ω_0, that is, the frequency where $u_T(t)$ lags $e(t)$ by 180°.

(c) Graph the step response of the closed-loop system.
(d) Repeat parts (a), (b), and (c) using $K_C = (K_C)_{max} = 2.62$.

4.79 The block diagram of a heading control system for a ship, presented in Section 4.4.4, is shown in Figure E4.79. The baseline parameter values are
$K_C = 10$ V/° (heading)
$K_P = 10°$ (rudder)/volt, $\tau_P = 0.2$ s,
$K_S = 0.5°$ (heading)/s/° (rudder), $\tau_S = 7.5$ s
(a) Find the closed-loop transfer functions

$$\frac{E(s)}{\theta_{com}(s)}, \quad \frac{U(s)}{\theta_{com}(s)}, \quad \frac{R(s)}{\theta_{com}(s)}, \quad \text{and} \quad \frac{\theta(s)}{\theta_{com}(s)}$$

(b) For a step input $\theta_{com} = 5°$, $t \geq 0$ graph $e(t)$, $u(t)$, $r(t)$, and $\theta(t)$.

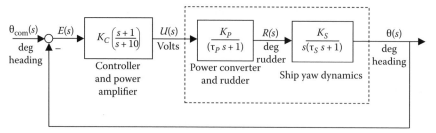

FIGURE E4.79

4.80 A system of two interacting tanks is shown in Figure E4.80a:

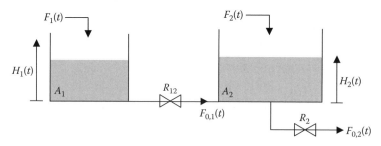

FIGURE E4.80a

The state equations are given as

$$
\begin{bmatrix} dH_1/dt \\ dH_2/dt \end{bmatrix} = \begin{bmatrix} -\dfrac{1}{A_1 R_{12}} & \dfrac{1}{A_1 R_{12}} \\ \dfrac{1}{A_2 R_{12}} & -\dfrac{1}{A_2 R_{12}} - \dfrac{1}{A_2 R_{12}} \end{bmatrix} \begin{bmatrix} H_1 \\ H_2 \end{bmatrix} \begin{bmatrix} \dfrac{1}{A_1} & 0 \\ 0 & \dfrac{1}{A_2} \end{bmatrix} \begin{bmatrix} F_1 \\ F_2 \end{bmatrix}
$$

$$
\begin{bmatrix} H_1 \\ H_2 \\ H_3 \end{bmatrix} = \begin{bmatrix} 1 & 0 \\ 0 & 1 \\ A_1 & A_2 \end{bmatrix} \begin{bmatrix} H_1 \\ H_2 \end{bmatrix}
$$

The parameter values are

$$A_1 = 25 \text{ ft}^2, \quad A_2 = 100 \text{ ft}^2, \quad R_{12} = 0.1 \text{ ft/ft}^3/\text{min}, \quad R_2 = 0.4 \text{ ft/ft}^3/\text{min}$$

(a) Find the transfer functions $V_T(s)/F_1(s)$ and $V_T(s)/F_2(s)$.
(b) With both tanks initially empty, find and graph $H_1(t)$ and $H_2(t)$ in response to
   (i) $F_1(t) = 12 \text{ ft}^3/\text{min}, F_2(t) = 0 \text{ ft}^3/\text{min}$
   (ii) $F_1(t) = 0 \text{ ft}^3/\text{min}, F_2(t) = 12 \text{ ft}^3/\text{min}$
   (iii) $F_1(t) = 12 \text{ ft}^3/\text{min}, F_2(t) = 12 \text{ ft}^3/\text{min}$
   (iv) $F_1(t)$ in Figure E4.80b

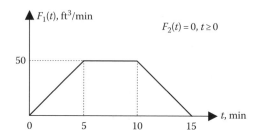

FIGURE E4.80b

4.81 The transfer function for the circuit in Figure E4.81 is (see Equation 4.183)

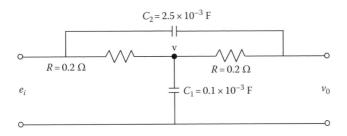

FIGURE E4.81

$$\frac{V_0(s)}{E_i(s)} = \frac{R^2 C_1 C_2 s^2 + 2RC_2 s + 1}{R^2 C_1 C_2 s^2 + R(C_1 + 2C_2)s + 1}$$

(a) Convert the system transfer function to a state variable model with output v_0.
(b) Use the state variable model to find and plot the impulse response.
(c) Find the unit step response of the circuit by inverse Laplace transforming $V_0(s)$.
(d) Repeat part (c) using the control system toolbox to find the unit step response. Compare the results from parts (c) and (d).
(e) Approximate the continuous-time transfer function with a discrete-time z-domain transfer function based on Tustin's method. Choose an appropriate integration step size.
(f) Find and plot the unit step response of the discrete-time system. Compare the step responses of the continuous-time and discrete-time systems.

4.82 Use "BODE" instead of "BODEMAG" to plot the magnitude and phase plots for the filters with transfer functions in Equations 4.719 through 4.721.

4.83 Compare the phase characteristics of the continuous- and discrete-time band-pass filters introduced in this section.

4.84 A simple control system block diagram is shown in Figure E4.84:

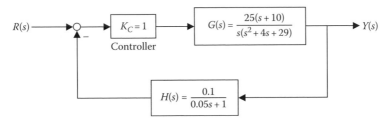

FIGURE E4.84

(a) Find the closed-loop transfer function of the system using block diagram reduction.
(b) Check your answer to part (a) using the control system toolbox.
(c) Draw a simulation diagram of the system.
(d) Represent the system in state variable form based on your simulation diagram.
(e) Use the control system toolbox to find a state variable model for the system.
(f) Compare the eigenvalues (characteristic poles) of the coefficient matrix A in parts (d) and (e).
(g) Use "BODE" to plot the frequency response of the open-loop system transfer function. Find the gain and phase margins of the system.

(h) Compute the maximum gain $(K_C)_{crit}$ which makes the system marginally stable. Redraw the Bode plot for $K_C = (K_C)_{crit}$.

(i) Check your answer to part (h) using a root-locus plot and identifying the value of gain K_C where the locus is on the Imaginary axis.

4.85 A continuous-time system is modeled by the differential equation

$$\frac{d^3y}{dt^3} + 5\frac{d^2y}{dt^3} + 33\frac{dy}{dt} + 29y = u$$

(a) Find the transfer function $H(s) = Y(s)/U(s)$ of the system.

(b) Create a model object "`sys`" to represent $H(s)$.

(c) Use the control system toolbox to plot the impulse and step response of the system.

(d) Approximate the continuous-time transfer function $H(s)$ with a discrete-time z-domain transfer function $H(z) = Y(z)/U(z)$ using Tustin's method with appropriate sample time T.

(e) Find the difference equation for the discrete-time system approximation.

(f) Write a MATLAB M-file to find and plot the step response of the discrete-time system.

(g) Use the control system toolbox to plot the step response of the discrete-time system, and compare the result with your answer in part (f).

4.11   CASE STUDY: LONGITUDINAL CONTROL OF AN AIRCRAFT

The equations of motion for an aircraft are derived using a moving coordinate system fixed to the aircraft as shown in Figure 4.100. The x–y–z axes are referred to as body axes. The x-axis is aligned with the longitudinal axis of the airplane. The equations are based on Newton's laws of motion for a rigid body in translation and rotation. The result is a system of six coupled nonlinear differential equations. Three of the six equations express accelerations $\dot{u}, \dot{v}, \dot{w}$ in terms of body axis velocities u, v, w, angular velocities p, q, r, and external, aerodynamic, and gravitational forces acting on the plane. The remaining three equations relate the angular accelerations $\dot{p}, \dot{q}, \dot{r}$ to p, q, r and moments produced by the external and aerodynamic forces about the plane's center of mass.

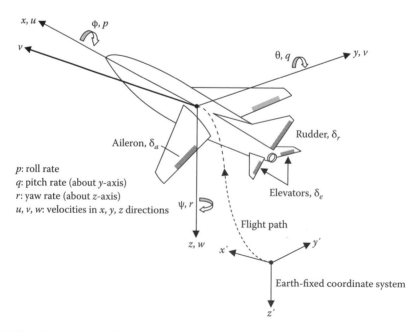

FIGURE 4.100   Body axis coordinates (x, y, z) and Euler angles (ψ, θ, ϕ).

The position and orientation of the airplane are referenced to an inertial (earth-fixed) coordinate system $x'-y'-z'$ also shown in Figure 4.100. The horizontal x'-axis is in the vertical plane containing the initial velocity vector, and the plane's center of mass is located at the origin of the $x'-y'-z'$ system at $t = 0$. The plane's attitude is fixed by three rotations of the $x-y-z$ axes starting from an orientation initially aligned with the $x'-y'-z'$ axes of the inertial coordinate system. The angular rotations ψ, θ, and φ are called Euler angles and denote the roll, pitch, and yaw of the plane, respectively.

Solution to the flight dynamics equations yields u, v, w in the $x-y-z$ body axis coordinate system. The velocity vector v is converted from body axis components u, v, w to inertial components $\dot{x}'$, $\dot{y}'$, $\dot{z}'$ by a transformation matrix C_e^b (Etkin 1982),

$$\begin{bmatrix} \dot{x}' \\ \dot{y}' \\ \dot{z}' \end{bmatrix} = C_e^b \begin{bmatrix} u \\ v \\ w \end{bmatrix} \tag{4.726}$$

$$C_e^b = \begin{bmatrix} \cos\theta\cos\psi & \sin\phi\sin\theta\cos\psi - \cos\phi\sin\psi & \cos\phi\sin\theta\cos\psi + \sin\phi\sin\psi \\ \cos\theta\sin\psi & \sin\phi\sin\theta\sin\psi + \cos\phi\cos\psi & \cos\phi\sin\theta\sin\psi - \sin\phi\cos\psi \\ -\sin\theta & \sin\phi\cos\theta & \cos\phi\cos\theta \end{bmatrix} \tag{4.727}$$

The position of the plane's center of mass in inertial coordinates x', y', z' is obtained by integration of the respective velocities in Equation 4.726.

Solving the equations of motion also yields the angular velocities p, q, r, which are transformed into $\dot{\psi}, \dot{\theta}, \dot{\phi}$ by

$$\begin{bmatrix} \dot{\psi} \\ \dot{\theta} \\ \dot{\phi} \end{bmatrix} = \begin{bmatrix} 0 & \sin\phi\sec\theta & \cos\phi\sec\theta \\ 0 & \cos\phi & -\sin\phi \\ 1 & \sin\phi\tan\theta & \cos\phi\tan\theta \end{bmatrix} \begin{bmatrix} p \\ q \\ r \end{bmatrix} \tag{4.728}$$

The Euler angles ψ, θ, and ϕ are obtained by integration of the respective velocities in Equation 4.728.

Solution of the nonlinear flight dynamics equations is complicated by the dependency of the aerodynamic forces and moments on the variable flight conditions, for example, altitude, cruising speed, weight, angle of attack, side slip, and control surface positions. A simpler approach is based on a linearized model that describes the aircraft's motion provided the excursions in flight from a known steady state are small. The subject of linearization is treated in some detail in Chapter 7.

When the conditions for linearization of the flight equations are satisfied, the linearized model can be decoupled into two sets of equations. One set describes the longitudinal dynamics of the aircraft, and the remaining equations apply to the lateral dynamics. The longitudinal dynamics involve changes in u and w, the plane's velocity in the x- and z-directions, and the pitch rate q about the y-axis. Lateral dynamics involve changes in side velocity v and the yaw and roll rates r and p about the z- and x-axes, respectively.

Figure 4.100 shows the velocity vector v aligned differently from the x-axis. The projection of v in the $x-z$ plane is v_{xz} shown in Figure 4.101. The angle between v_{xz} and the x-axis (longitudinal axis of plane) is called the angle of attack. Note that when the lateral dynamics of the plane are zero, the flight path is confined to the $x-z$ plane, $v = v_{xz}$, and the instantaneous direction of flight is given by γ in Figure 4.101, the angle between the velocity vector and the horizontal direction. The thrust (δ_T) from the engine, the aerodynamic forces, lift (L) and drag (D), and the gravitational force (W) are also shown in Figure 4.101.

The primary control surfaces for controlling the aircraft's position and attitude are the elevators, ailerons, and rudder. The longitudinal dynamics respond to changes in elevator deflection δ_e and

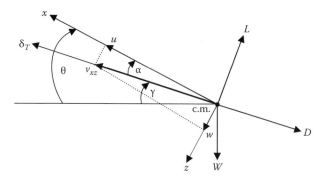

FIGURE 4.101   Illustration of angle of attack (α) and forces influencing flight dynamics.

thrust δ_T. Elevator deflection and thrust result from changes to the yoke and throttle by the pilot (or autopilot). The rudder and ailerons are used primarily to control the lateral response for banking and turning maneuvers.

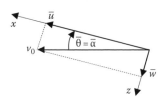

Our interest is solely in the longitudinal dynamics, specifically pitch and altitude response of the aircraft to changes in elevator deflection when the plane is flying at a constant cruising speed in horizontal flight under steady-state conditions. From Figure 4.101, for the plane to be in level flight, the velocity vector v must be horizontal, the flight angle $\gamma = 0$, and the pitch is equal to the angle

FIGURE 4.102   Initial steady-state conditions of aircraft.

of attack. The plane is pitched slightly in order for the wings to develop sufficient lift to overcome gravity. The steady-state conditions are shown in Figure 4.102 with v_0 (horizontal cruising speed), $\bar{u}$ (longitudinal speed), $\bar{w}$ (speed in z-direction), $\bar{\theta}$ (pitch), and $\bar{\alpha}$ (angle of attack). The elevator input and engine thrust necessary to maintain these conditions are $\bar{\delta}_e$ and $\bar{\delta}_T$, respectively.

The deviations in u, α, θ, w, and q from their steady-state operating levels are

$$\Delta u = u - \bar{u}, \quad \Delta w = w - \bar{w} = w, \quad \Delta\alpha = \alpha - \bar{\alpha}, \quad \Delta\theta = \theta - \bar{\theta}, \quad \Delta q = q - \bar{q} = q \quad (4.729)$$

Since we are considering only changes in elevator deflection,

$$\Delta\delta_e = \delta_e - \bar{\delta}_e, \quad \Delta\delta_T = \delta_T - \bar{\delta}_T = 0 \quad (4.730)$$

The state vector $\Delta\underline{x}$ in a linearized model of the longitudinal dynamics can be chosen as either $[\Delta u \quad \Delta w \quad \Delta q \quad \Delta\theta]^T$ or $[\Delta u \quad \Delta\alpha \quad \Delta q \quad \Delta\theta]^T$. The relationship between u, w, and α is (see Figure 4.101)

$$\tan\alpha = \frac{w}{u} \quad (4.731)$$

For small angles of attack, $\tan\alpha = \sin\alpha/\cos\alpha \approx \alpha$. Replacing $\tan\alpha$ in Equation 4.731 with α and solving for w give

$$w = u\alpha \quad (4.732)$$

Solving for u, α, and w in Equation 4.729 and substituting the results into Equation 4.732,

$$\bar{w} + \Delta w = (\bar{u} + \Delta u)(\bar{\alpha} + \Delta\alpha) = \bar{u}\bar{\alpha} + \bar{u}\Delta\alpha + \bar{\alpha}\Delta u + \Delta u\Delta w \quad (4.733)$$

Recognizing that $\overline{w} = \overline{u}\,\overline{\alpha}$ and ignoring the high-order term $\Delta u \Delta w$ lead to

$$\Delta w = \overline{u}\Delta\alpha + \overline{\alpha}\Delta u \tag{4.734}$$

Suppose the linearized model of an aircraft cruising in level flight under steady-state conditions with $v_0 = 500$ ft/s and $\overline{\alpha} = \overline{\theta} = 0.05$ rad (2.86°) is

$$\frac{d}{dt}\begin{bmatrix} \Delta u \\ \Delta\alpha \\ \Delta q \\ \Delta\theta \end{bmatrix} = \begin{bmatrix} -0.04 & 11.59 & 0 & -32.2 \\ -0.00073 & -0.65 & 1 & 0 \\ 0.000048 & -0.49 & -0.58 & 0 \\ 0 & 0 & 1 & 0 \end{bmatrix} \begin{bmatrix} \Delta u \\ \Delta\alpha \\ \Delta q \\ \Delta\theta \end{bmatrix} + \begin{bmatrix} 0 & 0.1 \\ 0 & 0 \\ -0.014 & 0 \\ 0 & 0 \end{bmatrix} \begin{bmatrix} \Delta\delta_e \\ \Delta\delta_T \end{bmatrix} \tag{4.735}$$

where
Δu has units of ft/s
$\Delta\alpha$, $\Delta\theta$ are in rad
Δq is in rad/s
$\Delta\delta_e$ is in degree of elevator deflection
$\Delta\delta_T$ is in lb of thrust

Choosing the output $\Delta y = \Delta x = [\Delta u\ \Delta\alpha\ \Delta q\ \Delta\theta]^T$ leads to the system of state equations $\Delta\dot{x} = A\Delta x + B\Delta u$, $\Delta y = C\Delta x + D\Delta u$ with A and B the matrices in Equation 4.735, C equal to the 4×4 identity matrix and D is a 4×2 matrix of zeros. Note that $\Delta u = [\Delta\delta_e\ \Delta\delta_T]^T$ is the input vector, not to be confused with Δu, the first component of the state vector.

The linearized equations in state variable form can be converted to a transfer function matrix relating the four outputs $\Delta u(s)$, $\Delta\alpha(s)$, $q(s)$, and $\Delta\theta(s)$ to the two inputs $\Delta\delta_e(s)$ and $\Delta\delta_T(s)$. The transfer function matrix can be found using Equation 4.231, repeated again for convenience in Equation 4.736.

$$G(s) = \begin{bmatrix} \dfrac{\Delta u(s)}{\Delta\delta_e(s)} & \dfrac{\Delta u(s)}{\Delta\delta_T(s)} \\[2ex] \dfrac{\Delta\alpha(s)}{\Delta\delta_e(s)} & \dfrac{\Delta\alpha(s)}{\Delta\delta_T(s)} \\[2ex] \dfrac{q(s)}{\Delta\delta_e(s)} & \dfrac{q(s)}{\Delta\delta_T(s)} \\[2ex] \dfrac{\Delta\theta(s)}{\Delta\delta_e(s)} & \dfrac{\Delta\theta(s)}{\Delta\delta_T(s)} \end{bmatrix} = C(sI - A)^{-1}B + D \tag{4.736}$$

The control system toolbox in MATLAB contains a function ``ss2tf'' for expediting the process of converting from the state-space model to the transfer function description of an LTI system. Calling this function with arguments (A, B, C, D, i), where $i = 1$ designates the first input $\Delta\delta_e$ and $i = 2$ specifies the second input $\Delta\delta_T$, generates the eight transfer functions in Equation 4.736. The MATLAB statement "[numG denG] = ss2tf (A, B, C, D, 1)" returns

```
numG = 0    0.0000   -0.0000    0.2906    0.2951
       0    0.0000   -0.0141   -0.0006   -0.0003
       0   -0.0141   -0.0097   -0.0005    0.0000
       0    0.0000   -0.0141   -0.0097    0.0125

denG = 1.0000  1.2700  0.9247  0.0406  0.0125
```

The transfer function relating elevator input to aircraft pitch is therefore

$$G_{\Delta\delta e}^{\Delta\theta}(s) = \frac{\Delta\theta(s)}{\Delta\delta e(s)} = \frac{-0.0141s^2 - 0.0097s - 0.0005}{s^4 + 1.2700s^3 + 0.9247s^2 + 0.0406s + 0.0125} \quad (4.737)$$

Factoring the numerator and denominator gives

$$G_{\Delta\delta e}^{\Delta\theta}(s) = \frac{\Delta\theta(s)}{\Delta\delta e(s)} = \frac{K_\theta(s + c_1)(s + c_2)}{(s^2 + a_1 s + b_1)(s^2 + a_2 s + b_2)} \quad (4.738)$$

The constants in Equation 4.738, computed in M-file "*Chap4_CaseStudy1.m*," are

$$K_\theta = -0.0141, \quad c_1 = 0.6358, \quad c_2 = 0.0542, \quad a_1 = 1.2440, \quad a_2 = 0.0260,$$
$$b_1 = 0.8780, \quad b_2 = 0.0143$$

The quadratic factors in the denominator of Equation 4.738 are both underdamped, regardless of whether the aircraft is a small passenger plane, a commercial jet, or a high-performance military aircraft. However, as we shall soon learn, the natural frequencies and damping ratios of each quadratic are quite different.

We begin by finding the pitch response to a step change in elevator input of "A" deg. The Laplace transform of the response is

$$\Delta\theta(s) = \frac{K_\theta(s + c_1)(s + c_2)}{(s^2 + a_1 s + b_1)(s^2 + a_2 s + b_2)} \cdot \frac{A}{s} \quad (4.739)$$

Using partial fraction expansion, Equation 4.739 is written as

$$\Delta\theta(s) = K_\theta A \left[\frac{R_1}{s - p_1} + \frac{R_2}{s - p_2} + \frac{R_3}{s - p_3} + \frac{R_4}{s - p_4} + \frac{R_5}{s} \right] \quad (4.740)$$

where p_1 and p_2 are the poles from the quadratic $s^2 + a_1 s + b_1$, and p_3 and p_4 are the poles associated with the quadratic $s^2 + a_2 s + b_2$. R_1, R_2, R_3, R_4, and R_5 are the constants (residues) in the partial fraction expansion. Letting $p_1 = \alpha_1 + j\beta_1, p_3 = \alpha_3 + j\beta_3$ and recognizing that $p_2 = \bar{p}_1 = \alpha_1 - j\beta_1$, $p_4 = \bar{p}_3 = \alpha_3 - j\beta_3$ as well as $R_2 = \bar{R}_1, R_4 = \bar{R}_3$ give

$$\Delta\theta(t) = \mathcal{L}^{-1}\{\theta(s)\} = K_\theta A[R_1 \, e^{p_1 t} + \bar{R}_1 \, e^{\bar{p}_1 t} + R_3 \, e^{p_3 t} + \bar{R}_3 \, e^{\bar{p}_3 t} + R_5], \quad t \geq 0 \quad (4.741)$$

It is left as an exercise to show that

$$Re^{pt} + \bar{R}e^{\bar{p}t} = 2e^{\alpha t}[\text{Re}(R) \cos \beta t - \text{Im}(R) \sin \beta t] \quad (4.742)$$

where

$$p = \alpha + j\beta, \quad \bar{p} = \alpha - j\beta, \quad R = \text{Re}(R) + j\text{Im}(R), \quad \bar{R} = \text{Re}(R) - j\text{Im}(R)$$

The pitch response (in rad) to an $A = 1°$ elevator deflection is given by

$$\Delta\theta(t) = K_\theta\{2e^{\alpha_1 t}[\text{Re}(R_1) \cos \beta_1 t - \text{Im}(R_1) \sin \beta_1 t]$$
$$+ 2e^{\alpha_3 t}[\text{Re}(R_3) \cos \beta_3 t - \text{Im}(R_3) \sin \beta_3 t] + R_5\} \quad (4.743)$$

Assuming the aircraft's natural dynamics are stable, the poles are located in the left-half plane, that is, $\alpha_1 < 0$ and $\alpha_3 < 0$. From Equation 4.739 and the final value theorem and Equation 4.743 with $t \to \infty$, the steady-state pitch response to a unit step input is

$$\Delta\theta_{ss} = \frac{K_\theta c_1 c_2}{b_1 b_2} = K_\theta R_5 \qquad (4.744)$$

The poles and residues are obtained in "*Chap4_CaseStudy1.m.*"

$$p_{1,2} = -0.6220 \pm j0.7008, \quad p_{3,4} = -0.0130 \pm j0.1187$$
$$R_{1,2} = -0.0331 \pm j0.5586, \quad R_{3,4} = -1.3429 \pm j2.9777, \quad R_5 = 2.7519$$

From Equation 4.743, the pitch step response is

$$\Delta\theta(t) = -0.0141\{2e^{-0.6220t}[-0.0331\cos 0.7008t - 0.5586\sin 0.7008t]$$
$$+ 2e^{-0.0130t}[-1.3429\cos 0.1187t - 2.9777\sin 0.1187t] + 2.7519\} \qquad (4.745)$$

The two damped oscillatory components are referred to as the short period and phugoid modes. The natural frequencies, damping ratios, and exponential envelope time constants are given in Table 4.12.

The complete step response is shown in Figure 4.103. The steady-state pitch is from Equation 4.744, $\theta_{ss} = -0.0388$ rad ($-2.2232°$).

The short period and phugoid mode oscillation components of the step response are shown in Figure 4.104.

TABLE 4.12
Short Period and Phugoid Mode Parameters

Mode	ω_n (rad/s)	z	$\tau_{envelope} = 1/\zeta\omega_n$ (s)
Short period	0.9370	0.6638	1.6077
Phugoid	0.1194	0.1089	76.9042

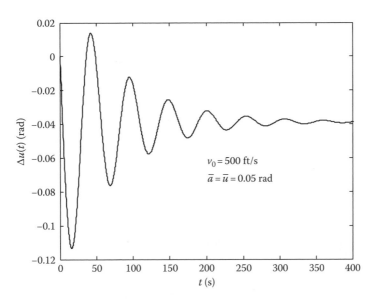

FIGURE 4.103 Linearized aircraft pitch response due to 1° step change in elevator deflection.

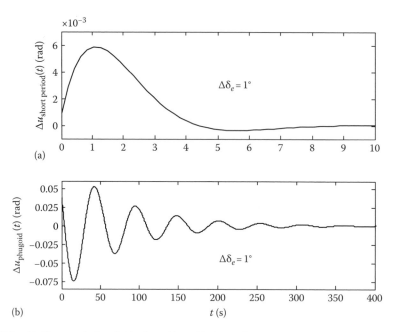

FIGURE 4.104 (a) Short period and (b) phugoid oscillations of elevator unit step response.

Shortly, we will look at the design of an autopilot to control the plane's altitude. Before doing so, a way of determining altitude is needed. From Equations 4.726 and 4.727,

$$\dot{z}' = (-\sin\theta)u + (\sin\varphi\cos\theta)v + (\cos\varphi\cos\theta)w \tag{4.746}$$

where $\dot{z}'$ is the rate of change of altitude, a positive value indicating that the plane is descending. For small values of θ and motion in the longitudinal direction only, $v = 0$, $\phi = 0$, $\sin\theta \approx \theta$, $\cos\theta \approx 1$, $\sin\phi = 0$, $\cos\phi = 1$ and Equation 4.746 simplifies to

$$\dot{z}' = -\theta u + w \tag{4.747}$$

In terms of steady-state values and deviations, Equation 4.747 becomes

$$\frac{d}{dt}(\bar{z} + \Delta\bar{z}') = -(\bar{\theta} + \Delta\theta)(\bar{u} + \Delta u) + \bar{w} + \Delta w \tag{4.748}$$

$$\Rightarrow \frac{d}{dt}(\bar{z}') + \frac{d}{dt}(\Delta z') = -(\bar{\theta}\bar{u} + \bar{u}\Delta\theta + \bar{\theta}\Delta u + \Delta\theta\Delta u) + \bar{w} + \Delta w \tag{4.749}$$

$$\Rightarrow \frac{d}{dt}(\bar{z}') + \frac{d}{dt}(\Delta\bar{z}') = -(-\bar{\theta}\bar{u} + \bar{w}) - \bar{u}\Delta\theta - \bar{\theta}\Delta u - \Delta\theta\Delta u + \Delta w \tag{4.750}$$

Equation 4.747 evaluated at steady state is

$$\frac{d}{dt}(\bar{z}') = -(\bar{\theta}\bar{u} + \bar{w}) \tag{4.751}$$

Subtracting Equation 4.751 from Equation 4.750, ignoring the higher order term $\Delta\theta\,\Delta u$, and recognizing that $d\Delta z'/dt = d(z' - \bar{z}')/dt = dz'/dt$ yield

$$\frac{dz'}{dt} = -\bar{u}\Delta\theta - \bar{\theta}\Delta u + \Delta w \tag{4.752}$$

Substituting Δw in Equation 4.734 into Equation 4.752 gives

$$\frac{dz'}{dt} = -\bar{u}\Delta\theta - \bar{\theta}\Delta u + (\bar{u}\Delta\alpha + \bar{\alpha}\Delta u) \qquad (4.753)$$

$$= -(\bar{\theta} - \bar{\alpha})\Delta u - \bar{u}(\Delta\theta - \Delta\alpha) \qquad (4.754)$$

$$= -\bar{u}(\Delta\theta - \Delta\alpha) \qquad (4.755)$$

Laplace transforming Equation 4.755,

$$\dot{z}'(s) = -\bar{u}[\Delta\theta(s) - \Delta\alpha(s)] \qquad (4.756)$$

The transfer function from elevator input $\Delta\delta_e(t)$ to output $\dot{z}'(t)$ is

$$G_{\dot{z}'}(s) = \frac{\dot{z}'(s)}{\Delta\delta_e(s)} = -\bar{u}\left[\frac{\Delta\theta(s)}{\Delta\delta_e(s)} - \frac{\Delta\alpha(s)}{\Delta\delta_e(s)}\right] \qquad (4.757)$$

The transfer function $\Delta\alpha(s)/\Delta\delta_e(s)$ is obtained in the same way we found $\Delta\theta(s)/\delta_e(s)$ in Equation 4.738. The result is

$$\frac{\Delta\alpha(s)}{\Delta\delta_e(s)} = \frac{K_\alpha(s^2 + d_1 s + d_0)}{(s^2 + a_1 s + b_1)(s^2 + a_2 s + b_2)} \qquad (4.758)$$

where $K_\alpha = -0.141$, $d_1 = 0.0400$, and $d_0 = 0.0235$ are from "*Chap4_CaseStudy1.m*."
Substituting Equations 4.738 and 4.758 into Equation 4.757 gives

$$G_{\dot{z}'}(s) = -\bar{u}\left[\frac{K_\theta(s + c_1)(s + c_2)}{(s^2 + a_1 s + b_1)(s^2 + a_2 s + b_2)} - \frac{K_\alpha(s^2 + d_1 s + d_0)}{(s^2 + a_1 s + b_1)(s^2 + a_2 s + b_2)}\right] \qquad (4.759)$$

$$\Rightarrow G_{\dot{z}'}(s) = \frac{-\bar{u}[(K_\theta - K_\alpha)s^2 + \{K_\theta(c_1 + c_2) - K_\alpha d_1\}s + K_\theta c_1 c_2 - K_\alpha d_0]}{(s^2 + a_1 s + b_1)(s^2 + a_2 s + b_2)} \qquad (4.760)$$

$$\Rightarrow G_{\dot{z}'}(s) = \frac{\lambda_2 s^2 + \lambda_1 s + \lambda_0}{(s^2 + a_1 s + b_1)(s^2 + a_2 s + b_2)} \qquad (4.761)$$

$$\lambda_2 = -\bar{u}(K_\theta - K_\alpha), \quad \lambda_1 = -\bar{u}[K_\theta(c_1 + c_2) - K_\alpha d_1], \quad \lambda_0 = -\bar{u}(K_\theta c_1 c_2 - K_\alpha d_0) \qquad (4.762)$$

From "*Chap4_CaseStudy1.m*," $\lambda_2 = 0$, $\lambda_1 = 4.5768$, and $\lambda_0 = 0.0771$.
For a step input in elevator deflection of $A°$, Equation 4.761 and $\lambda_2 = 0$ give

$$\dot{z}'(s) = \frac{\lambda_1 s + \lambda_0}{(s^2 + a_1 s + b_1)(s^2 + a_2 s + b_2)}\left(\frac{A}{s}\right) \qquad (4.763)$$

The partial fraction expansion of $\dot{z}'(s)$ is

$$\dot{z}'(s) = A\left[\frac{R_1}{s - p_1} + \frac{R_2}{s - p_2} + \frac{R_3}{s - p_3} + \frac{R_4}{s - p_4} + \frac{R_5}{s}\right] \qquad (4.764)$$

where the residues, evaluated in "*Chap4_CaseStudy1.m*," are

$$R_{1,2} = 3.7283 \pm j0.4124, \quad R_{3,4} = -6.8081 \pm j21.2231, \quad R_5 = 6.1596$$

From Equations 4.763 and 4.764, the final value of $\dot{z}'$ is given by

$$\dot{z}'_{ss} = \frac{A\lambda_0}{b_1 b_2} = AR_5 \tag{4.765}$$

The step response is from Equation 4.764,

$$\dot{z}'(t) = A[R_1 e^{p_1 t} + R_2 e^{p_2 t} + R_3 e^{p_3 t} + R_4 e^{p_4 t} + R_5] \tag{4.766}$$

Equation 4.766 is converted to a trigonometric form with real coefficients and real exponents similar to Equation 4.743 for $\Delta\theta(t)$. The unit step response is graphed in Figure 4.105. According to Equation 4.765, the steady-state value $\dot{z}'_{ss} = AR_5 = 1 \times 6.1596$ ft/s.

The change in altitude $\Delta z(t)$ resulting from a step change in elevator input is obtained by integration of $\dot{z}'(t)$. From Equation 4.763,

$$\Delta z'(s) = \frac{1}{s} \dot{z}'(s) = \frac{1}{s} \left[\frac{\lambda_1 s + \lambda_0}{(s^2 + a_1 s + b_1)(s^2 + a_2 s + b_2)} \right] \frac{A}{s} \tag{4.767}$$

$$= \frac{A(\lambda_1 s + \lambda_0)}{s^2(s^2 + a_1 s + b_1)(s^2 + a_2 s + b_2)} \tag{4.768}$$

The inverse transform of Equation 4.768 is left as an exercise problem. The change in altitude $\Delta z'(t)$ is graphed in Figure 4.105 below the derivative $d\dot{z}'/dt$.

The phugoid mode is an undesirable fact of life when it comes to control of an aircraft. In the previous example, it takes 300–400 s for the plane to establish a new steady-state pitch and rate of descent following a step change in the elevator position.

Consider a scenario where the plane is required to decrease its cruising altitude by some amount. One approach is for the pilot to pull back on the yoke to increase the elevator deflection from its neutral position, which produces steady-state level flight conditions. The plane will begin a descent similar to the one shown in Figure 4.105. The actual descent will depend on the magnitude of the elevator deflection. Some time later, the yoke is returned to the neutral position, and the plane

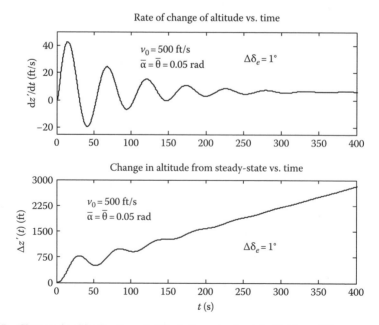

FIGURE 4.105 Changes in altitude rate and altitude from steady-state flight conditions.

returns to level flight conditions at a reduced altitude. To illustrate, suppose the pilot's action results in an elevator deflection of $\Delta\hat{\delta}_e$ degree for a period of T_{pulse} s. The aircraft's altitude response to the pulse input in elevator deflection is obtained as the difference between the step response and the delayed step response, that is,

$$\Delta z_p(t) = \Delta\hat{\delta}_e \Delta z_1(t) - \Delta\hat{\delta}_e \Delta z_1(t - T_{pulse})\hat{u}(t - T_{pulse}) \tag{4.769}$$

where
$\Delta z_1(t)$ is the change in altitude response to a unit step elevator deflection
$\hat{u}(t - T_{pulse})$ is the unit step function starting at $t = T_{pulse}$
$\Delta z_p(t)$ is the change in altitude response to a pulse elevator deflection of $\Delta\hat{\delta}_e$ deg lasting T_{pulse} s

For a 5° elevator pulse input of 30 s, the aircraft's descent is computed according to Equation 4.769 in "*Chap4_CaseStudy1.m*" and shown in Figure 4.106. The label "open-loop" refers to the lack of feedback used to determine the control surface deflection $\Delta\delta_e(t)$.

The open-loop response settles at a value of approximately 927.5 ft once the phugoid oscillations have disappeared. Some form of corrective action is necessary to dampen the excessive phugoid mode oscillations. A feedback control system or autopilot can automate the process without relying on human input.

Figure 4.107 is a simplified block diagram of a control system for regulating an aircraft's altitude. Sensors convert the plane's altitude and rate of descent (or ascent) to voltages, which are transmitted

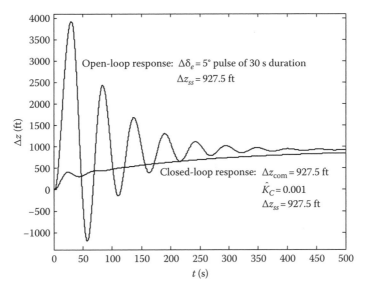

FIGURE 4.106 Open- and closed-loop altitude response vs. time.

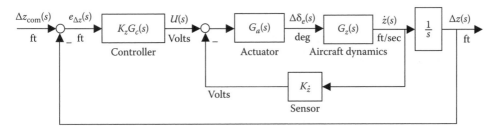

FIGURE 4.107 Block diagram for altitude control system.

to the autopilot. In Figure 4.107, the gain of the altitude sensor K_z is shown combined with the controller transfer function $G_C(s)$, allowing the command signal Δz_{com} to be in ft rather than volts. (Note that the ' symbol designating inertial coordinates is dropped from here on.)

The inner loop provides feedback of the altitude rate, which improves the damping and speed of the outer altitude control loop. There are several ways of obtaining the closed-loop transfer function $\Delta z(s)/\Delta z_{com}(s)$. The inner loop can be reduced to

$$\frac{\dot{z}(s)}{U(s)} = \frac{G_a(s)G_{\dot{z}}(s)}{1 + K_{\dot{z}}G_a(s)G_{\dot{z}}(s)} \tag{4.770}$$

Using the same block diagram reduction formula for the outer loop gives

$$\frac{\Delta z(s)}{\Delta z_{com}(s)} = \frac{K_z G_c(s)[\dot{z}(s)/U(s)]1/s}{1 + K_z G_c(s)[\dot{z}(s)/U(s)]1/s} \tag{4.771}$$

$$= \frac{K_z G_c(s)[G_a(s)G_{\dot{z}}(s)/(1 + K_{\dot{z}}G_a(s)G_{\dot{z}}(s))]1/s}{1 + K_z G_c(s)[G_a(s)G_{\dot{z}}(s)/(1 + K_{\dot{z}}G_a(s)G_{\dot{z}}(s))]1/s} \tag{4.772}$$

$$= \frac{K_z G_c(s)G_a(s)G_{\dot{z}}(s)}{[1 + K_{\dot{z}}G_a(s)G_{\dot{z}}(s)]s + K_z G_c(s)G_a(s)G_{\dot{z}}(s)} \tag{4.773}$$

To start with, a proportional controller $G_c(s) = K_c$ is considered. The product of the gain K_z and controller gain K_c is $\hat{K}_C$, that is, $\hat{K}_C = K_z K_c$ is the effective controller gain for design purposes. For now, we ignore the actuator dynamics and let $G_a(s) = K_a$ measured in deg/volt. Equation 4.773 becomes

$$\frac{\Delta z(s)}{\Delta z_{com}(s)} = \frac{\hat{K}_C K_a G_{\dot{z}}(s)}{[1 + K_{\dot{z}} K_a G_{\dot{z}}(s)]s + \hat{K}_C K_a G_{\dot{z}}(s)} \tag{4.774}$$

The DC gain of the autopilot is

$$\lim_{s \to 0} \frac{\Delta z(s)}{\Delta z_{com}(s)} = \lim_{s \to 0} \frac{\hat{K}_C K_a G_{\dot{z}}(s)}{[1 + K_{\dot{z}} K_a G_{\dot{z}}(s)]s + \hat{K}_C K_a G_{\dot{z}}(s)} = 1 \tag{4.775}$$

Substituting Equation 4.761 with $\lambda_2 = 0$ for $G_{\dot{z}}(s)$ into Equation 4.774 gives

$$\frac{\Delta z(s)}{\Delta z_{com}(s)} = \frac{\hat{K}_C K_a(\lambda_1 s + \lambda_0)}{s^5 + \mu_4 s^4 + \mu_3 s^3 + \mu_2 s^2 + \mu_1 s + \mu_0} \tag{4.776}$$

$$\left.\begin{array}{l} \mu_4 = a_1 + a_2 \\ \mu_3 = a_1 a_2 + b_1 + b_2 \\ \mu_2 = a_1 b_2 + a_2 b_1 + K_{\dot{z}} K_a \lambda_1 \\ \mu_1 = b_1 b_2 + K_{\dot{z}} K_a \lambda_0 + \hat{K}_C K_a \lambda_1 \\ \mu_0 = \hat{K}_C K_a \lambda_0 \end{array}\right\} \tag{4.777}$$

"*Chap4_CaseStudy1.m*" creates a system object for the control system transfer function in Equation 4.776 and then issues the MATLAB ``step'' command to acquire the unit step response values, which are multiplied by Δz_{com} and then plotted. The statements are

```
num_cs_z = Kc_hat*Ka*[lambda1 lambda0];
den_cs_z = [1 mu4 mu3 mu2 mu1 mu0];
sys_cs_z = tf(num_cs_z, den_cs_z)
```

```
T=linspace(0, 500, 1000); % t array for step response
[Y,T] =step(sys_cs_z,T); %Y is unit step response of control system
z_com=927.5; % command input (ft)
z_cs=z_com*Y; % control system response to z_com
plot(T,z_cs,'r')
```

Numerical values used to obtain the closed-loop response in Figure 4.106 were $\Delta z_{\text{com}} = 927.5$ ft, $K_a = 1°/\text{V}$, $K_{\dot{z}} = 0.1$ volt/ft/s, and $\hat{K}_C = 0.001$. The closed-loop transfer function corresponding to those values is

$$\frac{\Delta z(s)}{\Delta z_{\text{com}}(s)} = \frac{0.004577s + 0.00007713}{s^5 + 1.27s^4 + 0.9247s^3 + 0.08634s^2 + 0.01787s + 0.00007713} \quad (4.778)$$

Both responses in Figure 4.106 approach 927.5 ft; however, the closed-loop response is far superior to the open-loop pulse response. The elevator deflection in the closed-loop system response must be small enough to justify the use of the linearized model in Equation 4.735, which assumes small deviations in u, α, q, and θ. The small angle approximations and omission of high-order terms, key to the linearized model's accuracy, may not hold if there are sizable changes in any of the responses. We must look at a graph of $\Delta \delta_e(t)$ responsible for the closed-loop response in Figure 4.106.

$\Delta \delta_e(s)/\Delta z_{\text{com}}(s)$ can be obtained by observing from Figure 4.107 that

$$\Delta z(s) = \frac{1}{s} G_{\dot{z}}(s) \Delta \delta_e(s) \quad (4.779)$$

Solving Equation 4.779 for $\Delta \delta_e(s)$ and then dividing both sides by $\Delta z_{\text{com}}(s)$ lead to

$$\frac{\Delta \delta_e(s)}{\Delta z_{\text{com}}(s)} = \frac{s}{G_{\dot{z}}(s)} \frac{\Delta z(s)}{\Delta z_{\text{com}}(s)} \quad (4.780)$$

Substituting for $G_{\dot{z}}(s)$ the expression in Equation 4.761 gives

$$\frac{\Delta \delta_e(s)}{\Delta z_{\text{com}}(s)} = \frac{\hat{K}_C K_a s(s^2 + a_1 s + b_1)(s^2 + a_2 s + b_2)}{s^5 + \mu_4 s^4 + \mu_3 s^3 + \mu_2 s^2 + \mu_1 s + \mu_0} \quad (4.781)$$

The closed-loop elevator and altitude step responses for $\hat{K}_C = 0.001, 0.003$, and 0.005 along with the open-loop response are shown in Figures 4.108 and 4.109.

Looking at Figure 4.108, it is clear that the closed-loop system elevator input $\Delta \delta_e(t)$, $t \geq 0$ remains less than the 5° pulse amplitude in the open-loop system. It is left as an exercise problem to investigate the deviations Δu, $\Delta \alpha$, q, and $\Delta \theta$ as well.

The proportional gain compensator for the autopilot is far too simplistic; however, the results are fairly dramatic even for this simple design. One of the problems with this design is related to stability. The sluggish response ($\hat{K}_C = 0.001$) in Figure 4.109 is the most stable, yet the location of the closed-loop system poles, which determine the transient response, is far from optimal. Table 4.13 lists the location of the closed-loop system poles corresponding to the values of $\hat{K}_C$.

The reader should consult one of the numerous control system texts for a discussion of more sophisticated compensators to achieve superior dynamic response with increased stability margins.

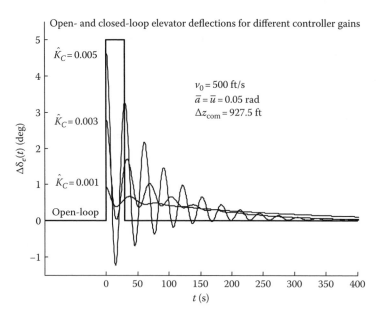

FIGURE 4.108 Elevator response for open- and closed-loop control of altitude.

The gain (magnitude in db) of the open- and closed-loop frequency response functions is shown in Figure 4.110. The open-loop $|\dot{z}(j\omega)/\Delta\delta_e(j\omega)|$ is obtained from the transfer function in Equation 4.761, (recall $\lambda_2 = 0$). The open-loop $|\Delta z(j\omega)/\Delta\delta_e(j\omega)|$ comes from the transfer function

$$G_{\Delta z}(s) = \frac{\Delta z(s)}{\Delta\delta_e(s)} = \frac{\lambda_1 s + \lambda_0}{s(s^2 + a_1 s + b_1)(s^2 + a_2 s + b_2)} \tag{4.782}$$

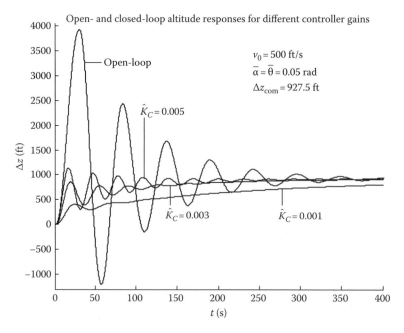

FIGURE 4.109 Altitude response for open- and closed-loop control.

TABLE 4.13
Closed-Loop System Poles for Autopilot
with Proportional Control

$\hat{K}_C$	Closed-Loop Poles
0.001	$-0.5981 \pm j0.6759, \ -0.0347 \pm j0.1424, \ -0.0044$
0.003	$-0.6066 \pm j0.6748, \ -0.0240 \pm j0.1772, \ -0.0088$
0.005	$-0.6150 \pm j0.6742, \ -0.0145 \pm j0.2055, \ -0.0109$

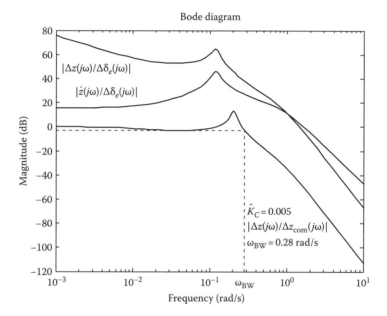

FIGURE 4.110 Open- and closed-loop magnitude functions.

The closed-loop $|\Delta z(j\omega)/\Delta z_{\text{com}}(j\omega)|$ is based on the transfer function in Equation 4.776 with $\hat{K}_C = 0.005$.

Note that the resonant frequency in the open-loop functions at the natural frequency of the phugoid $\omega_n = 0.1194$ rad/s (see Table 4.12). The closed-loop system gain is close to 0 db from DC to somewhat less than the resonant frequency. The bandwidth of the control system is approximately 0.28 rad/s.

4.11.1 DIGITAL SIMULATION OF AIRCRAFT LONGITUDINAL DYNAMICS

A digital simulation of longitudinal dynamics requires z-domain transfer functions to approximate the corresponding continuous-time transfer functions. A z-domain transfer function to approximate the continuous-time transfer function in Equation 4.776 based on explicit Euler integration is

$$\frac{\Delta z(z)}{\Delta z_{\text{com}}(z)} = \left. \frac{\hat{K}_C K_a(\lambda_1 s + \lambda_0)}{s^5 + \mu_4 s^4 + \mu_3 s^3 + \mu_2 s^2 + \mu_1 s + \mu_0} \right|_{s=(z-1)/T} \qquad (4.783)$$

Substituting $(z - 1)/T$ for s in Equation 4.783 leads to

$$\frac{\Delta z(z)}{\Delta z_{com}(z)} = \hat{K}_C K_a T^4 \left[\frac{\lambda_1 z - (\lambda_1 - \lambda_0 T)}{z^5 + \gamma_4 z^4 + \gamma_3 z^3 + \gamma_2 z^2 + \gamma_1 z + \gamma_0} \right] \tag{4.784}$$

where

$$\left. \begin{array}{l} \gamma_4 = -5 + \mu_4 T \\ \gamma_3 = -10 - 4\mu_4 T + \mu_3 T^2 \\ \gamma_2 = -10 + 6\mu_4 T - 3\mu_3 T^2 + \mu_2 T^3 \\ \gamma_1 = 5 - 4\mu_4 T + 3\mu_3 T^2 - 2\mu_2 T^3 + \mu_1 T^4 \\ \gamma_0 = -1 + \mu_4 T - \mu_3 T^2 + \mu_2 T^3 - \mu_1 T^4 + \mu_0 T^5 \end{array} \right\} \tag{4.785}$$

To simulate the altitude response to a step input command of magnitude $\Delta z_{com} = A$, we need the difference equation relating Δz_k and $(\Delta z_{com})_k$. Cross multiplying Equation 4.784 after multiplying numerator and denominator by z^{-5} gives

$$(1 + \gamma_4 z^{-1} + \gamma_3 z^{-2} + \gamma_2 z^{-3} + \gamma_1 z^{-4} + \gamma_0 z^{-5})\Delta z(z)$$
$$= \hat{K}_C K_a T^4 [\lambda_1 z^{-4} - (\lambda_1 - \lambda_0 T) z^{-5}]\Delta z_{com}(z) \tag{4.786}$$

Invert z-transforming both sides of Equation 4.786 and solving for Δz_k give

$$\Delta z_k = -\gamma_4 \Delta z_{k-1} - \gamma_3 \Delta z_{k-2} - \gamma_2 \Delta z_{k-3} - \gamma_1 \Delta z_{k-4} - \gamma_0 \Delta z_{k-5}$$
$$+ \hat{K}_C K_a T^4 [\lambda_1 (\Delta z_{com})_{k-4} - (\lambda_1 - \lambda_0 T)(\Delta z_{com})_{k-5}] \tag{4.787}$$

The first several values of Δz_k are evaluated sequentially from Equation 4.787 as

$$k = 0, 1, 2, 3: \quad \Delta z_k = 0 \tag{4.788}$$

$$k = 4: \quad \Delta z_4 = \hat{K}_C K_a T^4 \lambda_1 (\Delta z_{com})_0 = \hat{K}_C K_a T^4 \lambda_1 A \tag{4.789}$$

$$k = 5: \quad \Delta z_5 = -\gamma_4 \Delta z_4 + \hat{K}_C K_a T^4 [\lambda_1 (\Delta z_{com})_1 - (\lambda_1 - \lambda_0 T)(\Delta z_{com})_0] \tag{4.790}$$

$$= -\gamma_4 (\hat{K}_C K_a T^4 \lambda_1 A) + \hat{K}_C K_a T^4 [\lambda_1 A - (\lambda_1 - \lambda_0 T)A] \tag{4.791}$$

$$= \hat{K}_C K_a T^4 A(-\gamma_4 \lambda_1 + \lambda_0 T) \tag{4.792}$$

$\Delta z_k, k = 6, 7, 8, \ldots$ is computed by recursion according to

$$\Delta z_k = -\gamma_4 \Delta z_{k-1} - \gamma_3 \Delta z_{k-2} - \gamma_2 \Delta z_{k-3} - \gamma_1 \Delta z_{k-4} - \gamma_0 \Delta z_{k-5} + \hat{K}_C K_a T^5 A \lambda_0 \tag{4.793}$$

"*Chap4_CaseStudy1.m*" contains statements to implement Equations 4.788, 4.789, 4.792, and 4.793. The simulated altitude response of the closed-loop system with $\hat{K}_C = 0.003$ to the altitude command previously considered ($\Delta z_{com} = 927.5$ ft) is shown in Figure 4.111. The analytical solution previously plotted in Figure 4.109 is also presented. For purposes of clarity, the simulated points are plotted 1 s apart, that is, every 10th point is plotted. The exact and simulated responses are in close agreement.

Analytical and simulated (Euler $T = 0.1$ s) closed-loop system altitude response

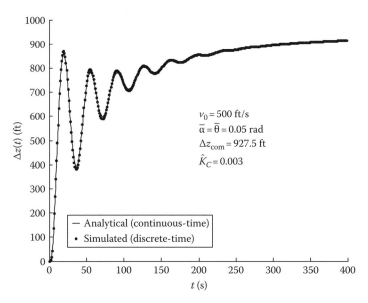

$v_0 = 500$ ft/s
$\bar{\alpha} = \bar{\theta} = 0.05$ rad
$\Delta z_{com} = 927.5$ ft
$\hat{K}_C = 0.003$

— Analytical (continuous-time)
• Simulated (discrete-time)

FIGURE 4.111   Altitude step responses of analytical and simulated closed-loop system.

4.11.2 SIMULATION OF STATE VARIABLE MODEL

The linearized model describing the longitudinal dynamics of an aircraft was given in state variable form in Equation 4.735. Subsequent analysis of dynamic response, however, was done using transfer function descriptions relating a specific input, namely, $\Delta\delta_e(t)$, and a certain output, for example, $\Delta\theta(t)$, $\dot{z}(t)$, and $\Delta z(t)$. The conversion from a state-space description to input–output models is accomplished using Equation 4.736 or the MATLAB function "ss2tf" available in the control system toolbox. The remainder of this section is devoted to simulation of the aircraft dynamics based on the continuous-time state-space model

$$\Delta\underline{x} = A\Delta\underline{x} + B\Delta\underline{u}, \quad \Delta\underline{y} = C\Delta\underline{x} + D\Delta\underline{u} \tag{4.794}$$

where
  A, B, $\Delta\underline{x}$, $\Delta\underline{u}$ are evident from Equation 4.735
  $\Delta\underline{y}$ is the output vector, which determines C and D

Suppose a simulation of the state equations using trapezoidal integration is required. Equation 3.121 is the difference equation for updating the discrete-time state based on trapezoidal integration. It is repeated below (using the deviation variable notation) along with the equation for computing the output vector.

$$\Delta\underline{x}(n+1) = \left(I - \frac{1}{2}TA\right)^{-1}\left(I + \frac{1}{2}TA\right)\Delta\underline{x}(n) + \frac{1}{2}\left(I - \frac{1}{2}TA\right)^{-1}TB[\Delta\underline{u}(n) + \Delta\underline{u}(n+1)] \tag{4.795}$$

$$\Delta\underline{y}(n) = C\Delta\underline{x}(n) + D\Delta\underline{u}(n) \tag{4.796}$$

Equations 4.795 and 4.796 represent a straightforward approach to simulation of the state equations using trapezoidal integration. The equations are implemented in the script file "*Chap4_CaseStudy1.m*"

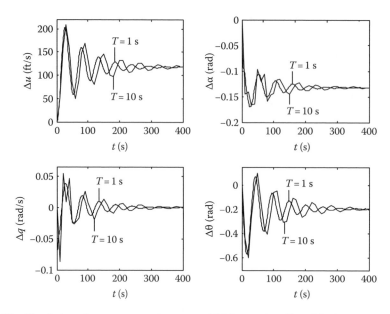

FIGURE 4.112   Simulation of state vector using trapezoidal integration ($\delta_e = 5$ deg).

for the case where $\Delta\underline{u} = [\Delta\delta_e\ \Delta\delta_T]^T = [5°\ 0\ \mathrm{lb}]^T$, $\Delta\underline{y} = [\Delta u\ \ \Delta\alpha\ \ \Delta q\ \ \Delta\theta]^T$. Accordingly, C is the 4×4 identity matrix and D is a 4×2 matrix of zeros. The simulated output $\Delta\underline{y}(n) = [\Delta u(n)\ \ \Delta\alpha$ $(n)\ \ \Delta q(n)\ \ \Delta\theta(n)]^T$ was recorded for $T = 1, 5, 10$ s and the results graphed for $T = 1$ and 10 s in Figure 4.112.

There was very little difference in the outputs for $T = 1$ and 5 s suggesting that the higher value is appropriate for further simulation studies using trapezoidal integration.

Setting $\Delta\underline{\dot{x}} = \underline{0}$ in Equation 4.794 and solving for $\Delta\underline{x}$ at steady state give

$$\Delta\underline{x}_{ss} = -A^{-1}B\Delta\underline{u} \tag{4.797}$$

$$\Rightarrow \begin{bmatrix} \Delta u_{ss} \\ \Delta\alpha_{ss} \\ \Delta q_{ss} \\ \Delta\theta_{ss} \end{bmatrix} = - \begin{bmatrix} -0.04 & 11.59 & 0 & -32.2 \\ -0.00073 & -0.65 & 1 & 0 \\ 0.000048 & -0.49 & -0.58 & 0 \\ 0 & 0 & 1 & 0 \end{bmatrix}^{-1} \begin{bmatrix} 0 & 0.1 \\ 0 & 0 \\ -0.014 & 0 \\ 0 & 0 \end{bmatrix} \begin{bmatrix} 5 \\ 0 \end{bmatrix}$$

$$= \begin{bmatrix} 117.83\ \mathrm{ft/s} \\ -0.13\ \mathrm{rad} \\ 0\ \mathrm{rad/s} \\ -0.19\ \mathrm{rad} \end{bmatrix} \tag{4.798}$$

Setting $\Delta\underline{x}(n+1) = \Delta\underline{x}(n) = \Delta\underline{x}(\infty)$ in Equation 4.795,

$$\Delta\underline{x}(\infty) = \left(I - \frac{1}{2}TA\right)^{-1} \left(I + \frac{1}{2}TA\right)\Delta\underline{x}(\infty)$$

$$+ \frac{1}{2}\left(I - \frac{1}{2}TA\right)^{-1} TB[\Delta\underline{u}(\infty) + \Delta\underline{u}(\infty)] \tag{4.799}$$

Solving for the steady-state vector $\Delta\underline{x}(\infty)$ gives

$$\Delta\underline{x}(\infty) = \left[I - \left(I - \frac{1}{2}TA\right)^{-1}\left(I + \frac{1}{2}TA\right)\right]^{-1}\left(I - \frac{1}{2}TA\right)^{-1} TB\Delta\underline{u}(\infty)$$

$$= [117.83 \text{ ft/s} \quad -0.13 \text{ rad} \quad 0 \text{ rad/s} \quad -0.19 \text{ rad}]^T \qquad (4.800)$$

The continuous-time $\Delta\underline{x}_{ss}$ and discrete-time (simulated) $\Delta\underline{x}(\infty)$ are identical, in agreement with the values observed in Figure 4.112.

EXERCISES

4.86 Prove the relationship in Equation 4.742 involving complex numbers.

4.87 Use the control system toolbox to
   (a) Find the transfer functions $\Delta u(s)/\Delta\delta_T(s)$, $\Delta\alpha(s)/\Delta\delta_T(s)$, $q(s)/\Delta\delta_T(s)$, $\Delta\theta(s)/\Delta\delta_T(s)$.
   (b) Plot the unit step responses for the linearized model in Equation 4.735 with $\Delta y = \Delta x$.

4.88 Find $\Delta z'(t)$ by inversion of $\Delta z'(s)$ in Equation 4.768.

4.89 (a) Use a similar approach to the one for finding $\Delta z'(s)/\Delta\delta_e(s)$ to determine $\Delta x'(s)/\Delta\delta_e(s)$.
   (b) Use the control system toolbox to plot $\Delta x'(t)$ in response to a step change in elevator input of $5°$.
   (c) Find the response $\Delta z'(t)$ to the same input.
   (d) Plot the aircraft's flight trajectory $\Delta z'$ vs. $\Delta x'$ for $(0 \leq \Delta x' \leq 25{,}000$ ft).

4.90 Find the time duration of a $-5°$ elevator pulse input required to increase the plane's elevation by 1500 ft.

4.91 The actuator that controls elevator deflection was assumed to exhibit negligible dynamics in the typical range of frequencies encountered. The actuator transfer function is first-order with gain $K_a = 1°/V$ and time constant $\tau_a = 0.4$ s.
   (a) Find the closed-loop transfer functions $\Delta z(s)/\Delta z_{com}(s)$ and $\Delta\delta_e(s)/\Delta z_{com}(s)$ with the actuator dynamics included. Express both transfer functions as a ratio of polynomials similar to Equations 4.776 and 4.781.
   (b) Find the closed-loop system ($\hat{K}_C = 0.005$) poles with and without the actuator dynamics. Comment on the results.
   With $\hat{K}_C = 0.005$, verify the assumption of negligible actuator dynamics by
   (c) Plotting the frequency response of the open-loop transfer function with and without actuator dynamics.
   (d) Comparing the elevator deflection response when $\Delta z_{com} = 500$ ft with and without the actuator dynamics.
   (e) Comparing the aircraft altitude response when $\Delta z_{com} = 500$ ft with and without the actuator dynamics.

4.92 For the conditions in Figure 4.111, find the maximum deviation between the analytical and simulated altitude responses when using Euler integration with the step sizes shown in the table below. Fill in the table.

Step Size	$T=0.025$ s	$T=0.05$ s	$T=0.1$ s	$T=0.25$ s
$\text{Max}|\Delta z_{anal} - \Delta z_{sim}|$				

4.93 Starting with the open-loop transfer function $G_{\Delta\delta_e}^{\Delta\theta}(s) = \Delta\theta(s)/\Delta\delta_e(s)$ in Equation 4.738,
   (a) Use Tustin's method with a sample time of $T=1$ s to obtain a discrete-time system approximation $G_{\Delta\delta_e}^{\Delta\theta}(z)$. Use the control system toolbox function "c2d" if available, otherwise be prepared for some tedious algebraic work.

(b) Use the pulse transfer function $G^{\Delta\theta}_{\Delta\delta_e}(z)$ to find the difference equation relating $\Delta\theta_k$ and $(\Delta\delta_e)_k$.

(c) Find the aircraft's pitch response to a unit step change in elevator position by recursive solution of the difference equation.

(d) Compare the simulated pitch step response in part (c) to the continuous-time pitch step response shown in Figure 4.103.

4.12 CASE STUDY: NOTCH FILTER FOR ELECTROCARDIOGRAPH WAVEFORM

An electrocardiograph (ECG) signal is corrupted with 60 Hz noise from an electrical power source. A portion of the noisy signal, sampled regularly at 0.004 s intervals, is shown in Figure 4.113.

A notch filter is needed to remove the noise. One realization of a second-order filter transfer function is given by (Orfanidis 1996)

$$H(z) = \frac{Y(z)}{U(z)} = b\left[\frac{1 - 2(\cos\omega_0 T)z^{-1} + z^{-2}}{1 - 2b(\cos\omega_0 T) + (2b-1)z^{-2}}\right] \tag{4.801}$$

where ω_0 is the notch frequency (in rad/s). The filter parameter Q relates the notch frequency ω_0 to the width of the 3 db interval $\Delta\omega$ on a plot of $|H(e^{j\omega T})|^2$ vs. ω.

$$Q = \frac{\omega_0}{\Delta\omega} \tag{4.802}$$

The higher Q is, the narrower is the 3 db interval $\Delta\omega$. The filter parameter b is obtained from

$$b = \frac{1}{1 + \tan(\omega_0 T/2Q)} \tag{4.803}$$

Two notch filters will be investigated. One with $Q = 10$ and the other with $Q = 50$. The M-file "*Chap4_CaseStudy2.m*" computes the filter coefficients and plots both $|H(e^{j\omega T})|^2$ vs. ω and the magnitude function (in db), $|H(e^{j\omega T})|$ vs. ω (see Figures 4.114 through 4.117).

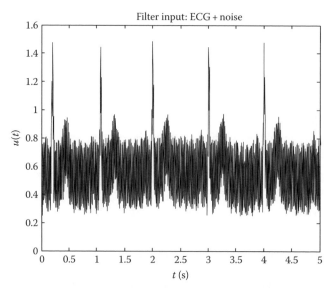

FIGURE 4.113 ECG signal corrupted with 60 Hz noise sampled at $T = 0.004$ s intervals.

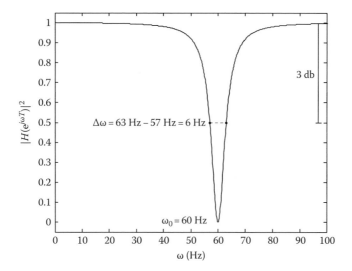

FIGURE 4.114 Magnitude squared function for notch filter ($Q = 10$).

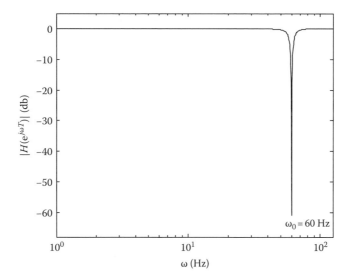

FIGURE 4.115 Magnitude function (in db) for notch filter ($Q = 10$).

Note, when $|H(e^{j\omega T})|^2 = 0.5$ it is 3 db below the DC value $|H(e^{j0T})|^2 = 1$.

The filtered outputs are shown in Figures 4.118 and 4.119. There is little difference in the outputs of the two filters except for the longer transient period of the filter with $Q = 50$.

4.12.1 MULTINOTCH FILTERS

When more than one notch frequency exists, a multinotch filter design is required. The previous reference includes several methods of designing a multinotch filter. One approach is to simply use the singlenotch design for each notch frequency and cascade the respective filters. To illustrate, suppose the ECG signal contains a 25 Hz square wave noise signal like the one shown in Figure 4.120.

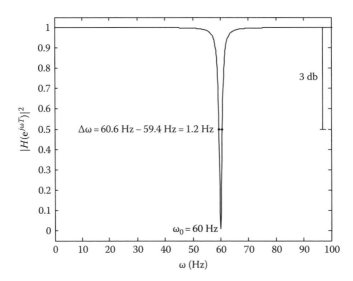

FIGURE 4.116 Magnitude squared function for notch filter ($Q = 50$).

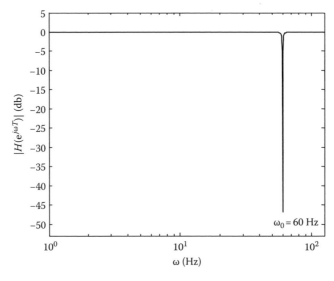

FIGURE 4.117 Magnitude function (in db) for notch filter ($Q = 50$).

The noise $n(t)$ contains harmonics at multiples of the fundamental frequency $\omega_0 = 2\pi f_0 = 50\pi$ rad/s. The Fourier Series expansion of $n(t)$ is given by (see Exercise 4.95)

$$n(t) = \frac{1}{\pi} \sin \omega_0 t + \frac{1}{3\pi} \sin 3\omega_0 t + \frac{1}{5\pi} \sin 5\omega_0 t + \cdots \tag{4.804}$$

Example 4.37

A clean ECG signal, 10 s in duration, is sampled every $T_s = 0.004$ s and stored in the data file "*clean_ecg_10sec.mat*." The time and signal data are stored in arrays "t" and "s."

(a) Sample the square wave noise shown in Figure 4.120 at the sampling frequency $\omega_s = 1/T_s$ and plot the sampled noise $n(t)$ and the noisy ECG signal $s(t) + n(t)$.

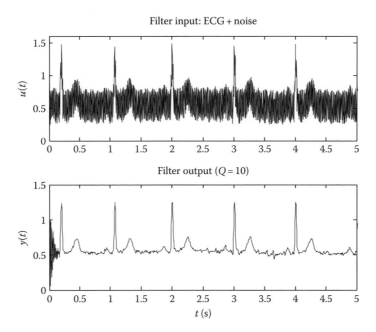

FIGURE 4.118   Output of notch filter ($Q = 10$).

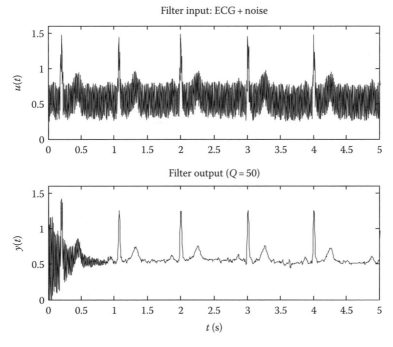

FIGURE 4.119   Output of notch filter ($Q = 50$).

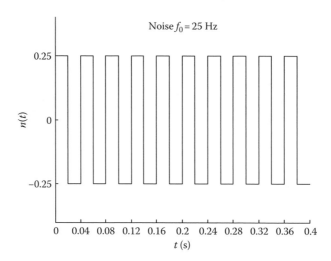

FIGURE 4.120 Square wave noise component of ECG signal.

(b) Design notch filters:
   (i) $H\omega_0(z)$ to remove the fundamental frequency
   (ii) $H_{3\omega_0}(z)$ to remove the first nonzero harmonic term
   (iii) $H_{5\omega_0}(z)$ to remove the second nonzero harmonic term
   Choose the Q values such that the 3 db width $\Delta\omega$ for $|H(e^{j\omega T})|^2$ vs. ω is the same for each filter.
(c) Draw the magnitude function (in db) for the following filters:
   (i) $H_{\omega_0}(z)$ (ii) $H_{\omega_0}(z)H_{3\omega_0}(z)$ (iii) $H_{\omega_0}(z)H_{3\omega_0}(z)H_{5\omega_0}(z)$
(d) Filter the noisy ECG signal in part (a) using the three filters in part (c) and graph the results.

(a) Figure 4.121 shows 5 s of the noise square wave $n(t)$ and the combined signal plus noise $s(t) + n(t)$.

(b) The filter parameter Q was chosen as 10 for the first filter. From Equation 4.802 the 3 db width $\Delta\omega = 2.5$ Hz. Using this value for notch frequencies $3\omega_0 = 75$ Hz and $5\omega_0 = 125$ Hz in Equation 4.802 gives

$$Q = \frac{3\omega_0}{\Delta\omega} = \frac{3(25)}{2.5} = 30, \quad Q = \frac{5\omega_0}{\Delta\omega} = \frac{5(25)}{2.5} = 50 \tag{4.805}$$

The M-file "Chap4_Ex12_1.m" computes the filter coefficients for the three notch filters with Q values 10, 30, and 50 using Equations 4.801 and 4.803. The results are

$$H_{\omega_0}(z) = 0.9695\left(\frac{1 - 1.6180z^{-1} + z^{-2}}{1 - 1.5687z^{-1} + 0.9391}\right) \quad (Q = 10) \tag{4.806}$$

$$H_{3\omega_0}(z) = 0.9695\left(\frac{1 + 0.6180z^{-1} + z^{-2}}{1 + 0.5992z^{-1} + 0.9391}\right) \quad (Q = 30) \tag{4.807}$$

$$H_{5\omega_0}(z) = 0.9695\left(\frac{1 + 2z^{-1} + z^{-2}}{1 + 1.9391z^{-1} + 0.9391}\right) \quad (Q = 50) \tag{4.808}$$

(c) "Chap4_Ex12_1.m" includes statements to plot the magnitude functions of $H_{\omega_0}(z)$ and the cascaded filters $H_{\omega_0}(z)H_{3\omega_0}(z)$ and $H_{\omega_0}(z)H_{3\omega_0}(z)H_{5\omega_0}(z)$. The results are shown in Figures 4.122 through 4.124.

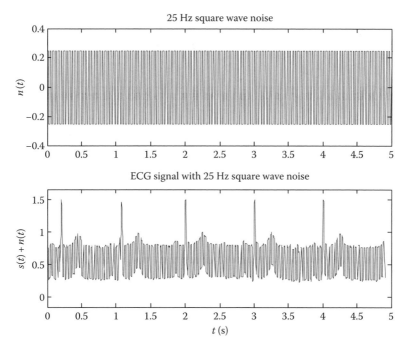

FIGURE 4.121   Square wave noise and noise-corrupted ECG signal.

(d) The three filters are shown in Figure 4.125 with their corresponding inputs and outputs. Output of the filter with transfer function $H_{\omega_0}(z)$ in Equation 4.806 is shown in Figure 4.126. The simple notch filter was designed to remove the fundamental frequency term in Equation 4.804.

Output of the first filter $y_1(k)$ is passed to the notch filter with transfer function $H_{3\omega_0}(z)$ in Equation 4.807. Output $y_2(k)$ of the multinotch filter $H_{\omega_0}(z)H_{3\omega_0}(z)$ is shown in Figure 4.127. Finally, the output of the middle filter in Figure 4.125 is the input to the third filter in the series of cascaded filters. The output of the last filter $y_3(k)$ is plotted as $y_3(t)$ in Figure 4.128.

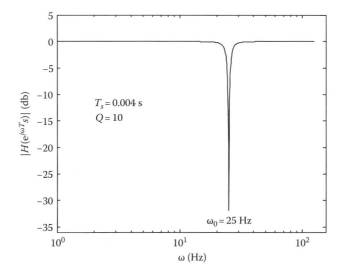

FIGURE 4.122   Magnitude function (db) for notch filter $H_{\omega_0}(z)$.

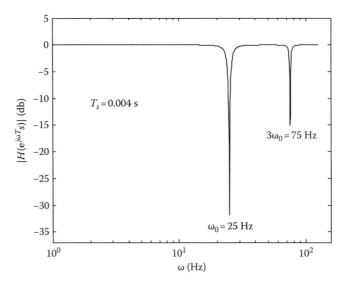

FIGURE 4.123 Magnitude function (db) for multinotch filter $H_{\omega_0}(z)H_{3\omega_0}(z)$.

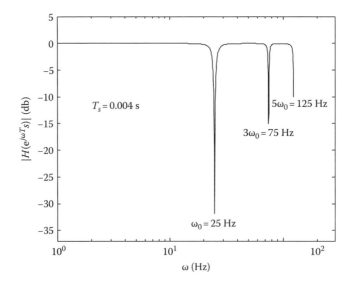

FIGURE 4.124 Magnitude function (db) for multinotch filter $H_{\omega_0}(z)H_{3\omega_0}(z)H_{5\omega_0}(z)$.

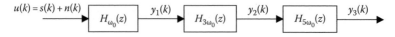

FIGURE 4.125 Multinotch filter for removing fundamental frequency and first two nonzero harmonics.

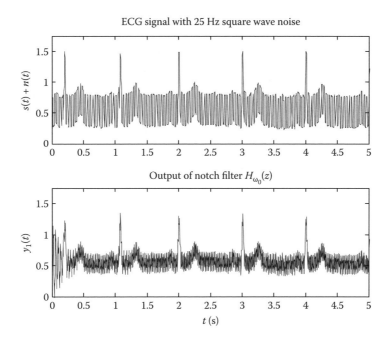

FIGURE 4.126 Input and output of notch filter $H_{\omega_0}(z)$.

The multinotch filter output in Figure 4.128 is similar in appearance to the single notch filter outputs shown in Figures 4.118 and 4.119 (after the transient response has vanished) when the noise was a pure sinusoid at 60 Hz. Even though the square wave noise contains an infinite number of harmonics, that is, odd multiples of the fundamental frequency (see Equation 4.804), all but the first two nonzero harmonics $3\omega_0 = 75$ Hz and $5\omega_0 = 125$ Hz are above the Nyquist

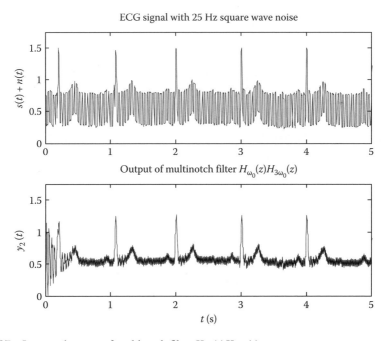

FIGURE 4.127 Input and output of multinotch filter $H_{\omega_0}(z)H_{3\omega_0}(z)$.

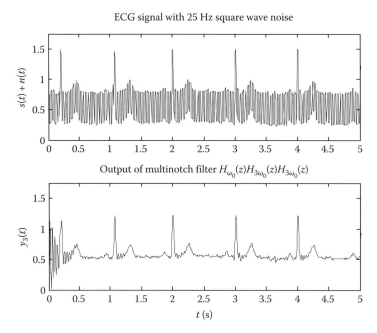

FIGURE 4.128 Input and output of multinotch filter $H_{\omega_0}(z)H_{3\omega_0}(z)H_{5\omega_0}(z)$.

frequency $\omega_{nyq} = 0.5\omega_s = 0.5 \times (1/T_s) = 125$ Hz. Consequently, the harmonics at $7\omega_0 = 175$ Hz, $9\omega_0 = 225$ Hz, and so forth, are aliased back to the lower frequencies which are effectively removed by the multinotch filter in Figure 4.125.

EXERCISES

4.94 Create a noisy ECG signal $u(t_k)$ by starting with the clean signal $s(t_k)$, where $t_k = kT_s$, $k = 0, 1, 2,\ldots$ ($T_s = 0.004$ s) in "*clean_ecg_10sec.mat.*" Add a 50 Hz sinusoidal noise $n(t_k)$ with amplitude of 0.75.

(a) Design and implement an appropriate notch filter to remove the noise.

(b) Graph the filter input $u(t_k)$ and its output $y(t_k)$ below it.

(c) Compare the clean ECG signal $s(t_k)$ and the filter output $y(t_k)$.

4.95 The clean ECG signal described in Exercise 4.94 is corrupted by the periodic noise $n(t)$ shown in Figure E4.95. The period $P = 1/30$ s ($\omega_0 = 30$ Hz) and amplitude $A = 1$.

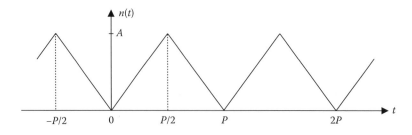

FIGURE E4.95

(a) Sample the noise at the frequency $\omega_s = 250$ Hz ($T_s = 0.004$ s), and add it to the clean ECG signal. Denote the corrupted signal by $u(t_k)$, where $t_k = kT_s$, $k = 0, 1, 2, 3,\ldots$.

(b) Expand the noise in a Fourier series expansion,

$$n(t) = \frac{a_0}{2} + \sum_{k=1,2,\ldots} (a_k \cos k\omega_0 t + b_k \sin k\omega_0 t),$$

$$\omega_0 = \frac{2\pi}{P} = \frac{2\pi}{1/30} = 60\pi \text{ rad/s}$$

$$a_k = \frac{2}{P} \int_{-P/2}^{P/2} n(t) \cos k\omega_0 t \, dt, \quad k = 0, 1, 2, \cdots$$

$$b_k = \frac{2}{P} \int_{-P/2}^{P/2} n(t) \sin k\omega_0 t \, dt, \quad k = 1, 2, \cdots$$

(c) Design and implement a multinotch filter to remove all the frequency components (except DC) below the Nyquist frequency $\omega_{nyq} = 0.5\omega_s = 125$ Hz.

(d) Graph the filter input $u(t_k)$ and its output $y(t_k)$ below it.

(e) Compare the clean ECG signal $s(t_k)$ and the filter output $y(t_k)$.

5 Simulink®

5.1 INTRODUCTION

This chapter serves as an introduction to the continuous simulation program, Simulink®. It is similar in many ways to its predecessors such as CSMP (Continuous System Modeling Program), ACSL (Advanced Continuous Simulation Language), TUTSIM (Twente University of Technology Simulator), MATRIX-X, STELLA, and EASY5. The major advantage of Simulink stems from its tight integration with MATLAB®, the data analysis and visualization program with its own structured programming language. The numerous (37 at the time of this printing) MATLAB toolboxes in diverse areas of engineering, science, and business extend the capabilities of Simulink.

In addition to the toolboxes, there are a number of Simulink blocksets that extend Simulink into various disciplines such as aerospace, communications, signal processing, image processing, and so forth. A complete list of toolboxes and blocksets with descriptions of each can be found at http://www.mathworks.com/products/product_listing/index.html.

Chapters 1 through 4 cover some basic essentials of linear continuous- and discrete-time systems. Elementary simulation techniques based on numerical integration are also introduced. In all but the simplest cases, the simulated solutions were programmed in MATLAB M-files.

The early continuous-time system simulation languages (CSSLs) consisted of individual sections, for example, "Initial," "Dynamic," "Derivative," and "Terminal" with special demarcation headers for inputting constants and system parameters, calculating new parameters, setting initial conditions for the states, evaluating inputs over time, numerically integrating the state derivative vector, and computing the system outputs (Korn 1978). The continuous-time system dynamics were confined to a section containing expressions for the state derivatives. Lookup tables (in one or more dimensions) were often included in the section to evaluate the state derivatives. Crucial savings in simulation development time resulted from the built-in numerical integration routines and graphing capabilities.

Despite minor variations among the CSSLs, they were classified as "equation-oriented" because expressions for the state derivatives, difference equations, and outputs were entered on one or more lines in equation format. Later, general-purpose, block-oriented simulation programs emerged with powerful graphical user interfaces (GUIs). Dragging and dropping blocks from libraries containing blocks of similar functionality is the most intuitive way for creating a simulation model. Even more so than equation-oriented CSSLs, block-oriented simulation programs such as Simulink free the simulationist from the tedious grunt work required to develop a model structure, implement numerical integration, and produce useful output.

Our initial exploration of Simulink in this chapter is merely the "tip of the iceberg." Later chapters will delve further into the world of Simulink and its capabilities.

5.2 BUILDING A SIMULINK® MODEL

To begin our introduction to Simulink, we will demonstrate the procedure for creating a model of a simple system and run the model to obtain useful information about its dynamic response. Our purpose here is to get comfortable with the Simulink user interface at a macroscopic level. *Mastering Simulink* (Dabney 2001) and The Math Works Web page http://www.mathworks.com/access/helpdesk/help/toolbox/simulink/ug/ug.html are excellent references for the beginner interested in getting started with Simulink. The Simulink models in this text were developed using Simulink Version 6.

5.2.1 SIMULINK® LIBRARY

The Simulink library contains blocks for representing the mathematical models of commonly occurring components in dynamic systems. The blocks are grouped in sublibraries according to function. The standard Simulink sublibraries are shown in the left pane of Figure 5.1. The blocks residing in the selected "Continuous" sublibrary are shown in the right pane. The "Integrator" block is selected, and there is a brief description of it in the top pane. The transfer function, $1/s$, is used to designate the integrator.

Building a Simulink model of a system consists of selecting the appropriate blocks and connecting them in a way that represents the mathematical model. Inputs, when present, are implemented using blocks from the "Sources" sublibrary, which can generate a host of input signals. Simulation output is saved and displayed using various blocks such as "Scopes," "XY Graphs," and "Displays" from the "Sinks" sublibrary.

Our first Simulink model will simulate the dynamics of the linear second-order system model introduced in Chapter 2. The differential equation is

$$\frac{d^2}{dt^2}y(t) + 2\zeta\omega_n\frac{d}{dt}y(t) + \omega_n^2 y(t) = K\omega_n^2 u(t) \tag{5.1}$$

Assuming for the moment that the second derivative term d^2y/dt^2 is present in a new model window, it can be twice integrated as shown in Figure 5.2 where "ydd," "yd," and "y" are the Simulink variable names. The "Integrator" blocks are dragged or copied from the "Continuous" sublibrary into the model window.

By inspection of Equation 5.1, the second derivative term is a linear combination of the input $u(t)$, the output $y(t)$, and its first derivative dy/dt. The Simulink library browser allows us to search the standard sublibraries for the blocks needed to "build" the second derivative and, thus, complete the Simulink model.

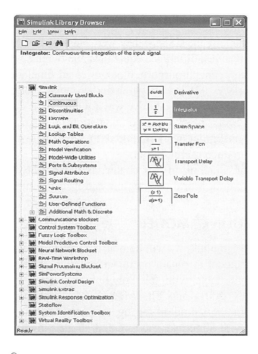

FIGURE 5.1 The Simulink® Library Browser.

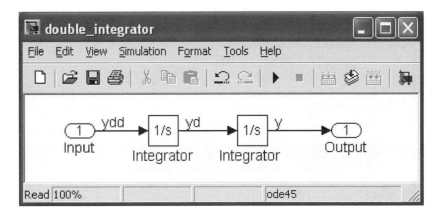

FIGURE 5.2   Integrating the second derivative "ydd" twice to obtain the first derivative "yd" and output "y."

The system parameters K, ω_n, and ζ and the literal constant "2" are generated using a "Constant" block found in the "Sources" sublibrary. The "Math" sublibrary provides the additional blocks for addition and multiplication of the signals.

We have yet to specify an input or forcing function, assuming there is one. For now, let us pick a simple step input applied at $t = 0$. Looking in the "Sources" sublibrary, the step input can be implemented with a "Constant" or "Step" block; however, the latter is more flexible should we later decide to delay the time at which the step is applied.

Numerical values of the system parameters are set by selecting the individual blocks and typing in the appropriate values in a properties dialog box. Some Simulink blocks contain several parameters, all of which should be specified or else the default values will be used. For example, the "Step" block generally requires values for "Step time," "Initial value," and "Final value" as shown in Figure 5.3, and the "Integrator" block requires an "Initial condition."

Figure 5.4 shows a Simulink diagram for simulation of the unit step response of the second-order system. The choice of Simulink blocks and their location in a Simulink diagram is not unique. The appearance or layout of blocks depends to a large extent on individual user preferences. Some prefer

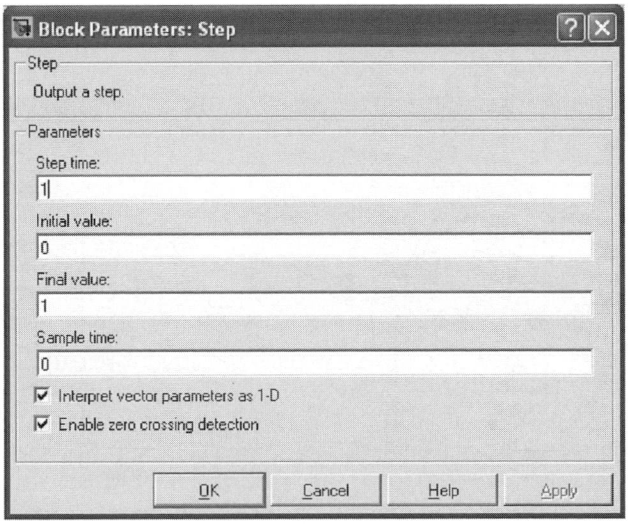

FIGURE 5.3   Dialog box for specifying input step parameter values.

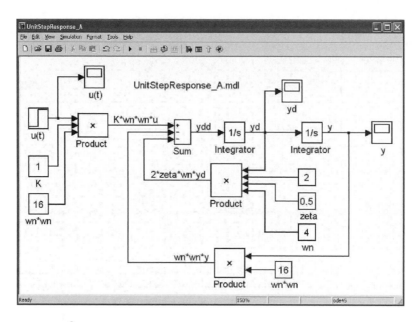

FIGURE 5.4 Simulink® diagram for step response of a second-order system.

that the diagram be the most economical in terms of Simulink blocks used. Others are more concerned with layout style, striving to make the diagram visually appealing.

Oftentimes, the mathematical model of the system is available in block diagram form, as in the case of a control system. A Simulink diagram of the system will be strikingly similar, especially when Simulink blocks for modeling actual system components are available.

An alternate Simulink diagram for the second-order system in Equation 5.1 is shown in Figure 5.5. A "Gain" block with a parameter value equal to the product $2\zeta\omega_n$ replaces the "Product" block in the inner feedback loop and the three constant blocks feeding it. Another "Gain" block is inserted in the outer feedback loop with a parameter value numerically equal to ω_n^2 replacing the "Product" and "Constant" blocks in Figure 5.4. The third "Gain" block is employed to multiply the input $u(t)$ by $K\omega_n^2$, further reducing the number of blocks required.

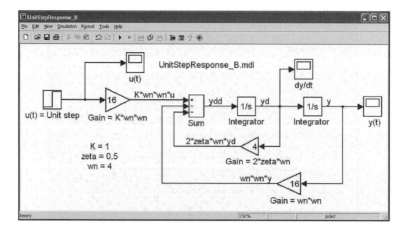

FIGURE 5.5 Alternate Simulink® diagram for a second-order system step response.

Note the similarity between the Simulink diagram in Figure 5.5 and the simulation diagram of the system in Figure 2.13. In fact, the thought process for preparing a simulation diagram of a system is nearly identical to the steps required to arrive at a Simulink diagram.

Before we delve further into the Simulink library, let us run one of the Simulink models for simulating the step response of the second-order system.

5.2.2 RUNNING A SIMULINK® MODEL

The Simulink model is similar to a conventional block diagram of a system. For a system with analog components, it embodies the algebraic and differential equations of the continuous-time math model. For inherently discrete-time systems, the Simulink model encapsulates algebraic and difference equations governing the system's behavior. Simulink models of hybrid systems containing analog and discrete-time components implement solutions to algebraic, differential, and difference equations.

A computer program is created from the Simulink model to solve the equations that comprise the mathematical model of the system. Some of its functions include initialization of state variables, calculation of state derivatives, solution of algebraic equations, updating the state variables, and calculation of the system's outputs. Simulink offers a variety of numerical integrators to advance the continuous-time state vector over an integration step. The user has the option of choosing a particular integrator and step size (applicable for fixed-step size algorithms), tolerances for satisfying accuracy requirements, the simulation start and stop times, and exchanging simulation data with MATLAB via The MATLAB Workspace.

Clicking on "Simulation" in the model window menu followed by "Configuration Parameters" leads to a dialog box like the one shown in Figure 5.6 where the simulation is configured according to the user's preferences as previously described. The improved Euler integrator (Heun's method) with a fixed-step size of 0.01 s and simulation time of 5 s has been selected.

After configuring the simulation, the "Simulation" pull-down menu is reopened and "Start" is selected. The simulation terminates when the simulation time reaches the selected "stop time" of 5 s.

The simplest way to view simulation output is to select one of the scopes and observe the time history of its input. The output of the second integrator "y" is displayed in several ways, as shown in Figures 5.7 through 5.9. Figure 5.7 is a screen capture of the scope labeled "y(t)" after running the simulation and viewing the scope output by double clicking on it.

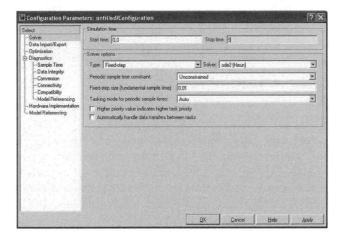

FIGURE 5.6 Dialog box for configuring simulation.

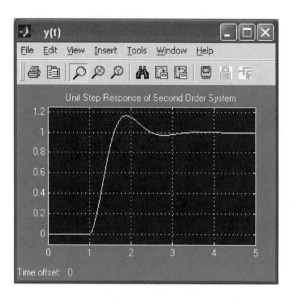

FIGURE 5.7   Screen capture of scope output.

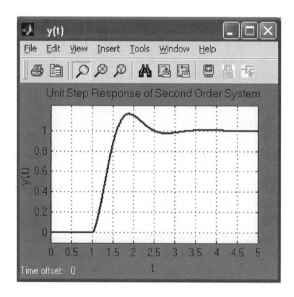

FIGURE 5.8   Screen capture of edited scope output.

Figure 5.8 is a screen shot of the edited scope output made possible by running an M-file "*SimScopeControl.m*," which brings up the MATLAB Property Editor for editing graphs.

Figure 5.9 is the result of copying the edited scope output to the clipboard and pasted into the text. Scope outputs throughout the text will be shown in one of the three formats.

As expected, the step response reflects a moderately underdamped second-order system.

The simulation results can also be imported to the MATLAB Workspace several different ways. In this example, the scopes were configured to communicate the results in named arrays specified in the parameters dialog box, which opens after clicking on the icon shown in Figure 5.10.

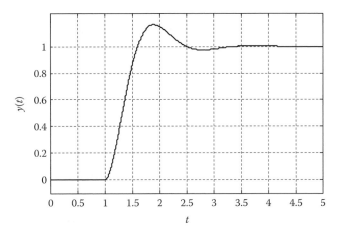

FIGURE 5.9 Simulink® plot of unit step response of a second-order system.

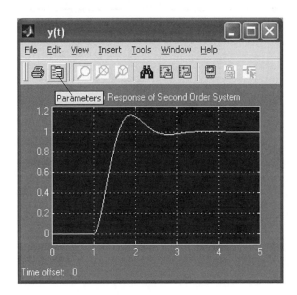

FIGURE 5.10 Icon to open the parameter dialog box of "$y(t)$" scope.

The "Data history" tab in Figure 5.11 was used to save the second integrator's output in the array "t_y," which consists of two columns. The first consists of the time values for the simulation, and the second column contains the associated $y(t)$ values output from the integrator.

Once in the MATLAB Workspace, the various signals can be graphed as shown in Figure 5.12.

EXERCISES

5.1 For the first-order system modeled by

$$\tau \frac{dy}{dt} + y = Ku, \quad y(0) = 0 \quad (\tau = 3 \text{ s}, K = 0.1)$$

$$u(t) = \begin{cases} 0, & t \le 0 \\ A, & t > 0 \quad (A = 5) \end{cases}$$

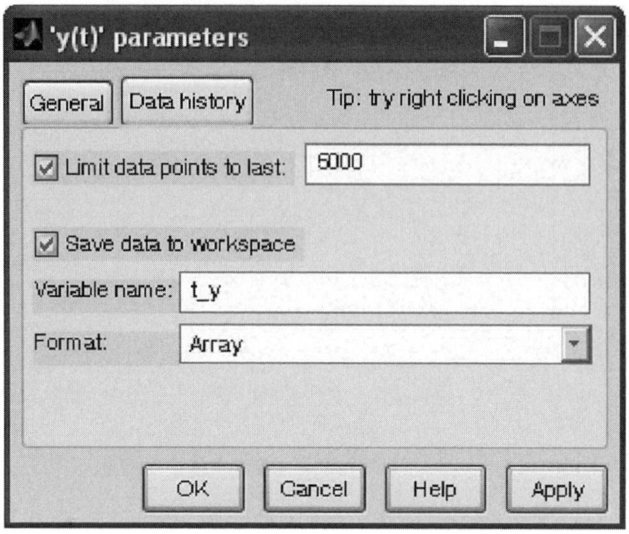

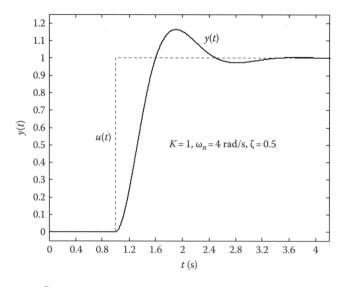

FIGURE 5.11 Parameter dialog box for saving output to the MATLAB® Workspace.

FIGURE 5.12 MATLAB® plot of a second-order system unit step response.

with initial condition $y(0) = 0$,

(a) Prepare a Simulink diagram for simulating the response.

(b) Plot $y(t)/y_{ss}$ vs. t where $y_{ss} = \lim_{t \to \infty} y(t)$ is the steady-state response.

(c) Compare the results from part (b) with the exact solution.

(d) Repeat parts (b) and (c) for $y(0) = 0.5y_{ss}$, y_{ss}, and $1.5y_{ss}$.

5.2 Simulate the second-order system unit step response for $\omega_n = 25$ rad/s and $\zeta = 0.1$ and

(a) Prepare a MATLAB plot of $y(t)$ vs. the dimensionless independent variable $\omega_n t$.

(b) Repeat part (a) for $\zeta = 0.7, 1, 2$.

5.3 The temperature $T(t)$, in °F, of a turkey baking in an oven is approximately governed by the differential equation

$$C \frac{d}{dt} T(t) = Q_i(t) - Q_0(t)$$

where
    C is the thermal capacity in (Btu/°F) of the turkey
    $Q_i(t)$ is the heat input to the turkey
    $Q_0(t)$ is the heat loss due to conduction and convection from the oven, both in Btu/h

Expressions for $Q_i(t)$ and $Q_0(t)$ are as follows:

$$Q_i(t) = \bar{Q}, \quad t \geq 0$$

$$Q_0(t) = \frac{1}{R}[T(t) - T_0(t)]$$

R is the overall thermal resistance (°F/Btu/h) of the oven, and $T_0(t)$ is the room temperature surrounding the oven. Simulate the baking of a 15 lb turkey in an oven with thermal resistance $R = 0.025$°F/Btu/h and constant heat input $\bar{Q} = 4000$ Btu/h. The room temperature is a constant 75°F.

    Note that the specific heat of turkey is $c = 1.25$ Btu/lb/°F, and the thermal capacity of the turkey is given by $C = mc$ where m is the mass (in lb) of the turkey.

    Assume the initial temperature of the turkey is 40°F.

(a) Plot the temperature $T(t)$ on one graph and heat flows $Q_i(t)$ and $Q_0(t)$ on separate graphs. Be sure to run the simulation for a period of time sufficient to examine the complete transient response.
(b) Estimate the final temperature of the turkey if left unattended in the oven.
(c) Estimate the time required to heat the turkey to 160°F.
(d) Compare the results in parts (a), (b), and (c) with results obtained using the solution to the continuous-time differential equation model for $T(t)$.
(e) What size turkey can be heated to 150°F in 2 h?

5.3 SIMULATION OF LINEAR SYSTEMS

Simulink offers the user a variety of approaches when it comes to simulation of linear continuous-time systems. The form of the system model generally dictates the choice of blocks from the "`Continuous`" sublibrary to be used in the Simulink model. For example, a linear second-order system comprising two first-order systems in series like that shown in Figure 5.13 suggests an overall Simulink model constructed using Simulink models of the individual first-order systems.

    The Simulink diagram of the system is shown in Figure 5.14. Note that the two integrators are not in series like they were when the system model was a second-order differential equation. The state variables are x and y.

    A Simulink model of the cascaded first-order systems employing consecutive "`Integrator`" blocks is easily obtained once the variable x is eliminated from the coupled first-order differential equations in Figure 5.13. The resulting second-order differential equation in y and the Simulink diagram is left as an exercise.

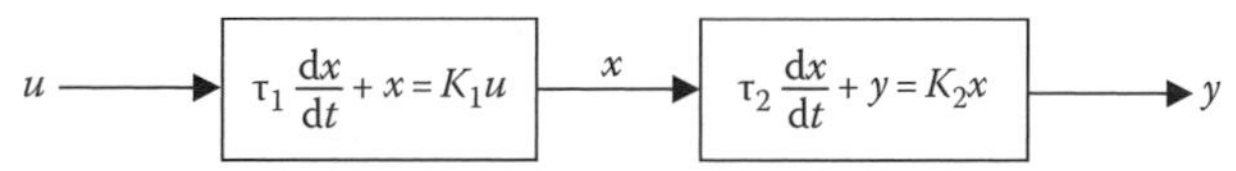

$$u \longrightarrow \boxed{\tau_1 \frac{dx}{dt} + x = K_1 u} \xrightarrow{x} \boxed{\tau_2 \frac{dx}{dt} + y = K_2 x} \longrightarrow y$$

FIGURE 5.13 A second-order system comprised of two cascaded first-order systems.

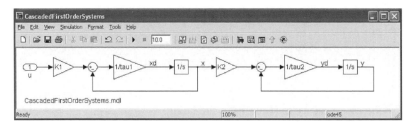

FIGURE 5.14 Simulink® diagram of a second-order system shown in Figure 5.13.

5.3.1 TRANSFER FCN BLOCK

A glimpse of the Simulink blocks in the "Continuous" sublibrary reveals additional options for simulation of linear system models. The "Transfer Fcn" and "Zero-Pole" blocks provide alternative representations for the dynamics of a linear continuous-time component. The n individual integrators and arithmetic blocks for a system component with nth-order dynamics are collapsed into a single block, incorporating the higher-order dynamics. The "Transfer Fcn" and "Zero-Pole" blocks correspond to transfer function models in polynomial and factored form, respectively.

To illustrate the use of the "Transfer Fcn" block, consider a variation of the case study in Section 2.8 for the submarine depth control system. The reference signal $v_{com}(t)$ for the control loop is the command depth rate and the controlled variable is the actual depth rate $v(t)$ as shown in Figure 5.15. The depth $y(t)$ is obtained by integrating the depth rate $v(t)$.

The submarine is assumed initially to be in steady state at the surface when the command depth rate is suddenly increased to 25 ft/s and held constant for 30 s and then returned to zero. The transfer function for the controller and stern plane actuator is

$$G_C(s) = \frac{\theta(s)}{E(s)} = \frac{K_C s + K_I}{s} \quad (K_C = 0.6, K_I = 0.1) \tag{5.2}$$

and the submarine dynamics is modeled by the transfer function

$$G_P(s) = \frac{V(s)}{\theta(s)} = \frac{K_{\dot\theta} s + K_\theta}{\tau s + 1} \quad (K_{\dot\theta} = 20, K_\theta = 10, \tau = 10) \tag{5.3}$$

The Simulink diagram is shown in Figure 5.16. A "Transfer Fcn" block was used to model the controller and submarine dynamics.

Note the use of two step blocks with the same amplitude (25 ft/s), the first commencing at $t = 0$ and the second starting at $t = 30$ s along with the summation block to implement the overall command depth rate signal. The command and actual depth rates are multiplexed and fed to the scope in the upper right corner of the diagram. The submarine depth is captured by the scope directly below. The simulation was configured using Simulink's fixed-step "ode4" numerical integrator with step size 0.01 s. The "ode4" numerical integrator belongs to a family of numerical

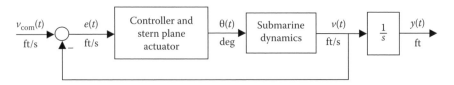

FIGURE 5.15 Submarine depth rate control system.

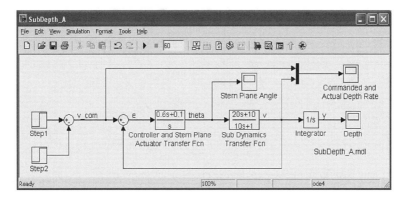

FIGURE 5.16   Simulink® diagram for sub depth control using transfer function blocks.

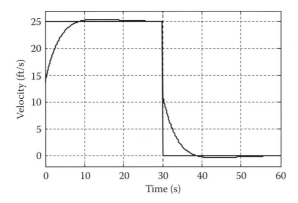

FIGURE 5.17   Command and actual submarine depth rates.

integrators collectively referred to as Runga–Kutta. Chapter 6 includes a discussion of Runga–Kutta integration.

The command and actual depth rate signals are shown in Figure 5.17. Note the discontinuity in the actual depth rate at $t = 0$ and $t = 30$ s. This implies the existence of a direct path from the command depth rate v_{com} to the actual depth rate v without integrators present. The direct path is not apparent in Figure 5.16; however, it would be evident on a simulation diagram of the system.

The stern plane angle (°) and the actual submarine depth (ft) are shown in Figure 5.18.

The presence of a direct path with only algebraic blocks from command input v_{com} to the actual submarine depth rate v is easier to visualize if we express the transfer functions in Figure 5.16 differently, that is,

$$G_C(s) = \frac{\theta(s)}{E(s)} = K_C + \frac{K_I}{s} \tag{5.4}$$

$$G_P(s) = \frac{V(s)}{\theta(s)} = \frac{K_{\dot\theta}s + K_\theta}{\tau s + 1} \tag{5.5}$$

$$= \frac{K_{\dot\theta}}{\tau} + \frac{(K_\theta - (K_{\dot\theta}/\tau))}{\tau s + 1} \tag{5.6}$$

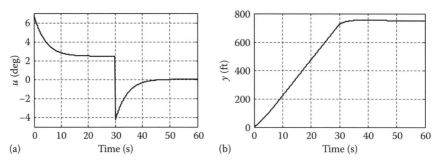

FIGURE 5.18 Simulated (a) stern plane angle (deg) and (b) sub depth (ft).

Hence, the direct path starts from "v_com" through the summer and on through constant blocks with gains "KC" and "Kthd/tau" to the output "v." The Simulink diagram in Figure 5.16 can be modified to implement the controller and submarine dynamics transfer functions as given in Equations 5.4 and 5.6 (see Exercise 5.5).

The submarine depth, shown in Figure 5.16, is continuous at $t=0$ due to the presence of the integrator between "v" and "y."

Referring to Figure 5.15, the closed-loop transfer function is

$$\frac{V(s)}{V_{com(s)}} = \frac{G_C(s)G_P(s)}{1 + G_C(s)G_P(s)} \tag{5.7}$$

$$= \frac{((K_C s + K_I)/s)((K_{\dot\theta}s + K_\theta)/(\tau s + 1))}{1 + ((K_C s + K_I)/s)((K_{\dot\theta}s + K_\theta)/(\tau s + 1))} \tag{5.8}$$

$$= \frac{(K_C s + K_I)(K_{\dot\theta}s + K_\theta)}{s(\tau s + 1) + (K_C s + K_I)(K_{\dot\theta}s + K_\theta)} \tag{5.9}$$

The steady-state value $v(\infty)$ resulting from the step input $\theta_{com}(t) = 25$, $t \geq 0$ is obtained from the final value theorem (Section 4.2),

$$v(\infty) = \lim_{s \to 0} sV(s) = \lim_{s \to 0} s \left[\frac{(K_C s + K_I)(K_{\dot\theta}s + K_\theta)}{s(\tau s + 1) + (K_C s + K_I)(K_{\dot\theta}s + K_\theta)} \right] \frac{25}{s} = 25 \tag{5.10}$$

confirmed in Figure 5.17, which shows the depth rate $v(t)$ approaching the commanded 25 ft/s once the transient response has vanished.

The discontinuity in depth rate at $t=0$ shown in Figure 5.17 can also be verified. According to the initial value; theorem,

$$v(0^+) = \lim_{s \to \infty} sV(s) = \lim_{s \to \infty} s \left[\frac{(K_C s + K_I)(K_{\dot\theta}s + k_\theta)}{s(\tau s + 1) + (K_C s + K_I)(K_{\dot\theta}s + K_\theta)} \right] \frac{25}{s} \tag{5.11}$$

$$= \lim_{s \to \infty} \left[\frac{(K_C + (K_I/s))(K_{\dot\theta} + (K_\theta/s))}{(\tau + (1/s)) + (K_C + (K_I/s))(K_{\dot\theta} + (K_\theta/s))} \right] 25 \tag{5.12}$$

$$= \frac{25 K_C K_{\dot\theta}}{\tau + K_C K_{\dot\theta}} = \frac{25(0.6)(20)}{10 + (0.6)(20)} = 13.64 \, \text{ft/s} \tag{5.13}$$

is in agreement with the graph of $v(t)$ shown in Figure 5.17.

The following example further illustrates the use of the "Transfer fcn" block.

Example 5.1

For the submarine depth rate control system shown in Figure 5.15,

(a) Find the analytical solution for the submarine depth rate $v(t)$, $0 < t \leq 30$ in response to the command input $v_{com}(t) = 25$, $t \geq 0$.

(b) Model the closed-loop control system dynamics using a "Transfer fcn" block for $V(s)/V_{com}(s)$ and use it to simulate the depth rate response to the command depth rate shown in Figure 5.17. Compare the simulated and analytical depth rate responses for $v(t)$, $0 < t \leq 30$.

(a) From Equation 5.9,

$$V(s) = \left[\frac{(K_C s + K_I)(K_{\dot{\theta}} s + K_{\theta})}{s(\tau s + 1) + (K_C s + K_I)(K_{\dot{\theta}} s + K_{\theta})} \right] V_{com}(s) \tag{5.14}$$

$$= \left[\frac{K_C K_{\dot{\theta}} s^2 + (K_C K_{\theta} + K_I K_{\dot{\theta}})s + K_I K_{\theta}}{(\tau + K_C K_{\dot{\theta}})s^2 + (1 + K_C K_{\theta} + K_I K_{\dot{\theta}})s + K_I K_{\theta}} \right] \frac{25}{s} \tag{5.15}$$

$$= \left[\frac{12s^2 + 8s + 1}{22s^2 + 9s + 1} \right] \frac{25}{s} \tag{5.16}$$

$$= \frac{25}{22} \left[\frac{12s^2 + 8s + 1}{s\{s^2 + (9/22)s + 1/22\}} \right] \tag{5.17}$$

Using partial fraction expansion of the right-hand side of Equation 5.17 followed by inverse Laplace transformation, the solution for $v(t)$, $0 < t \leq 30$ becomes

$$v(t) = 25 - \frac{25}{22} e^{-9t/44} \left[10 \cos\left(\frac{\sqrt{7}}{44} t \right) - \frac{47}{\sqrt{7}} \sin\left(\frac{\sqrt{7}}{44} t \right) \right], \quad 0 < t \leq 30 \tag{5.18}$$

(b) The analytical solution for $v(t)$ in Equation 5.18 is incorporated in Simulink using a "Sine Wave" block from the "Sources" sublibrary and a "Math Function" block from the "Math" sublibrary for the exponential term. The Simulink diagram appears in Figure 5.19.

The "Sine Wave" parameters dialog box for the cosine term $\cos(\sqrt{7}/44 \, t)$ in the analytical solution, Equation 5.18, is shown in Figure 5.20. Note that the phase angle is $\pi/2$ rad to produce the cosine function.

The control system loop with input $v_{com}(t)$ and output $v(t)$ in Figure 5.15 is replaced with the equivalent closed-loop transfer function $V(s)/V_{com}(s)$ in Figure 5.21.

The Simulink diagram shown in Figure 5.22 includes a "Transfer fcn" block for implementing the closed-loop transfer function. The simulated and analytical depth rates for a time period $0 < t \leq 12$ s are shown in Figure 5.23. The graphs were generated in the MATLAB M-file "Chap5_Fig3_11.m" by saving the data in the scope shown with the heavy line multiplexed input in Figure 5.22. The complete set of time values along with the simulated and analytical results is saved in the MATLAB Workspace in a named array set in the scope dialog box. Also shown is the difference between the two depth rates. It is clear from looking at the difference that the simulated depth rate is nearly identical to the analytical solution.

5.3.2 STATE-SPACE BLOCK

The process of transforming models consisting of linear algebraic and differential equations into state variable form was demonstrated in Section 2.6. Conversion of SISO (single input–single output) or MIMO (multiple input–multiple output) system transfer functions to state-space

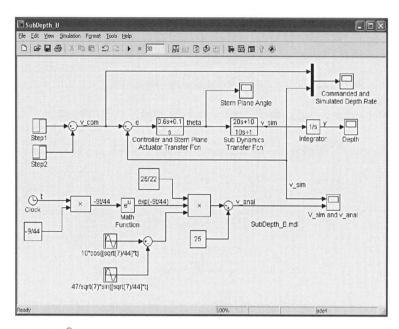

FIGURE 5.19 Simulink® diagram with simulated and analytical submarine depth rate.

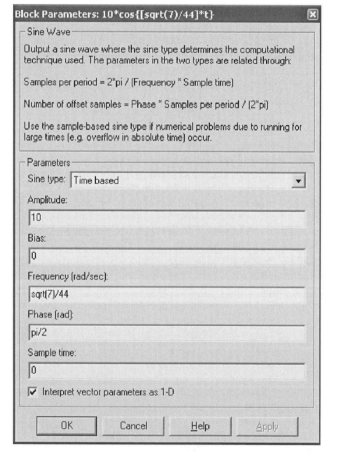

FIGURE 5.20 "Sine Wave" parameter box to generate cosine term in analytical solution.

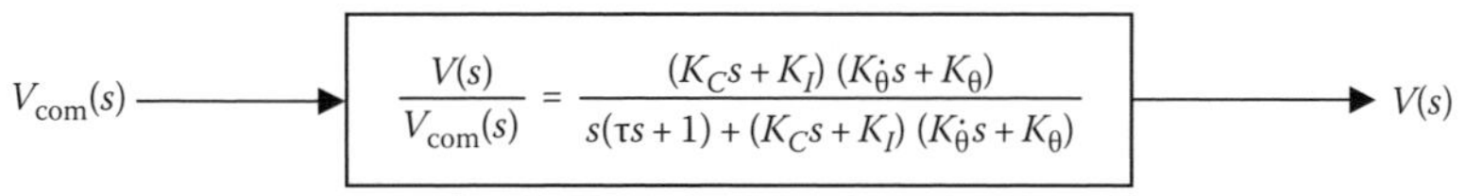

$$V_{com}(s) \longrightarrow \boxed{\frac{V(s)}{V_{com}(s)} = \frac{(K_C s + K_I)(K_{\dot{\theta}} s + K_{\theta})}{s(\tau s + 1) + (K_C s + K_I)(K_{\dot{\theta}} s + K_{\theta})}} \longrightarrow V(s)$$

FIGURE 5.21 Closed-loop transfer function of submarine depth rate control system.

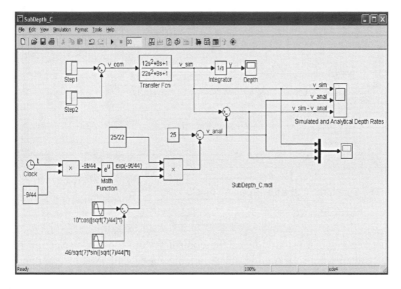

FIGURE 5.22 Simulink® diagram using "Transfer fcn" block for submarine closed-loop depth rate control system dynamics.

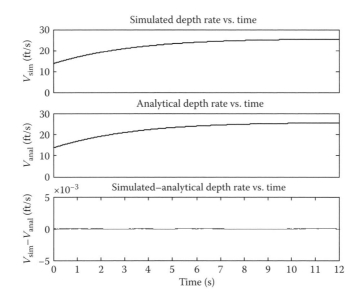

FIGURE 5.23 Analytical and simulated depth rate using "Transfer fcn" for $V(s)/V_{com}(s)$.

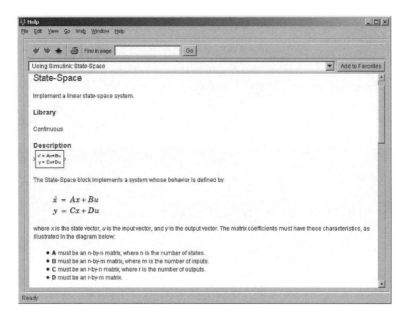

FIGURE 5.24   The Simulink® "State-Space" block.

models and vice versa was illustrated using the control system toolbox in Section 4.10. Simulink provides a mechanism for incorporating state variable models of system components using the "State-Space" block located in the "Continuous" sublibrary. A partial description of the "State-Space" block is shown in Figure 5.24. The next example illustrates its use.

Example 5.2

An automobile traveling along a level road at a constant speed v_0 encounters a speed bump shown in Figure 5.25. The vehicle's suspension system (front and rear springs and shock absorbers) is modeled by linear springs and dampers, and the compliance of the tires is modeled by front and rear springs. The vehicle cab motion is limited to heave in the y-direction and a small amount of pitch θ of the vehicle's longitudinal axis. The tires are assumed to remain in contact with the road surface at all times.

The road profile is responsible for the system's input $\underline{u} = [u_f\, u_r]^T$, where u_f and u_r are the height of the road (with respect to some reference) underneath the front and rear tires, respectively. The system has three translational degrees of freedom, y, y_f, y_r, which are the vertical displacements of the vehicle cab and both front and rear axles from their equilibrium positions. The lone rotational degree of freedom is the pitch angle θ.

Three of the four model equations are obtained by equating the sum of suspension and tire forces acting on the three masses to the appropriate acceleration term, $M\ddot{y}$, $M_f\ddot{y}_f$, and $M_r\ddot{y}_r$. The fourth equation sets the torques about the vehicle cab's center of gravity created by the suspension forces equal to the inertial acceleration $I\ddot{\theta}$.

The model equations are listed as follows:

$$M\ddot{y} = K_{fs}[y_f - (y + L_f\theta)] + B_f[\dot{y}_f - (\dot{y} + L_f\dot{\theta})] + K_{rs}[y_r - (y - L_r\theta)] + B_r[\dot{y}_r - (\dot{y} - L_r\dot{\theta})] \tag{5.19}$$

$$= -(K_{fs} + K_{rs})y - (B_f + B_r)\dot{y} + K_{fs}y_f + B_f\dot{y}_f + K_{rs}y_r + B_r\dot{y}_r$$

$$+ (K_{rs}L_r - K_{fs}L_f)\theta + (B_rL_r - B_fL_f)\dot{\theta} \tag{5.20}$$

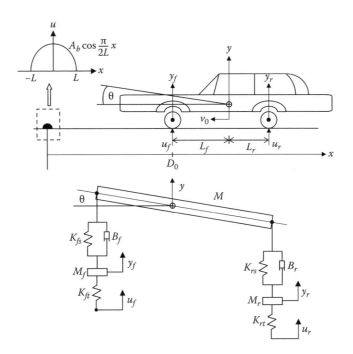

FIGURE 5.25 Moving vehicle and suspension system model.

$$M_f \ddot{y}_f = -K_{fs}[y_f - (y + L_f\theta)] - B_f[\dot{y}_f - (\dot{y} + L_f\dot{\theta})] + K_{ft}(u_f - y_f) \tag{5.21}$$

$$= -(K_{fs} + K_{ft})y_f - B_f\dot{y}_f + K_{fs}y + B_f\dot{y} + K_{fs}L_f\theta + B_fL_f\dot{\theta} + K_{ft}u_f \tag{5.22}$$

$$M_r \ddot{y}_r = -K_{rs}[y_r - (y - L_r\theta)] - B_r[\dot{y}_r - (\dot{y} - L_r\dot{\theta})] + K_{rt}(u_r - y_r) \tag{5.23}$$

$$= -(K_{rs} + K_{rt})y_r - B_r\dot{y}_r + K_{rs}y + B_r\dot{y} - K_{rs}L_r\theta - B_rL_r\dot{\theta} + K_{rt}u_r \tag{5.24}$$

$$I\ddot{\theta} = \{K_{fs}[y_f - (y + L_f\theta)] + B_f[\dot{y}_f - (\dot{y} + L_f\dot{\theta})]\}L_f$$

$$\quad - \{K_{rs}[y_r - (y - L_r\theta)] + B_r[\dot{y}_r - (\dot{y} - L_r\dot{\theta})]\}L_r \tag{5.25}$$

$$= -(K_{fs}L_f^2 + K_{rs}L_r^2)\theta - (B_fL_f^2 + B_rL_r^2)\dot{\theta} + (K_{rs}L_r - K_{fs}L_f)y$$

$$\quad + (B_rL_r - B_fL_f)\dot{y} + K_{fs}L_fy_f - K_{rs}L_ry_r + B_fL_f\dot{y}_f - B_rL_r\dot{y}_r \tag{5.26}$$

Note that the equations are linear as a result of assuming small pitch angles, allowing the approximations $\sin\theta \approx \theta$ and $\cos\theta \approx 1$.

(a) Introduce state variables

$$x_1 = y, \quad x_3 = y_f, \quad x_5 = y_r, \quad x_7 = \theta,$$

$$x_2 = \dot{y}, \quad x_4 = \dot{y}_f, \quad x_6 = \dot{y}_r, \quad x_8 = \dot{\theta},$$

and solve for the state derivatives, that is, find the matrices A and B in $\underline{\dot{x}} = A\underline{x} + B\underline{u}$.

(b) Define the outputs as $y_1 = y$, $y_2 = y_f$, $y_3 = y_r$, and $y_4 = \theta$ and find matrices C and D in $\underline{y} = C\underline{x} + D\underline{u}$.

(c) Simulate and plot the vehicle dynamics using the following values for the weight of the vehicle and tires, suspension parameters, forward speed, and speed bump profile.

$$W = 4{,}200\,\text{lb}, \quad W_f = 125\,\text{lb}, \quad W_r = 125\,\text{lb}, \quad K_{fs} = 120\,\text{lb/in}, \quad K_{rs} = 180\,\text{lb/in},$$

$$B_f = 25\,\text{lb s/in}, \quad B_r = 35\,\text{lb s/in}, \quad K_{ft} = 1{,}100\,\text{lb/in}, \quad K_{rt} = 1{,}100\,\text{lb/in},$$

$$I = 40{,}000\,\text{in. lb s}^2, \quad L_f = 55\,\text{in.}, \quad L_r = 65\,\text{in.}, \quad v_0 = 20\,\text{mph}, \quad A_b = 4\,\text{in.}, \quad L = 1\,\text{ft}$$

(a) Using the definition of the state variables and solving for the state derivatives in Equations 5.19 through 5.26 give

$$\dot{x}_1 = x_2 \tag{5.27}$$

$$\dot{x}_2 = \frac{-(K_{fs} + K_{rs})}{M}x_1 - \frac{(B_f + B_r)}{M}x_2 + \frac{K_{fs}}{M}x_3 + \frac{B_f}{M}x_4 + \frac{K_{rs}}{M}x_5 + \frac{B_r}{M}x_6$$

$$+ \frac{(K_{rs}L_r - K_{fs}L_f)}{M}x_7 + \frac{(B_r L_r - B_f L_f)}{M}x_8 \tag{5.28}$$

$$\dot{x}_3 = x_4 \tag{5.29}$$

$$\dot{x}_4 = \frac{K_{fs}}{M_f}x_1 + \frac{B_f}{M_f}x_2 - \frac{(K_{fs} + K_{ft})}{M_f}x_3 - \frac{B_f}{M_f}x_4 + \frac{K_{fs}L_f}{M_f}x_7 + \frac{B_f L_f}{M_f}x_8 + \frac{K_{ft}}{M_f}u_f \tag{5.30}$$

$$\dot{x}_5 = x_6 \tag{5.31}$$

$$\dot{x}_6 = \frac{K_{rs}}{M_r}x_1 + \frac{B_r}{M_r}x_2 - \frac{(K_{rs} + K_{rt})}{M_r}x_5 - \frac{B_r}{M_r}x_6 - \frac{K_{rs}L_r}{M_r}x_7 - \frac{B_r L_r}{M_r}x_8 + \frac{K_{rt}}{M_r}u_r \tag{5.32}$$

$$\dot{x}_7 = x_8 \tag{5.33}$$

$$\dot{x}_8 = \frac{(K_{rs}L_r - K_{fs}L_f)}{I}x_1 + \frac{(B_r L_r - B_f L_f)}{I}x_2 + \frac{K_{fs}L_f}{I}x_3 + \frac{B_f L_f}{I}x_4$$

$$- \frac{K_{rs}L_r}{I}x_5 - \frac{B_r L_r}{I}x_6 - \frac{(K_{fs}L_f^2 + K_{rs}L_r^2)}{I}x_7 - \frac{(B_f L_f^2 + B_r L_r^2)}{I}x_8 \tag{5.34}$$

The system matrix A and input matrix B are

$$A = \begin{bmatrix}
0 & 1 & 0 & 0 \\
\dfrac{-(K_{fs} + K_{rs})}{M} & \dfrac{-(B_f + B_r)}{M} & \dfrac{K_{fs}}{M} & \dfrac{B_f}{M} \\
0 & 0 & 0 & 1 \\
\dfrac{K_{fs}}{M_f} & \dfrac{B_f}{M_f} & \dfrac{-(K_{fs} + K_{ft})}{M_f} & \dfrac{-B_f}{M_f} \\
0 & 0 & 0 & 0 \\
\dfrac{K_{rs}}{M_r} & \dfrac{B_r}{M_r} & 0 & 0 \\
0 & 0 & 0 & 0 \\
\dfrac{K_{rs}L_r - K_{fs}L_f}{1} & \dfrac{B_r L_r - B_f L_f}{1} & \dfrac{K_{fs}L_f}{1} & \dfrac{B_f L_f}{1} \\
0 & 0 & 0 & 0 \\
\dfrac{K_{rs}}{M} & \dfrac{B_r}{M} & \dfrac{K_{rs}L_r - K_{fs}L_f}{M} & \dfrac{B_r L_r - B_f L_f}{M} \\
0 & 0 & 0 & 0 \\
0 & 0 & \dfrac{K_{fs}L_f}{M_f} & \dfrac{B_f L_f}{M_f} \\
0 & 1 & 0 & 1 \\
\dfrac{-(K_{rs} + K_{rt})}{M_r} & \dfrac{-B_r}{M_r} & \dfrac{-K_{rs}L_r}{M_r} & \dfrac{-B_r L_r}{M_r} \\
0 & 0 & 0 & 1 \\
\dfrac{-K_{rs}L_r}{I} & \dfrac{-B_r L_r}{I} & \dfrac{-(K_{fs}L_f^2 + K_{rs}L_r^2)}{I} & \dfrac{-(B_f L_f^2 + B_r L_r^2)}{I}
\end{bmatrix} \tag{5.35}$$

$$B = \begin{bmatrix} 0 & 0 \\ 0 & 0 \\ 0 & 0 \\ \dfrac{K_{ft}}{M_f} & 0 \\ 0 & 0 \\ 0 & \dfrac{K_{ft}}{M_r} \\ 0 & 0 \\ 0 & 0 \end{bmatrix}$$

(5.36)

(b) The output matrix C and direct transmission matrix D are given by

$$C = \begin{bmatrix} 1 & 0 & 0 & 0 & 0 & 0 & 0 & 0 \\ 0 & 0 & 1 & 0 & 0 & 0 & 0 & 0 \\ 0 & 0 & 0 & 0 & 1 & 0 & 0 & 0 \\ 0 & 0 & 0 & 0 & 0 & 0 & 1 & 0 \end{bmatrix}, \quad D = \begin{bmatrix} 0 & 0 \\ 0 & 0 \\ 0 & 0 \\ 0 & 0 \end{bmatrix}$$

(5.37)

The direct transmission matrix D is all zeros, since the system inputs u_f and u_r are not directly coupled to the outputs, that is, step changes in either input are integrated before influencing the outputs, and, hence, the outputs are continuous at the time the step input(s) is applied.

(c) The Simulink diagram for simulating the vehicle's response as it travels over the speed bump is shown in Figure 5.26. The "State-Space" block parameters are the matrices A, B, C, and D of Equations 5.35 through 5.37, which have been defined in a MATLAB M-file "Chap5_VehParams. m" for convenience.

The input displacements u_f and u_r are based on the speed bump profile shown in Figure 5.25 and the forward speed of the car. The front tire displacement is given by

$$u_f = \begin{cases} 0, & t < \dfrac{D_0 - L}{v_0} \\ A_b \cos \dfrac{\pi}{2L}(D_0 - v_0 t), & \dfrac{D_0 - L}{v_0} \le t \le \dfrac{D_0 + L}{v_0} \\ 0, & t > \dfrac{D_0 + L}{v_0} \end{cases}$$

(5.38)

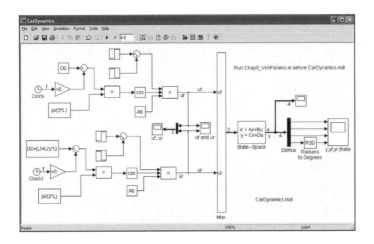

FIGURE 5.26 Simulink® diagram for vehicle response traveling over a speed bump.

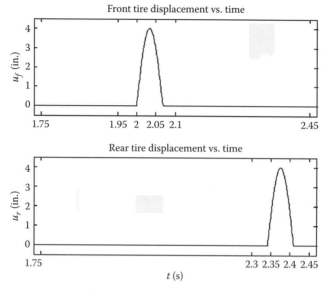

FIGURE 5.27   Inputs u_f and u_r for vehicle traveling at constant speed v_0.

The Simulink blocks to implement u_f (and u_r) are shown in the top left (and lower left) corner of Figure 5.26. Note the use of the "Clock" from the "Sources" sublibrary to generate the simulation time variable "t." Also, the wider (and heavier) arrows in and out of the "State-Space" block designate the presence of nonscalar signals, and the "2" and "4" indicate the number of components in each.

The inputs u_f and u_r are captured in a scope and plotted for $1.75 < t \le 2.5$ s in the M-file "*Chap5_Figs3_15and3_16.m*" (see Figure 5.27).

The output vector "y" of the "State-Space" block is decomposed in a "Demux" block and sent to a scope with four input channels (one for each output). It is also saved for use by the M-file "*Chap5_Figs3_15and3_16.m*." The results are plotted for the interval $1.5 \le t \le 3.5$ s in Figure 5.28.

The vehicle cab displacement varies from -0.189 to 0.627 in. despite the 4 in. height of the speed bump. Also, the pitch of the vehicle is constrained to $-0.403° \le \theta \le 0.358°$.

The "Data history" tab in the "Scope" with multiplexed input containing "uf" and "ur" is shown in Figure 5.29. Simulation time values and front and rear tire displacements are saved to the MATLAB Workspace in array "uf_ur."

The following MATLAB statements placed at the beginning of M-file "*Chap5_Figs3_15and3_16.m*" store the saved values of the time array and tire displacements in arrays "t," "uf," and "ur" and produce the graph shown in Figure 5.27.

```
Chap5_VehParams
sim('CarDynamics')
t=uf_ur(:,1);
uf=uf_ur(:,2);
ur=uf_ur(:,3);
figure(1) % begin Figure 5.27
subplot(2,1,1)
plot(t,uf)
ylabel('uf (in)', 'Font Size',11)
title('Front Tire Displacement vs. Time', 'FontSize',11)
subplot(2,1,2)
plot(t,ur)
```

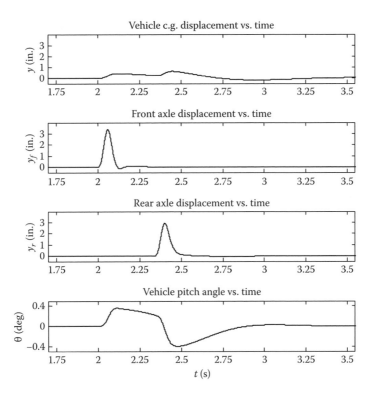

FIGURE 5.28 Outputs y, y_f, y_r, and θ of vehicle suspension system for $1.5 \leq t \leq 4$ s.

FIGURE 5.29 Saving "`uf`" and "`ur`" for plotting in M-file "*Chap5_Figs3_15and3_16.m.*"

```
ylabel('ur (in)','FontSize',11)
xlabel('\itt \rm(sec)','FontSize',11)
title('Rear Tire Displacement vs. Time','FontSize',11)
```

The first statement runs another M-file "Chap5_VehParams.m," which loads the parameter values. The next command sim("CarDynamics") causes execution of the Simulink model "CarDynamics.mdl."

EXERCISES

5.4 For the second-order system shown in Figure 5.13,
    (a) Find the second-order differential equation relating the output y and input u.
    (b) Draw a Simulink diagram of the system with two integrators in series.

5.5 For the submarine depth control system shown in Figure 5.15,
    (a) Draw a simulation diagram. Is there a direct connection from v_{com} to v?
    (b) Redraw the Simulink diagram in Figure 5.16 using the alternate expressions for the controller and submarine dynamics transfer functions in Equations 5.4 and 5.6.
    (c) Run the Simulink model and compare the responses for $v(t)$, $y(t)$, and $\theta(t)$ with those shown in the text.

5.6 Two linear tanks are arranged in series as shown in Figure E5.6:

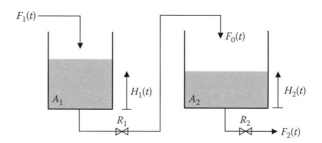

FIGURE E5.6

    (a) Write the differential equation models for the tanks.
    (b) The system parameters are
        $A_1 = 50$ ft^2, $A_2 = 100$ ft^2, $R_1 = 0.2$ ft/ft^3/min, and $R_2 = 0.3$ ft/ft^3/min
        Prepare a Simulink diagram of the system, and simulate the response of both tank levels under the following conditions:
        (i) $H_1(0) = 0$, $H_2(0) = 0$, $F_1(t) = 40$ ft^3/min, $t \geq 0$
        (ii) $H_1(0) = 0$, $H_2(0) = 10$, $F_1(t) = 0$, $t \geq 0$

$$(iii)\ H_1(0) = 0,\ H_2(0) = 0,\ F_1(t) = \begin{cases} 5t, & 0 \leq t \leq 5 \\ -5t + 50, & 5 < t \leq 10 \\ 0, & t > 0 \end{cases}$$

        Obtain one graph with time histories of $H_1(t)$ and $H_2(t)$ and a second graph with $F_0(t)$, $F_1(t)$, and $F_2(t)$.
    (c) Eliminate $H_1(t)$ from the two first-order differential equations in part (a) to obtain a second-order differential equation relating $H_2(t)$ and $F_1(t)$.
    (d) Prepare a Simulink diagram based on the continuous-time model in part (c).
    (e) Run the Simulink model for the same conditions in part (b), and compare the response for $H_2(t)$ with the one obtained in part (b).
    (f) Find the analytical solution $[H_2(t)]_{anal}$ when both tanks are initially empty and $F_1(t) = 40$ ft^3/min, $t \geq 0$. Compare the analytical solution $[H_2(t)]_{anal}$ with the simulated solution $[H_2(t)]_{sim}$ obtained in part (b).

    Hint: Use Simulink to implement the analytical solution and feed both $[H_2(t)]_{sim}$ and $[H_2(t)]_{anal}$ into a summer to obtain the difference.

5.7 Solve Exercise 2.3 using Simulink.
5.8 Solve Exercise 2.4 using Simulink.

5.4 ALGEBRAIC LOOPS

Execution of the Simulink model in this chapter, Figure 5.16, poses a dilemma often encountered when simulating dynamic systems with feedback loops. A runtime warning (default) or error appears in the MATLAB Command Window stating

```
Warning:Block diagram 'SubDepth_A' contains 1 algebraic loop(s).
Found algebraic loop containing block(s):
'SubDepth_A/Controller and Stern Plane Actuator Transfer Fcn'
'SubDepth_A/Sub Dynamics Transfer Fcn'
'SubDepth_A/Sum1' (algebraic variable)
```

An algebraic loop is any closed loop appearing in the Simulink diagram composed of strictly algebraic and implicit blocks such as the implicit discrete-time numerical integrators (discussed in Section 5.6). Consequently, the output of any block in an algebraic loop is ultimately an implicit function of itself. In large scale simulations with several 100 blocks, it is nearly impossible to identify the presence of an algebraic loop by visual inspection. The Simulink diagrams of even relatively simple simulations with only a handful of blocks may contain algebraic loops, which escape detection. Simulink (and other block-oriented continuous simulation languages) detects the presence of an algebraic loop and reports the blocks comprising it.

Before we discuss its implications, let us confirm the existence of an algebraic loop in the Simulink model "*SubDepth_A.mdl*" consisting of the two "Transfer fcn" blocks and the "Sum" block. Referring to Figure 5.16, the controller and stern plane actuator transfer function can be rewritten as follows:

$$G_C(s) = \frac{0.6s + 0.1}{s} = 0.6 + \frac{0.1}{s} \tag{5.39}$$

and the submarine dynamics transfer function is expressible as

$$G_P(s) = \frac{20s + 10}{10s + 1} = 2 + \frac{8}{10s + 1} \tag{5.40}$$

leading to an equivalent block diagram shown in Figure 5.30.

The algebraic loop is shown in bold, and a similar algebraic loop is present in the Simulink diagram for "*SubDepth_A.mdl*." Note that if the controller and stern plane actuator transfer function were replaced by a pure gain, the diagram would still have an algebraic loop due to the direct path from the input to the output in the submarine dynamics transfer function.

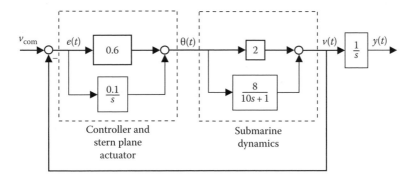

FIGURE 5.30 Block diagram for submarine depth control showing algebraic loop.

The dilemma posed by algebraic loops can be demonstrated by looking at the equations the Simulink program is attempting to solve in the submarine example at the time $t = 0$. After initializing the state $\theta(0)$ and $v(0)$ and evaluating the input $v_{\text{com}}(0)$, Simulink calculates $e(0)$ according to

$$e(0) = v_{\text{com}}(0) - v(0) \tag{5.41}$$

Existence of direct paths, that is, pure gain (zero-order dynamics), from $e(t)$ to $\theta(t)$ and $\theta(t)$ to $v(t)$ implies

$$v(0) = 2\theta(0) \tag{5.42}$$

$$\theta(0) = 0.6e(0) \tag{5.43}$$

Substituting $\theta(0)$ in Equation 5.43 into Equation 5.42 gives

$$v(0) = 2[0.6e(0)] = 1.2e(0) \tag{5.44}$$

Replacing $e(0)$ in Equation 5.44 with $e(0)$ in Equation 5.41 results in

$$v(0) = 1.2[v_{\text{com}}(0) - v(0)] \tag{5.45}$$

The circular nature of algebraic loops is demonstrated by Equation 5.45, an implicit equation with $v(0)$ on both sides. In the general case, the implicit equation is nonlinear. Simulink attempts to solve the implicit equations associated with an algebraic loop using the iterative Newton–Raphson method (Chapra 2002). Solving implicit equation(s) at each iteration, especially nonlinear ones, can dramatically decrease the simulation execution speed. Further, the method can fail to converge to a solution.

The initial depth rate value, more precisely, the value at $t = 0^+$ in the submarine example, is easily verified from Equation 5.45.

$$v(0^+) = 1.2[v_{\text{com}}(0) - v(0^+)] \tag{5.46}$$

$$\Rightarrow v(0^+) = \frac{1.2}{2.2} v_{\text{com}}(0) = \frac{1.2}{2.2}(25) = 13.64 \tag{5.47}$$

in agreement with the value given in Equation 5.13 as well as the graph for $v(t)$ shown in Figure 5.17.

5.4.1 ELIMINATING ALGEBRAIC LOOPS

The most desirable method for eliminating an algebraic loop is by means of algebraic manipulation of the loop equations to produce an equivalent system explicit in nature. It is up to the user to obtain an explicit solution, if one exists, and modify the Simulink diagram accordingly. Simulink does not perform the symbolic math operations necessary to obtain the solution shown in Equation 5.47.

To illustrate, consider the block diagram of a system shown in Figure 5.31. The algebraic loop is shown in bold.

By algebraic manipulation or similar block diagram reduction techniques, the transfer function $Y(s)/R(s)$ is obtained as

$$\frac{Y(s)}{R(s)} = \frac{K + (1 + K)G(s)}{(1 + K) + (2 + K)G(s)} \tag{5.48}$$

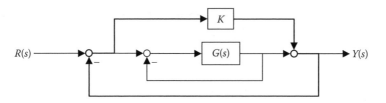

FIGURE 5.31 Block diagram of system with algebraic loop.

Suppose the constant $K = 1$ and the transfer function $G(s) = 1/(s + 10)$. The transfer function $Y(s)/R(s)$ reduces to

$$\frac{Y(s)}{R(s)} = \frac{0.5(s + 12)}{s + 11.5} \tag{5.49}$$

It is left as an exercise to demonstrate that a Simulink diagram based on the block diagram in Figure 5.31 and one with a single "`Transfer Fcn`" to implement Equation 5.49 produce identical outputs.

Unfortunately, the dynamic model equations rarely permit this approach. In most cases, the algebraic loop entails nonlinear blocks, making it difficult or impossible to reformulate the equations to produce a new block diagram with the algebraic loop removed. Several algebraic loops with shared blocks may exist, complicating matters even further.

A second approach to dealing with algebraic loops consists of inserting a "`Memory`" block into the loop. A "`Memory`" block is equivalent to a one-integration step delay. Its output is the input from the previous time step. This allows Simulink to calculate outputs of all the blocks in the algebraic loop in the proper sequence.

The system shown in Figure 5.32 consists of a cart with an inverted pendulum. The position of the cart $x(t)$ and the angle of the pendulum from the vertical $\theta(t)$ are of interest. The pendulum is free to rotate without friction in a plane, and the cart moves along a frictionless surface. The input is a horizontal force u. The outputs are x and θ.

From Newton's second law (translation and rotation), the equations of motion are

$$(M + m)\ddot{x} - ml\dot{\theta}^2 \sin \theta + ml\ddot{\theta} \cos \theta = u \tag{5.50}$$

$$m\ddot{x} \cos \theta + ml\ddot{\theta} = mg \sin \theta \tag{5.51}$$

where
   l is the length of the pendulum
   m is the pendulum mass (assumed to be concentrated at the end)
   M is the mass of the cart
   g is the gravitational constant

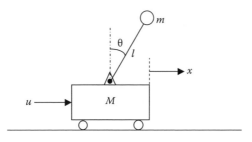

FIGURE 5.32 Inverted pendulum.

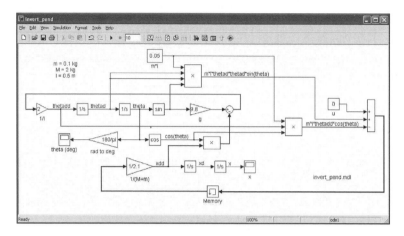

FIGURE 5.33 Simulink® model of inverted pendulum with "Memory" block.

Later, in Section 5.6, Equations 5.50 and 5.51 will be converted into a pair of equations, one for $\ddot{x}$ and the other with $\ddot{\theta}$ where both are explicit functions of the state variables θ and $\dot{\theta}$.

A Simulink diagram of the system is shown in Figure 5.33.

The algebraic loop shown in bold is broken by the insertion of a "Memory" block, eliminating the need for the Newton–Raphson iterative root solving at each integration step.

A simulation of the inverted pendulum when $u(t) = 0$, $t \geq 0$ was run for a period of 10 s using a fixed-step numerical integrator. All initial conditions are zero except the initial pendulum deflection, $\theta(0) = \pi/2$ rad. The output $\theta(t)$ is shown in Figure 5.34.

It is important to verify the results obtained when "Memory" blocks are employed to break algebraic loops. The delay introduced by the "Memory" block adversely affects the numerical accuracy and stability of the simulation. A considerable reduction in the time required to execute a simulation is hardly a suitable trade-off for inaccurate results. In other words, if the integration step size has to be reduced significantly to combat the existence of the "Memory" block, then the overall savings in execution time may be insignificant, or worse yet, the net result might be an overall increase in time of execution. A "Memory" block is worth considering when Simulink reports difficulty in converging to a solution of the implicit equations arising from an algebraic loop.

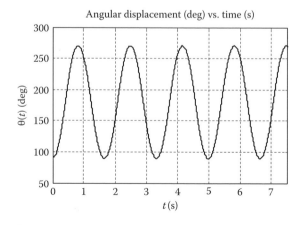

FIGURE 5.34 Simulink® output for $\theta(t)$, $t \geq 0$ using a "Memory" block.

5.4.2 ALGEBRAIC EQUATIONS

While Simulink is generally used for simulating dynamic systems described by ordinary differential equations, it can also be used to solve a system of algebraic equations. For example, the algebraic equations

$$\left. \begin{array}{l} y = f(x) \\ x = g(y) \end{array} \right\}$$ (5.52)

comprise an algebraic loop. Consider the dynamic system modeled by

$$\left. \begin{array}{l} \dfrac{dy}{dt} = F(x, y) = f(x) - y \\ x = g(y) \end{array} \right\}$$ (5.53)

The two parts of Equation 5.53 represent the model of a first-order autonomous system, that is,

$$\frac{dy}{dt} + y - f[g(y)] = 0$$ (5.54)

Suppose we are able to find an equilibrium point y_0 of the system described by Equation 5.54. Then (x_0, y_0), where $x_0 = g(y_0)$, constitutes a solution to the system of algebraic equations in Equation 5.52. To illustrate, let us attempt to find a point that lies on the circle $x^2 + y^2 = 100$ and the curve $x = y^2/5$. In this case,

$$\begin{array}{l} y = f(x) = (100 - x^2)^{1/2} \\ x = g(y) = \dfrac{y^2}{5} \end{array}$$ (5.55)

The Simulink diagram in Figure 5.35 incorporates an integrator for solution to

$$\frac{dy}{dt} = f(x) - y = (100 - x^2)^{1/2} - y$$ (5.56)

along with the block to generate x from the second of the two equations in Equation 5.55.

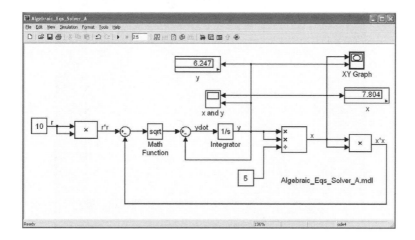

FIGURE 5.35 Simulink® diagram for solving algebraic equations in Equation 5.55.

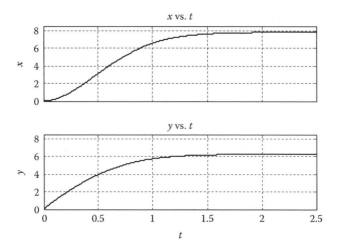

FIGURE 5.36   Graph of $x(t)$ and $y(t)$ from Simulink® scope block.

The search for the solution to the algebraic equations in Equation 5.55 begins at $(x(0),\ y(0))$ where $y(0)$ is the initial condition of the integrator and $x(0) = g[y(0)]$. Starting from the point $(0, 0)$, the approach to the equilibrium point y_0 and corresponding value of x_0 is viewable by clicking on the "Scope" block. The edited output is shown in Figure 5.36.

The solution $x_0 = 7.804$, $y_0 = 6.247$ is visible in the respective "Display" blocks shown in the Simulink diagram. The "XY Graph" block allows a view of the trajectory $x = g(y) = y^2/5$ from $(0, 0)$ up to the solution (x_0, y_0), as shown in Figure 5.37.

If the simulation fails to converge to an equilibrium point, restarting from a new point $(x(0),\ y(0))$ may help. Only stable equilibrium points of Equation 5.53 can be discovered. Keep in mind that nonlinear algebraic equations may possess none, one, or several equilibrium points, and the number of such points may not be known beforehand.

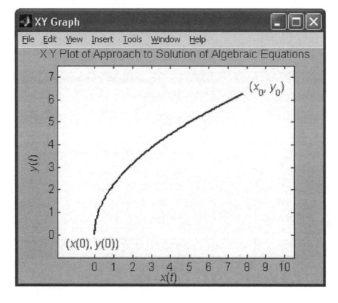

FIGURE 5.37   Trajectory from initial point $x(0) = 0$, $y(0) = 0$ to solution (x_0, y_0).

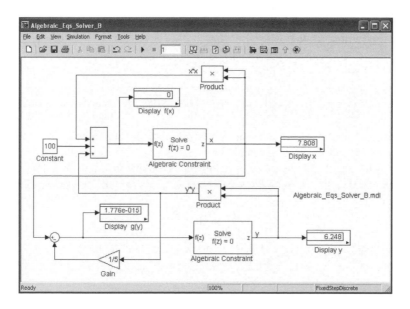

FIGURE 5.38 Using algebraic constraint blocks to solve algebraic equations in Equation 5.55.

A more direct approach to solving nonlinear algebraic equations with Simulink involves the use of an "Algebraic Constraint" block. This block changes its output in an iterative manner until its input approaches zero indicating that the algebraic constraint equation is satisfied, that is, the existence of a solution. Note that a feedback path must exist from the output to the input.

The previous system of algebraic equations is solved using "Algebraic Constraint" blocks as shown in Figure 5.38. Initial guesses for the variables x and y are required. Note that the inputs to both "Algebraic Constraint" blocks have converged to zero and the algebraic states x and y are in agreement with the previous solution.

The "Algebraic Constraint" block is an effective tool for locating the equilibrium points of a nonlinear dynamic system.

EXERCISES

5.9 Run the Simulink model in Figure 5.33 using the "ode1" Euler integrator, and determine the largest step size possible for simulating the inverted pendulum dynamics with $u(t) = 0$, $t \geq 0$, and $\theta(0) = \pi/2$ for a period of 10 s. Repeat without the "Memory" block.

5.10 Starting with Equations 5.50 and 5.51 for the inverted pendulum,
   (a) Find explicit functions $f(\theta, \dot{\theta}, u)$ and $g(\theta, \dot{\theta}, u)$ where

$$\ddot{x} = f(\theta, \dot{\theta}, u) \quad \text{and} \quad \ddot{\theta} = g(\theta, \dot{\theta}, u)$$

   (b) Introduce state variables x_1, x_2, x_3, and x_4 where $x_1 = x$, $x_2 = \dot{x}$, $x_3 = \theta$, and $x_4 = \dot{\theta}$, and find the state derivative functions $f_1(x_1, x_2, x_3, x_4, u), f_2(x_1, x_2, x_3, x_4, u), f_3(x_1, x_2, x_3, x_4, u)$, and $f_4(x_1, x_2, x_3, x_4, u)$, where

$$\dot{x}_1 = f_1(x_1, x_2, x_3, x_4, u)$$

$$\dot{x}_2 = f_2(x_1, x_2, x_3, x_4, u)$$

$$\dot{x}_3 = f_3(x_1, x_2, x_3, x_4, u)$$

$$\dot{x}_4 = f_4(x_1, x_2, x_3, x_4, u)$$

(c) The outputs are $y_1 = x$ and $y_2 = \theta$. Find the output functions $g_1(x_1, x_2, x_3, x_4, u)$ and $g_2(x_1, x_2, x_3, x_4, u)$, that is,

$$y_1 = g_1(x_1, x_2, x_3, x_4, u)$$

$$y_2 = g_2(x_1, x_2, x_3, x_4, u)$$

(d) Prepare a Simulink diagram of the system based on the nonlinear state equations obtained in parts (b) and (c). Is an algebraic loop present?

(e) Compare outputs for $\theta(t)$, $t \geq 0$ using the Simulink diagram from Figure 5.33 and a Simulink diagram based on the state equations $\dot{x} = f(x, u)$, $y = g(x, u)$ for the following cases:

    (i) $u(t) = 0$, $t \geq 0$ and $x_1(0) = x_2(0) = 0$, $x_3(0) = 1°$, $x_4(0) = 0$
    (ii) $u(t) = 0$, $t \geq 0$ and $x_1(0) = x_2(0) = x_3(0) = 0$, $x_4(0) = 10°/s$

5.11 Rework the example designed to find the first quadrant solution to

$$y = f(x) = (100 - x^2)^{1/2} \quad \text{and} \quad x = g(y) = \frac{y^2}{5}$$

by looking for an equilibrium point of

$$\frac{dx}{dt} = G(x, y) = g(y) - x$$

$$y = f(x)$$

5.12 Find both solutions to the algebraic equations

$$y = e^x - 1, \quad y = 5 - (x - 1)^2$$

using "Algebraic Constraint" blocks.

5.13 Consider the system represented in block diagram form in Figure 5.31 and the equivalent closed-loop transfer function in Equation 5.48.

(a) Find the differential equation relating the output $y(t)$ and input $r(t)$ when

$$G(s) = \frac{K_1}{\tau s + 1} \tag{i}$$

$$G(s) = \frac{K_1(\tau_1 s + 1)}{\tau_2 s + 1} \tag{ii}$$

$$G(s) = \frac{K_1}{s^2 + 2\zeta\omega_n s + 1} \tag{iii}$$

(b) Prepare Simulink diagrams to simulate the block diagram and transfer function representations of the system when $G(s) = 2/(0.5s + 1)$ and $K = 10$. Find and plot the responses to the following inputs:

    (i) $r(t) = \hat{u}(t)$, the unit step input
    (ii) $r(t) = e^{-t/2}$, $t \geq 0$
    (iii) See graph of $r(t)$ in Figure E5.13

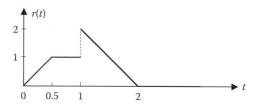

FIGURE E5.13

5.5 MORE SIMULINK® BLOCKS

In this section, we introduce additional Simulink blocks to extend the simulation capabilities developed so far. The next example is a common one from the field of traffic engineering. The objective is to formulate a mathematical model suitable for describing the characteristics of a driver/vehicle attempting to follow a lead vehicle in a single lane of traffic. The result is referred to as a microscopic car-following model. Car-following models are an essential component of traffic simulation software used to predict traffic flows in tunnels and other roads where passing is restricted.

The basic situation is illustrated in Figure 5.39, which shows a lead vehicle $(n-1)$ and a following vehicle (n), each of length L.

The system, comprised of the lead and following vehicle, is driven (no pun intended) by the speed of the lead vehicle $\dot{x}_{n-1}$, and the outputs include $\{x_{n-1}, x_n, \dot{x}_n, \ddot{x}_n\}$ in addition to the following quantities, which relate directly to the combination of lead and following vehicle movements.

$$\text{Vehicle spacing: } s_n = x_{n-1} - x_n \tag{5.57}$$

$$\text{Vehicle following distance: } d_n = (x_{n-1} - L) - x_n \tag{5.58}$$

$$\text{Speed difference: } \Delta\dot{x}_n = \dot{x}_{n-1} - \dot{x}_n \tag{5.59}$$

$$\text{Vehicle gap: } g_n = \frac{x_{n-1} - x_n}{\dot{x}_n} \tag{5.60}$$

The subscripts "$n-1$" and "n" are used, so that we can model a platoon consisting of a lead vehicle and several following vehicles. Except for the platoon leader and the last vehicle in the platoon, each vehicle operates in a following and lead vehicle mode as depicted in Figure 5.39. Platoon dynamics is considered in the next section.

We have yet to formulate a mathematical model that governs the motion of the following vehicle in the case of small-to-moderate vehicle spacing. Note that car-following models are not applicable at low traffic densities since each vehicle is essentially a leader moving independently of the preceding vehicle.

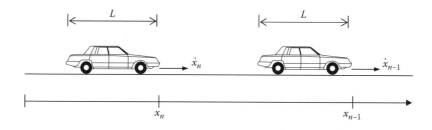

FIGURE 5.39 Diagram showing lead and following vehicles.

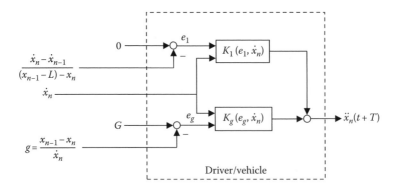

FIGURE 5.40 Block diagram of a car-following model.

Standard practice is to postulate an equation for the acceleration of the following vehicle in response to certain stimuli that are based on the relative movements of the two vehicles, that is,

$$\ddot{x}_n(t+T) = f(x_{n-1}(t), x_n(t), \dot{x}_{n-1}(t), \dot{x}_n(t)) \tag{5.61}$$

The acceleration response is delayed by an amount T, which represents the sum of the driver's cognition and reaction times in addition to the vehicle response time. The literature is replete with articles and chapters in books describing suitable candidates for the function "$f()$" in Equation 5.61 (Bender and Fenton 1966; Haberman 1977; Mesterton-Gibbons 1988; Aycin and Benekohal 2001).

The block diagram in Figure 5.40 represents a specific function developed by the author used to simulate realistic traffic in a driving simulator.

The driver/vehicle combination behaves like a regulatory controller with output $\ddot{x}_n(t+T)$, a function of two error terms e_1, e_g and the following vehicle's speed $\dot{x}_n$. The first error term e_1 is the difference between 0 and $\dot{x}_n - \dot{x}_{n-1}$ weighted by the reciprocal of the spacing $(x_{n-1} - L) - x_n$. The second term e_g represents a gap error, that is, the difference between some desirable gap G and the actual gap g. The driver/vehicle controller attempts to drive both errors to zero by implementation of the control law

$$\ddot{x}_n(t+T) = K_1(e_1, \dot{x}_n) \cdot e_1 + K_g(e_g, \dot{x}_n) \cdot e_g \tag{5.62}$$

Note that when $\dot{x}_{n-1}$ is constant and both errors are zero, the following vehicle is traveling at the same speed with a separation $x_{n-1} - x_n = G\dot{x}_{n-1}$.

The functions $K_1(e_1, \dot{x}_n)$ and $K_g(e_g, \dot{x}_n)$ are implemented as shown in Tables 5.1 and 5.2. The constants $K_{1,a}$, $K_{1,d}$, $K_{g,d}$, and $K_{g,a}$ are gain parameters reflecting driver aggressiveness, SL is the speed limit, and Δ is a threshold above the speed limit.

A block diagram of the system is shown in Figure 5.41. The blocks to limit the acceleration and speed are self-explanatory. The spacing limiter assures that the minimum vehicle separation $x_{n-1} - x_n$ is greater than one car length at all times.

A Simulink diagram of the system is shown in Figure 5.42. The M-file "*Chap5_cfparams1.m*" assigns values to the system parameters referenced in a number of the Simulink blocks. Accordingly, it must be run prior to executing the simulation model file "*car_following.mdl*." The new Simulink blocks in Figure 5.42 and their function are described briefly as follows.

TABLE 5.1
Function $K_1(e_1, \dot{x}_n)$

$\dot{x}_n$		
e_1	$\leq SL + \Delta$	$> SL + \Delta$
>0	$K_{1,a}$	0
≤ 0	$K_{1,d}$	$K_{1,d}$

TABLE 5.2
Function $K_g(e_g, \dot{x}_n)$

$\dot{x}_n$		
e_g	$\leq SL + \Delta$	$> SL + \Delta$
>0	$K_{g,d}$	$K_{g,d}$
≤ 0	$K_{g,a}$	0

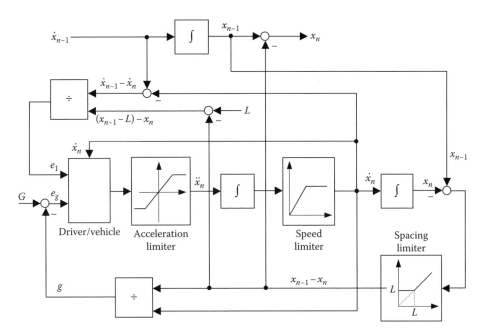

FIGURE 5.41 Block diagram of a car-following system.

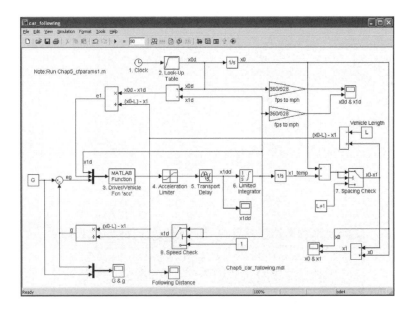

FIGURE 5.42 Simulink® diagram for a car-following system.

1. "Clock": Outputs the simulation time variable "t" for the "Lookup Table" block.
2. "Lookup Table": Linearly interpolates between specified data points to generate the lead car speed profile.
3. "MATLAB fcn": Passes the inputs "x1d," "e1," and "eg" to the MATLAB function "$acc.m$," which computes the vehicle's acceleration response.
4. "Saturation": Sets limits for minimum and maximum vehicle acceleration.
5. "Transport Delay": Delays vehicle acceleration by T, the driver/vehicle reaction time.

6. "Limited Integrator": An integrator configured to limit vehicle speed between zero and a maximum value of "vmax."

7. "Switch": Logical blocks that limit the spacing "x0-x1" to at least $L + 1$ ft and the speed "x1d" to at least 1 ft/s for calculation of the gap g.

Access to the MATLAB Workspace during execution allows Simulink block parameters to be variables specified in MATLAB script files. For example, The "Lookup Table" block parameters "T0," "T1," "T2," "A1," and "A2" shown in Figure 5.43 are set in the M-file "Chap5_cfparams1.m."

The "MATLAB Function" block is a powerful feature of Simulink, which exploits the tight integration between MATLAB and Simulink. The Simulink block outputs "x1d," "e1," and "eg" are accessible as inputs to the MATLAB function M-file "acc.m," which implements the car-following algorithm in Equation 5.62. The computed output is sent to the "Acceleration Limiter" block in Figure 5.42.

The M-file "acc.m" is listed as follows.

```
% function acc.m computes the temporary commanded acceleration
  function y = acc(x1d,e1,eg,K1d,K1a,Kgd,Kga,SL,delta)
  if e1<=0
    y1 = K1d*e1;
  elseif x1d<=SL+delta
    y1 = K1a*e1;
  else
    y1 = 0
  end
  if eg>0
    yg = Kgd*eg;
  elseif x1d<=SL+delta
    yg = Kga*eg;
  else
    yg = 0;
  end
y = y1+yg;
```

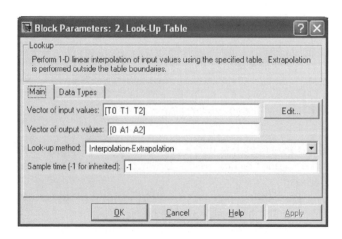

FIGURE 5.43 "Lookup Table" block parameters.

The results of simulating a pair of initially stopped vehicles, with one car length separation, followed by the lead vehicle accelerating (with constant acceleration) to 60 mph in 30 s are obtained by running the M-file "*Chap5_Figs5_6thru5_10.m*" and shown in Figures 5.44 through 5.48. The commanded gap G is 2 s, the value recommended for highway driving by The American Automobile Association.

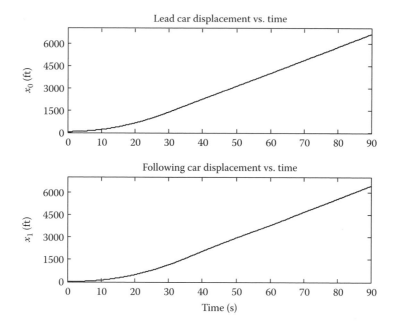

FIGURE 5.44   Lead and following vehicle positions.

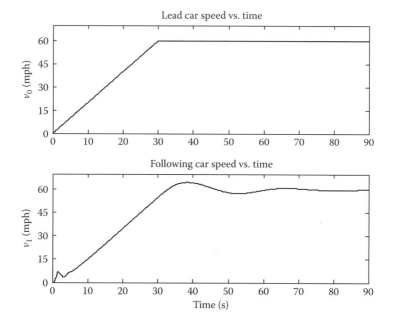

FIGURE 5.45   Lead and following vehicle speeds.

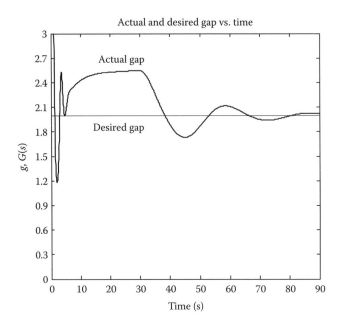

FIGURE 5.46   Desired and actual gaps.

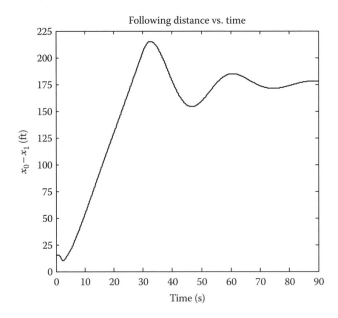

FIGURE 5.47   Following distance.

The initial blip in speed of the following vehicle (see Figure 5.45) is due to the excessive gap g that results whenever the following vehicle is moving at very low speeds. The car-following model, Equation 5.62, implemented in the M-file "*acc.m*" is not robust, that is, it is not valid at following vehicle speeds close to zero, which occurs when the simulation begins. Similar artifacts are present in the gap (Figure 5.46) and acceleration (Figure 5.48) plots. One of the exercise problems addresses this point further.

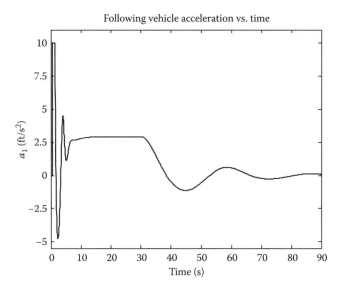

FIGURE 5.48 Simulink® output of following vehicle acceleration.

5.5.1 DISCONTINUITIES

Each of the nonlinear elements presented in Section 2.7 are available as blocks in Simulink. From within the Simulink Library Browser, click on "Discontinuities" to display the element blocks as shown in Figure 5.49.

In the right-hand column are nonlinear blocks for friction, dead zone, saturation, backlash, hysteresis (relay), and quantization.

5.5.2 FRICTION

Figure 5.50 shows the "Coulomb and Viscous Friction" parameter dialog box.

While the default conditions are shown in Figure 5.50, a more practical way to use the block is to assign a scalar value to the Coulomb friction value (Offset). This would represent the coefficient of *static* friction as in the case of initiating the motion of a sliding mass. Of course, the Coefficient of viscous friction (Gain) corresponds to the kinetic friction as the coefficient of the velocity term in the dynamic equations of motion.

A detailed description of the "Coulomb and Viscous Friction" block can found by clicking on Help from the dialog box.

5.5.3 DEAD ZONE AND SATURATION

Figure 5.51 shows the "Dead Zone" parameter dialog box.

The parameter dialog box for the dead zone block is rather intuitive. The user simply sets the beginning and the end of the dead zone according to the input being sent to the block. In the default example, the output is zero if the input signal is between −0.5 and 0.5. Otherwise, the output tracks the input.

A detailed description of the "Dead Zone" block can found by clicking on Help from the dialog box.

Figure 5.52 shows the "Saturation" parameter dialog box.

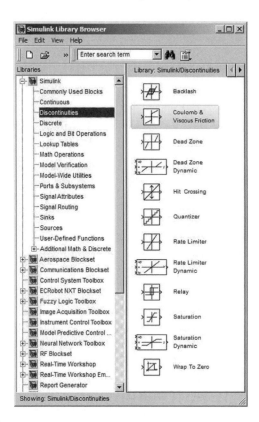

FIGURE 5.49 Simulink® Library Browser—Discontinuities.

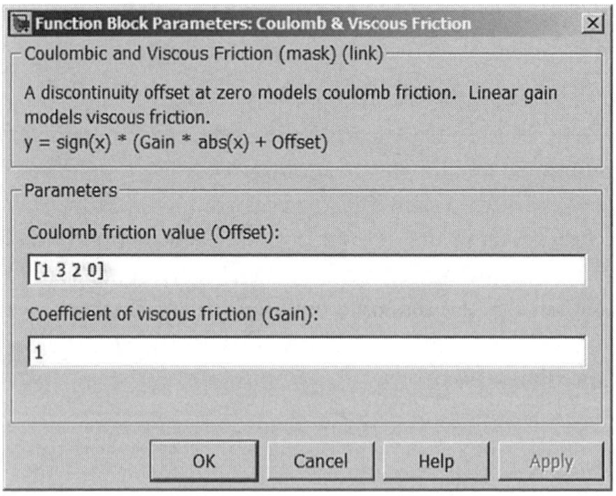

FIGURE 5.50 "Coulomb and Viscous Friction" parameter dialog box.

The parameter dialog box for the saturation block is also intuitive. The user simply sets the beginning and the end of the saturation limits according to the input being sent to the block. In the default example, the output is -0.5 for input values less than -0.5, the output tracks the input between -0.5 and 0.5, and the output is 0.5 for input values greater than 0.5.

A detailed description of the "Saturation" block can found by clicking on Help from the dialog box.

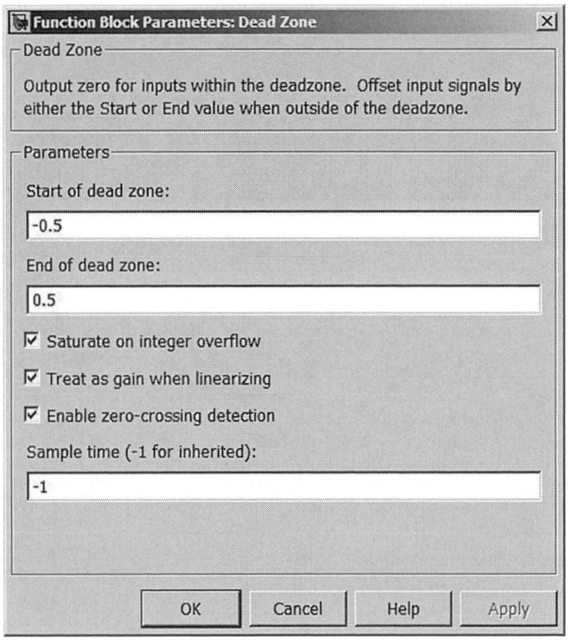

FIGURE 5.51 "Dead Zone" parameter dialog box.

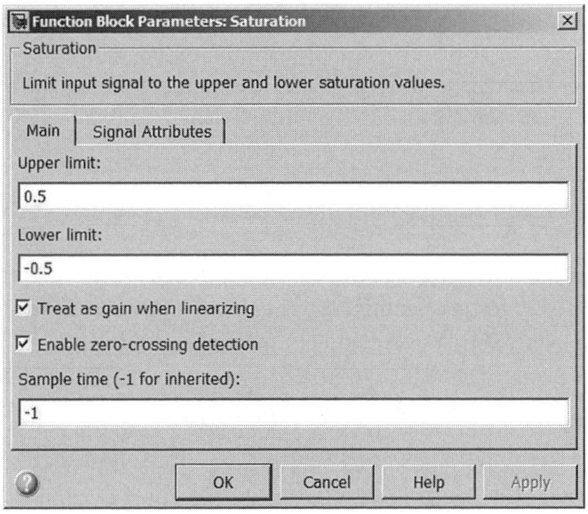

FIGURE 5.52 "Saturation" parameter dialog box.

5.5.4 BACKLASH

Figure 5.53 shows the "Backlash" parameter dialog box.

For the backlash block, the user sets the Deadband width and the Initial output. If the defaults are taken, the output of the backlash block is split evenly between upper and lower values of the input. For example, if the input is a square wave with an upper limit of +1 and a lower limit of −1, the deadband width is centered on zero (the Initial output default), and half of the Deadband width (0.5) is taken from the upper limit while the other half of the Deadband width is taken from the lower limit yielding an output of +0.5 (when the input is +1) and an output of −0.5 (when the input is −1).

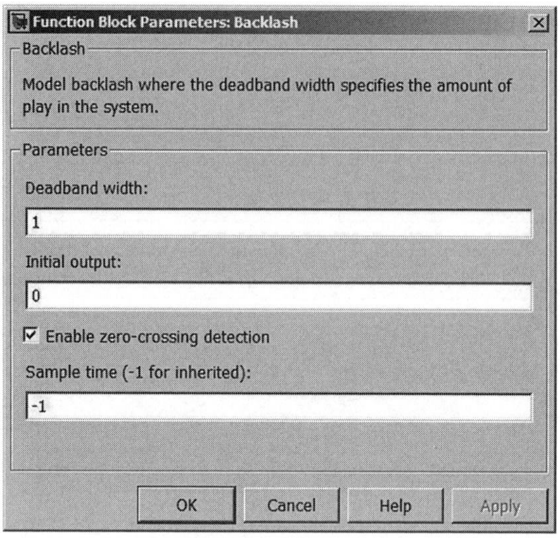

FIGURE 5.53 "Backlash" parameter dialog box.

Clarifying, if the Deadband width is 0.4, then 0.2 will be taken from each of the input values, that is, the output is a square wave between +0.8 and −0.8 (using the input square wave between +1 and −1).

If the Initial output is nonzero and exceeds the input value, then the backlash block can be used to simulate gears that have yet to be engaged. Continuing with the same square wave input between +1 and −1, if the Initial output is set to 2, then the output is +1 plus half of the Deadband width or +1.2. It is only when the gears engage that the output returns to the limits of +0.8 and −0.8.

A detailed description of the "Backlash" block can found by clicking on Help from the dialog box.

5.5.5 Hysteresis

One of the examples in Section 2.7 on nonlinear systems dealt with maintaining the temperature inside a building using a thermostat to control the heat from a furnace. The building temperature and thermostat control are governed by the equations repeated as follows.

$$\tau \frac{dT}{dt} + T = RQ + T_0 \tag{5.63}$$

$$Q = \begin{cases} \bar{Q}, \ T \leq T_d - \Delta \quad \text{or} \quad T_d - \Delta < T < T_d + \Delta \quad \text{and} \quad \dfrac{dT}{dt} > 0 \\[2ex] 0, \ T > T_d + \Delta \quad \text{or} \quad T_d - \Delta < T < T_d + \Delta \quad \text{and} \quad \dfrac{dT}{dt} < 0 \end{cases} \tag{5.64}$$

where
   T is the building temperature (°F)
   Q is the heat input from furnace (Btu/h)
   T_0 is the outside temperature (°F)
   R is the thermal resistance of building (°F/Btu/h)
   τ is the time constant of building temperature response (h)
   $\bar{Q}$ is the rating of furnace (Btu/h)
   T_d is the thermostat setting (°F)
   Δ is the dead zone parameter for thermostat

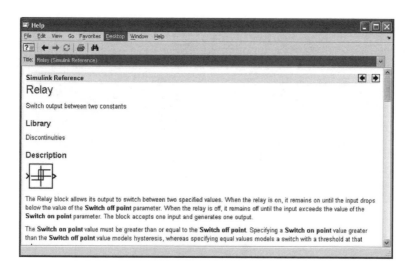

FIGURE 5.54 Description of the Simulink® "Relay" block.

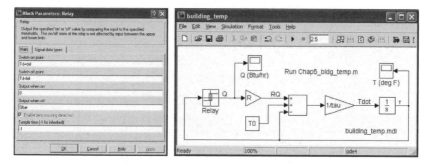

FIGURE 5.55 Simulink® diagram for a simulation of building temperature using the "Relay" block for thermostat.

The hysteresis effect associated with the thermostat, Equation 5.64, is illustrated graphically in Figure 2.38. This type of nonlinear behavior is readily simulated using a "Relay" block from the Simulink "Discontinuities" sublibrary. A description of the "Relay" block can be found in the online Simulink Reference, which contains detailed documentation for each block (see Figure 5.54).

A Simulink diagram for simulating building temperature with conditions as described in Example 2.11 is shown in Figure 5.55. The "Relay" block parameter box is also shown.

The furnace output and building temperature are shown in Figure 5.56. The building temperature increases from 50°F, the constant outside temperature. The hysteresis kicks in once the temperature exceeds "Td-del" = 75 − 3 = 72 °F. The initial portion of the building temperature response in Figure 5.56 is identical to the temperature response graphed in Section 2.7.

5.5.6 Quantization

Figure 5.57 shows the "Quantization" parameter dialog box.

To demonstrate the use of the quantization block, use the default value given for the Quantization interval, that is, 0.5, where the input is a ramp with a slope of 0.5. When the input is between 0 and 0.5, the quantized output is 0; between 0.5 and 1, the quantized output is 0.5; et cetera.

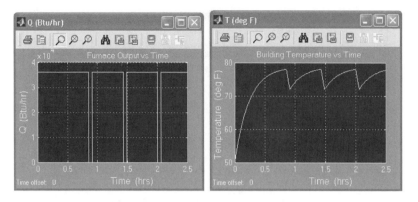

FIGURE 5.56   Furnace output and building temperature.

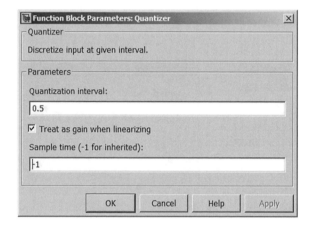

FIGURE 5.57   "Quantization" parameter dialog box.

Following the example from Section 2.7, the Quantization interval is set to 10/256. This corresponds to the analog input range of 0–10 V for an 8-bit microprocessor with $2^8 = 256$ states, 0–255.

A detailed description of the "Quantization" block can be found by clicking on Help from the dialog box.

EXERCISES

5.14 Simulate the motion of a lead vehicle and following vehicle for a period of time sufficient to reach steady state for the conditions given in Table E5.14. The lead vehicle speed is

$$\dot{x}_0(t) = \begin{cases} v_0, & 0 \leq t \leq P/4 \\ v_0 + A\,\sin\dfrac{2\pi t}{P}, & P/4 < t \leq 5P/4 \\ v_0, & t > 5P/4 \end{cases}$$

Plot graphs similar to the ones shown in Figures 5.44 through 5.48.

TABLE E5.14

Parameters	Case I	Case II	Case III
v_0	50 mph	60 mph	70 mph
A	5 mph	5 mph	10 mph
P	30 s	30 s	60 s
G	2 s	2.5 s	3 s
$x_0(0)$	Gv_0	Gv_0	Gv_0
$\dot{x}_1(0)$	v_0	v_0	v_0
$x_1(0)$	0 ft	0 ft	0 ft
SL	v_0	v_0	v_0
Δ	5 mph	10 mph	55 mph
a_{min}	-10 ft/s^2	-12 ft/s^2	-15 ft/s^2
a_{max}	10 ft/s^2	12 ft/s^2	15 ft/s^2
v_{max}	90 mph	90 mph	90 mph
$K_{1,a}$	3 ft/s	4 ft/s	5 ft/s
$K_{1,d}$	3 ft/s	4 ft/s	5 ft/s
$K_{g,a}$	-5 ft/s^3	-5 ft/s^3	-5 ft/s^3
$K_{g,d}$	-4 ft/s^3	-4 ft/s^3	-4 ft/s^3
T (delay)	0.5 s	0.75 s	1 s
L	15 ft	15 ft	15 ft

5.15 Consider the second column of Table E5.14 as baseline numerical values for simulation of a pair of vehicles. Perform a simulation study to analyze the effect of the desirable gap G on the following vehicle's ability to follow at or near the desirable gap. Run the simulation for $G = \{1, 1.5, 2, 2.5, 3, 3.5, 4\}$ for a duration of $3P$ s, and record the value of the average absolute gap error, that is,

$$\left| e_g \right|_{ave} = \frac{1}{2P} \int_P^{3P} |g(t) - G| \, \mathrm{d}t$$

Plot $\left| e_g \right|_{ave}$ vs. G

5.16 Improve the robustness of the car-following simulation to make the output more realistic at very low vehicle speeds. Specifically, modify the code in "*acc.m*" and add additional blocks as necessary to the Simulink diagram in Figure 5.42. Use the baseline values from Table E5.14 to simulate the following vehicle's response to a lead car with speed profile shown in Figure E5.16.

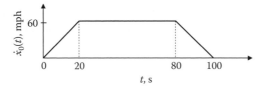

FIGURE E5.16

5.17 The Simulink diagram in Figure 5.42 contains a "Switch" block to maintain the vehicle separation $x_n - x_{n-1}$ greater than $L + 1$. This effectively eliminates the possibility of a rear end collision.

(a) Remove the "Switch" block and add the necessary Simulink blocks to detect the existence of a collision and halt the simulation.

(b) Use the lead car profile in Figure E5.16 and adjust the parameters Δ and $K_{1,a}, K_{1,d}, K_{g,a}, K_{g,d}$ to force a rear-end collision.

5.18 Flow into the tank shown in Figure E5.18a is either on or off. It turns on when the level falls below 20 ft and remains on when the tank is filling until there is 25 ft of liquid in the tank. It remains off as the tank empties until the level falls below 20 ft. In the on condition, the flow rate is 18 ft³/min. The tank dynamics are described by

$$A\frac{dH}{dt} + F_0 = F_1, \quad F_0 = \alpha H^{1/2}$$

where $A = 50\,\text{ft}^2$, $\alpha = 3\,\text{ft}^3/\text{min}/\text{ft}^{1/2}$, and $H(0) = 0$ ft.

(a) Develop your own Simulink diagram or use the one shown in Figure E5.18b to simulate the tank dynamics for a period of time sufficient to see several cycles of filling and emptying. Plot the tank level and the two flows vs. time.

(b) What is the minimum flow necessary to assure the tank is capable of filling to a level of 25 ft? Find the answer analytically and verify with Simulink.

(c) Supplement your Simulink diagram with additional blocks to measure the percentage of time during a cycle in which the tank is filling up.

(d) Instead of switching the flow off and on immediately when the level reaches 25 and 20 ft, respectively, suppose the flow switches off 30 s after the tank level reaches 25 ft and switches on 30 s after the tank level falls to 20 ft. The tank is 25 ft tall. Add Simulink blocks to account for spillover from the tank. Plot the flow in, flow out, spillover, and tank level vs. time.

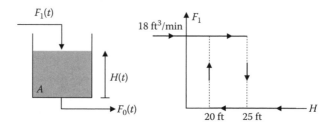

FIGURE E5.18a

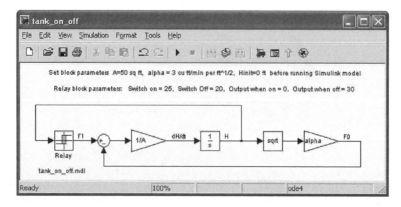

FIGURE E5.18b

5.19 Cascading the dead zone and saturation blocks, model the valve described in Section 2.7.2. Set the dead zone block's parameters to model opening currents of −0.5 and 0.5 amp (default settings) and set the saturation block's parameters to model saturation currents of −1.0 and 1.0 amp. Verify that the valve is modeled correctly by using a ramp input and observing the characteristics shown in Figure 2.39.

5.6 SUBSYSTEMS

As the physical systems we model become progressively more complex, the Simulink representation increases in size, that is, the number of blocks required to model the systems' dynamics grows significantly. A Simulink diagram with hundreds of blocks makes it difficult, if not impossible, to understand the interactions among the systems' components. A more instructive approach consists of grouping specific blocks associated with various subsystems into single entities. At the highest level, the system is viewed in terms of the interactions between these entities.

This hierarchical approach is illustrated for the case of modeling the dynamics of an automobile. Figure 5.58 shows a block diagram of the top level description for modeling the dynamics. At this level, the important interconnections between individual subsystems are identified.

The next step requires the development of concrete descriptions of the individual subsystems, either in mathematical or block diagram form. The mathematical models are transformed into Simulink models that are reusable, much in the same way a procedural function is used in high-level programming languages. Multiple levels are possible in this modeling hierarchy. Moving down one or more levels from the top subsystem level provides more of a microscopic, that is, detailed, description involving low-level components.

An advantage of this approach is the distribution of the modeling effort to individuals with expertise necessary for modeling the individual subsystems. For example, the "Tire Model" subsystem is a critical component in modeling vehicle response. A person knowledgeable in tire/road surface interaction phenomena and the properties of specific tires is needed to develop models that will produce correct tire forces required by the equations of motion.

Suppose a question arises concerning the handling characteristics of a vehicle with different classes of tires. The existing vehicle subsystem models are already in place and can be reused with a "Tire Model" developed specifically for the class of tire under consideration.

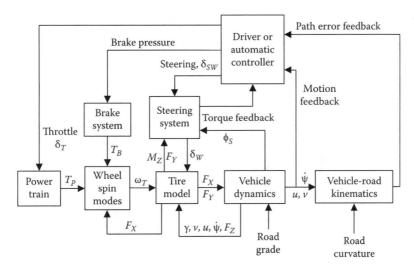

FIGURE 5.58 Top-level description of vehicle dynamics model. (From Allen, R.W. and Rosenthal, T. Systems technology/requirements for vehicle dynamics simulation models, Society of Automotive Engineers, SAE 941075, 1994.)

5.6.1 PHYSBE

PHYSBE is a benchmark simulation of the human circulatory system. It was first introduced by John Mcleod in 1966 in an article titled "PHYSBE...a Physiological Simulation Benchmark Experiment." Over the years, it has appeared in numerous references involving modeling and simulation. The underlying dynamics have been simulated using the popular continuous simulation programs including Simulink.

The human circulatory system is represented by three main components: the lungs (pulmonary circulation), heart (coronary circulation), and the rest of the body (systemic circulation). Coronary and systemic circulations were further divided into subsystems as shown in the Simulink diagram in Figure 5.59 (provided by The Mathworks, Inc.).

The simulation computes pressures, blood flows, volumes, temperatures, and heat flows after a number of parameters describing the physical nature of each subsystem have been specified. The dynamics of each subsystem are hidden in the macroscopic view of the human circulatory system in Figure 5.59. A detailed description of the individual subsystem models is accessible by "looking inside" each of the blocks. For example, the LUNGS subsystem opens up to reveal the components used in modeling the blood flow, blood temperature, heat content, and heat dissipation within the lungs (see Figure 5.60).

The modular structure of the overall system makes it relatively simple to simulate, for example, the effects of partial blockages in the blood vessels of the systemic regions or the effect of changes in vascular compliance on blood pressure.

5.6.2 CAR-FOLLOWING SUBSYSTEM

The next example of Simulink subsystems involves the dynamic behavior of a platoon of vehicles, that is, a lead car (platoon leader) followed by several vehicles whose motion is governed by the dynamics of the preceding vehicle. The Simulink diagram in Figure 5.42 was used to simulate car-following behavior. Deleting the "Scope" blocks, the "Clock" and "Lookup" blocks for generating the lead vehicle speed profile, and the integrator block for creating the lead vehicle position leaves the essential blocks for defining a Simulink car-following subsystem.

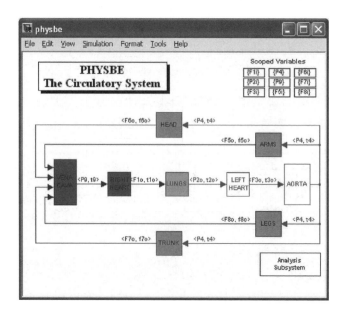

FIGURE 5.59 Simulink® diagram of PHYSBE model.

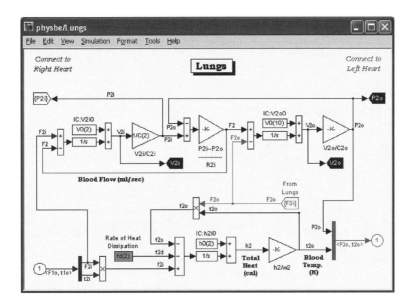

FIGURE 5.60 Subsystem model description of LUNGS.

Input and output ports to the subsystem are created using Simulink "In" and "Out" blocks from the "Ports and Subsystems" sublibrary. A certain amount of discretion is possible when choosing subsystem inputs and outputs. For example, "In" blocks can be connected to the lead vehicle speed "x0d" and position "x0," while the following vehicle speed "x1d" and position "x1" are selected as outputs by connecting them to "Out" blocks. Alternatively, the subsystem could be described in terms of a single input, namely, vehicle speed "x0d," and a single output such as vehicle acceleration "x1dd."

A "car-following" subsystem is created by enclosing selected blocks in the Simulink diagram with a bounding box and choosing "Edit: Create Subsystem" from the menu. The selected blocks collapse into the "car-following" subsystem with renamed inputs and outputs as shown in Figure 5.61. Opening (double clicking) the subsystem reveals the underlying Simulink blocks that can be edited at any time.

The "car-following" subsystem constituent blocks are shown in Figure 5.62. Note that conversion factors from mph to fps (3600/5280) and vice versa were added (see "Gain" blocks in Figure 5.62) to maintain the vehicle speeds in and out of the subsystem in mph, while internal to the subsystem, vehicle speeds are in fps. Simulation of a platoon of vehicles is accomplished by repeated use of the "car-following" subsystem block.

Suppose we wish to simulate the dynamics of a five-vehicle platoon in response to a lead vehicle that decelerates and then accelerates back to a constant steady-state speed. In particular, our interest will focus on the induced perturbations in the stream of traffic. At the beginning of the simulation, each vehicle is traveling at the speed "*SL*," separated in time by the desired gap *G*, as shown in Figure 5.63.

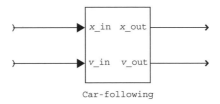

FIGURE 5.61 "Car-following" subsystem with lead vehicle inputs "x_in," "v_in" and following vehicle outputs "x_out," "v_out."

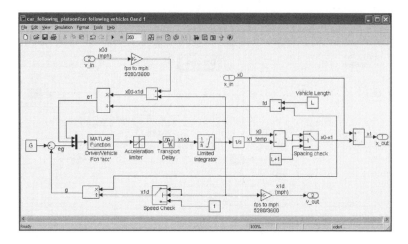

FIGURE 5.62 Simulink® blocks comprising the car-following subsystem.

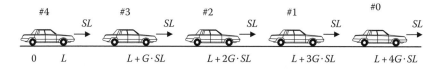

FIGURE 5.63 Initial conditions of the platoon vehicles.

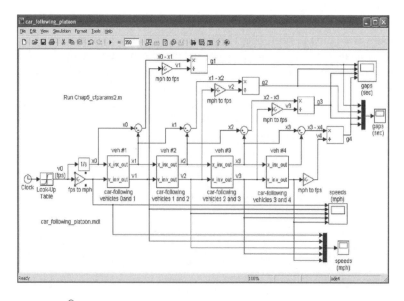

FIGURE 5.64 Simulink® diagram with multiple instances of the "car-following" subsystem.

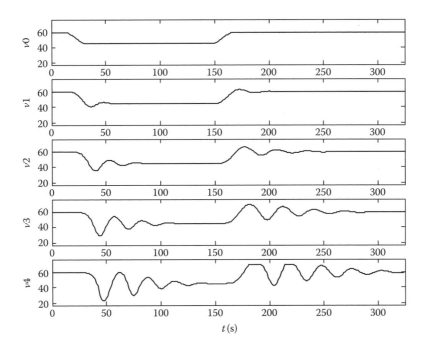

FIGURE 5.65   Speeds (mph) of platoon leader and following vehicles vs. time (s).

The file "*Chap5_cfparams2.m*" loads the parameters required by the Simulink subsystems and "Lookup Table" block for setting the lead vehicle speed. A top-level view of the model "car_following_platoon.mdl" is shown in Figure 5.64.

Initial conditions, namely, speeds and positions, of the trailing vehicles are set in the integrators of the appropriate subsystem blocks. The initial position of the lead vehicle is determined by the parameter of the integrator block feeding the first subsystem. The "Mux" block (lower right) multiplexes the five vehicle speeds on a single line for input to a "Scope" block that draws the five plots on a single set of axes. The heavy arrow emanating from the "Mux" indicates the presence of multiple signals.

Speeds of the lead vehicle and four following vehicles are shown in Figure 5.65. Figure 5.66 is a graph of the successive gaps between the vehicles of the platoon. The responses in Figures 5.65 and 5.66 indicate that the platoon achieves a new steady state identical to the initial one after the perturbations die out.

The "Car-following" subsystem can be daisy-chained as shown in Figure 5.64 to simulate the response of platoons of vehicles of any size. Furthermore, the following vehicles can be individualized by including a vector of randomly chosen driver/vehicle parameters as an additional input to each "Car-following" subsystem block.

5.6.3   SUBSYSTEM USING FCN BLOCKS

The "Fcn" block is a convenient time saver when the mathematical model of the system consists primarily of algebraic and differential equations. Lengthy expressions are evaluated in equation form instead of being constructed from Simulink blocks. To illustrate, the frictionless inverted pendulum introduced in Section 5.4 can be treated as a subsystem with the governing equations for $\ddot{x}$ and $\ddot{\theta}$ implemented by the use of "Fcn" blocks. Equations 5.50 and 5.51 are implicit in nature as a result of $\ddot{x}$ and $\ddot{\theta}$ appearing in both equations. (Recall the presence of an algebraic loop in the Simulink diagram.) This can be overcome by solving for the second derivative terms explicitly leading to Equations 5.65 and 5.66.

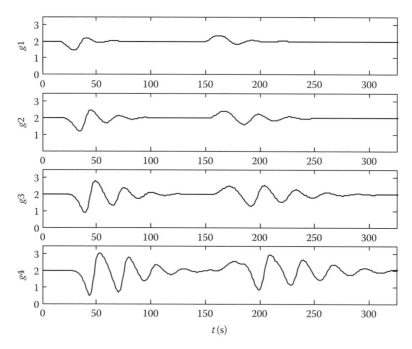

FIGURE 5.66 Gaps (s) of following vehicles vs. time (s).

$$\ddot{x} = \frac{ml\dot{\theta}^2 \sin\theta - mg\cos\theta\sin\theta + u}{M + m\sin^2\theta} \tag{5.65}$$

$$\ddot{\theta} = \frac{-ml\dot{\theta}^2 \cos\theta\sin\theta + (m+M)g\sin\theta - u\cos\theta}{l(M + m\sin^2\theta)} \tag{5.66}$$

Figure 5.67 shows the top layer with the "cart model" subsystem.

It includes Simulink blocks to generate the input u, decompose the state vector $[x, \dot{x}, \theta, \dot{\theta}]$ into its components, and feed the components to individual "scope" blocks. Note the use of the Simulink supplied "R2D" block for converting from radians to degrees. It is found in the "Simulink Extras" sublibrary under the "Transformations" heading. A number of useful coordinate transformation blocks are available there.

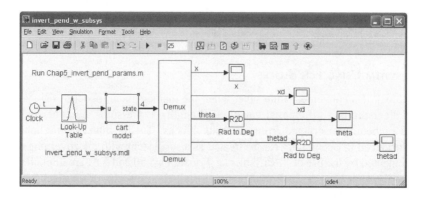

FIGURE 5.67 Top layer of a Simulink® diagram for simulating an inverted pendulum.

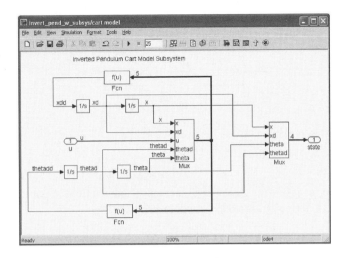

FIGURE 5.68   Cart subsystem using "Fcn" blocks.

Opening the "cart model" subsystem reveals the blocks shown in Figure 5.68. Note that the "Display option" of the "Mux" parameter blocks is set to "signals" in order to identify its inputs. The parameters of the two "Fcn" blocks are expressions relating the accelerations "xdd" and "thetadd" to the inputs "x," "xd," "u," "thetad," and "theta" (from the "mux" block). The "Fcn" block input notation is u[1], u[2], . . . ,u[5] where u[1] is the first input "x," u[2] is "xd," and so forth.

From Equation 5.65, the "Fcn" block parameter expression for "xdd" is

```
(m*l*u[4]^2*sin(u[5])-m*g*cos(u[5])*sin(u[5])+u[3])/
   (M+m*sin(u[5])^2)
```

Referring to Equation 5.66, the "Fcn" block parameter expression for "thetadd" is

```
(-m*l*u[4]^2*cos(u[5])*sin(u[5])+(m+M)*g*sin(u[5])-u[3]*
   cos(u[5])/(l*(M+m*sin(u[5])^2))
```

The angular position of the pendulum $\theta(t)$ is plotted in Figure 5.69 for the case when the cart and pendulum are initially at rest, that is, $x(0) = \dot{x}(0) = \theta(0) = \dot{\theta}(0) = 0$, and the input $u(t)$ is the triangular pulse shown in Figure 5.70. The numerical values of the system parameters are $M = 2$ kg, $m = 0.1$ kg, and $l = 0.5$ m.

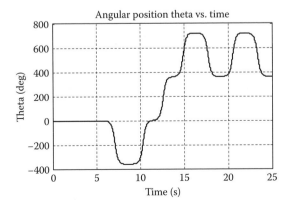

FIGURE 5.69 $\theta(t)$ vs. t for $u(t)$ shown in Figure 5.70.

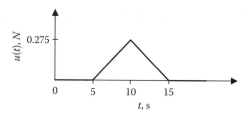

FIGURE 5.70 Force $u(t)$ applied to cart.

EXERCISES

5.20 Randomize the car-following behavior by assuming that the desired gap G and driver/vehicle delay T are both normally distributed random variables, that is,

$$T \sim N(\mu_T, \sigma_T), \quad \mu_T = 0.75\,\text{s}, \quad \sigma_T = 0.15\,\text{s}$$

$$G \sim N(\mu_G, \sigma_G), \quad \mu_G = 2.5\,\text{s}, \quad \sigma_G = 0.3\,\text{s}$$

Use MATLAB to generate $\{G_i, T_i\}$, $i = 1, 2, 3, 4$ for four following vehicles, and repeat the simulation of the five vehicles shown in Figure 5.63.

5.21 Six vehicles are stopped at a traffic light with a distance of L feet from the rear bumper of the car in front to the front bumper of the following vehicle. The lead car accelerates uniformly from zero mph to the speed limit $SL = 45$ mph in 30 s and continues traveling at the speed limit. Use the robust car-following model developed in Exercise 5.16 to simulate the transient response of the platoon. Use the baseline conditions in the second column of Table E5.14 for the parameters, or choose a new set of appropriate values. Obtain time history plots of
  (a) Vehicle positions
  (b) Vehicle speeds
  (c) Vehicle gaps
  (d) Vehicle-following distances

5.22 A total of 11 cars are traveling at the speed limit SL with initial spacing similar to those in Figure 5.63. At $t = 0$, the lead car speed begins to vary sinusoidally with amplitude of 3 mph and period of 20 s. Determine the peak amplitude in speed of the following vehicles for the nine combinations: $SL = 30, 45, 60$ mph and $G = 1.5, 2, 3$ s.

5.23 Starting with Equations 5.65 and 5.66 for the cart and inverted pendulum,
  (a) Develop a state variable model of the system, that is, $\dot{x} = f(x, u)$ and $y = g(x, u)$ where the state $x = [x, \dot{x}, \theta, \dot{\theta}]^T$ and output $y = [x, \theta]^T$.
  (b) Find the state equations for updating the discrete-time state $x_A(n)$ and computing $y_A(n)$ using forward Euler integration with step size T.
  (c) Solve the equations in part (b) recursively to find $x_A(n)$ and $y_A(n)$, $n = 1, 2, \dots, n_f$ where $T = 0.05$ s, $T_{\text{final}} = n_f T = 5$ s, $u(t) = 0$, $t \geq 0$, and $x(0) = [0, 0, \pi/6, 0]$.
  (d) Plot the discrete-time state vector $x_A(n)$, $n = 0, 5, 10, \dots, n_f$.
  (e) Simulate the response with Simulink for the same conditions in part (c) using the ode1 (Euler) integrator. Plot the state vector and compare the results to part (d).

5.24 Show that the frictionless cart and pendulum have two equilibrium points when the input $u(t) = 0$, $t \geq 0$, namely,

$$x_{1,e} = 0\,\text{m}, \quad x_{2,e} = 0\,\text{m/s}, \quad x_{3,e} = 0\,\text{rad}, \quad x_{4,e} = 0\,\text{rad/s}$$

$$x_{1,e} = 0\,\text{m}, \quad x_{2,e} = 0\,\text{m/s}, \quad x_{3,e} = \pi\,\text{rad}, \quad x_{4,e} = 0\,\text{rad/s}$$

and verify by using Simulink that the first equilibrium point is unstable and the second one is stable. Is the second equilibrium point asymptotically stable?

5.25 Develop a subsystem model of the cart and pendulum where the pendulum rotation is opposed by a damping torque $T_D = c\dot{\theta}$ and the cart motion is subject to a constant friction force

$$f_\mu = \begin{cases} -\mu(m+M)g \cdot \mathrm{sgn}(\dot{x}), & \dot{x} \neq 0 \\ 0, & \dot{x} = 0 \end{cases}$$

Simulate the response of the cart and pendulum, starting from the stable equilibrium point, to the input

$$u(t) = \begin{cases} 0, & 0 \leq t < 1 \\ U_0, & 1 \leq t < 3 \\ 0, & 3 \leq t \end{cases}$$

Numerical values of the system parameters are $m = 0.25$ kg, $M = 10$ kg, $l = 1$ m, $c = 0.5$ N·m/rad/s, $\mu = 0.015$, and $U_0 = 10$ N.

5.7 DISCRETE-TIME SYSTEMS

Up to this point, we have focused on using Simulink for simulation of systems with continuous-time mathematical models. Discrete-time systems evolve as approximate representations of continuous-time systems at specific points in time. The numerical integrators already considered as well as those to come are predicated on some form of discrete-time approximation to the derivative function. Digital processors that manipulate streams of sampled numerical data are likewise discrete time in nature. Indeed, much of the first part of this book deals with methods for obtaining discrete-time model approximations of continuous-time systems. In the case of linear time-invariant (LTI) discrete-time systems, methods for finding solutions to specific inputs were presented as well.

Alternatively, some discrete-time systems process information, which by its very nature is allowed to change only at discrete instants of time. In that case, the systems are inherently discrete time. The difference equations are solved either recursively or by the use of a general solution (if one exists), resulting in an output sequence of numbers defined solely at discrete times 0, T, $2T$, $3T$, $\ldots$.

Simulink is well suited to obtain solutions of discrete-time system models regardless of whether they are approximations of continuous-time systems or inherently discrete time to begin with. The procedure for obtaining a Simulink diagram of a discrete-time system is similar to the way Simulink diagrams of continuous-time systems were developed. With discrete-time systems, the goal is to express the highest order difference term as an explicit function of the lower order terms. For example, suppose an nth-order discrete-time system with output $y(k)$ and input $u(k)$ is modeled by the nth-order difference equation,

$$g[y(k+n), y(k+n-1), \ldots, y(k+1), y(k), u(k+p), \ldots, u(k+1), u(k)] = 0 \tag{5.67}$$

with initial conditions $y(0)$, $y(1)$, $\ldots$, $y(n-1)$. Oftentimes it is possible to solve Equation 5.67 explicitly for $y(k+n)$, giving

$$y(k+n) = f[y(k+n-1), \ldots, y(k+1), y(k), u(k+p), \ldots, u(k+1), u(k)] \tag{5.68}$$

Starting with $y(k+n)$ and $u(k+p)$, delayed signals $y(k+n-1), \ldots$, $y(k+1)$, $y(k)$ and $u(k+p-1), \ldots$, $u(k+1)$, $u(k)$ are generated using the "Unit Delay" block and combined according to Equation 5.68 to complete the simulation diagram of the discrete-time system.

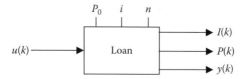

FIGURE 5.71 Repayment and amortization of a loan.

5.7.1 Simulation of an Inherently Discrete-Time System

We begin with an inherently discrete-time system most of us are familiar with, namely, a fixed interest loan with constant periodic payments. An amount of money is borrowed for a specified period of time, and equally spaced installments are paid to the lender until the loan is completely repaid. The interest rate on the loan is established at the time of the loan. Furthermore, each payment consists of a portion that reduces the loan principal and the remaining portion that is interest on the outstanding balance. The situation is illustrated in Figure 5.71.
The system parameters consist of

P_0: Loan amount
i: Interest rate per period (fixed for the duration of the loan)
n: Number of interest periods for duration of loan

The discrete-time input

$$u(k) = \begin{cases} 0, & k = 0 \\ A, & k = 1, 2, 3, \ldots, n \end{cases} \tag{5.69}$$

is the constant payment A made at the end of the kth interest period. The discrete-time outputs are

$y(k)$: Outstanding balance of loan immediately following the kth payment
$P(k)$: Portion of kth payment used to reduce the outstanding balance
$I(k)$: Interest portion of kth payment

The unpaid balance after the $(k+1)$st payment is simply the unpaid balance following the kth payment plus the interest accrued for one period on the unpaid balance minus the amount of the $(k+1)$st payment. Thus,

$$y(k+1) = y(k) + iy(k) - u(k+1), \quad k = 0, 1, 2, \ldots, n-1 \tag{5.70}$$

$$= (1+i)y(k) - u(k+1), \quad k = 0, 1, 2, \ldots, n-1 \tag{5.71}$$

$P(k+1)$, the portion of $u(k+1)$ used for loan principal reduction, is equal to the reduction in outstanding balance from the kth to the $(k+1)$st payment, that is,

$$P(k+1) = y(k) - y(k+1), \quad k = 0, 1, 2, \ldots, n-1 \tag{5.72}$$

$I(k)$, the interest portion of $u(k)$, is obtained from

$$P(k) + I(k) = u(k) = A, \quad k = 1, 2, \ldots, n \tag{5.73}$$

$$\Rightarrow I(k) = A - P(k), \quad k = 1, 2, \ldots, n \tag{5.74}$$

It can be shown (Thuesen 1971) that the constant payment A necessary to fully repay the loan in n periods, that is, make $y(n) = 0$, is given by

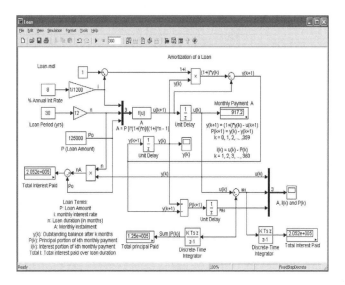

FIGURE 5.72   Simulink® diagram for loan repayment.

$$A = P_0 \left[\frac{i(1+i)^n}{(1+i)^n - 1} \right] \tag{5.75}$$

Equations 5.71, 5.72, and 5.74 are the difference equations for the first-order discrete-time system in Figure 5.71. A Simulink diagram of the system is shown in Figure 5.72. Note the use of a single "Unit Delay" block to generate the signal $y(k)$ and the sum block in the upper right corner producing $y(k+1)$ as the difference of $(1+i)y(k)$ and the payment amount $u(k+1)$ according to Equation 5.71.

The "Simulation Parameters" dialog box is shown in Figure 5.73. A "Fixed-step" integrator with "Fixed-step size" of 1 is selected to force the simulation to step through integer values of discrete time. Since there is no continuous-time integration present in an inherently discrete-time system, the "discrete (no continuous states)" option is chosen from the drop-down menu of integrators.

The numerical values shown in Figure 5.72 correspond to a $125,000 loan at 8% interest per annum repaid over 30 years. The monthly payment A is calculated inside the "Fcn" block and appears in the "Display" as $917.20. The unpaid balance $y(k)$ is shown in Figure 5.74. As expected, the loan balance is zero following the 360th monthly payment.

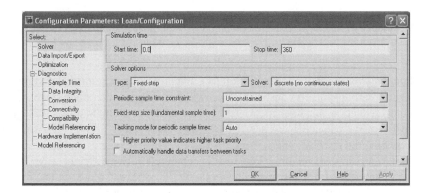

FIGURE 5.73   Simulation parameters dialog box for loan simulation.

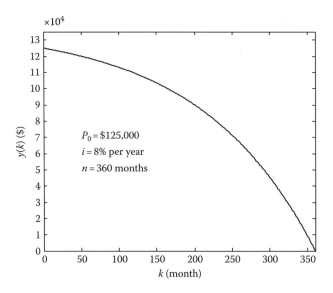

FIGURE 5.74 Unpaid balance $y(k)$ vs. interest period k.

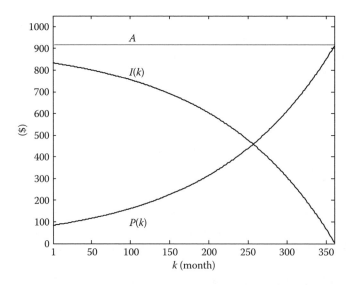

FIGURE 5.75 Monthly installment A, interest portion $I(k)$, and principal portion $P(k)$ vs. k.

The total monthly payment A, interest portion $I(k)$, and principal portion $P(k)$ are shown in Figure 5.75. Note that the early payments consist almost entirely of interest with only a small amount going towards principal reduction. As the loan progresses, the portion of each monthly installment used to reduce the outstanding balance increases. Conversely, the interest portion of each subsequent payment is less than the previous one.

The total interest paid over the life of the loan is computed in two different ways. The simplest approach is to compute $nA - P_0$, the result shown in the "Display" block on the left side of the Simulink diagram. The second method employs a "Discrete-Time Integrator" situated in the lower right corner of Figure 5.72.

FIGURE 5.76   Simulink® discrete-time integrators.

5.7.2 DISCRETE-TIME INTEGRATOR

A "Discrete-Time Integrator" is a numerical integrator from the "Discrete" sublibrary reserved for discrete-time systems. The "Discrete-Time Integrator" can be configured as a forward (explicit) Euler, backward (implicit) Euler, or trapezoidal integrator. The z-domain transfer functions derived in Section 4.7 for each of the discrete-time integrators are placed inside the appropriate block shown in Figure 5.76.

The two discrete-time integrators in Figure 5.72 function as summing devices, one for the total interest and the other for the computation of the total of all the principal payments. To understand why, consider a discrete-time backward (implicit) Euler integrator with input signal $I(k)$ and output $I_T(k)$. The difference equation is

$$I_T(k + 1) = I_T(k) + T \cdot I(k + 1) \tag{5.76}$$

With $T = 1$ and $I_T(0) = 0$, it follows that $I_T(1) = I(1)$, $I_T(2) = I(1) + I(2), \ldots$ and

$$I_T(n) = \sum_{k=1}^{n} I(k) \tag{5.77}$$

Simulink's fixed-step integrators "ode1" (Euler), "ode2" (Heun) through "ode5" are all explicit. Thus, if we elect to use a fixed-step implicit integrator, our choice is limited to either "Backward Euler" or "Trapezoidal" from the "Discrete" sublibrary. In this case, the Simulink diagram is similar to the simulation diagram representation of the continuous-time system with the exception that the continuous-time integrators ($1/s$ blocks) are replaced by the preferred implicit discrete-time integrator.

To illustrate, consider the vehicle of weight W lb in Figure 5.77 rolling backwards down an incline (θ) subject to an aerodynamic drag force F_D, a rolling friction force F_μ ($F_\mu/4$ on each tire), and a gravitational component of weight F_W. The vehicle travels a length L along the inclined section and then continues on a level section of road.

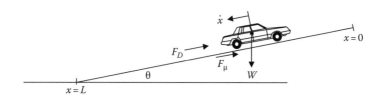

FIGURE 5.77   Vehicle rolling down an incline.

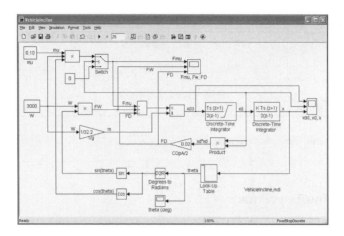

FIGURE 5.78 Simulink® diagram for vehicle rolling down incline.

Summing the forces on the vehicle in the direction of travel gives

$$m\ddot{x} = -F_D - F_\mu + F_W \tag{5.78}$$

$$= \begin{cases} -0.5C_D\rho A\dot{x}^2 - \mu W\cos\theta + W\sin\theta, & x \le L \\ -0.5C_D\rho A\dot{x}^2 - \mu W, & x > L \end{cases} \tag{5.79}$$

where
m is the mass of the vehicle
C_D is the aerodynamic drag coefficient
ρ is the density of air
A is the exposed vehicle front area
μ is the coefficient of rolling friction between the tires and the road

A Simulink diagram using trapezoidal integration for both integrators is shown in Figure 5.78. The first is a limited integrator with the lower limit set to zero. The "Switch" block guarantees that the friction force is zero when the vehicle is stopped.

Numerical values of the system parameters are shown in the "Con" and "Gain" blocks except for $L = 200$ ft and $\theta = 10°$, which appear in the "Lookup table" parameters. Results are shown in Figures 5.79 through 5.81 for the case when the vehicle is released from a stopped position and starts rolling down the incline.

Implicit integrators like the "Backward Euler" and "Trapezoidal" may lead to algebraic loops, which require additional computational effort to resolve at each discrete-time step. In fact, an "Algebraic Loop Warning" appears in executing the simulation corresponding to the Simulink diagram in Figure 5.78. Algebraic loops never include a continuous-time integrator because all of Simulink's continuous-time integrators are implemented by explicit numerical integration algorithms.

Typical of all Simulink blocks in the "Discrete" sublibrary, the discrete-time integrator outputs are clamped or held constant for the duration of the sample time (integration step), 0.01 s, in this example (see Figure 5.82).

5.7.3 CENTRALIZED INTEGRATION

The simulation diagram of an nth-order continuous-time system model will contain n distinct integrators. The Simulink diagram will contain one integrator for each continuous-time state or equivalently a "State-Space," "Transfer Fcn," or "Zero-Pole" block from the "Continuous"

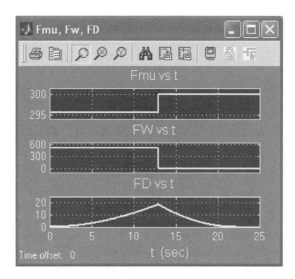

FIGURE 5.79 F_μ, F_W, F_D vs. t.

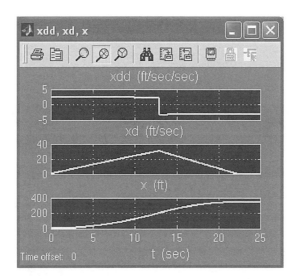

FIGURE 5.80 $\ddot{x}, \dot{x}, x$ vs. t.

sublibrary to model components with one or more continuous-time states. In either case, the numerical integrator selected from the choice of fixed-step and variable-step integrators in the "Simulation Parameters" dialog box will be applied to all the continuous-time state derivatives.

For "One-Step" numerical integration algorithms (discussed in Chapter 6), which includes the explicit Euler integrator, all state derivatives are calculated at $t_n = nT$ prior to updating a single state at $t_{n+1} = (n+1)T$. The entire collection of continuous-time states are updated at t_{n+1} based on the state derivatives at t_n, which in turn depend on the values of the states and inputs (if present) at $t_n = nT$. This is referred to as centralized integration.

There are situations when centralized integration is not the most advantageous approach when it comes to updating the states. For example, when the state derivatives are themselves states, some of the states can be updated at time $t_{n+1} = (n+1)T$ based on calculated values of other states at t_{n+1}.

Figure 5.83 is a simulation diagram of a mechanical system with inertia. It contains an acceleration term that is twice integrated to produce velocity and position.

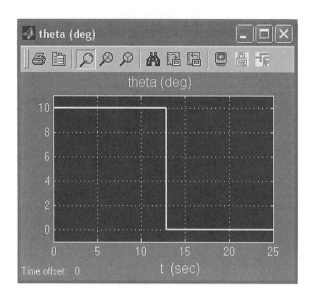

FIGURE 5.81   θ vs. *t*.

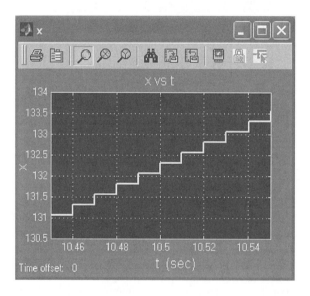

FIGURE 5.82   Close-up of discrete-time integrator output illustrating discrete-time nature (sample time equal 0.01 s).

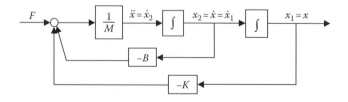

FIGURE 5.83   Simulation diagram of second-order system with sequential integrators.

With centralized explicit Euler integration, the discrete-time states $x_{1,A}(n)$ and $x_{2,A}(n)$ are updated at time $t_{n+1} = (n+1)T$ in the sequence of steps as follows:

$$\dot{x}_2(n) = \ddot{x}(n) = \frac{1}{M}[F(n) - Kx_{1,A}(n) - Bx_{2,A}(n)] \tag{5.80}$$

$$\dot{x}_1(n) = x_{2,A}(n) \tag{5.81}$$

$$x_{2,A}(n+1) = x_{2,A}(n) + T\dot{x}_2(n) \tag{5.82}$$

$$\Rightarrow x_{2,A}(n+1) = x_{2,A}(n) + \frac{T}{M}[F(n) - Kx_{1,A}(n) - Bx_{2,A}(n)] \tag{5.83}$$

$$x_{1,A}(n+1) = x_{1,A}(n) + T\dot{x}_1(n) \tag{5.84}$$

$$\Rightarrow x_{1,A}(n+1) = x_{1,A}(n) + Tx_{2,A}(n) \tag{5.85}$$

Instead of starting with Equation 5.84 to update $x_{1,A}(n)$, the implicit Euler form can be used, that is, the updated state $x_{1,A}(n+1)$ is obtained from

$$x_{1,A}(n+1) = x_{1,A}(n) + T\ddot{x}_1(n+1) \tag{5.86}$$

$$\Rightarrow x_{1,A}(n+1) = x_{1,A}(n) + Tx_{2,A}(n+1) \tag{5.87}$$

Equations 5.83 and 5.87 form the basis of a noncentralized integration scheme, which uses explicit Euler integration to update $x_{2,A}(n)$ and implicit Euler integration to update $x_{1,A}(n)$ at $t_{n+1} = (n+1)T$. The explicit Euler/implicit Euler combination is superior to explicit Euler/explicit Euler integration by virtue of its using the updated velocity $x_{2,A}(n+1)$ in the computation for the new state $x_{1,A}(n+1)$ in Equation 5.87.

With Simulink, explicit Euler/implicit Euler (or explicit Euler/trapezoidal) integration of a second-order component is straightforward. Figure 5.84 is a Simulink diagram for simulating the unit step response of the second-order system

$$\ddot{x} + 2\zeta\omega_n\dot{x} + \omega_n^2 x = \omega_n^2 u \tag{5.88}$$

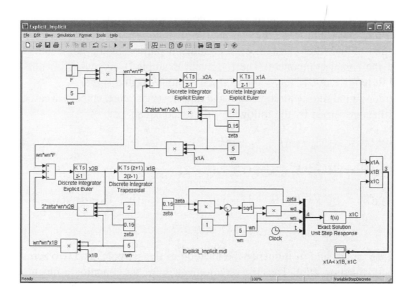

FIGURE 5.84 Simulink® diagram with explicit Euler/explicit Euler integration, explicit Euler/trapezoidal integration and exact solution of second-order system unit step response ($\zeta = 0.15$, $\omega_n = 5$ rad/s), sample time $T = 0.02$ s.

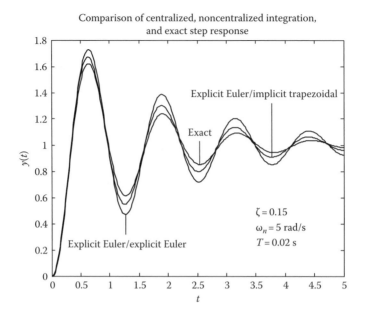

FIGURE 5.85 Results of centralized integration and noncentralized integration.

The explicit Euler/explicit Euler and explicit Euler/trapezoidal integration routines are implemented, and the exact solution is generated for comparison. The lightly damped system step responses are shown in Figure 5.85. The response obtained using noncentralized integration (explicit Euler/ trapezoidal) is closer to the exact solution.

5.7.4 DIGITAL FILTERS

Digital filters were introduced in this chapter. A digital filter is a discrete-time system designed to process discrete-time data for the purpose of extracting useful information. In many cases, the data is comprised of a useful signal and an unwanted component such as noise. When the frequency components in the signal and noise are confined to distinct regions in the frequency spectrum, a properly designed digital filter can remove a significant portion of the noise without appreciable degradation of the signal component. The following example (Cadzow 1973) illustrates a notch filter designed to remove 60 Hz ($\omega_0 = 2\pi \times 60$ rad/s) noise from a signal.

The sixth-order digital filter with input $u(k)$ and output $y(k)$ is represented as a cascaded system of second-order filters governed by the following difference equations:

$$y_1(k) = u(k) + b_1 u(k-1) + u(k-2) - a_1 y_1(k-1) - a_2 y_1(k-2) \tag{5.89}$$

$$y_2(k) = y_1(k) + b_1 y_1(k-1) + y_1(k-2) - a_3 y_2(k-1) - a_4 y_2(k-2) \tag{5.90}$$

$$y_3(k) = y_2(k) + b_1 y_2(k-1) + y_2(k-2) - a_5 y_3(k-1) - a_6 y_3(k-2) \tag{5.91}$$

$$y(k) = b y_3(k) \tag{5.92}$$

Parameters $a_1, a_2, \ldots, a_6$ and b_1 influence the location of the six poles and two zeros of the filter's z-domain transfer function $H(z)$. The first requirement is for the magnitude function at the noise frequency, $|H(e^{j\omega_0 T})| = 0$.

The constant b is selected to make the magnitude function approximately one at other frequencies. Numerical values of the filter's constants are listed in the Simulink diagram shown in Figure 5.86. A sampling period of $T = 0.001$ s is used.

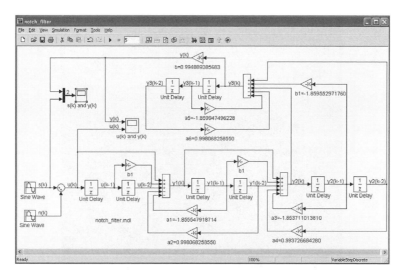

FIGURE 5.86 Simulink® diagram for a notch filter (notch frequency = 60 Hz).

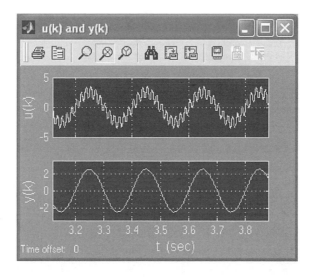

FIGURE 5.87 $u(k)$, $y(k)$, $\omega = 10\pi$ rad/s.

The continuous-time signal $u(t)$ consists of a sinusoidal component $s(t) = 2.5 \sin \omega t$ and 60 Hz noise component $n(t) = \sin 120\pi t$. Figure 5.87 shows the sampled input signal $u(k)$ with a low frequency 5 Hz ($\omega = 10\pi$ rad/s) signal component. Also shown is the filter output $y(k)$ at steady state following the transient period. Figure 5.88 shows that the signal component $s(k)$ and the discrete-time filter output $y(k)$ are nearly identical.

Figures 5.89 and 5.90 show the steady-state response of the filter when the frequency of the signal component is closer to the noise frequency, that is, 57.5 Hz ($\omega = 115\pi$ rad/s).

The reader should try running the simulation for the case when the signal frequency is higher than the notch frequency (but less than the Nyquist frequency $\pi/T = 1000\pi$ rad/s) to verify similar results.

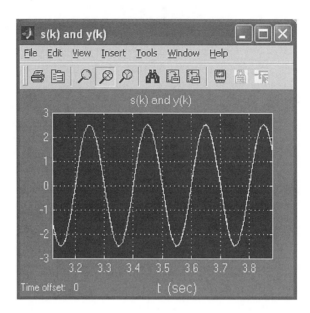

FIGURE 5.88 $s(k)$, $y(k)$, $\omega = 10\pi$ rad/s.

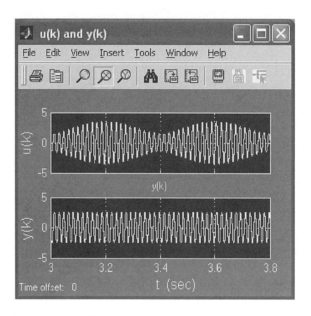

FIGURE 5.89 $u(k)$, $y(k)$, $\omega = 115\pi$ rad/s.

5.7.5 DISCRETE-TIME TRANSFER FUNCTION

Simulink is capable of simulating the response of linear discrete-time systems based on the knowledge of the system's discrete-time transfer function. Similar to continuous-time systems with transfer function $H(s)$ in pole-zero form or a ratio of polynomials, the response of a discrete-time system with transfer function $H(z)$ to a discrete-time input is obtained by using one of the blocks shown in Figure 5.91.

The "`Discrete Filter`" block is used primarily for digital filters where the numerator and denominator are polynomials in z^{-1}. It is easily obtained from either of the other two forms by

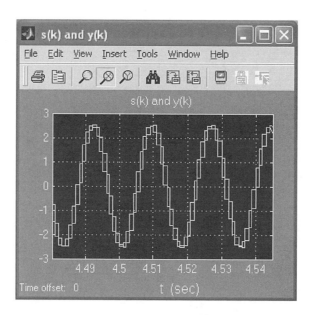

FIGURE 5.90 $s(k)$, $y(k)$, $\omega = 115\pi$ rad/s.

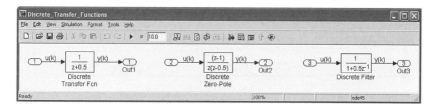

FIGURE 5.91 Simulink® discrete-time transfer function blocks.

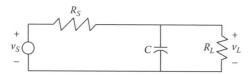

FIGURE 5.92 Circuit for a low-pass filter.

multiplying numerator and denominator by z^{-n} where n is the order of the denominator polynomial. The following example illustrates the use of the "Discrete Zero-Pole" block for a low-pass digital filter obtained by approximating the dynamics of a continuous-time filter.

The RC circuit in Figure 5.92 was shown to exhibit the characteristics of a low-pass filter in Chapter 4.

The continuous-time transfer function $H(s)$ is given by

$$H(s) = \frac{K\omega_c}{s + \omega_c} \tag{5.93}$$

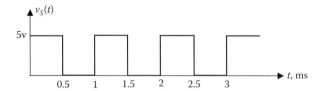

FIGURE 5.93 Input signal $v_S(t)$, $t \geq 0$.

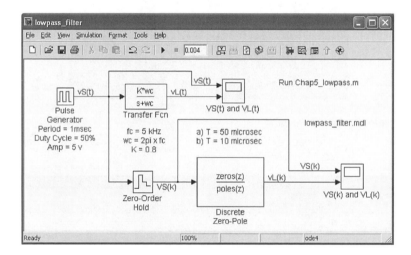

FIGURE 5.94 Simulink® diagram for continuous- and discrete-time low-pass filters.

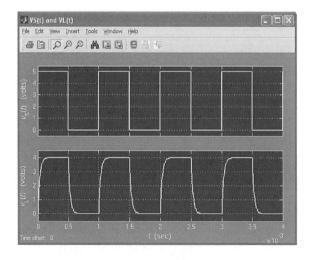

FIGURE 5.95 Low-pass continuous-time filter input and output.

where ω_c, the cutoff frequency, and DC gain K are related to circuit parameters R_S, R_L, and C according to

$$\omega_c = \frac{(1/R_S) + (1/R_L)}{C} \tag{5.94}$$

$$K = \frac{R_L}{R_S + R_L} \tag{5.95}$$

Suppose the signal $v_S(t)$ is the square wave shown in Figure 5.93.

The capacitor in the circuit of Figure 5.92 removes high-frequency components from $v_S(t)$, resulting in smoother pulse transitions in $v_L(t)$.

For values of $R_S = 50\ \Omega$ and $R_L = 200\ \Omega$, the capacitance C is selected to make the cutoff frequency 5 kHz ($\omega_c = 5 \times 10^3 \times 2\pi$ rad/s). The initial design calls for the Nyquist frequency π/T to be twice the cutoff frequency making the sample time $T = 50$ μs (20 kHz sampling rate).

The discrete-time filter transfer function $H(z)$ is to be synthesized from the continuous-time filter transfer function $H(s)$ using the bilinear transform (see Chapter 4).

$$H(z) = H(s)|_{s=(2/T)((z-1)/(z+1))} \tag{5.96}$$

$$= \frac{K\omega_c}{(2/T)((z-1)/(z+1)) + \omega_c} \tag{5.97}$$

$$\Rightarrow H(z) = \frac{K\omega_c T}{2 + \omega_c T}\left[\frac{z+1}{z - ((2-\omega_c T)/(2+\omega_c T))}\right] \tag{5.98}$$

Figure 5.94 shows the Simulink diagram for simulating the response of the continuous-time filter when the input is $v_S(t)$, $t \geq 0$ and the digital filter when the discrete-time input is $v_S(kT)$, $k = 0, 1, 2,\ldots$. The "Zero-Order Hold" block is required when the discrete-time system sample time T exceeds the integration step size (10 μs) for the continuous-time "Transfer Fcn" block. The continuous-time signals are shown in Figure 5.95.

The discrete-time signals are shown in Figure 5.96 ($T = 50$ μs) and Figure 5.97 ($T = 10$ μs). As expected, the digital filter response $v_L(K)$ is closer to the output $v_L(t)$of the analog circuit at the higher sampling rate.

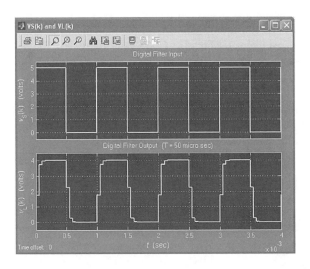

FIGURE 5.96   Digital filter input and output ($T = 50$ μs).

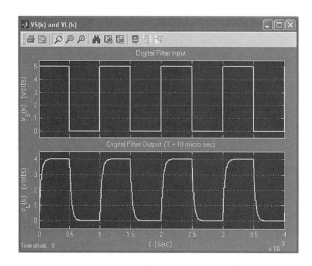

FIGURE 5.97　Digital filter input and output ($T = 10$ μs).

EXERCISES

5.26 A car loan in the amount of \$25,000 is to be paid off in 5 years with an annual interest rate of 8%. Use the Simulink loan simulation to find
  (a) The monthly installment
  (b) The unpaid balance after the 30th payment
  (c) The principal portion of the 12th payment
  (d) The total interest paid over the life of the loan
  (e) The time required for the unpaid balance to equal \$12,500

5.27 A prospective home buyer is considering purchasing a \$200,000 house with a 10% down payment and financing the balance over 30 years. He is able to afford monthly payments of \$1,450.
  (a) What is the maximum interest rate per annum on the mortgage for which the house is affordable?
  (b) Repeat part (a) for a 15 year mortgage.

5.28 A college savings account is created on January 1, 2000 with a deposit of \$1000. The account earns 4% per year. End-of-month deposits in the amount of \$150 are made for a period of 18 years with the last deposit scheduled for December 31, 2017.
  (a) Write a difference equation for $y(k)$, $k = 1, 2, 3, \ldots$ the account balance after the kth deposit. Note that $y(0) = 1000$ and $u(k) = 150$, $k = 1, 2, 3, \ldots, 216$.
  (b) Find the account balance after the last deposit.

5.29 Numerical differentiation is a procedure for approximating the derivatives of a mathematical function based on sampled values from it. The following backward difference formulas estimate the first three derivatives of a signal $f(t)$ at time t_i:

$$f'(t_i) = \frac{1}{2T}[f(t_i - 2T) - 4f(t_i - T) + 3f(t_i)]$$

$$f''(t_i) = \frac{1}{T^2}[-f(t_i - 3T) + 4f(t_i - 2T) - 5f(t_i - T) + 2f(t_i)]$$

$$f'''(t_i) = \frac{1}{2T^3}[3f(t_i - 4T) - 14f(t_i - 3T) + 24f(t_i - 2T) - 18f(t_i - T) + 5f(t_i)]$$

Develop a Simulink program to approximate
(a) $f'(t_i), t_i = 2T, 3T, 4T, \ldots$ given $f(0), f(T)$
(b) $f''(t_i), t_i = 3T, 4T, 5T, \ldots$ given $f(0), f(T), f(2T)$
(c) $f'''(t_i), t_i = 4T, 5T, 6T, \ldots$ given $f(0), f(T), f(2T), f(3T)$
where the function $f(t)$ is obtained from an "Fcn" block
as shown in Figure E5.29:

FIGURE E5.29

(d) Run the simulations for approximating the derivatives for the following cases:

$$f(t) = A \sin \omega t, \quad A = 1, \quad \omega = 2\pi, \quad T = 0.01 \frac{2\pi}{\omega} \tag{i}$$

$$f(t) = Ke^{-t/\tau}, \quad K = 10, \quad \tau = 1, \quad T = 0.01\tau \tag{ii}$$

$$f(t) = K\left[1 - \frac{\omega_n}{\omega_d}e^{-\zeta\omega_n t}\sin(\omega_d t + \varphi)\right], \quad K = 1, \quad \zeta = 0.25, \quad \omega_n = 4, \quad T = 0.01\left(\frac{1}{\zeta\omega_n}\right) \tag{iii}$$

where
$$\omega_d = \sqrt{1 - \zeta^2}\,\omega_n$$
$$\varphi = \tan^{-1}(\omega_d/\zeta\omega_n)$$

Run the simulations for a period of time sufficient to include two cycles of the sine function and the transient periods of the second and third functions.

(e) Compare the approximate and exact values of $f'(t_i), f''(t_i),$ and $f'''(t_i)$ for each function at 10 equally spaced points.

5.30 A second-order system is governed by the differential equation

$$\ddot{y}(t) + 2\zeta\omega_n\dot{y}(t) + \omega_n^2 y(t) = K\omega_n^2 u(t) + b_1\dot{u}(t) + b_2\ddot{u}(t)$$

(a) Draw a simulation diagram of the system and label the states x_1 and x_2.
(b) Draw a Simulink diagram using "Forward Euler" and "Trapezoidal" discrete-time integrators to calculate $x_{2,A}(n)$ and $x_{1,A}(n)$, respectively.
(c) Use the values $\zeta = 0.5, \omega_n = 4, K = 2, b_1 = 0,$ and $b_2 = 1$ to simulate $y(t)$ in response to the input

$$u(t) = \begin{cases} 9 - t^2, & 0 \le t < 3 \\ 0, & t \ge 3 \end{cases}$$

with $y(0) = \dot{y}(0) = 0$.
(d) Find the analytical solution for $y(t)$. Use an "Fcn" block in the Simulink diagram with input t and output the analytical solution for $y(t)$. Compare the simulated and analytical solutions.

5.31 For the notch filter given by Equations 5.89 through 5.92, find an expression for the constant b in Equation 5.92 if the DC gain of the filter is one.

5.32 An electronic sensor is used to measure ambient temperature $T_{amb}(t)$, which varies over a 24 h period in sinusoidal fashion about 45°F with amplitude of 15°F (Figure E5.32a). The sensor output $T_s(t)$ is corrupted with an additive white noise component (power $= 0.01$). A Simulink diagram for generating $T_{amb}(t)$ and $T_s(t)$ is shown in Figure E5.32b, along with time histories of each signal.

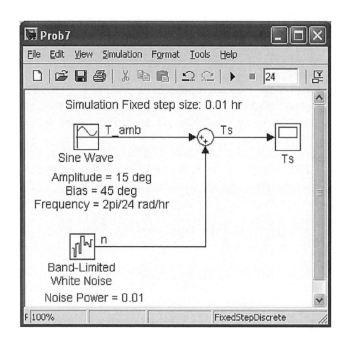

FIGURE E5.32a

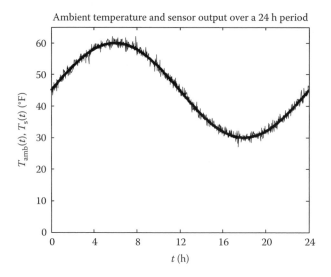

FIGURE E5.32b

A simple low-pass digital filter is required to smooth the output signal without seriously degrading the signal component. The first-order filter difference equation is

$$T_f(k+1) = (1 - \alpha)T_s(k+1) + \alpha T_f(k), \quad k = 0, 1, 2, \ldots$$

where
   $T_s(k)$ is the sampled output of the sensor
   $T_f(k)$ is the filter output

Note that the initial filter output $T_f(0) = T_{\text{amb}}(0) = 45°\text{F}$.

The filter sample time is $T = 0.01$ h. The parameter α is related to the bandwidth of the filter ω_0 by

$$\alpha = 2 - \cos\omega_0 T - \sqrt{(3 - \cos\omega_0 T)(1 - \cos\omega_0 T)}$$

Choose ω_0 as the frequency of the ambient temperature (in rad/h), and simulate the filter's response over a 24 h period. Plot $T_{amb}(t)$, $T_s(t)$, and $T_f(k)$ on the same graph.

5.33 Differentiation of noisy analog signals in continuous-time systems is often accomplished by first removing the high-frequency components. To illustrate, suppose the signal $x(t) = s(t) + n(t)$ where $s(t) = 2t$, $t \geq 0$, and $n(t) = 0.25 \sin 120\pi t$ is fed to a differentiator as shown in Figure E5.33. A series of low-pass filters with $H(z) = (1 - a)z/(z - a)$ like the ones shown in Figure E5.33 are inserted between the signal $x(t)$ and another differentiator. Vary the number of low-pass filters and the constant "a" and compare

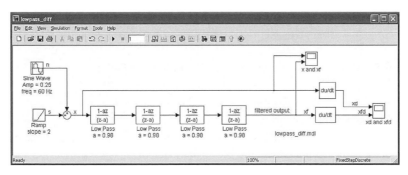

FIGURE E5.33

(a) The noisy analog signal "x" and the filtered signal "xf"
(b) The outputs "xd" and "xfd" of the two differentiators

5.8   MATLAB® AND SIMULINK® INTERFACE

While it is possible to work with signals and systems exclusively within the MATLAB environment, it is far more efficient to utilize Simulink and the MATLAB toolboxes to solve problems in specific disciplines. Sharing of data between MATLAB and Simulink is a seamless process, enabling MATLAB's extensive capabilities in data analysis and visualization to be utilized.

The following example illustrates how to effectively exploit the MATLAB and Simulink interface. It deals with the Fourier Series and its application to frequency response of linear systems.

A periodic signal $u(t)$ is shown in Figure 5.98. Equation 5.99 describes the signal over one period from $-T/2$ to $T/2$. It is periodic as a result of $u(t + T) = u(T)$, $-\infty < t < \infty$.

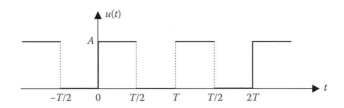

FIGURE 5.98   Periodic signal $u(t)$.

$$u(t) = \begin{cases} 0, & -\dfrac{T}{2} \le t < 0 \\[2mm] A, & 0 \le t < \dfrac{T}{2} \end{cases} \tag{5.99}$$

Its Fourier Series expansion (O'Neil 1983) is

$$u(t) = a_0 + \sum_{n=1,3,5,\dots}^{\infty} a_n \sin n\omega_0 t \tag{5.100}$$

where

$\omega_0 = 2\pi/T$ is called the fundamental frequency

$u_n(t) = a_n \sin n\omega_0 t$, $n = 1, 3, 5, \dots$ is the nth harmonic

the Fourier coefficients are

$$a_0 = \frac{A}{2}, \quad a_n = \frac{2}{n\pi}, \quad n = 1, 3, 5, \dots \tag{5.101}$$

Suppose $u(t)$ is the input to a second-order system with transfer function

$$G(s) = \frac{Y(s)}{U(s)} = \frac{\omega_n^2}{s^2 + 2\zeta\omega_n s + \omega_n^2} \tag{5.102}$$

By the principle of superposition, $y(t)$ is equal to the sum of the second-order system response to the constant a_0 and the responses to the harmonic components $u_n(t)$, $n = 1, 3, 5, \dots$ Fourier coefficients of the truncated series $a_0 + \sum_{n=1,3,5,\dots}^{19} a_n \sin n\omega_0 t$ are evaluated in the M-file "*Chap5_Fourier_Series.m*," a portion of which is listed as follows:

```
% MATLAB Script File Chap5_Fourier_Series.m
% Fourier Series of periodic function u(t)
% f(t) = A, 0 <= t < T/2
% = 0, T/2 <= t < T
n=19; % order of truncated Fourier Series of u(t)
k=1:2:n; % harmonics of u(t)
A=10; % amplitude of u(t)
a=2*A./(k.*pi); % Fourier Series coefficients a(k), k=1,3,5,...,n
a0=0.5*A; % ave value of u(t)
T=0.1; % period of u(t)
w0=2*pi/T; % input frequency
wh=k*w0; % harmonic frequencies
wr=wh(5); % Set resonant frequency equal to freq of 9th harmonic
wn=1.01*wr; % calculate natural frequency
zeta=sqrt((1-(wr/wn)^2)/2); % calculate damping ratio
w=linspace(0,1500,500); % range of freq for |G(jw)| plot
s=j*w; % complex freqs
magG_w=(wn.^2)./abs(s.^2+2*zeta*wn*s+wn^2); % |G(jw)|
plot(w,magG_w)
hold on, s=j*wh;
magG_wh=(wn.^2)./abs(s.^2+2*zeta*wn*s+wn^2); % |G(jwh)|
plot(wh,magG_wh,''.'',''MarkerSize'',12)
num=[wn^2]; % numerator of G(s)
denom=[1 2*zeta*wn wn^2]; % denominator of G(s)
```

```
SYS = TF (num, denom); % transfer function of G(s)
Figure, bode(SYS,{50, 1500})
[MAG, PHASE, wh] = BODE (SYS, wh); % Evaluate |G(jwh)| and Angle(G(jwh))
sim(''resonance'') % call Simulink model ''resonance.mdl''
subplot(4, 2, 1); plot(t, harmonics (:,1))
subplot (4, 2, 3); plot(t, harmonics (:,2))
```

The truncated series expansion of $u(t)$ is evaluated and compared to the input $u(t)$ as part of a Simulink simulation shown in Figure 5.99. The Fourier coefficients and harmonic frequencies are used to set the parameters in the "Sine Wave" blocks. A comparison of the signal $u(t)$ and the truncated Fourier Series is shown in Figure 5.100.

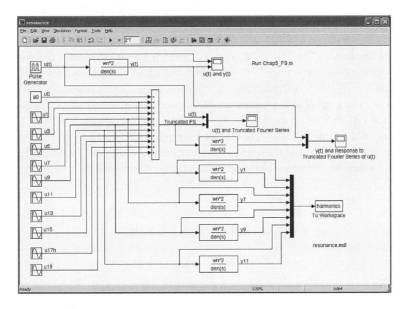

FIGURE 5.99 Simulink® diagram for finding the response to $u(t)$ and truncated Fourier Series of $u(t)$.

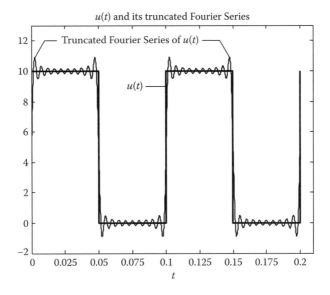

FIGURE 5.100 Periodic signal $u(t)$ ($t = 0.1$ s) and 19th-order truncated series.

The resonant frequency ω_r of the second-order system with transfer function $G(s)$ is set equal to the frequency of the ninth harmonic $9\omega_0 = 9(2\pi/T) = 565.49$ rad/s. The natural frequency ω_n is chosen slightly higher, that is, $\omega_n = 1.01\omega_r$ producing a lightly damped system with damping ratio of approximately 0.1 calculated from (Ogata 1988)

$$\zeta = \sqrt{\frac{1 - (\omega_r/\omega_n)^2}{2}} \tag{5.103}$$

The second-order system response to $u(t)$ and its response to the truncated Fourier Series representation of $u(t)$ are shown in Figure 5.101.

Clearly, enough harmonics of $u(t)$ have been retained in the truncated Fourier Series to accurately predict the response of the second-order system under consideration.

Next, we discuss how MATLAB and Simulink can be used effectively to demonstrate the phenomenon of resonance. The M-file "*Chap5_Fourier_Series.m*" evaluates the magnitude function $|G(j\omega)|$ over the frequency range $0 \leq \omega \leq 1500$ rad/s and plots the results with the harmonic frequencies shown in Figure 5.102. Note that the resonant frequency ω_r where the peak amplitude of $|G(j\omega)|$ occurs is in fact equal to the frequency of the ninth harmonic. A similar finding is possible using the control system toolbox to specify the transfer function and draw a Bode plot or merely compute the magnitude function with "MAG" at selected frequencies and plot the results.

The Fourier coefficients of the truncated series expansion of $u(t)$ are shown in Table 5.3. Also listed are the frequency response characteristics of the second-order system at the harmonic frequencies.

The peak magnitude at the resonant frequency is

$$\underset{\omega \geq 0}{\text{Max}} |G(j\omega)| = |G(j\omega_r)| = 5.0624 \tag{5.104}$$

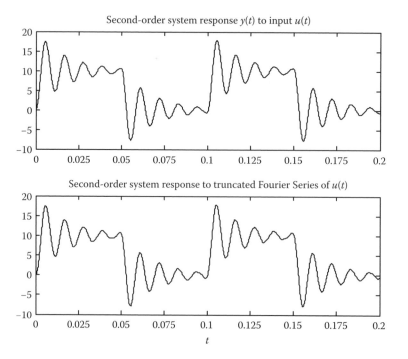

FIGURE 5.101 Response of a second-order system to $u(t)$ and its truncated Fourier Series.

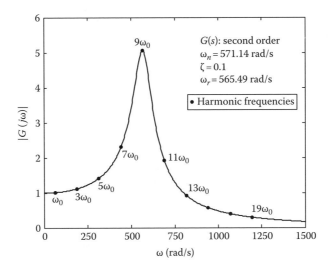

FIGURE 5.102 Magnitude function of a second-order system.

TABLE 5.3
**Fourier Coefficients and Magnitude Function
at Selected Frequencies**

n	$n\omega_0$ (rad/s)	a_n	$|G(jn\omega_0)|$	$\angle G(jn\omega_0)$ (rad)
0	0	5	1	0
1	$\omega_0 = 62.8$	6.3662	1.0120	−0.0221
3	$3\omega_0 = 188.5$	2.1221	1.1192	−0.0734
5	$5\omega_0 = 314.2$	1.2732	1.4166	−0.1553
7	$7\omega_0 = 439.8$	0.9095	2.3002	−0.3593
9	$9\omega_0 = 565.5$	0.7074	5.0624	−1.4709
11	$11\omega_0 = 691.2$	0.5787	1.9126	−2.6642
13	$13\omega_0 = 816.8$	0.4897	0.9232	−2.8764
15	$15\omega_0 = 942.5$	0.4244	0.5702	−2.9537
17	$17\omega_0 = 1068.1$	0.3745	0.3960	−2.9940
19	$19\omega_0 = 1193.8$	0.3351	0.2946	−3.0190

The harmonic components $u_n(t)$, $n = 1, 7, 9, 11$ in the truncated series expansion of $u(t)$ in Figure 5.98 are given by

$$u_1(t) = a_1 \sin \omega_0 t = 6.3662 \sin 62.8t \tag{5.105}$$

$$u_7(t) = a_7 \sin 7\omega_0 t = 0.9095 \sin 439.8t \tag{5.106}$$

$$u_9(t) = a_9 \sin 9\omega_0 t = 0.7074 \sin 565.5t \tag{5.107}$$

$$u_{11}(t) = a_{11} \sin 11\omega_0 t = 0.5787 \sin 691.2t \tag{5.108}$$

The second-order system response to the above components is

$$y_1(t) = |G(j\omega_0)|a_1 \sin[\omega_0 t + \angle G(j\omega_0)] \tag{5.109}$$

$$= 1.0120(6.3662)\sin(62.8t - 0.0221) \tag{5.110}$$

$$= 6.4426\sin(62.8t - 0.0221) \tag{5.111}$$

$$y_7(t) = |G(j7\omega_0)|a_7 \sin[7\omega_0 t + \angle G(j7\omega_0)] \tag{5.112}$$

$$= 2.3002(0.9095)\sin(439.8t - 0.3593) \tag{5.113}$$

$$= 2.0920\sin(439.8t - 0.3593) \tag{5.114}$$

$$y_9(t) = |G(j9\omega_0)|a_9 \sin[9\omega_0 t + \angle G(j9\omega_0)] \tag{5.115}$$

$$= 5.0624(0.7074)\sin(565.5t - 1.4709) \tag{5.116}$$

$$= 3.5811\sin(565.5t - 1.4709) \tag{5.117}$$

$$y_{11}(t) = |G(j11\omega_0)|a_{11} \sin[11\omega_0 t + \angle G(j11\omega_0)] \tag{5.118}$$

$$= 1.9126(0.5787)\sin(691.2t - 2.6642) \tag{5.119}$$

$$= 1.1068\sin(691.2t - 2.6642) \tag{5.120}$$

The Simulink data for the signals in Equations 5.105 through 5.108, 5.111, 5.114, 5.117, and 5.120 are returned to the MATLAB Workspace (see Figure 5.99) for use by M-file "*Chap5_Fourier_Series.m*" in preparing the graph shown in Figure 5.103. The input and output components can be identified by referring to the amplitudes in Equations 5.105 through 5.120.

Note that $y_1(t)$ has a larger amplitude than $y_9(t)$, the system response to the harmonic component at the resonant frequency. This results from $|G(j\omega_0)|a_1 = 6.4426$ being larger than $|G(j9\omega_0)|a_9 = 3.5811$.

As a final comment, the line in "*Chap5_Fourier_Series.m*"

```
sim('resonance') % call Simulink model 'resonance.mdl'
```

enables the MATLAB script file "*Chap5_Fourier_Series.m*" to initiate execution of the Simulink model file "*resonance.mdl*." With additional parameters in the "sim" command, the user has control of many of the settings entered in the Simulink "Simulation Parameters" dialog box.

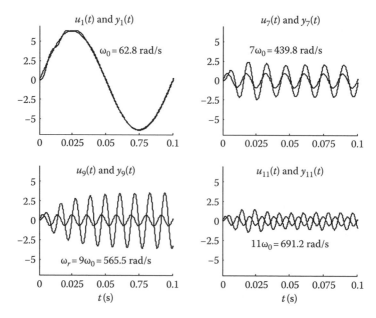

FIGURE 5.103 Several harmonic components of $u(t)$ and response of second-order system.

EXERCISES

5.34 A spring mass system described by $m\ddot{y} + ky = F$ is subject to an external periodic force $F(t)$ shown in Figure E5.34:

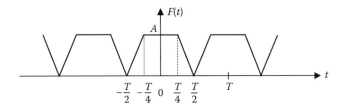

FIGURE E5.34

(a) The Fourier series expansion of $F(t)$ is

$$f(t) = \frac{a_0}{2} + \sum_{n=1}^{\infty} a_n \cos n\omega_0 t, \quad \omega_0 = \frac{2\pi}{T}$$

$$a_0 = \frac{2}{T} \int_{-T/2}^{T/2} F(t)dt, \quad a_n = \frac{2}{T} \int_{-T/2}^{T/2} F(t)\cos\left(\frac{2n\pi t}{T}\right)dt, \quad n = 1, 2, 3, \ldots$$

Find expressions for the Fourier coefficients a_n, $n = 0, 1, 2, 3, \ldots$.

(b) The mass $m = 1$ slug and the natural frequency of the system is $\omega_n = 25$ rad/s. The period of the forcing function T is related to the natural frequency according to $T = 2N\pi/c\omega_n$ where N is a positive integer and c is a constant. Write a MATLAB script file that reads values of c and N and computes the period T and Fourier coefficients a_n, $n = 0, 1, 2, 3, \ldots, 3N$.

For $A = 1$, $c = 1$, and $N = 5$, use the MATLAB script file to

(c) Plot on the same graph $F(t)$ and the truncated Fourier Series

$$f_{FS}(t) = \frac{a_0}{2} + \sum_{n=1}^{3N} a_n \cos n\omega_0 t$$

for $0 \le t \le 3T$. Comment on the results.

(d) Prepare a Simulink diagram for simulating the response of the system with zero initial conditions. Call the simulation from the script file using the same values for c and N. Return the values of $\{t, y(t)\}$ to the MATLAB Workspace and plot the response. (The simulation should run long enough to recognize the steady-state response.) Comment on the results.

5.35 The dynamic interaction of rabbit and fox populations in a forest is under investigation. The predator–prey ecosystem is illustrated in block diagram form in Figure E5.35a:

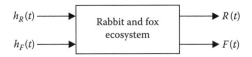

FIGURE E5.35a

$R(t) =$ Population of rabbits after "t" weeks
$F(t) =$ Population of foxes after "t" weeks
$h_R(t) =$ Rate of rabbit hunting (rabbits/week)
$h_F(t) =$ Rate of fox hunting (fox/week)

A Simulink diagram of the system is shown in Figure E5.35b:

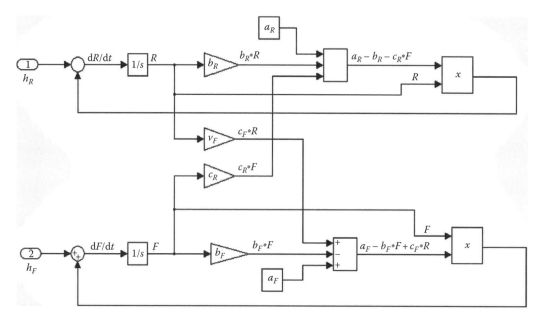

FIGURE E5.35b

a_R, b_R, $c_R =$ constant parameters defining the growth rate of rabbits
a_F, b_F, $c_F =$ constant parameters defining the growth rate of foxes

(a) Find the mathematical model governing the system dynamics.
(b) Find the nontrivial equilibrium points (R_e, F_e) when $h_R(t) = 0$, $t \geq 0$ and $h_F(t) = 0$, $t \geq 0$.
(c) Write a MATLAB script file to set the following baseline parameter values:

$$a_R = 0.05 \; \frac{\text{rabbits/week}}{\text{rabbit}^2}, \quad b_R = 5 \times 10^{-7} \; \frac{\text{rabbits/week}}{\text{rabbit}^3}, \quad c_R = 1.25 \times 10^{-5} \; \frac{\text{rabbits/week}}{\text{rabbit}^2 - \text{fox}}$$

$$a_F = 0.04 \; \frac{\text{foxes/week}}{\text{fox}^2}, \quad b_F = 2 \times 10^{-5} \; \frac{\text{foxes/week}}{\text{fox}^3}, \quad c_F = 8 \times 10^{-7} \; \frac{\text{foxes/week}}{\text{foxes}^2 - \text{rabbit}}$$

$$h_R(t) = 0, t \geq 0 \quad h_F(t) = 0, t \geq 0$$

$$R(0) = 50,000 \quad F(0) = 1,000$$

(d) Run the simulation and plot $R(t)$ and $F(t)$ vs. t until the system reaches equilibrium.
(e) Obtain a solution trajectory R vs. F. Place a vertical line at $F = F_e$ and a horizontal line at $R = R_e$. This will allow you to verify by inspection if the solution trajectory approaches the theoretical equilibrium.

(f) Investigate the effect of changes in c_R, a parameter that measures the interaction between foxes and rabbits. Plot families of appropriate responses corresponding to 0%–50% change in c_R.

(g) Establish a policy for hunting rabbits that makes the number of foxes equal to approximately 2500 at equilibrium.

(h) Establish a policy for hunting foxes that makes the number of rabbits equal to approximately 35,000 at equilibrium.

5.36 The tank shown in Figure E5.36 has a brine solution flowing into it. The solution is stirred well enough, so that the concentration of salt in the tank is uniform.

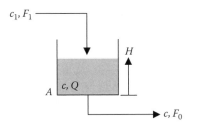

c_1: Brine concentration (lbs/gal)
F_1: Brine flow (gal/min)
c: Salt concentration in tank (lbs/gal)
Q: Quantity of salt in tank (lbs)
H: Liquid level in tank (ft)
V: Volume of liquid in tank (gal)
F_0: Flow rate from tank (gal/min)

FIGURE E5.36

The mathematical model consists of the following equations:

$$\frac{dQ}{dt} = c_1 F_1 - c F_0$$

$$c = \frac{Q}{V}, \quad V = AH$$

$$A\frac{dH}{dt} + F_0 = F_1, \quad F_0 = \alpha H^{1/2}$$

The system baseline parameter values are $A = 20$ ft^2 and $\alpha = 6$ gal/min per ft$^{1/2}$.

Note: 1 ft^3 of water is roughly 8.3 gal.

(a) Draw a simulation diagram of the system.

(b) Choose the state variables as $x_1 = Q$ and $x_2 = H$ and the outputs $y_1 = c$, $y_2 = Q$, and $y_3 = V$. Write the state equations in the form

$$\dot{x}_1 = f_1(x_1, x_2, c_1, F_1), \quad y_1 = g_1(x_1, x_2, c_1, F_1)$$

$$\dot{x}_2 = f_2(x_1, x_2, c_1, F_1), \quad y_2 = g_2(x_1, x_2, c_1, F_1)$$

$$y_3 = g_3(x_1, x_2, c_1, F_1)$$

(c) Find expressions for the steady-state values of the states $x_1(\infty)$ and $x_2(\infty)$ and the outputs $y_1(\infty)$, $y_2(\infty)$, and $y_3(\infty)$ assuming c_1 and F_1 are constant.

(d) The tank is initially filled with 100 gal of water (no salt). Brine starts flowing into the tank at the rate of 12 gal/min. The salt concentration of the brine is 0.25 lb/gal. Both the flow rate and salt concentration of the brine flow remain constant. Using explicit Euler integration, find the discrete-time state equations

$$\underline{x}_A(n+1) = \underline{f}[(\underline{x}_A(n), \underline{u}(n)]$$

$$\underline{y}_A(n) = g[(\underline{x}_A(n), \underline{u}(n)]$$

used to obtain an approximate solution for the continuous-time states and outputs.

(e) Solve the discrete-time state equations recursively for the discrete-time states $x_{1,A}(n)$ and $x_{2,A}(n)$ and the outputs $y_{1,A}(n)$, $y_{2,A}(n)$, and $y_{3,A}(n)$. Graph the transient responses. Comment on the value of T used for the numerical integrator.

(f) Compare the steady-state results obtained in part (e) with the predicted values from part (c). Comment on the results.

(g) Use Simulink to verify the responses obtained in part (e).

5.9 HYBRID SYSTEMS: CONTINUOUS- AND DISCRETE-TIME COMPONENTS

Hybrid systems consist of continuous- and discrete-time components and the interfaces bridging the gap between them. A good example is a digital controller (microprocessor or general-purpose digital computer) determining discrete-time input(s) to a continuous-time process.

Figure 5.104 shows a digital controller used to regulate the temperature inside a chamber. The DC voltage input to the heater $v(t)$ is determined by a digital control algorithm represented by discrete-time transfer function $D(z)$. The heat input to the chamber is assumed proportional to the square of the heater voltage. A temperature sensor with gain K_S produces a voltage signal $v_S(t)$ for comparison with a reference voltage $v_R(t)$. The reference voltage is based on the commanded temperature $T_R(t)$ (not shown in Figure 5.104).

The error signal $e(t)$ is sampled every T s in an analog-to-digital (A/D) converter. The A/D converter functions as an interface between the continuous-time inputs (sensor and reference voltage) and the discrete-time digital controller. The error signal $e(k)$ is processed by the digital controller, resulting in an output $v(k)$, the intended voltage to the heater. A digital-to-analog (D/A) converter, operating synchronously with the A/D, produces the voltage. Internal circuitry in the D/A latches the discrete-time input for the duration of the sampling period, resulting in a stepwise constant voltage $v(t)$ applied to the heater. The D/A converter serves as an interface between the discrete-time and continuous-time components. It is modeled by a zero-order hold (ZOH) in Figure 5.104.

The digital controller implements a linear difference equation for $v(k)$ in terms of past values $v(k-1)$, $v(k-2)$, ..., $v(k-n)$ as well as present and past values $e(k)$, $e(k-1)$, ..., $e(k-p)$. Digital controllers are often synthesized by approximating continuous-time controllers. For example, the transfer function of a continuous-time proportional-integral-derivative (PID) controller is

$$\frac{V(s)}{E(s)} = K_P + \frac{K_I}{s} + K_D s \tag{5.121}$$

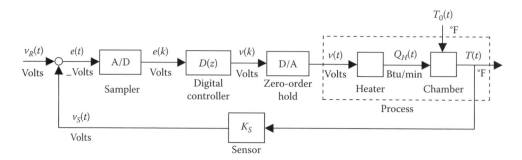

FIGURE 5.104 Digital control of chamber temperature.

Approximating the integral by trapezoidal integration and the derivative by a backward difference equation leads to the equivalent digital transfer function (Jacquot 1981)

$$D(z) = \frac{c_1 z^2 + c_2 z + c_3}{z^2 - z} \tag{5.122}$$

where

$$c_1 = K_P + \frac{K_I T}{2} + \frac{K_D}{T}, \quad c_2 = \frac{K_I T}{2} - K_P - \frac{2K_D}{T}, \quad c_3 = \frac{K_D}{T} \tag{5.123}$$

The continuous-time process model is (see Section 2.7)

$$C \frac{dT}{dt} + \frac{1}{R} T = \frac{1}{R} T_0 + Q_H \tag{5.124}$$

where
C is the thermal capacitance of the interior space (objects and volume of air assumed to be at the same temperature)
R is the effective thermal resistance of the material separating the inside and outside of the chamber

The heater output is

$$Q_H(t) = \frac{v^2(t)}{R_e} \tag{5.125}$$

where R_e is the electrical resistance of the heater coil. A temperature sensor produces a voltage proportional to the interior temperature

$$v_s(t) = K_s T(t) \tag{5.126}$$

A Simulink diagram of the system is shown in Figure 5.105. The commanded reference temperature T_Y is converted to a reference voltage input with a Units Converter, that is, a "Gain" block with

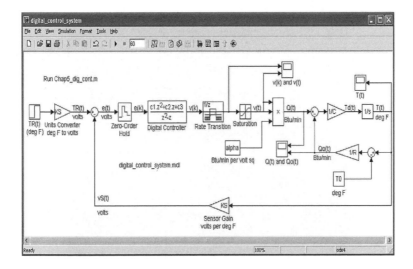

FIGURE 5.105 Simulink® diagram of a digital control system for chamber temperature.

parameter equal to K_S. A "Saturation" block limits the actual voltage $v(t)$ to the heater. The "Zero-Order Hold" and "Rate Transition" blocks are included to resolve timing issues related to the faster simulation execution rate (based on the integration step) and the slower sampling rate of the digital controller.

System parameters were set in the MATLAB script file "*Chap5_dig_cont.m*."
Chamber: $R = 0.175°$F/Btu/min, $C = 50$ Btu/°F
Sensor: $K_S = 0.25$ V/°F
Controller: $K_P = 2$, $K_I = 2$, $K_D = 0.25$
Heater: $R_e = 1.25$ Ω, $v_{max} = 100$ V
Inputs: $T_R(t) = 125°$F, $t \geq 5$, $T_0(t) = 75°$F, $t \geq 0$
Timing: $T = 0.02$ min (sample time), $\Delta t = 0.002$ min (integration step size)

Figure 5.106 shows the voltage $v(k)$ computed from the digital control algorithm and the actual voltage $v(t)$ to the heater. Note the initial spike due to the presence of the proportional control and derivative action in the controller. The initial continuous-time voltage to the heater is "maxed out" at a 100 V, the upper limit of the saturation block.

Figure 5.107 shows the heat flows to and from the chamber. Note the constant heat flow to the chamber when the heater is at saturation. At the end of the transient response period, the heat flows have equalized, and the chamber interior is in thermal equilibrium with its surroundings.

Figure 5.108 is a graph of the chamber temperature increasing from its initial value of 75°F to the commanded value of 125°F. The step response is typical of a slightly underdamped second-order system with a settling time between 50 and 60 min.

The thermal time constant of the chamber is

$$\tau = RC = 0.175 \; \frac{°\text{F}}{\text{Btu}/\text{min}} \times 50 \; \frac{\text{Btu}}{°\text{F}} = 8.75 \; \text{min}$$

The sampling time $T = 0.02$ min of the A/D converter is chosen several orders of magnitude less than the process time constant in order to capture the transient behavior of the chamber temperature. A more precise way of determining the sampling rate will be discussed in a subsequent chapter.

The control system is nonlinear as a consequence of Equation 5.125. Laplace transforms cannot be used to find an analytical solution for the system variables. Simulation is the only viable approach to examining the system dynamics.

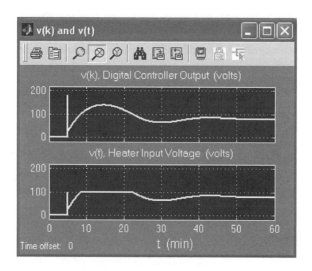

FIGURE 5.106 Digital controller output $v(k)$ and heater input $v(t)$.

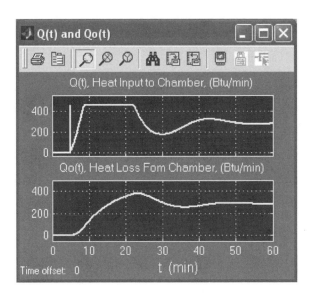

FIGURE 5.107   Heater input $Q_H(t)$ and heat loss $Q_0(t)$ from chamber to surroundings.

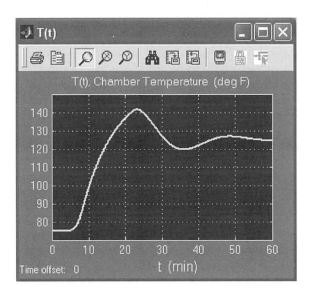

FIGURE 5.108   Chamber temperature response $T(t)$ to reference input $T_R(t) = 125°F$, $t \geq 5$.

EXERCISES

In Exercises 5.37 through 5.40, use baseline values for the system parameters found in "*Chap5_dig_cont.m*" unless otherwise stated.

5.37 Plot the simulated chamber temperature responses (on the same graph) corresponding to a range of sampling intervals from 0.01 to 0.25 min. Comment on the results.

5.38 The maximum output from the chamber heater in watts is $(QH)_{max} = v_{max}^2 / R_e$.
(a) Find T_{max}, the maximum temperature achievable in the chamber.
(b) *Note:* 1 kW = 56.896 Btu/min.

(c) Simulate the chamber temperature when the commanded temperature is set to
   (i) T_{max}
   (ii) 10% higher than T_{max}
   (iii) 25% higher than T_{max}

5.39 Simulate the temperature response of the control system with proportional control only, that is, $K_I = 0$ and $K_D = 0$. The set point temperature is 200°F. Vary K_P from 1 to 10 and plot the responses on the same graph.

5.40 Suppose the chamber temperature has been constant at $T_R = 125°F$ for some time. Simulate the chamber temperature $T(t)$ when
   (a) the heater is turned off
   (b) the reference temperature is set to 150°F

5.41 Simulate the chamber temperature using a digital controller obtained by approximating the continuous controller in Equation 5.121 using Tustin's method (trapezoidal integration). Compare the results with those shown in Figures 5.106 through 5.108.

5.10 MONTE CARLO SIMULATION

The dynamic systems, which have been simulated to this point, were all deterministic, that is, there have been no random components associated with either the system's parameters or inputs. In reality, knowledge of the values of a system's parameters is inexact for a number of reasons. Precise measurement or observation of the parameters may be difficult, or it is possible that the numerical values drift over time as the components age. Quantitative descriptions of the input signals a priori may be probabilistic in nature. The existence of random inputs and uncertain system parameter values leads to stochastic differential equation models with solutions in the form of stochastic processes.

An alternate approach is based on the technique of Monte Carlo simulation. An empirical rather than analytical method, its name stems from the random nature of gambling and associated probabilities. The underlying premise in Monte Carlo simulation is that by repeatedly sampling from known probability distributions, the probabilities of events or probability distributions of functions of a random variable(s) can be approximated. Sampling from the probability distribution of a random variable (or random variables) to generate random deviates is substituted for the process of making observations of the random variable(s) from the real world or physical process itself. In other words, random samples obtained by actual measurements or observations of a random variable are replaced by simulated random samples based on random number generators and known probability distributions.

Consider a simple mechanical system with mass M, spring constant K, and damping coefficient B described by

$$M\ddot{y} + B\dot{y} + Ky = f(t) \tag{5.127}$$

where
   y is the displacement of the mass from equilibrium
   $f(t)$ is a force acting on the mass

Suppose M, B, and K are continuous random variables with known probability density functions (pdf's) $f_M(u)$, $f_B(u)$, and $f_K(u)$, respectively. The damping ratio ζ

$$\zeta = \zeta(M, B, K) = \frac{B}{2\sqrt{MK}} \tag{5.128}$$

is a new random variable, which, along with the natural frequency, characterizes the system's natural dynamics. Finding the theoretical probability distribution of ζ, that is, its pdf $f_\zeta(u)$, is a formidable task despite the relative simplicity of Equation 5.128. The following example demonstrates a Monte Carlo simulation to obtain what we shall refer to as an empirical probability density function denoted $\hat{f}_\zeta(u)$ to distinguish it from the true pdf $f_\zeta(u)$. The empirical pdf can be used to approximate probability distributions of other random variables functionally related to the damping ratio such as the overshoot in the step response of underdamped second-order systems.

The parameters M, B, and K are each assumed to vary uniformly between specified limits. The pdf for random variable M is the uniform pdf, denoted $U(M_l, M_u)$ where M_l and M_u are the lower and upper limits of M, respectively. In mathematical terms, the pdf is given by

$$f_M(u) = \begin{cases} \dfrac{1}{M_u - M_l}, & M_l \le u \le M_u \\ 0, & \text{elsewhere} \end{cases} \tag{5.129}$$

Similar expressions apply for the pdfs of random variables B and K, that is,

$$f_B(u) = \begin{cases} \dfrac{1}{B_u - B_l}, & B_l \le u \le B_u \\ 0, & \text{elsewhere} \end{cases} \tag{5.130}$$

$$f_K(u) = \begin{cases} \dfrac{1}{K_u - K_l}, & K_l \le u \le K_u \\ 0, & \text{elsewhere} \end{cases} \tag{5.131}$$

A random variable, uniformly distributed between 0 and 1, also referred to as a random number, is generated by the MATLAB function "rand." To be more precise, the generated numbers are actually pseudo random numbers, which depend on the specific algorithm implemented for generation. A random number R_i uniformly distributed $U(0, 1)$ is transformed to a new random variable X_i with pdf $U(A, B)$ by

$$X_i = A + (B - A)R_i \tag{5.132}$$

The MATLAB M-file "Chap5_MonteCarlo_damping_ratio.m" generates 100,000 random vectors (M_i, B_i, K_i), $i = 1, 2, \ldots, 100,000$ using lower and upper limits $M_l = 0.9$, $M_u = 1.1$, $B_l = 1.75$, $B_u = 2.25$, $K_l = 3.8$, and $K_u = 4.2$. The corresponding 100,000 damping ratios ζ_i, $i = 1, 2, \ldots,$ 100,000 computed from Equation 5.128 are segregated into equal intervals of width 0.005, several of which are shown in Table 5.4.

TABLE 5.4
Monte Carlo Simulation Results for Damping Ratio

Interval $(\zeta_{i-1} \le \zeta \le \zeta_i)$	Center of Interval $\bar{\zeta}_i$	Frequency of Occurrence n_i	Normalized Frequency of Occurrence f_i
(0.3975, 0.4025)	0.4000	0	0
(0.4025, 0.4075)	0.4050	0	0
(0.4075, 0.4125)	0.4100	33	0.0660
(0.4875, 0.4925)	0.4900	4053	8.1060
(0.4925, 0.4975)	0.4950	4098	8.1960
(0.4975, 0.5025)	0.5000	4062	8.1240
(0.5025, 0.5075)	0.5050	4033	8.0660
(0.5075, 0.5125)	0.5100	3986	7.9720
(0.5875, 0.5925)	0.5900	341	0.6820
(0.5925, 0.5975)	0.5950	164	0.3280
(0.5975, 0.6025)	0.6000	59	0.1180

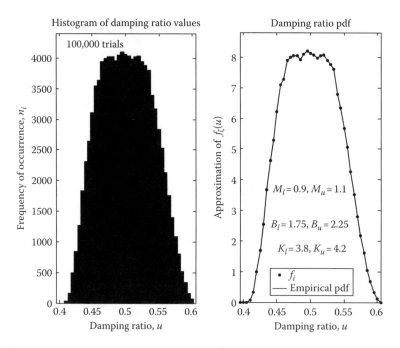

FIGURE 5.109 Histogram of ζ values and empirical pdf $\hat{f}_\zeta(u)$.

A histogram based on the first and third columns of the complete table is shown in the left graph of Figure 5.109. The empirical probability density function $\hat{f}_\zeta(u)$ is obtained by connecting the points $(\bar{\zeta}_i, n_i)$ and rescaling the ordinate values to f_i using Equation 5.133 to make the area under the resulting curve equal to 1.

$$f_i = \frac{n_i}{\text{Number of trials} \times \text{width of interval}} = \frac{n_i}{100,000 \times 0.005} = \frac{n_i}{500} \qquad (5.133)$$

Finally, a data point is added at $\bar{\zeta}_i = 0.6050$, $f_i = 0$ to assure the pdf $\hat{f}_\zeta(u)$ returns to zero at the upper tail. The result is shown in the right graph of Figure 5.109.

The theoretical probability of ζ falling in a certain interval is the area under $f_\zeta(u)$ for that interval. It is approximated by the area under the empirical pdf $\hat{f}_\zeta(u)$ for the same interval. For example, the estimate of $\Pr(0.45 \le \zeta \le 0.5)$ is computed in the M-file "*Chap5_MonteCarlo_damping_ratio.m*" to be 0.4105.

The empirical pdf $\hat{f}_\zeta(u)$ can be used to approximate probabilities involving various performance measures related to the damping ratio. For example, the percent overshoot in the unit step response and the peak amplitude of the frequency response

$$P.O. = f_1(\zeta) = 100e^{-\zeta\pi/\sqrt{1-\zeta^2}} \qquad (5.134)$$

$$M_{p\omega} = f_2(\zeta) = \frac{1}{2\zeta\sqrt{1-\zeta^2}} \qquad (5.135)$$

How shall we go about determining the empirical pdf $\hat{f}_{M_{p\omega}}(u)$? A table similar to Table 5.4 with equally spaced intervals of $M_{p\omega}$ and frequencies of occurrence is needed. The first step is to generate a random sample from a population with pdf $\hat{f}_\zeta(u)$. The random sample $(\zeta_1, \zeta_2, \dots, \zeta_n)$ and Equation 5.135 are used to generate the sample $[(M_{p\omega})_1, (M_{p\omega})_2, \dots, (M_{p\omega})_n]$ needed for the new table.

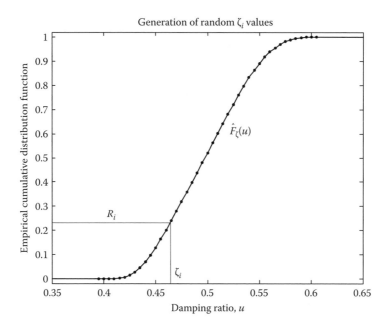

FIGURE 5.110   Illustration of method for generating ζ_i using cdf $\hat{F}_\zeta(u)$.

The random sample $(\zeta_1, \zeta_2, \ldots, \zeta_n)$ can be generated in several ways. One method relies on the use of random numbers $(R_1, R_2, \ldots, R_n)$ and the cumulative probability distribution function (cdf), $\hat{F}_\zeta(u)$ given by

$$\hat{F}_\zeta(u) = \int_{-\infty}^{u} \hat{f}_\zeta(x)dx, \quad -\infty < u < \infty \tag{5.136}$$

The empirical pdf $\hat{f}_\zeta(u)$ is numerically integrated in "*Chap5_MonteCarlo_damping_ratio.m*," resulting in $\hat{F}_\zeta(u)$ shown in Figure 5.110. A random damping ratio ζ_i is obtained as the solution to the equation

$$R_i = \hat{F}_\zeta(\zeta_i) \tag{5.137}$$

where R_i is a random number uniformly distributed between 0 and 1. That is, ζ_i is obtained from

$$\zeta_i = \hat{F}_\zeta^{-1}(R_i) \tag{5.138}$$

where $\hat{F}_\zeta^{-1}(R_i)$ is the inverse function. The Inverse Transformation Method (Gordon 1978) based on Equation 5.138 is illustrated in Figure 5.110.

After the random sample $[(M_{p\omega})_1, (M_{p\omega})_2, \ldots, (M_{p\omega})_n]$ is generated from $(\zeta_1, \zeta_2, \ldots, \zeta_n)$ and Equation 5.135, the empirical pdf $\hat{f}_{M_{p\omega}}(u)$ is obtained in the same way $\hat{f}_\zeta(u)$ was determined.

Suppose we have reason to estimate $\Pr[1.1 \leq M_{p\omega} \leq 1.3]$. The area under $\hat{f}_{M_{p\omega}}(u)$ between 1.1 and 1.3 is easily computed. Alternatively, we could numerically integrate $\hat{f}_{M_{p\omega}}(u)$ to obtain $\hat{F}_{M_{p\omega}}(u)$ and estimate the required probability from

$$\Pr[1.1 \leq M_{p\omega} \leq 1.3] = \hat{F}_{M_{p\omega}}(1.3) - \hat{F}_{M_{p\omega}}(1.1) \tag{5.139}$$

The details are left for an exercise at the end of the section.

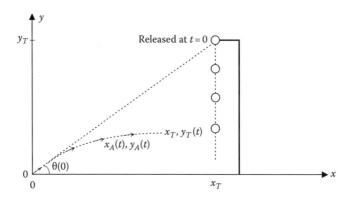

FIGURE 5.111   Arrow fired at a falling target.

5.10.1 Monte Carlo Simulation Requiring Solution of a Mathematical Model

In the previous example, the parameters M, B, and K of a second-order system were random variables with known probability density functions $f_M(u)$, $f_B(u)$, and $f_K(u)$. Additional random variables were introduced, namely, ζ and $M_{p\omega}$ in Equations 5.128 and 5.135. Using Monte Carlo simulation, the theoretical probability density functions $f_\zeta(u)$ and $f_{M_{p\omega}}(u)$ were approximated by empirical pdfs $\hat{f}_\zeta(u)$ and $\hat{f}_{M_{p\omega}}(u)$ without ever solving the differential equation model, Equation 5.127. In the next example, simulation of the mathematical model is an integral component of the overall Monte Carlo simulation study.

Suppose an archer is attempting to hit a falling target as shown in Figure 5.111.

Considering the aerodynamic drag forces on the arrow and target, the differential equations governing the motions of each are

$$m_A \ddot{x}_A = -\alpha_A \dot{x}_A^{n_A} \tag{5.140}$$

$$m_A \ddot{y}_A = -m_A g - \mathrm{sgn}(\dot{y}_A) \cdot \alpha_A |\dot{y}_A|^{n_A} \tag{5.141}$$

$$m_T \ddot{y}_T = -m_T g + \alpha_T |\dot{y}_T|^{n_T} \tag{5.142}$$

where
  m_A and m_T are masses of the arrow and target
  α_A, n_A, α_T and n_T are parameters for modeling arrow and target drag forces
  $x_A(t)$, $y_A(t)$, x_T and $y_T(t)$ are the x–y coordinates of the center of the arrow and center of the
    circular target

The sgn $(\dot{y}_A)$ function

$$\mathrm{sgn}(\dot{y}_A) = \frac{\dot{y}_A}{|(\dot{y}_A)|} = \begin{cases} 1, & \dot{y}_A > 0 \\ -1, & \dot{y}_A < 0 \end{cases} \tag{5.143}$$

is required to produce the proper sign on the arrow drag term for both upward and downward flights. Absolute values appear in Equations 5.141 and 5.142 to avoid raising negative speeds to noninteger powers.

The velocity of the arrow is uniquely determined by its speed $v_A(t)$ and direction $\theta(t)$, that is, the angle between the arrow and the horizontal axis. The speed is calculated from

$$v_A(t) = \left[\dot{x}_A^2(t) + \dot{y}_A^2(t) \right]^{1/2} \tag{5.144}$$

and the angle $\theta(t)$ is determined from

$$\tan \theta(t) = \frac{\dot{y}_A(t)}{\dot{x}_A(t)} \tag{5.145}$$

$$\Rightarrow \theta(t) = \tan^{-1}\left[\frac{\dot{y}_A(t)}{\dot{x}_A(t)}\right] \tag{5.146}$$

Baseline parameter values for the system are

$$m_A = 0.125/g \text{ slugs}, \quad \alpha_A = 4.75 \times 10^{-6}, \quad n_A = 1.85$$

$$m_T = 1/g \text{ slugs}, \quad \alpha_T = 3 \times 10^{-6}, \quad n_T = 2.3$$

$$x_A(0) = 0 \text{ ft}, \quad y_A(0) = 0 \text{ ft}, \quad v_A(0) = 80 \text{ ft/s}$$

$$x_T = 100 \text{ ft}, \quad y_T(0) = 150 \text{ ft}$$

The arrow is 2.4 ft in length and the target is 3 ft in diameter.

A Simulink diagram for simulating the trajectories of the arrow and target is shown in Figure 5.112. The Simulink model "*arrow.mdl*" is called from the M-file "*Chap5_MonteCarlo_arrow.m*."

Initially, the aerodynamic drag forces were zeroed out ($\alpha_A = \alpha_T = 0$) and the arrow's angle of departure $\theta(0)$ was set to the angle of the line of sight to the target because the archer knows, from Physics, that the target will be struck (in the absence of aerodynamic drag forces) under those conditions. The flight path of the arrow and its position at 0.25 s increments is shown in Figure 5.113. The target is captured at 0.25 s increments starting at 2 s and shown as well. Figure 5.113 confirms that the arrow and target appear to be at the same point after approximately 2.25 s.

Figure 5.114 is a close-up snapshot of the arrow and target when the arrow has traveled the horizontal distance to the target. The arrow strikes the target after 2.25 s have elapsed. The arrow and target coordinates at impact are ($x_A = 100$ ft, $y_A = 68.32$ ft) and ($x_T = 100$ ft, $y_T = 68.39$ ft), respectively. How do you explain the slight difference between y_A and y_T? Note that the simulation is halted when the arrow strikes the ground after 4.15 s (see "`display`" in Figure 5.112).

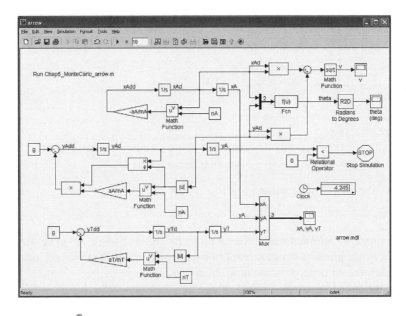

FIGURE 5.112 Simulink® diagram of arrow and target simulation.

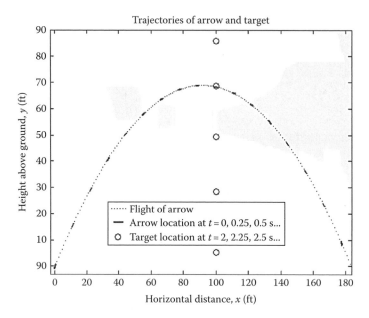

FIGURE 5.113   Path of the arrow and target at selected times.

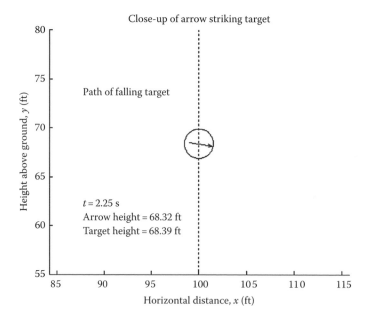

FIGURE 5.114   Close-up of the arrow and target.

The reader should experiment with different target heights and arrow initial velocities to confirm the result whereby the arrow always intercepts the target. Naturally, the time and location where the arrow hits the target will depend on the new initial conditions.

A natural question to ask is "What happens when the drag forces are accounted for and an element of randomness associated with the archer's aim and initial speed of the arrow is introduced?" To answer the question, let us assume that the arrow's angle of departure $\theta(0) = \theta_0$ and its initial velocity $v(0) = v_0$ off the bow are random variables with probability density functions

$$f_{\theta_0}(u) \sim N(\mu_{\theta_0}, \sigma_{\theta_0}), \quad -\infty < u < \infty \tag{5.147}$$

$$f_{v_0}(u) \sim U[(v_0)_L, (v_0)_U], \quad (v_0)_L \le v \le (v_0)_U \tag{5.148}$$

where

μ_{θ_0} and σ_{θ_0} are the mean and standard deviation of the Normal population
$(v_0)_L$ and $(v_0)_U$ are the lower and upper limits of the Uniform population

A Monte Carlo experiment can be designed to estimate the probability of hitting the target. "*Chap5_MonteCarlo_arrow.m*" uses the MATLAB functions "rand" and "randn" to generate random deviates $R_i \sim U(0, 1)$ and $z_i \sim N(0, 1)$. R_i and z_i are transformed to random deviates from the desired populations, Equations 5.147 and 5.148 by

$$(\theta_0)_i = \mu_{\theta 0} + z_i \sigma_{\theta_0} \tag{5.149}$$

$$(v_0)_i = (v_0)_L + [(v_0)_U - (v_0)_L]R_i \tag{5.150}$$

For now, let us assume that the mean angle of departure of the arrow is equal to the sight angle to the target and the standard deviation is 1°. Further, assume that the initial velocities are uniformly distributed between 75 and 85 ft/s. Two Monte Carlo experiments were performed, each with a total of 50,000 random vectors $[(\theta_0)_i, (v_0)_i]$ generated and 50,000 Simulink simulation runs executed. During each run, the occurrence of a "hit" or "miss" is determined and recorded. A "hit" occurs at time $t = \hat{t}$ when the arrow has traveled a horizontal distance x_T, that is, $x_A(\hat{t}) = x_T$ provided

$$y_T(\hat{t}) - r_T \le y_A(\hat{t}) \le y_T(\hat{t}) + r_T \tag{5.151}$$

where $r_T = 1.5$ ft is the radius of the target. The vertical separation between the arrow and target at $t = \hat{t}$ is the distance Δ,

$$\Delta = y_A(\hat{t}) - y_T(\hat{t}) \tag{5.152}$$

Results of both experiments are saved in MATLAB data files "arrowdata1.mat" and "arrowdata2.mat." Histograms of the separations Δ_i, $i = 1, 2, \ldots 50{,}000$ for both Monte Carlo runs are plotted in M-file "*Chap5_plot_arrow_histogram.m*" and shown in Figure 5.115 along with the estimated probability of hitting the target.

The histograms suggest that the separation Δ is approximately normally distributed with mean zero. Since more than 99% of the total area under a Normal pdf lies within the mean plus and minus three standard deviations, the standard deviation of Δ is approximately 5 ft.

A question that naturally arises with Monte Carlo simulation is "How many random trials are needed to accurately estimate an unknown theoretical probability?" Figure 5.116 is a plot of the estimated probability of hitting the target computed after 100, 200, ..., 1000 trials from the first data file "arrowdata1.mat."

After 1000 trials, an estimate of the true (unknown) probability of hitting the target, under the conditions given for $\theta(0)$ and $v(0)$, is accurate to one place after the decimal point. Figure 5.117 is a similar plot showing the estimated probability of hitting the target computed after every 1000 trials. The estimated probability of a hit based on 50,000 trials is now accurate to three places after the decimal point.

Figure 5.118 shows arrow and target locations from four of the random trials. Note the correlation between the height of the arrow and the initial speed $v(0)$. The arrow is located at a higher elevation when the initial speed is greater. Also, Δ, the separation between the arrow and

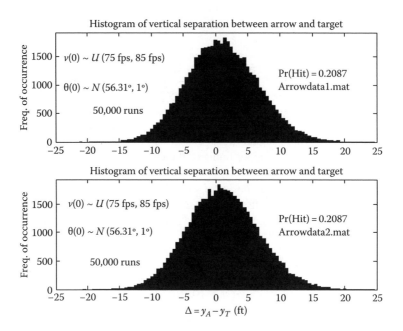

FIGURE 5.115 Estimated Pr(hit) and histogram of separations.

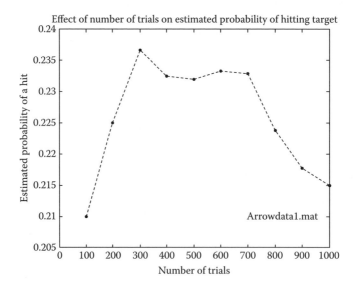

FIGURE 5.116 Estimated probability of a hit after 100, 200, . . . , 1000 trials.

target, should be dependent on the angle of departure $\theta(0)$, specifically its relationship to the line of sight angle θ_{LS} (see Figure 5.111).

$$\theta_{LS} = \tan^{-1}\left(\frac{y_T(0)}{x_T}\right) = \tan^{-1}\left(\frac{150}{100}\right) = 0.9828\,\text{rad}\,(56.31°) \qquad (5.153)$$

In the first run (upper left graph in Figure 5.118), the angle of departure (55.99°) is slightly less than the line of sight angle, and the arrow strikes the target just below its center. In the second run

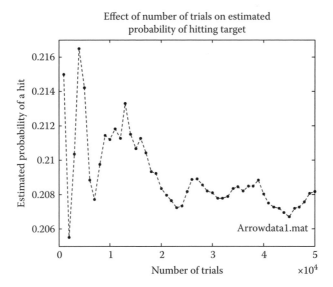

FIGURE 5.117   Estimated probability of a hit after 1000, 2000, ..., 50,000 trials.

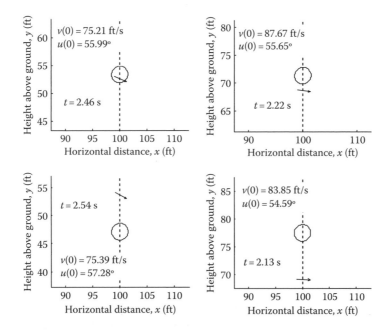

FIGURE 5.118   Arrow and target positions from four runs.

(upper right), the initial angle $\theta(0)$ is even less and the arrow passes under the target. In the last two runs, the angle of departure is significantly greater (lower left) and significantly less (lower right) than θ_{LS}, and the corresponding separations are greater and in the expected direction.

The arrow speed $v(t)$ and pitch angle $\theta(t)$ for the last case (lower right corner) are shown in Figure 5.119. When the arrow is directly below the target at $t = 2.13$ s, the speed is 45.22 ft/s and the pitch angle is $-2.93°$.

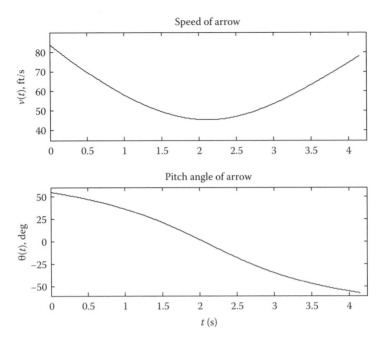

FIGURE 5.119 Arrow speed $v(t)$ and pitch $\theta(t)$ for $v(0) = 83.85$ ft/s, $\theta(0) = 54.59°$.

EXERCISES

In Exercises 5.42 and 5.43, M, B, and K are randomly distributed according to Equations 5.129 through 5.131 with the same limits given in the text.

5.42 Use "*Chap5_MonteCarlo_damping_ratio.m*" or write your own program to find and graph the approximate pdf $\hat{f}_{M_{p\omega}}(u)$ and cdf $\hat{F}_{M_{p\omega}}(u)$. Find $\Pr[1.1 \leq M_{p\omega} \leq 1.3]$.

5.43 The resonant frequency of a second-order system depends on the damping ratio and natural frequency according to

$$\omega_r = \omega_n \sqrt{1 - 2\zeta^2}, \quad \zeta \leq 0.707$$

(a) Use Monte Carlo simulation to approximate the true pdf and cdf for ω_r.
(b) Graph $\hat{f}_{\omega_r}(u)$ and $\hat{F}_{\omega_r}(u)$.
(c) Estimate $\Pr[\omega_r > \sqrt{2}$ rad/s].

5.44 Repeat Exercise 5.43 if the mass M is normally distributed with mean $\mu_M = 1$ slug and standard deviation $\sigma_M = 0.25$ slugs. Assume B and K are no longer random, instead $B = 2$ lb s/ft and $K = 4$ lb/ft.

5.45 Suppose the arrow and target with mass and aerodynamic properties given in the text are dropped from an airplane in level flight at a cruising speed of $v_{cr} = 600$ ft/s.
(a) Find expressions for the terminal velocities of both.
(b) Simulate their descent from an altitude of 10,000 ft with zero initial velocity.
(c) Plot the acceleration of each during their descent.

5.46 Neglecting aerodynamic damping forces and assuming that the initial firing angle of the arrow is equal to the sight angle to the target, perform a simulation study to produce the missing graphs in Figure E5.46:

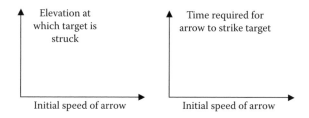

FIGURE E5.46

5.47 A boy is throwing rocks, aiming at a circular target with diameter D. The center of the target is x_T ft down range from where he is located (see Figure E5.47). The aerodynamic drag force is proportional to the speed of the rock with drag constant α. The rocks are launched from a height of y_0 at an angle $\varphi(0)$ and initial speed $v(0)$. The weight of the rock is W. The distance downrange where the rock lands is R.

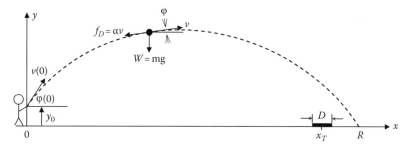

FIGURE E5.47

Baseline system parameter values are
$y_0 = 6$ ft, $x_T = 160$ ft, $D = 4$ ft, $\alpha = 9 \times 10^{-4}$ lb/ft/s, $W_0 = 0.5$ lb, $\varphi(0) = 45°$, and $v(0) = 75$ ft/s

(a) Write the equations comprising the mathematical model of the system in state variable form $\dot{\underline{x}} = f(\underline{x}, \underline{u})$ where the state vector $\underline{x} = [x\ \dot{x}\ y\ \dot{y}]$.

(b) Use Simulink to simulate the system under baseline conditions, and verify the stone trajectory shown in Figure E5.47:

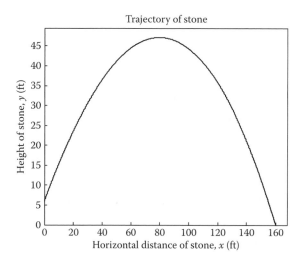

FIGURE E5.47

(c) The boy picks up a rock, the weight of which is uniformly distributed between 0.25 and 0.75 lb, and throws it with initial speed and angle given by the baseline values. Find the probability of the rock landing on the target.

(d) Prepare a histogram for the random variable $\Delta = |R - x_T|$, and use it to find the empirical probability density function $\hat{f}_\Delta(u)$, $\Delta \geq 0$.

(e) Repeat parts (d) and (e) if $W = W_0 = 0.5$ lb and $\theta(0) \sim U(40°, 50°)$.

5.48 A particle slides without friction along a path given by $y = f(x) = x^{1/2}$ under the influence of gravity as shown in Figure E5.48:

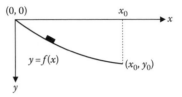

FIGURE E5.48

The time required for the particle to slide down the curve starting from the origin to the point (x_0, y_0) is (Speckhart 1976)

$$t_0 = \frac{1}{\sqrt{2g}} \int_0^{x_0} \sqrt{\frac{1 + (dy/dx)^2}{y}} \, dx$$

The termination value x_0 is a random variable uniformly distributed between 1 and 5 along the curve. Implement a Monte Carlo experiment culminating in a histogram for the random variable t_0.

5.49 Consider the second-order system $\ddot{y} + 2\zeta\omega_n\dot{y} + \omega_n^2 y = 0$ with initial conditions $y(0) = y_0$, $\dot{y}(0) = 0$. Introduce state variables $x_1 = y$, $x_2 = \dot{y}$. Phase plots for an underdamped ($\zeta = 0.25$), critically damped ($\zeta = 1$), and overdamped ($\zeta = 2$) case with $\omega_n = 1$ rad/s and $y_0 = 1$ are shown in Figure E5.49:

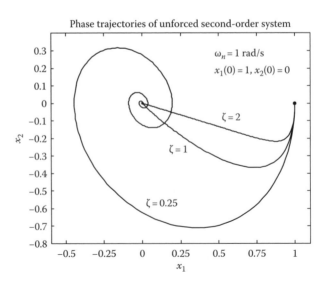

FIGURE E5.49

(a) Plot a histogram for the distance from the initial point $x_1(0) = 1$, $x_2(0) = 0$ to the steady-state equilibrium point $x_1(\infty) = 0$, $x_2(\infty) = 0$ along the trajectories in state space if the damping ratio is uniformly distributed between 0 and 2.

Note that the distance from the initial point $(1,0)$ to the point $[x_1(t), x_2(t)]$ along the trajectory is given by

$$s(t) = \int_0^t \left(\dot{x}_1^2 + \dot{x}_2^2 \right)^{1/2} dt$$

(i) Repeat part (a) for the case where $\zeta = 0.25$, and the natural frequency ω_n is uniformly distributed between 0 and 100 rad/s.

(ii) Repeat part (a) for the case where $\zeta = 1$, and the natural frequency ω_n is uniformly distributed between 0 and 12.5 rad/s.

(iii) Repeat part (a) for the case where $\zeta \sim U(0, 2)$, $\omega_n \sim U(0, 100)$, and $y_0 \sim U(0, 1)$.

5.11   CASE STUDY: PILOT EJECTION

Several benchmark applications of continuous-time simulation using analog and digital computers have been around for decades. Simulation of a pilot and seat ejected from a fighter aircraft falls in this category (Korn 1978). The system is shown in Figure 5.120.

When forced to eject, the combination of pilot and seat trajectory is controlled by a set of guide rails until it is clear of the plane. The ejection velocity v_E is constant along a direction θ_E from the y axis of the plane. Ejection occurs when the pilot and seat have traveled a vertical distance y_1.

After ejection from the aircraft, the pilot and seat follow a ballistic trajectory subject to an aerodynamic drag force and its own weight. The equations of motion can be developed in the x–y coordinate system or n–t coordinate system, where n and t refer to directions normal and tangential to the flight of the pilot and seat as shown in Figure 5.121. Summing forces in the n and t directions,

$$\sum F_t = ma_t \tag{5.154}$$

$$\Rightarrow -F_D - W \sin \theta = m\dot{v} \tag{5.155}$$

$$\sum F_n = ma_n \tag{5.156}$$

$$\Rightarrow -W \cos \theta = m\frac{v^2}{R} \tag{5.157}$$

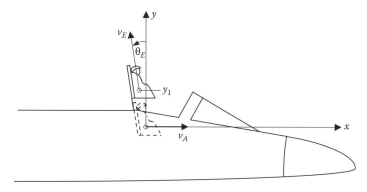

FIGURE 5.120   Diagram of pilot ejection.

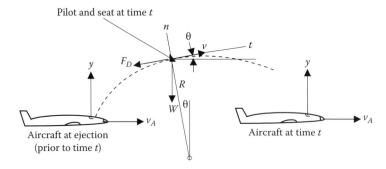

FIGURE 5.121   Trajectory of pilot and seat after ejection.

where R is the instantaneous radius of curvature of the pilot and seat trajectory. The plane is assumed to be traveling in a horizontal direction at constant speed v_A.

The forward velocity v and angular velocity $\dot{\theta}$ are related by

$$v = R\dot{\theta} \tag{5.158}$$

Solving for R in Equation 5.158 and substituting the result in Equation 5.157 give

$$-W \cos \theta = mv\dot{\theta} \tag{5.159}$$

With $W = mg$ and state variables v and θ, the state derivatives are obtained from Equations 5.155 and 5.159 as

$$\dot{v} = \begin{cases} 0, & 0 \le y < y_1 \\ -\dfrac{F_D}{m} - g \sin \theta, & y \ge y_1 \end{cases} \tag{5.160}$$

$$\dot{\theta} = \begin{cases} 0, & 0 \le y < y_1 \\ -\dfrac{g \cos \theta}{v}, & y \ge y_1 \end{cases} \tag{5.161}$$

The intervals $0 \le y < y_1$ and $y \ge y_1$ correspond to before and after ejection.

Additional state variables x and y, the relative coordinates of the pilot and seat with respect to the moving aircraft, are needed to view its trajectory with respect to the plane in order to determine if it safely clears the plane's rear vertical stabilizer. The state derivatives are expressed as (see Figure 5.121)

$$\dot{x} = v \cos \theta - v_A \tag{5.162}$$

$$\dot{y} = v \sin \theta \tag{5.163}$$

It is convenient to start the simulation, that is, integrating the state derivatives, at the moment of ejection. The initial conditions are obtained with the help of Figure 5.122.

The initial states $v(0)$ and $\theta(0)$ are computed from

$$v(0) = \left[v_x^2(0) + v_y^2(0)\right]^{1/2} \tag{5.164}$$

$$\Rightarrow v(0) = [(v_A - v_E \sin \theta_E)^2 + (v_E \cos \theta_E)^2]^{1/2} \tag{5.165}$$

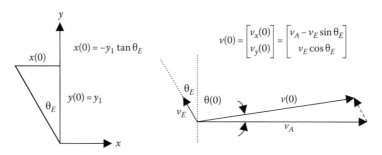

FIGURE 5.122   Initial states $x(0)$, $y(0)$, $v(0)$, and $\theta(0)$ at ejection ($t = 0$).

$$\theta(0) = \tan^{-1}\left[\frac{v_y(0)}{v_x(0)}\right] \qquad (5.166)$$

$$\Rightarrow \theta(0) = \tan^{-1}\left(\frac{v_E \cos\theta_E}{v_A - v_E \sin\theta_E}\right) \qquad (5.167)$$

Finally, the drag force F_D is obtained from

$$F_D = \frac{1}{2}C_D\rho A v^2 \qquad (5.168)$$

where
   C_D is the drag coefficient
   ρ is the density of air
   A is the surface area of the pilot and seat normal to the velocity vector

A simulation study is required to investigate the combinations of aircraft speed v_A and altitude h associated with safe ejection, that is, pilot and seat clear the rear vertical stabilizer by a predetermined amount. First, we shall simulate a single case where $v_A = 500$ ft/s and $h = 0$ (sea level). A Simulink diagram is shown in Figure 5.123.

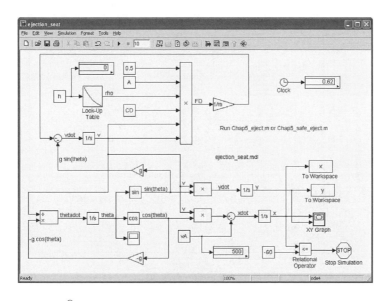

FIGURE 5.123   Simulink® diagram of pilot ejection.

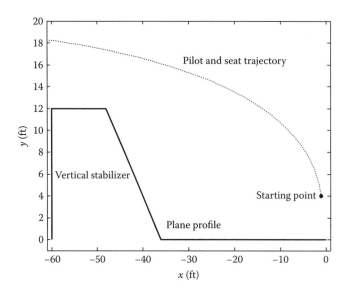

FIGURE 5.124 Plot of pilot and seat trajectory relative to the aircraft ($h = 0$ ft, $v_A = 500$ ft/s).

Baseline numerical values of the system parameters are $\theta_E = 15°$, $v_E = 40$ ft/s, $m = 8$ slugs, $A = 10$ ft^2, $C_D = 1$, and $y_1 = 4$ ft. The "Lookup Table" contains air density ρ (slug ft^2) vs. altitude h (ft) data points from sea level to 60,000 ft.

The pilot and seat trajectory relative to the aircraft is obtained by calling the Simulink model "*ejection_seat.mdl*" from the M-file "*Chap5_eject.m*" using the command "sim('ejection_seat')." Figure 5.124 illustrates the relative separation between the pilot and seat combination and the plane during the time when the pilot and seat are located above the plane. The pilot and seat safely clear the vertical stabilizer.

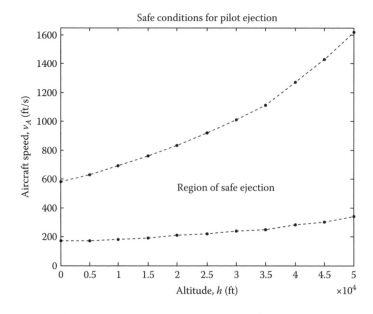

FIGURE 5.125 Lower and upper aircraft speeds at a given altitude for safe ejection.

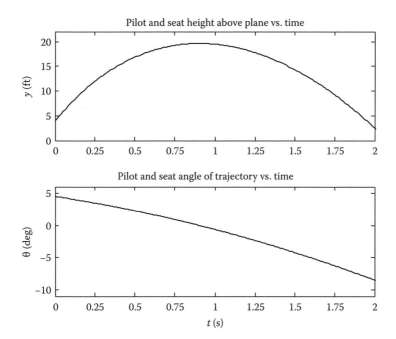

FIGURE 5.126 Pilot and seat height above plane and trajectory after ejection.

At a given altitude h, the pilot and seat trajectory will safely clear the stabilizer provided the aircraft cruising speed v_A falls within a range of values. At slow speeds, the exit velocity is insufficient to propel the pilot and seat safely over the stabilizer, while at very high speeds, the excessive drag force and backward velocity (relative to the plane) produce a similar outcome.

A simulation study was performed to determine a region of safe ejection conditions, that is, altitude and speed combinations resulting in a clearance of 5 ft when the pilot and seat are directly over the back part of the rear stabilizer. The M-file "*Chap5_safe_eject.m*" calls the simulation model for altitudes from zero to 50,000 ft (in increments of 5,000 ft) and finds the range of aircraft speeds for a safe ejection. The result is shown in Figure 5.125.

Figure 5.126 shows a plot of $y(t)$, the height of the pilot and seat combination above the plane, corresponding to the safe ejection trajectory shown in Figure 5.124. The lower graph shows $\theta(t)$, the angle between the velocity vector and the horizontal. Can you locate the point on each plot where the pilot and seat are located at the rear of the plane?

EXERCISES

5.50 With respect to the ballistic trajectory of the pilot and seat,
  (a) Develop an alternate mathematical model using x, y coordinates. The states are x, $\dot{x}$, y, and $\dot{y}$.
  (b) Prepare a Simulink diagram for simulating the trajectory following ejection.
  (c) Run the simulation for the same conditions as in Figure 5.124 and compare results.
  (d) Suppose the aircraft is cruising at 30,000 ft in level flight when ejection occurs. Simulate pilot and seat trajectories corresponding to $v_A = 500, 600, \ldots, 1200$ ft/s. Plot the entire set of trajectories (with respect to the plane) on the same axes with the plane profile similar to Figure 5.124. Are the results consistent with the safe ejection conditions portrayed in Figure 5.125?

5.51 Use either n–t or x–y coordinate systems to model the pilot and seat trajectory and obtain plots of

(a) x vs. t

(b) y vs. t

(c) θ vs. t

when ejection occurs from 50,000 ft at a speed of 900 ft/s.

5.52 Reexamine the limiting plane speeds for a safe ejection from 25,000 ft as the mass of the pilot and seat varies from 8 slugs to 12 slugs. How important is the combined mass of the pilot and seat with respect to the limiting plane speeds at 25,000 ft?

5.53 Obtain new curves for lower and upper safe ejection speeds in terms of altitude if the criterion for a safe ejection is that the pilot and seat simply clear the rear vertical stabilizer. Use the baseline value for $m = 8$ slugs.

5.54 Modify the code in M-file "*Chap5_safe_eject.m*" to check whether the pilot and seat have cleared the rear stabilizer over its entire length of 48–60 ft back from the point of ejection. How does this affect the curves in Figure 5.125?

5.12   CASE STUDY: KALMAN FILTERING

Estimations of the Moon and planetary orbits were performed by early pioneers such as Kepler, Legendre, and Gauss. More recent estimation algorithms have been developed in an effort to obtain the optimal estimate of a dynamic object, the Kalman filter being the most popular. In this case study, the continuous-time Kalman filter, the steady-state Kalman filter, and the discrete-time Kalman filter (Simon 2006) are applied to the trajectory of an asteroid. First, the algorithms of the different filters will be presented in summary form, and then simulations will be run in Simulink for comparison.

5.12.1   CONTINUOUS-TIME KALMAN FILTER

The state equations of a continuous dynamic system are given by

$$\dot{x} = Ax + Bu + w$$
$$y = Cx + v$$

(5.169)

where

x is the state vector

u is the input vector

y is the output vector

A is the system matrix

B is the input matrix

C is the output matrix

In the state equations, w and v are zero-mean, uncorrelated, continuous-time, white noise with process covariance matrix Q_c and measurement covariance matrix R_c, respectively. Mathematically,

$$w \sim (0, Q_c)$$
$$v \sim (0, R_c)$$
$$E[ww^T] = Q_c \delta_{ij}$$
$$E[vv^T] = R_c \delta_{ij}$$
$$E[vw^T] = 0$$

(5.170)

The algorithm of the continuous-time Kalman filter is given by

$$K = PC^T R_c^{-1}$$

$$\dot{\hat{x}} = A\hat{x} + Bu + K(y - C\hat{x}) \tag{5.171}$$

$$\dot{P} = -PC^T R_c^{-1} CP + AP + PA^T + Q_c$$

where the last equation in 5.171 is referred to as the Riccati equation. The algorithm is initialized with the expectation values of the state and state covariance

$$\hat{x}(0) = E[x(0)]$$

$$P(0) = E[(x(0) - \hat{x}(0))(x(0) - \hat{x}(0))^T] \tag{5.172}$$

5.12.2 Steady-State Kalman Filter

In the case of the steady-state Kalman filter, the system dynamics do not change with respect to time; therefore, $\dot{P} = 0$, so that the Riccati equation of 5.171 becomes

$$0 = -PC^T R_c^{-1} CP + AP + PA^T + Q_c \tag{5.173}$$

5.12.3 Discrete-Time Kalman Filter

The state equations of a discrete dynamic system are given by

$$x_k = F_{k-1} x_{k-1} + G_{k-1} u_{k-1} + w_{k-1}$$

$$y_k = H_{k-1} x_{k-1} + v_{k-1} \tag{5.174}$$

where
  F_{k-1} is the system matrix
  G_{k-1} is the input matrix
  H_{k-1} is the output matrix

In this case, w_{k-1} and v_{k-1} are zero-mean, uncorrelated, discrete-time, white noise with process covariance matrix Q_k and measurement covariance matrix R_k, respectively. Mathematically,

$$w_k \sim (0, Q_k)$$

$$v_k \sim (0, R_k)$$

$$E\left[w_k w_j^T\right] = Q_k \delta_{k-j}$$

$$E\left[v_k v_j^T\right] = R_k \delta_{k-j} \tag{5.175}$$

$$E\left[w_k v_j^T\right] = 0$$

The algorithm of the discrete-time Kalman filter is given by

$$
\begin{aligned}
\hat{x}_k^- &= F_{k-1}\hat{x}_{k-1}^+ + G_{k-1}u_{k-1} \\
P_k^- &= F_{k-1}P_{k-1}^+ F_{k-1}^T + Q_{k-1} \\
K_k &= P_k^- H_k^T \left(H_k P_k^- H_k^T + R_k \right)^{-1} \\
\hat{x}_k^+ &= \hat{x}_k^- + K_k \left(y_k - H_k\hat{x}_k^- \right) \\
P_k^+ &= (I - K_k H_k)P_k^- (I - K_k H_k)^T + K_k R_k K_k^T
\end{aligned}
\tag{5.176}
$$

and is initialized with the expectation values of the state and state covariance

$$
\begin{aligned}
\hat{x}_0^+ &= E[x_0] \\
P_0^+ &= E\left[\left(x_0 - \hat{x}_0^+\right)\left(x_0 - \hat{x}_0^+\right)^T \right]
\end{aligned}
\tag{5.177}
$$

5.12.4 Simulink® Simulations

The three different Kalman filters (continuous, steady-state, and discrete) are used to estimate the kinematics (position and velocity) of an incoming meteorite. It is assumed that the meteorite is tracked with a radar system that picks up the object at a range of 200,000 m with a velocity of 5,000 m/s. The measurement error R of the radar tracking station is 100 m. The process noise statistics Q in range, velocity, and acceleration are 1 m, 0.1 m/s, and 0.1 m/s², respectively. Since the initial conditions of the meteorite are unknown, the diagonal elements of the state covariance matrix P are large. The meteorite is tracked for 30 s at a frequency of 10 Hz.

Figure 5.127 shows a Simulink diagram for estimating the range of the meteorite with a continuous-time Kalman filter. (In most cases, element blocks retained their default names for ease of locating them in the Simulink library. A few subsystem names were changed to reflect their contents.) At the top of the continuous-time Kalman filter hierarchy, two major subsystems are shown: (1) the actual range of the meteorite corrupted by noise and (2) the estimated range containing the continuous-time Kalman filter elements. To run this model, execute the MATLAB M-file CTKF_Model_Data.m.

By double clicking on the "Actual" subsystem, Figure 5.128 shows the elemental blocks that calculate the kinematics of the meteorite $y = y_0 + v_0 t + 1/2at^2$ and $v = v_0 + at$ where the initial conditions are represented by *xhat*0, a vector defined in the MATLAB M-file.

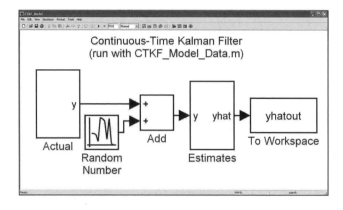

FIGURE 5.127 Top view of the continuous-time Kalman filter.

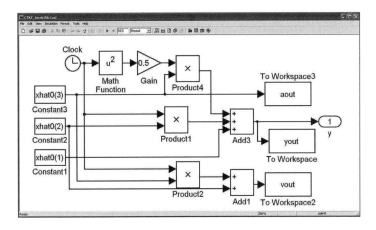

FIGURE 5.128   The "Actual" subsystem.

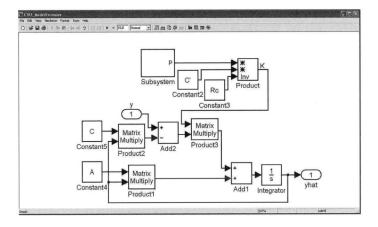

FIGURE 5.129   The continuous-time Kalman filter algorithm.

Returning to the top-level view and then double clicking on the "Estimates" subsystem, Figure 5.129 shows the elemental blocks of the continuous-time Kalman filter algorithm, Equation 5.171. The integrator block requires the initial conditions $xhat0$ defined in the MATLAB M-file. For legibility, the computation of the state covariance matrix P is placed into its own subsystem.

By double clicking on the "P" subsystem, Figure 5.130 shows the elemental blocks that update the state covariance matrix P, Equation 5.171. The integrator in this subsystem requires the initial conditions $P0$ defined in the M-file.

Simulating the model by executing the MATLAB M-file CTKF_Model_Data.m created the following plots. Figure 5.131 shows the actual range R and the estimated range $Rhat$ of the meteorite vs. time. The meteorite is picked up at a range of 200,000 m and tracked for 30 s. Over this time period, the meteorite traveled approximately 150,000 m. The continuous-time Kalman filter performs very well, such that it is difficult to see any differences between the actual range and the estimated range.

Figure 5.132 shows the actual velocity V and the estimated velocity $Vhat$ of the meteorite vs. time. The continuous-time Kalman filter takes approximately 10 s for transients to settle before obtaining reasonable velocity estimates.

Figure 5.133 shows the actual acceleration A and the estimated acceleration $Ahat$ of the meteorite vs. time. It is unnecessary to estimate the acceleration of gravity, but it is shown here for

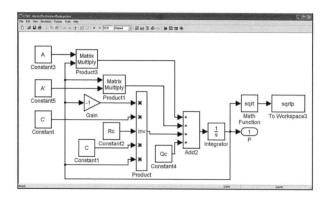

FIGURE 5.130 Simulink® diagram of the continuous-time Kalman filter.

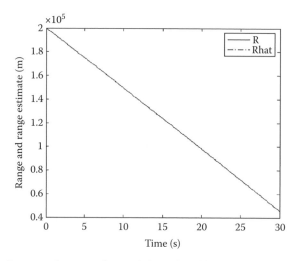

FIGURE 5.131 Plot of range and range estimates (m) vs. time (s).

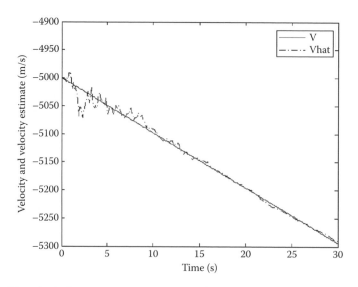

FIGURE 5.132 Plot of velocity and velocity estimates (m/s) vs. time (s).

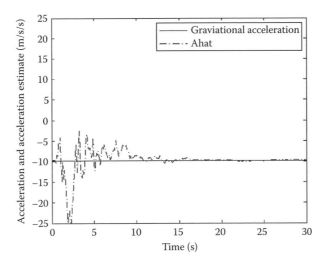

FIGURE 5.133   Plot of acceleration and acceleration estimates (m/s/s) vs. time (s).

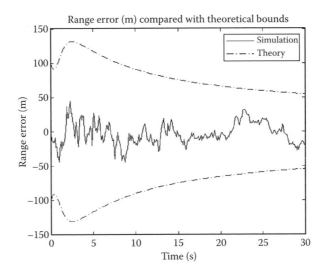

FIGURE 5.134   Plot of range error vs. time.

completeness. Again, the transients take approximately 10 s to settle before obtaining reasonable estimates.

Figure 5.134 shows the range error, the difference between the actual range and the estimated range, vs. time. In theory, the range error should be bounded by the standard deviation of the 1,1 element of the state covariance matrix, which it is. It appears as if the maximum range error at any given time is about 50 m. Recall (Figure 5.131) that the meteorite traveled roughly 150,000 m over 30 s. An error of 50 m, even at the end of the 30 s when the meteorite is at a range of 50,000 m, is 0.1%.

Figure 5.135 shows the velocity error, the difference between the actual velocity and the estimated velocity, vs. time. In this case, the velocity error should be bounded by the standard deviation of the 2,2 element of the state covariance matrix, which it is. After the filter transients settle out, the maximum velocity error appears to be less than 10 m/s. Recall (Figure 5.132) that the meteorite obtained a speed of roughly 5300 m/s over 30 s. An error of 10 m/s is less than 0.2%.

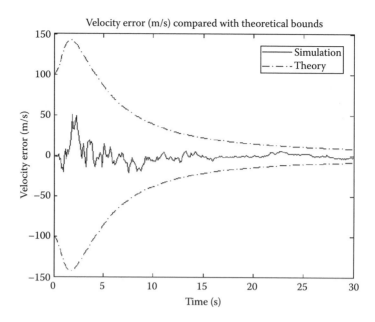

FIGURE 5.135 Plot of velocity error vs. time.

This concludes the implementation and analysis of the continuous-time Kalman filter as applied to the range and velocity estimates of an incoming meteorite.

Next, the steady-state Kalman filter is applied to the same problem for comparison with the continuous-time Kalman filter. The only difference between the two models is the calculation of the state covariance matrix P. In the continuous-time algorithm, the Riccati equation is time dependent; for the steady-state algorithm, the Riccati equation is independent of time, Equation 5.173. With regard to model structure, the top-level diagram and "Actual" subsystem diagram are the same for the steady-state Kalman filter as they were for the continuous-time Kalman filter. However, the "Estimates" subsystem reflects the difference with regard to the Riccati equation, which is represented by a constant element block called "SSP" seen in Figure 5.136.

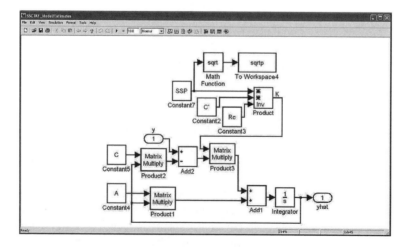

FIGURE 5.136 The steady-state Kalman filter algorithm.

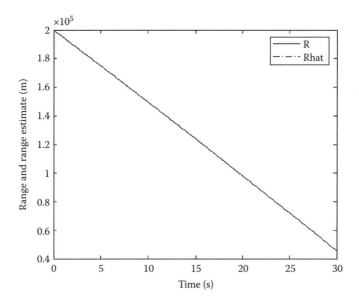

FIGURE 5.137   Plot of range and range estimates (m) vs. time (s).

Simulating the model by executing the MATLAB M-file SSCTKF_Model_Data.m created the following plots. Figure 5.137 shows the actual range *R* and the estimated range *Rhat* of the meteorite vs. time. From this plot, it appears that the steady-state Kalman filter performs just as well as the continuous-time Kalman filter. As before, it is difficult to see any differences between the actual range and the estimated range.

Figure 5.138 shows the actual velocity *V* and the estimated velocity *Vhat* of the meteorite vs. time. From this plot, it can be seen that the steady-state Kalman filter performs better than the continuous filter in estimating the velocity of the meteorite. Obviously missing from this plot are the transients associated with the time-dependent state covariance updates. The steady-state Kalman

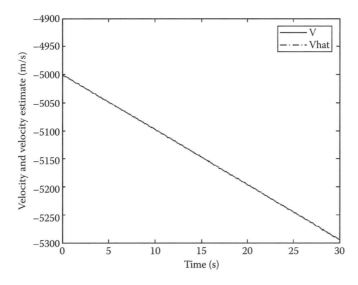

FIGURE 5.138   Plot of velocity and velocity estimates (m/s) vs. time (s).

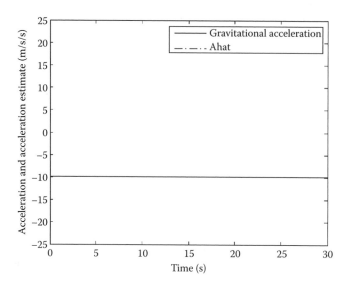

FIGURE 5.139   Plot of acceleration and acceleration estimates (m/s/s) vs. time (s).

filter eliminates the need to perform this calculation—which may be significant for an application where real-time processing is limited.

Figure 5.139 shows the actual acceleration *A* and the estimated acceleration *Ahat* of the meteorite vs. time. As mentioned before, it is unnecessary to estimate the acceleration of gravity, but it is shown for completeness. Again, there are no transients with the steady-state Kalman filter.

Figure 5.140 shows the range error, the difference between the actual range and the estimated range, vs. time. Again, the range error is bounded by the standard deviation of the 1,1 element of the state covariance matrix, which is constant. The maximum range error at any given time is negligible for the steady-state Kalman filter.

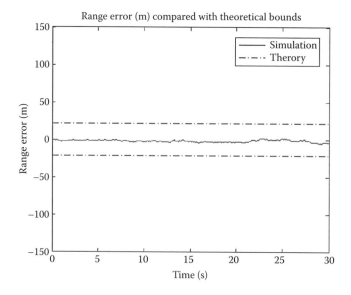

FIGURE 5.140   Plot of range error vs. time.

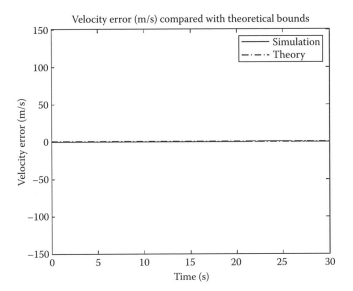

FIGURE 5.141   Plot of velocity error vs. time.

Figure 5.141 shows the velocity error, the difference between the actual velocity and the estimated velocity, vs. time. The velocity error is bounded by the standard deviation of the 2,2 element of the state covariance matrix, which is constant. Here, too, the maximum velocity error at any given time is negligible for the steady-state Kalman filter.

This concludes the implementation and analysis of the steady-state Kalman filter as applied to the range and velocity estimates of an incoming meteorite.

Next, the discrete-time Kalman filter is applied to the same problem for comparison with the continuous-time Kalman filter. The dynamic system of the meteorite kinematics are discretized, Equation 5.174, and then simulated with the discrete-time Kalman filter algorithm, Equation 5.176. At this time, a few comments regarding the algorithm are in order. The first two equations of the algorithm $\hat{x}_k^-$ and P_k^- are known as the a priori state and state covariance estimates, respectively. They take the name "a priori" because the calculations are performed *before* the meteorite's state is measured. The third equation of the algorithm K_k is the Kalman gain. The last two equations of the algorithm $\hat{x}_k^+$ and P_k^+ are known as the a posteriori state and state covariance estimates, respectively. They take the name "a posteriori" because the calculations are performed *after* the meteorite's state is measured.

As in the previous two cases, the top-level diagram and "Actual" subsystem diagram are the same for the discrete-time Kalman filter. However, the "Estimates" subsystem, shown in Figure 5.142, shows the Simulink diagram for the discrete-time Kalman filter algorithm. From this view, the a priori state and state covariance, the Kalman gain, and the a posteriori state and state covariance subsystems are clearly represented.

By double clicking on the "a priori state" subsystem, Figure 5.143 shows the elemental blocks that calculate the a priori state estimate of the algorithm. The initial conditions are represented by *xm*0, a vector defined in the corresponding MATLAB M-file.

Returning to the top-level view and then double clicking on the "a priori covariance" subsystem, Figure 5.144 shows the elemental blocks that calculate the a priori state covariance estimate of the algorithm. The initial conditions are represented by *Pm*0, a matrix defined in the corresponding MATLAB M-file.

Returning to the top-level view and then double clicking on the "Kalman gain" subsystem, Figure 5.145 shows the elemental blocks that calculate the Kalman gain of the algorithm.

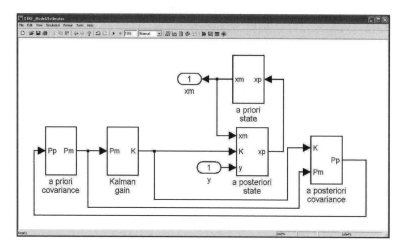

FIGURE 5.142 The discrete-time Kalman filter algorithm.

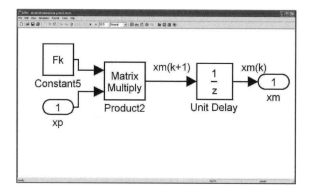

FIGURE 5.143 The "a priori state" subsystem.

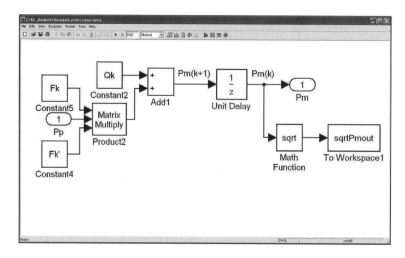

FIGURE 5.144 The "a priori covariance" subsystem.

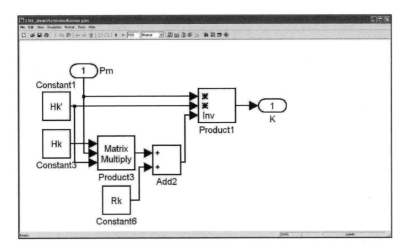

FIGURE 5.145 The "Kalman gain" subsystem.

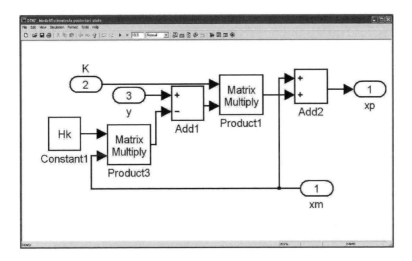

FIGURE 5.146 The "a posteriori state" subsystem.

By double clicking on the "a posteriori state" subsystem, Figure 5.146 shows the elemental blocks that calculate the a posteriori state estimate of the algorithm.

Returning to the top-level view and then double clicking on the "a posteriori covariance" subsystem, Figure 5.147 shows the elemental blocks that calculate the a posteriori state covariance estimate of the algorithm.

Simulating the model by executing the MATLAB M-file DTKF_Model_Data.m created the following plots. Figure 5.148 shows the actual range R and the estimated range $Rhat$ of the meteorite vs. time. The meteorite is picked up at a range of 200,000 m and tracked for 30 s. Over this time period, the meteorite traveled approximately 150,000 m. Like the previous two filters, the discrete-time Kalman filter performs very well. Indeed, it is difficult to see any differences between the actual range and the estimated range.

Figure 5.149 shows the actual velocity V and the estimated velocity $Vhat$ of the meteorite vs. time. The discrete-time Kalman filter takes approximately 10 s for transients to settle before obtaining reasonable velocity estimates. This is similar to the behavior of the continuous-time Kalman filter.

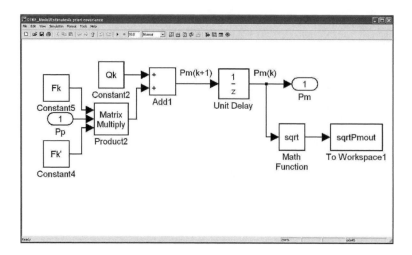

FIGURE 5.147   The "a posteriori covariance" subsystem.

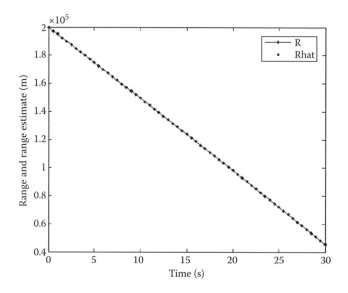

FIGURE 5.148   Plot of range and range estimates (m) vs. time (s).

Figure 5.150 shows the actual acceleration A and the estimated acceleration $Ahat$ of the meteorite vs. time. The transients take approximately 15 s to settle before obtaining reasonable estimates, 5 s more than the continuous-time Kalman filter.

Figure 5.151 shows the range error, the difference between the actual range and the estimated range, vs. time. In theory, the range error should be bounded by the standard deviation of the 1,1 element of the state covariance matrix. For the discrete-time Kalman filter, a few data points lie outside this theoretical limit, but only marginally. Recall (Figure 5.149) that the meteorite traveled roughly 150,000 m over 30 s. An error of 100 m, even at the end of the 30 s when the meteorite is at a range of 50,000 m, is 0.2%.

Figure 5.152 shows the velocity error, the difference between the actual velocity and the estimated velocity, vs. time. Again, in theory, the velocity error should be bounded by the standard deviation of the 2,2 element of the state covariance matrix. After the discrete-time Kalman filter transients settle

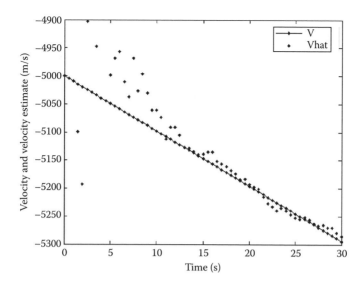

FIGURE 5.149 Plot of velocity and velocity estimates (m/s) vs. time (s).

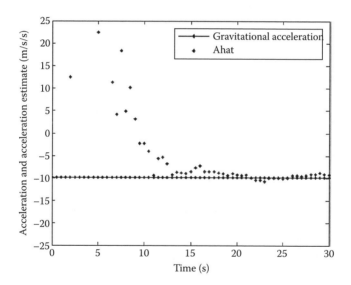

FIGURE 5.150 Plot of acceleration and acceleration estimates (m/s/s) vs. time (s).

out, the maximum velocity error appears to be less than 10 m/s. Recall (Figure 5.149) that the meteorite obtained a speed of roughly 5300 m/s over 30 s. An error of 10 m/s is less than 0.2%.

5.12.5 SUMMARY

Three different Kalman filters (continuous, steady-state, and discrete) were used to estimate the kinematics (position and velocity) of an incoming meteorite. Once filter transients settled out, both the continuous-time and discrete-time Kalman filters provided acceptable results with regard to meteorite range and velocity estimation as evidenced by comparing the range and velocity errors with actual range and velocity magnitudes. If real-time processing poses limitations, it is recommended to use the steady-state Kalman filter.

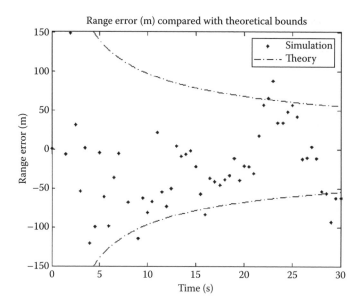

FIGURE 5.151   Plot of range error vs. time.

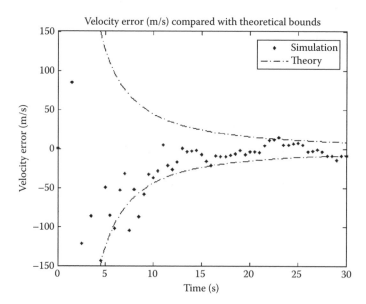

FIGURE 5.152   Plot of velocity error vs. time.

EXERCISE

5.55 Develop the steady-state Kalman filter for the discrete model. Hint: Combine the a priori and the a posteriori equations into a single equation and note in the steady-state, $\hat{x}_k^- = \hat{x}_{k-1}^- = \hat{x}^-$ $P_k^- = P_{k-1}^- = P^-$ in the a priori case or $\hat{x}_k^+ = \hat{x}_{k-1}^+ = \hat{x}^+$ $P_k^+ = P_{k-1}^+ = P^+$ in the a posteriori case.

6 Intermediate Numerical Integration

6.1 INTRODUCTION

We continue our exposition of numerical integration introduced in Chapter 3. Additional algorithms to approximate the solution of differential equation models of continuous-time systems will be examined. In previous chapters, there was no mention of how to quantify the degree of accuracy one could expect with the simple Euler and trapezoidal integrators. Truncation errors are introduced in this chapter as a way of remedying this omission.

This chapter introduces two broad classifications of numerical integrators known as one-step methods and multistep formulas and presents a case for when to use each type. Adaptive techniques for changing the integration step size when using one-step methods are discussed.

Later on, a property of system models referred to as "stiffness" is explored along with ways of dealing with it to make sure accurate and stable simulations result. Numerical stability is mentioned only briefly near the end of the chapter; however, more will be mentioned about this important property when we revisit numerical integration in Chapter 8.

This chapter concludes with a case study that relies on one of the numerical integration methods introduced earlier in the chapter.

6.2 RUNGE–KUTTA (RK) (ONE-STEP METHODS)

One-step methods refer to a family of numerical integration algorithms designed to update the current state across an interval of time, called the integration step, in such a way that the state derivative function is evaluated at one or more points of the interval. In contrast, multistep methods incorporate computed state values from previous intervals in the process of updating the state.

Our discussion of one-step methods begins with an autonomous system involving a single state variable $x = x(t)$ with state derivative function $f(t, x)$.

$$\frac{dx}{dt} = f(t, x) \tag{6.1}$$

The state derivative function could be written $f(t, x, u)$ when there are external inputs present. The reason for choosing a first-order system is simple. Dynamic system models are typically higher than first order; however, the differential equations comprising an nth-order model can be recast as a set of coupled first-order differential equations for the state derivatives $\dot{x}_1(t), \dot{x}_2(t), \ldots, \dot{x}_n(t)$ in terms of the state variables $x_1(t), x_2(t), \ldots, x_n(t)$ and when present, inputs $u_1(t), u_2(t), \ldots, u_r(t)$. The algorithms derived for numerical integration of Equation 6.1 are easily extended to the case of more than one state variable.

Suppose $x(t_i)$, the solution to Equation 6.1 at time $t = t_i$, were known and denoted x_i for short. A way of approximating $x_{i+1} = x(t_{i+1})$, the state $x(t)$ at $t = t_{i+1} = t_i + T$, is needed. The approximation is written as $x_A(i + 1)$ (see Figure 6.1).

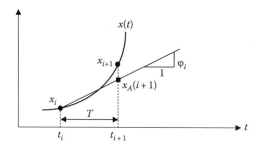

FIGURE 6.1   Graphical representation of calculation for new state $x_A(i+1)$.

We can proceed along a line whose slope is φ_i (see Figure 6.1) starting from the point (t_i, x_i) on the solution $x(t)$ and terminating when $t = t_{i+1}$. This leads to

$$x_A(i+1) = x_i + T\varphi_i \tag{6.2}$$

The slope φ_i is a suitably chosen approximation to the state derivative function $f(t, x)$ over the interval $t_i \leq t \leq t_{i+1}$. We shall return to this notion of a line with slope φ_i from (t_i, x_i) to $[t_{i+1}, x_A(i+1)]$ momentarily.

6.2.1 TAYLOR SERIES METHOD

Consider the Taylor Series expansion of the function $x(t)$ shown in Figure 6.1.

Expanding the function $x(t)$ in a Taylor Series about the point t_i,

$$x_{i+1} = x_i + \frac{d}{dt}x(t_i)T + \frac{1}{2!}\frac{d^2}{dt^2}x(t_i)T^2 + \frac{1}{3!}\frac{d^3}{dt^3}x(t_i)T^3 + \cdots \tag{6.3}$$

Equation 6.3 can be expressed in terms of the state derivative function,

$$f(t, x) = \frac{d}{dt}x(t) \tag{6.4}$$

$$\Rightarrow x_{i+1} = x_i + f(t_i, x_i)T + \frac{1}{2!}\frac{d}{dt}f(t_i, x_i)T^2 + \frac{1}{3!}\frac{d^2}{dt^2}f(t_i, x_i)T^3 + \cdots \tag{6.5}$$

The derivatives $(d/dt)f(t_i, x_i)$, $(d^2/dt^2)f(t_i, x_i)$, and so forth can be obtained from the chain rule. For example, the first derivative is

$$\frac{d}{dt}f(t_i, x_i) = \frac{\partial}{\partial t}f(t_i, x_i) + \frac{\partial}{\partial x}f(t_i, x_i)\frac{d}{dt}x(t_i) \tag{6.6}$$

$$= f_t(t_i, x_i) + f_x(t_i, x_i)f(t_i, x_i) \tag{6.7}$$

where

$$f_t(t_i, x_i) = \frac{\partial}{\partial t}f(t_i, x_i), \quad f_x(t_i, x_i) = \frac{\partial}{\partial x}f(t_i, x_i) \tag{6.8}$$

Substituting Equation 6.7 into Equation 6.5 yields

$$x_{i+1} = x_i + Tf(t_i, x_i) + \frac{T^2}{2}[f_t(t_i, x_i) + f_x(t_i, x_i)f(t_i, x_i)] + \cdots \qquad (6.9)$$

Truncating Equation 6.9 after the second term produces the explicit Euler integrator

$$x_A(i+1) = x_i + Tf(t_i, x_i) \qquad (6.10)$$

which would normally be written as

$$x_A(i+1) = x_A(i) + Tf[t_i, x_A(i)] \qquad (6.11)$$

since x_i is known only at the initial point $(0, x_0)$.

Truncating Equation 6.9 after the third term results in a more accurate approximation of the true value x_{i+1}, namely,

$$x_A(i+1) = x_A(i) + Tf[t_i, x_A(i)] + \frac{T^2}{2}\{f_t[t_i, x_A(i)] + f_x[t_i, x_A(i)]f[t_i, x_A(i)]\} \qquad (6.12)$$

The Taylor Series method can be used to obtain difference equations such as Equations 6.11 and 6.12 for updating the discrete-time state $x_A(i)$. However, it is rarely attempted because expressions for the higher-order derivatives of $f(t, x)$ are often complex functions involving higher-order partial derivatives of $f(t, x)$. What is needed is an algorithm for computing $x_A(i+1)$ with comparable accuracy to the truncated Taylor Series without requiring partial derivatives of $f(t, x)$.

6.2.2 SECOND-ORDER RUNGE–KUTTA METHOD

Recalling our previous discussion of φ_i, the slope of the line from the point (t_i, x_i) to $[t_{i+1}, x_A(i+1)]$ in Figure 6.1, suppose we choose it to be a weighted sum of the state derivative $f(t, x)$ evaluated at several points on the interval. In particular, if φ_i is a weighted average of $f(t, x)$ at two points on the interval $t_i \leq t \leq t_{i+1}$, the result is

$$\varphi_i = a_1 k_1 + a_2 k_2 \quad (0 \leq a_1 \leq 1, 0 \leq a_2 \leq 1, a_1 + a_2 = 1) \qquad (6.13)$$

where k_1 is the state derivative function $f(t, x)$ at (t_i, x_i), that is,

$$k_1 = f(t_i, x_i) \qquad (6.14)$$

and k_2 is the state derivative function $f(t, x)$ at $[t_i + pT, x_i + qTf(t_i, x_i)]$, that is,

$$k_2 = f[t_i + pT, x_i + qTf(t_i, x_i)], \quad (0 \leq p \leq 1, 0 \leq q \leq 1) \qquad (6.15)$$

Lines with slopes k_1 and k_2 are shown in Figure 6.2.

From Equations 6.14 and 6.15,

$$k_2 = f[t_i + pT, x_i + qTk_1] \qquad (6.16)$$

indicating that k_2 can be determined once k_1 is known. The weights a_1 and a_2 as well as the constants p and q are to be determined.

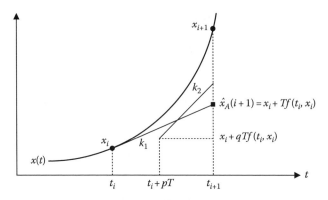

FIGURE 6.2 Representation of $\varphi_i = a_1 k_1 + a_2 k_2$ as weighted sum of $f(t, x)$ at two points.

Substituting Equation 6.13 into Equation 6.2 gives

$$x_A(i + 1) = x_i + T(a_1 k_1 + a_2 k_2) \tag{6.17}$$

The derivative function $f(t, x)$ can be expanded in a two-dimensional Taylor Series about the point (t_i, x_i) as follows:

$$f(t_i + \Delta t, x_i + \Delta x) = f(t_i, x_i) + f_t(t_i, x_i)\Delta t + f_x(t_i, x_i)\Delta x$$
$$+ \frac{1}{2}\left[f_{tt}(t_i, x_i)\Delta t^2 + 2f_{tx}(t_i, x_i)\Delta t \Delta x + f_{xx}(t_i, x_i)\Delta x^2\right] + \cdots \tag{6.18}$$

Letting $\Delta t = pT$, $\Delta x = qTf(t_i, x_i)$ in Equation 6.18 makes k_2 in Equation 6.16 equal to

$$k_2 = f(t_i, x_i) + f_t(t_i, x_i)pT + f_x(t_i, x_i)qTf(t_i, x_i)$$
$$+ \frac{1}{2}\left\{f_{tt}(t_i, x_i)(pT)^2 + 2f_{tx}(t_i, x_i)(pT)[qTf(t_i, x_i)] + f_{xx}(t_i, x_i)[qTf(t_i, x_i)]^2\right\} + \cdots \tag{6.19}$$

Substituting Equation 6.14 for k_1 and Equation 6.19 for k_2 into Equation 6.17 results in

$$x_A(i + 1) = x_A(i) + Ta_1 f(t_i, x_i) + Ta_2[f(t_i, x_i) + f_t(t_i, x_i)pT + f_x(t_i, x_i)qTf(t_i, x_i)]$$
$$+ \frac{1}{2}Ta_2\left\{f_{tt}(t_i, x_i)(pT)^2 + 2f_{tx}(t_i, x_i)(pT)[qTf(t_i, x_i)] + f_{xx}(t_i, x_i)[qTf(t_i, x_i)]^2\right\} + \cdots \tag{6.20}$$

Simplifying Equation 6.20 by collecting terms involving powers of T leads to

$$x_A(i + 1) = x_i + (a_1 + a_2)Tf(t_i, x_i) + a_2 T^2[pf_t(t_i, x_i) + qf_x(t_i, x_i)f(t_i, x_i)] + \cdots \tag{6.21}$$

Equating the right-hand sides of Equations 6.9 and 6.21 gives

$$a_1 + a_2 = 1, \quad a_2 p = \frac{1}{2}, \quad a_2 q = \frac{1}{2} \tag{6.22}$$

The first three terms in Equation 6.3 comprise the second-order truncated Taylor Series expansion of $x(t)$ about the point t_i, that is,

$$x_2(t_i + T) = x(t_i) + \frac{\mathrm{d}}{\mathrm{d}t}x(t_i)T + \frac{1}{2!}\frac{\mathrm{d}^2}{\mathrm{d}t^2}x(t_i)T^2 \tag{6.23}$$

where the subscript "2" indicates that the Taylor Series is truncated after the term containing T^2. Hence, by choosing the constants a_1, a_2, p, and q according to Equation 6.22, we can be certain that the computed state $x_A(i+1)$ in Equation 6.21 achieves comparable accuracy as the second-order truncated Taylor Series.

There are, however, an infinite number of solutions to the three equations in four unknowns in Equation 6.22. Numerical integrators based on the use of Equation 6.17 with a_1, a_2, p, and q satisfying the constraints in Equation 6.22 are referred to as second-order RK or RK-2 integrators.

6.2.3 TRUNCATION ERRORS

The local truncation error ε_T is the difference between the exact solution $x(t_i + T)$ and the approximate solution $x_A(i+1)$ obtained by the Taylor Series method or some other numerical approximation technique such as the RK-2 integrators. Hence,

$$\varepsilon_T = x(t_i + T) - x_A(i+1) \tag{6.24}$$

For the approximation based on the second-order truncated Taylor Series method, Equation 6.24 becomes

$$\varepsilon_T = x(t_i + T) - \left[x_i + f(t_i, x_i)T + \frac{1}{2!}\frac{d}{dt}f(t_i, x_i)T^2 \right] \tag{6.25}$$

Thus, the local truncation error reduces to the sum of all the terms in the Taylor Series expansion for $x(t_i + T)$ beginning with the term containing T^3. That is,

$$\varepsilon_T = \frac{1}{3!}\frac{d^3}{dt^3}x(t_i)T^3 + \frac{1}{4!}\frac{d^4}{dt^4}x(t_i)T^4 + \cdots \tag{6.26}$$

Since the first term on the right-hand side of Equation 6.26 is generally the dominant term (magnitude-wise), the local truncation error is proportional to T^3 and is said to be of order T^3, denoted $\varepsilon_T \sim O(T^3)$. The global truncation error E_T is the accumulation of individual truncation errors incurred in the process of numerically integrating over several intervals. It turns out that E_T is proportional to T^2 or equivalently $E_T \sim O(T^2)$.

It is important to distinguish between the order of the local truncation error and its actual value for a particular numerical integrator. We should not expect to find the numerical value of ε_T in the process of computing $x_A(i), i = 0, 1, 2,\ldots$. Were that possible, the exact solution $x(t_i), i = 0, 1, 2,\ldots$ could be computed from Equation 6.24.

We have seen that RK-2 integrators achieve comparable accuracy to the second-order truncated Taylor Series method and, as a result, are referred to as second-order accurate. The local truncation error $\varepsilon_T \sim O(T^3)$ regardless of how we solve for a_1, a_2, p, and q in Equation 6.22. The numerical value of ε_T will, however, be sensitive to the particular RK-2 integrator.

Knowing $\varepsilon_T \sim O(T^3)$ and $E_T \sim O(T^2)$ for RK-2 integrators makes the consequence of adjusting the integration step size predictable. For example, halving the step size reduces the local and global truncation errors by a factor of $\frac{1}{8}$ and $\frac{1}{4}$, respectively. For the explicit Euler integrator (RK-1), $\varepsilon_T \sim O(T^2)$ and $E_T \sim O(T)$ implying the local truncation error are reduced by $\frac{1}{4}$ while the global truncation is approximately $\frac{1}{2}$ as large when the step size is halved.

We now investigate two possible choices for the set of constants a_1, a_2, p, and q.

Solution I: $a_1 = a_2 = 1/2$ and $p = q = 1$

From Equations 6.2 and 6.13, the RK-2 integrator becomes

$$x_A(i+1) = x_i + \frac{T}{2}(k_1 + k_2) \tag{6.27}$$

Since x_i is unknown after the initial step, it must be replaced by $x_A(i)$ in Equation 6.27 to yield the difference equation for a numerical integrator. Using the definitions for k_1 and k_2 in Equations 6.14 and 6.15 and remembering that $p = q = 1$ give

$$x_A(i + 1) = x_A(i) + \frac{T}{2}\{f[t_i, x_A(i)] + f[t_i + T, x_A(i) + Tf[t_i, x_A(i)]]\} \tag{6.28}$$

Denoting $x_A(i) + Tf[t_i, x_A(i)]$ by $\hat{x}_A(i + 1)$ in Equation 6.28 gives

$$x_A(i + 1) = x_A(i) + \frac{T}{2}\{f[t_i, x_A(i)] + f[t_i + T, \hat{x}_A(i + 1)]\} \tag{6.29}$$

You should recognize $\hat{x}_A(i + 1)$ as the explicit Euler estimate of x_{i+1} in Equation 6.11 (see Figure 6.2). Hence, the explicit Euler (an RK-1 integrator) establishes the second point $[t_i + T, \hat{x}_A(i + 1)]$ for evaluating the derivative function, and the average derivative function or slope is then used to update the state according to Equation 6.29.

The RK-2 integrator of Equation 6.29 is the improved Euler or Heun's method introduced in Section 3.6. At that time, it was developed using a geometrical argument instead of the formal approach presented here.

The second solution for the constants a_1, a_2, p, and q will also look familiar.

Solution II: $a_1 = 0$, $a_2 = 1$ and $p = q = 1/2$.

From Equations 6.2 and 6.13, the RK-2 integrator is

$$x_A(i + 1) = x_i + Tk_2 \tag{6.30}$$

As in the case of the improved Euler integrator, the difference equation for $x_A(i)$ results from replacing x_i by $x_A(i)$ in Equation 6.30 giving

$$x_A(i + 1) = x_A(i) + Tf\left[t_i + \frac{T}{2}, x_A(i) + \frac{T}{2}f[t_i, x_A(i)]\right] \tag{6.31}$$

Introducing the notation

$$x_A\left(i + \frac{1}{2}\right) = x_A(i) + \frac{T}{2}f[t_i, x_A(i)] \tag{6.32}$$

implies the new state $x_A(i + 1)$ is calculated according to

$$x_A(i + 1) = x_A(i) + Tf\left[t_i + \frac{T}{2}, x_A\left(i + \frac{1}{2}\right)\right] \tag{6.33}$$

Equation 6.33 is identical to the modified Euler integrator in Section 3.6.

In summary, the Taylor Series method (second order and higher) for approximating $x(t_i + T)$ requires the derivative function $f(t_i, x_i)$ as well as its derivatives (see Equation 6.5). RK-2 integrators produce estimates of x_{i+1} to the same accuracy as the first three terms in Equation 6.5 without requiring the total derivative $(d/dt)f(t, x)$. The price is an extra derivative function evaluation $f(t, x)$.

The following example illustrates use of the Taylor Series method and the RK-2 integrators. Results are compared with the first-order explicit Euler (RK-1) integrator and the exact solution.

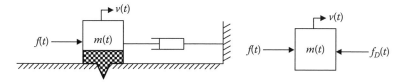

FIGURE 6.3 Moving object with decreasing mass.

Example 6.1

The object shown in Figure 6.3 is initially at rest and then subjected to a constant force $f(t) = \bar{F}, t \geq 0$. The motion of the object is opposed by the damper force $f_D(t) = \alpha v(t)$. The contents of the object are leaking so that the object's mass diminishes from its initial value m_0 to a final mass m_f.

At a given time t, the mass of the object is given by

$$m(t) = \begin{cases} m_0 - ct, & 0 \leq t \leq \dfrac{(m_0 - m_f)}{c} \\ \\ m_f, & t > \dfrac{(m_0 - m_f)}{c} \end{cases} \tag{6.34}$$

(a) Find an expression for the state derivative function $f(t, v)$ while the mass of the object is still decreasing.

(b) Find the difference equation for updating the state $v_A(i)$ using the second-order Taylor Series method.

(c) Find the difference equation for updating the state $v_A(i)$ using the RK-1 explicit Euler integrator.

(d) Find the difference equation for updating the state $v_A(i)$ using the RK-2 improved Euler integrator.

(e) Find the difference equation for updating the state $v_A(i)$ using the RK-2 modified Euler integrator.

(f) Find the exact solution for the state $v(t)$.

(g) Numerical values of the system parameters are $m_0 = 1$ slug, $m_f = 0.2$ slugs, $c = 0.05$ slugs/min, and $\alpha = 0.25$ lb/ft/min and the external force is $\bar{F} = 10$ lb. Tabulate and graph the results when $T = 0.5$ min.

(a) The differential equation model for the system is

$$m(t) \frac{dv}{dt} = \bar{F} - \alpha v \tag{6.35}$$

Solving for the derivative function,

$$\frac{dv}{dt} = f(t, v) = \frac{\bar{F} - \alpha v}{m_0 - ct}, \quad 0 \leq t \leq \frac{(m_0 - m_f)}{c} \tag{6.36}$$

(b) From Equation 6.9,

$$v_{i+1} = v_i + T f(t_i, v_i) + \frac{T^2}{2} [f_t(t_i, v_i) + f v(t_i, v_i) f(t_i, v_i)] + \cdots \tag{6.37}$$

Partial differentiation of Equation 6.36 gives

$$f_t(t_i, v_i) = (\overline{F} - \alpha v_i) \frac{c}{(m_0 - ct_i)^2} \tag{6.38}$$

$$f_v(t_i, v_i) = \frac{-\alpha}{m_0 - ct_i} \tag{6.39}$$

Substituting Equations 6.36, 6.38 and 6.39 into Equation 6.37 yields

$$v_{i+1} = v_i + T\left[\frac{\overline{F} - \alpha v_i}{m_0 - ct_i}\right] + \frac{T^2}{2}\left[\frac{c(\overline{F} - \alpha v_i)}{(m_0 - ct_i)^2} - \frac{\alpha}{(m_0 - ct_i)}\frac{\overline{F} - \alpha v_i}{m_0 - ct_i}\right] + \cdots \tag{6.40}$$

Truncating Equation 6.40 after the T^2 term, replacing v_i by $v_A(i)$, v_{i+1} by $v_A(i+1)$, and setting $t = iT$ lead to the difference equation

$$v_A(i+1) = v_A(i) + \left[\frac{\overline{F} - \alpha v_A(i)}{m_0 - ciT}\right]T + \frac{(c - \alpha)}{2}\left[\frac{\overline{F} - \alpha v_A(i)}{(m_0 - ciT)^2}\right]T^2 \tag{6.41}$$

(c) The RK-1 explicit Euler integrator is

$$\hat{v}_A(i+1) = \hat{v}_A(i) + Tf[t_i, \hat{v}_A(i)] \tag{6.42}$$

$$= \hat{v}_A(i) + T\left[\frac{\overline{F} - \alpha \hat{v}_A(i)}{m_0 - ciT}\right] \tag{6.43}$$

(d) The RK-2 improved Euler integrator, Equation 6.29, is

$$v_A(i+1) = v_A(i) + \frac{T}{2}\{f[t_i, v_A(i)] + f[t_i + T, \hat{v}_A(i+1)]\} \tag{6.44}$$

$$= v_A(i) + \frac{T}{2}\left[\frac{\overline{F} - \alpha v_A(i)}{m_0 - ciT} + \frac{\overline{F} - \alpha \hat{v}_A(i+1)}{m_0 - c(i+1)T}\right] \tag{6.45}$$

(e) The RK-2 modified Euler integrator, Equations 6.32 and 6.33, is

$$v_A\left(i + \frac{1}{2}\right) = v_A(i) + \frac{T}{2}f[t_i, v_A(i)] \tag{6.46}$$

$$= v_A(i) + \frac{T}{2}\left[\frac{\overline{F} - \alpha v_A(i)}{m_0 - ciT}\right] \tag{6.47}$$

$$v_A(i+1) = v_A(i) + Tf\left[t_{i+1/2}, v_A\left(i + \frac{1}{2}\right)\right] \tag{6.48}$$

$$= v_A(i) + T\left[\frac{\overline{F} - \alpha v_A\left(i + \frac{1}{2}\right)}{m_0 - c\left(i + \frac{1}{2}T\right)}\right] \tag{6.49}$$

(f) The exact solution for $v(t)$ is obtained from Equation 6.36 by integration.

$$\int_{v(0)}^{v} \frac{dv'}{\overline{F} - \alpha v'} = \int_0^t \frac{dt'}{m_0 - ct'} \tag{6.50}$$

$$\Rightarrow v(t) = \frac{\overline{F}}{\alpha} - \left[\frac{\overline{F}}{\alpha} - v(0)\right]\left[1 - \frac{ct}{m_0}\right]^{\alpha/c}, \quad 0 \le t \le \frac{(m_0 - m_f)}{c} \tag{6.51}$$

TABLE 6.1

Taylor Series Method, RK-1 (Explicit Euler), RK-2 (Improved Euler), RK-2 (Modified Euler) with $T = 0.5$ min, and Exact Solution

i	$t_i = iT$	Taylor Series Method $v_A(i)$	RK-1 Explicit Euler $\hat{v}_A(i)$	RK-2 Improved Euler $v_A(i)$	RK-2 Modified Euler $v_A(i)$	Exact Solution
0	0	0	0	0	0	0
1	0.5	4.75	5.0	4.7436	4.7468	4.7562
2	1	9.0375	9.4872	9.0257	9.0317	9.0488
4	2	16.3617	17.0828	16.3421	16.3520	16.3804
6	3	22.2287	23.0849	22.2045	22.2168	22.2518
8	4	26.8677	27.7584	26.8415	26.8548	26.8928
10	5	30.4826	31.3371	30.4562	30.4696	30.5078
12	6	33.2532	34.0256	33.2280	33.2408	33.2772
14	7	35.3370	36.0013	35.3139	35.3256	35.3588
16	8	36.8704	37.4162	36.8501	36.8604	36.8896
18	9	37.9706	38.3992	37.9534	37.9622	37.9869
20	10	38.7368	39.0575	38.7226	38.7298	38.7500
22	11	39.2515	39.4792	39.2403	39.2460	39.2619
24	12	39.5826	39.7345	39.5741	39.5785	39.5904
26	13	39.7843	39.8783	39.7782	39.7814	39.7899
28	14	39.8991	39.9519	39.8948	39.8970	39.9028
30	15	39.9586	39.9847	39.9559	39.9573	39.9609

(g) For the numerical values given, results from the Taylor Series method, the three numerical integrators, and the exact solution are tabulated in Table 6.1 at 1 min intervals after the first two steps.

Figure 6.4 contains a graph of the four numerical integrators and the exact solution. Both the table and figure confirm the improved accuracy possible with the use of the Taylor Series method and RK-2 integration compared to the explicit Euler (RK-1) integrator.

Knowing the exact solution, we can check the results obtained from the Taylor Series method. For the numerical values given, the exact solution in Equation 6.51 becomes

$$v(t) = 40 - 40(1 - 0.05t)^5, \quad 0 \le t \le 16 \tag{6.52}$$

The second-order truncated Taylor Series $v_2(t)$ about the point $t = 0$ is

$$v_2(T) = v(0) + \frac{d}{dt}v(0)T + \frac{1}{2}\frac{d^2}{dt^2}v(0)T^2 \tag{6.53}$$

Setting $v(0)$ to zero, differentiating Equation 6.52 to find the first two derivatives and substituting the results into Equation 6.53 give

$$v_2(T) = 10T + \frac{1}{2}(-2)T^2 \tag{6.54}$$

$$\Rightarrow v_2(0.5) = 10(0.5) - (0.5)^2 = 4.75$$

which agrees with the value in Table 6.1.

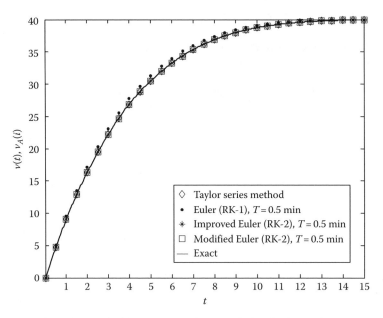

FIGURE 6.4   Comparison of numerical integrators and exact solution for Example 6.1.

6.2.4 HIGH-ORDER RUNGE–KUTTA METHODS

Higher-order RK formulas are derived in the same manner as the RK-2 integrators. For RK-3 integration, the formula for updating the state $x_A(i)$, is

$$x_A(i+1) = x_A(i) + T(a_1k_1 + a_2k_2 + a_3k_3) \tag{6.55}$$

where k_1, k_2, and k_3 are derivative function evaluations at specific points. There are now three constants p, q, and r, which determine the points at which the derivatives are to be evaluated. Matching coefficients of powers of T in the expression for $x_A(i+1)$ using Equation 6.55 with the truncated Taylor Series for $x(t)$ through the T^3 term generates four equations in the six unknowns a_1, a_2, and a_3 and p, q, and r.

One particular solution leads to the frequently used RK-3 integration formula

$$x_A(i+1) = x_A(i) + \frac{T}{6}(k_1 + 4k_2 + k_3) \tag{6.56}$$

where

$$k_1 = f[t_i, x_A(i)] \tag{6.57}$$

$$k_2 = f\left[t_i + \frac{1}{2}T, x_A(i) + \frac{1}{2}k_1T\right] \tag{6.58}$$

$$k_3 = f[t_i + T, x_A(i) - k_1T + 2k_2T] \tag{6.59}$$

The local truncation error of an RK-3 integrator $\varepsilon_T \sim O(T^4)$ and the global truncation error $E_T \sim O(T^3)$.

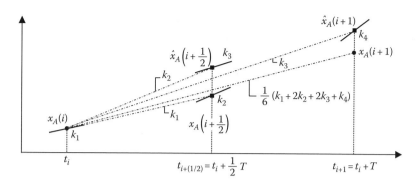

FIGURE 6.5 Illustration of an RK-4 integrator.

Fourth-order RK formulas are the most common of all the RK numerical integrators for reasons we shall discuss shortly. The derivation is patterned after the approach used for the lower-order RK methods. Flexibility in the choice of several parameters results in a family of RK-4 integrators. A popular RK-4 integrator is illustrated in Figure 6.5.

The derivative function evaluations are computed according to

$$k_1 = f[t_i, x_A(i)], \quad x_A\left(i+\frac{1}{2}\right) = x_A(i) + \frac{T}{2}k_1 \tag{6.60}$$

$$k_2 = f\left[t_{i+1/2}, x_A\left(i+\frac{1}{2}\right)\right], \quad \hat{x}_A\left(i+\frac{1}{2}\right) = x_A(i) + \frac{T}{2}k_2 \tag{6.61}$$

$$k_3 = f\left[t_{i+1/2}, \hat{x}_A\left(i+\frac{1}{2}\right)\right], \quad \hat{x}_A(i+1) = x_A(i) + Tk_3 \tag{6.62}$$

$$k_4 = f[t_{i+1}, \hat{x}_A(i+1)] \tag{6.63}$$

and the updated state $x_A(i+1)$ is obtained from

$$x_A(i+1) = x_A(i) + \frac{T}{6}(k_1 + 2k_2 + 2k_3 + k_4) \tag{6.64}$$

Note that of the four required derivative evaluations, one is at the beginning of the interval, two occur at the midpoint, and the last one takes place at the end of the interval. The algorithm is straightforward to program because of the sequential nature in the calculations of k_1, k_2, k_3, and k_4.

RK-1 through RK-4 (and higher) integrators are incorporated in simulation and numerical analysis software packages. MATLAB® and Simulink® offer a choice of RK-1 through RK-5 integrators.

6.2.5 LINEAR SYSTEMS: APPROXIMATE SOLUTIONS USING RK INTEGRATION

The special case of linear system models is worth looking at in some detail. Suppose the derivative function in Equation 6.1 is linear in x, that is,

$$\frac{dx}{dt} = f(t, x) = ax \tag{6.65}$$

Applying RK-1, RK-2, RK-3, and RK-4 integrators to the linear system in Equation 6.65 produces the following difference equations for updating the state $x_A(i)$:

$$\text{RK-1: } x_A(i+1) = (1+aT)x_A(i) \tag{6.66}$$

$$\text{RK-2: } x_A(i+1) = \left[1 + aT + \frac{1}{2!}(aT)^2\right] x_A(i) \tag{6.67}$$

$$\text{RK-3: } x_A(i+1) = \left[1 + aT + \frac{1}{2!}(aT)^2 + \frac{1}{3!}(aT)^3\right] x_A(i) \tag{6.68}$$

$$\text{RK-4: } x_A(i+1) = \left[1 + aT + \frac{1}{2!}(aT)^2 + \frac{1}{3!}(aT)^3 + \frac{1}{4!}(aT)^4\right] x_A(i) \tag{6.69}$$

The general solutions to Equations 6.66 through 6.69 are easily obtained by recursion. The results are

$$\text{RK-1: } x_A(i) = (1+aT)^i x(0) \tag{6.70}$$

$$\text{RK-2: } x_A(i) = \left[1 + aT + \frac{1}{2!}(aT)^2\right]^i x(0) \tag{6.71}$$

$$\text{RK-3: } x_A(i) = \left[1 + aT + \frac{1}{2}(aT)^2 + \frac{1}{2!}(aT)^3\right]^i x(0) \tag{6.72}$$

$$\text{RK-4: } x_A(i) = \left[1 + aT + \frac{1}{2}(aT)^2 + \frac{1}{3!}(aT)^3 + \frac{1}{4!}(aT)^4\right]^i x(0) \tag{6.73}$$

where $x(0)$ is the initial condition. In general, an RK-m integrator applied to the linear system model, Equation 6.65, results in

$$x_A(i) = \left[\sum_{k=0}^{m} \frac{(aT)^k}{k!}\right]^i x(0) \tag{6.74}$$

$$= \left[1 + aT + \frac{1}{2!}(aT)^2 + \frac{1}{3!}(aT)^3 + \cdots + \frac{1}{m!}(aT)^m\right]^i x(0) \tag{6.75}$$

In the case of more than a single state variable, that is, $\underline{\dot{x}} = A\underline{x}$, a similar result applies for RK-m integrators.

$$\underline{x}_A(i) = \left[\sum_{k=0}^{m} \frac{(TA)^k}{k!}\right]^i \underline{x}(0) \tag{6.76}$$

$$= \left[I + TA + \frac{1}{2!}(TA)^2 + \frac{1}{3!}(TA)^3 + \cdots + \frac{1}{m!}(TA)^m\right]^i \underline{x}(0) \tag{6.77}$$

Equation 6.76 for the explicit Euler integrator ($m=1$) as well as the improved and modified Euler integrators ($m=2$) was first introduced in Section 3.6.

The discrete-time signal $x_A(i)$, $i=0, 1, 2, 3, \ldots$ is intended to approximate the continuous-time state $x(t)|_{t=iT}$, $i=0, 1, 2, 3, \ldots$ The solution $x(t)$ to Equation 6.65 is

$$x(t) = x(0)e^{at}, \quad t \geq 0 \tag{6.78}$$

$$\Rightarrow x(iT) = x(0)e^{aiT} = x(0)(e^{aT})^i \tag{6.79}$$

Expanding e^{aT} in a Taylor Series about zero, Equation 6.79 becomes

$$x(iT) = \left[1 + aT + \frac{1}{2!}(aT)^2 + \frac{1}{3!}(aT)^3 + \cdots + \frac{1}{m!}(aT)^m + \cdots\right]^i x(0) \qquad (6.80)$$

From Equations 6.74 and 6.80 with $i = 1$, the $m + 1$ terms in the approximate value $x_A(1)$ are identical to the first $m + 1$ terms of the infinite series expression for $x(T)$.

After one step, the local truncation error of an RK integrator is

$$\varepsilon_T = x(T) - x_A(1) \qquad (6.81)$$

For an RK-m integrator,

$$\varepsilon_T = e^{aT}x(0) - \left[1 + aT + \frac{1}{2!}(aT)^2 + \frac{1}{3!}(aT)^3 + \cdots + \frac{1}{m!}(aT)^m\right]x(0) \qquad (6.82)$$

Replacing e^{aT} in Equation 6.82 by its Taylor Series expansion leads to

$$\varepsilon_T = \left[\frac{1}{(m+1)!}(aT)^{m+1} + \frac{1}{(m+2)!}(aT)^{m+2}\cdots\right]x(0) \qquad (6.83)$$

and, therefore, $\varepsilon_T \sim O(T^{m+1})$ as expected. All RK-m integrators are said to be of mth order, not to be confused with their local truncation error, which is of order $m + 1$, that is, $\varepsilon_T \sim O(T^{m+1})$. The mth order reference stems from the high-order term in the truncated Taylor Series. For an RK-m integrator, the global truncation error $E_T \sim O(T^m)$.

RK-1 through RK-4 integrators require one to four derivative function evaluations per step. RK integrators of order higher than four are not as efficient. For example, an RK-5 integrator requires six derivative function evaluations per step for comparable agreement with the fifth-order Taylor Series expansion of the solution. A penalty of one additional derivative function evaluation per step is the price incurred in moving from an RK-4 integrator with $\varepsilon_T \sim O(T^5)$ to an RK-5 integrator with $\varepsilon_T \sim O(T^6)$. The computational effort during each integration step results primarily from evaluating the derivative function. Hence, the penalty is nontrivial.

Worse yet, RK-6 integrators require eight derivative function evaluations to achieve a local truncation error $\varepsilon_T \sim O(T^7)$. RK-4 methods are popular because they are the highest order one-step integrators that do not require more derivative function evaluations than their order.

6.2.6 Continuous-Time Models with Polynomial Solutions

The Taylor Series method for finding $x_A(i + 1)$ starting from the point (t_i, x_i) on the solution $x(t)$ is

$$x_A(i + 1) = x_i + \frac{d}{dt}x(t_i)T + \frac{1}{2!}\frac{d^2}{dt^2}x(t_i)T^2 + \cdots + \frac{1}{m!}\frac{d^m}{dt^m}x(t_i)T^m \qquad (6.84)$$

where the total derivatives $(d^2/dt^2)x(t_i)$, $(d^3/dt^3)x(t_i), \ldots, (d^m/dt^m)x(t_i)$ are computed from partial derivatives of the derivative function $f(t_i, x_i)$.

Suppose the exact solution is the mth-order polynomial

$$x(t) = a_0 + a_1 t + a_2 t^2 + \cdots + a_m t^m \qquad (6.85)$$

The exact solution at $t = t_{i+1}$ is

$$x(t_{i+1}) = a_0 + a_1 t_{i+1} + a_2 t_{i+1}^2 + \cdots + a_m t_{i+1}^m \qquad (6.86)$$

With $x_A(0)$ set equal to $x(0)$, Equations 6.84 and 6.86 produce identical results at the discrete points $0, T, 2T, \ldots$ In other words,

$$x_A(i+1) = x(t_{i+1}), \quad i = 0, 1, 2, \ldots \tag{6.87}$$

Proof for the case when $m = 2$ follows. Starting with

$$x_A(i+1) = x_i + \frac{d}{dt}x(t_i)T + \frac{1}{2!}\frac{d^2}{dt^2}x(t_i)T^2 \tag{6.88}$$

The two derivatives in Equation 6.88 are obtained from the exact solution for $x(t)$ in Equation 6.85 with $m = 2$. Substituting them into Equation 6.88 and simplifying give

$$x_A(i+1) = x_i + (a_1 + 2a_2t_i)T + \frac{1}{2}(2a_2)T^2 \tag{6.89}$$

$$= (a_0 + a_1t_i + a_2t_i^2) + a_1T + 2a_2t_iT + a_2T^2 \tag{6.90}$$

$$= a_0 + a_1(t_i + T) + a_2(t_i + T)^2 \tag{6.91}$$

$$= x(t_{i+1}) \tag{6.92}$$

The proof is similar for higher-order polynomial solutions.

In Example 6.1, the exact solution $v(t)$ in Equation 6.52 is a fifth-order polynomial. Hence, the Taylor Series method using the fifth-order truncated Taylor Series would agree with the exact solution at $0, T, 2T, \ldots$ However, in Example 6.1, a second-order Taylor Series was used to generate the discrete-time values $v_A(1)$, $v_A(2)$, $v_A(4)$, $v_A(6)$, $\ldots$, $v_A(30)$ shown in Table 6.1. This explains the discrepancy between the discrete-time values and the exact solution $v(t_1)$, $v(t_2)$, $v(t_4)$, $v(t_6)$, $\ldots$, $v(t_{30})$ shown in the last column of the table.

In general, when $x(t)$ is an mth-order polynomial, unlike the mth-order Taylor Series method, RK-m integrators will not generate the true solution values $x(t_1)$, $x(t_2)$, $x(t_3)$, $\ldots$ Different RK-m integrators will produce different discrete-time solutions; however, they achieve comparable accuracy with the Taylor Series method in the sense that the local truncation errors are the same order of magnitude. A similar result holds for RK integrators and the truncated Taylor Series method when both are the same order and less than m. In that case, the Taylor Series method will no longer be exact. The following example illustrates this point.

Example 6.2

In Example 6.1, if we change the value of α from 0.25 to 0.1, the exact solution to Equation 6.35 becomes

$$v(t) = 100 - \frac{1}{4}(20 - t)^2 \tag{6.93}$$

Approximate the solution for $v(t)$ using the second-order Taylor Series, RK-1 integration, and both RK-2 integrators with a step size of $T = 0.5$. Compare results with the exact solution.

The state derivative function, Equation 6.36, is given by

$$f(t_i, v_i) = \frac{\bar{F} - \alpha v_i}{m_0 - ct_i} = \frac{10 - 0.1v_i}{1 - 0.05t_i} = 2\left(\frac{100 - v_i}{20 - t_i}\right) \tag{6.94}$$

The first partials in Equations 6.38 and 6.39 become

$$f_t(t_i, v_i) = 2\left[\frac{100 - v_i}{(20 - t_i)^2}\right] \tag{6.95}$$

TABLE 6.2
Comparison of Taylor Series Method, RK-1, RK-2, and Exact Solution

i	t_i	Taylor Series $v_A(i)$	RK-1 Explicit $v_A(i)$	RK-2 Improved $v_A(i)$	RK-2 Modified $v_A(i)$	Exact $v(t_i)$
0	0	0	0	0	0	0
1	0.5	4.9375	5.0000	4.9359	4.9367	4.9375
2	1.0	9.7500	9.8718	9.7468	9.7484	9.7500
4	2.0	19.0000	19.2308	18.9938	18.9970	19.0000
6	3.0	27.7500	28.0769	27.7410	27.7455	27.7500
8	4.0	36.0000	36.4103	35.9883	35.9942	36.0000
10	5.0	43.7500	44.2308	43.7357	43.7430	43.7500

$$f_v(t_i, v_i) = \frac{-2}{(20 - t_i)} \tag{6.96}$$

and the Taylor Series method for calculating the approximation to $v(t + T)$ is

$$v_A(i + 1) = v_A(i) + Tf[t_i, v_A(i)] + \frac{T^2}{2}\{f_t[t_i, v_A(i)] + f_v[t_i, v_A(i)]f[t_i, v_A(i)]\} \tag{6.97}$$

From Equations 6.94 through 6.97 with $i = 0$, $t_i = t_0 = 0$ and $v_A(i) = v_A(0) = v(0) = 0$,

$$v_A(1) = 10T - \frac{1}{4}T^2 \tag{6.98}$$

For a step size of $T = 0.5$, $v_A(1) = 4.9375$. The exact solution $v(T)|_{T=0.5}$ is computed from

$$v(t)|_{t=T=0.5} = 100 - \frac{1}{4}(20 - t)^2\bigg|_{t=T=0.5} = 4.9375$$

which agrees with the result from the Taylor Series method.

Results for the Taylor Series method, RK-1, both RK-2 integrators, and the exact solution are tabulated in Table 6.2.

6.2.7 HIGHER-ORDER SYSTEMS

The application of RK numerical integration to higher-order systems is straightforward. The differential equations of an nth-order system model are expressed as a system of first-order differential equations as shown in Equations 6.99 through 6.101.

$$\frac{dx_1}{dt} = f_1(t, x_1, x_2, \ldots, x_n) \tag{6.99}$$

$$\frac{dx_2}{dt} = f_2(t, x_1, x_2, \ldots, x_n) \tag{6.100}$$

$$\vdots$$

$$\frac{dx_n}{dt} = f_n(t, x_1, x_2, \ldots, x_n) \tag{6.101}$$

Updating the current discrete-time state vector $[x_{1,A}(i), x_{2,A}(i), \ldots, x_{n,A}(i)]$ to the new vector $[x_{1,A}(i+1), x_{2,A}(i+1), \ldots, x_{n,A}(i+1)]$ with RK-m integration consists of determining, in the proper sequence, the derivatives $k_{j,p}$, $j = 1, 2, \ldots, m$ and $p = 1, 2, 3, \ldots, n$. By the proper sequence, we mean $k_{1,1}, k_{1,2}, \ldots, k_{1,n}$, followed by $k_{2,1}, k_{2,2}, \ldots, k_{2,n}$ up through by $k_{m,1}, k_{m,2}, \ldots, k_{m,n}$.

To illustrate, suppose we are dealing with a third-order ($n = 3$) system and choose to implement a fourth-order ($m = 4$) RK-4 integrator to update the discrete-time state. The three derivative functions are each calculated four times in the following order:

$$\left.\begin{aligned}
k_{1,1} &= f_1[t_i, x_{1,A}(i), x_{2,A}(i), x_{3,A}(i)] \\
k_{1,2} &= f_2[t_i, x_{1,A}(i), x_{2,A}(i), x_{3,A}(i)] \\
k_{1,3} &= f_3[t_i, x_{1,A}(i), x_{2,A}(i), x_{3,A}(i)]
\end{aligned}\right\} \tag{6.102}$$

$$\left.\begin{aligned}
k_{2,1} &= f_1[t_i + 0.5T, x_{1,A}(i) + 0.5Tk_{1,1}, x_{2,A}(i) + 0.5Tk_{1,2}, x_{3,A}(i) + 0.5Tk_{1,3}] \\
k_{2,2} &= f_2[t_i + 0.5T, x_{1,A}(i) + 0.5Tk_{1,1}, x_{2,A}(i) + 0.5Tk_{1,2}, x_{3,A}(i) + 0.5Tk_{1,3}] \\
k_{2,3} &= f_3[t_i + 0.5T, x_{1,A}(i) + 0.5Tk_{1,1}, x_{2,A}(i) + 0.5Tk_{1,2}, x_{3,A}(i) + 0.5Tk_{1,3}]
\end{aligned}\right\} \tag{6.103}$$

$$\left.\begin{aligned}
k_{3,1} &= f_1[t_i + 0.5T, x_{1,A}(i) + 0.5Tk_{2,1}, x_{2,A}(i) + 0.5Tk_{2,2}, x_{3,A}(i) + 0.5Tk_{2,3}] \\
k_{3,2} &= f_2[t_i + 0.5T, x_{1,A}(i) + 0.5Tk_{2,1}, x_{2,A}(i) + 0.5Tk_{2,2}, x_{3,A}(i) + 0.5Tk_{2,3}] \\
k_{3,3} &= f_3[t_i + 0.5T, x_{1,A}(i) + 0.5Tk_{2,1}, x_{2,A}(i) + 0.5Tk_{2,2}, x_{3,A}(i) + 0.5Tk_{2,3}]
\end{aligned}\right\} \tag{6.104}$$

$$\left.\begin{aligned}
k_{4,1} &= f_1[t_i + T, x_{1,A}(i) + Tk_{3,1}, x_{2,A}(i) + Tk_{3,2}, x_{3,A}(i) + Tk_{3,3}] \\
k_{4,2} &= f_2[t_i + T, x_{1,A}(i) + Tk_{3,1}, x_{2,A}(i) + Tk_{3,2}, x_{3,A}(i) + Tk_{3,3}] \\
k_{4,3} &= f_3[t_i + T, x_{1,A}(i) + Tk_{3,1}, x_{2,A}(i) + Tk_{3,2}, x_{3,A}(i) + Tk_{3,3}]
\end{aligned}\right\} \tag{6.105}$$

The components of the state are updated according to

$$x_{1,A}(i+1) = x_{1,A}(i) + \frac{T}{6}(k_{1,1} + 2k_{2,1} + 2k_{3,1} + k_{4,1}) \tag{6.106}$$

$$x_{2,A}(i+1) = x_{2,A}(i) + \frac{T}{6}(k_{1,2} + 2k_{2,2} + 2k_{3,2} + k_{4,2}) \tag{6.107}$$

$$x_{3,A}(i+1) = x_{3,A}(i) + \frac{T}{6}(k_{1,3} + 2k_{2,3} + 2k_{3,3} + k_{4,3}) \tag{6.108}$$

An example of a second-order system model using RK-4 integration is now presented. The standard form of a linear second-order system is

$$\frac{d^2x}{dt^2} + 2\zeta\omega_n\frac{dx}{dt} + \omega_n^2 x = K\omega_n^2 u \tag{6.109}$$

Letting $x_1 = x$ and $x_2 = dx/dt$ leads to the state equation model

$$\frac{dx_1}{dt} = f_1(t, x_1, x_2, u) = x_2 \tag{6.110}$$

$$\frac{dx_2}{dt} = f_2(t, x_1, x_2, u) = -\omega_n^2 x_1 - 2\zeta\omega_n x_2 + K\omega_n^2 u \tag{6.111}$$

where the last argument u of $f_1(t, x_1, x_2, u)$ and $f_2(t, x_1, x_2, u)$ refers to the system input.

Expressions for the derivatives k_1, k_2, k_3, and k_4 associated with states x_1 and x_2 are

$$k_{1,1} = f_1[t_i, x_{1,A}(i), x_{2,A}(i), u(t_i)] \tag{6.112}$$

$$= x_{2,A}(i) \tag{6.113}$$

$$k_{1,2} = f_2[t_i, x_{1,A}(i), x_{2,A}(i), u(t_i)] \tag{6.114}$$

$$= -\omega_n^2 x_{1,A}(i) - 2\zeta\omega_n x_{2,A}(i) + K\omega_n^2 u(t_i) \tag{6.115}$$

$$k_{2,1} = f_1[t_i + 0.5T, x_{1,A}(i) + 0.5Tk_{1,1}, x_{2,A}(i) + 0.5Tk_{1,2}, u(t_i + 0.5T)] \tag{6.116}$$

$$= x_{2,A}(i) + 0.5Tk_{1,2} \tag{6.117}$$

$$k_{2,2} = f_2[t_i + 0.5T, x_{1,A}(i) + 0.5Tk_{1,1}, x_{2,A}(i) + 0.5Tk_{1,2}, u(t_i + 0.5T)] \tag{6.118}$$

$$= -\omega_n^2[x_{1,A}(i) + 0.5Tk_{1,1}] - 2\zeta\omega_n[x_{2,A}(i) + 0.5Tk_{1,2}] + K\omega_n^2 u(t_i + 0.5T) \tag{6.119}$$

$$k_{3,1} = f_1[t_i + 0.5T, x_{1,A}(i) + 0.5Tk_{2,1}, x_{2,A}(i) + 0.5Tk_{2,2}, u(t_i + 0.5T)] \tag{6.120}$$

$$= x_{2,A}(i) + 0.5Tk_{2,2} \tag{6.121}$$

$$k_{3,2} = f_2[t_i + 0.5T, x_{1,A}(i) + 0.5Tk_{2,1}, x_{2,A}(i) + 0.5Tk_{2,2}, u(t_i + 0.5T)] \tag{6.122}$$

$$= -\omega_n^2[x_{1,A}(i) + 0.5Tk_{2,1}] - 2\zeta\omega_n[x_{2,A}(i) + 0.5Tk_{2,2}] + K\omega_n^2 u(t_i + 0.5T) \tag{6.123}$$

$$k_{4,1} = f_1[t_i + T, x_{1,A}(i) + Tk_{3,1}, x_{2,A}(i) + Tk_{3,2}, u(t_i + T)] \tag{6.124}$$

$$= x_{2,A}(i) + Tk_{3,2} \tag{6.125}$$

$$k_{4,2} = f_2[t_i + T, x_{1,A}(i) + Tk_{3,1}, x_{2,A}(i) + Tk_{3,2}, u(t_i + T)] \tag{6.126}$$

$$= -\omega_n^2[x_{1,A}(i) + Tk_{3,1}] - 2\zeta\omega_n[x_{2,A}(i) + Tk_{3,2}] + K\omega_n^2 u(t_i + T) \tag{6.127}$$

The updated state $[x_{1,A}(i), x_{2,A}(i)]$ is obtained from

$$x_{1,A}(i+1) = x_{1,A}(i) + \frac{T}{6}(k_{1,1} + 2k_{2,1} + 2k_{3,1} + k_{4,1}) \tag{6.128}$$

$$x_{2,A}(i+1) = x_{2,A}(i) + \frac{T}{6}(k_{1,2} + 2k_{2,2} + 2k_{3,2} + k_{4,2}) \tag{6.129}$$

An example illustrating the application of Equations 6.110 through 6.129 follows.

Example 6.3

A simplified model to predict the levels of a drug in an individual is accomplished using compartmental analysis similar to the iodine model in Section 4.3. After oral ingestion, a drug enters the gastrointestinal tract (compartment 1) and is then distributed into the bloodstream (compartment 2) where it is metabolized and eliminated. State equations describing the drug dynamics in each compartment are

$$\text{Gastrointestinal tract: } \frac{dm_1}{dt} = -c_1 m_1 + u \tag{6.130}$$

$$\text{Bloodstream: } \frac{dm_2}{dt} = -c_1 m_1 - c_2 m_2 \tag{6.131}$$

where
  m_1 and m_2 are the amounts of drug in each compartment (mg)
  u is the ingestion rate of the drug (mg/min)
  c_1 and c_2 are drug distribution and elimination constants of the individual (min^{-1})

The output y is the amount of drug in the bloodstream, that is, m_2.

(a) Convert the state equations into a single second-order differential equation relating the output y and input u. Find ζ, ω_n, and K for the second-order system.
(b) Define $x_1 = y = m_2$ and $x_2 = dy/dt = dm_2/dt$ and simulate the response using classic RK-4 integration with a step size $T = 1$ min for the following conditions:

$$m_1(0) = 0 \text{ mg}, \quad m_2(0) = 0 \text{ mg}, \quad c_1 = 0.06 \text{ min}^{-1}, \quad \text{and} \quad c_2 = 0.015 \text{ min}^{-1}$$

Assume the drug ingestion rate is given by

$$u(t) = Me^{-t/\tau}, \quad t \geq 0 \ (M = 5 \text{ mg/min}, \ \tau = 4 \text{ min}) \tag{6.132}$$

(c) Find the exact solution for $x_1(t)$.
(d) Plot the simulated response $x_{1,A}(i)$ and the exact solution $x_1(t)$ on the same graph.

(a) Elimination of m_1 from Equations 6.130 and 6.131 is easily accomplished by Laplace transforming the equations and algebraically solving for $M_2(s)$, which is also $Y(s)$.

$$(s + c_1)M_1(s) = U(s) \tag{6.133}$$

$$(s + c_2)M_2(s) = c_1 M_1(s) \tag{6.134}$$

$$Y(s) = M_2(s) = \frac{c_1}{s + c_2} M_1(s) \tag{6.135}$$

$$= \frac{c_1}{s + c_2}\left[\frac{1}{s + c_1} U(s)\right] \tag{6.136}$$

Inverse Laplace transformation of $Y(s)$ leads to the differential equation

$$\frac{d^2y}{dt^2} + (c_1 + c_2)\frac{dy}{dt} + c_1 c_2 y = c_1 u \tag{6.137}$$

Comparing Equation 6.137 with the standard form of Equation 6.109 yields

$$\omega_n = (c_1 c_2)^{1/2}, \quad \zeta = \frac{c_1 + c_2}{2(c_1 c_2)^{1/2}}, \quad K = \frac{1}{c_2} \tag{6.138}$$

Solving for the second-order system parameters,

$$\omega_n = (c_1 c_2)^{1/2} = [(0.06)(0.015)]^{1/2} = 0.03 \text{ rad/min}$$

$$\zeta = \frac{c_1 + c_2}{2(c_1 c_2)^{1/2}} = \frac{0.06 + 0.015}{2(0.03)} = 1.25$$

$$K = \frac{1}{c_2} = \frac{1}{0.015} = 66.6\bar{6} \text{ min}^{-1}$$

(b) The RK-4 calculations follow the procedure outlined in Equations 6.112 through 6.129. The integration step begins with the k_1 derivative evaluation for each state, that is,

$$k_{1,1} = x_{2,A}(0) = x_2(0) = 0$$
$$k_{1,2} = -\omega_n^2 x_{1,A}(0) - 2\zeta\omega_n x_{2,A}(0) + K\omega_n^2 u(0)$$
$$= -\omega_n^2 x_1(0) - 2\zeta\omega_n x_2(0) + K\omega_n^2 M$$
$$= -0.0009(0) - 2(1.25)(0.03)(0) + \left(\frac{1}{0.015}\right)(0.0009)(5)$$
$$= 0.3$$

followed by the k_2 derivative evaluation for each state, that is,

$$k_{2,1} = x_{2,A}(0) + \frac{1}{2}k_{1,2}T$$

$$= 0 + \frac{1}{2}(0.3)(1)$$

$$= 0.15$$

$$k_{2,2} = -\omega_n^2 \left[x_{1,A}(i) + \frac{1}{2}k_{1,1}T \right] - 2\zeta\omega_n \left[x_{2,A}(i) + \frac{1}{2}k_{1,2}T \right] + K\omega_n^2 u\left(t_i + \frac{1}{2}T \right)$$

$$= -0.0009 \left[0 + \frac{1}{2}(0)(1) \right] - 2(1.25)(0.03) \left[0 + \frac{1}{2}(0.3)(1) \right]$$

$$+ \left(\frac{1}{0.015} \right)(0.0009)5e^{-0.5/4}$$

$$= 0.2535$$

The remaining derivative evaluations are obtained in a similar fashion.

$$k_{3,1} = 0.1267, \quad k_{3,2} = 0.2552, \quad k_{4,1} = 0.2552, \quad k_{4,2} = 0.2144$$

M-file "Chap6_Ex2_3.m" recursively solves the RK-4 difference equations for the discrete-time states $x_{1,A}(i)$ and $x_{2,A}(i)$. Table 6.3 contains selected values of $x_{1,A}(i)$.

(c) The exact solution for $x_1(t) = y(t)$ can be determined by substituting the Laplace transform of $u(t)$ into Equation 6.136,

$$X_1(s) = \frac{c_1}{s + c_2} \left[\frac{1}{s + c_1} \left(\frac{M}{s + (1/\tau)} \right) \right] \tag{6.139}$$

TABLE 6.3
Approximate (RK-4, $T = 1$ min) and Exact Solutions

i	t_i	$x_{1,A}(i)$	$x_1(t_i)$
0	0	0.0	0.0
1	1	0.13477907	0.13477243
2	2	0.48560007	0.48558842
3	3	0.98658006	0.98656465
4	4	1.58751933	1.58750114
5	5	2.25036682	2.25034661
25	25	11.68143041	11.68141034
50	50	11.65359364	11.65357994
100	100	6.24296647	6.24295986
150	150	2.98572189	2.98571876
200	200	1.41218500	1.41218352
250	250	0.66716006	0.66715936
300	300	0.31514864	0.31514831

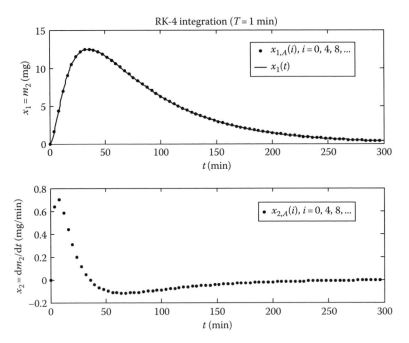

FIGURE 6.6   Discrete-time signals $x_{1,A}(i)$ and $x_{2,A}(i)$ and continuous-time signal $x_1(t)$.

Inverse Laplace transforming Equation 6.139 gives the exact solution,

$$x_1(t) = \frac{Mc_1\tau}{(c_1 - c_2)(1 - c_1\tau)(1 - c_2\tau)} \left[(1 - c_1\tau)e^{-c_2t} - (1 - c_2\tau)e^{c_1 t} + (c_1 - c_2)\tau e^{-t/\tau} \right] \quad (6.140)$$

(d) Table 6.3 contains values of the discrete-time response $x_{1,A}(i)$ and the exact solution $x_1(t)$ at different times. The approximate and exact solutions agree to four places after the decimal point.

Figure 6.6 contains plots of the discrete-time states $x_{1,A}(i)$ and $x_{2,A}(i)$ and the exact solution $x_1(t)$. Every fourth point of the discrete-time signals is plotted for the sake of clarity.

Before we proceed further, it would be useful to know the total amount of drug ingested by the individual. Integrating the rate of drug ingestion over time,

$$M_T = \int_0^\infty u(t)dt = \int_0^\infty Me^{-t/\tau}dt = M\tau \quad (6.141)$$

For a continuous-time integrator, the derivative function $f(t, x)$ is equal to the input $u(t)$ to the integrator. Hence, M_T in Equation 6.141 can be found for an arbitrary input function $u(t)$ using any of the numerical integrators we have studied. Of course, the upper limit must be finite, presumably the time required for the drug to be fully ingested.

We conclude this section with a simple nonlinear system model.

Example 6.4

The cooling of a high-temperature oven is governed by the following differential equation (McClamroch 1980):

$$C\frac{d\tilde{T}}{dt} = -K_c(\tilde{T} - T_0) - K_r(\tilde{T}^4 - T_0^4) \quad (6.142)$$

where

$\tilde{T} = \tilde{T}(t)$ is the oven temperature (°R)

T_0 is the surrounding temperature (°R)

C is the thermal capacity of the oven

K_c and K_r are convective and radiation heat loss coefficients

Simulate the oven's cooling from an initial temperature of 1000°R if the surrounding temperature is 500°R. Numerical values of the thermal parameters are

$$C = 24\,\text{Btu/°R}, \quad K_c = 8\,\text{Btu/h/°R}, \quad K_r = 2 \times 10^{-8}\,\text{Btu/h/°R}^4$$

RK-1 through RK-4 integrators were used to numerically integrate the derivative function

$$f(\tilde{T}) = \frac{d\tilde{T}}{dt} = -\frac{K_c}{C}(\tilde{T} - T_0) - \frac{K_r}{C}(\tilde{T}^4 - T_0^4) \qquad (6.143)$$

The results are shown in Table 6.4. The integration step size was chosen for each integration method to make the total number of derivative function evaluations and, hence, the computational effort, roughly the same.

The last column is labeled "Exact"; however, the exact solution is not easily obtained. The numbers in the last column were obtained using RK-4 integration with a small enough step size ($T = 0.005$ h) to generate approximate values in agreement with the exact solution values to at least one place after the decimal point. How can we check this assumption?

Figure 6.7 shows the results of using RK-1 and RK-4 to integrate the derivative function, Equation 6.143, with step sizes $T = 0.1, 0.2, 0.3$ h and $T = 0.3, 0.6, 0.9$ h, respectively. RK-1 produces reasonably accurate results only with $T = 0.1$ h whereas RK-4 integration generates accurate approximations to the "exact" solution for $T = 0.4$ h and $T = 0.8$ h.

TABLE 6.4

Comparison of RK-1, 2, 3, 4 Integrators with Different Step Sizes and Exact Solution

		RK-1			RK-2			RK-3			RK-4	"Exact"
		$T=0.1$			$T=0.2$			$T=0.3$			$T=0.4$	
i	t_i	$\tilde{T}A(i)$	i	t_i	$\tilde{T}A(i)$	i	t_i	$\tilde{T}A(i)$	i	t_i	$\tilde{T}A(i)$	$\tilde{T}A(i)$
0	0	1000.0	0	0	1000.0	0	0	1000.0	0	0	1000.0	1000.0
2	0.2	841.0	1	0.2	864.1							859.9
3	0.3	793.1				1	0.3	806.2				813.2
4	0.4	755.6	2	0.4	784.9				1	0.4	774.3	775.5
6	0.6	699.8	3	0.6	730.8	2	0.6	713.5				718.0
8	0.8	660.0	4	0.8	691.1				2	0.8	675.6	676.2
9	0.9	644.0				3	0.9	656.1				659.3
10	1.0	630.1	5	1.0	660.6							644.4
12	1.2	607.0	6	1.2	636.4	4	1.2	617.2	3	1.2	619.1	619.5
16	1.6	573.9	8	1.6	600.8				4	1.6	583.4	583.6
20	2.0	552.1	10	2.0	576.3				5	2.0	559.4	559.6
30	3.0	522.7	15	3.0	540.7	10	3.0	526.2				526.7
40	4.0	510.2	20	4.0	523.3				10	4.0	512.3	512.3
48	4.8	505.4	24	4.8	515.4	16	4.8	506.6	12	4.8	506.7	506.7

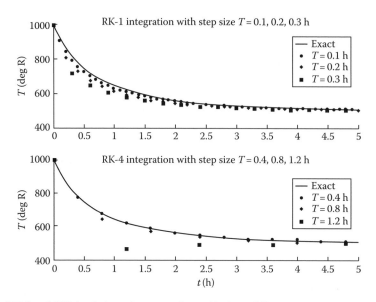

FIGURE 6.7 RK-1 and RK-4 solution of oven cooling with three different step sizes.

EXERCISES

6.1 Show that the difference equation resulting from using RK-m integration to approximate the behavior of $dx/dt = f(t, x) = ax$ is

$$x_A(i + 1) = \left[1 + aT + \frac{1}{2!}(aT)^2 + \frac{1}{3!}(aT)^3 + \cdots + \frac{1}{m!}(aT)^m\right] x_A(i)$$

6.2 The mass m of a radioactive material in a container decays according to the equation

$$\frac{dm}{dt} = -km \quad \text{subject to } m(0) = m_0$$

where

$m = m(t)$

k is a constant for the specific radioactive material

m_0 is the initial mass of radioactive material in the container

The half-life of a radioactive material, $T_{1/2}$, is related to the decay constant k by $T_{1/2} = \ln 2/k$.

Suppose the half-life of a radioactive material is 2 years and there is initially 1 kg of material present in the container.

(a) Use RK-1 through RK-4 integration to obtain approximations of the mass of radioactive material present in the container every month until less than 0.25 kg of material remains.

(b) Compare the results from part (a) with the exact solution for $m(t)$.

6.3 The amount of fish in a lake at any time is assumed to obey the following logistic growth model:

$$\frac{dx}{dt} = Kx(M - x) - u$$

where

$x = x(t)$ is the number of fish present

$u = u(t)$ is the rate at which fish are harvested

Nominal values of the system parameters are $K=2.5 \times 10^{-7}$ (fish-day)$^{-1}$ and $M=200,000$ fish. The lake is initially stocked with 50,000 fish.

(a) Use RK-4 integration with appropriate step size T to approximate the fish population in the absence of harvesting. Plot the results.

(b) Plot the exact solution $x(t)=M/(1-[1-M/x(0)]e^{-KMt})$ on the same graph.

(c) Repeat part (a) for a constant harvesting rate $u=2750$ fish/day.

(d) Repeat part (a) for a constant harvesting rate $u=2250$ fish/day.

6.4 For the system in Example 6.1, let $\alpha=0.15$, $c=0.05$, $m_0=1$, $\bar{F}=0$, and $v(0)=2$. The derivative function is

$$\frac{dv}{dt} = f(t, v) = \frac{-3v}{20-t}, \quad 0 \le t \le 16$$

and the exact solution is

$$v(t) = 2(1-0.05t)^3, \quad 0 \le t \le 16$$

Show that $v(0.5)$ and $v_A(1)$ are identical, where $v_A(1)$ is the result of using RK-3 integration with a step size $T=0.5$.

6.5 Repeat Example 6.3 using RK-1 and RK-2 integrator with a step size $T=1$ min. Compare the accuracy of each with the RK-4 method results shown in Table 6.3.

6.6 In Example 6.3, choose the states $x_1=m_1$ and $x_2=m_2$ and the outputs $y_1=x_1$ and $y_2=x_2$.

(a) Find the matrices A, B, C, and D in the continuous-time state variable model of the system.

(b) Apply RK-4 integration with $T=1$ min to obtain approximate solutions for the amount of drug in the gastrointestinal tract and the bloodstream. Compare the results of drug amounts in the bloodstream with results in Table 6.3.

(c) Use RK-4 integration with $T=1$ min to find an approximate solution for the case where $m_1(0)=20$ mg, $m_2(0)=0$ mg, and $u(t)=0$, $t \ge 0$.

(d) Verify the results in part (c) by using

$$\underline{x}_A(i) = \left[I + TA + \frac{1}{2!}(TA)^2 + \frac{1}{3!}(TA)^3 + \frac{1}{4!}(TA)^4 \right]^i \underline{x}(0)$$

6.7 Approximate the amount of drug ingested by an individual using RK-1 through RK-4 integration (using an appropriate step size T) when the drug ingestion rate is

(a) $u(t)=Me^{-t/\tau}$, $t \ge 0$ ($M=5$ mg/min, $\tau=4$ min)

(b) $u(t)=Me^{-t/\tau}$, $t \ge 0$ ($M=1$ mg/min, $\tau=45$ min)

(c) $u(t)=A \sin(2\pi t/P)$, $0 \le t \le P/2$ ($A=2$ mg/min, $P=30$ min)

(d) $u(t)$ is available in tabular form in the following table:

t (min)	0	0.5	1.0	1.5	2.0	2.5	3.0	3.5	4.0	4.5	5.0
$u(t)$ (mg/min)	0.0	0.4	1.0	3.0	2.2	1.4	0.8	0.4	0.2	0.1	0.0

6.8 Since RK-4 integrators require four times the number of derivative function evaluations as RK-1 integrators and RK-2 integrators require twice the number of derivative function evaluations as RK-1 integrators, it is reasonable to compare the three integrators when the computational effort is roughly the same for all three. In other words, if the step size for the RK-1 integrator is T, then the RK-2 and RK-4 integrators should be run with step sizes $2T$ and $4T$, respectively. Simulate the response of the system in Example 6.1 where

$$\frac{dv}{dt} = f(t, v) = 5\left[\frac{2 - v}{20 - t}\right], \quad v(0) = 0$$

using RK-1, RK-2 (improved or modified Euler), and the classic RK-4 integrator using step sizes of 0.25, 0.5, and 1 s, respectively. Enter the results in the following table. Comment on the results.

RK-1 ($T=0.25$)				RK-2 ($T=0.5$)				RK-4 ($T=1$)		
i	t_i	$v_A(i)$	$v(t_i)$	i	t_i	$v_A(i)$	$v(t_i)$	i	t_i	$v_A(i)$	$v(t_i)$
0	0	0.00000	0.00000	0	0	0.00000	0.00000	0	0	0.00000	0.00000
4	1			2	1			1	1		
8	2			4	2			2	2		
12	3			6	3			3	3		
16	4			8	4			4	4		
20	5			10	5			5	5		
24	6			12	6			6	6		
28	7			14	7			7	7		
32	8			16	8			8	8		

6.9 The model for finding the temperature in the oven of Example 6.4 when there is an internal heat source is

$$C\frac{d\tilde{T}}{dt} = -K_c(\tilde{T} - T_0) - K_r(\tilde{T}^4 - T_0^4) + u$$

where $u = u(t)$ is the heat source. Suppose the oven and its surroundings are in equilibrium at a temperature of 600°R.
(a) Simulate the transient response of the oven temperature using an RK-2 integrator with step size $T = 0.25$ h when the heat transferred to the oven is as shown in Figure E6.9:

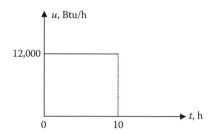

FIGURE E6.9

(b) Find the exact solution for $\tilde{T}(t)$ to one place after the decimal point by solving for $\tilde{T}_A(i) \approx \tilde{T}(t_i)$, $i = 0, 1, 2, 3, \ldots$ using RK-4 integration with step size $T = 0.001$ h. Compare the results with those from part (a) and (b).

6.3 ADAPTIVE TECHNIQUES

The computational efficiency of RK methods can be improved if the step size is allowed to vary during a simulation. A reasonable criterion must be established for determining when it is appropriate to modify the step size and by how much. The criterion is usually based on an estimate of the local truncation error as the simulation progresses with time. If an estimate of the local truncation error is outside an acceptable tolerance, then it is possible to either reduce the step size when the estimated error is too large or quite possibly increase the step size if it appears that the error is unnecessarily small. Techniques for estimating the local truncation error and modifying the step size, if warranted, are referred to as adaptive step size control.

6.3.1 REPEATED RK WITH INTERVAL HALVING

If we use the local truncation error as the basis for determining when the step size needs adjustment, then a method is needed for approximating it. One approach requires that we obtain two estimates of the updated state from an RK integrator and use the difference to estimate the local truncation error. Interval halving refers to the case where the step sizes differ by a factor of 2.

Refer to Figure 6.8 to understand how the method works. Let $x_A(i+1|T)$ be the approximate solution to $\dot{x} = f(t,x)$ at t_{i+1} obtained using a step size of T. Similarly, let $x_A(i+1|T/2)$ be the approximate solution to $\dot{x} = f(t,x)$ at t_{i+1} obtained after two steps using a step size of $T/2$.

Assume $x_A(i)$ is exact, that is, $x_A(i) = x(t_i)$. It follows that

$$x(t_{i+1}) = x_A(i+1|T) + \varepsilon_T \tag{6.144}$$

$$x(t_{i+1}) = x_A\left(i+1\left|\frac{T}{2}\right.\right) + \varepsilon_{T/2} \tag{6.145}$$

where ε_T and $\varepsilon_{T/2}$ are the local truncation errors in $x_A(i+1|T)$ and $x_A(i+1|T/2)$, respectively.

From Equations 6.144 and 6.145,

$$x_A(i+1|T) + \varepsilon_T = x_A\left(i+1\left|\frac{T}{2}\right.\right) + \varepsilon_{T/2} \tag{6.146}$$

Suppose the numerical integrator is an RK-4 with local truncation error $\varepsilon_T \sim O(T^5)$. Then ε_T can be expressed as

$$\varepsilon_T = cT^5 \tag{6.147}$$

and $\varepsilon_{T/2}$, which is the sum of local truncation errors for the two half-intervals, is given by

$$\varepsilon_{T/2} = c\left(\frac{T}{2}\right)^5 + c\left(\frac{T}{2}\right)^5 = 2c\left(\frac{T}{2}\right)^5 = \frac{1}{16}cT^5 = \frac{1}{16}\varepsilon_T \tag{6.148}$$

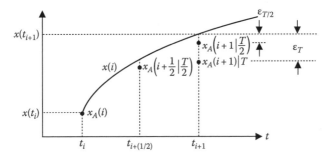

FIGURE 6.8 Illustration of interval halving for estimation of local truncation error.

TABLE 6.5
Step Size Adjustment Based on Outcome of $|\varepsilon_{T/2}|$

Outcome	Action (Next Integration Step)
$\varepsilon_L > |\varepsilon_{T/2}|$	Double current step size
$\varepsilon_L \leq |\varepsilon_{T/2}| \leq \varepsilon_U$	Keep current step size
$|\varepsilon_{T/2}| > \varepsilon_U$	Halve current step size

In reality, c in Equation 6.147 and the two occurrences of c in Equation 6.148 are different and depend on the derivative function and the intervals; however, for suitably small T, the differences are negligible. Eliminating ε_T from Equations 6.146 and 6.148 gives

$$x_A(i+1|T) + 16\varepsilon_{T/2} = x_A\left(i+1\left|\frac{T}{2}\right.\right) + \varepsilon_{T/2} \tag{6.149}$$

Solving for $\varepsilon_{T/2}$ in Equation 6.149 gives

$$\varepsilon_{T/2} = \frac{x_A\left(i+1\left|\frac{T}{2}\right.\right) - x_A(i+1|T)}{15} \tag{6.150}$$

$\varepsilon_{T/2}$ in Equation 6.150 is an estimate of the local truncation error of the RK-4 integrator when the step size is $T/2$. It can be used to adjust the step size in subsequent calculations. For example, Table 6.5 shows a possible approach to step size adjustment using predetermined tolerance limits ε_L and ε_U.

The truncation error $\varepsilon_{T/2}$ can be added to $x_A(i+1|(T/2))$ to obtain a fifth-order accurate estimate of $x(t_i + 1)$, that is, a new estimate $x_A(i+1)$ with local truncation error $\varepsilon_T \sim O(T^5)$. This gives

$$\varepsilon_L \leq |\varepsilon_{T/2}| \leq \varepsilon_U \tag{6.151}$$

$$= \frac{16x_A\left(i+1\left|\frac{T}{2}\right.\right) - x_A(i+1|T)}{15} \tag{6.152}$$

The following example includes a one-step numerical integrator with the step size determined by the interval halving method previously described.

Example 6.5

A cone-shaped tank is filling with water at a constant rate $F_1(t) = \bar{F}$ as shown in Figure 6.9. The initial level is h_0. Water evaporates from the tank at a rate proportional to the surface area of liquid. The constant of proportionality is α.

(a) Find the derivative function in the continuous-time model of the tank.
(b) For $\bar{F} = \pi\,\text{ft}^3/\text{min}$, $\alpha = 0.01$ ft/min, and $h_0 = 10$ ft, estimate the local truncation error in the RK-4 estimate $h_A(1)$ when $T = 1$ in using interval halving, that is, find $\varepsilon_{T/2}$.

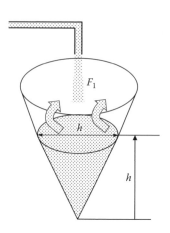

FIGURE 6.9 Conical tank with evaporation.

(c) Use $\varepsilon_{T/2}$ to obtain a fifth-order accurate estimate of $h(T)$.
(d) Simulate the tank dynamics for a period of time sufficient to allow the tank level to increase by 90% of the ultimate change in level. Use an adaptive step size based on the algorithm in Table 6.5 with $\varepsilon_L = 10^{-13}$ and $\varepsilon_U = 10^{-11}$.
(e) Find the analytical solution of the continuous-time model and plot it along with the simulated solution.

(a) The continuous-time model of the tank is

$$\frac{dV}{dt} = F_1(t) - \alpha S \tag{6.153}$$

where
  V is the volume of water in the tank
  S is the surface area of water at the top where evaporation occurs

For the conical shape in Figure 6.9,

$$V = \frac{\pi h^3}{12}, \quad S = \frac{\pi h^2}{4} \tag{6.154}$$

Substituting Equation 6.154 into Equation 6.153 and solving for the derivative function give

$$\frac{dh}{dt} = \frac{4\bar{F}}{\pi h^2} - \alpha \tag{6.155}$$

(b) Using the given values for α and $\bar{F}$ gives

$$\frac{dh}{dt} = \frac{4}{h^2} - 0.01 \tag{6.156}$$

Starting from the initial point $(0, h_0) = (0, 10)$, the results from interval halving after one integration step are given in Table 6.6.

The second column contains the results from a single-step RK-4 integrator with step size $T = 1$ min. The last two columns list the results from two consecutive steps of an RK-4 integrator with step size $T = 0.5$ min.

From Equation 6.150, an estimate of the local truncation error in $h_A(1|(T/2))$ is given by

$$\varepsilon_{T/2} = \frac{h_A\left(1\left|\left(\frac{T}{2}\right)\right.\right) - h_A(1|T)}{15}$$
$$= \frac{10.02988067543399 - 10.02988067543460}{15}$$
$$= -0.40619359727619 \times 10^{-13}$$

TABLE 6.6
Results after One Step of Interval Halving Using RK-4 ($T = 1$ min)

	$t_1 = T$	$t_{1/2} = T/2$	$t_1 = T/2 + T/2$
k_1	0.03000000000000	0.03000000000000	0.02988050771064
k_2	0.02988026946101	0.02994006743256	0.02982108077964
k_3	0.02988074623874	0.02994018702867	0.02982119883721
k_4	0.02976202120809	0.02988050764055	0.02976202170051
	$h_A(1|T) = 10.02988067543460$	$h_A\left(\frac{1}{2}\left|\frac{T}{2}\right.\right) = 10.01497008471359$	$h_A\left(1\left|\frac{T}{2}\right.\right) = 10.02988067543399$

(c) From Equation 6.152, the fifth-order accurate estimate of $h(T)$ is

$$h_A(1) = \frac{16h_A\left(1\left|\left(\frac{T}{2}\right)\right.\right) - h_A(1|T)}{15}$$

$$= \frac{16(10.02988067543399) - 10.02988067543460}{15}$$

$$= 10.02988067543395$$

(d) The steady-state tank level is easily obtained by setting the derivative function to zero in Equation 6.155 and solving for $h = h(\infty)$.

$$\Rightarrow h(\infty) = \sqrt{\frac{4\bar{F}}{\pi\alpha}} = \sqrt{\frac{4\pi}{\pi(0.01)}} = 20\,\text{ft} \tag{6.157}$$

The tank dynamics were simulated using RK-4 integration with interval halving for step size control in "Chap6_Ex3_1.m." The ultimate change in tank level is $h(\infty) - h(0) = 20 - 10 = 10$ ft. The simulation terminates when $h_A(i)$ exceeds the level $h(0) + 0.9[h(\infty) - h(0)] = 10 + 0.9(10) = 19$ ft.

Table 6.7 summarizes how the step size was changed in accordance with the given tolerances on the estimated local truncation error. Note the significant increase in step size from the starting value of $T = 1$ min as the simulation progresses.

(e) The analytical solution is an implicit function for $h(t)$. The derivation is left as an exercise. The result is

$$100\left[10 - h(t) - 10\ln\left(\frac{60 - 3h(t)}{h(t) + 20}\right)\right] = t \tag{6.158}$$

Data points were obtained by increasing $h(t)$ from 10 to 19 ft in small increments, solving for the corresponding value of t, and plotted in Figure 6.10 with t values along the abscissa. The simulated results (every fourth point) are also plotted demonstrating the close agreement with the exact solution. Notice that the step size is progressively increased as the slope, that is, derivative of the solution, gradually decreases.

The average of the estimated local truncation errors is

$$\bar{\varepsilon}_{T/2} = \sum_{i=1}^{169}(\varepsilon_{T/2})_i = 1.6653 \times 10^{-12} \tag{6.159}$$

TABLE 6.7
Simulation Time Interval and the Constant Step Size T Using Interval Halving for Adaptive RK-4 Integration

Time Interval	Step Size, T
$0 \le t \le 1$	1
$1 \le t < 69$	2
$69 \le t < 209$	4
$209 \le t < 313$	8
$313 \le t < 1673$	16

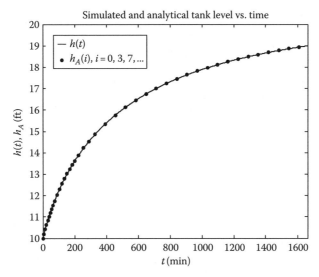

FIGURE 6.10 Results of RK-4 integration with adaptive step size control.

In this example, the simulated time was approximately 1670 min, which would have required 1670 integration steps if the step size had remained constant at $T = 1$ min. With adaptive step size control, the number of integration steps was 169, a nearly 90% reduction. Of course, each of the 169 integration steps requires two passes, one using a full step size and the other using two half-steps. The number of derivative function evaluations for each method is summarized below.

6.3.2 CONSTANT STEP SIZE ($T = 1$ min)

$$\text{Total number of function evaluations} = \frac{1670 \text{ min}}{1 \text{ min/step}} \times 4 \frac{\text{function evaluations}}{\text{step}} = 6680$$

6.3.3 ADAPTIVE STEP SIZE (INITIAL $T = 1$ min)

1. Number of function evaluations (first pass)

$$169 \text{ steps} \times 4 \frac{\text{function evaluations}}{\text{step}} = 676$$

2. Number of new function evaluations (second pass)

$$169 \text{ steps} \times 3 \frac{\text{function evaluations}}{\text{first half interval}} = 507$$
$$169 \text{ steps} \times 4 \frac{\text{function evaluations}}{\text{second half interval}} = 676$$

Total number of function evaluations $= 676 + 507 + 676 = 1859$.

The number of derivative function evaluations has been reduced by over 72%. This comparison is clearly sensitive to the order of the RK integrator used as well as the constant step size T and total simulation time. For example, halving the value of T from 1 to 0.5 min doubles the number of derivative function evaluations in the first case where the step size remains constant. With interval halving and adaptive step size control, the total number of steps would remain nearly the same regardless of the initial step size. Consequently, the total number of derivative function evaluations would remain about the same in either case.

The step size control logic is also significant. The adaptive step size control is typically more complex (Borse 1997; Chapra and Canalel 2002) than the simple approach presented here where the new step size is either one half, the same, or twice the current step size.

Since an implicit solution for $h(t)$ is known (Equation 6.158), it is possible to compare the estimated local truncation error with the actual local truncation error, although not in a straightforward manner due to the implicit nature of the solution.

6.3.4 RK–FEHLBERG

In the case of RK-4 integration, the interval halving method requires seven additional derivative function evaluations for the second pass over the two half-intervals. A total of 11 function evaluations are required for each interval, independent of the interval size. An alternative method for estimating the local truncation error is based on the difference of two different order RK integrators over the same integration time step. By choosing two RK integrators with several common points for the derivative function evaluations, efficiency is improved significantly compared to the interval halving method.

The RK–Fehlberg method employs RK-4 and RK-5 integrators where the four function evaluations k_1, k_2, k_3, and k_4 of the RK-4 integrator are used in the RK-5 integrator as well. Recall that RK-5 integration methods require six function evaluations per step. Therefore, RK–Fehlberg methods combining RK-4 and RK-5 integrators employ a total of six derivative function evaluations per interval.

A common RK–Fehlberg integrator is given as follows (Rao 2002):

$$k_1 = f[t_i, x_A(i)] \tag{6.160}$$

$$k_2 = f\left[t_i + \frac{1}{4}T, x_A(i) + \frac{1}{4}Tk_1\right] \tag{6.161}$$

$$k_3 = f\left[t_i + \frac{3}{8}T, x_A(i) + \frac{3}{32}Tk_1 + \frac{9}{32}Tk_2\right] \tag{6.162}$$

$$k_4 = f\left[t_i + \frac{12}{13}T, x_A(i) + \frac{1932}{2197}Tk_1 - \frac{7200}{2197}Tk_2 + \frac{7296}{2197}Tk_3\right] \tag{6.163}$$

$$k_5 = f\left[t_i + T, x_A(i) + \frac{439}{216}Tk_1 - 8Tk_2 + \frac{3680}{513}Tk_3 - \frac{845}{4104}Tk_4\right] \tag{6.164}$$

$$k_6 = f\left[t_i + \frac{1}{2}T, x_A(i) - \frac{8}{27}Tk_1 + 2Tk_2 - \frac{3544}{2565}Tk_3 + \frac{1859}{4104}Tk_4 - \frac{11}{40}Tk_5\right] \tag{6.165}$$

The estimate of $x[(i+1)T]$ using RK-4 integration is

$$x_A(i+1) = x_A(i) + T\left[\frac{25}{216}k_1 + \frac{1408}{2565}k_3 + \frac{2197}{4104}k_4 - \frac{1}{5}k_5\right] \tag{6.166}$$

The RK-5 estimate of $x[(i+1)T]$ and eventual updated state is

$$x_A(i+1) = x_A(i) + T\left[\frac{16}{135}k_1 + \frac{6656}{12825}k_3 + \frac{28561}{56430}k_4 - \frac{9}{50}k_5 + \frac{2}{55}k_6\right] \tag{6.167}$$

The local truncation error incurred in the ith integration interval is estimated from the difference of Equations 6.167 and 6.166. Thus,

$$\text{Estimate of } (\varepsilon_T)_i = T\left[\frac{1}{360}k_1 - \frac{128}{4275}k_3 - \frac{2197}{75240}k_4 + \frac{1}{50}k_5 + \frac{2}{55}k_6\right] \tag{6.168}$$

Example 6.6 illustrates the use of RK–Fehlberg integration, specifically, the RK-4 and RK-5 methods previously described.

Example 6.6

A motor boat is being driven across a river L ft wide to the opposite side as shown in Figure 6.11. The boat departs from point 0, the origin of an x–y coordinate system, attempting to reach a point H ft upstream. The boat travels at a constant speed v_b mph relative to the water that flows downstream at a speed of v_r mph. The boat is continuously steered in the direction of its intended destination. The boat's heading is given by the angle θ as shown in the diagram. Numerical values of the system parameters are $L = 1000$ ft, $H = 5000$ ft, $v_b = 15$ mph, and $v_r = 5$ mph.

(a) Choose the state variables as $x(t)$ and $y(t)$ and obtain expressions for the state derivative functions in terms of x and y and the system parameters L, H, v_b, and v_r.
(b) Use the "ode45" numerical integrator in Simulink and simulate the boat's x and y position as a function of time. Plot x vs. t, y vs. t, and θ vs. t.
(c) Plot the steering angle θ vs. horizontal position x.
(d) Find an expression for the derivative dy/dx in terms of x and y and the system parameters L, H, v_b, and v_r.
(e) Write a program to implement the RK–Fehlberg method to numerically integrate dy/dx. Adjust the step size when the estimated local truncation error falls outside an acceptable tolerance range. Choose the initial integration step to be $T = 1$ ft.
(f) Find the exact solution for $y(x)$ and compare it with the simulated results.

(a) From the diagram, the state derivatives are

$$\frac{dx}{dt} = v_b \cos\theta = v_b \frac{L - x}{[(L - x)^2 + (H - y)^2]^{1/2}} \tag{6.169}$$

$$\frac{dy}{dt} = -v_r + v_b \sin\theta \tag{6.170}$$

$$= -v_r + v_b \frac{H - y}{[(L - x)^2 + (H - y)^2]^{1/2}} \tag{6.171}$$

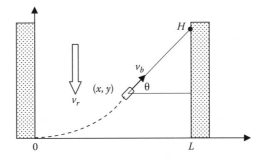

FIGURE 6.11 Boat trajectory crossing the river.

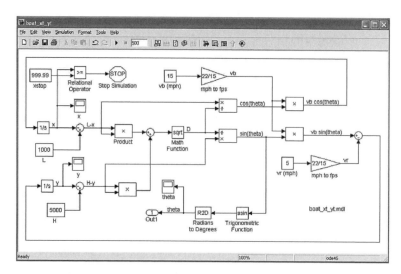

FIGURE 6.12   Simulink® diagram of boat crossing.

(b) A Simulink diagram of the system is shown in Figure 6.12.

Selecting the "ode45" integrator with maximum step size set to 20 and relative tolerance equal to 10^{-6} produces graphs of the state variables $x(t)$ and $y(t)$ and the additional output $\theta(t)$ shown in Figure 6.13.

The exact solutions for $x(t)$, $y(t)$, and $\theta(t)$ were approximated using Simulink's RK-4 integrator with a very small step size, namely, $T = 0.01$ s. The adaptive step size control quickly adjusts the step size to its maximum value and maintains it at the upper limit until the simulation is nearly complete. Comparison with the "exact" ($T = 0.01$ s) solution shows that the truncation errors are minimal.

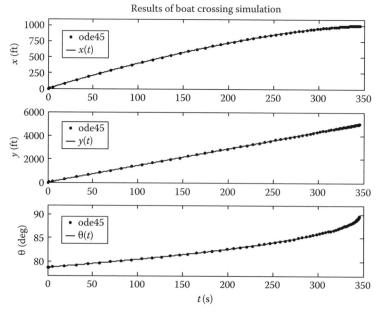

FIGURE 6.13   Time histories of state variables $x(t)$ and $y(t)$ and output $\theta(t)$ using Simulink® variable-step integrator "ode45" and "exact" solutions (RK-4 with $T = 0.01$).

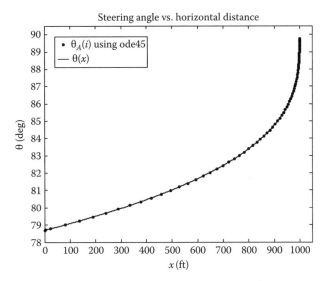

FIGURE 6.14 Steering output θ vs. horizontal location x with Simulink® variable-step integrator "ode45" and "exact" solution.

(c) A plot of steering angle θ vs. horizontal position x is shown in Figure 6.14. Note that the river current causes the boat to be steered at increasingly greater angles as it approaches the right bank of the river.

(d) States $x(t)$ and $y(t)$ represent a parametric description of the boat's trajectory $y = y(x)$. The trajectory can be found in one of two ways. First, the parameter t can be eliminated from equations for the states $x(t)$ and $y(t)$. Since we have not developed the solutions for $x(t)$ and $y(t)$, we resort to the second approach, namely, integration of the derivative dy/dx.

Dividing Equation 6.170 by Equation 6.169 gives

$$\frac{dy}{dx} = \frac{-v_r + v_b \sin\theta}{v_b \cos\theta} \tag{6.172}$$

Expressing $\sin\theta$ and $\cos\theta$ in terms of the distances x, y, L, and H (see Figure 6.11) and simplifying the result yield

$$\frac{dy}{dx} = \frac{H-y}{L-x} - \frac{v_r}{v_b}\left[1 + \left(\frac{H-y}{L-x}\right)^2\right]^{1/2} \tag{6.173}$$

(e) The RK–Fehlberg Equations 6.160 through 6.168 are solved in "Chap6_Ex3_2.m." A simulation summary is shown in Table 6.8.

TABLE 6.8
Summary of RK–Fehlberg Simulation Results for Boat Crossing

Minimum tolerance	10^{-7} ft
Minimum tolerance	10^{-5} ft
Minimum step size	0.1 ft
Minimum step size	25 ft
Number of integration steps	87
Average step size	11.49 ft
Average estimated local truncation error	1.6427×10^{-4}

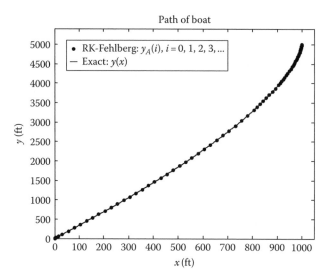

FIGURE 6.15   Simulated (RK–Fehlberg) and exact solutions for $y(x)$ in boat crossing.

(f) Derivation of the exact solution to Equation 6.173 is left as an exercise. The result is

$$y(x) = H - \frac{L-x}{2}\left[c(L-x)^{-k} - \frac{(L-x)^k}{c} \right] \tag{6.174}$$

where
$$c = L^{k-1}(H + \sqrt{L^2 + H^2})$$
$$k = v_r/v_b$$

Plots of the approximate solution $y_A(i)$ obtained in part (e) and the exact solution, Equation 6.174, are shown in Figure 6.15.

EXERCISES

6.10 In Example 6.5,
- (a) Find the analytical solution to the continuous-time model of Equation 6.155.
- (b) Write a program to numerically integrate the derivative function using RK-4 with adaptive control of step size. Compare your results with those in Tables 6.6 and 6.7.
- (c) Experiment with the tolerances used to establish the step size, and plot the results on the same graph with the analytical solution.
- (d) Is the tank initially in equilibrium? Explain. Find the constant flow in $\overline{F}1$ for which the tank is in equilibrium when the height of water is 15 ft.
- (e) With 15 ft of water in the tank and equilibrium conditions established, the flow in is decreased by 50%. Simulate the response using RK-4 with step size control.
- (f) For the same conditions as in part (e), find the estimated and true local truncation errors resulting from the use of RK-4 numerical integration when the water level is 14.9 ft.

Hint: Find the actual time required for the tank level to reach 14.9 ft, and use that value as the initial integration step size.

6.11 In Example 6.6,

(a) Find the analytical solution $y = y(x)$ to Equation 6.173 repeated as follows:

$$\frac{dy}{dx} = \frac{H-y}{L-x} - \frac{v_r}{v_b}\left[1 + \left(\frac{H-y}{L-x}\right)^2\right]^{1/2}$$

Hint: Let $\hat{x} = L - x, \hat{y} = H - y$, introduce $u = \hat{y}/\hat{x}$, and obtain an implicit solution of the separable differential equation in u.

(b) Compute the estimated and actual local truncation errors using RK-4 with interval halving to adjust the step size. Choose the initial step size $T = 1$ ft.

6.12 Find the trajectory of the boat in Example 6.6, assuming it is steered continuously at the destination point (L, H), if the river current varies sinusoidally according to

$$v_r(x) = A\sin\frac{\pi}{L}x, \quad 0 \le x \le L$$

where $A = 10$ mph.

6.13 In the boat-crossing problem of Example 6.6, suppose the boat steering angle θ is an input to the continuous-time model.

(a) Find the state derivative functions in $dx/dt = f_1(x, y, \theta)$ and $dy/dt = f_2(x, y, \theta)$.

For parts (b) through (d), find the analytical solution and check your answer using Simulink with the "ode45" solver.

(b) The steering angle θ is held constant at the initial heading of the destination point, that is,

$$\theta(t) = \bar{\theta} = \tan^{-1}\left(\frac{H}{L}\right), \quad t \ge 0$$

Find the location of the boat when it reaches the other side. Use the values of v_r, v_b, L, and H from Example 6.6.

(c) The captain wishes to cross the river and reach the opposite shore line at $x = L$, $y = 0$, which is directly across from where he started. Find the constant heading $\bar{\theta}$, which allows him to accomplish this. Assume the river current is constant at $v_r = 6$ mph and the boat moves at a constant speed of $v_b = 24$ mph. Plot the boat's trajectory.

(d) Make a plot of $y(L)$ vs. $\bar{\theta}$ for $0 \le \bar{\theta} < \pi/2$ where $y(L)$ is the y coordinate of the location where the boat reaches the opposite side of the river. Assume $v_r = 10$ mph and $v_b = 25$ mph.

(e) The captain observes a large fish swimming upstream at a constant speed of $v_f = 6$ mph in the middle of the river $(x = L/2)$. Starting from $(0, 0)$, he begins to steer directly at the fish when it is directly across from him, that is, located at $(L/2, 0)$. Find and plot the boat's trajectory until it catches up with the fish if the river current is 0 mph and the boat speed is 10 mph.

6.14 A hydraulic accumulator is shown in Figure E6.14. Its purpose is to damp fluctuations in the input flow rate $f_1(t)$ caused by pressure peaks upstream. The flow exits downstream of the accumulator through a linear resistance. The continuous-time model for the pressure $p(t)$ in the accumulator section is (Palm 1983)

$$\frac{A^2}{k}\frac{dp}{dt} = f_1 - \frac{1}{R}(p - p_0)$$

where
  A is the area of the accumulator plate
  k is the spring constant
  R is the fluid resistance

The input flow rate is given by

$$f_1(t) = \begin{cases} 0.01 \text{ ft}^3/\text{s}, & t \le 0 \text{ s} \\ 0.05 \text{ ft}^3/\text{s}, & 0 < t \le 0.01 \text{ s} \\ 0.01 \text{ ft}^3/\text{s}, & t > 0.01 \text{ s} \end{cases}$$

Numerical values of the system parameters are

$$A = 0.0055 \text{ ft}^2, \quad k = 30 \text{ lb/ft}, \quad R = 10^5 \text{ lb s/ft}^2, \quad p_0 = 14.7 \text{ lb/in.}^2$$

The system is at steady state prior to the pulse input in flow.
(a) Use the RK–Fehlberg integrator to simulate the transient response of $p(t)$.
(b) Find the analytical solution for $p(t)$.
(c) Find the solution for $p(t)$ without the accumulator present.
(d) Plot the responses from parts (a), (b), and (c) on the same graph.
(e) Simulate the response for $p(t)$ with Simulink using the "ode45" integrator, and compare the results with those in parts (a) and (b).

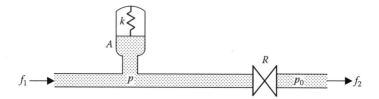

FIGURE E6.14

6.4 MULTISTEP METHODS

RK integrators were classified as one-step methods. The calculations for determining $x_A(i+1)$, the approximate solution to the continuous-time model

$$\frac{dx}{dt} = f(t, x) \tag{6.175}$$

at $t = t_{i+1}$, relies on the previous estimate $x_A(i)$ and one or more derivative function evaluations on the interval $t_i \le t \le t_{i+1}$. The previous state estimate $x_A(i)$ is ignored once $x_A(i+1)$ has been computed. In contrast, multistep methods exploit knowledge of previous state estimates because they provide information about the local behavior of $x(t)$ that can be used to advance the state.

Formulas for multistep methods are derived by integrating Equation 6.175 from t_i to t_{i+1},

$$\int_{x(t_i)}^{x(t_{i+1})} dx = \int_{t_i}^{t_{i+1}} f[t, x(t)] dt \tag{6.176}$$

$$\Rightarrow x(t_{i+1}) = x(t_i) + \int_{t_i}^{t_{i+1}} f[t, x(t)] dt \tag{6.177}$$

The integrand $f[t, x(t)]$ is unknown since $x(t)$ is the solution to Equation 6.175.

6.4.1 EXPLICIT METHODS

An mth-order interpolating polynomial $P_m(t)$ that passes through the current derivative $f[t_i, x_A(i)]$ and previous m derivatives $f[t_{i-1}, x_A(i-1)], f[t_{i-2}, x_A(i-2)], \ldots, f[t_{i-m}, x_A(i-m)]$ can be used to obtain an approximation of the integral in Equation 6.177 (see Figure 6.16). Replacing the integrand in Equation 6.177 by the interpolating polynomial $P_m(t)$ gives

$$x_A(i+1) = x_A(i) + \int_{t_i}^{t_{i+1}} P_m(t)dt \qquad (6.178)$$

where the approximations $x_A(i)$ and $x_A(i+1)$ are used instead of $x(t_i)$ and $x(t_{i+1})$, the actual points on the solution $x(t)$. The integral in Equation 6.178 is equal to the shaded area under the polynomial $P_m(t)$, which has been extrapolated over the current integration interval (t_i, t_{i+1}).

To illustrate, suppose the polynomial is the linear function passing through $\{t_{i-1}, f[x_A(i-1)]\}$ and $\{t_i, f[x_A(i)]\}$. Then $m = 1$ and

$$P_1(t) = f[t_i, x_A(i)] + \left\{ \frac{f[t_i, x_A(i)] - f[t_{i-1}, x_A(i-1)]}{t_i - t_{i-1}} \right\}(t - t_i) \qquad (6.179)$$

Integrating $P_1(t)$ and substituting the result in Equation 6.178 yield after simplifying

$$x_A(i+1) = x_A(i) + \frac{T}{2}\{3f[t_i, x_A(i)] - f[t_{i-1}, x_A(i-1)]\} \qquad (6.180)$$

The formula in Equation 6.180 is known as the two-step Adams–Bashforth (AB-2) method. "Two-step" refers to the use of two intervals, (t_{i-1}, t_i) and (t_i, t_{i+1}), to compute the new state $x_A(i+1)$. Note that the method is explicit since $x_A(i+1)$ does not appear on the right-hand side of Equation 6.180.

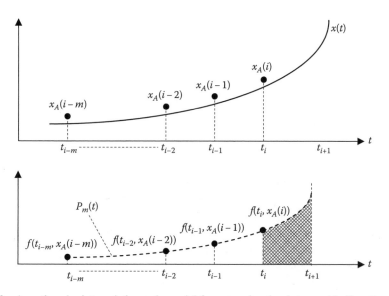

FIGURE 6.16 An mth-order interpolating polynomial for approximating integrand in Equation 6.177.

The Taylor Series expansion of the derivative function $f(t, x)$ leads to an alternative derivation of Equation 6.180, which also provides an expression for the local truncation error of the AB-2 integrator. The Taylor Series expansion of $x(t)$ about t_i evaluated at t_{i+1} is given by

$$x(t_{i+1}) = x(t_i) + \frac{d}{dt}x(t_i)(t_{i+1} - t_i) + \frac{1}{2}\frac{d^2}{dt^2}x(t_i)(t_{i+1} - t_i)^2 + \cdots \tag{6.181}$$

$$= x(t_i) + Tf[t_i, x(t_i)] + \frac{T^2}{2}\frac{d}{dt}f[t_i, x(t_i)] + \cdots \tag{6.182}$$

where T is the fixed-step size, that is, $T = t_{i+1} - t_i = t_i - t_{i-1} = \cdots$. The Taylor Series expansion of the derivative function $f(t, x)$ about t_i evaluated at t_{i-1} is

$$f[t_{i-1}, x(t_{i-1})] = f[t_i, x(t_i)] + \frac{d}{dt}f[t_i, x(t_i)](t_{i-1} - t_i) + \frac{1}{2}\frac{d^2}{dt^2}f[t_i, x(t_i)](t_{i-1} - t_i)^2 + \cdots \tag{6.183}$$

$$= f[t_i, x(t_i)] - T\frac{d}{dt}f[t_i, x(t_i)] + \frac{T^2}{2}\frac{d^2}{dt^2}f[t_i, x(t_i)] + \cdots \tag{6.184}$$

Solving for $T(d/dt)f[t_i, x(t_i)]$ in Equation 6.184 and substituting into Equation 6.182 give

$$x(t_{i+1}) = x(t_i) + \frac{T}{2}\{3f[t_i, x(t_i)] - f[t_{i-1}, x(t_{i-1})]\} + \frac{5}{12}T^3\frac{d^2}{dt^2}f[t_i, x(t_i)] + \cdots \tag{6.185}$$

Truncating Equation 6.185 after the linear term and replacing $x(t_{i-1})$, $x(t_i)$, $x(t_{i+1})$ with $x_A(i-1)$, $x_A(i)$, $x_A(i+1)$ lead to the AB-2 formula in Equation 6.180. Furthermore, since the first term omitted in Equation 6.185 is of order T^3, the local truncation error $\varepsilon_T \sim O(T^3)$. The global truncation error $E_T \sim O(T^2)$ and AB-2 is said to be second-order accurate.

More accurate Adams–Bashforth integration formulas exist. It is simply a question of the number of points, that is, $m + 1$ in Figure 6.16, used to establish the interpolating polynomial $P_m(t)$. Several higher-order AB integrators are listed as follows using the simpler notation $f_A(i) = f[t_i, x_A(i)]$, $f_A(i-1) = f[t_{i-1}, x_A(i-1)]$, etc.

$$\text{AB-3: } x_A(i+1) = x_A(i) + \frac{T}{12}[23f_A(i) - 16f_A(i-1) + 5f_A(i-2)] \tag{6.186}$$

$$\text{AB-4: } x_A(i+1) = x_A(i) + \frac{T}{24}[55f_A(i) - 59f_A(i-1) + 37f_A(i-2) - 9f_A(i-3)] \tag{6.187}$$

$$\text{AB-5: } x_A(i+1) = x_A(i) + \frac{T}{720}[1901f_A(i) - 2774f_A(i-1) + 2616f_A(i-2)$$

$$- 1274f_A(i-3) + 251f_A(i-4)] \tag{6.188}$$

Local truncation errors for AB integrators are obtained using the Taylor Series expansion approach illustrated for deriving the AB-2 formula. Truncating the respective series to obtain Equations 6.186 through 6.188 results in $(3/8)T^4(d^3/dt^3)f[t_i, x(t_i)]$, $(251/720)T^5(d^4/dt^4)f[t_i, x(t_i)]$, and $(475/1440)$ $T^6(d^5/dt^5)f[t_i, x(t_i)]$ as the first omitted terms in the AB-3, AB-4, and AB-5 formulas. The local and global truncation errors for the third-order accurate AB-3 integrator are $\varepsilon_T \sim O(T^4)$ and $E_T \sim O(T^3)$, respectively. An mth-order accurate AB-m integrator has a local truncation error $\varepsilon_T \sim O(T^{m+1})$ and global truncation error $E_T \sim O(T^m)$.

Both AB and RK integrators rely on a weighted sum of derivative function evaluations. In the case of one-step RK integration, the derivative function is evaluated numerous times over a single interval in contrast to the multistep AB integrators, which rely on derivative evaluations from previous intervals.

The mth-order accurate multistep integration formulas are more efficient than one-step methods of identical order because the same derivative function $f_A(i)$ is utilized m times for updating the state over m consecutive intervals. Another way of looking at it is only a single new derivative function evaluation $f_A(i)$ is required to advance the state from $x_A(i)$ to $x_A(i+1)$. For example, suppose we have just determined the state $x_A(i)$ using AB-3 integration. Since $f_A(i-1)$ and $f_A(i-2)$ are still in memory, only $f_A(i)=f[t_i, x_A(i)]$ is needed to compute the new state $x_A(i+1)$ in Equation 6.186.

Multistep methods are not self-starting. One approach is to utilize a one-step method for the first several integration steps before transitioning to a multistep formula. Alternatively, a one-step method followed by lower-order multistep methods can be used prior to implementing a specific multistep method. Once again, let us choose the AB-3 integrator for illustration purposes. From Equation 6.186 with $i=0, 1$

$$x_A(1) = x_A(0) + \frac{T}{12}[23f_A(0) - 16f_A(-1) + 5f_A(-2)] \tag{6.189}$$

$$x_A(2) = x_A(1) + \frac{T}{12}[23f_A(1) - 16f_A(0) + 5f_A(-1)] \tag{6.190}$$

It is impossible to know $f_A(-1)$ and $f_A(-2)$ without knowing $x(-T)$ and $x(-2T)$. Hence, two integrations are performed using a one-step method starting from the known initial point $[0, x(0)]$ to determine $x_A(1)$ and $x_A(2)$. Subsequent state estimates $x_A(3), x_A(4), \ldots$ are computed from the AB-3 formula. The "weakest link in the chain" argument dictates the choice of an appropriate one-step method to initiate the numerical solution. In other words, for a third-order accurate AB-3 integrator with local truncation error $\varepsilon_T \sim O(T^4)$, a third-order accurate RK-3 integrator with comparable local truncation error $\varepsilon_T \sim O(T^4)$ is used.

In the second approach, a third-order accurate one-step method can be used to find $x_A(1)$ followed by the second-order accurate multistep AB-2 integrator to determine x_A (2) before switching to AB-3 integration. The first approach is preferred since the AB-2 integrator degrades the accuracy of the numerical solution.

The difference equations for AB-2, AB-3, and so forth are higher order than the first-order differential equation of the continuous-time system given in Equation 6.175. In other words, the resulting discrete-time systems for approximating the first-order continuous-time system dynamics have two or more discrete-time states depending on the order of the AB integrator used. Later, in Chapter 8, we shall see that there is a penalty for implementing higher order (and hence more accurate) multistep integrators to simulate linear continuous-time systems. The penalty takes the form of a constraint imposed on the integration step size in order to assure a stable simulation.

6.4.2 Implicit Methods

Equations 6.180 and 6.186 through 6.188 are explicit methods since all the terms on the right-hand side have already been computed. There are, however, compelling reasons for using the derivatives $f[t_{i+1}, x_A(i+1)], f[t_i, x_A(i)], f[t_{i-1}, x_A(i-1)], \ldots, f[t_{i-m+1}, x_A(i-m+1)]$ instead of $f[t_i, x_A(i)], f[t_{i-1}, x_A(i-1)], f[t_{i-2}, x_A(i-2)], \ldots, f[t_{i-m}, x_A(i-m)]$ (see Figure 6.16) to determine the mth-order interpolating polynomial $P_m(t)$. Since our objective is to compute $x_A(i+1)$, the eventual difference equation will be implicit, that is, $x_A(i+1)$ will appear on both sides of Equation 6.178.

TABLE 6.9
Local Truncation Errors for AB-m, AM-m Integrators ($m = 2, 3, 4, 5$)

	Local Truncation Error, ϵ_T		Local Truncation Error, ϵ_T
AB-2	$\dfrac{5}{12}T^3\dfrac{d^2}{dt^2}f[\hat{t}_i, x(\hat{t}_i)]$	AM-2	$-\dfrac{1}{12}T^3\dfrac{d^2}{dt^2}f[\hat{t}_i, x(\hat{t}_i)]$
AB-3	$\dfrac{3}{8}T^4\dfrac{d^3}{dt^3}f[\hat{t}_i, x(\hat{t}_i)]$	AM-3	$-\dfrac{1}{24}T^4\dfrac{d^3}{dt^3}f[\hat{t}_i, x(\hat{t}_i)]$
AB-4	$\dfrac{251}{720}T^5\dfrac{d^4}{dt^4}f[\hat{t}_i, x(\hat{t}_i)]$	AM-4	$-\dfrac{19}{720}T^5\dfrac{d^4}{dt^4}f[\hat{t}_i, x(\hat{t}_i)]$
AB-5	$\dfrac{475}{1440}T^6\dfrac{d^5}{dt^5}f[\hat{t}_i, x(\hat{t}_i)]$	AM-5	$-\dfrac{27}{1440}T^6\dfrac{d^5}{dt^5}f[\hat{t}_i, x(\hat{t}_i)]$

Using the implicit form in Equation 6.178 yields formulas for the Adams–Moulton implicit numerical integrators given in Equations 6.191 through 6.194.

$$\text{AM-2: } x_A(i+1) = x_A(i) + \frac{T}{2}[f_A(i+1) + f_A(i)] \tag{6.191}$$

$$\text{AM-3: } x_A(i+1) = x_A(i) + \frac{T}{12}[5f_A(i+1) + 8f_A(i) - f_A(i-1)] \tag{6.192}$$

$$\text{AM-4: } x_A(i+1) = x_A(i) + \frac{T}{24}[9f_A(i+1) + 19f_A(i) - 5f_A(i-1) + f_A(i-2)] \tag{6.193}$$

$$\text{AM-5: } x_A(i+1) = x_A(i) + \frac{T}{720}[251f_A(i+1) + 646f_A(i) - 246f_A(i-1)$$
$$+ 106f_A(i-2) - 19f_A(i-3)] \tag{6.194}$$

Note that the AM-2 integration formula is the implicit trapezoidal integrator introduced in Section 3.4. If the system model is linear, $f_A(i+1)$ is a linear function of $x_A(i+1)$, and an explicit solution for $x_A(i+1)$ in Equations 6.191 through 6.194 is possible. In general, implicit equations are solved in iterative fashion by numerical methods.

AB-m and AM-m integrators are both mth-order accurate. However, the local truncation error ε_T for the implicit AM-m integrator is less than the comparable explicit AB-m integrator (see Table 6.9). The local truncation errors cannot actually be calculated because the value of $\hat{t}_i$ is unknown except for $t_i \leq \hat{t}_i \leq t_{i+1}a$.

Multistep methods are not well suited for adaptively changing the step size based on the estimated local truncation error. With a change in step size from $x_A(i)$ to $x_A(i+1)$, some or all of the past values $[x_A(i), f_A(i)], [x_A(i-1), f_A(i-1)], \ldots, [x_A(i-m), f_A(i-m)]$ can no longer be used, defeating the essential reason for using a multistep method in the first place. The use of multistep integration methods is demonstrated in Example 6.7.

Example 6.7

The dynamics of a tumor growth is described by the first-order differential equation as follows (Braun 1978):

$$\frac{d}{dt}V(t) = \lambda e^{-\alpha t}V(t) \tag{6.195}$$

(a) Find the difference equations for approximate tumor growth $V_A(i)$, $i = 1, 2, 3, \ldots$ using AB-2 and AM-3 integrators.
(b) Find the analytical solution $V(t)$ to Equation 6.195.

The model parameters are $\lambda = 0.2$ new cells per cell per week and $\alpha = 0.02$ per week. A tumor initially contains one thousand cells.

(c) Compare results from the exact solution and approximate solutions with a step size of $T = 0.25$ week. Plot the approximate and exact solutions on the same graph.

(a) Combining the derivative function

$$f_A(i) = f[t_i, V_A(i)] = \lambda e^{-\alpha t_i} V_A(i) \qquad (6.196)$$

with the AB-2 integrator of Equation 6.180, that is,

$$V_A(i+1) = V_A(i) + \frac{T}{2}[3f_A(i) - f_A(i-1)]$$

yields the second-order difference equation

$$V_A(i+1) = V_A(i) + \frac{T}{2}[3\lambda e^{-\alpha t_i} V_A(i) - \lambda e^{-\alpha t_{i-1}} V_A(i-1)] \qquad (6.197)$$

$$= \left(1 + \frac{3}{2}\lambda T e^{-\alpha i T}\right) V_A(i) - \frac{1}{2}\lambda T e^{-\alpha(i-1)T} V_A(i-1), \quad i = 1, 2, 3, \ldots \qquad (6.198)$$

Repeating the steps for the AM-2 integrator, Equation 6.191 yields the implicit form of the difference equation, that is,

$$V_A(i+1) = \frac{5}{12}\lambda T e^{-\alpha(i+1)T} V_A(i+1) + \left(1 + \frac{8}{12}\lambda T e^{-\alpha i T}\right) V_A(i) - \lambda T e^{-\alpha(i-1)T} V_A(i) \qquad (6.199)$$

Solving for $V_A(i+1)$ in Equation 6.199 produces the explicit form,

$$V_A(i+1) = \left[\frac{1 + (2/3)\lambda T e^{-\alpha i T}}{1 - (5/12)\lambda T e^{-\alpha(i+1)T}}\right] V_A(i) - \left[\frac{(1/12)\lambda T e^{-\alpha(i-1)T}}{1 - (5/12)\lambda T e^{-\alpha(i+1)T}}\right] V_A(i-1), \quad i = 1, 2, \ldots \quad (6.200)$$

Note that the discrete-time system models, Equations 6.198 and 6.200, are time-varying due to the appearance of the discrete-time variable "i" in the coefficients of $V_A(i)$ and $V_A(i-1)$. This is expected since the continuous-time model, Equation 6.195 is time-varying as a result of the $e^{-\alpha t}$ term in the coefficient of $V(t)$.

(b) The exact solution is obtained by separating the differential equation, Equation 6.195, and integrating from $t = 0$, $V = V_0$ where V_0 is the initial volume of cells.

$$\int_{V_0}^{v} \frac{dV}{V} = \int_{0}^{t} \lambda e^{-\alpha t} dt \qquad (6.201)$$

$$\Rightarrow V(t) = V_0 e^{(\lambda/\alpha)(1 - e^{-at})} \qquad (6.202)$$

(c) The AB-2 and AM-3 integrators require a single integration step using a one-step method to provide a starting value for $V_A(1)$. Ordinarily, an RK-2 one-step integrator would be used for the first step with the AB-2 and AM-3 multistep integrators. In lieu of that, we shall use the exact solution to generate $V_A(1)$ and leave the use of one-step methods to start the solution process as an exercise.

From the initial condition and Equation 6.202, the starting values for the AB-2 and AM-3 integrators are $V_A(0) = 1000$, $V_A(1) = 1051.14$.

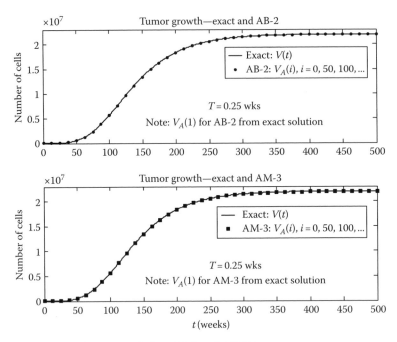

FIGURE 6.17 Tumor growth—exact solution, AB-2 and AM-3 integrators.

Plots of the exact solution for tumor growth and every 50th discrete-time output of the AB-2 and AM-3 integrators are shown in Figure 6.17. Based on comparison with the exact solution of the continuous-time model, both integrators appear to predict cell growth exceptionally well (see "Chap6_Ex4_1.m").

The limiting value of tumor size $V(\infty)$ occurs when the growth rate $(1/V)(dV/dt)$ approaches zero. This limit cannot be obtained from the continuous-time model by setting the derivative to zero as in the case of logistic growth (see Section 1.5). However, it is possible to compute $V(\infty)$ as the limiting value of the exact solution, that is,

$$V(\infty) = \lim_{t \to \infty} V_0 e^{(\lambda/\alpha)(1-e^{-\alpha t})} = V_0 e^{\lambda/\alpha} \qquad (6.203)$$

Substituting the value of $V_A(0)$ for V_0 along with the given values for λ and α gives $V(\infty) = 1000e^{0.2/0.002} = 2.203 \times 10^7$ in agreement with the plots in Figure 6.17.

6.4.3 Predictor–Corrector Methods

Generally speaking, implicit methods are more accurate than explicit methods of the same order. In all but the simplest cases, the solution requires an iterative root-solving scheme, which can wreak havoc on the computational efficiency of the implicit integrator. Fortunately, a solution to the problem exists, although with a slight trade-off in the number of required derivative function evaluations.

The alternative approach is to employ an explicit method to predict the new state followed by an implicit method using the predicted state on the right-hand side of the equation. This eliminates the primary obstacle of implicit methods, namely, a nonlinear algebraic equation with the unknown updated state on both sides. The combination of explicit and implicit numerical integration is called a predictor–corrector method.

If this sounds familiar, it is because we have already implemented a simple predictor–corrector method in Section 3.6, namely, the improved Euler or Heun's method. In that case, the predictor is the first-order explicit Euler integrator, and the corrector is the second-order trapezoidal integrator. The common practice is to combine explicit Adams–Bashforth and implicit Adams–Moulton integrators of the same order. Integration formulas for several of these predictor–corrector combinations are

$$\text{AB-2 predictor: } \hat{x}_A(i+1) = x_A(i) + \frac{T}{2}\{3f[t_i, x_A(i)] - f[t_{i-1}, x_A(i-1)]\} \tag{6.204}$$

$$\text{AM-2 corrector: } x_A(i+1) = x_A(i) + \frac{T}{2}[\hat{f}_A(i+1) + f_A(i)] \tag{6.205}$$

where $\hat{f}_A(i+1) = f[t_{i+1}, \hat{x}_A(i+1)]$ is the derivative based on the predicted state $\hat{x}_A(i+1)$.

$$\text{AB-3 predictor: } \hat{x}_A(i+1) = x_A(i) + \frac{T}{12}[23f_A(i) - 16f_A(i-1) + 5f_A(i-2)] \tag{6.206}$$

$$\text{AM-3 corrector: } x_A(i+1) = x_A(i) + \frac{T}{12}[5\hat{f}_A(i+1) + 8f_A(i) - f_A(i-1)] \tag{6.207}$$

$$\text{AB-4 predictor: } \hat{x}_A(i+1) = x_A(i) + \frac{T}{24}[55f_A(i) - 59f_A(i-1) + 37f_A(i-2) - 9f_A(i-3)] \tag{6.208}$$

$$\text{AM-4 corrector: } x_A(i+1) = x_A(i) + \frac{T}{24}[9\hat{f}_A(i+1) + 19f_A(i) + 5f_A(i-1) + f_A(i-2)] \tag{6.209}$$

It should be noted that some authors refer to the implicit numerical integrators in Equations 6.191 through 6.194 as Adams integrators and the predictor–corrector formulas in Equations 6.204 through 6.209 as Adams–Moulton integration formulas.

In certain applications, it may be desirable to execute several iterations of the corrector equation before advancing to the next integration step. In other words, corrected values are continually inserted on the right-hand side of the corrector equation until some threshold or tolerance is attained, resulting in improved estimates of the new state. In general, it is inadvisable to execute the corrector equation more than once or twice due to the additional derivative function calculations required. When the corrector equation is implemented only once, predictor–corrector integration formulas are examples of a two-pass (one for the predictor and one for the corrector) approach to updating the discrete-time state. There are no implicit equations to solve.

When the order of the predictor and corrector is the same, the combined predictor–corrector integration formula is also of that order. Furthermore, the truncation errors (local and global) are the same as those of the more accurate implicit corrector (see Table 6.9). Combining same order predictor and corrector makes it possible to estimate the local truncation error after each step (Ralston and Wilf 1965) based on the predicted and corrected states with virtually no computational overhead. This permits the step size to be changed in an adaptive fashion. Of course, repeatedly changing the step size with a multistep integration method is counterproductive.

The stability of numerical integration methods refers to the sequence of numerical values computed for the discrete-time states when simulating a stable continuous-time system. We shall learn in Chapter 8 that explicit multistep methods exhibit poorer stability characteristics compared with implicit methods. Suffice it to say for now that the higher-order AB multistep integrators are prone to instability. This is mitigated to some extent by the choice of step size. However, reducing the step size to combat the problem adversely impacts computational efficiency reflected in the total number of derivative function evaluations required to simulate the system.

Example 6.8

A manufacturer of high-end luxury automobiles has determined that the monthly demand for its cars follows an inverse price relationship, that is,

$$d(p) = a\left(\frac{1}{p}\right), \quad p > 0 \tag{6.210}$$

where
    p is the base price of a single vehicle
    d is the monthly demand
    a is a constant

The number of vehicles produced by the manufacturer is based on the fluctuating price. Suppose the monthly supply of vehicles (up to some limit) is related to price by

$$s(p) = bp^{1/2}, \quad p > 0 \tag{6.211}$$

where
    s is the monthly production
    b is another constant

Furthermore, assume the actual price is governed by supply and demand according to

$$\frac{dp}{dt} = K[d(p) - s(p)], \quad p > 0 \tag{6.212}$$

$$= K\left[a\left(\frac{1}{p}\right) - bp^{1/2}\right] \tag{6.213}$$

where K is also a constant.
    Several months ago when the price was $200,000, 16 cars were sold. The car maker would produce 25 vehicles per month if the vehicle price were $250,000. The current price is $180,000. The numerical value of K is $2,000 per vehicle.

(a) Use an AB-4/AM-4 predictor–corrector with step size $T = 0.5$ month to find the response of the price. Generate the required starting values from an RK-4 integrator.
(b) Simulate the response in part (a) using RK-4 with step size sufficiently small to approximate the exact response. Graph the simulated and "exact" response.

(a) The MATLAB file to compute the RK-4 starting values and implement AB-4/AM-4 predictor–corrector integration is "Chap6_Ex4_2.m." Using the classic RK-4 integrator with step size $T = 0.5$ month, the following values were obtained to start the AB-4/AM-4 predictor–corrector:

$$p_A(0) = \$180,000, \quad p_A(1) = 176,823.92$$

$$p_A(2) = \$174,122.06, \quad p_A(3) = \$171,832.02$$

(b) Basing the exact solution $p(t)$ on RK-4 with $T = 0.01$ months produced the results in Table 6.10 and plotted in Figure 6.18. Values from the simulated response $p_A(i)$ are also tabulated in

TABLE 6.10
$p_A(i)$ from AB-4/AM-4 Integration Using RK-4 Starting Values with $T = 0.5$ Months and Exact Solution $p(t)$ Approximated by RK-4 with $T = 0.01$ Months

i	t_i, Months	$p_A(i)$, $	$p(t_i)$, $	i	t_i, Months	$p_A(i)$, $	$p(t_i)$, $
0	0	180,000.00	180,000.00	16	8	161,086.76	161,086.85
2	1	174,122.06	174,122.02	18	9	160,748.08	160,748.17
4	2	169,897.30	169,897.25	20	10	160,514.70	160,514.77
6	3	166,897.64	166,897.62	22	11	160,354.00	160,354.06
8	4	164,787.43	164,787.46	24	12	160,243.42	160,243.47
10	5	163,313.01	163,313.07	26	13	160,167.36	160,167.39
12	6	162,287.89	162,287.97	28	14	160,115.05	160,115.08
14	7	161,577.64	161,577.73	30	15	160,079.08	160,079.10

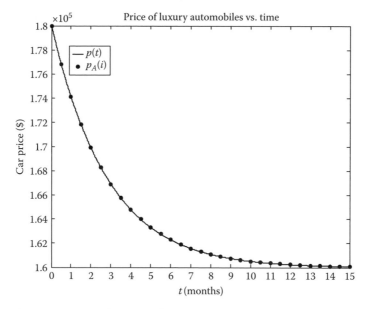

FIGURE 6.18 Price response $p_A(i)$ from AB-4/AM-4 ($T = 0.25$ months) and "Exact" $p(t)$ based on RK-4 ($T = 0.01$ months).

Table 6.10 and plotted in Figure 6.18. According to the graphs, the transient period for the price to reach equilibrium is approximately 15 months. The equilibrium point is easily obtained from Equation 6.213, that is,

$$0 = K\left[a\left(\frac{1}{p(\infty)}\right) - bp^{1/2}(\infty)\right] \tag{6.214}$$

$$\Rightarrow p(\infty) = \left(\frac{a}{b}\right)^{2/3} = \left(\frac{3{,}200{,}000}{0.05}\right)^{2/3} = \$160{,}000 \tag{6.215}$$

in agreement with the value shown in Figure 6.18.

EXERCISES

6.15 Rework Example 6.7 using RK-2 to find the starting value $V_A(1)$ for the AB-2 and AM-3 integrators, respectively. Comment on the results.

6.16 Rework Example 6.7 with step size $T=2$ weeks using an RK-4 method to find the starting values $V_A(1)$, $V_A(2)$, and $V_A(3)$ for the AB-4 and AM-4 integrators. Comment on the results.

6.17 Show that the equilibrium price in Example 6.8 is stable by choosing initial prices slightly less and slightly greater than $p(\infty)$ and observing the transient price responses. Use a suitable numerical integrator to obtain the transient response.

6.18 An unforced continuous-time system is described by the first-order differential equation $(t+1)\,(dx/dt)+x=0$. An AB-2 numerical integrator with step size T is used to simulate the response of the system with initial condition $x(0)=1$.

(a) The difference equation for updating the discrete-time state $x_A(n)$ is

$$x_A(n+1) = \alpha_0 x_A(n) + \alpha_1 x_A(n-1), \quad n=1,2,3,\ldots$$

Express α_0 and α_1 in terms of the step size T and discrete-time variable n.

(b) Use an RK-2 integrator with step size $T=0.1$ s to find $x_A(1)$, the starting value needed for the AB-2 integration.

(c) Use the AB-2 integrator to find $x_A(2)$.

(d) Compare the approximate values $x_A(2)$, $x_A(3)$,..., $x_A(10)$ with the exact values $x(0.2)$, $x(0.3)$,..., $x(1)$. Note that the exact solution is given by $x(t)=1/(t+1)$.

6.19 A double integrator is shown in Figure E6.19. Initial conditions are $x(0)=y(0)=0$.

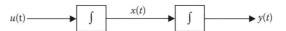

FIGURE E6.19

(a) Find the discrete-time system approximation (difference equation) of the first integrator using explicit Euler integration with step size T. Denote the input as $u(n)$ and the output as $x_A(n)$.

(b) The input is a unit step $u(t)=1$, $t \geq 0$. Find $x_A(1)$, $x_A(2)$, and $x_A(3)$. Leave your answers in terms of T.

(c) Find the general solution for $x_A(n)$.

(d) Show that the local truncation error $(\varepsilon_T)_n = x_A(n) - x(nT) = 0$, $n=0$, 1, 2, 3,... Comment on the result, that is, explain why the discrete-time output $x_A(n)$ is identical with the continuous-time output at the end of each integration step.

(e) Find the discrete-time system approximation (difference equation) of the second integrator using explicit Euler integration with step size T. Denote the input as $x_A(n)$ and the output as $y_A(n)$.

(f) Find $y_A(1)$, $y_A(2)$, $y_A(3)$, $y_A(4)$, and $y_A(5)$. Leave your answers in terms of T.

(g) The general solution for the output $y_A(n)$ is

$$y_A(n) = (an^2 + bn + c)T^2, \quad n=0,1,2,3,\ldots$$

Find the numerical values of the constants a, b, and c.

(h) Find the differential equation relating the output $y(t)$ and input $u(t)$.

(i) Find the local truncation error $(\varepsilon_T)_n = y_A(n) - y(nT)$.

6.5 STIFF SYSTEMS

Linear time-invariant models of dynamic systems are termed "stiff" when the time constants, or more specifically the characteristic roots (eigenvalues of the system matrix A), vary significantly in magnitude. For nonlinear system models, the concept applies to the characteristic roots of a linearized model that represents the dynamics of the nonlinear system in some operating region. Linearization of nonlinear systems is discussed in Chapter 7.

Systems tend to be stiff for a number of reasons. Mechanical systems composed of stiff and soft components exhibit resonant frequencies that differ greatly in magnitude. The natural response of certain electrical networks contains spikes, which die out rapidly in comparison to terms with far slower dynamics. Control system components such as controllers, actuators, and sensors oftentimes respond much quicker than the plant or process being controlled.

Figure 6.19a and b contain s-plane pole plots corresponding to stiff systems, and Figure 6.19c is a pole plot of a fast system, but not stiff. The stiffness can be quantified by the ratio of the largest (in magnitude) to the smallest characteristic root.

Stiff systems impose requirements on numerical integrators, in particular explicit methods, which can result in exceedingly small integration steps to assure the result is a stable solution. Numerical stability is considered in some detail in Chapter 8. For the present time, we can think of numerical stability as a property of numerical integration, which implies that a stable discrete-time system will result whenever the continuous-time system model is stable.

Suppose the fast pole in Figure 6.19a, the pole furthest from the imaginary axis, is designated s_1 and the slower poles are s_2 and s_3, $s_4 = -\zeta\omega_n \pm j\omega_d$. The natural response consists of a linear combination of the real modes $e^{s_1 t}$, $e^{s_2 t}$ and the oscillatory modes $e^{-\zeta\omega_n t}\sin\omega_d t$, $e^{-\zeta\omega_n t}\cos\omega_d t$. That is,

$$x_{nat}(t) = c_1 e^{s_1 t} + c_2 e^{s_2 t} + e^{-\zeta\omega_n t}[A_1 \sin\omega_d t + A_2 \cos\omega_d t] \tag{6.216}$$

The transient response of the system with poles in Figure 6.19a is of the same form as the natural response in Equation 6.216. Due to the inherent stiffness of the system, the time constant $\tau_1 = -1/s_1$ is considerably shorter than either $\tau_2 = -1/s_2$ or $\tau = 1/\zeta\omega_n$, the effective time constant of the damped oscillations. Hence, the fast component $c_1 e^{s_1 t}$ vanishes well before the remaining terms. However, the numerical stability of fixed-step explicit integrators is controlled by the fast mode, requiring the use of a far smaller integration step than would be necessary in the absence of the fast characteristic root s_1.

Integration formulas have been developed specifically for stiff systems. References by Gear (1971) and Hartley (1994) contain excellent descriptions of specific "stiff" integrators. MATLAB and Simulink offer a choice of one-step and multistep integrators designed for efficient simulation of stiff systems.

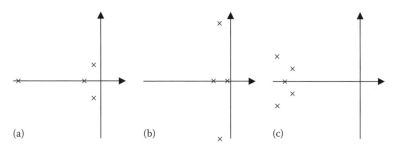

FIGURE 6.19 s-Plane location of characteristic roots for stiff system (a), (b), and nonstiff system (c).

6.5.1 STIFFNESS PROPERTY IN FIRST-ORDER SYSTEM

Before we illustrate an example of a stiff system, it should be mentioned that the "stiffness" property can be present in a forced system with only a single state variable, that is, a system with a single characteristic root or eigenvalue modeled by a linear first-order differential equation. The basic requirement is merely the existence of two or more terms in the response with markedly different time constants. Consider the simple forced mechanical system shown in Figure 6.20.

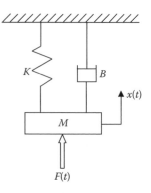

Assuming that the mass M is negligible leads to the continuous-time model,

FIGURE 6.20 Simple mechanical system.

$$B\frac{d}{dt}x(t) + Kx(t) = F(t) \qquad (6.217)$$

The state derivative function is

$$f(x) = \frac{d}{dt}x(t) = \frac{1}{\tau}\left[\frac{1}{K}F - x\right] \qquad (6.218)$$

where $\tau = B/K$ is the first-order system time constant.

Suppose the forcing function $F(t)$ is an ideal step input whose amplitude is numerically equal to the spring constant K, that is, $F(t) = K$, $t \geq 0$. Because a step input is physically impossible, it is approximated by $\hat{F}(t)$

$$\hat{F}(t) = K(1 - e^{-t/\tau_F}), \quad t \geq 0 \qquad (6.219)$$

where τ_F is the time constant of the exponential rise. From the system's perspective, $\hat{F}(t)$ will look like a step input provided its rise time is several orders of magnitude less than τ, the system time constant.

Analytical solutions for the state $x(t)$ based on the ideal step input $F(t)$ of magnitude K and the approximation in Equation 6.219 are easily obtained by the use of Laplace transforms. Laplace transforming Equation 6.217 and solving for $X(s)$ give

$$X(s) = \frac{1}{\tau s + 1}\left[\frac{1}{K}F(s)\right] \qquad (6.220)$$

The Laplace transform of the state response to an ideal step input of magnitude K is therefore

$$X(s) = \frac{1}{\tau s + 1}\left(\frac{1}{K} \cdot \frac{K}{s}\right) = \frac{1}{\tau s + 1}\left(\frac{1}{s}\right) \qquad (6.221)$$

When the input $\hat{F}(t)$ is used, $\hat{F}(s)$ replaces $F(s)$ in Equation 6.220, making the Laplace transform of the state response, denoted $\hat{x}(t)$, equal to

$$\hat{X}(s) = \frac{1}{\tau s + 1}\left[\frac{1}{K} \cdot K\left(\frac{1}{s} - \frac{\tau_F}{\tau_F s + 1}\right)\right] = \frac{1}{\tau s + 1}\left[\frac{1}{s(\tau_F s + 1)}\right] \qquad (6.222)$$

Inverse Laplace transformation of Equations 6.221 and 6.222 gives

$$x(t) = 1 - e^{-t/\tau}, \quad t \geq 0 \tag{6.223}$$

$$\hat{x}(t) = 1 - \frac{1}{\tau - \tau_F}\left(\tau e^{-t/\tau} - \tau_F e^{-t/\tau_F}\right), \quad t \geq 0 \ (\tau \gg \tau_F) \tag{6.224}$$

Note that the response $\hat{x}(t)$ consists of a fast and a slow component, that is,

$$\hat{x}(t) = \hat{x}_F(t) + \hat{x}_S(t) \tag{6.225}$$

where

$$\hat{x}_F(t) = \frac{\tau_F}{\tau - \tau_F} e^{-t/\tau_F} \tag{6.226}$$

and

$$\hat{x}_S(t) = 1 - \frac{\tau}{\tau - \tau_F} e^{-t/\tau} \tag{6.227}$$

Simulation of the first-order system response to $\hat{F}(t)$ poses problems not previously encountered. To illustrate, consider the case when $B = 1$ and $K = 10$. The system time constant $\tau = B/K = 0.1$ s. Suppose the fast component time constant τ_F in Equation 6.219 is chosen two orders of magnitude less than the system time constant, that is, $\tau_F = \tau/100 = 0.001$ s.

Dividing $\hat{F}(t)$ by K produces the exponential rise approximation to a unit step input shown in the upper left corner of Figure 6.21. The ideal unit step input and unit step response in Equation 6.223 are shown in the lower left quadrant of Figure 6.21.

RK-4 integration was used to generate the simulated responses shown on the right side of Figure 6.21. In the top right quadrant, the integration time step T was chosen to be an order of magnitude

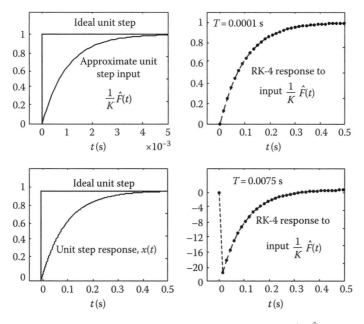

FIGURE 6.21 Unit step response and simulated (RK-4) response to input $(1/K)\hat{F}(t)$.

less than τ_F, that is, $T = \tau_F/10 = 0.0001$ s, to guarantee accuracy of the simulation. Every 150th point of the simulated response is plotted. The unit step response $x(t)$ and the simulated step response are nearly identical at $t_i = iT$, $i = 0, 1, 2, \ldots$.

The integration step size $T = 0.0001$ s is a great deal smaller than would seem necessary for RK-4 integration of a first-order system with time constant $\tau = 0.1$ s. Since the fast component $\hat{x}_F(t)$ decays in $5\tau_F = 5 \times 0.001 = 0.005$ s, an adaptive procedure can be employed, which increases the step size after the transient period of the fast component has elapsed.

What do you suppose would happen if we tried a fixed-step RK-4 integrator with T, an order of magnitude smaller than the system time constant, that is, $T = \tau/10 = 0.01$ s? To answer that question, the simulation was rerun using RK-4 with T a little less than 0.01 s, namely, $T = 0.0075$ s. The simulated response (every other point) is shown in the lower right quadrant of Figure 6.21. It bears no resemblance to either $x(t)$ or $\hat{x}(t)$.

Despite the gross inaccuracy, the numerical integrator is nonetheless stable as evidenced by the limiting value approaching the correct steady-state value of unity. Further increases in T will eventually result in an unstable response of the discrete-time system. The integration step size is therefore limited by the fast time constant τ_F.

This example illustrates how a first-order system appears to be stiff, despite the fact there is only a single state. The fast input component ($\tau_F = 0.001$ s) in conjunction with the slower system natural mode ($\tau = 0.1$ s) is responsible for this happening.

6.5.2 STIFF SECOND-ORDER SYSTEM

A second-order system is stiff if it contains a "fast" and a "slow" natural mode. Consequently, for a second-order system to be inherently stiff, it must be overdamped. The second-order circuit shown in Figure 6.22 is stiff provided the circuit parameters produce a pair of real characteristic roots several orders of magnitude apart.

Compared with fixed-step-size numerical integrators, stiff integrators increase the step size after the fast transients decay to zero, reducing execution time significantly. The following example illustrates the use of one of Simulink's stiff integrators.

Example 6.9

In the circuit shown in Figure 6.22, after the capacitor has fully charged to the battery voltage v_0, the switch disconnects the battery at $t = 0$, and the capacitor discharges its stored energy to the RLC circuit. The current $i(t)$ satisfies the differential equation

$$L\frac{d^2 i}{dt^2} + R\frac{di}{dt} + \frac{1}{C}i = 0 \tag{6.228}$$

$$v_C(0) = v_0, \quad i(0) = 0$$

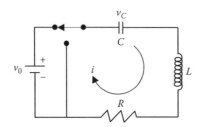

FIGURE 6.22 A second-order RLC circuit.

(a) Represent the circuit in state variable form where $x_1 = i$ and $x_2 = di/dt$.

(b) Show that the system is stiff when the circuit parameter values are $R = 25\ \Omega$, $L = 20$ mH, $C = 200$ mF, and $v_0 = 12$ V.

(c) Simulate the transient response using a fixed-step RK-2 integrator, and determine the largest step size T, which yields a stable and accurate solution.

(d) Use one of the stiff numerical integrators available in Simulink to simulate the transient response.

(e) Find the analytical solution for the transient response, and compare the results of parts (c) and (d) with the exact solution.

(a) Derivation of the state equations is straightforward.

$$\dot{x}_1 = \frac{di}{dt} = x_2 \tag{6.229}$$

$$\dot{x}_2 = \ddot{x}_1 = \frac{d^2 i}{dt^2} \tag{6.230}$$

$$= \frac{1}{L}\left[-\frac{1}{C}i - R\frac{di}{dt}\right] \tag{6.231}$$

$$= -\frac{1}{LC}x_1 - \frac{R}{L}x_2 \tag{6.232}$$

(b) The characteristic equation is $|sI - A| = 0$ where A is the system matrix in the state representation $\dot{\underline{x}} = A\underline{x}$. Thus,

$$|sI - A| = \left| s\begin{pmatrix} 1 & 0 \\ 0 & 1 \end{pmatrix} - \begin{pmatrix} 0 & 1 \\ -\frac{1}{LC} & -\frac{R}{L} \end{pmatrix} \right| = 0 \tag{6.233}$$

$$s^2 + \frac{R}{L}s + \frac{1}{LC} = 0 \tag{6.234}$$

The characteristic roots are

$$s_{1,2} = \frac{-R/L \pm \sqrt{(R/L)^2 - 4(1/LC)}}{2} \tag{6.235}$$

Substituting the given values for R, L, and C in Equation 6.235 yields a stiff system with characteristic roots $s_1 = -1249.8$ rad/s and $s_2 = -0.2$ rad/s.

(c) The Simulink model for the system is shown in Figure 6.23.

The natural modes are $e^{s_1 t} = e^{-1249.8t}$ and $e^{s_2 t} = e^{-0.2t}$. Using Simulink's RK-2 integrator with different step sizes eventually produces a stable and accurate simulation with $T = 0.0015$ s. The discrete-time state $x_{1,A}(i)$ is plotted in the upper left graph of Figure 6.24. It requires 16,663 steps to simulate the transient response, lasting approximately 25 s. The first 41 points $x_{1,A}(i)$, $i = 0, 1, 2, \ldots, 40$ are shown in Figure 6.24 and every 500th point thereafter.

Increasing the step size T from 0.0015 s to 0.0016 s with RK-2 produces the graph of $x_{1,A}(i)$ in the lower left corner of Figure 6.24. Every 500th point is plotted. While the response is stable, that is, $\lim_{i \to \infty} x_{1,A}(i) = 0$, it is clearly inaccurate. The graph in the lower right quadrant contains the first 0.04 s of the discrete-time state $x_{1,A}(i)$ when the RK-2 integration step size is 0.002 s. In this case, the discrete-time system is unstable with the simulated response becoming increasingly more negative (approaching $-\infty$) as time increases.

(d) Choosing the "ode23s" stiff integrator produces the response shown in the upper right corner of Figure 6.24. It is similar in appearance to the graph obtained with RK-2 integration and step size $T = 0.0015$ s; however, the entire simulation required a total of 72 steps. The improvement in

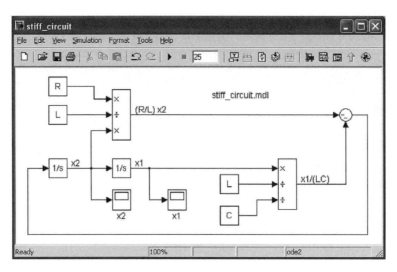

FIGURE 6.23 Simulink® model for RLC circuit.

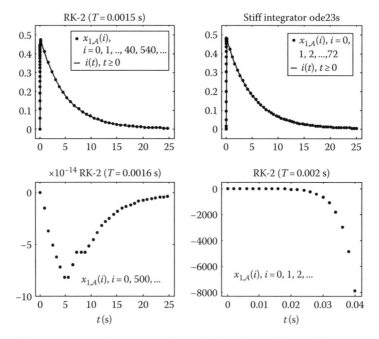

FIGURE 6.24 RK-2 integration with step sizes $T = 0.0015$, 0.0016, 0.002 s, stiff integrator "ode23s" and exact solution for current $i(t)$.

efficiency compared to the RK-2 integrator is dramatic, that is, an average step size of $25/72 = 0.3472$ s compared to 0.0015 s.

(e) The exact solution for $i(t)$ is obtained by Laplace transformation of Equation 6.228, that is,

$$L\left[s^2 I(s) - si(0) - \frac{di}{dt}(0)\right] + R[sI(s) - i(0)] + \frac{1}{C}I(s) = 0 \tag{6.236}$$

$$I(s) = \frac{di}{dt}(0)\left[\frac{1}{s^2 + (R/L)s + 1/LC}\right] \tag{6.237}$$

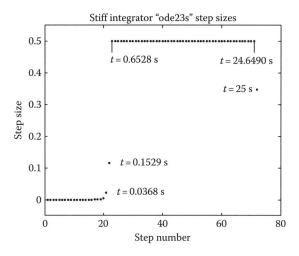

FIGURE 6.25 Step size vs. step number for "ode23s" integrator in Example 6.9.

Replacing the initial derivative $(di/dt)(0)$ by $v_C(0)/L$ gives

$$i(t) = \frac{v_C(0)}{L}\left[\frac{1}{s_1 - s_2}\right](e^{s_1 t} - e^{s_2 t}) \tag{6.238}$$

where s_1 and s_2 are the characteristic roots found in Equation 6.235. Using the values for $v_C(0) = v_0$, L and the characteristic roots s_1 and s_2, the exact solution for $i(t)$ is

$$i(t) = -0.4802(e^{-1249.8t} - e^{-0.2t}) \tag{6.239}$$

$$= -0.4802(e^{-t/0.0008} - e^{-t/5}) \tag{6.240}$$

The exact solution for $i(t)$ is plotted on the graphs with the RK-2. ($T = 0.0015$ s) and "ode23s" responses. Both are in excellent agreement with the exact solution. Note the initial spike in $i(t)$ from zero to approximately 0.48 amp. This results from the rapid decay of the fast mode $e^{-1249.8t}$ in the first $5 \times 0.0008 = 0.004$ s. After 0.004 s have elapsed, the response is essentially the slow component $0.4802e^{-0.2t}$, which lasts for approximately $5 \times 5 = 25$ s.

Stiff integrators are designed to take smaller steps while the fast component of the transient response is decaying and then accelerate after the fast component has vanished. Figure 6.25 illustrates how the integrator "ode23s" creeps along for the first 20 or so steps and then ramps up for the last 52 integration steps. Indeed, after the first 21 steps, the simulation has progressed to 0.03675 s with an average step size of 0.00175 s. The average step size over the final 51 steps is 0.4894 s.

6.5.3 APPROXIMATING STIFF SYSTEMS WITH LOWER-ORDER NONSTIFF SYSTEM MODELS

Stiff systems typically consist of components or subsystems that operate at significantly different speeds. For example, consider the control system shown in Figure 6.26 comprising a proportional controller, a second-order system, and a first-order sensor in the feedback loop. An additive disturbance or load component combines with the second-order system output to produce the complete output signal $y(t)$.

The output $Y(s)$ is expressed in terms of two transfer functions $G_R(s)$ and $G_D(s)$

$$Y(s) = G_R(s)R(s) + G_D(s)D(s) \tag{6.241}$$

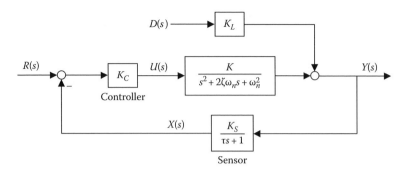

FIGURE 6.26 A stiff system with fast and slow components.

where

$$G_R(s) = \frac{Y(s)}{R(s)}\bigg|_{D(s)=0} = \frac{K_C K(\tau s + 1)}{\tau s^3 + (1 + 2\zeta\omega_n\tau)s^2 + \left(2\zeta\omega_n + \omega_n^2\tau\right)s + \omega_n^2 + K_C K K_S} \qquad (6.242)$$

and

$$G_D(s) = \frac{Y(s)}{D(s)}\bigg|_{R(s)=0} = \frac{K_L\left[\tau s^3 + (1 + 2\zeta\omega_n\tau)s^2 + \left(2\zeta\omega_n + \omega_n^2\tau\right)s + \omega_n^2\right]}{\tau s^3 + (1 + 2\zeta\omega_n\tau)s^2 + \left(2\zeta\omega_n + \omega_n^2\tau\right)s + \omega_n^2 + K_C K K_S} \qquad (6.243)$$

The sensor dynamics are considerably faster than those of the second-order plant, a common situation in control systems. Suppose the numerical values of the system parameters are $K_C = 2$, $K = 5$, $\zeta = 0.7$, $\omega_n = 1.5$ rad/s, $K_S = 0.75$, $\tau = 0.00125$ s, and $K_L = 3$. The characteristic polynomial of the third-order system is

$$\Delta(s) = \tau s^3 + (1 + 2\zeta\omega_n\tau)s^2 + \left(2\zeta\omega_n + \omega_n^2\tau\right)s + \omega_n^2 + K_C K K_S \qquad (6.244)$$

Substituting the given values of the system parameters into Equation 6.244 and using the MATLAB function "roots" to find the characteristic roots (poles) of the closed-loop control system result in $p_1 = -800.01$, $p_{2,3} = -1.0453 \pm j2.9423$. The stiffness ratio is

$$\text{stiffness} = \frac{|p_1|}{|p_2|} = \frac{|-800.01|}{|-1.0453 + j2.9423|} = 256.21 \qquad (6.245)$$

indicating a moderately stiff system. A Simulink diagram of the system is shown in Figure 6.27. Both reference input and disturbance inputs are accounted for.

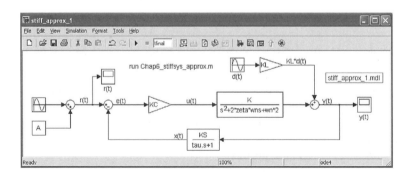

FIGURE 6.27 Simulink® diagram for simulating stiff control system dynamics.

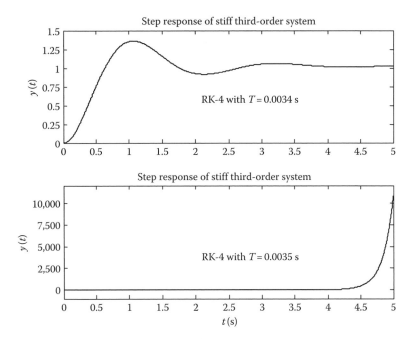

FIGURE 6.28   Stable and unstable simulated responses using RK-4 integration.

The simulated response to a unit step input $r(t) = 1$, $t \geq 0$ is to be obtained using RK-4 integration. Analytical methods exist to compute the largest value of step size T, which results in a stable simulation; however, they are deferred until Chapter 8. Trial and error with different values of T produced the responses shown in Figure 6.28.

The correct step response is shown on top, whereas the one on the bottom is the result of numerical instability of the RK-4 integrator at the larger step size of $T = 0.0035$ s. The results are typical of what happens when a numerical integrator becomes unstable, that is, the simulated results may be quite accurate and suddenly become useless as the integration step size is increased by a slight amount. Try modifying the Simulink model "*stiff_approx_1.mdl*" to allow a disturbance step input or simply make one of the initial conditions nonzero and look at the natural response. In either case, a step size of $T = 0.0034$ s produces a stable output and $T = 0.0035$ s does not.

The stiffness is attributable to the disparity in the time constant of the sensor and the effective time constant of the second-order system. The question that naturally arises is "What happens if the sensor dynamics are ignored, that is, the sensor responds instantaneously to its inputs?" The characteristic polynomial in Equation 6.244 becomes second order when the sensor time constant τ is set to zero. The control system is underdamped with a pair of complex poles, $-1.50 \pm j2.9407$, nearly identical to the complex poles of the third-order control system with sensor time constant included. The system is no longer stiff and a larger value of T can be used for RK-4 simulation.

Step responses of the original third-order control system and the reduced second-order system are generated in the MATLAB script file "*Chap6_stiffsys_approx.m*," which calls the Simulink model "*stiff_approx_2.mdl*" shown in Figure 6.29. Both systems are simulated concurrently using RK-4 integration with step size $T = 0.001$ s.

The plant output $y(t)$ and sensor output $x(t)$ for the third-order control system with the sensor dynamics included and second-order control system with sensor approximated as a pure gain are shown in Figure 6.30. There is no noticeable difference in $y(t)$ or $x(t)$ for the second- and third-order systems.

The second-order system was simulated to determine how large the step size could be without concern for numerical instability of the RK-4 integrator. The reader should verify that step sizes up

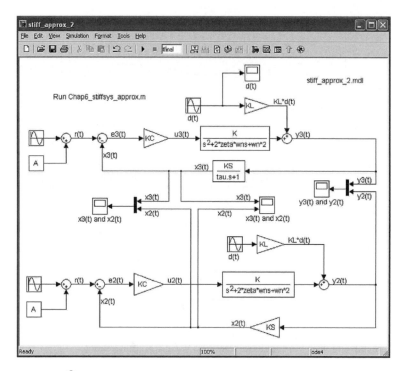

FIGURE 6.29 Simulink® diagram for third-order and second-order control systems.

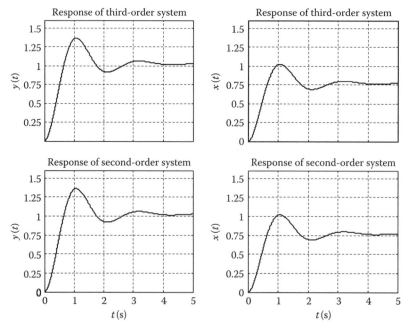

FIGURE 6.30 Step response of stiff and nonstiff control system models.

to approximately $T = 0.2$ s produce accurate (and therefore stable) results. This represents a sizable reduction in execution time, a speedup of roughly $0.2/0.0034 \approx 59$ times. Chapter 8 includes a discussion on how to find the limiting value of T precisely.

Consider the load transfer function $G_D(s)$ in Equation 6.243 for the case when $\tau = 0.00125$ s and when $\tau = 0$. Putting $G_D(s)$ in pole-zero form,

$$G_D(s) = \frac{b_3 s^3 + b_2 s^2 + b_1 s + b_0}{a_3 s^3 + a_2 s^2 + a_1 s + a_0} \tag{6.246}$$

$$= \left(\frac{b_3}{a_3}\right) \frac{(s + z_1)(s + z_2)(s + z_3)}{(s + p_1)(s + p_2)(s + p_3)} \tag{6.247}$$

From M-file "*Chap6_stiffsys_approx.m*," the results are

$\tau = 0.00125$ s:

$b_0 = 6.75$	$a_0 = 9.75$	$z_1 = -800$	$p_1 = -800.0094$
$b_1 = 6.3084$	$a_1 = 2.1028$	$z_2 = -1.05 + j1.0712$	$p_2 = -1.0453 + j2.9423$
$b_2 = 3.0079$	$a_2 = 1.0026$	$z_3 = -1.05 - j1.0712$	$p_3 = -1.0453 - j2.9423$
$b_3 = 0.0039$	$a_3 = 0.0013$		

$$G_D(s) = \frac{3(s + 800)(s^2 + 2.1s + 2.25)}{(s + 800.0094)(s^2 + 2.0906s + 9.7499)} \tag{6.248}$$

$\tau = 0$:

$b_0 = 6.75$	$a_0 = 9.75$	$z_1 = -1.05 + j1.0712$	$p_1 = -1.05 + j2.9407$
$b_1 = 6.3$	$a_1 = 2.1$	$z_2 = -1.05 + j1.0712$	$p_2 = -1.05 + j2.9407$
$b_2 = 3$	$a_2 = 1$		
$b_3 = 0$	$a_3 = 0$		

$$G_D(s) = \frac{3(s^2 + 2.1s + 2.25)}{s^2 + 2.1s + 9.75} \tag{6.249}$$

Canceling the real pole and real zero in Equation 6.248 results in a nonstiff second-order system, which accurately represents the dynamics of the stiff third-order system.

Canceling factors from the numerator and denominator in a transfer function when the pole-zero plot indicates that a pole and zero are close to each other is valid under most conditions. In fact, one of the goals of control system design based on "pole placement" is to mitigate or eliminate entirely the effect of undesirable modes in the open-loop system natural response. A controller with a combination zero and pole is inserted in the loop with the zero located near the undesirable open-loop pole.

Another example of approximating a stiff system model with a lower-order dynamics model is now given. In this case, the order of the approximate system is reduced by ignoring a fast mode and retaining the slower dominant mode as opposed to canceling nearly equivalent numerator and denominator factors.

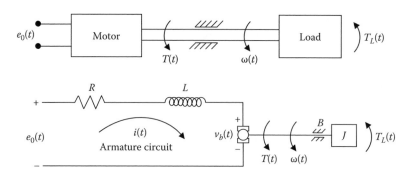

FIGURE 6.31   Armature-controlled DC motor and load.

Example 6.10

An armature-controlled DC motor with a load inertia mounted on its shaft is shown in Figure 6.31. The inputs are the armature voltage $e_0(t)$ and the load torque $T_L(t)$. The outputs are the motor torque $T(t)$ and angular speed of the motor $\omega(t)$. Dependent variables (in addition to the outputs) are the armature current $i(t)$ and back emf of the motor $v_b(t)$. R and L are the electrical resistance and inductance of the armature circuit while B and J are the viscous damping coefficient and load inertia. K_b and K_T are the back emf and torque constants of the motor.

The following equations govern the dynamics of this electromechanical system:

$$e_0(t) = Ri(t) + L\frac{d}{dt}i(t) + v_b(t) \tag{6.250}$$

$$v_b(t) = K_b\omega(t) \tag{6.251}$$

$$T(t) = K_T i(t) \tag{6.252}$$

$$J\frac{d}{dt}\omega(t) + B\omega(t) = T(t) + T_L(t) \tag{6.253}$$

(a) Draw a block diagram of the system and find the transfer functions $I(s)/E_0(s)$ and $\Omega(s)/E_0(s)$ where $E_0(s) = \mathcal{L}\{e_0(t)\}$, $I(s) = \mathcal{L}\{i(t)\}$, and $\Omega(s) = \mathcal{L}\{\omega(t)\}$.

(b) Find the steady-state gain (from armature voltage to angular speed), natural frequency, and damping ratio of the motor as a function of the motor parameters.

(c) Find expressions for the motor time constants in terms of the motor parameters.

(d) The motor constants and load inertia are

$$R = 0.2\ \Omega, \quad L = 0.1\ \text{mH}, \quad K_T = 8 \times 10^{-3}\ \text{ft lb}_f/\text{A}$$

$$K_b = 0.05\ \frac{V}{\text{rad/s}}, \quad B = 0.01\ \frac{\text{ft lb}_f}{\text{rad/s}}, \quad J = 4.5 \times 10^{-3}\ \frac{\text{ft lb}_f}{\text{rad/s}^2}$$

Compute the second-order system parameters, characteristic roots, time constants, and stiffness ratio.

(e) Find expressions for the time constants when the armature inductance is assumed to be negligible. Find the reduced order transfer functions $I(s)/E_0(s)$ and $\Omega(s)/E_0(s)$ when $L \approx 0$.

(f) Simulate the response $\omega(t)$, $t \ge 0$ of the first- and second-order models to a unit step input in armature voltage using Simulink's Euler integrator. Compare the results and comment on the step size required to achieve a stable response in each case.

(g) Use one of Simulink's stiff integrators to obtain the step response of the DC motor second-order system model. Compare the number of steps and execution time required for the stiff integrator and the RK-1 Euler integrator with step size $T = 0.0005$ s.

(h) Compare the frequency response function $G_\Omega(j\omega) = \Omega(j\omega)/E_0(j\omega)$ when $L = 0.1$ and 0 mH. Comment on the results.

(i) Compare the outputs $i(t)$, $t \geq 0$ and $\omega(t)$, $t \geq 0$ in response to a load torque $T_L(t) = \sin \omega_L t$, $t \geq 0$ for the following cases shown in Table 6.11.

TABLE 6.11
Motor Inductance and Load Torque Frequency Values

	ω_L	
L (mH)	2π rad/s	200 rad/s
0		
1		
0.1		

(a) Laplace transforming Equations 6.250 through 6.253 with initial conditions zero provides the basis for constructing the block diagram shown in Figure 6.32.

$$G_\Omega(s) = \left.\frac{\Omega(s)}{E_0(s)}\right|_{T_L(s)=0} = \frac{(1/(Ls+R))K_T(1/(Js+B))}{1 + K_b(1/(Ls+R))K_T(1/(Js+B))} \tag{6.254}$$

$$= \frac{K_T}{(Ls+R)(Js+B) + K_bK_T} \tag{6.255}$$

$$G_I(s) = \left.\frac{I(s)}{E_0(s)}\right|_{T_L(s)=0} = \frac{\Omega(s)}{E_0(s)} \bigg/ \frac{\Omega(s)}{I(s)} \tag{6.256}$$

$$= \frac{K_T/[(Ls+R)(Js+B) + K_bK_T]}{K_T/(Js+B)} \tag{6.257}$$

$$= \frac{(Js+B)}{(Ls+R)(Js+B) + K_bK_T} \tag{6.258}$$

(b) Dividing $G_\Omega(s)$ in Equation 6.255 by JL and equating the result to the standard form of a second-order system,

$$G_\Omega(s) = \frac{K_T/JL}{[s + (R/L)][s + (B/J)] + K_bK_T/JL} = \frac{K_m\omega_n^2}{s^2 + 2\zeta\omega_n s + \omega_n^2} \tag{6.259}$$

Solving for the steady-state gain K_m, the natural frequency ω_n, and the damping ratio ζ in terms of the motor parameters results in

$$K_m = \frac{K_T}{BR + K_bK_T} \tag{6.260}$$

$$\omega_n = \left(\frac{BR + K_bK_T}{JL}\right)^{1/2} \tag{6.261}$$

$$\zeta = \frac{BL + JR}{2[JL(BR + K_bK_T)]^{1/2}} \tag{6.262}$$

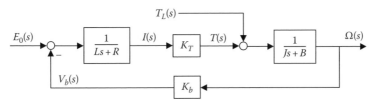

FIGURE 6.32 Block diagram of armature-controlled DC motor.

(c) It is possible to show that the motor is overdamped ($\zeta > 1$), and, therefore, the transfer function in Equation 6.259 is expressible as

$$G_{\Omega}(s) = \frac{K_m \omega_n^2}{s^2 + 2\zeta\omega_n s + \omega_n^2} = \frac{K_m \omega_n^2 \tau_1 \tau_2}{(\tau_1 s + 1)(\tau_2 s + 1)} \tag{6.263}$$

The denominator of $G_{\Omega}(s)$ is the characteristic polynomial $\Delta(s)$ whose roots are

$$s_1, s_2 = -\zeta\omega_n \pm \sqrt{\zeta^2 - 1}\,\omega_n \tag{6.264}$$

The motor time constants in Equation 6.263 are related to the characteristic roots according to $\tau_1 = -1/s_1$, $\tau_2 = -1/s_2$. Substituting Equations 6.261 and 6.262 into Equation 6.264 produces an expression for the characteristic roots,

$$s_1, s_2 = -\frac{1}{2JL}[(BL + JR) \pm \{(BL + JR)^2 - 4JL(BR + K_b K_T)\}^{1/2}] \tag{6.265}$$

Taking the negative reciprocals of s_1 and s_2 gives

$$\tau_1, \tau_2 = \frac{2JL}{(BL + JR) \pm \{(BL + JR)^2 - 4JL(BR + K_b K_T)\}^{1/2}} \tag{6.266}$$

(d) The second-order system parameters are computed using Equations 6.260 through 6.262. The results are $K_m = 3.33$ rad/s/V, $\omega_n = 73.03$ rad/s, and $\zeta = 13.71$. The characteristic polynomial of the second-order system model is

$$\Delta(s) = (Ls + R)(Js + B) + K_b K_T \tag{6.267}$$

$$= LJs^2 + (LB + RJ)s + RB + K_b K_T \tag{6.268}$$

Substituting the numerical values of the motor constants gives

$$\Delta(s) = 4.5 \times 10^{-7} s^2 + 9.0 \times 10^{-4} s + 2.4 \times 10^{-3} \tag{6.269}$$

The characteristic roots s_1 and s_2 can be found directly from Equation 6.265 or by solving for the roots of $\Delta(s)$ in Equation 6.269. The result is $s_1 = -1999.6$ rad/s and $s_2 = -2.67$ rad/s. The motor time constants are $\tau_1 = -1/s_1 = 0.0005$ s and $\tau_2 = -1/s_2 = 0.375$ s. The stiffness ratio is $s_1/s_2 = 749.7$.

(e) Ignoring terms involving L in the denominator of Equation 6.266 gives

$$\tau_1 \approx \left[\frac{2JL}{(BL + JR) + \{(BL + JR)^2 - 4JL(BR + K_b K_T)\}^{1/2}}\right]_{L \approx 0} = \frac{L}{R} \tag{6.270}$$

$$\tau_2 \approx \lim_{L \to 0} \left[\frac{2JL}{(BL + JR) - \{(BL + JR)^2 - 4JL(BR + K_b K_T)\}^{1/2}}\right] \tag{6.271}$$

Application of L'Hospital's rule in Equation 6.271 results in

$$\tau_2 \approx \frac{JR}{BR + K_b K_T} \tag{6.272}$$

Ignoring the effect of armature inductance, that is, assuming $L = 0$ in Equations 6.255 and 6.258, yields a first-order model of the motor with transfer functions

$$\frac{\Omega(s)}{E_0(s)} = \frac{K_T}{JRs + RB + K_b K_T} \tag{6.273}$$

$$\frac{I(s)}{E_0(s)} = \frac{Js + B}{JRs + RB + K_b K_T} \tag{6.274}$$

Hence, the motor can be modeled as a first-order component

$$\frac{\Omega(s)}{E_0(s)} = \frac{K_m}{\tau_m s + 1} \tag{6.275}$$

with time constant $\tau_m = \tau_2 = 0.375$ s and $K_m = 3.33$ rad/s/V.

(f) The Simulink diagram for the step responses of the first- and second-order system models using Euler integration is shown in Figure 6.33.

The simulated responses of the motor to a unit step input in armature voltage occurring at $t = 0.25$ s are shown in Figure 6.34. Euler integration at $T = 0.001$ s is stable for both cases, $L = 0.1$ and 0 mH. Note that both responses approach the predicted steady-state value $\omega_{ss} = K_m \times 1 = 3.33$ rad/s/V × 1V = 3.33$\overline{3}$ rad/s in roughly $5 \times \tau_m = 5 \times 0.375 = 1.875$ s after the unit step is applied.

Figure 6.35 shows the simulated response of the second-order system model ($L = 0.1$ mH) with Euler integration for step sizes of $T = 0.001001$ and 0.001002 s. The first plot indicates the onset of numerical instability, while the second shows clear instability at the larger step size. By trial and error, the upper limit for stable Euler integration of the stiff system model, the second-order system with $L = 0.1$ mH, is approximately $T = 0.001$ s.

The first-order system model obtained by ignoring the fast pole at $s_1 = -1999.6$ rad/s leaving only the dominant pole at $s_2 = -2.7$ rad/s can be simulated with Euler integration using a far greater integration step. Figure 6.36 shows what to expect with step sizes of $T = 0.1$, 0.25, 0.75, and 1 s, respectively. The lowest value of T results in a step response nearly identical to the analytical solution (not shown). The result is still quite acceptable for $T = 0.25$ s. The integrator appears to be marginally stable (and grossly inaccurate) when T is equal to 0.75 s. The response in the lower right is clearly unstable.

(g) The Simulink model in Figure 6.33 was called from M-file "Chap6_Ex5_2.m" with the "ode1" and "ode15s" integrators selected to simulate the motor angular speed and current. Simulated outputs of the second-order system are plotted in Figure 6.37. "ode1" is Euler and "ode15s" is one of the stiff integrators available in MATLAB.

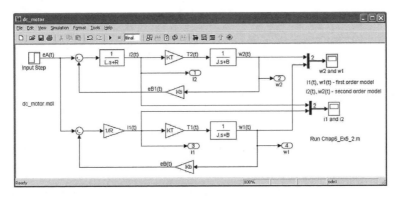

FIGURE 6.33 Simulink® diagram for step responses of first- and second-order models.

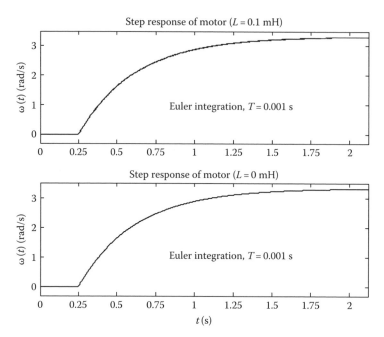

FIGURE 6.34 Unit step responses of first- and second-order system models.

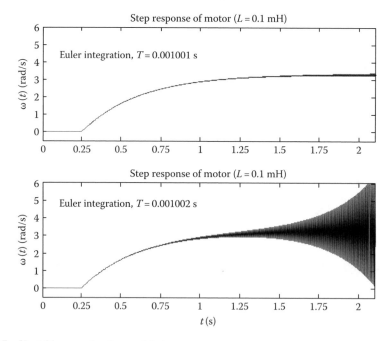

FIGURE 6.35 Unstable second-order model step responses.

The y-labels are written as $\omega(t)$ and $i(t)$ even though the plots are actually of the discrete-time (simulated) system outputs. The armature voltage $e_0(t)$ was applied at $t = 0.25$ s, and the simulation ran for $0.25 + 5\tau_m = 0.25 + 5(0.375) = 2.125$ s. The analytical solutions for $\omega(t)$ and $i(t)$ are considered in Exercise 6.24.

Euler simulation required $(0.25 + 5\tau_m)/T = 4250$ integration steps. The stiff integrator needed only 79 steps to produce comparably accurate results. The execution times for each were obtained

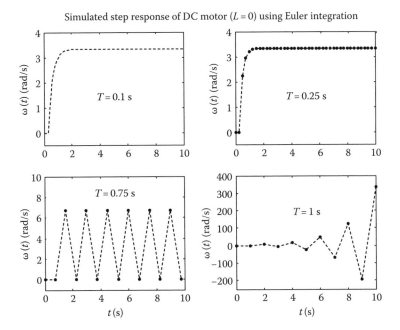

FIGURE 6.36   Simulated response using Euler integration with four different step sizes.

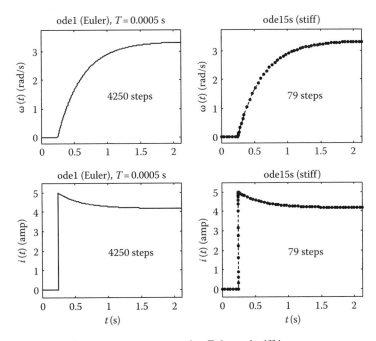

FIGURE 6.37   Simulated DC motor step response using Euler and stiff integrator.

using the MATLAB command "`cputime`," which returns the CPU time used by MATLAB from the time it is first loaded. Execution times for the Euler and stiff integrator were 63 and 47 ms, respectively.

(h) The frequency response functions $G_\Omega(j\omega)$ for the first-order model ($L = 0$) and second-order model ($L = 0.1$ mH) are shown in Figure 6.38. The magnitude function $|G_\Omega(j\omega)|$ for $L = 0.1$ and 0 mH is nearly identical up to 1000 rad/s well beyond the cutoff frequency or bandwidth of the

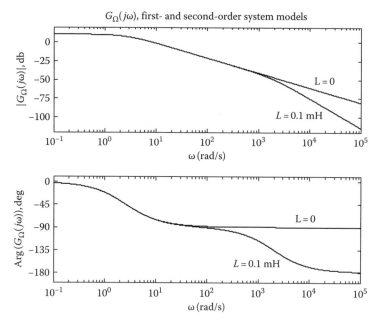

FIGURE 6.38 Frequency response function $G_\Omega(j\omega)$ for $L=0$ and 0.1 mH.

motor. The DC gain $G_\Omega(j0)$ is the same as the motor gain $K_m = 3.3\bar{3}$ rad/s/V (10.46 db). At $\omega = 2000$ rad/s, the magnitudes are 0.0031 rad/s/V (-50.05 db) with $L = 0.1$ mH and 0.0044 rad/s/V (-47.04 db) with $L = 0$.

Figure 6.38 suggests that the dynamic response of the motor to changes in armature voltage be accurately predicted by the first-order (nonstiff) model.

(i) The six cases in Table 6.11 were simulated using the Simulink model "*dc_motor_2.mdl*" shown in Figure 6.39. "*Chap6_Ex5_2.m*" calls "*dc_motor_2.mdl*" twice, once with $\omega_L = 2\pi$ rad/s and

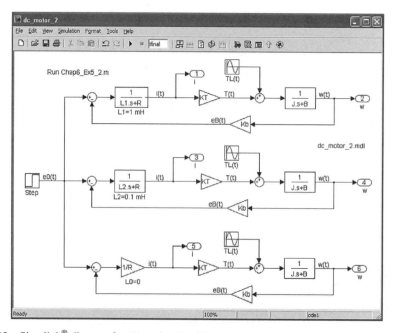

FIGURE 6.39 Simulink® diagram for $i(t)$ and $\omega(t)$ with $L = 0$, 0.1, and 1 mH.

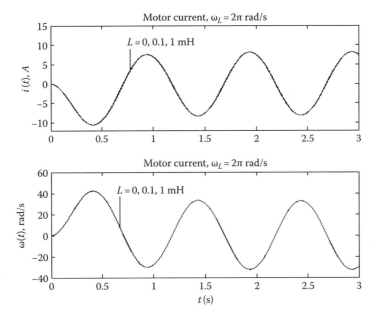

FIGURE 6.40 Motor current and speed for $L = 0, 0.1, 1$ mH, $\omega_L = 2\pi$ rad/s.

the second time with $\omega_L = 2000$ rad/s. The armature voltage $e_0(t)$ is zero for both calls. RK-4 integration with step size 0.0001 s was specified in *"Chap6_Ex5_2.m."*

The motor current and angular speeds for $L = 0$, 0.1, and 1 mH are indistinguishable from each other when the load torque frequency is 2π rad/s (see Figure 6.40). Figure 6.41 shows angular speed and current of the motor when the load torque frequency $\omega_L = 200$ rad/s. The angular speeds are nearly identical; however, there is a noticeable difference in current when $L = 1$ mH. Hence, for an accurate simulation of motor current for the case when $L = 1$ mH and the load torque frequency is 200 rad/s (or greater), the stiff second-order system model is required.

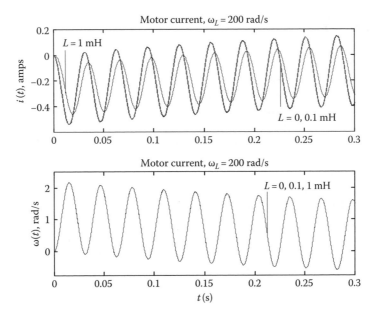

FIGURE 6.41 Motor current and speed for $L = 0, 0.1, 1$ mH, $\omega_L = 200$ rad/s.

EXERCISES

6.20 In Example 6.9,
- (a) Find the largest integration time step T, which yields stable and accurate approximations of the current $i(t)$ using RK-1, RK-3, and RK-4 integrators.
- (b) Find the analytical solution for the current $i(t)$.
- (c) Simulate the transient response of the circuit using the remaining stiff integrators available with Simulink and compare the number of integration steps required for each one. Calculate

$$|\bar{e}| = \frac{1}{N} \sum_{k=1,2,\ldots,N} |i(t_k) - x_{1,A}(t_k)|$$

where t_k, $k=1, 2,\ldots, N$ are the discrete times used by the stiff integration method to approximate the exact solution $i(t_k)$.

6.21 Figure E6.21 shows a thermal second-order system with input $u(t)$ and output $y(t)$. The temperature output is converted by a transducer, modeled as a first-order lag, to an electronic signal $v(t)$.

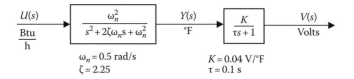

$$\omega_n = 0.5 \text{ rad/s} \qquad K = 0.04 \text{ V/°F}$$
$$\zeta = 2.25 \qquad\qquad \tau = 0.1 \text{ s}$$

FIGURE E6.21

- (a) Find the exact solution for the unit step response of $v(t)$.
- (b) Find the stiffness ratio relating the ratio of the largest to the smallest (in magnitude) characteristic root of the system. Is the system stiff?
- (c) Simulate the unit step response with a fixed-step RK integrator. What is the largest integration step size that can be used to obtain a stable solution?
- (d) Repeat part (c) using one of Simulink's stiff integrators, and compare the number of steps used by the RK and stiff integrator.
- (e) Compare the frequency response function $V(j\omega)/U(j\omega)$ with and without the sensor dynamics by generating a Bode plot for each on the same graph. Comment on the results.

6.22 The liquid level in the tank shown in Figure E6.22 is regulated by controlling the flow in F_1 using an electronically actuated control valve. A level transmitter provides a voltage signal v_T to the controller. The set point level H_{com} is converted to a voltage v_{com} inside the controller. The actuating signal $e_v = v_{com} - v_T$ is input to the controller that outputs the voltage signal v that determines the valve opening. The valve dynamics are described by a gain K_v and time constant τ_v as shown in the block diagram of the control system. The outflow from the tank F_0 is assumed to be proportional to the level, that is, $F_0 = cH$.

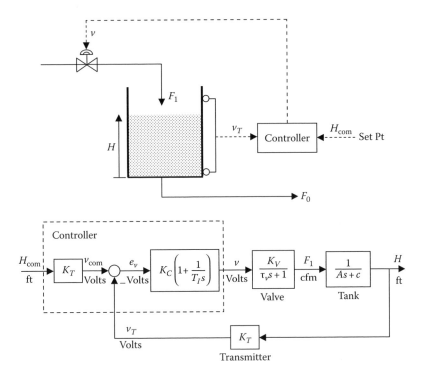

FIGURE E6.22

(a) The characteristic equation of the closed-loop control system is

$$1 + K_T K_C \left(1 + \frac{1}{T_I s} \right) \left(\frac{K_V}{\tau_v s + 1} \right) \left(\frac{1}{As + c} \right) = 0$$

The numerical values of the system parameters are

$$K_T = 0.25 \, \text{V/ft}, \quad K_C = 2, \quad K_I = 10 \, \text{min}, \quad K_V = 4 \, \text{cfm/V}, \quad \tau_v = 0.01 \, \text{min},$$
$$A = 100 \, \text{ft}^2, \quad c = 3 \, \text{cfm/ft}.$$

(b) Find the characteristic roots and the stiffness ratio.
(c) The system is initially at steady state with the tank empty. The set point input is a step function $H_{com}(t) = 3$ ft, $t > 0$. The step response $H(t)$, $0 \le t \le 180$ min is simulated using Simulink's fixed-step integrators "ode1" through "ode4." Use trial and error to estimate the integration step size T (to eight places after the decimal point), resulting in a marginally stable simulated response. Enter the values in the second column of the following table.
(d) Obtain plots of $H(t)$ and $F_1(t)$ with each integrator when the step size is one half the limiting values found in part (b). Enter the number of integration steps used to simulate the tank level response in the third column of the following table.
(e) Obtain plots of $H(t)$ and $F_1(t)$ using Simulink's stiff integrator "ode23s."

(f) Find the number of integration steps in part (d) and enter the value in the following table.

Integrator	T (Marginally Stable Response)	Number of Steps (Step Size $T/2$)
Ode1		
Ode2		
Ode3		
Ode4		
Ode23s	n/a	

6.23 Consider a third-order system with transfer function in Equation 6.248 and second-order system approximation with transfer function in Equation 6.249. Denote the transfer functions by $G_3(s)$ and $G_2(s)$, respectively. Suppose the input to both systems is $u(t) = 100e^{at}$, $t \geq 0$.

(a) Simulate the responses of each system and plot them on the same graph for the following cases:

   (i) $a = 0$ (ii) $a = -100$ (iii) $a = -800$ (iv) $a = -800.0094$ (v) $a = -5000$

6.24 Find analytical solutions for $\omega(t)$ and $i(t)$ in response to a step input $e_0(t)$ in Example 6.10. Compare the exact solutions with the simulated results obtained using "ode1" and "ode15s" integrators.

6.25 For the DC motor in Example 6.10 with armature voltage zero,

(a) Find the transfer functions

$$G_\Omega(s)|_{E_0(s)=0} = \left.\frac{\Omega(s)}{T_L(s)}\right|_{E_0(s)=0}, \quad G_I(s)|_{E_0(s)=0} = \left.\frac{I(s)}{T_L(s)}\right|_{E_0(s)=0}$$

(b) Draw Bode plots for $G_\Omega(j\omega)|_{E_0(s)=0}$ and $G_I(j\omega)|_{E_0(s)=0}$ for $L = 0$, 0.1, 1 mH.

(c) Are the motor current and speed profiles in Figures 6.40 and 6.41 consistent with the results in part (b)?

6.26 An angular speed control system is shown in Figure E6.26a:

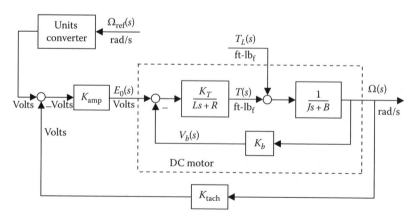

FIGURE E6.26a

The motor constants and load inertia are

$$R = 1\,\Omega, \quad L = 0.1\,\text{mH}, \quad K_T = 0.8\,\text{ft lb}_f/A, \quad K_b = 0.05\,\text{V/rad/s},$$

$$B = 0.01\,\text{ft lb}_f/\text{rad/s}, \quad J = 0.045\,\text{ft lb}_f/\text{rad/s}^2.$$

The tachometer gain in the feedback path is $K_{\text{tach}} = 0.0475$ V/rad/s and the amplifier gain $K_{\text{amp}} = 50$. A units converter is inserted before the first summer to convert the reference input from rad/s to volts. The gain of the units converter is the same as K_{tach}.

(a) Find the stiffness of the DC motor.

(b) Find the stiffness of the closed-loop control system.

(c) Prepare a Simulink diagram for simulating the control system. The reference input and load torque profiles are shown in Figure E6.26b.

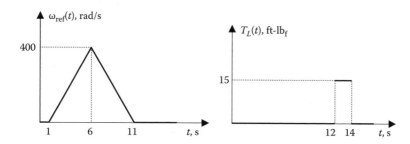

FIGURE E6.26b

(d) Use trial and error to find the maximum step size for stable integration of the model using RK-1 through RK-4 integration.

(e) Simulate the control system using RK-1 through RK-4 integration with step sizes equal to one half the values found in part (d). Repeat using the stiff integrators "ode15s," "ode23s," "ode23t," and "ode23tb." Compare the execution times and number of integration steps required with each.

6.27 The block diagram of a control system is shown in Figure E6.27.

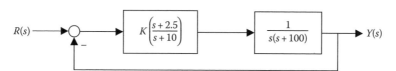

FIGURE E6.27

(a) Find the closed-loop system transfer function $Y(s)/R(s)$.

(b) Find the closed-loop system poles (characteristic roots) and the stiffness ratio for controller gains of $K = 1$, 100, 1000.

(c) Find the analytical solutions for the unit step responses when $K = 1$, 100, 1000.

(d) Select any order RK integrator and find the step size for each value of K where the integrator is on the verge of becoming unstable.

(e) Simulate the step responses using the selected RK integrator with a step size of one half the value found in part (d) for each value of K.

(f) Plot the analytical and simulated step responses on the same graphs.

(g) Approximate the stiff closed-loop system dynamics when $K = 100$ with a second-order transfer function obtained by ignoring the fast pole of the third-order closed-loop transfer function. Introduce a gain in the numerator of the second-order transfer function that makes the DC gain of the second- and third-order system transfer functions identical. Compare the third-order system analytical and simulated step responses to the second-order system analytical and simulated step responses. Compare the step sizes, number of integration steps, and execution times used to simulate the original system and the reduced order system approximation.

6.6 LUMPED PARAMETER APPROXIMATION OF DISTRIBUTED PARAMETER SYSTEMS

Dynamic systems involving variables that exhibit both spatial and temporal variations are modeled by partial differential equations and referred to as distributed parameter systems. The introductory section in Chapter 1 cited the example of a room temperature $T(x, y, z, t)$ that varies as a function of the point coordinates (x, y, z) as well as time t. Analytical solutions of partial differential equation models subject to various boundary conditions are rare in all but the simplest of examples. Numerical solutions are based on a partitioning of the entire volume and surface areas within the system into meshes comprising finite-sized triangular elements with interior and exterior nodes at the vertices. Difference equations, sometimes numbering in the hundreds of thousands depending on the size and shape of the finite elements, are written for the dependent variable(s) at a subset of the nodes. Accurate approximations to the continuous solutions of the partial differential equation models are possible using this "finite element analysis" approach. Examples include the temperature distribution and heat flows from irregular-shaped cooling surfaces, structural analysis, fluid dynamics, and so forth.

In dynamic systems with regular-shaped geometries, a continuously varying spatial parameter can be discretized into a finite number of values associated with discrete geometric regions. For example, consider a long, thin cylindrical rod with perfect insulation along its length and top face like the one shown in Figure 6.42.

Suppose one end of the rod is immersed in a liquid bath of constant temperature $\overline{T}$. Assuming negligible heat flow in the x and y directions, temperature gradients exist solely in the longitudinal direction, that is, along the z-axis of the cylinder. The temperature is described by $T(t, z)$. The initial temperature distribution $T_0(z)$ is known as well.

Derivation of the equation governing the cylinder's temperature $T(t, z)$ is straightforward (Miller 1975). The result is the partial differential equation

$$\frac{\partial}{\partial t} T(t, z) - \alpha \frac{\partial^2}{\partial z^2} T(t, z) = 0 \qquad (6.276)$$

subject to initial condition $T(0, z) = T_0(z)$, $0 \le z \le L$ along with the boundary conditions $T(t, 0) = \overline{T}$, $t \ge 0$ and $(\partial/\partial z) T(t, z)|_{z=L} = 0$, $t \ge 0$. L is the length of the cylinder, and α is a parameter related to the physical and thermal properties of the cylinder material.

A lumped parameter model consisting of coupled ordinary differential equations is obtained by dividing the cylinder into n equal segments of length $\Delta z = L/n$ (see Figure 6.42). Each segment has, associated with it, a thermal capacitance C_i and is assigned a node temperature T_i. Energy balances for each segment relate the net heat flow to the accumulation of thermal energy, that is,

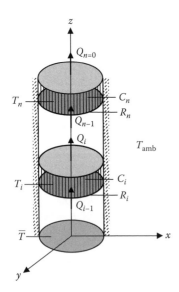

FIGURE 6.42 Lumped parameter depiction of rod with discrete thermal capacitances.

$$C_i \frac{d}{dt} T_i(t) = Q_{i-1} - Q_i, \quad i = 1, 2, 3, \dots, n \qquad (6.277)$$

Heat flows across the boundaries of each segment along the z-axis by conduction. Fourier's law of heat conduction states that the conductive heat flow per unit area is negatively proportional to the temperature gradient in the direction of flow. The heat flow from the constant temperature source at the bottom to the first segment with temperature T_1 is

$$Q_0 = -kA\left(\frac{T_i - \overline{T}}{\Delta z/2}\right) = \frac{\overline{T} - T_1}{R_1} \tag{6.278}$$

The term in parenthesis is the temperature gradient, and k is the thermal conductivity of the material. R_1 represents the thermal resistance at the lower boundary and is computed from

$$R_1 = \frac{\Delta z}{2kA} \tag{6.279}$$

The internal heat flows are described by

$$Q_i = -kA\left(\frac{T_{i+1} - T_i}{\Delta z}\right) = \frac{T_i - T_{i+1}}{R_{i+1}}, \quad i = 1, 2, \ldots, n-1 \tag{6.280}$$

where

$$R_{i+1} = \frac{\Delta z}{kA}, \quad i = 1, 2, \ldots, n-1 \tag{6.281}$$

Heat flow between the top segment and its surroundings is zero as a result of assuming that the top face is perfectly insulated. Consequently,

$$Q_n = 0 \tag{6.282}$$

The cylindrical rod with $n = 5$ segments is illustrated in Figure 6.43.

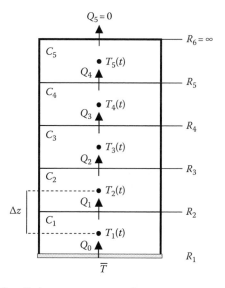

FIGURE 6.43 Cylinder with five distinct temperature nodes.

Combining Equations 6.277, 6.278, and 6.280 through 6.282 leads to the linear system of differential equations

$$\frac{d}{dt}\begin{bmatrix} T_1(t) \\ T_2(t) \\ T_3(t) \\ T_4(t) \\ T_5(t) \end{bmatrix} = A \begin{bmatrix} T_1(t) \\ T_2(t) \\ T_3(t) \\ T_4(t) \\ T_5(t) \end{bmatrix} + B\bar{T} \tag{6.283}$$

The coefficient matrix A and input matrix B are given by

$$A = \begin{bmatrix} -\left(\dfrac{1}{R_1}+\dfrac{1}{R_2}\right)\dfrac{1}{C_1} & \dfrac{1}{R_2 C_1} & 0 & 0 & 0 \\[2ex] \dfrac{1}{R_2 C_2} & -\left(\dfrac{1}{R_2}+\dfrac{1}{R_3}\right)\dfrac{1}{C_2} & \dfrac{1}{R_3 C_2} & 0 & 0 \\[2ex] 0 & \dfrac{1}{R_3 C_3} & -\left(\dfrac{1}{R_3}+\dfrac{1}{R_4}\right)\dfrac{1}{C_3} & \dfrac{1}{R_4 C_3} & 0 \\[2ex] 0 & 0 & \dfrac{1}{R_4 C_4} & -\left(\dfrac{1}{R_4}+\dfrac{1}{R_5}\right)\dfrac{1}{C_4} & \dfrac{1}{R_5 C_4} \\[2ex] 0 & 0 & 0 & \dfrac{1}{R_5 C_5} & -\dfrac{1}{R_5 C_5} \end{bmatrix} \tag{6.284}$$

$$B = \left[\dfrac{1}{R_1 C_1} \quad 0 \quad 0 \quad 0 \quad 0\right]^T \tag{6.285}$$

Example 6.11

The temperature of a 10 ft long, 2 ft diameter copper cylinder is initially 75°F throughout its entire length. One of its edges is placed in contact with a surface maintained at a constant temperature of 200°F. The cylinder is thermally insulated from its surroundings except for the edge surface in contact with the 200°F temperature. Assume heat flows in the longitudinal direction only.

The physical properties of copper are

thermal conductivity: $k = 224$ Btu/h/°F/ft
specific heat: $c = 2.93$ Btu/°F/slug
mass density: $\rho = 17.3$ slug/ft^3

Partition the cylinder into five equal-sized sections and

(a) Find the matrices A and B in the state equation $\dot{\underline{T}}(t) = A\underline{T}(t) + B\bar{T}$ where $\underline{T}(t) = [T_1(t)T_2(t)T_3(t)T_4(t)T_5(t)]^T$ is the state vector.
(b) Find the steady-state node temperatures.
(c) Simulate and plot the temperature responses of each section long enough for the transient response to die out.
(d) Plot the temperature profile along the bar at $t = 0, 2.5, 5, 10, 20, 30$ h.

(a) The volume of each section is

$$V_i = A_i \Delta z = \pi\left(\frac{D}{2}\right)^2 \Delta z = \pi\left(\frac{2}{2}\right)^2\left(\frac{10}{5}\right) = 2\pi \text{ ft}^3, \quad i = 1, 2, \ldots, 5 \tag{6.286}$$

The thermal capacitance of each section is

$$C_i = c_i \rho V_i = 2.93 \, \frac{\text{Btu}}{\text{°F} - \text{slug}} \times 17.3 \, \frac{\text{slug}}{\text{ft}^3} \times 2\pi \, \text{ft}^3 = 318.49 \, \frac{\text{Btu}}{\text{°F}}, \quad i = 1, 2, \ldots, 5 \qquad (6.287)$$

and the thermal resistances at the interfaces of each section are

$$R_1 = \frac{\Delta z_1}{2kA_1} = \frac{2 \, \text{ft}}{2 \times 224 \, (\text{Btu/h/°F ft}) \times \pi \, \text{ft}^2} = 0.0014 \, \frac{\text{°F}}{\text{Btu/h}} \qquad (6.288)$$

$$R_i = \frac{\Delta z_i}{kA_i} = 2 \times R_1 = 0.0028 \, \frac{\text{°F}}{\text{Btu/h}}, \quad i = 2, 3, 4, 5 \qquad (6.289)$$

Substituting the values for R_i and C_i into Equations 6.284 and 6.285 gives (see M-file "Chap6_Ex6_1.m")

$$A = \begin{bmatrix} -3.3143 & 1.1048 & 0 & 0 & 0 \\ 1.1048 & -2.2096 & 1.1048 & 0 & 0 \\ 0 & 1.1048 & -2.2096 & 1.1048 & 0 \\ 0 & 0 & 1.1048 & -2.2096 & 1.1048 \\ 0 & 0 & 0 & 1.1048 & -1.1048 \end{bmatrix}, \quad B = \begin{bmatrix} 2.2096 \\ 0 \\ 0 \\ 0 \\ 0 \end{bmatrix}$$

(b) The steady-state state vector $\underline{T}_{ss}$ is obtained from Equation 6.283 with the left-hand side equal to the zero vector. The result is

$$\underline{T}_{ss} = -A^{-1}B\overline{T}$$

$$= - \begin{bmatrix} -3.3143 & 1.1048 & 0 & 0 & 0 \\ 1.1048 & -2.2096 & 1.1048 & 0 & 0 \\ 0 & 1.1048 & -2.2096 & 1.1048 & 0 \\ 0 & 0 & 1.1048 & -2.2096 & 1.1048 \\ 0 & 0 & 0 & 1.1048 & -1.1048 \end{bmatrix}^{-1} \begin{bmatrix} 2.2096 \\ 0 \\ 0 \\ 0 \\ 0 \end{bmatrix} 200$$

$$= \begin{bmatrix} 200 \\ 200 \\ 200 \\ 200 \\ 200 \end{bmatrix} \qquad (6.290)$$

(c) A "Constant" block and the "state-space" block in Simulink are all that are needed to simulate the response of the lumped parameter system model. The output matrix C is chosen to be the 5×5 identity matrix, forcing the output vector to be identical to the state vector. The direct transmission matrix D is a 5×1 column vector of all zeros. The Simulink diagram is shown in Figure 6.44.

The "Workspace I/O" tab in the "Simulation Parameters" dialog box must have "Time" and "States" checked. The Simulink model file "temp_cylinder.mdl" is called from within "Chap6_Ex6_1.m." RK-4 integration with step size $T = 0.01$ h was used to generate the node temperature responses $T_1(t)$, $T_2(t)$, $\ldots$, $T_5(t)$ shown in Figure 6.45.

(d) The temperature profiles are approximated by linearly interpolating the node temperatures at the required times (see Figure 6.46).

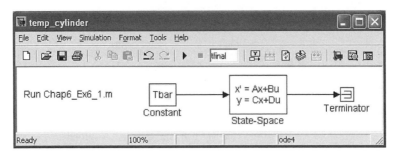

FIGURE 6.44   Simulink® diagram for simulation of lumped parameter system model.

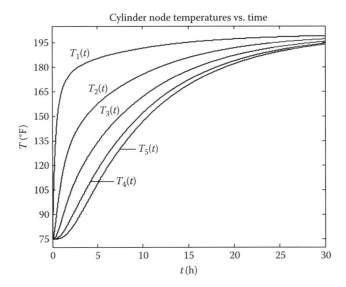

FIGURE 6.45   Time histories of cylinder node temperatures.

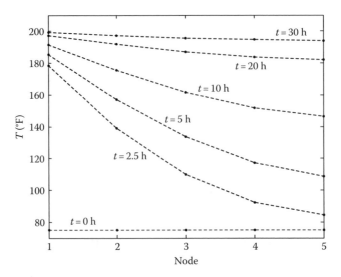

FIGURE 6.46   Temperature profiles along cylinder at $t = 0, 2.5, 5, 10, 20, 30$ h.

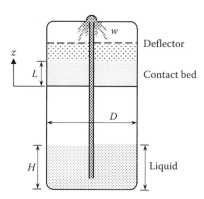

FIGURE 6.47 Coffee pot with liquid circulation.

6.6.1 NONLINEAR DISTRIBUTED PARAMETER SYSTEM

The next example illustrating the approximation of a distributed parameter system with a lumped parameter model is that of a coffee pot used for brewing coffee. In the coffee pot shown in Figure 6.47, liquid rises up through the riser, is distributed uniformly over the bed of coffee grounds, passes through the bed taking up coffee extract, and falls back to the bottom of the pot.

The following notation is used in the partial differential equation model, which governs the concentration of coffee in the liquid as it passes through the layer of coffee grounds, and the ordinary differential, which describes the concentration of coffee in the well-mixed reservoir at the bottom of the pot.

Notation:

A: cross-sectional area of the contact bed, ft^2
L: height of the contact bed, ft
H_L: holdup of liquid per unit height of contact bed, lb water/ft
H_t: holdup of liquid in reservoir of pot, lb water
a: mass transfer area per unit of volume of bed, ft^2/ft^3
k_m: mass transfer rate coefficient, lb/s coffee/($ft^2 \times$ (lb coffee/lb water))
c_s: saturated concentration of coffee, lb coffee/lb water
z: independent spatial variable measured from bottom to top of contact bed, ft
t: independent time variable, min
$w(t)$: circulation of liquid, lb water/s
$E_0(z, t)$: fraction of coffee not yet extracted at height z and time t
$c(z, t)$: concentration of coffee in liquid at height z and time t, lb coffee/lb water
$c_R(t)$: concentration of coffee in reservoir, lb coffee/lb water

Assuming no coffee concentration gradients in the radial direction of the contact bed and a well-mixed reservoir leads to a mathematical model based on conservation of coffee extract in the contact bed and reservoir (Huntsinger, personal notes)

$$\text{Contact bed: } -w\frac{\partial}{\partial z}c(z,t) + E_0(z,t)Aak_m[c_s - c(z,t)] = H_L\frac{\partial}{\partial t}c(z,t) \tag{6.291}$$

$$\text{subject to: } c(0,t) = c_R(t), \quad c(z,0) = 0 \tag{6.292}$$

$$\text{Reservoir: } H_t\frac{d}{dt}c_R(t) + wc_R(t) = wc(L,t), \quad c_R(0) = 0 \tag{6.293}$$

The lumped parameter model of the coffee pot is developed in a similar manner to the way it was obtained for the temperature distribution along the cylindrical rod. That is, the contact bed is divided

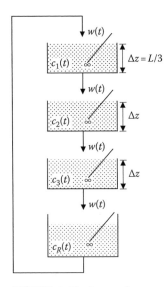

FIGURE 6.48 Lumped parameter view of coffee pot.

into a number of discrete layers with homogeneous properties throughout. The situation is illustrated in Figure 6.48 for the case of three sections with uniform liquid concentrations $c_1(t)$, $c_2(t)$, and $c_3(t)$. The liquid concentration in the reservoir is $c_R(t)$.

Equations expressing the conservation of coffee extract in each homogeneous section are

$$\text{Section 1:} \; w(t)c_R(t) - w(t)c_1(t) + (\Delta z)AaK_m[c_s - c_1(t)]E_{0,1}(t)$$

$$= H_L\Delta z\frac{d}{dt}c_1(t) \tag{6.294}$$

$$\text{Section 2:} \; w(t)c_1(t) - w(t)c_2(t) + (\Delta z)AaK_m[c_s - c_2(t)]E_{0,2}(t)$$

$$= H_L\Delta z\frac{d}{dt}c_2(t) \tag{6.295}$$

$$\text{Section 3:} \; w(t)c_2(t) - w(t)c_3(t) + (\Delta z)AaK_m[c_s - c_3(t)]E_{0,3}(t)$$

$$= H_L\Delta z\frac{d}{dt}c_3(t) \tag{6.296}$$

The third term in Equations 6.294 through 6.296 accounts for the mass transfer of coffee extracted from the coffee grounds to the liquid. $E_{0,i}(t)$, $i = 1, 2, 3$ represents the fraction of coffee not yet extracted from section "i" after time "t." The equation for $E_{0,i}(t)$ is

$$E_{0,i}(t) = \frac{B_0A\Delta z - K_{TE_i}(t)}{B_0A\Delta z}, \quad i = 1, 2, 3 \tag{6.297}$$

where

B_0 is the total coffee per volume of bed for fresh grounds
$K_{TE_i}(t)$ is the total coffee extracted from section i in time "t" obtained from

$$K_{TE_i}(t) = \int_0^t E_{0,i}(\Delta z)AaK_m[c_s - c_i(\tau)]d\tau \tag{6.298}$$

The final equation of the lumped parameter model is the mass balance on the coffee in and out of the reservoir.

$$\text{Reservoir:} \; w(t)c_3(t) - w(t)c_R(t) = H_t\frac{d}{dt}c_R(t) \tag{6.299}$$

A careful check of all terms in Equations 6.294 through 6.296 and 6.299 will reveal the units to be lb coffee/s.

The circulation of coffee is described by

$$w(t) = \begin{cases} \left(\dfrac{\bar{w}}{t_1}\right)t, & 0 \le t < t_1 \\ \bar{w}, & t \ge t_1 \end{cases} \tag{6.300}$$

The model equations are represented in the Simulink model file "*coffee.mdl*" shown in Figure 6.49. Numerical values of the system parameters are given in "*Chap6_coffee.m*" and listed as follows:

$$D = 6\,\text{in.}, \quad H = 5\,\text{in. of water}, \quad L = 2.5\,\text{in. of coffee}$$

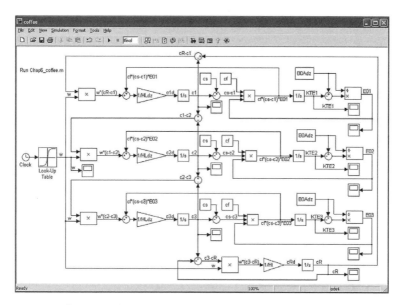

FIGURE 6.49 Simulink® model for simulating coffee pot.

$$a = 3000 \, \text{ft}^2 \text{ of bed/ft}^3 \text{ of bed}, \quad k_m = 0.00003 \, \frac{\text{lb/s coffee}}{\text{ft}^2 \times (\text{lb coffee/lb water})}$$

$$B_0 = 3 \, \text{lb coffee/ft}^3 \text{ of bed}, \quad c_s = 0.2 \, \text{lb coffee/lb water}$$

$$t_1 = 60 \, \text{s}, \quad \bar{w} = 0.05 \, \text{lb water/s}$$

Initial conditions: $c_1(0) = c_2(0) = c_3(0) = c_R(0) = 0 \, \text{lb of coffee/lb of water}$

Coffee concentration in the three sections and reservoir are shown in Figure 6.50.

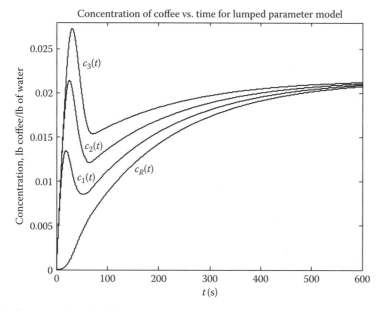

FIGURE 6.50 Concentration of coffee in lumped sections and reservoir.

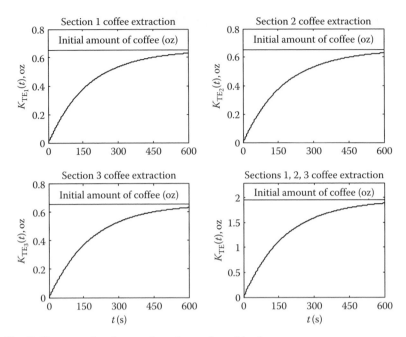

FIGURE 6.51 Coffee extraction from each section and combined.

The transient period is approximately 10 min (600 s). The steady-state concentration attained in each section and in the reservoir is slightly greater than 0.02 lb coffee/lb of water, well below the saturation limit of $c_s = 0.2$ lb coffee/lb water.

There is no analytical method for determining $(c_1)_{ss}$, $(c_2)_{ss}$, $(c_3)_{ss}$, and $(c_R)_{ss}$ from the model Equations 6.294 through 6.296 and 6.299 when all the coffee has been extracted from the coffee grounds, that is, $E_{0,1}(\infty) = E_{0,2}(\infty) = E_{0,3}(\infty) = 0$. When this occurs, the steady-state concentrations are an identical amount that depends on the quantity of coffee grounds initially placed in the coffee pot.

Figure 6.51 shows the amount of coffee extracted (in oz) from each section and the overall amount as a function of time. The initial amount of coffee in each section (0.6545 oz) and the total (1.9635 oz) are calculated from the initial volumes of coffee extracted in each section and B_0, the coffee density in lb coffee/cu ft of bed. After 10 min, the total amount of coffee extracted from sections 1, 2, and 3 is 1.8892 oz.

There is sufficient water for nearly ten 8 oz cups of coffee. Can you verify this? Figures 6.50 and 6.51 are plotted in M-file "*Chap6_coffee.m*."

EXERCISES

6.28 Rework Example 6.11 for the case where the top surface of the cylinder is no longer insulated. Instead, the top surface is maintained at 0°F.

6.29 Rework Example 6.11 using $n = 10$ and 20 segments, and compare the results with those shown in Figures 6.45 and 6.46.

6.30 Rework Example 6.11 for the case where the diameter of the cylinder is 1 ft instead of 2 ft. Compare the results to those shown in Figures 6.45 and 6.46.

6.31 Rework Example 6.11 for the case where the bottom face of the cylinder receives a constant supply of heat in the amount of 25,000 Btu/h and the top surface is maintained at 75°F, the same as the initial temperature of the cylinder.

6.7 SYSTEMS WITH DISCONTINUITIES

Mathematical models of dynamic systems sometimes exhibit discontinuities. Internal and external forces in mechanical systems and energy sources in electrical and thermal systems can change instantaneously as a result of infinitesimal displacements in the state of the these systems. Distinct regions exist in the state space where the system model is represented by different sets of algebraic and differential equations. The situation is illustrated in Figure 6.52 for the case of a discontinuous second-order system with state variables $x_1 \geq 0$, $x_2 \geq 0$ and 3 distinct regions S_1, S_2, and S_3.

For a second-order system without discontinuities, a suitable mathematical model assumes the form of a system of first-order differential equations

$$
\left.
\begin{aligned}
\frac{dx_1}{dt} &= f_1(t, x_1, x_2) \\[2mm]
\frac{dx_2}{dt} &= f_2(t, x_1, x_2)
\end{aligned}
\right\}
\tag{6.301}
$$

For the second-order system with discontinuities like the one shown in Figure 6.52,

$$
\left.
\begin{aligned}
\frac{dx_1}{dt} &= f_{11}(t, x_1, x_2), & (x_1, x_2) &\in S_1 \\[2mm]
\frac{dx_1}{dt} &= f_{12}(t, x_1, x_2), & (x_1, x_2) &\in S_2 \\[2mm]
\frac{dx_1}{dt} &= f_{13}(t, x_1, x_2), & (x_1, x_2) &\in S_3
\end{aligned}
\right\}
\tag{6.302}
$$

$$
\left.
\begin{aligned}
\frac{dx_2}{dt} &= f_{21}(t, x_1, x_2), & (x_1, x_2) &\in S_1 \\[2mm]
\frac{dx_2}{dt} &= f_{22}(t, x_1, x_2), & (x_1, x_2) &\in S_2 \\[2mm]
\frac{dx_2}{dt} &= f_{23}(t, x_1, x_2), & (x_1, x_2) &\in S_3
\end{aligned}
\right\}
\tag{6.303}
$$

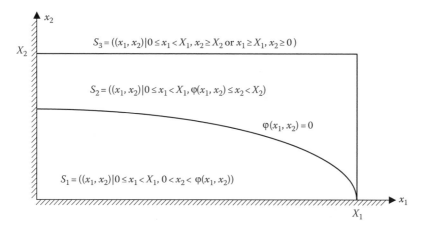

FIGURE 6.52 Discontinuous system with three distinct regions in state space.

In the general case of an nth-order system with m regions $S_1, S_2, \ldots, S_m$, we have

$$\frac{dx_i}{dt} = f_{ij}(t, x_1, x_2, \ldots, x_n), \quad i = 1, 2, \ldots, n \quad j = 1, 2, \ldots, m \qquad (6.304)$$

where the m regions are defined by a set of discontinuity functions $\phi_k(x_1, x_2, \ldots, x_n)$ such that a discontinuity occurs when one of the functions $\phi_k = 0$ (Hay 1973).

Simulation of a dynamic system modeled as in Equation 6.304 is not as straightforward as the systems previously encountered. The complication arises from the requirement of knowing which region the state resides in to assure numerical integration of the appropriate equations. With fixed-step as well as variable-step integration methods, the state $(x_1, x_2, \ldots, x_n)$ and the set of discontinuity functions ϕ_k are available only at discrete points in time corresponding to the end point of each integration step. The presence of a discontinuity (or several discontinuities) at an interior point of the step is sensed by a change in sign of one (or more) of the discontinuity functions.

Several approaches to the problem are possible. The simplest is to merely assume the discontinuity (or discontinuities) occurs at the end of the step in which it is detected. The appropriate model equations are numerically integrated, starting from the beginning of the next step. The shortcoming of this approach is apparent, namely, the creation of a cumulative error resulting from integration of the incorrect equations over a portion of the interval in which the discontinuity occurs. The error is minimized by choosing excessively small integration steps when using fixed-step integrators, not a very satisfactory solution, even impossible for certain applications.

The second approach is applicable for variable-step integration methods, which adjust the step size based on estimation of the local truncation error. Instead of waiting for the end of an integration step to check for the occurrence of a discontinuity, the discontinuity functions ϕ_k are evaluated after each derivative function evaluation within the interval. A change of sign in any ϕ_k triggers a switch in one of the derivative functions, eventually producing an artificially large estimate of the truncation error. The result is a self-correcting reduction in the current integration step leading up to the time of the discontinuity and slightly beyond.

The next approach is similar to the first in that the discontinuity functions ϕ_k are evaluated only at the end of each fixed-size integration step. When one or more discontinuities are found to have occurred in the current interval, some form of interpolation or possibly root finding is employed to locate their time(s) of occurrence to a prescribed accuracy. Once the time of occurrence is determined, the integration is repeated over the subinterval ending at the time of the first (earliest) discontinuity. Subsequent integrations proceed to the end of the fixed-size integration step using the state equations appropriate to the corresponding region in state space.

The last approach is best illustrated by a simple example. Figure 6.53 shows a pendulum swinging from a frictionless hinge with angular displacement confined to a single plane of motion. The bob at the end of the pendulum is immersed in a viscous fluid during a portion of its travel.

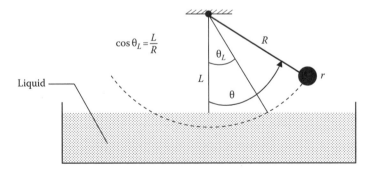

FIGURE 6.53 Pendulum traveling through air and liquid.

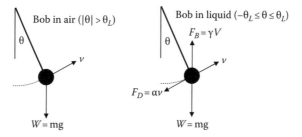

FIGURE 6.54 Diagram showing external forces acting on bob.

The pendulum rod is assumed to be of negligible mass as is the drag force on the bob when exposed to air.

The bob is subject to a gravitational force W at all times along with a drag force F_D and buoyant force F_B acting on it while it is submerged. The forces are shown in Figure 6.54 for both cases.

The pendulum dynamics are modeled by the differential equation

$$J\frac{d^2\theta}{dt^2} = \begin{cases} (-W + F_B)R\sin\theta - F_D R, & -\theta_L \leq \theta \leq \theta_L \\ -WR\sin\theta, & |\theta| > \theta_L \end{cases} \tag{6.305}$$

Expressions for the constant buoyant force and assumed linear drag force are

$$F_B = \gamma V = E\left(\frac{4}{3}\pi r^3\right) \tag{6.306}$$

$$F_D = \alpha v = \alpha R\frac{d\theta}{dt} \tag{6.307}$$

where
  γ is the specific weight of the liquid
  V is the volume of the bob
  α is the drag coefficient

Combining Equations 6.305 through 6.307 gives

$$J\frac{d^2\theta}{dt^2} = \begin{cases} \left(-mg + \frac{4}{3}\gamma\pi r^3\right)R\sin\theta - \alpha R^2\frac{d\theta}{dt}, & -\theta_L \leq \theta \leq \theta_L \\ -mg\,R\sin E, & |\theta| > \theta_L \end{cases} \tag{6.308}$$

Introducing state variables $x_1(t) = \theta(t)$ and $x_2(t) = \dot{\theta}(t)$ results in

$$\dot{x}_1 = \dot{\theta} = x_2, \quad \dot{x}_2 = \ddot{\theta} = \begin{cases} \frac{1}{J}\left\{\left(-mg + \frac{4}{3}\gamma\pi r^3\right)R\sin x_1 - \alpha R^2 x_2\right\} & -\theta_L \leq x_1 \leq \theta_L \\ \frac{1}{J}(-mg\,R\sin x_1), & |x_1| > \theta_L \end{cases} \tag{6.309}$$

Defining regions S_1 and S_2 in the state space according to

$$S_1 = \{(x_1, x_2),\ -\theta_L \leq x_1 \leq \theta_L\} \quad \text{and} \quad S_2 = \{(x_1, x_2), |x_1| > \theta_L\} \tag{6.310}$$

and using the notation in Equation 6.304, the state derivative functions become

$$f_{11}(x_1, x_2) = x_2, \quad (x_1, x_2) \in S_1 \tag{6.311}$$

$$f_{12}(x_1, x_2) = x_2, \quad (x_1, x_2) \in S_2 \tag{6.312}$$

$$f_{21}(x_1, x_2) = \frac{1}{J}\left[\left(-mg + \frac{4}{3}\gamma\pi r^3\right) R \sin x_1 - \alpha R^2 x_2\right], \quad (x_1, x_2) \in S_1 \tag{6.313}$$

$$f_{22}(x_1, x_2) = \frac{1}{J}(-mg R \sin x_1), \quad (x_1, x_2) \in S_2 \tag{6.314}$$

The discontinuity functions are

$$\phi_1(x_1, x_2) = x_1 - \theta_L \tag{6.315}$$

$$\phi_2(x_1, x_2) = -x_1 - \theta_L \tag{6.316}$$

Note that $\phi_1(x_1, x_2) = 0 \Rightarrow x_1 = \theta_L$ and $\phi_2(x_1, x_2) = 0 \Rightarrow x_1 = -\theta_L$. Hence, when either discontinuity function is zero, the pendulum is transitioning from region S_1 to S_2 or vice versa. Figure 6.55 shows the state vector (x_1, x_2) is inside region S_1 when the discontinuity functions satisfy the inequalities

$$\phi_1(x_1, x_2) \leq 0 \quad \text{and} \quad \phi_2(x_1, x_2) < 0 \tag{6.317}$$

Conversely, the state vector (x_1, x_2) is in region S_2 whenever

$$\phi_1(x_1, x_2) > 0 \quad \text{or} \quad \phi_2(x_1, x_2) \geq 0 \tag{6.318}$$

A flow chart is shown in Figure 6.56 for simulating the pendulum dynamics. MATLAB routines called by the main program "*Chap6_discont.m*" are listed followed by a brief explanation of their function.

```
function [phi_1, phi_2] = DFUNCT(x1,x2)
% Evaluates discontinuity functions given state components
% Inputs: x1,x2 - components of state
% Outputs: ph1,ph2 - discontinuity functions at (x1,x2)
global thetaL
phi_1 = x1-thetaL;
phi_2 = -x1-thetaL;
```

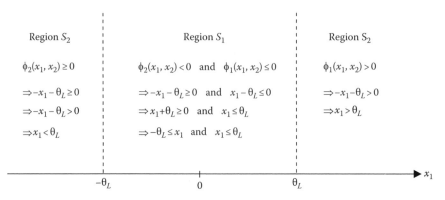

FIGURE 6.55 Definition of regions S_1 and S_2 in terms of discontinuity functions.

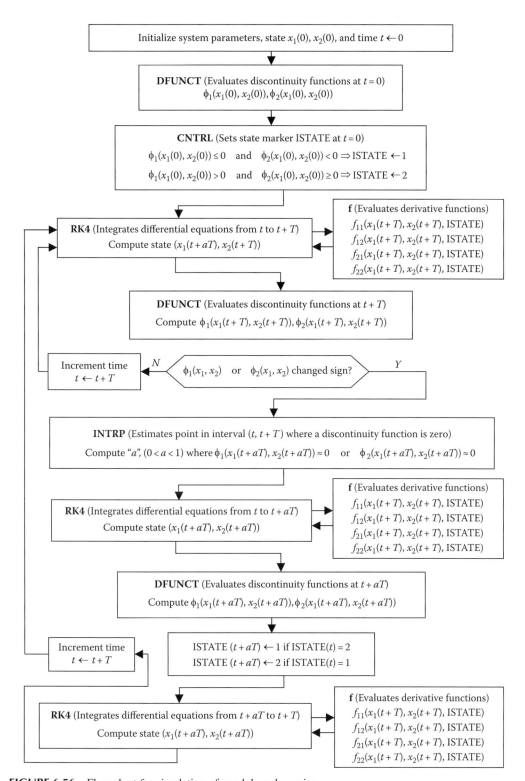

FIGURE 6.56 Flow chart for simulation of pendulum dynamics.

MATLAB function "DFUNCT.m" receives the coordinates (x_1, x_2) of the state vector and returns values of the two discontinuity functions $\phi_1(x_1, x_2)$ and $\phi_2(x_1, x_2)$.

```
function ISTATE = CNTRL(phi_1,phi_2)
% Determines whether state vector is in Region S1 or S2
% S1 - pendulum bob in liquid, i.e. |x1|<=theta_L
% S2 - pendulum bob in air, i.e. |x1|>theta_L
% Inputs: phi_1, phi_2 - discontinuity functions
% Output: ISTATE - marker indicating if state is in Region S1 or S2
if phi_1 <=0 & phi_2 <0
  ISTATE=1; % state is in Region S1
else
    ISTATE=2; % state is in Region S2
end
```

"CNTRL.m" accepts the values of the discontinuity functions $\phi_1(x_1, x_2)$ and $\phi_2(x_1, x_2)$ and checks which of the mutually exclusive conditions in Equation 6.317 or 6.318 are true. The marker "ISTATE" is set accordingly.

```
function [x1_new, x2_new] = RK4(T,x1_old,x2_old,ISTATE)
% RK-4 numerical integrator for updating state
% Inputs: T - integration step size
% x1_old,x2_old - starting values of state components
% ISTATE - marker indicating if state is in Region S1 or S2
% Outputs: x1_new,x2_new - updated state vector
global g R m J gamma r alpha
[k11 k12] = f(x1_old,x2_old,ISTATE);
x1_half=x1_old+(T/2)*k11;
x2_half=x2_old+(T/2)*k12;
[k21 k22] = f(x1_half,x2_half,ISTATE);
x1_half_hat=x1_old+(T/2)*k21;
x2_half_hat=x2_old+(T/2)*k22;
[k31 k32] = f(x1_half_hat,x2_half_hat,ISTATE);
x1_full_hat=x1_old+T*k31;
x2_full_hat=x2_old+T*k32;
[k41 k42] = f(x1_full_hat,x2_full_hat,ISTATE);
x1_new=x1_old+(T/6)*(k11+2*k21+2*k31+k41);
x2_new=x2_old+(T/6)*(k12+2*k22+2*k32+k42);
```

"RK4.m" implements the commonly used fourth-order RK integration algorithm presented in Equations 6.60 through 6.64. In addition to inputs specifying the integration step size and the current state vector, the last input "ISTATE" is passed to the function "f.m" to assure the appropriate state derivative equations are selected, that is, Equations 6.311 and 6.313 or 6.312 and 6.314.

```
function [f1, f2] = f(x1,x2,ISTATE)
% Inputs: x1,x2 - components of state
%     ISTATE - marker indicating if state is in Region S1 or S2
% Outputs: f1,f2 - state derivatives
   global g R m J gamma r alpha
f1=x2;
```

```
if ISTATE = = 1
   f2 = (R/J)*(-m*g+(4/3)*(gamma*pi*r^3))*sin(x1)-alpha*x2;
elseif ISTATE = = 2
   f2 = (R/J)*(-m*g*sin(x1));
end
```

"f.m" is called from "RK4.m" four times (once at the start, twice in the middle, and once at the end of the integration interval) in the process of updating the state. It returns the values of the state derivative functions.

```
function a = INTRP(ti,ph_old,ph_new,x11,x22,k)
% Interpolates to estimate pt ti+aT where one of the discontinuity
% functions is zero. Uses linear interpolation to find intermediate
% pt ti+bT followed by quadratic interpolation based on given two pts
   and intermediate pt.
% Inputs: ti - starting pt of interval to be interpolated
%      ph_old,ph_new - starting and ending value of discontinuity
%      function which changed sign over interval
%      x11,x22 - state vector at start of interval
%      k - index of discontinuity function which changed sign
% Outputs: a - decimal number between 0 and 1 indicating where
%      discontinuity function is estimated to be zero
   global T ISTATE
b=ph_old/(ph_old-ph_new); % zero crossing at ti+bT based on linear
      interpolation of (ti,ph_old) and (ti+T,ph_new)
t0 = ti+b*T;
t1 = ti;
t2 = ti+T;
y1 = ph_old;
   y2 = ph_new;
   [x11 x22] = RK4(b*T,x11,x22,ISTATE); % compute state at ti+bT
   [ph11 ph22] = DFUNCT(x11,x22); % compute ph1 and ph2 at ti+bT
   if k = = 1
     y0 = ph11;
   else
       y0 = ph22;
   end % if
   t = [t1 t0 t2];
   y = [y1 y0 y2];
   p = polyfit(t,y,2); % fit quadratic thru (ti,ph_old), (ti+T, ph_new)
       % and (ti+bT, y0)
   r = roots(p); % roots of quadratic
   if r(1) >= ti & r(1) <= ti+T % find root in interval (ti, ti+T)
     t_root = r(1);
else
     t_root = r(2);
end % if
a = (t_root-ti)/T; % normalizes ``a'' to between 0 and 1
```

"INTRP.m" is invoked when a change in sign of either discontinuity function is detected from one end of the integration interval to the other (see Figure 6.56). Several options are possible when it

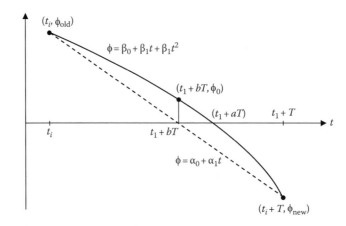

FIGURE 6.57 Quadratic interpolation to locate approximate pt of discontinuity.

comes to estimating the point in time where the discontinuity function is zero. One approach is illustrated in Figure 6.57.

The first step is to fit a linear function through the pts (t_i, ϕ_{old}) and $(t_i + T, \phi_{new})$, where t_i, ϕ_{old} and ϕ_{new} are provided as inputs to "INTRP.m." The root of the linear function occurs at $t_i + bT$, where

$$b = \frac{\phi_{old}}{\phi_{old} - \phi_{new}} \quad (0 < b < 1) \tag{6.319}$$

The time $t_i + bT$ can be treated as the pt where the discontinuity function is approximately zero. However, an improved estimate is possible if we determine ϕ_0, the value of the discontinuity function at $t_i + bT$, and generate the quadratic function through all three pts, namely, (t_i, ϕ_{old}), $(t_i + T, \phi_{new})$, and $(t_i + bT, \phi_0)$. The root of the quadratic interpolation polynomial that falls between t_i and $t_i + T$ is the desired time $t_i + aT$, $(0 < a < 1)$. "INTRP.m" returns the value of "a."

Once the pt $t_i + aT$ is identified, RK-4 integration is repeated for the interval $(t_i, t_i + T)$ by sequentially integrating from t_i to $t_i + aT$ and then from $t_i + aT$ to $t_i + T$. Note that since the state transitions between regions at points where either discontinuity function is zero, the state marker "ISTATE" is switched from 1 to 2 or vice versa in preparation of the RK-4 integration from $t_i + aT$ to $t_i + T$ (see Figure 6.56).

An alternative to the method described involves the use of an iterative root-solving technique (e.g., Bisection, False Position, and so forth) to locate the pt $t_i + aT$. The number of iterations is controlled by setting a tolerance on the magnitude of the discontinuity function at $t_i + aT$.

A numerical example for the pendulum shown in Figure 6.53 follows. Baseline system parameter values are

Radius of spherical pendulum bob: $r = 2.5$ in
Density of iron pendulum bob: $\gamma_{bob} = 491.32$ lb/ft^3
Length of negligible mass pendulum rod: $R = 3$ ft
Vertical distance from center of rotation to liquid surface: $L = 2.25$ ft
Density of liquid: $\gamma = 62.4$ lb/ft^3
Drag coefficient on pendulum bob in liquid: $\alpha = 0.15$ lb/ft/s

6.7.1 Physical Properties and Constant Forces Acting on the Pendulum BOB

weight: $W = \gamma_{\text{iron}} V = 491.32 \dfrac{\text{lb}}{\text{ft}^3} \times \dfrac{4}{3}\pi r^3 \text{ ft}^3 = 491.32 \dfrac{\text{lb}}{\text{ft}^3} \times \dfrac{4}{3}\pi \left(\dfrac{2.5}{12} \text{ ft}\right)^3 = 18.61 \text{ lb}$

mass: $m = \dfrac{W}{g} = \dfrac{18.61}{32.17} \text{ slug} = 0.5785 \text{ slug}$

moment of inertia about axis of rotation: $J = mR^2 = 0.5785 \text{ slug} \times (3 \text{ ft}^2) = 5.21 \text{ ft lb}_f \text{ s}^2$

buoyant force: $F_B = \gamma V = 62.4 \dfrac{\text{lb}}{\text{ft}^3} \times \dfrac{4}{3}\pi r^3 \text{ ft}^3 = 62.4 \dfrac{\text{lb}}{\text{ft}^3} \times \dfrac{4}{3}\pi \left(\dfrac{2.5}{12} \text{ ft}^3\right) = 2.36 \text{ lb}$

angle of pendulum at initial contact with liquid:

$$\theta_L = \cos^{-1}\left(\dfrac{L}{R}\right) = \cos^{-1}\left(\dfrac{2.25}{3}\right)$$
$$= 0.7227 \text{ rad } (41.41°)$$

In addition to the model system parameters, initial conditions must be specified. Choosing $\theta(0) = 75°$, $\dot\theta(0) = 0°/s$, the pendulum dynamics were simulated consistent with the logic outlined in the flow chart shown in Figure 6.56.

A Simulink diagram of the pendulum dynamics using fixed-step RK-4 integration, without searching for the precise time when a discontinuity occurs, is shown in Figure 6.58. A step size of $T = 0.1$ s was used in both cases.

Comparison of the simulation results for the pendulum angle $\theta(t)$ is shown in Figure 6.59. The MATLAB M-file "*Chap6_discont.m*" was executed with a time step of $T = 0.1$ s. A third plot intended to represent the exact solution for $\theta(t)$ is also shown. It was obtained by running the Simulink model with RK-4 and step size of $T = 0.001$ s. Using a time step of this magnitude negates

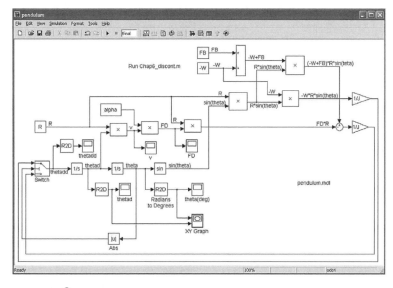

FIGURE 6.58 Simulink® diagram for pendulum dynamics.

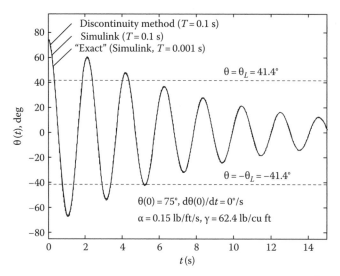

FIGURE 6.59 Simulated results using method for locating discontinuities, Simulink®, and approximation to "exact" solution.

almost entirely the adverse effect of a discontinuity occurring part way into the integration interval. The three responses are in close agreement resembling that of a lightly damped linear second-order system.

Useful information about the pendulum dynamics can be obtained from inspection of time histories and phase plots of additional system variables. Figure 6.60 is a phase portrait of the state trajectory evolving from the initial point $\theta(0) = 75°$, $\dot{\theta}(0) = 0°/s$ and lasting for a period of 15 s.

The points along the trajectory where $\theta(t) = \theta_L$ and $\theta(t) = -\theta_L$ indicates a transition from one region to the other, that is, the first marker corresponds to the pendulum entering the liquid for the first time on its way down.

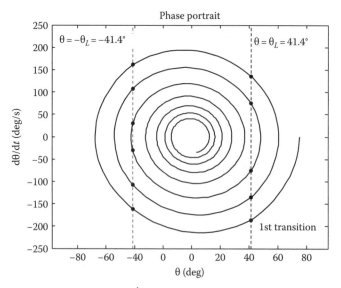

FIGURE 6.60 Plot of state trajectory $x_2(t) = \dot{\theta}(t)$ vs. $x_1(t) = \theta(t)$.

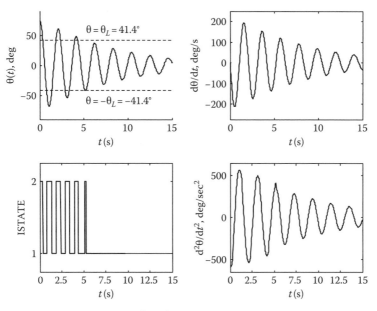

FIGURE 6.61 Time histories of θ, $d\theta/dt$, $d^2\theta/dt^2$ and state marker "ISTATE."

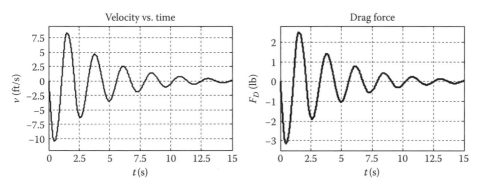

FIGURE 6.62 Velocity and drag force on pendulum bob.

Figure 6.61 includes time histories of $\theta(t)$, $\dot{\theta}(t)$, and $\ddot{\theta}(t)$. In addition, the marker "ISTATE" is shown fluctuating between 1 and 2 corresponding to transitions of the pendulum bob from air to water and vice versa.

The pendulum bob velocity and the drag force exerted by the liquid opposing its motion were captured in the Simulink model scopes and are shown in Figure 6.62.

The constant buoyant force of 2.36 lb opposes the motion of the pendulum bob on the way down and does the opposite while the bob is moving upward. The drag force never exceeds 2 lb in magnitude. The pendulum bob weighs 18.6 lb. From Figure 6.62, we notice that it continues to oscillate for a relatively long period of time due to minimal damping forces.

The discontinuous nature of the system is best illustrated by taking a closer look at the angular acceleration. Figure 6.63 shows the step changes that occur as the pendulum bob transitions between the two media. Note that the step changes in angular acceleration are greater at the moments when the pendulum bob is going from air to liquid compared with transitions from liquid to air. Can you explain why this happens? Exercise 6.34 addresses this point in greater detail.

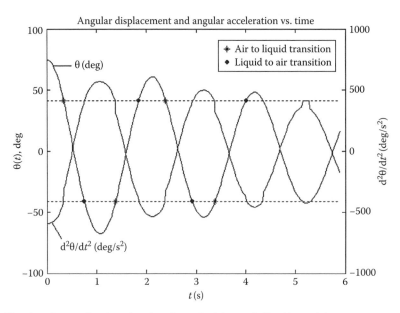

FIGURE 6.63 Angular acceleration showing discontinuities at air/liquid transitions.

Suppose we increase the damping effect of the liquid by replacing it with a heavier fluid. Instead of water, imagine a liquid with weight density of $\gamma = 150$ lb/ft³ responsible for producing a drag coefficient of $\alpha = 0.3$ lb/ft/s. Further, suppose the pendulum bob is released with an initial angular displacement $\theta(0) = 75°$ and initial velocity of $\dot{\theta}(0) = -90°/s$.

Figure 6.64 shows a portion of the transient responses obtained from the discontinuity method ($T = 0.1$ s), Simulink with RK-4 ($T = 0.1$ s) and Simulink with RK-4 ($T = 0.001$ s) as the approximation to the exact solution. Values obtained from the method based on locating the points of discontinuity within the integration interval are closer to the "exact" solution than the values obtained from conventional implementation of RK-4 integration.

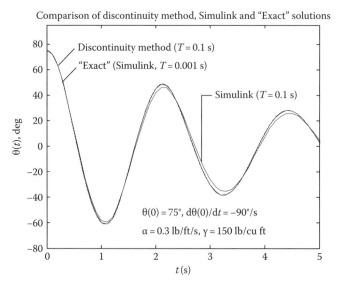

FIGURE 6.64 Comparison of solutions with new initial conditions and parameters.

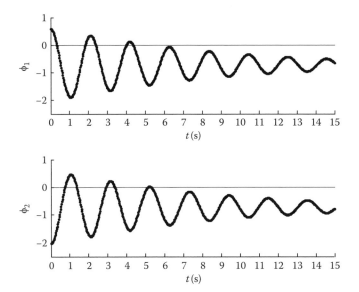

FIGURE 6.65   Plot of discontinuity functions.

It is instructive to look at graphs of the discontinuity functions $\phi_1(x_1, x_2)$ and $\phi_2(x_1, x_2)$. Figure 6.65 shows their time histories for the conditions listed in Figure 6.59.

The zero crossings of $\phi_1(x_1, x_2)$ and $\phi_2(x_1, x_2)$ correspond to the transitions of the system between regions S_1 and S_2. A close-up view of the discontinuity functions is shown in Figure 6.66.

Note how the quadratic interpolation function "INTRP" successfully locates the zero crossings, enabling the RK-4 integrator to stop at the correct point in time within the integration interval, reset the derivative functions, and then continue to integrate for the remainder of the interval as indicated in the flow chart in Figure 6.56.

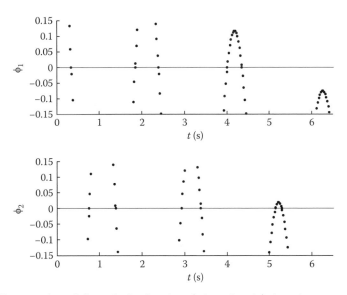

FIGURE 6.66   Close-up view of discontinuity functions $\phi_1(x_1, x_2)$ and $\phi_2(x_1, x_2)$.

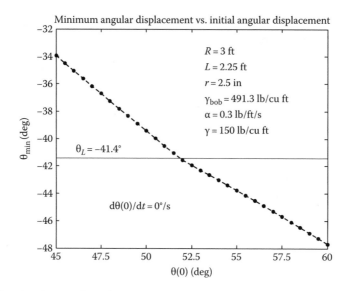

FIGURE 6.67   Results of search for $\theta(0)$ resulting in $\theta_{min} = -\theta_L$.

Example 6.12

Using the baseline conditions for the pendulum except for $\gamma = 150$ lb/ft^3 and $\alpha = 0.3$ lb/ft/s, determine the largest initial angle of the pendulum rod, so that when it is released with zero initial angular velocity, its fails to emerge from the liquid. Plot the angular rotation of the pendulum as a check.

The problem is to find the initial condition $\theta(0) \geq 0$, which satisfies

$$\underset{\theta(0) \geq 0}{\text{Min}}\ \theta(t) = -\theta_L \tag{6.320}$$

A simple search for the required initial condition was performed by varying $\theta(0)$ from 45° to 60° in increments of 0.5°. The results are shown in graphical form in Figure 6.67. The answer appears to be slightly less than 52°.

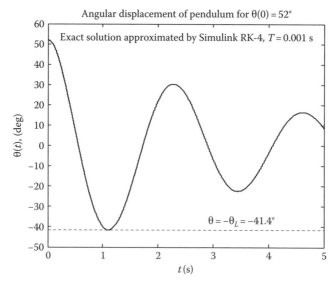

FIGURE 6.68   Simulated pendulum response with initial condition $\theta(0) = 52°$.

The pendulum response with $\theta(0) = 52°$ was generated for the conditions shown in Figure 6.68 using RK-4 integration with step size $T = 0.001$ s. The result is shown in Figure 6.68. As expected, the minimum angular response is approximately $-\theta_L = -41.4°$.

EXERCISES

6.32 The pendulum bob in Example 6.12 is released from the vertical position $\theta(0) = \pi$ rad with initial angular velocity $\dot{\theta}_0$. Find $\dot{\theta}_0$ if the bob makes a complete revolution and returns to the vertical position with zero angular velocity.

6.33 The pendulum bob shown in Figure E6.33 passes through two different nonmixing liquids. Physical parameter values are

Radius of spherical pendulum bob: $r = 3$ in
Density of pendulum bob: $\gamma_{bob} = 250$ lb/ft^3
Length of negligible mass pendulum rod: $R = 4$ ft
Vertical distance from center of rotation to liquid 1 surface: $L_1 = 2.5$ ft
Vertical distance from center of rotation to liquid 2 surface: $L_2 = 3.5$ ft
Density of liquid 1: $\gamma_1 = 62.4$ lb/ft^3
Density of liquid 2: $\gamma_1 = 175$ lb/ft^3
Drag coefficient on pendulum bob in liquid 1: $\alpha_1 = 0.10$ lb/ft/s
Drag coefficient on pendulum bob in liquid 2: $\alpha_2 = 0.65$ lb/ft/s

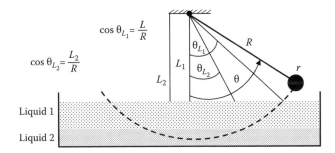

FIGURE E6.33

With state vector $(x_1, x_2) = (\theta, \dot{\theta})$, the state equations are

$$\dot{x}_1 = \begin{cases} f_{11}(x_1, x_2), & (x_1, x_2) \in S_1 \\ f_{12}(x_1, x_2), & (x_1, x_2) \in S_2 \\ f_{13}(x_1, x_2), & (x_1, x_2) \in S_3 \end{cases} \qquad \dot{x}_2 = \begin{cases} f_{21}(x_1, x_2), & (x_1, x_2) \in S_1 \\ f_{22}(x_1, x_2), & (x_1, x_2) \in S_2 \\ f_{23}(x_1, x_2), & (x_1, x_2) \in S_3 \end{cases}$$

where the regions S_1, S_2, and S_3 in state space are described by

S_1: $\{(x_1, x_2)|$pendulum bob in air$\}$
S_2: $\{(x_1, x_2)|$pendulum bob in liquid 1$\}$
S_3: $\{(x_1, x_2)|$pendulum bob in liquid 2$\}$

(a) Find expressions for S_1, S_2, and S_3 similar to those in Equation 6.310.
(b) Find the state derivative functions f_{ij}, $i = 1, 2, j = 1, 2, 3$.
(c) Find the discontinuity functions $\phi_1(x_1, x_2)$, $\phi_2(x_1, x_2)$, $\phi_3(x_1, x_2)$, and $\phi_4(x_1, x_2)$, where $\phi_i(x_1, x_2) = 0$, $i = 1, 2, 3, 4$ indicates the pendulum bob is passing from region S_1 to S_2, S_2 to S_3, S_3 to S_2, and S_2 to S_1, respectively.

(d) Use the method outlined in the flow chart in Figure 6.66 to simulate the angular position and angular velocity of the pendulum for initial conditions $\theta(0) = 90°$ and $\dot\theta(0) = 0°/s$. Choose any of the RK integrators with integration step size T selected on the basis of a trade-off between accuracy and computational effort. Plot time histories of $\theta(t)$ and $\dot\theta(t)$ as well as a phase portrait similar to the one in Figure 6.60 showing the points where the system transitions between regions.

(e) Simulate the same conditions in part (d) using Simulink with an excessively small integration step size T in order to obtain an approximation to the exact solution. Compare the results in parts (d) and (e).

6.34 According to the graphs in Figure 6.63, the angular acceleration appears to be continuous when the pendulum bob passes from liquid to air for the first time.

(a) Verify this by plotting $d^2\theta/dt^2$ vs. t shortly before to shortly after this occurs.

(b) For this to happen, the component of the buoyant force F_B in the direction of motion and the drag force F_D must effectively cancel each other out. On the same axes, plot both quantities and compare them at the moment the pendulum bob exits from the liquid for the first time.

6.35 Consider the pendulum in Figure 6.53 with physical properties

Radius of spherical pendulum bob: $r = 2$ in
Density of pendulum bob: $\gamma_{bob} = 400$ lb/ft^3
Length of negligible mass pendulum rod: $R = 5$ ft
Vertical distance from center of rotation to liquid surface: $L = 3$ ft

The drag coefficient α (lb/ft/s) is related to the density of the liquid γ (lb/ft^3) according to the relationship $\alpha = 0.05 + 0.02\gamma$, $50 \le \gamma \le 400$.

The pendulum is released from an almost vertical position $\theta(0) = 179.9°$ with zero angular velocity. Simulate the pendulum dynamics using any suitable method and prepare graphs of

(a) θ_{max} vs. $\gamma(50 \le \gamma \le 400)$ where θ_{max} is the total number of degrees the pendulum rotates through on its first swing.

(b) $t_{settling\ time}$ vs. $\gamma(50 \le \gamma \le 400)$ where $t_{settling\ time}$ is the time in seconds for the transient response to remain within 2% of its steady-state equilibrium value $\theta_{ss} = 0°$.

(c) $\dot\theta\ max$ vs. $\gamma(50 \le \gamma \le 400)$ where $|\dot\theta_{max}|$ is the absolute value of the maximum angular velocity in. $°/s$.

6.36 A rolling cart of mass m is connected to a stationary support located at $x = 0$ by a spring with stiffness k and damper with damping constant c as shown in Figure E6.36a:

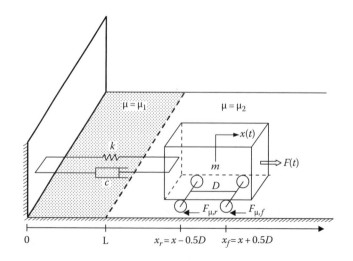

FIGURE E6.36a

The cart is subjected to an external force $F(t)$. The coefficient of rolling friction changes from surface 1 ($\mu = \mu_1$) to surface 2 ($\mu = \mu_2$) at $x = L$. The frictional force at each wheel is $F_\mu = \mu$ $(mg/4)$ where μ is either μ_1 or μ_2 depending on which surface the wheel is in contact with. A diagram of the cart and the forces acting on it is shown in Figure E6.36b. Note that the state definition $x_1 = x$ and $x_2 = \dot{x}$.

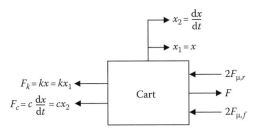

FIGURE E6.36b

Introduce regions S_1, S_2, and S_3 in state space $\{x_1, x_2\}$ according to the cart location, that is,

$S_1 = \{(x_1, x_2) \,|\, x_1 + 0.5D < L\}$—cart is located entirely on surface 1
$S_2 = \{(x_1, x_2) \,|\, x_1 - 0.5D < L \text{ and } x_1 + 0.5D > L\}$—cart is on surface 1 and surface 2
$S_3 = \{(x_1, x_2) \,|\, x_1 - 0.5D > L\}$—cart is located entirely on surface 2

(a) Find expressions for $f_{ij}(x_1, x_2)$, $i = 1, 2, j = 1, 2, 3$, the ith state derivative $\dot{x}_i$ when the state (x_1, x_2) is located in region S_j.

(b) Find the discontinuity functions $\phi_1(x_1, x_2)$ and $\phi_2(x_1, x_2)$ where
$\phi_1(x_1, x_2) = 0 \Rightarrow (x_1, x_2)$ is transitioning between S_1 and S_2
$\phi_2(x_1, x_2) = 0 \Rightarrow (x_1, x_2)$ is transitioning between S_2 and S_3

(c) Implement the method outlined in the flow chart of Figure 6.56 using RK-4 integration with integration step size T, based on a trade-off between accuracy and computational effort, to simulate the cart dynamics. Baseline conditions are
$\mu_1 = 0.4$, $\mu_2 = 0.05$
$m = 30$ slugs, $c = 5$ lb/ft/s, $k = 25$ lb$_f$/ft
$L = 25$ ft, $D = 5$ ft
$x(0) = L$, $\dot{x}(0) = 0$ ft/s
The applied force $F(t)$ is a step input of magnitude $F_0 = 250$ lb. Plot the cart position $x(t)$ vs. time and cart velocity $\dot{x}(t)$ vs. time.

(d) Simulate the cart dynamics with Simulink, and compare the results with those obtained in part (c).

6.37 A block diagram for a simple on–off tank level control system is shown in Figure E6.37. The flow in to the tank F_1 is either zero or $\overline{F}$ depending on the state of the on–off controller, that is,

$$F_1 = \overline{F}u, \quad u = \begin{cases} 0, & e \leq 0 \\ 1, & e > 0 \end{cases} \quad \text{where } e = H_{com} - H$$

The tank dynamics is modeled by

$$A\frac{dH}{dt} + F_0 = F_1, \quad F_0 = cH^{1/2}$$

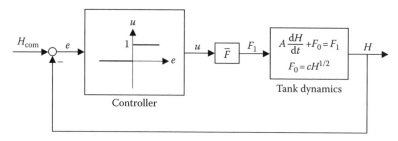

FIGURE E6.37

(a) The state derivative is

$$\frac{dH}{dt} = \begin{cases} f_1(H), & H \in S_1 \\ f_2(H), & H \in S_2 \end{cases}$$

Regions S_1 and S_2 are defined such that when H is in region S_1 of state space, the controller is off, and the opposite is true when H is in region S_2. Find expressions for S_1 and S_2 in terms of the state H.

(b) Find expressions for the state derivate functions $f_1(H)$ and $f_2(H)$.

(c) Find the discontinuity function $\phi(H)$ that specifies which region the state is in based on its sign, that is, $\phi(H) = 0$ implies the state H is transitioning between the two regions.

(d) Use the method that finds the time of the discontinuity to simulate the tank level. Choose any RK integrator with suitable integration step size based on accuracy and computation requirements.

The following conditions apply:

$$A = 20\,\text{ft}^2, \quad c = 0.4\,\text{ft}^3/\text{min}/\text{ft}^{1/2}, \quad \overline{F} = 10\,\text{ft}^3/\text{min}, \quad H(0) = 0\,\text{ft}$$

$$H_{\text{com}} = 15, \quad t \geq 0$$

Run the simulation for a period of time sufficient for the controller to cycle on and off several times and plot time histories of $H(t)$ and $\dot{H}(t)$.

(e) Plot a phase portrait $\dot{H}$ vs. H showing the points where the controller cycles between its two states.

(f) Simulate the system for the same conditions in part (d) with Simulink using RK-4 integration with an excessively small step size in order to approximate the exact solution. Compare the results with those in part (d).

6.8 CASE STUDY: SPREAD OF AN EPIDEMIC

Epidemic models for various fatal and nonfatal diseases in humans and animals have been postulated since the early 1900s (Kermack and McKendrick 1927; Hethcote 1976; Keen and Spain 1992; Brown and Rothery 1993). Modern-day epidemics such as the spread of AIDS have been studied with the help of simulation models (Isham 1988; Perelson 1993; Culshaw and Ruan 2000; Coutinho et al. 2001).

The formulation of a mathematical model in the field of epidemiology requires some basic information about disease and how it spreads among a population. To start with, symptoms of the disease may not appear at the time a host is infected, rather an incubation period may be necessary prior to appearance of the symptoms. A host infected with a pathogen may become infectious only after a period of latency. The infectious period is the duration of time during which the host is capable of transmitting the disease to others in the population. The incubation, latent, and infectious periods depend on the pathology of the disease.

For certain diseases, the host may experience an immune period where the infection has run its course, the host has recovered, and cannot be re-infected. However, the individual may still be a carrier and capable of transmitting the disease to susceptible individuals. As a means of preventing or limiting the scope of an epidemic, some infected individuals may be isolated from the population to prevent transmission of the disease to susceptible individuals. If the disease is potentially fatal, a number of infected individuals will die. If a vaccine exists, individuals receiving the vaccine pass from the class of susceptibles to the class of recovered individuals.

Early epidemic models concentrated on the movement of individuals through three stages, namely, (S)usceptible, (I)nfected carrier, and (R)ecovered. The so-called S-I-R models relate the state derivatives dS/dt, dI/dt, and dR/dt to the states S, I, and R using expressions formulated by epidemiologists to describe the interactions between individuals in each group. Inherent in the models are a number of parameters (rate constants) associated with infection, transmission, recovery, mortality, and so forth. Later on, more sophisticated models were developed to account for additional stages. Finally, partial differential equations evolved as modelers attempted to predict both temporal and spatial variations of the populations in each stage during the course of an epidemic.

The following information is postulated to provide a framework for studying the dynamics of an epidemic stemming from the spread of a fatal disease.

- The initial population consists entirely of susceptible individuals, that is, those at risk of contracting the disease.
- The disease is introduced by individuals immigrating from outside the area, a fraction of which are sick.
- A subset of the susceptible individuals contract the disease through contact with sick individuals.
- An outbreak of the disease is recognized after a specified period of time immediately followed by a cessation of immigration.
- After recognizing the existence of a possible epidemic, a segment of the susceptible individuals is inoculated with a vaccine making them immune to the disease.
- Starting at the same time inoculations begin, a portion of those who are sick or become sick later are separated from the general population by quarantine.
- Sick individuals either recover and become immune or die.

Members of the population exist in one of five states.

   $x_1(t)$: Number of susceptible people at time t
   $x_2(t)$: Number of sick people in population at time t
   $x_3(t)$: Number of immune people at time t
   $x_4(t)$: Number of deceased people at time t
   $x_5(t)$: Number of sick people quarantined from population at time t

Possible transitions between states are illustrated in Figure 6.69. Note that the $m = m(t)$ is the rate of immigration and $n = n(t)$ represents the rate of inoculation of susceptible individuals.

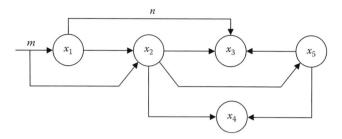

FIGURE 6.69 State transition diagram.

The state vector $\underline{x}$ is $[x_1 \ x_2 \ x_3 \ x_4 \ x_5]^T$. A mathematical model of the system requires knowledge of a vector function $\underline{f}(t, \underline{x}, m)$, describing the state derivatives. In this example, the system of coupled differential equations $\underline{\dot{x}} = \underline{f}(t, \underline{x}, m)$ is given by

$$\frac{dx_1}{dt} = f_1(t, \underline{x}, m) = -cx_1x_2 + \alpha m - n \tag{6.321}$$

$$\frac{dx_2}{dt} = f_2(t, \underline{x}, m) = cx_1x_2 - a_{23}x_2 - a_{24}x_2 - a_{25}x_2 + (1 - \alpha)m \tag{6.322}$$

$$\frac{dx_3}{dt} = f_3(t, \underline{x}, m) = a_{23}x_2 + a_{53}x_5 + n \tag{6.323}$$

$$\frac{dx_4}{dt} = f_4(t, \underline{x}, m) = a_{24}x_2 + a_{54}x_5 \tag{6.324}$$

$$\frac{dx_5}{dt} = f_5(t, \underline{x}, m) = \begin{cases} 0, & 0 \le t \le t_0 \\ a_{25}x_2 - a_{53}x_5 - a_{54}x_5, & t > t_0 \end{cases} \tag{6.325}$$

The constants $a_{23}, a_{24}, a_{25}, a_{53}, a_{54}, c$, and α are system parameters, which describe the transitions by individuals from one state to another. For example, the disease spreads by contact between susceptible and sick members of the population, and c is a transmission constant. The constant α is the fraction of immigrants who are susceptible. All terms on the right-hand side of Equations 6.321 through 6.325 are in units of individuals per unit of time, the same as the left-hand-side state derivatives.

The time t_0 in Equation 6.325 is the length of time it takes to recognize the outbreak of a possible epidemic. Quarantining of sick people, cessation of immigration, and inoculation of susceptible individuals begin at $t = t_0$. Immigration and inoculation profiles are shown in Figure 6.70.

The simple model ignores birth and deaths from other causes and does not account for emigration of individuals. The following values have been arbitrarily selected for conducting a baseline study.

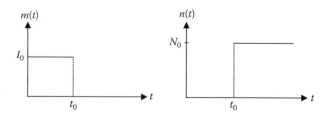

FIGURE 6.70 Immigration and inoculation profiles.

$a_{23} = 0.1$ per week, $a_{24} = 0.003$ per week, $a_{25} = 0.05$ per week

$a_{53} = 0.1$ per week, $a_{54} = 0.003$ per week

$\alpha = 0.9$, $c = 2.25 \times 10^{-8}$ per people/week

$t_0 = 8$ weeks, $I_0 = 2500$ people/week, $N_0 = 0$ inoculations/week

Note that the baseline conditions assume zero inoculations following the recognition of a possible epidemic. A number of interesting simulation studies are possible. First, we will investigate various inoculation policies and their mitigating effect on spreading of the disease in a population initially consisting of 10 million susceptible individuals.

The classic RK-4 numerical integrator introduced in Equations 8.60 through 8.64 was chosen for simulating the system response. After several trial runs with different integration step sizes, $T = 0.1$ weeks were selected. The results of a baseline and additional simulations using inoculation rates of 5,000, 10,000, and 15,000 people per week are shown in Figures 6.71 through 6.74. Refer to MATLAB M-file "Chap6_CaseStudy.m." For simplicity, the subscript "A" has been dropped from the notation for the discrete-time signals.

A summary of the results is listed in Table 6.12.

As expected, the highest inoculation level results in the fewest deaths. The third row shows the maximum number of sick people at any time in the 200 week study period. The peak is reduced from 585,834 sick at one time to 268,548 as a result of administering 15,000 vaccinations/week compared with none at all.

The discrete-time state variable $x_2(i)$ represents the number of infected individuals at the discrete times $t_i = iT$, $i = 0, 1, 2, \ldots$ The cumulative number of people who have been sick up through time t_i is denoted $s(i)$ (see Figures 6.71 through 6.74 and Table 6.12). It is computed by numerical integration of $\mathrm{d}s/\mathrm{d}t$, where

$$\frac{\mathrm{d}s}{\mathrm{d}t} = (1 - \alpha)m + cx_1 x_2 \tag{6.326}$$

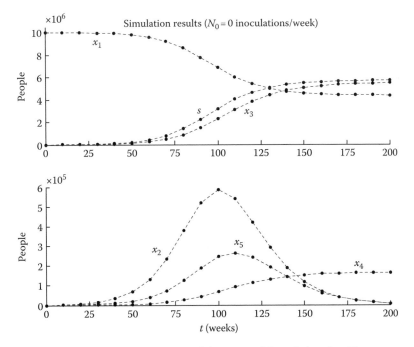

FIGURE 6.71 Epidemic response for baseline conditions ($N_0 = 0$ inoculations/week).

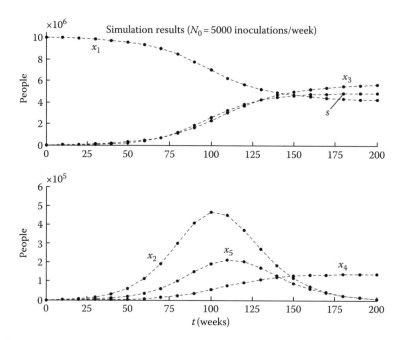

FIGURE 6.72 Epidemic response ($N_0 = 5000$ inoculations/week).

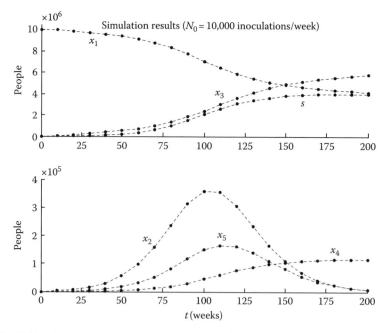

FIGURE 6.73 Epidemic response ($N_0 = 10,000$ inoculations/week).

Note the difference between ds/dt and dx_2/dt. The former is the rate of change of newly infected individuals, that is, those people entering state x_2. As a result, $s(t)$ is monotonically increasing. The state derivative dx_2/dt is the overall rate of change of infected people in the nonquarantined population. It is negative when more individuals are leaving state x_2 than entering, which results in $x_2(t)$ decreasing.

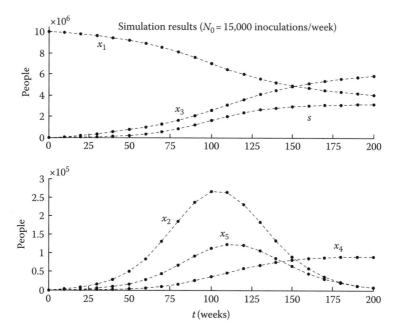

FIGURE 6.74 Epidemic response ($N_0 = 15,000$ inoculations/week).

TABLE 6.12
Summary of Epidemic Simulation Results after 200 Weeks

	$N_0 = 0$	$N_0 = 5,000$	$N_0 = 10,000$	$N_0 = 15,000$
$x_1(200)$	4,368,093	4,233,072	4,125,930	4,016,139
$x_2(200)$	8,089	8,038	7,614	6,395
Max x_2	585,834	469,690	362,637	268,548
$x_3(200)$	5,471,549	5,630,535	5,763,273	5,900,147
$x_4(200)$	164,146	140,116	115,298	90,604
$x_5(200)$	8,123	8,238	7,885	6,715
$s(200)$	5,700,412	4,867,698	4,007,144	3,149,525

The same RK-4 integration method and step size were used to numerically integrate the discrete-time signal $(1 - \alpha)m(i) + cx_1(i)x_2(i)$ to generate $s(i)$.

A valuable check on the accuracy of the simulation is possible. Conservation of individuals can be verified at every discrete point in time. In this case, the total number of individuals begins at 10 million and increases at a rate of 2500 per week for 8 weeks. Hence, after approximately 2 months, the total population consists of 10,020,000 people distributed among the five states x_1, x_2, x_3, x_4, and x_5. Summing $x_1(200)$, $x_2(200)$, $x_3(200)$, $x_4(200)$, and $x_5(200)$ in each column of Table 6.12 will show that all individuals are accounted for. This is crucial in the context of real-world simulations where analytical solutions of the continuous-time model are not available.

Sensitivity analyses with respect to each system parameter at baseline conditions offer insight into the dynamics of the epidemic. To illustrate this, suppose we are interested in relating the number of individuals who contract the disease with the system parameter that measures how contagious the disease is, that is, the transmission coefficient c that appears in Equations 6.321 and 6.322. The parameter was allowed to vary by 25% in both directions from the nominal or baseline value $c = 2.25 \times 10^{-8}$ per people/week, and the simulation is repeated with the remaining parameters

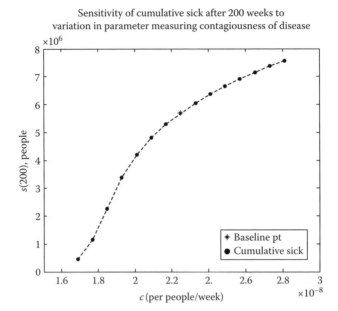

FIGURE 6.75   Sensitivity analysis: $s(200)$ vs. c.

fixed at their baseline values. Figure 6.75 shows that $s(200)$, the predicted number of sick people in the first 200 weeks, increases as the transmission coefficient parameter c increases, as one would expect.

A similar study was conducted to investigate the relationship between the cumulative number of deaths $x_4(200)$ and the transmission coefficient c. The graph in Figure 6.76 shows what can be expected in terms of the number of people dying over the 200 week period as the level of contagiousness varies about the baseline value.

Try running the simulation with the same baseline conditions to ascertain the numerical value of c that results in the epidemic spreading to every member of the population. A number of other studies are suggested in the exercise problems.

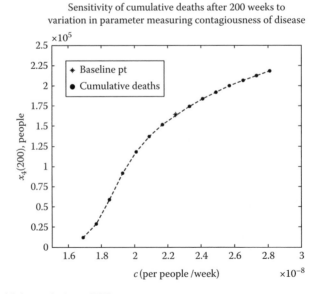

FIGURE 6.76   Sensitivity analysis: $x_4(200)$ vs. c.

EXERCISES

6.38 In the study that looked at the effect of various inoculation rates, prepare graphs of
(a) Cumulative sick vs. inoculation rate
(b) Number of deaths vs. inoculation rate
(c) Peak number of sick vs. inoculation rate

Use the classic RK-4 integrator with baseline values for all parameters (except inoculation rate), and run the simulations for a sufficient period of time to include the transient response. Consider inoculation rates from 0 to 50,000 per week.

Repeat parts (a), (b), and (c) using Simulink with the same numerical integrator.

6.39 In the inoculation study, find the cumulative number of individuals quarantined.

Hint: Let $q(t)$ represent the cumulative number of people quarantined through time t, and write the differential equation for $q(t)$ similar to the procedure for finding $s(t)$.

6.40 Investigate the duration of the epidemic transient period as a function of the parameter c. Does the epidemic last longer when the disease is more contagious?

7 Simulation Tools

7.1 INTRODUCTION

Mathematical models of dynamic systems are derived with their intended use in mind. For example, systems with fast internal dynamics driven by inputs that change infrequently (relative to the system time constants) reside in steady state the majority of the time. Accordingly, the model consists of a system of coupled, possibly nonlinear, algebraic equations. In this context, a solution (or solutions) defines an equilibrium state (or states) corresponding to fixed values of the system inputs. When one or more inputs change, a stable system transitions from one equilibrium state to another and the dynamics, that is, transient response, is ignored. Solving the steady-state algebraic equations for an equilibrium solution is rarely a straightforward task, particularly when dealing with nonlinear systems. The MATLAB® steady-state solver is introduced in Section 7.2. It is designed to locate equilibrium states of a Simulink® model.

Tuning a simulation model of a real, continuous-time system is an iterative process like the one shown in Figure 7.1. Assumptions about the structure of the mathematical model, that is, the state derivative vector $f(\underline{x}, \underline{u}, \underline{p})$ and parameter values $\underline{p}$, are tested and refined using observed data acquired from the system. An essential component of the validation process is minimization of an error function $\varepsilon(\underline{f}, \underline{p})$, a measure of the differences between the actual system and simulation model outputs.

A Simulink add-on called `Parameter Estimation` compares empirical data with data generated by a Simulink model. Using optimization techniques, Simulink `Parameter Estimation` estimates the parameter and (optionally) initial conditions of states such that a user-selected cost function is minimized. The cost function typically calculates a least-square error between the empirical and model data.

Once the mathematical model has been determined, some type of exploratory study can be performed to determine the "best" (in some sense) values for the controllable system parameters. Optimization theory is a broad area of study with roots in *Operations Research and Applied Mathematics*. It is the foundation for implementation of what are collectively called optimum seeking methods. The MATLAB optimization toolbox provides the simulationist access to a number of algorithms for locating points in the model's parameter space where the system performs at optimum or near optimum levels. Examples are presented in Section 7.3 of using optimization for both parameter identification and system optimization involving Simulink models.

Another Simulink add-on, `Response Optimization`, is a tool that helps you tune design parameters in Simulink models by optimizing time-based signals to meet user-defined constraints. It supports continuous-time, discrete-time, and multirate models accounting for model uncertainty by conducting Monte Carlo simulations. Simulink `Response Optimization` can be used to tune multiinput/multioutput and adaptive controllers in nonlinear systems and optimize physical parameters to minimize power consumption, reduce range of motion, and tune filter coefficients.

An equilibrium state is sometimes required to serve as the initial state for a simulation investigation of the system's dynamic response. After locating an equilibrium point of a nonlinear system model, the system's response to dynamically changing inputs can be approximated by linearizing the equations about the equilibrium point. Accuracy of the linearized model depends in part on the magnitude of the state vector's excursions from the equilibrium state. In general, if the changing input vector remains in close proximity to its equilibrium level, the state vector will do the same. Regulatory control systems are a good example of an application where linearization of process

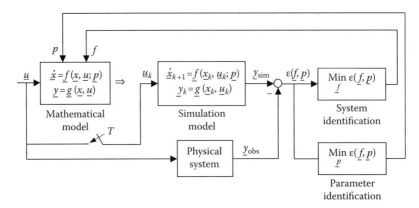

FIGURE 7.1   Iterative procedure for simulation model validation.

models has proven beneficial in both the design and analysis of the system. Section 7.4 illustrates the capabilities of MATLAB to linearize nonlinear models created with Simulink.

7.2   STEADY-STATE SOLVER

Unlike linear system models, it is possible for nonlinear systems to possess any number of equilibrium points, that is, points in state space where the state derivatives are all zero. Furthermore, there is no uniform approach guaranteed to determine the number or location of the equilibrium points.

Knowledge of a nonlinear system's equilibrium points is important for several reasons. The first relates to stability. Once an equilibrium point is located, stability can be determined by linearizing the model's state equations in the neighborhood of the equilibrium point. Second, the behavior of forced nonlinear systems is often approximated by "small signal" linearized models. The characteristic dynamics (time constants, poles, critical frequencies, eigenvalues, and so forth.) of the linearized system depend on the location of the equilibrium point. Linearization is discussed in a later section.

Consider a nonlinear state model

$$\dot{\underline{x}} = \underline{f}(\underline{x}, \underline{u}) \tag{7.1}$$

where
$\underline{x}$ and $\underline{u}$ are the state and input vectors, respectively
$\underline{f}(\underline{x}, \underline{u})$ is a vector of functions defining the state derivatives

Equilibrium points $\underline{x}_e$ corresponding to a constant input vector $\underline{u}_e$ are solutions to the nonlinear system of algebraic equations

$$\underline{f}(\underline{x}_e, \underline{u}_e) = \underline{0} \tag{7.2}$$

Some type of numerical method for finding the solutions $(\underline{x}_e)_1, (\underline{x}_e)_2, \ldots,$ given the input $\underline{u}_e$, is needed. Nonlinear autonomous systems described by

$$\dot{\underline{x}} = \underline{f}(\underline{x}) \tag{7.3}$$

may also possess a finite (or infinite) number of equilibrium points that satisfy

$$\underline{f}(\underline{x}_e) = \underline{0} \tag{7.4}$$

To illustrate the point, we focus on a nonlinear system model from the field of ecology. A predator–prey model for the population of fish (prey) and sharks (predator) in the ocean is (Haberman 1997)

$$\frac{dF}{dt} = F(a - bF - cS) \tag{7.5}$$

$$\frac{dS}{dt} = S\left(e - \lambda \frac{S}{F}\right) \tag{7.6}$$

where $F = F(t)$ and $S = S(t)$ are the instantaneous populations (or population densities) of fish and sharks in a fixed geographical area. The system is autonomous since there are no fish or sharks entering or leaving the region according to an external function of time t. Conversely, harvesting of either population according to some predetermined schedule, independent of the levels F and S, would require additional terms with explicit dependence on t resulting in a nonautonomous system model.

The model equations are based on the following observations:

1. The growth rate of fish $(1/F)dF/dt$ is reduced from a constant "a" by an amount proportional to the number of fish (which compete for the limited food supply) as well as an amount proportional to the number of sharks (for which the fish are the primary food source). Proportionality constants b and c reflect the level of competition among the fish for their food and the aggressiveness of the sharks.
2. Shark growth rate $(1/S)dS/dt$ is reduced from a constant e by an amount proportional to the ratio of sharks to fish. A higher S/F depletes the fish supply more rapidly.

Equilibrium points (F_e, S_e) satisfy the steady-state algebraic equations resulting from setting the state derivatives to zero. Thus,

$$0 = F_e(a - bF_e - cS_e) \tag{7.7}$$

$$0 = S_e\left(e - \lambda \frac{S_e}{F_e}\right) \tag{7.8}$$

There is more than one equilibrium point (see Exercise 7.2). Our interest is in the nontrivial equilibrium point where neither fish nor sharks vanish. From Equation 7.8 with the term in parenthesis equal to zero,

$$S_e = \frac{e}{\lambda} F_e \tag{7.9}$$

Substituting S_e from Equation 7.9 into Equation 7.7 gives (after simplification)

$$F_e = \frac{a\lambda}{b\lambda + ce} \tag{7.10}$$

Finally, S_e is obtained from Equations 7.9 and 7.10 as

$$S_e = \frac{ae}{b\lambda + ce} \tag{7.11}$$

We shall return to the predator–prey model to investigate the dynamic interaction between fish and sharks, particularly in the vicinity of the equilibrium point (F_e, S_e).

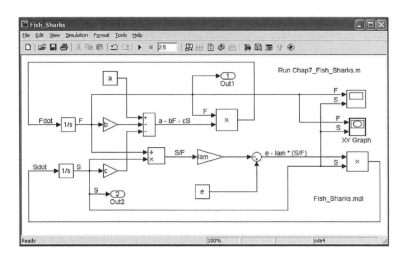

FIGURE 7.2 Simulink® diagram of predator–prey model.

7.2.1 TRIM FUNCTION

Figure 7.2 shows a Simulink diagram for simulating the predator–prey ecosystem modeled by Equations 7.5 and 7.6. The model name is "*Fish_Sharks.mdl*." Parameters "a," "b," "c," "e," and "lam" are assigned values in the MATLAB M-file "*Chap7_Fish_Sharks.m*." There are no inputs and the two states are designated as outputs.

A function "trim" is called from MATLAB to search for equilibrium points associated with a named Simulink model file. The "trim" function call is

```
[x, u, y, dx] = trim('Fish_sharks', x0)
```

The second parameter "x0" is the starting point in state space in the search for an equilibrium point. The output contains the equilibrium state "x," input and output vectors "u" and "y" at equilibrium, respectively, and the value of the state derivative vector at the equilibrium point. Empty vectors are returned when there are no inputs or outputs defined. Should the numerical search algorithm fail to converge to an equilibrium point, a different starting point will sometimes fix the problem.

Optional parameters are available for constraining selected components of the state, input, and output vectors. For example, instead of a true equilibrium point, we may look for points in the state space where a subset of the state derivative vector is zero.

Numerical values of the parameters in Equations 7.5 and 7.6 were arbitrarily chosen as $a = 50$, $b = 1$, $c = 5$, $e = 2$, and $\lambda = 10$. Running M-file "*Chap7_Fish_Sharks.m*" with starting point $x0 = (18.75; 3.75)$ results in

$$
\begin{array}{lll}
\mathtt{x = 25.0000} & \mathtt{u = Empty\ matrix:\ 0-by-1} & \mathtt{y = 25.0000} \\
\quad\ \ \mathtt{5.0000} & & \quad\ \ \mathtt{5.0000} \\
\mathtt{dx = 1.0e-011\ *} & \mathtt{-0.1954} & \\
& \quad\ \ \mathtt{0.0026} &
\end{array}
$$

According to Equations 7.10 and 7.11,

$$
F_e = \frac{a\lambda}{b\lambda + ce} = \frac{50(10)}{1(10) + 5(2)} = 25, \quad S_e = \frac{50(2)}{1(10) + 5(2)} = 5
$$

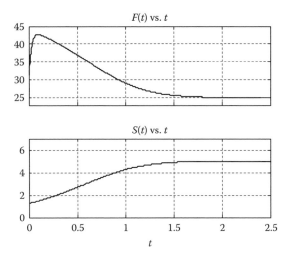

FIGURE 7.3 Transient response of ecosystem.

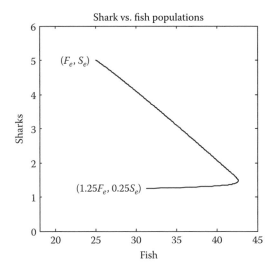

FIGURE 7.4 State trajectory of ecosystem.

in agreement with the results of the "trim" function call. Note that the ordering of the state vector "x" must be known. This will be addressed later in Section 7.4. The small "dx" values assure the accuracy of the located equilibrium point.

The transient response and state trajectory produced by the Simulink "scope" and "XY Graph" starting from the point $(1.25F_e, 0.25S_e) = (31.25, 1.25)$ are shown in Figures 7.3 and 7.4.

The behavior of the system starting from 100 randomly selected points in a region including the equilibrium point is shown in Figure 7.5. It appears that the equilibrium point is indeed stable since all trajectories terminate there.

7.2.2 EQUILIBRIUM POINT FOR A NONAUTONOMOUS SYSTEM

The "trim" function can be used to locate the steady state of a forced system subjected to a constant input(s). Figure 7.6 is a simplified diagram of a transducer that converts a low-level

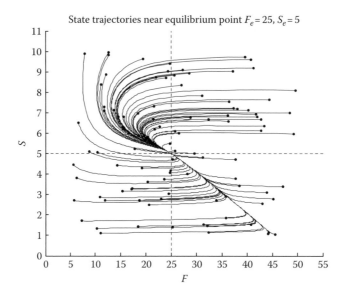

FIGURE 7.5 State trajectories demonstrating stability of equilibrium point.

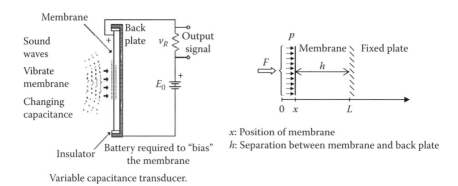

Variable capacitance transducer.

x: Position of membrane
h: Separation between membrane and back plate

FIGURE 7.6 Variable capacitance transducer.

acoustic pressure signal $p(t)$ to an output voltage $v_R(t)$. The movable membrane and fixed plate form a capacitor. The sound waves deflect the membrane changing the separation between it and the back plate and, therefore, the capacitance. A bias voltage is applied to produce an electrical charge on the membrane. The motion of the membrane is opposed by damping and elastic forces as well as an electrostatic force.

The mathematical model consists of the following differential and algebraic equations describing the circuit and the forces acting on the membrane:

$$R\frac{dQ}{dt} + v_C = E_0, \quad v_R = E_0 - v_C \tag{7.12}$$

$$Q = Cv_C \tag{7.13}$$

$$C = \frac{B}{h} = \frac{B}{L - x} \tag{7.14}$$

$$m\frac{d^2x}{dt^2} + \mu\frac{dx}{dt} + kx = -F_e + F \tag{7.15}$$

$$F_e = \frac{Q^2}{2B}, \quad F = pA \tag{7.16}$$

where
$Q = Q(t)$ is the electric charge on the capacitor (Cs)
$v_C = v_C(t)$ is the voltage across the capacitor (Vs)
$v_R = v_R(t)$ is the output voltage across the resistor (Vs)
$C = C(t)$ is the variable capacitance of the capacitor (Fs)
$h = h(t)$ is the separation between the movable membrane and back plate (mm)
$x = x(t)$ is the membrane displacement from equilibrium, that is, when $p = E_0 = 0$ (mm)
$F_e = F_e(t)$ is the electrostatic force on the membrane (N)
$F = F(t)$ is the force acting on the membrane due to pressure p (N)
$p = p(t)$ is the input acoustic pressure acting uniformly on the membrane (psi)
E_0 is the bias voltage on the capacitor (Vs)
m is the mass of membrane (gs)
μ is the damping coefficient (N/(mm/s))
k is the elastic constant (N/mm)

Choosing the states and outputs

$$x_1 = x, \quad x_2 = Q, \quad x_3 = \dot{x}$$

$$y_1 = x, \quad y_2 = h, \quad y_3 = C, \quad y_4 = F_e, \quad y_5 = F, \quad y_6 = v_C, \quad y_7 = Q, \quad y_8 = v_R$$

leads to the state equations (see Exercise 7.4)

$$\dot{x}_1 = x_3, \quad \dot{x}_2 = \frac{1}{BR}[-x_2(L - x_1) + BE_0], \quad \dot{x}_3 = \frac{1}{m}\left[-kx_1 - \frac{x_2^3}{2B} - \mu x_3 + Ap\right] \tag{7.17}$$

$$y_1 = x_1, \quad y_2 = L - x_1, \quad y_3 = \frac{B}{L - x_1}, \quad y_4 = \frac{x_2^2}{2B} \tag{7.18}$$

$$y_5 = Ap, \quad y_6 = \frac{x_2(L - x_1)}{B}, \quad y_7 = x_2, \quad y_8 = E_0 - \frac{x_2(L - x_1)}{B} \tag{7.19}$$

For constant inputs E_0 and $p(t) = p_0$, the equilibrium states are found by setting the state derivatives in Equation 7.17 to zero resulting in

$$x_{3,e} = 0, \quad -x_{2,e}(L - x_{1,e}) + BE_0 = 0, \quad -kx_{1,e} - \frac{x_{2,e}^2}{2B} + Ap_0 = 0 \tag{7.20}$$

Eliminating $x_{2,e}$ from the two equations yields a third-order polynomial in $x_{1,e}$.

$$kx_{1,e}^3 - (2kL + Ap_0)x_{1,e}^2 + L(kL + 2Ap_0)x_{1,e} + 0.5BE_0^2 - AL^2p_0 = 0 \tag{7.21}$$

Equation 7.21 is solved in the M-file "*chap7_cap.m*" using the following baseline parameter values:

$$A = \pi (20\,\text{mm})^2, \quad L = 10\,\text{mm}, \quad m = 5\,\text{g}, \quad \mu = 0.01\,\text{N}/(\text{mm/s}),$$

$$k = 0.5\,\text{N/mm}, \quad R = 100\,\Omega, \quad B = 5 \times 10^{-5}\,\text{F-mm},$$

$$E_0 = 48\,\text{V}, \quad p_0 = 0.01\,\text{psi}$$

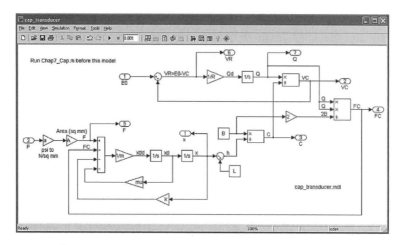

FIGURE 7.7 Simulink® diagram of capacitive transducer for use by "Trim" function.

The single real root of Equation 7.21 is $x_{1,e} = x_e = 0.17208282239091$ mm. From the second of the equations in Equation 7.20, $x_{2,e} = Q_e = 2.442023021386376 \times 10^{-4}$ C.

A Simulink diagram of the system is shown in Figure 7.7. For reference by the "trim" function, the inputs are "E0" and "p0," the states are "x," "Q," and "xd," and the eight outputs are designated as shown. The "trim" function call in the M-file "Chap7_Cap.m" is

```
[x, u, y, dx] = trim('cap_transducer', x0, u0, y0 ix, iu iy)
```

where the outputs "x," "u," and "y" are the computed equilibrium values of the state, input, and output vectors, respectively. The last argument "dx" is the state derivative vector that is identically zero at true equilibrium conditions. The input parameters "x0, u0, and y0" are used to set initial guesses for the equilibrium state, input, and output while the remaining arguments "ix, iu, and iy" serve to constrain selected components of the state, input, and output vectors at equilibrium.

Running script file "Chap7_Cap.m" produces the results shown in Table 7.1.

Note that the second input $u_{2,e} = 0.00006894413789$ is the equivalent of $p_0 = 0.001$ psi converted to N/m². The equilibrium state from the "trim" function call is in agreement with the solution to the equilibrium equations in Equation 7.20 obtained by running the M-file "Chap7_Cap.m."

A common use of the "trim" function is in applications where a subset of the equilibrium state and/or output vector is specified and the goal is to determine the input conditions resulting in the partially or fully specified equilibrium state. Figure 7.8 portrays a block diagram of a system with four inputs, six states, and three outputs.

Instead of specifying constants for inputs $u_1(t)$, $u_2(t)$, $u_3(t)$, and $u_4(t)$, to establish an equilibrium state $\underline{x}_e = [x_{1,e} \quad x_{2,e} \quad x_{3,e} \quad x_{4,e} \quad x_{5,e} \quad x_{6,e}]^T$ and output $\underline{y}_e = [y_{1,e} \quad y_{2,e} \quad y_{3,e}]^T$, only input u_2 is fixed. Equilibrium levels of x_2, x_6, and y_1 are also fixed, and the steady-state equations of the system

---

TABLE 7.1
"Trim" Function Results

$x_{1,e} = 0.17208282239053$,	$x_{2,e} = 0.00024420230214$,	$x_{3,e} = -0.00000000000000$	
$u_{1,e} = 48.00000000000000$,	$u_{2,e} = 0.00006894413789$		
$y_{1,e} = 0.17208282239053$,	$y_{2,e} = 9.82791717760947$,	$y_{3,e} = 0.00000508754796$,	$y_{4,e} = 0.00059634764370$,
$y_{5,e} = 0.08663775883916$,	$y_{6,e} = 48.00000000001074$,	$y_{7,e} = 0.00024420230214$,	$y_{8,e} = -0.00000000001074$
$dx_1 = 10^{-7} \times -0.00000000000000$,	$dx_2 = 10^{-7} \times -0.00000107363007$,	$dx_3 = 10^{-7} \times -0.38882785879935$	

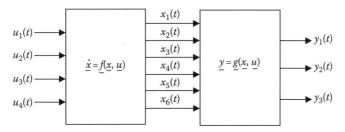

FIGURE 7.8 Dynamic system with equilibrium conditions specified for one input, two states, and one output.

must be solved subject to these constraints. The solution will include the values for the nonconstrained inputs u_1, u_3, and u_4 along with the equilibrium values for the nonconstrained state and output components. Keep in mind that there may be no feasible solution or several solutions depending on the values assigned to the constrained variables.

To be more specific, consider an aircraft flying in level flight at a given altitude with constant speed, heading, and angle of attack. Certain constraints are imposed on the state vector of translational (longitudinal, lateral, and vertical) velocities (u, v, and w) and angular (roll, pitch, and yaw) velocities (r, p, and q) in a body reference coordinate system. The pilot wishes to know the throttle position and input settings that control the orientation of the control surfaces in order for the plane to achieve steady-state "trim" flight conditions.

The following example (Beltrami 1993) illustrates the point for an ecological system.

Example 7.1

The growth rate of fish in a confined space at a fishery is modeled by

$$g(x) = \frac{u_R}{x} + rx\left(1 - \frac{x}{k}\right) - \varepsilon u_E \tag{7.22}$$

where
    $x = x(t)$ is the density of fish measured in tons per square mile
    Parameters r and k determine the natural growth rate function $rx(1 - x/k)$ of fish in the absence of external inputs related to harvesting and restocking
    Input u_E is a measure of the effort (ships, gear, manpower, etc.) per year expended in harvesting
    The parameter ε represents the efficiency of catching fish, measured as a fraction of each ton of fish caught per unit of effort

Finally, the first term accounts for restocking of fish with u_R, the restocking rate measured in tons of fish per square mile per year.

(a) Find the state derivative function and verify whether the given units for parameters and variables are consistent. In particular, determine the units of r, k, and ε.
(b) Find the equation relating the equilibrium state x_e and the constant inputs $\bar{u}_R$ and $\bar{u}_E$ where $u_R(t) = \bar{u}_R$, $t \geq 0$ and $u_E(t) - \bar{u}_E$, $t \geq 0$.
    Baseline numerical values of the system parameters are $k = 4$, $r = 2$, and $\varepsilon = 0.1$.
(c) Find x_e when $\bar{u}_R = 0.3$ tons/mi^2/year and $\bar{u}_E = 10$ effort units/year.
(d) Repeat part (c) using the "`trim`" function. In addition, use x_e found in part (c) and fix $\bar{u}_R = 0.3$ to find the equilibrium value of u_E. Repeat using x_e found in part (c) and fix $\bar{u}_E = 10$ to find the equilibrium value of u_R.
(e) Change the numerical value of $\bar{u}_E$ to 20 and show that there exist three real solutions for x_e.

(f) Verify the results in part (e) using the "trim" function with initial states $x_0 = 0$, 2, and 5.

(g) Show that the middle equilibrium point is unstable and the remaining two are stable. Verify the nature of the equilibrium points by simulation.

(a) The state derivative function is obtained from

$$g(x) = \frac{1}{x}\frac{dx}{dt} = \frac{u_R}{x} + rx\left(1 - \frac{x}{k}\right) - \varepsilon u_E \tag{7.23}$$

$$\Rightarrow f(x, u_R, u_E) = \frac{dx}{dt} = u_R + rx^2\left(1 - \frac{x}{k}\right) - \varepsilon u_E x \tag{7.24}$$

The units for each term in Equation 7.24 are tons/mi²/year, that is,

$$\frac{tons/mi^2}{year} = \frac{tons/mi^2}{year} + \left(\frac{1/year}{tons/mi^2}\right)\left(\frac{tons}{mi^2}\right)^2$$

$$- \left(\frac{1}{effort}\right)\left(\frac{effort}{year}\right)(tons/mi^2)$$

The units for r, k, and ε are (1/year)/(tons/mi²), tons/(mi²), and 1/effort, respectively.

(b) Setting the state derivative function in Equation 7.24 to zero gives

$$\bar{u}_R + rx_e^2\left(1 - \frac{x_e}{k}\right) - \varepsilon\bar{u}_E x_e = 0 \tag{7.25}$$

(c) Substituting the given values for r, k, ε, $\bar{u}_R$, and $\bar{u}_E$ into Equation 7.25 produces a cubic polynomial in x_e. The M-file "Chap7_Ex2_1.m" employs the "roots" function to find the roots. The results are 3.4740, $0.2630 \pm j0.3218$.

(d) "Chap7_Ex2_1.m" contains the statements

```
% A.1 Given uR_bar and uE_bar, find xe
uR_bar=0.3; uE_bar=10;
x0=10; ix=[];
u0=[uR_bar;uE_bar];
iu=[1,2]; y0=0; iy=[];
[x, u, y, dx]=trim('fishery_1', x0, u0, y0, ix, iu, iy)
```

where "fishery_1" is the Simulink file name of the model shown in Figure 7.9. Note that "iu=[1,2]" constrains the inputs to $\bar{u}_R = 3$ and $\bar{u}_E = 10$. The results are given in Table 7.2. The second part of part (d) is implemented using the following statements (see Table 7.2 for results):

```
% A.2 Given uR_bar and xe, find uE_bar
uR_bar=0.3;
x0=3.4740;
ix=1;
u0=[uR_bar;0];
iu=1;
y0=0; iy=[];
[x, u, y, dx]=trim(''fishery_1'', x0, u0, y0, ix, iu, iy)
```

The third part of part (d) is accomplished in a similar fashion except that $\bar{u}_E$ and x_e are fixed and $\bar{u}_R$ is returned by the "trim" function. In this case, the assignments "ix=1" and "iu=1" fix $x_e = 3.4740$ and $u_R = \bar{u}_R$

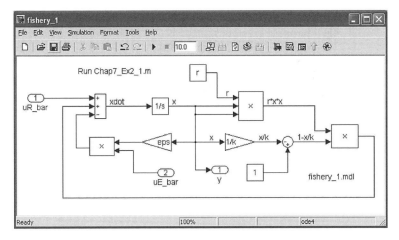

FIGURE 7.9   Simulink® diagram of fishery system dynamics.

TABLE 7.2
Results of Using "trim" Function for Different Conditions

Case	Given	Given	Initial Guess	Result	"xd"
A.1	$\bar{u}_R = 0.3$	$\bar{u}_E = 10$	$x_0 = 10$	$x_e = 3.4740$	-2.8234×10^{-8}
A.2	$\bar{u}_R = 0.3$	$x_e = 3.4740$	$x_0 = 10$	$\bar{u}_E = 10$	5.1090×10^{-11}
A.3	$\bar{u}_E = 10$	$x_e = 3.4740$	$x_0 = 10$	$\bar{u}_R = 0.3$	1.2261×10^{-12}
B.1	$\bar{u}_R = 0.3$	$\bar{u}_E = 20$	$x_0 = 0$	$x_e = 0.1814$	2.1407×10^{-12}
B.2	$\bar{u}_R = 0.3$	$\bar{u}_E = 20$	$x_0 = 2$	$x_e = 1.3278$	0
B.3	$\bar{u}_R = 0.3$	$\bar{u}_E = 20$	$x_0 = 5$	$x_e = 2.4908$	-3.3323×10^{-10}

(e) When $\bar{u}_E = 20$, the "roots" function in "Chap7_Ex2_1.m" yields three solutions to Equation 7.25, namely, 0.1814, 1.3278, and 2.4908.

(f) Using the "trim" function with initial state guesses of 0, 2, and 5 produces the identical equilibrium states (see Cases B.1, B.2, and B.3 in Table 7.2).

(g) The stability of each equilibrium point can be ascertained by looking at a graph of the growth rate function shown in Figure 7.10. Note that the fish density increases wherever $g(x)$ is positive as indicated by right-pointing arrows and conversely decreases in regions where $g(x)$ is negative, shown with left-pointing arrows. Fish densities initially located in the region $(x_e)_1 < x < (x_e)_2$ will move towards $(x_e)_1 = 0.1814$ whereas initial densities satisfying $(x_e)_2 < x < (x_e)_3$ eventually approach $(x_e)_3 = 2.4908$. Consequently, the equilibrium point $(x_e)_2 = 1.3278$ is unstable.

The system is simulated using "fishery_2.mdl" (not shown) that is identical to "fishery_1.mdl" except for the input blocks that are replaced by "constant" blocks and the "output" block is removed. Results are shown in Figure 7.11.

Note that the middle responses in the second graph starting at $x_0 = 1.3$, slightly less than $(x_e)_2 = 1.3278$ and $x_0 = 1.35$ and slightly more than $(x_e)_2 = 1.3278$, diverge from the neighborhood of $(x_e)_2$, the unstable equilibrium point.

Before we proceed further, it is interesting to consider the market place's influence on the fish supply. If we adopt a very rudimentary model for the harvesting effort u_E, one that says the rate of change of harvesting depends solely on net profit as measured by the difference between revenue and cost, then $u_E(t)$ is governed by the first-order differential equation

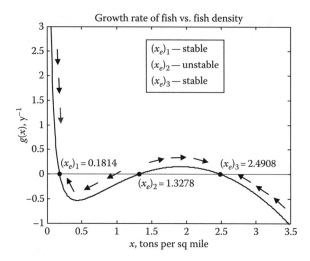

FIGURE 7.10 Graph of fish growth rate and equilibrium points.

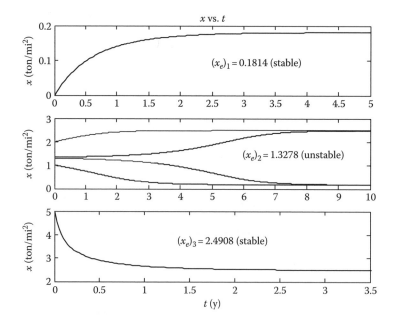

FIGURE 7.11 State responses starting from several initial points.

$$\frac{du_E}{dx} = \alpha(R - C) \tag{7.26}$$

where
  R and C are the revenue and cost, respectively, in \$/year/mi²
  α is a constant

Assuming revenue depends on harvesting $\varepsilon u_E x$ and selling price p leads to

$$R = \varepsilon u_E x \cdot p \tag{7.27}$$

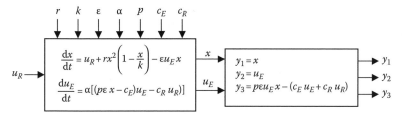

FIGURE 7.12 Block diagram of system showing parameters, input, states, and output.

where p is the selling price in \$/ton. The cost is a function of effort and restocking, that is,

$$C = c_E u_E + c_R u_R \qquad (7.28)$$

where
  c_E is in \$/effort/year/mi^2
  c_R is in \$/ton

Equations 7.26 through 7.28 give

$$\frac{du_E}{dt} = \alpha[p\varepsilon u_E x - (c_E u_E + c_R u_R)] \qquad (7.29)$$

The expanded system dynamics are now modeled by the coupled nonlinear differential equations given in Equations 7.24 and 7.29. A block diagram of the system displaying the system parameters, input, states, and outputs is shown in Figure 7.12.

System parameters $r, k, \ldots, c_R$ can be thought of as inputs to the system. However, they are distinguished from the system input u_R because they generally remain fixed at assigned values. When they do vary, fluctuations (often unpredictable) occur with less frequency than the input.

Suppose the set of parameters in Figure 7.12 are fixed at baseline values except for the selling price p. Viewing the system as being "driven" by the conventional input restocking rate u_R as well as p, we can employ the "trim" function to search for the equilibrium state and output vectors. The M-file "Chap7_fishery_w_economics.m" uses $k = 4$, $r = 2$, $\varepsilon = 0.1$, $\alpha = 0.75$, $c_E = 2$, and $c_R = 1$ for the fixed parameters and varies u_R and p as inputs in the "trim" function call. The results are shown in Table 7.3.

Certain combinations of $\bar{u}_R$ and $\bar{p}$ produce no solution, which raises the question of whether there may in fact be other equilibrium states in addition to the one given in Table 7.3 for the combinations considered. There is no certainty when using a search algorithm. All we can do is

TABLE 7.3
Equilibrium States x_e and $(u_E)_e$ as a Function of Input $(\bar{u}_R, \bar{p})$

$\bar{u}_R$	$\bar{p}$		
	6	7	8
0.3	$x_e = 3.3772$	$x_e = 2.8821$	$x_e = 2.5189$
	$(u_E)_e = 11.4053$	$(u_E)_e = 17.1501$	$(u_E)_e = 19.8447$
0.5	$x_e = 3.4052$	$x_e = 2.8975$	$x_e = 2.5304$
	$(u_E)_e = 11.5954$	$(u_E)_e = 17.6981$	$(u_E)_e = 20.5694$
1	$x_e = 3.4716$	$x_e = 2.9321$	$x_e = 2.5559$
	$(u_E)_e = 12.0522$	$(u_E)_e = 19.0668$	$(u_E)_e = 22.3675$

begin the search from different starting points in the hope of finding additional equilibrium states, should they exist. The results in Table 7.3 were obtained using a starting guess of $x = 10$ and $u_E = 20$.

We now explore the possibility of the existence of an analytical solution for the equilibrium state. The algebraic equations resulting from setting the state derivative functions in Equations 7.24 and 7.29 to zero are

$$0 = \bar{u}_R + rx_e^2\left(1 - \frac{x_e}{k}\right) - \varepsilon(u_E)_e x_e \tag{7.30}$$

$$0 = \bar{p}\varepsilon x_e - c_E(u_E)_e - c_R\bar{u}_R \tag{7.31}$$

Solving for x_e in Equation 7.31 gives

$$x_e = \frac{1}{\bar{p}\varepsilon(u_E)_e}[c_E(u_E)_e + c_R\bar{u}_R] \tag{7.32}$$

and substituting the result for x_e in Equation 7.30 produces a fourth-order polynomial in $(u_E)_e$. The details are left for an exercise problem; however, the result is

$$\beta_4(u_E)_e^4 + \beta_3(u_E)_e^3 + \beta_2(u_E)_e^2 + \beta_1(u_E)_e + \beta_0 = 0 \tag{7.33}$$

where

$$\begin{aligned}
\beta_4 &= -k\bar{p}^2\varepsilon^3 c_E \\
\beta_3 &= k\bar{p}^2\varepsilon^3\bar{u}_R(\bar{p} - c_R) + rc_E^2(k\bar{p}\varepsilon - c_E) \\
\beta_2 &= rc_E c_R\bar{u}_R(2k\bar{p}\varepsilon - 3c_E) \\
\beta_1 &= -r(c_R\bar{u}_R)^2(k\bar{p}\varepsilon - 3c_E) \\
\beta_0 &= -r(c_R\bar{u}_R)^3
\end{aligned} \tag{7.34}$$

For $\bar{u}_R = 0.3$ and $\bar{p} = 6$, the solutions from "Chap7_fishery_w_economics.m" are

$$(uE)_e = 11.4053, \quad 0.7500, \quad -0.1471 \pm j0.0168$$

$$x_e = 3.3772, \quad 4.0000, \quad -0.0219 \pm j0.3842$$

There are two feasible solutions, namely, $x_e = 3.3772$, $(u_E)_e = 11.4053$ and $x_e = 4.0000$, $(u_E)_e = 0.7500$. The "trim" function has converged to the first solution (see Table 7.3).

The values shown in Table 7.3 can be verified by simulation. For example, Figure 7.13 is a simulation of the system initially at equilibrium with inputs $u_R = \bar{u}_R = 0.3$, $p = \bar{p} = 6$, and $x_e = 3.3772$, $(u_E)_e = 11.4053$. Step changes in u_R and p occur at $t = 1$ year. The new inputs correspond to the lower right corner of Table 7.3, namely, $u_R = \bar{u}_R = 1$, $p = \bar{p} = 8$. The new equilibrium state agrees with the values shown in the table. Refer to M-file "Chap7_Fig2_12.m."

EXERCISES

7.1 An alternate predator–prey model for fish and sharks is

$$\frac{dF}{dt} = aF - bSF$$

$$\frac{dS}{dt} = -cS + dFS$$

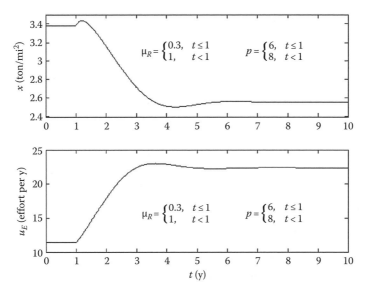

FIGURE 7.13 Simulated transient response to step changes in u_R and p.

(a) Find the nontrivial equilibrium point (F_e, S_e) in terms of parameters a, b, c, and d.
(b) Use the "`trim`" function to find the equilibrium point when the system parameters are $a = 0.1$, $b = 0.03$, $c = 0.02$, and $d = 0.0025$, and compare the answer with the value obtained using the result in part (a).
(c) Verify by simulation that the equilibrium point (F_e, S_e) is "neutrally stable," which means that sustained oscillations in fish and shark populations occur regardless of the initial conditions $F(0) \neq 0$ and $S(0) \neq 0$. Plot time histories $F(t)$ and $S(t)$, $t \geq 0$ and a phase plot S vs. F.

7.2 For the predator–prey system governed by Equations 7.5 and 7.6,
(a) Show that the points (F_e, S_e) in the following table are equilibrium points.

F_e	S_e
0	0
$\dfrac{a}{b}$	0
$\dfrac{a\lambda}{b\lambda + ce}$	$\dfrac{ae}{b\lambda + ce}$

(b) Investigate the local stability of each equilibrium point by simulation of the system with initial conditions in the neighborhood of each point. Draw the phase trajectories for each case.
(c) The system parameters are $a = 50$, $b = 2.5$, $c = 4$, $e = 2$, and $\lambda = 8$. Use the "`trim`" function starting at different points in the F–S plane to try and locate the last two equilibrium points.
(d) The system parameters are $a = 40$, $b = 4$, $c = 3$, $e = 2$, and $\lambda = 5$. Use the "`trim`" function with S constrained to zero to find the equilibrium point $(a/b, 0)$.
(e) The system parameters are $a = 50$, $b = 2.5$, $c = 0$, $e = 2$, and $\lambda = 8$. Simulate the system and obtain time histories of $F(t)$ and $S(t)$ along with a phase trajectory when the initial populations are $F(0) = 5$ and $S(0) = 2$.

7.3 A three-species predator–prey model (Edelstein-Keshet 1988) is

$$\frac{dx}{dt} = axz + \beta xy - yx$$

$$\frac{dy}{dt} = \delta y - \varepsilon xy$$

$$\frac{dz}{dt} = \mu z(v - z) - \lambda xy$$

where
   x is a predator
   y and z are its prey

(a) Express the nontrivial equilibrium pt (x_e, y_e, z_e) in terms of the system parameters.
(b) Use the "`trim`" function to find the equilibrium pt when the system parameters are $\alpha = 0.075$, $\beta = 0.009$, $\gamma = 0.2$, $\delta = 0.1$, $\varepsilon = 0.025$, $\mu = 0.0015$, $v = 10$, and $\lambda = 0.003$. Compare the answer with the value obtained using the result in part (a).
(c) The equilibrium point is asymptotically stable. Verify by simulating the response of the autonomous system starting from various randomly selected points in the neighborhood of the equilibrium point.

7.4 Derive the state Equations 7.17 through 7.19.

7.3 OPTIMIZATION OF SIMULINK® MODELS

System designers often resort to simulation to verify whether a newly designed system performs in a manner consistent with a set of predefined requirements and constraints. Generally speaking, multiple simulations are necessary to "observe" how the system responds to a range of inputs and parameter variations. For some systems, a subset of the inputs and parameters that affect its performance are controllable. For example, ground vehicle performance can be characterized by fuel economy, vehicle handling, ride comfort, acceleration, emergency braking, and so forth. Given a single unambiguous measure of system performance, the design objective reduces to a determination of numerical values for the controllable system parameters (wheel base, springs and shocks, carburetor design, weight, steering ratio, and so forth) resulting in optimal performance. In contrast, a simulation model used to predict weather relies on knowledge of atmospheric conditions to forecast future weather patterns. Neither inputs (at least not yet) nor the resulting weather is controllable.

Inherent in the process of optimizing system performance is the ability to observe or, somehow, measure performance, that is, acquire data about the system as the parameters are varied. Experimenting with the real system is oftentimes impractical for reasons of expense and time consumption or even dangerous depending on the levels of the system parameters. Herein lies the value of simulation in optimizing a system's performance. The simulation model can be "exercised" in a systematic way to achieve optimum or near optimum results without the previously cited pitfalls of dealing with the physical system. A simple example of system optimization follows.

Figure 7.14 portrays a pair of objects, one designated the target and the other object intent on destroying it by firing a projectile weapon at it. The target is assumed to be a point moving at constant velocity v_T in a circular trajectory of radius L with the attacker permanently positioned at the center of the coordinate system. The attacker fires its weapon along a fixed direction denoted by the azimuth angle θ. The projectile is subjected to a linear drag force. Both objects are assumed to be

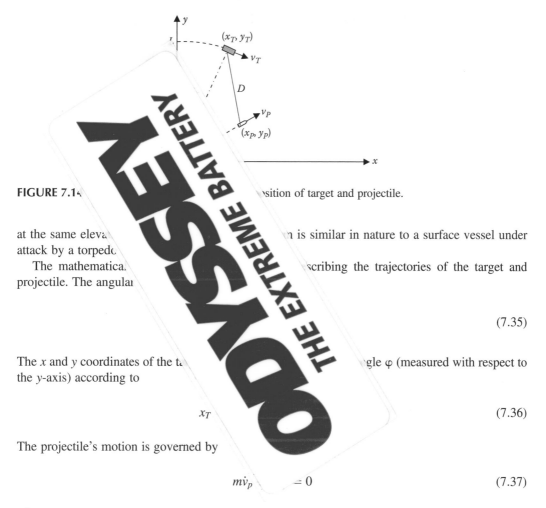

FIGURE 7.1 ...sition of target and projectile.

at the same eleva... ...n is similar in nature to a surface vessel under attack by a torpedo...

The mathematica... ...scribing the trajectories of the target and projectile. The angular

$$(7.35)$$

The x and y coordinates of the t... ...gle φ (measured with respect to the y-axis) according to

$$x_T \qquad (7.36)$$

The projectile's motion is governed by

$$m\dot{v}_p \qquad = 0 \qquad (7.37)$$

where
  m is the projectile mass
  μ *is* the drag coefficient for determining the linear drag force acting on the projectile

Resolving the projectile's velocity into x and y coordinates,

$$\dot{x}_p = v_P \cos\theta, \quad \dot{y}_P = v_P \sin\theta \qquad (7.38)$$

The distance separating the target and projectile is given by

$$D = [(x_T - x_P)^2 + (y_T - y_P)^2]^{1/2} \qquad (7.39)$$

A Simulink diagram incorporating Equations 7.35 through 7.39 is shown in Figure 7.15. Since the intended purpose of firing the projectile is to intercept the target, the performance measure of the system is taken as the separation between the target and projectile at the moment the projectile has traveled a distance L. This distance is denoted as D_{final}. Note the presence of a "Relational Operator" block for terminating the simulation when the projectile's distance "rP" exceeds "L." The simulation final time is chosen as some arbitrarily large number ensuring that the simulation is halted at the appropriate time, which incidentally is monitored in the "Display" block.

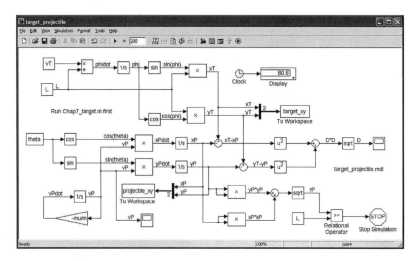

FIGURE 7.15   Simulink® diagram of target and projectile system.

The firing angle θ is treated as a controllable parameter. Our objective is to find θ_{opt}, that is, the projectile firing angle that minimizes the performance measure D_{final} (ideally to zero). A number of calls are made from the M-file "*Chap7_target.m*" to the simulation model "*target_projectile.mdl*" to explore the relationship between D_{final} and θ. The result is shown in Figure 7.16.

The function $D_{final}(\theta)$ is seen to possess a single minimum in the neighborhood of 70° when the remaining system parameter values are as shown in Figure 7.16. We must perform a search for θ_{opt} where

$$D_{final}(\theta_{opt}) = \underset{\theta \geq 0}{\text{Min}}\, D_{final}(\theta) \tag{7.40}$$

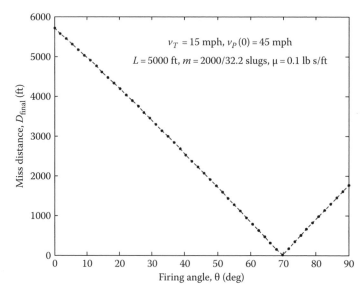

FIGURE 7.16   Graph of miss distance D_{final} vs. projectile firing angle θ.

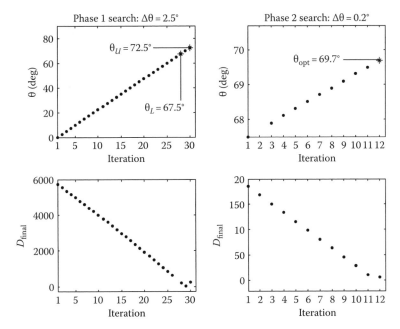

FIGURE 7.17   Two-phase search for optimum firing angle.

The M-file "*Chap7_opt_search.m*" performs a very rudimentary search for the optimum angle θ_{opt}. It begins by incrementing θ (starting from zero) until it finds an angle θ_U where $D_{final}(\theta_U)$ is greater than D_{final} at the previous firing angle. Since the previous point could be to the right of the minimum, the angle prior to the previous one is designated θ_L and the interval ($\theta_L \leq \theta \leq \theta_U$) is guaranteed to contain θ_{opt}. A second sweep, with a finer gradation of θ values, is initiated, beginning at θ_L. It continues until θ_{opt} is found or the entire interval ($\theta_L \leq \theta \leq \theta_U$) is traversed. θ_{opt} is detected when $D_{final}(\theta)$ is below some threshold, 10 ft in this case. A second sweep is tried with even finer divisions if the first one is unsuccessful. It must be borne in mind that each value of θ requires a simulation run to find $D_{final}(\theta)$. The two search phases are illustrated in Figure 7.17.

The optimization toolbox includes a number of algorithms for iteratively searching parameter spaces to locate local minima and maxima of a function that depends on the parameters. Optimum seeking methods are available for both unconstrained and constrained optimization. The optimization toolbox and Simulink complement each other when the performance measure (objective function in optimization terminology) at some point in the parameter space depends on the dynamic response of a system. That is, the actual system response must be observed or simulated to obtain a numerical value of the objective function. This could be a final value of some output (dependent variable) or perhaps a certain function of several dependent variables. A common situation is where the objective function is evaluated as the integral of an appropriate function of the system's outputs.

In situations where the objective function dependence on the system's parameters is expressible in analytical or tabular form, a dynamic simulation is unnecessary and Simulink is not required. In either case, a unique value for the objective function at different locations in the parameter space must be available to the optimization routine.

Before we delve more into the practical aspects of optimization, let us take a look at how the MATLAB optimization toolbox can be used to find the optimum firing angle θ_{opt} in the previous example. The first step is the creation of a MATLAB function file to evaluate the objective function D_{final} for a given value of firing angle θ. The function M-file "*obj_fcn_D*" is listed as follows:

```
function f = obj_fcn_D(angle, L, m, mu, vT, vP)
% Objective function for finding D_final
T = 0.05; % integration step (sec) for RK-4
tfinal = 200; % final sim time (sec)
opts = simset('SrcWorkSpace', 'current', 'DstWorkSpace', 'current');
theta = angle; % firing angle (rad) for 'CON' Simulink model block
sim('target_projectile', tfinal, opts); %run sim and return array D
f = D(end); % objective function: D_final (ft)
```

The first argument of "*obj_fcn_D.m*" is "angle" (the optimization parameter θ), and the remaining arguments are simply parameters passed to the function from the main program "*Chap7_Toolbox_opt_search.m*." The main program initializes the starting value of θ in the variable "angle_init" and then calls the optimization toolbox function "fminunc" to start the search $θ_{opt}$. The preferred way of calling "fminunc" depends on the version of MATLAB in use. Prior to MATLAB 6.0 (R12), the correct syntax was

```
[opt_angle_rad, FVAL] = fminunc('obj_fcn_D',angle_init, [], L, m, mu,
    vT, vP_initial) % optimum angle (rad)
```

For MATLAB 6.0 (R12) and later, the string 'obj_fcn_D' was replaced by the function handle '@obj_fcn_D' for faster calls to the objective function.

$θ_{opt}$ and the minimum D_{final} are returned in "opt_angle_rad" and "FVAL" if the search algorithm converges to a solution. "*Chap7_Toolbox_opt_search.m*" contains additional statements to simulate the system using the optimum firing angles returned by "fminunc" when the target speed is 15 and 75 mph. Target and projectile trajectories are shown in Figures 7.18 and 7.19. The projectile's position is plotted at 2.5 s intervals. The elapsed time in both cases is 80.8 s.

It is not surprising that elapsed time t_f is the same for both target speeds. The time is easily found by recognizing that the solution to Equation 7.37 is given by

$$v_P(t) = v_P(0)e^{-(\mu/m)t}, \quad t \geq 0 \tag{7.41}$$

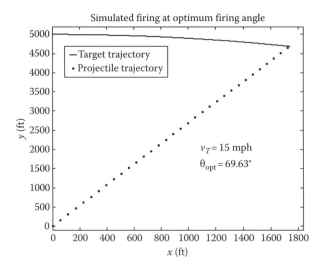

FIGURE 7.18 Target and projectile motion for optimum firing angle ($v_T = 15$ mph).

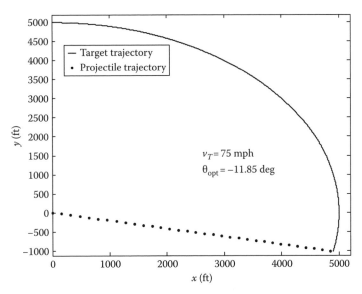

FIGURE 7.19 Target and projectile motion for optimum firing angle ($v_T = 75$ mph).

and then integrating to obtain $s_p(t)$, the distance traveled by the projectile

$$s_p(t) = \int_0^t v_p(0)e^{-(\mu/m)\tau}d\tau \tag{7.42}$$

$$= \frac{m}{\mu}v_p(0)[1 - e^{-(\mu/m)t}], \quad t \geq 0 \tag{7.43}$$

Setting $s_p(t_f) = L$ and solving for t_f give

$$t_f = -\frac{m}{\mu}\ln\left[1 - \frac{\mu L}{mv_p(0)}\right]$$

$$= -\frac{2000/32.2}{0.1}\ln\left[1 - \frac{0.1(5000)}{(2000/32.2)45(5280/3600)}\right] = 80.79 \text{ s} \tag{7.44}$$

There are a number of system parameters that can be varied to study their effect on the optimum firing angle. Suppose we wish to investigate the relationship between θ_{opt} and the target's speed v_T. The M-file "*Chap7_opt_theta_vT.m*" sequences through a range of target speeds $v_T = 5, 15, 25, \ldots,$ 65, 75 mph and finds the optimum firing angle for each speed. The result is shown in Figure 7.20 where it is apparent that the relationship is linear. Could this have been predicted?

Referring to Figure 7.14, at time t_f, when the projectile has struck the target,

$$\theta = \frac{\pi}{2} - j(t_f) = \frac{\pi}{2} - \left(\frac{v_T}{L}\right)t_f \quad (v_T \text{ in ft/s}, \theta \text{ in rad}) \tag{7.45}$$

$$\Rightarrow \theta = 90 - \left[\frac{v_T}{5000}\left(\frac{5280}{3600}\right)\left(\frac{180}{\pi}\right)\right]80.79 \quad (v_T \text{ in mph}, \theta \text{ in deg}) \tag{7.46}$$

$$\Rightarrow \theta = 90 - 1.3579v_T \tag{7.47}$$

which is the equation of the line shown in Figure 7.20.

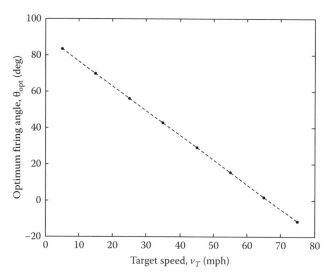

FIGURE 7.20 Results of target speed sensitivity analysis.

The search algorithm used to find the optimum firing angle, illustrated in Figure 7.17, is rather simple. More sophisticated algorithms rely on the local topography of the objective function to guide the search for the local optimum point. The gradient vector (to be defined shortly) is computed at a point in multidimensional parameter space and used to arrive at a new direction and distance for continuing the search. The gradient vector reduces to the first derivative for one-dimensional searches.

Following is an example of a one-dimensional parameter search using the slope, that is, first derivative to locate the minimum of an objective function. In the target–projectile system, the projectile decelerates with time due to the linear drag force. Suppose the target attempts to "outrun" the projectile by traveling in the y-direction starting from the point $(0, L)$, (see Figure 7.14) at constant speed v_T. The target is in the clear if the pursuing projectile is moving slower than the target, that is, $v_P(t) < v_T$ at some point in time and $y_T(t) > y_P(t)$ have been true up to that time.

We focus on the minimum separation between the target and projectile to see how close the two come. For a set of fixed parameters L, m, μ, v_T, and $v_P(0) > v_T$, the time at which the minimum separation occurs is required. Position of the target is given by

$$y_T(t) = L + v_T t, \quad t \geq 0 \tag{7.48}$$

From Equation 7.42, the distance traveled and, hence, position of the projectile are

$$y_P(t) = \tau v_P(0)(1 - e^{-t/\tau}), \, t \geq 0 \quad \text{where } \tau = \frac{m}{\mu} \tag{7.49}$$

The separation between the target and projectile as a function of time is

$$D(t) = y_T(t) - y_P(t) = L + v_T t - \tau v_P(0)(1 - e^{-t/\tau}), \quad t \geq 0 \tag{7.50}$$

A graph of Equation 7.50 with the nominal parameter values is shown in Figure 7.21. Differentiating $D(t)$ in Equation 7.50 gives

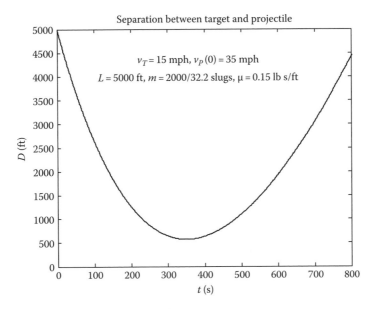

FIGURE 7.21   Graph of separation distance vs. time.

$$\frac{\mathrm{d}}{\mathrm{d}t}D(t) = v_T - v_P(0)e^{-t/\tau}, \quad t \geq 0 \tag{7.51}$$

A search algorithm is implemented in "*Chap7_min_sep_search.m*," which sequences through values of t based on the first derivative. Specifically,

$$t_{i+1} = t_i - \frac{\mathrm{d}}{\mathrm{d}t}D(t_i) \cdot \Delta, \quad i = 0, 1, 2, \ldots \tag{7.52}$$

The search terminates when the magnitude of the derivative falls below a threshold that was set to 0.1 and Δ was fixed at a value of 10.

Figure 7.22 shows the results of two searches for the minimum separation. The one on the left starts at $t_0 = 0$ s and the other begins at $t_0 = 800$ s. The two searches quickly locate the same minimum separation of 572.4 ft at $t = 349.7$ s. Note the derivative function approaching zero from opposite directions as the two searches progress.

This example illustrates the power of using the derivative to scale the step size between sequential points in the search for the optimum value of the objective function.

From elementary calculus, the local minima and maxima of a continuously differentiable function occur at critical points where the first derivative is equal to zero. Hence, from Equation 7.51, we can find $t_{\min}$ as follows

$$\frac{\mathrm{d}}{\mathrm{d}t}D(t)\bigg|_{t=t_{\min}} = v_T - v_P(0)e^{-t/\tau}\big|_{t=t_{\min}} = 0$$

$$\Rightarrow t_{\min} = -\tau \ln\left(\frac{v_T}{v_P(0)}\right) = -\frac{2000/32.2}{0.15}\ln\left(\frac{15}{35}\right) = 350.8\,\mathrm{s} \tag{7.53}$$

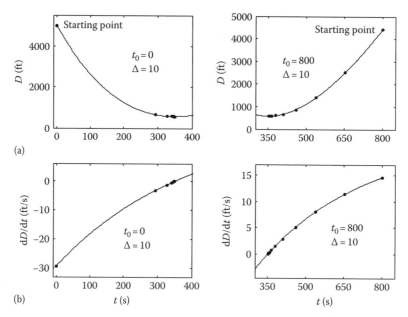

(a)

(b)

FIGURE 7.22 Results of two searches for minimum separation. (a) Sequence of points in search for minimum. (b) Sequence of derivative evaluations in search for minimum.

Substituting t_{min} in Equation 7.53 for t into Equation 7.50 gives (after simplification),

$$D_{min} = L - \tau \left[v_P(0) - v_T + v_T \ln \left(\frac{v_T}{v_P(0)} \right) \right] \tag{7.54}$$

$$= 5000 - \frac{2000/32.2}{0.15} \left[35 - 15 + 15 \ln \left(\frac{15}{35} \right) \right] \left(\frac{5280}{3600} \right) = 572.3 \, \text{ft} \tag{7.55}$$

The numerical values obtained analytically in Equations 7.53 and 7.55 are in agreement with the values obtained from the iterative search for the minimum separation.

Analytical solutions for finding the optimum point are seldom possible. When the system dynamics are modeled by nonlinear equations, iterative searches using simulation to obtain the objective function and numerical derivative approximations are often the only recourse. For example, the existence of nonlinear damping functions in the target–projectile system would necessitate a simulation-based approach to finding the minimum separation.

The optimization toolbox employs a different search method when the objective function and its derivative (partial derivatives in the multivariable case) are expressible in analytic form. The M-file "*obj_fcn_D_sep.m*" includes definitions of both the objective function and its first derivative. The essential statements are

```
function [f,g] =obj_fcn_D_sep(t,L,tau,vT,vP_initial)
f =L+vT*t-tau*vP_initial*(1-exp(-t/tau));
g =vT-vP_initial*exp(-t/tau);
```

and the calling program "*Chap7_Toolbox_opt_sep_search.m*" references the function file "*obj_fcn_D_sep.m*" using

```
options =optimset ('GradObj','on');
t_min =fminunc (@obj_fcn_D_sep, t_init, options, L, tau, vT, vP_
    initial)
```

The "`options`" declaration is required to enable the gradient search method, which uses the first derivative of the objective function given in "*obj_fcn_D_sep.m*," when the call to "`fminunc`" is made. The results obtained from running the M-file "*Chap7_Toolbox_opt_sep_search.m*" are identical with the analytical values given in Equations 7.53 and 7.55.

7.3.1 GRADIENT VECTOR

Our experience in the previous example taught us that knowledge of the slope, that is, first derivative of the objective function, could be used to reduce the number of iterations required to locate a local optimum. The same holds for objective functions involving several parameters. Instead of a single derivative, a gradient vector with components equal to the partial derivatives of the objective function with respect to each parameter is computed. The gradient vector of a multivariable function at a point in parameter space points in the direction of maximum increase of the function. Furthermore, the magnitude of the gradient vector is a measure of the rate of increase in the objective function in the direction of the gradient.

Consider the function

$$f(x_1, x_2) = c_1(x_1 - h)^2 + c_2(x_2 - k)^2, \quad -\infty < x_1 < \infty, \quad -\infty < x_2 < \infty \tag{7.56}$$

The gradient vector at the point (x_1, x_2) is

$$\nabla f(x_1, x_2) = \begin{bmatrix} \dfrac{\partial f(x_1, x_2)}{\partial x_1} \\ \dfrac{\partial f(x_1, x_2)}{\partial x_2} \end{bmatrix} = \begin{bmatrix} 2c_1(x_1 - h) \\ 2c_2(x_2 - k) \end{bmatrix} \tag{7.57}$$

Figure 7.23 portrays the objective function as a surface $z = f(x_1, x_2)$ for the case where $h = 5$, $k = 10$, $c_1 = 1$, and $c_2 = 4$. Several contours that are projections of constant z in the x_1–x_2 plane are also shown. The global minimum occurs at $x_1 = h = 5$ and $x_2 = k = 10$ and the minimum value is $f(h, k) = f(5, 10) = 0$.

Since the gradient vector at (x_1, x_2) points in the direction of maximum increase of $f(x_1, x_2)$, the orthogonal direction that coincides with the tangent to the contour at (x_1, x_2) represents the direction of zero change in $f(x_1, x_2)$. The negative of the gradient vector is drawn at several points in the x_1–x_2

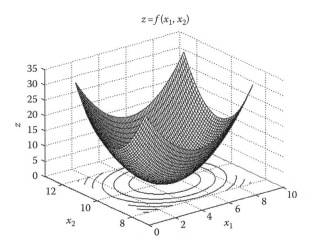

FIGURE 7.23 Graph of surface $z = f(x_1, x_2)$ and several contours $z = $ const.

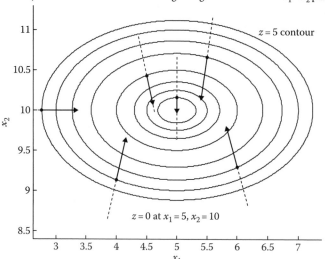

FIGURE 7.24 Contours of the objective function $f(x_1, x_2) = (x_1 - 5)^2 + 4(x_2 - 10)^2$.

parameter space in Figure 7.24 because $-\nabla f(x_1, x_2)$ points in the direction of maximum decrease of $f(x_1, x_2)$, and we are looking at minimizing the objective function.

Table 7.4 includes the points shown in Figure 7.24, the value of z for the contour, which the points lie on, the negative gradient vector, and its magnitude. The lengths of the negative gradient vectors are drawn proportional to their magnitudes given in the table.

TABLE 7.4
Contour and Gradient Data for Points Shown in Figure 7.24

(x_1, x_2)	Contour	$-\nabla f(x_1, x_2)$	$\|\nabla f(x_1, x_2)\|$
(2.7639, 10)	5	$\begin{bmatrix} 4.4721 \\ 0 \end{bmatrix}$	4.4721
(4, 9.1340)	4	$\begin{bmatrix} 2 \\ 6.9282 \end{bmatrix}$	7.2111
(6, 9.2929)	3	$\begin{bmatrix} -2 \\ 5.6569 \end{bmatrix}$	6.0000
(5.5, 10.6614)	2	$\begin{bmatrix} -1 \\ -5.2915 \end{bmatrix}$	5.3852
(4.5, 10.4330)	1	$\begin{bmatrix} 1 \\ -3.4641 \end{bmatrix}$	3.6056
(5, 10.1581)	0.1	$\begin{bmatrix} 0 \\ -1.2649 \end{bmatrix}$	1.2649

A multivariable function like $f(x_1, x_2)$ is expandable about a point $(\bar{x}_1, \bar{x}_2)$ using a two-dimensional Taylor Series, that is,

$$f(x_1, x_2) = f(\bar{x}_1, \bar{x}_2) + \frac{\partial f(\bar{x}_1, \bar{x}_2)}{\partial x_1}(x_1 - \bar{x}_1) + \frac{\partial f(\bar{x}_1, \bar{x}_2)}{\partial x_2}(x_2 - \bar{x}_2) + h.o.t. \qquad (7.58)$$

where *h.o.t.* represents higher order terms involving powers and products of $(x_1 - \bar{x}_1)$ and $(x_2 - \bar{x}_2)$. To a first-order approximation, the change in $f(x_1, x_2)$ about the point $(\bar{x}_1, \bar{x}_2)$ is

$$f(x_1, x_2) - f(\bar{x}_1, \bar{x}_2) \approx \frac{\partial f(\bar{x}_1, \bar{x}_2)}{\partial x_1} \Delta x_1 + \frac{\partial f(\bar{x}_1, \bar{x}_2)}{\partial x_2} \Delta x_2 \tag{7.59}$$

$$\Rightarrow \Delta f(\bar{x}_1, \bar{x}_2) \approx \frac{\partial f(\bar{x}_1, \bar{x}_2)}{\partial x_1} \Delta x_1 + \frac{\partial f(\bar{x}_1, \bar{x}_2)}{\partial x_2} \Delta x_2 \tag{7.60}$$

$$\Rightarrow \Delta f(\bar{x}_1, \bar{x}_2) \approx \begin{bmatrix} \dfrac{\partial f(\bar{x}_1, \bar{x}_2)}{\partial x_1} \\[2ex] \dfrac{\partial f(\bar{x}_1, \bar{x}_2)}{\partial x_2} \end{bmatrix}^T \begin{bmatrix} \Delta x_1 \\[1ex] \Delta x_2 \end{bmatrix} = \nabla f(\bar{x}_1, \bar{x}_2)^T \Delta \tag{7.61}$$

And, therefore, $\Delta f(\bar{x}_1, \bar{x}_2) \approx 0$ provided the gradient vector is identically zero at $(\bar{x}_1, \bar{x}_2)$. Quite understandably, the search for local extremes (minima and maxima) of the objective function $f(x_1, x_2)$ is based on finding points where the gradient vector $\nabla f(x_1, x_2) = [0 \ 0]^T$. The gradient vector also vanishes at a saddle point, which is neither a local minimum nor maximum. A test involving the matrix of second partials at points where the gradient is zero can distinguish between local extrema and saddle points.

Optimum seeking methods search for extreme values (minima and maxima) of an objective function using the gradient vector in some way (Converse 1970; Miller 1975, 2000; Hasdorff 1976; Daniels 1978; Bryson 1999). The references include both constrained and unconstrained optimization problems. In constrained optimization, a subset of the parameters are constrained in some fashion limiting the region of feasible solutions for finding the optimum. Typically, the constraints are inequalities reflecting limitation of system resources or existence of physical boundaries for safe operation.

7.3.2 Optimizing Multiparameter Objective Functions Requiring Simulink® Models

We now focus on multiparameter objective functions, which require execution of a Simulink model to evaluate. The following example is one of a control system where the objective is to minimize a performance measure by choosing two parameters associated with the controller. The performance measure is obtained from simulation and the optimization toolbox is used to find the optimum control settings.

A block diagram of a heading control system for a ship is shown in Figure 7.25. The ship's autopilot and power amplifier are an ideal proportional-derivative (PD) controller that converts an error signal to an amplified voltage for driving the steering gear connected to the ship's rudder. The steering gear, rudder, and hull dynamics are combined into a single ship dynamics transfer function. A gyro compass in the feedback loops senses the ship's heading and sends a voltage to the autopilot. A saturation block is inserted between the controller and ship transfer function to account for the

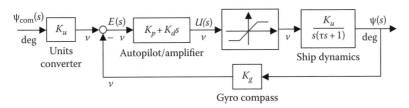

FIGURE 7.25 Block diagram of ship heading control system.

limited power available to the steering system. The units converter transforms the commanded heading from degree to volts for compatibility with the autopilot's electronics.

The control parameters K_p and K_d are to be selected to optimize the system response to a step input in command heading. There are numerous measures that can be used to characterize the step response. Five specific measures are enumerated as follows:

1. Rise time t_r—Time required for response to go from 10% to 90% of its final heading
2. Maximum overshoot, OS_{max}—Difference between maximum heading and final heading in underdamped systems
3. Maximum heading rate, $|\dot{\psi}_{max}|$—Maximum rate of change in ship's heading
4. Integral squared error, ISE—Integral of squared error from time zero to infinity
5. Integral absolute error, IAE—Integral of absolute value of error from zero to infinity

The objective function f is assumed to be a function of these measures, that is,

$$f = f(t_r, OS_{max}, |\dot{\psi}_{max}|, ISE, IAE) = F(K_P, K_d) \tag{7.62}$$

Note that the objective function is implicitly dependent on K_p and K_d because each of the measures t_r, OS_{max}, $|\dot{\psi}_{max}|$, ISE, and IAE depends on these parameters. The goal is to find the optimum value f_{opt} where

$$f_{opt} = \underset{K_P > 0, K_d > 0}{Min} F(K_P, K_d) \tag{7.63}$$

In this example, f is set to a linear combination of the five measures. Hence,

$$f(t_r, OS_{max}, |\dot{\psi}_{max}|, ISE, IAE) = c_1 t_r + c_2 OS_{max} + c_3 |\dot{\psi}_{max}| + c_4 ISE + c_5 IAE \tag{7.64}$$

The constants c_1, c_2, c_3, c_4, and c_5 determine the weights of each measure. For example, if the goal is to minimize the integral squared error (ISE),

$$ISE = \int_0^\infty e^2(t)dt = \int_0^\infty [\psi_{com} - \psi(t)]^2 dt \tag{7.65}$$

the weights are set to $c_1 = c_2 = c_3 = c_5 = 0$ and $c_4 = 1$.

The constrained optimization routine "fmincon" in the optimization toolbox implements a search for f_{opt} subject to parameter constraints. The statement

```
[opt_Kp_Kd,FVAL,EXITFLAG,OUTPUT] = fmincon(@obj_fcn_ship,Kp_Kd_
init,A,B,Aeq,Beq,LB,UB,NONLCON,OPTIONS,Kg,L,Ks,tau,t1,theta_com,c)
```

in "*Chap7_Toolbox_opt_ship.m*" invokes a constrained search for the optimum values of parameters K_p and K_d. The arguments "A, B, Aeq, Beq, LB, UB, NONLCON" define the constraints. "LB" and "UB" are used to set lower and upper bounds on the parameters, and the remaining arguments are empty arrays not applicable in this example.

Before we look at the results, it is instructive to visualize the objective function surface with respect to the K_p–K_d plane. The objective function in this example is

$$f = t_r + OS_{max} + |\dot{\psi}_{max}| = F(K_p, K_d) \tag{7.66}$$

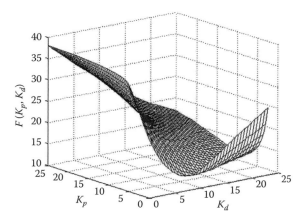

FIGURE 7.26 Objective function surface $f = t_r + OS_{max} + |\dot{\psi}_{max}| = F(K_p, K_d)$.

It is shown in Figure 7.26 for the region $0 \leq K_p \leq 25$, $0 \leq K_d \leq 25$. The data points for drawing the surface were obtained by repeated calls to the Simulink model "*ship.mdl*" from the M-file "*Chap7_ship_control.m*." The simulated step responses were executed for 100 s, a period of time sufficient to allow the transient response to vanish, except for heavily damped cases (low K_p, high K_d). Numerical values of the system parameters are $K_u = K_g = 10$ V/rad, $K_s = 0.04$ rad/s/V, and $\tau = 10$ s, and the autopilot/amplifier saturates at 25 V. The commanded heading ψ_{com} was set to 30°.

In runs where the ship's heading had yet to reach 90% of the final heading (which did not occur for the points shown in Figure 7.26), the rise time was set to 100 s. When the ship's heading failed to reach the final heading, the overshoot was set to zero. The final heading is the commanded heading for all combinations of K_p and K_d resulting in a stable response.

The Simulink block diagram for the model "*ship.mdl*" is shown in Figure 7.27.

The "PID" block in Figure 7.27 is present in the "Simulink Extras" library. It is an ideal PID controller with parameters P, I, and D in the transfer function

$$G(s) = P + \frac{I}{s} + D_s \tag{7.67}$$

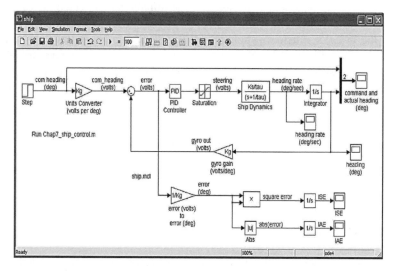

FIGURE 7.27 Simulink® block diagram for ship heading step response.

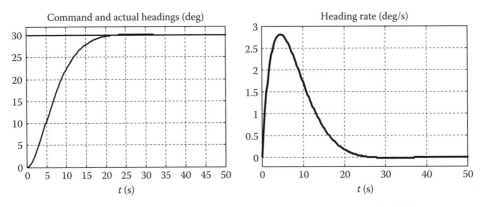

FIGURE 7.28   Results with optimal parameter settings: $(K_p)_{opt} = 1.3011$, $(K_d)_{opt} = 7.1913$.

For simulation runs, P assumed the value of K_p, I was zero, and D assumed the value of K_d. The optimization toolbox search algorithm started from the point (1, 1) in the K_p–K_d plane. A gradient search is not used since the gradient of the objective function is not available in analytic form. A gradient-based search would require numerical approximations to the gradient at a number of points along the way to finding the optimum. The number of objective function evaluations would likely increase significantly depending upon the algorithm's efficiency in locating the optimum. In this example, a "medium-scale, SQP, Quasi-Newton, and line-search" algorithm (see optimization toolbox reference manual), was used successfully to find the optimum solution, namely, $(K_p)_{opt} = 1.3011$, $(K_d)_{opt} = 7.1913$, and $f_{opt} = 14.4395$. A total of 344 function evaluations and, hence, the same number of simulation runs were required.

The simulation was run with the optimal parameter settings to verify the objective function value. The ship's heading and heading rate are shown in Figure 7.28. The rise time, maximum overshoot, and maximum heading rate are 11.5000 s, 0.1154 deg, and 2.8241 deg/s, respectively. From Equation 7.66, $f = 11.5000 + 0.1154 + 2.8241 = 14.4395$.

7.3.3   Parameter Identification

Knowledge of the system parameters appearing in the differential and algebraic equations used to model continuous-time dynamic systems is often imperfect. When the simulationist is reasonably confident that the model's structure is suitable for its intended purpose, the subject of parameter identification arises (see Figure 7.1). A simple approach predicated on minimizing the differences between observed and simulated responses is presented in the following example.

The decreasing concentration of a chemical in solution follows a law from reaction kinetics that states

$$\frac{dx}{dt} = -kx^n \quad (k > 0,\ n > 0) \tag{7.68}$$

where
     $x = x(t)$ is the concentration
     k is a rate constant
     n is the order of the reaction

Suppose the concentration of a chemical in solution was measured and recorded once a minute for 60 min. The values at 5 min intervals are shown in Table 7.5.

TABLE 7.5
Measured Concentration of Chemical in Solution at the End of 5 min Intervals

t (min)	0	5	10	15	20	25	30
$\hat{x}$ (mol/L)	0.5000	0.4015	0.3386	0.2945	0.2617	0.2363	0.2159
t (min)		35	40	45	50	55	60
$\hat{x}$ (mol/L)		0.1991	0.1851	0.1731	0.1628	0.1538	0.1459

The problem before us is to estimate the reaction constant k and reaction order n. We will do this by simulating the response for the chemical concentration starting with guessed values for k and n. The observed and simulated responses are used to compute the sum of squared errors, that is,

$$\text{SSE} = f(k, n) = \sum_{i=0}^{60} [\hat{x}_i - x_i]^2 \tag{7.69}$$

where

$x_i = x(t_i)$, $i = 0, 1, 2, \ldots, 60$ are simulated concentrations a minute apart
$\hat{x}_i = \hat{x}(t_i)$, $i = 0, 1, 2, \ldots, 60$ are values of concentration measured at one-minute intervals, some of which are shown in Table 7.5

Minimizing the objective function $f(k, n)$ yields the optimal estimates of the reaction parameters.

Observed concentrations $\hat{x}_i$, $i = 0, 1, 2, \ldots, 60$ are obtained by running the M-file "*Chap7_reaction_kinetics.m*," which calls the Simulink model "*chemical.mdl*" with $k = 0.125$ and $n = 2.3$, representative of the true reaction. The Simulink block diagram is shown in Figure 7.29. A search constrained to the first quadrant of the k–n plane is performed using one of the routines from the optimization toolbox. The search concludes with $k_{opt} = 0.1256$, $n_{opt} = 2.3037$ and $\text{SSE} = f(k_{opt}, n_{opt}) = 3.2303 \times 10^{-7}$.

A graph of the simulated concentration response with the optimal parameter values is shown in Figure 7.30. As expected, the observed concentrations fall on the simulated concentration response curve.

7.3.4 EXAMPLE OF A SIMPLE GRADIENT SEARCH

The common feature of all gradient search algorithms is their reliance on calculation of the gradient vector at a point in the parameter space. The logic for choosing a direction and step size leading to

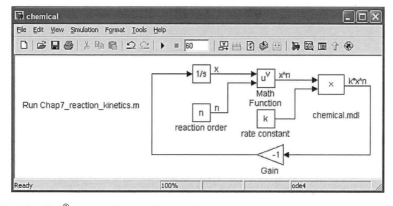

FIGURE 7.29 Simulink® diagram for chemical reaction.

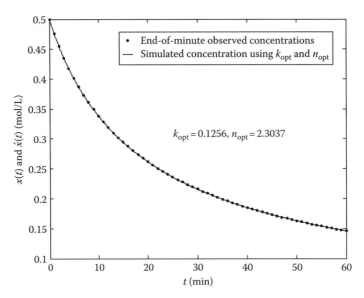

FIGURE 7.30 Graph of observed and simulated ($k = k_{opt}$, $n = n_{opt}$) concentration with optimal parameters.

the next point along with the frequency of gradient calculations is what distinguishes one gradient search algorithm from another. The gradient search presented in this section is intended to demonstrate how to exploit the property of the gradient vector to find a local minimum of an objective function. It is less efficient in comparison with established gradient search algorithms reported in the literature.

The focus of our attention is a bowl-shaped tank shown in Figure 7.31. The bowl is the lower half of a sphere of radius R. Water flow into and out of the tank is controlled by the valves located in the inflow and exiting pipes. The inflow $F_1(t)$ is maintained at a constant value F_1. The outflow $F_2(t)$ is a function of water level $H(t)$ and the opening of the valve in the discharge line, which effectively determines the constant c in the equation

$$F_2(t) = c[H(t)]^{1/2} \tag{7.70}$$

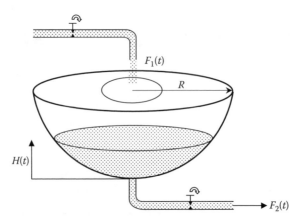

FIGURE 7.31 Hemispherical bowl with flows in and out.

Conservation of mass requires

$$\frac{d}{dt}V(t) = F_1(t) - F_2(t) \tag{7.71}$$

where the volume $V(t)$ is related to the water level by

$$V(t) = \frac{1}{3}\pi H^2(t)[3R - H(t)] \tag{7.72}$$

Differentiating Equation 7.72 with respect to t and simplifying yield

$$\frac{d}{dt}V(t) = \pi[2R - H(t)]H(t)\frac{d}{dt}H(t) \tag{7.73}$$

Combining Equations 7.70, 7.71, and 7.73 results in the differential equation model

$$\pi[2R - H(t)]H(t)\frac{d}{dt}H(t) = F_1(t) - c[H(t)]^{1/2} \tag{7.74}$$

The term $\pi[2R - H(t)]H(t)$ is equal to the cross-sectional area of the bowl at the water level $H(t)$, that is,

$$A(H) = \pi(2R - H)H \tag{7.75}$$

And, therefore, Equation 7.74 is expressible as

$$A(H)\frac{dH}{dt} = F_1 - cH^{1/2} \tag{7.76}$$

The objective is to fill the tank in a specified period of time. The inflow F_1 and discharge constant c are the controllable parameters at our disposal. Before we discuss the gradient search, the objective function must be defined. Since the goal is to fill the tank in a given period of time, say T_{des}, the objective function is defined as

$$F(t_{fill}) = \begin{cases} A\left(\dfrac{t_{fill}}{T_L} - 1\right)^2, & 0 \le t_{fill} < T_L \\ 0, & T_L \le t_{fill} \le T_H \\ B\left(\dfrac{t_{fill} - T_H}{T_{max} - T_H}\right)^2, & T_H < t_{fill} \le T_{max} \\ B, & T_{max} < t_{fill} \end{cases} \tag{7.77}$$

$F(t_{fill})$ is zero whenever the time to fill the tank t_{fill} falls between $T_L = T_{des} - \Delta/2$ and $T_H = T_{des} + \Delta/2$ where Δ is the width of the interval centered at T_{des}. The constant T_{max} is an arbitrarily chosen upper limit. A and B determine the objective function at the points $t_{fill} = 0$ and $t_{fill} = T_{max}$. A graph of $F(t_{fill})$ is shown in Figure 7.32.

It is helpful to visualize the objective function surface over the F_1–c plane. For convenience, let the maximum inflow be $(F_1)_{max} = 10$ ft^3/min when the inlet valve is wide open. Furthermore, a maximum value of $c_{max} = 2$ ft^3/min/ft$^{1/2}$ is assumed, corresponding to a wide-open valve in the discharge line.

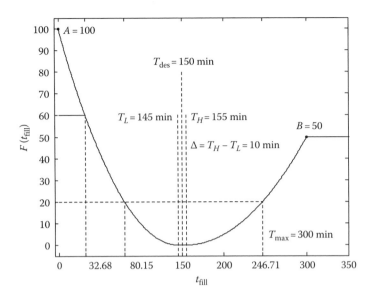

FIGURE 7.32   Graph of objective function $F(t_{fill})$.

The objective function surface is shown in Figure 7.33. It is plotted in the M-file "*Chap7_globe_fill_surface.m*," which loops through a range of F_1 and c values, calling the Simulink model file "*globe.mdl*" to determine the fill time t_{fill}. The simulation terminates when the tank is full, that is, $H(t) = R = 5$ ft, or failing that when the simulated time reaches $T_{max} = 300$ min.

It appears from looking at Figure 7.33 that the surface contains a ridge extending from $c = 0$ to $c = c_{max} = 2$ (with corresponding F_1 values) for which the objective function is zero. Indeed, this is consistent with our intuition, which suggests the likelihood of numerous combinations of F_1 and c yielding a tank fill time between $T_L = 145$ min and $T_H = 155$ min and, thus, $F(F_1, c) = 0$.

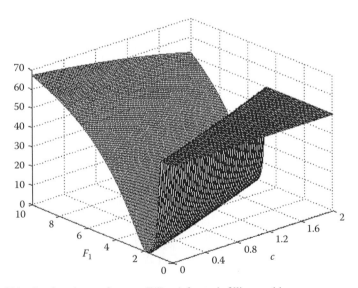

FIGURE 7.33   Objective function surface $z = F(F_1, c)$ for tank-filling problem.

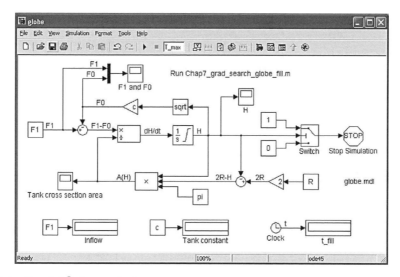

FIGURE 7.34 Simulink® diagram for hemispherical tank-filling simulation.

Another distinguishing characteristic in the surface's topology is the plateau at an elevation of 50 (the value of B) corresponding to points (F_1, c) for which the tank fill time is greater than or equal to 300 min or else the tank never fills. The challenge will be for the gradient search algorithm to find points in parameter space along the aforementioned ridge where the objective function is a minimum, that is, zero.

The Simulink block diagram is shown in Figure 7.34.

The parameters F_1 and c and the tank fill time t_{fill} are visible in Simulink "`display`" blocks. Note the limited integrator with upper limit set to $R = 5$ ft, which also happens to be identical to the threshold parameter of the "`Switch`" block. Consequently, the simulation is halted when the tank is full, that is, the level $H(t) \geq R$.

A variable-step "`ode45 Dormand Prince`" numerical integrator with default tolerance settings is used to control the truncation error. Execution times are reduced by a significant amount compared to one of the RK fixed-step integrators with suitably chosen integration step (see Exercise 7.10).

The gradient search implemented in "*Chap7_grad_search_globe_fill.m*" is outlined in flow chart form in Figure 7.35. It begins with a user-selected starting point (F_1, c) in F_1–c parameter space. Prior to the calculation of the gradient, the point is checked to verify the possibility of the tank filling up. At steady state, we know from Equation 7.76 that

$$(F_1)_{ss} - c(H_{ss})^{1/2} = 0 \tag{7.78}$$

and, therefore, imposing the constraint

$$F_1 > cR^{1/2} \tag{7.79}$$

guarantees that the water will eventually attain a level of $R = 5$ ft, although not necessarily in less than $T_{\text{max}} = 300$ min. If the initial point fails to satisfy the inequality in Equation 7.79, the initial inflow F_1 is adjusted according to

$$F_1 = \min \{1.5cR^{1/2}, (F_1)_{\text{max}}\} \tag{7.80}$$

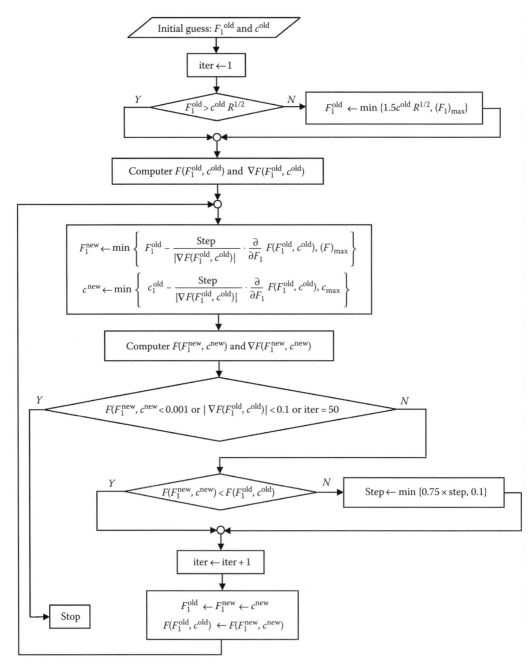

FIGURE 7.35 Flow chart for gradient search algorithm.

The remaining blocks in the flow chart are for computing the objective function, the gradient vector, and for determining how big a step to take in the negative gradient direction in searching for a minimum, that is, points where $F(F_1, c) = 0$.

The so-called steepest descent gradient searches (Wilde 1964) look for the optimum distance to travel in the negative gradient direction before changing directions. The optimum distance is determined by the local minimum of the objective function along the negative gradient direction. When the local minimum is reached, the gradient vector is recalculated, and the search proceeds in the new direction that happens to be orthogonal to the previous search direction. Hence, with steepest descent as described, the search consists of a sequence of orthogonal moves from point to point. The distance between consecutive points varies, generally decreasing as the optimum is approached.

The gradient search illustrated in Figure 7.35 is not of the steepest descent type; rather, it consistently takes a single step in the negative gradient direction from one point to the next and then recomputes the gradient vector.

The magnitude of the step is altered based on a comparison of the objective function at neighboring points, that is, after taking a full step, if the new objective function is greater than the previous value, the step size is reduced by 25% next time around. A lower threshold on step size is imposed to prevent the search from "slowing down to a crawl." Compared to steepest descent, the steps are either too small or too large, and the search will require more gradient calculations. Even worse, the new gradient direction may steer the search away from the minimum altogether, and the method fails to converge.

The search is terminated using a stop condition based on the magnitude of the gradient vector, the value of the objective function, and the number of steps taken. After considerable experimentation, the tolerances were chosen to stop the search if

$$|\nabla F(F_1, c)| = \left\| \begin{bmatrix} \dfrac{\partial}{\partial F_1} F(F_1, c) \\[2mm] \dfrac{\partial}{\partial c_1} F(F_1, c) \end{bmatrix} \right\| \leq 0.1 \quad \text{or} \quad F(F_1, c) \leq 0.001 \quad \text{or} \quad \# \text{ steps} = 50 \qquad (7.81)$$

The gradient vector $\nabla F(F_1, c)$ is calculated numerically using a central difference approximation formula, namely,

$$\frac{\partial}{\partial F_1} F(\overline{F}_1, \overline{c}) = \frac{F(\overline{F}_1 + \Delta F_1, \overline{c}) - F(\overline{F}_1 - \Delta F_1, \overline{c})}{2\Delta F_1} \qquad (7.82)$$

$$\frac{\partial}{\partial c} F(\overline{F}_1, \overline{c}) = \frac{F(\overline{F}_1, \overline{c} + \Delta c) - F(\overline{F}_1, \overline{c} - \Delta c)}{2\Delta c} \qquad (7.83)$$

where the deviations ΔF_1 and Δc are 0.005 and 0.01, respectively. The gradient vector is computed by calling the MATLAB function "gradF_globe.m" from the M-file "Chap7_grad_search_globe_fill.m" with arguments F_1 and c. The components in Equations 7.82 and 7.83 are returned as outputs.

Results of successful gradient searches starting from randomly chosen starting points in the region $0 \leq F_1 \leq (F_1)_{\max} = 10$, $0 \leq c \leq c_{\max} = 2$ of F_1–c parameter space are shown in Table 7.6. The search failed to locate the minimum on a few occasions.

A different approach to finding the optimum points located along the ridge in Figure 7.33 is to plot the $F(F_1, c) = 0$ contour. Other contours $F(F_1, c) = F_0$, (F_0 constant) can be plotted as well by searching for points in the F_1–c plane, which result in filling times corresponding to the required contour values. Figure 7.32 shows the two filling times that result in $F(F_1, c) = 20$ and the single filling time that leads to $F(F_1, c) = 60$.

With $F_0 = 20$, the next step is fixing the parameter c and varying F_1 until the two values that lead to $t_{\min} = 80.15$ min and $t_{\text{fill}} = 246.71$ min are found. The search for F_1 is constrained to the interval $(F_1)_{\min} \leq F_1 \leq (F_1)_{\max}$, where $(F_1)_{\min}$ is the minimum flow needed to fill the hemispherical tank and is given by

$$(F_1)_{\min} = cR^{1/2} \qquad (7.84)$$

where c is the current fixed value. In other words, points $\{(c, F_1)|F_1 < (F_1)_{\min}\}$ are infeasible and not searched. The process is repeated for c ranging from $c_{\min} = 0$ to $c_{\max} = 2$ ft^3/min/ft$^{1/2}$.

TABLE 7.6
Summary of Gradient Search Results for Five Starting Points in F_1–c Plane

	#1	#2	#3	#4	#5
$(F_1)_{start}$	2.026	9.218	1.762	0.578	8.131
c_{start}	1.344	1.476	0.811	0.705	0.019
$F[(F_1)_{start}, c_{start}]$	1.756	52.922	6.776	17.013	60.377
$\nabla F[(F_1)_{start}, c_{start}]$	$\begin{bmatrix} 11.293 \\ -21.173 \end{bmatrix}$	$\begin{bmatrix} 6.038 \\ -10.780 \end{bmatrix}$	$\begin{bmatrix} -46.318 \\ 86.859 \end{bmatrix}$	$\begin{bmatrix} -96.765 \\ 81.472 \end{bmatrix}$	$\begin{bmatrix} 4.288 \\ -7.400 \end{bmatrix}$
$\lvert \nabla F[(F_1)_{start}, c_{start}] \rvert$	23.996	12.356	98.437	205.659	8.553
$(F_1)_{opt}$	5.407	5.279	2.777	2.381	5.393
c_{opt}	1.995	1.898	0.558	0.345	1.995
t_{fill}	153.5	147.8	146.3	147.5	155.0
$F[(F_1)_{opt}, c_{opt}]$	0	0	0	0	0
$\nabla F[(F_1)_{opt}, c_{opt}]$	$\begin{bmatrix} -0.039 \\ 0.073 \end{bmatrix}$	$\begin{bmatrix} 0 \\ 0 \end{bmatrix}$	$\begin{bmatrix} 0 \\ 0 \end{bmatrix}$	$\begin{bmatrix} 0 \\ 0 \end{bmatrix}$	$\begin{bmatrix} -0.530 \\ 1.040 \end{bmatrix}$
$\lvert \nabla F[(F_1)_{opt}, c_{opt}] \rvert$	0.083	0	0	0	1.167
Iterations	8	17	3	9	10

A similar process occurs when $B \leq F_0 \leq A$, except in this case, there is a single value of fill time corresponding to F_0 (see Figure 7.32). For $F(F_1, c) = F_0 = 60$, the fill time is 32.68 min. The $F_0 = 20$ and $F_0 = 60$ contours are shown in Figure 7.36. The $F_0 = 0$ contour is also shown. The upper portion corresponds to fill times of $t_{fill} = T_L = 145$ min, and the lower segment is for $t_{fill} = T_H = 155$ min. (see M-file "*Chap7_Fig3_23.m*.") Note that only three values of the parameter c were used, namely, c_{min}, $(c_{min} + c_{max})/2$, and c_{max}, when searching for the corresponding value of F_1 because the contours appear to be linear.

The M-file "*Chap7_globe_contours.m*" can be used to draw the contours ranging from $F_0 = 0$ up to a maximum value $(F_0)_{max}$ corresponding to $c = c_{min} = 0$ and $F_1 = (F_1)_{max} = 10$ ft³/min. The contour

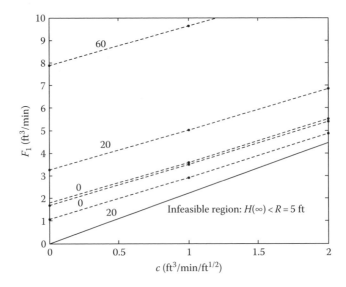

FIGURE 7.36 Objective function contours $F_0 = 0, 20, 60$.

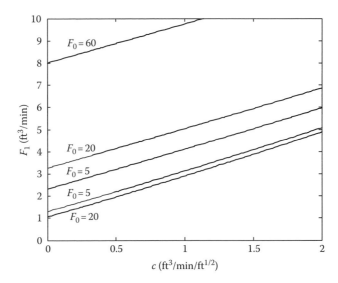

FIGURE 7.37 Graph of several contours of objective function $F(F_1, c)$.

for $F_0 = (F_0)_{max}$ is a single point located at $(0,10)$ in Figure 7.36. "*Chap7_globe_contours.m*" reports the value of $(F_0)_{max}$ along with the corresponding fill time, which happens to be the shortest time in which the tank can be filled. From Figure 7.33, $(F_0)_{max}$ appears to be approximately 68.

MATLAB will also draw the objective function contours. The statements

```
v = [5 20 60];
contour(cc,F11,z,v)
```

in "*Chap7_Fig3_24.m*" produce the contours corresponding to objective function values of 5, 20, and 60 shown in Figure 7.37. There is substantial agreement between the contours in Figures 7.36 and 7.37.

7.3.5 Optimization of Simulink® Discrete-Time System Models

We conclude this section with a simplified model of hospital–patient occupancy (McClamroch 1980) using Simulink to simulate the dynamics. The goal will be to investigate the relationship between the average number of scheduled patients per day on the hospital's utilization of existing capacity. Stochastic systems of this nature, where entities arrive in nondeterministic fashion requiring services of random duration at different stages, are typically studied using discrete-event simulation (Banks 2005). Popular programs for simulating systems of this nature are Process Model (Evans) and ARENA (Kelton 1997).

While Simulink may not be the ideal program to simulate the dynamics of patients flowing through a hospital's facilities, a macroscopic discrete-time system model that captures some of the important features is still possible. In the model to be formulated, the basic unit of discrete-time is a day.

The types of daily arrivals and departures from the hospital are accounted for by

e_i = number of emergency arrivals on $(i + 1)st$ day
s_i = number of scheduled arrivals on $(i + 1)st$ day
d_i = number of departures on $(i + 1)st$ day
m_i = number of deaths on $(i + 1)st$ day

Letting x_i denote the number of occupied beds at the end of the ith day and L the total number of beds, a simple model describing the hospital's daily occupancy is

$$x_{i+1} = \text{Min}\{L, x_i + u_i\}, \quad i = 0, 1, 2, 3, \ldots \tag{7.85}$$

where $u_i = s_i + e_i - d_i - m_i$. The components of u_i in Equation 7.85 are assumed to be normally distributed, that is,

$$s_i \sim N(\mu_S, \sigma_S^2), \quad e_i \sim N(\mu_E, \sigma_E^2), \quad d_i \sim N(\mu_D, \sigma_D^2), \quad m_i \sim N(\mu_M, \sigma_M^2)$$

where μ_S, μ_E, μ_D, μ_M and $\sigma_S^2, \sigma_E^2, \sigma_D^2, \sigma_M^2$ are the respective means and variances.

Typical sequences of u_i and x_i are shown in Figure 7.38. Note that u_i represents a summation of input components (arrivals and departures) during the $(i + 1)st$ day.

A Simulink block diagram of the nonlinear, first-order, discrete-time system is shown in Figure 7.39.

In addition to generating the input components and implementation of the difference equation, additional blocks are used to decompose the input "u(i)" into two series, called "u(i) > 0" and "u(i) < 0." The first series "u(i) > 0" is the subset of positive values in "u(i)" corresponding to days where the number of new patients exceeds the number of patients discharged or who have died. At the end of those days, the hospital's occupancy either increases (relative to the previous day) or else remains constant at its capacity.

The second series "u(i) < 0" is the subset of negative values in "u(i)" corresponding to days when the number of discharged and dying patients surpasses the number of arrivals and the hospital's occupancy at the end of the day is diminished from the previous day.

Note also the presence of two Simulink "switch" blocks feeding "scopes" labeled "delta (i) > 0" and "delta(i) < 0." The former outputs a time series showing on which days and by how much the demand for beds exceeds the hospital's capacity. The numerical values represent the overflow demand, that is, the amount of additional beds required to accommodate the influx of additional patients. The scope labeled "delta(i) < 0" shows the days when the hospital is operating at less than capacity and by how much.

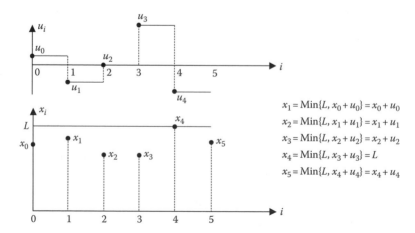

FIGURE 7.38 Illustration of discrete-time input and output relationship in Equation 7.85.

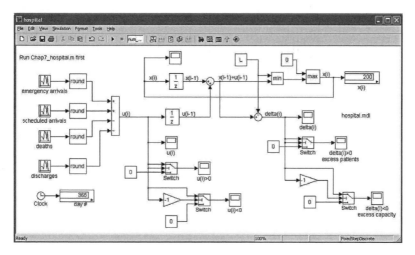

FIGURE 7.39 Simulink® diagram of hospital occupancy.

Typical profiles for "u(i) > 0," "u(i) < 0," "delta(i) > 0," and "delta(i) < 0" are shown in Figures 7.40 through 7.43 for the case where the average number of admissions exceeds the average number of discharges plus deaths.

Figures 7.44 and 7.45 show results of a single run for 100 days under the following conditions:

$$\mu_S = 21, \quad \sigma_S^2 = 4, \quad \mu_E = 5, \quad \sigma_E^2 = 2$$
$$\mu_D = 23, \quad \sigma_D^2 = 9, \quad \mu_M = 2, \quad \sigma_M^2 = 0.25$$
$$L = 200, \quad x_0 = 200$$

Hospital occupancy fluctuates between 195 and 200 corresponding to occupancy rates ranging from 97.5% to 100%. The high occupancy rates are consistent with the condition $\mu_S + \mu_E > \mu_D + \mu_M$, that is, the average daily arrival of new patients is greater than the average number of patients leaving the hospital. From Figure 7.45, it is clear there are a number of days when patients may have been scheduled for admittance but were not admitted. (Keep in mind the simplistic nature of the model that does not account for the hospital's ability to accommodate excess patients.)

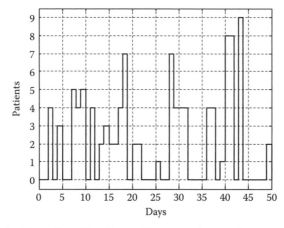

FIGURE 7.40 Typical "u(i) >0" profile—Days with excess of new patient.

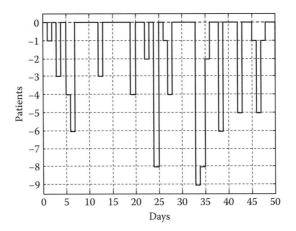

FIGURE 7.41 "$u(i) < 0$" profile—Excess discharged and dying patients.

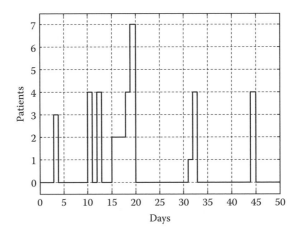

FIGURE 7.42 Typical "$delta(i) > 0$" profile—Days when capacity exceeded.

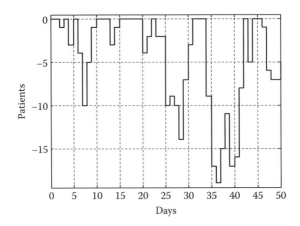

FIGURE 7.43 Typical "$delta(i) < 0$" profile—Days at less than capacity.

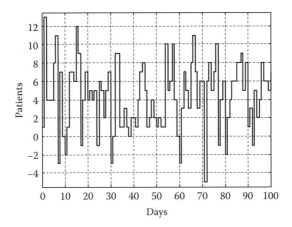

FIGURE 7.44 Daily net patient input.

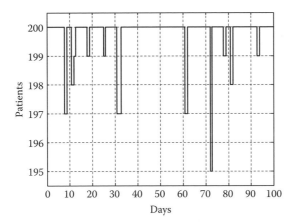

FIGURE 7.45 Hospital occupancy for 100 days.

A Monte Carlo simulation can be performed to investigate the effect of scheduled arrivals on hospital utilization (occupancy rate) and the number of patients turned away due to lack of beds. "*Chap7_hospital.m*" is an M-file, which varies μ_S, the mean number of scheduled arrivals, and computes a number of performance measures based on 10 simulated records, each containing 1 year (365 days) of information. That is, for a given value of μ_S, 365 days of operation are simulated. The initial occupancy is reset to $x_0 = L$, and the process repeated nine more times. The remaining system parameters are fixed at the baseline values previously given.

A number of performance measures are computed for each value of μ_S:

1. An objective function that accounts for the days when the hospital is unable to accept new patients due to excess demand, that is, "delta(i)>0," and other days when the hospital operates at less than capacity, that is, "delta(i)<0." It is a weighted average over all 10 records given by

$$F(\mu_S) = c_1 \left\{ \frac{1}{10} \sum_{j=1}^{10} \left[\frac{1}{365} \sum_{\substack{i=1 \\ \Delta(i)>0}}^{365} \Delta(i) \right]_j \right\} + c_2 \left\{ \frac{1}{10} \sum_{j=1}^{10} \left[\frac{1}{365} \sum_{\substack{i=1 \\ \Delta(i)<0}}^{365} |\Delta(i)| \right]_j \right\} \qquad (7.86)$$

where c_1 and c_2 are the weights applied to the average number of excess patients per day and the average number of unused beds per day, respectively.

2. The percent occupancy averaged over all 10 records (10×365 days).
3. The average number of excess patients per day averaged over all 10×365 days.
4. The average excess capacity (unused beds) per day averaged over all 10×365 days.

Results are shown in Figure 7.46 for $c_1 = c_2 = 1$.

Note the steep decline in objective function until $\mu_S = 20$, which represents an equilibrium condition in the sense that new arrivals and departures are balanced (on average), that is $\mu_S + \mu_E = \mu_D + \mu_M$. Choosing $c_1 = c_2 = 1$ implies that an unused bed and a nonadmitted patient have equal importance.

A few points to consider as we conclude this section are as follows:

1. Can you explain why the graphs of the objective function and the average excess capacity are nearly identical?
2. What should the hospital's admitting policy with respect to scheduled number of arrivals be to assure 100% occupancy rates?
3. What will happen to the objective function as the mean number of arrivals continues to increase beyond 25 as shown in Figure 7.46.
4. What is the implication of changing the weight c_2 from 1 to 5 and what effect will it have on the objective function?
5. What is the significance of setting $\sigma_S^2 = 0$?
6. Is there a difference between simulating ten 1 year periods and one 10 year period as far as the Monte Carlo simulation is concerned?

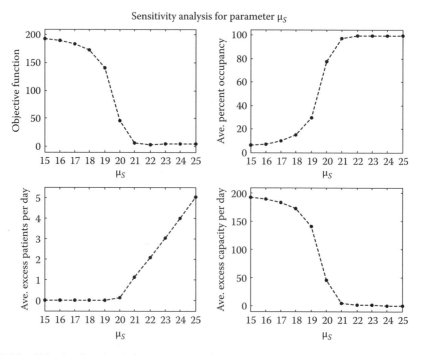

FIGURE 7.46 Objective function ($c_1 = c_2 = 1$) and other performance measures.

EXERCISES

7.5 Suppose the movement of the target in Figure 7.14 is along the circular path of radius $R = 2.5$ mi with speed given by

$$v(s) = V + v_0 \sin\left(\frac{2\pi s}{s_0}\right), \quad s \geq 0$$

where s is the distance traveled along the circular trajectory. The mean speed V and amplitude v_0 are uniformly distributed according to

$$V \sim U(20, 40\,\text{mph}), \quad v_0 \sim U(0, 10\,\text{mph})$$

and the period $s_0 = 2000$ ft. The projectile's dynamics are defined by the parameters $m = 4000/32.2$ slugs, $\mu = 0.15$ lb s/ft, and $v_p(0) = 60$ mph.

Use the MATLAB random number generator to generate values for V and v_0.

(a) Simulate single firings of the projectile corresponding to firing angles of $\theta = 0°, 5°, 10°, \ldots,$ 90°. Halt the simulation when the projectile has traveled a distance greater than R mi. Plot the miss distance (minimum separation between target and projectile) vs. the firing angle.

(b) From the graph in part (a), estimate the optimum firing angle, that is, the one that results in the projectile striking the target.

(c) Using the same values of V and v_0, write an M-file to find the optimum firing angle. The use of MATLAB's optimization toolbox is optional.

(d) Verify the result in part (c) by simulation.

7.6 A projectile of mass m is fired with initial velocity v_0 at an angle α_0 from the horizontal direction. Its position, while in flight, is given by coordinates (x, y), and its velocity is represented by v as shown in Figure E7.6a. The projectile is subject to a linear drag force f_D in the tangential direction and a constant gravitational force in the vertical direction.

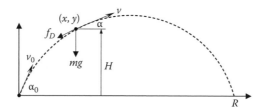

FIGURE E7.6a

The equations of motion are

$$\frac{d^2x}{dt^2} = -\frac{f_D}{m}\cos\alpha, \quad \frac{d^2y}{dt^2} = -\frac{f_D}{m}\sin\alpha - g$$

$$f_D = \mu|v| = \mu\left[\left(\frac{dx}{dt}\right)^2 + \left(\frac{dy}{dt}\right)^2\right]^{1/2}$$

$$\tan\alpha = \frac{dy/dt}{dx/dt}$$

Baseline values of the system parameters are

$$m = 0.25 \text{ slugs}, \quad v_0 = 500 \text{ ft/s}, \quad c = 0.015 \text{ lb s/ft}, \quad \alpha_0 = 45°$$

Use the Simulink diagram as shown in Figure E7.6b or construct your own Simulink model to answer the following questions.

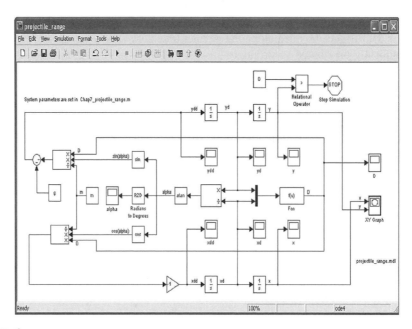

FIGURE E7.6b

(a) The projectile is fired in the vertical direction. Find
   (i) The analytical solution $y(t)$, $t \geq 0$ and plot it for the time period when $y \geq 0$
   (ii) The peak altitude H attained by the projectile
   (iii) The time t_p when the peak altitude is reached

(b) Find the maximum altitude H by optimization, that is, search for the time when $-y(t)$ is a minimum. Compare the result with your answer in part (a).

(c) Find the horizontal distance R corresponding to firing angles $\alpha_0 = 0°, 5°, 10°, \ldots, 90°$ and plot the results. Estimate the firing angle α_0 that maximizes R.

(d) Find the value $(\alpha_0)^{\text{opt}}$ that maximizes $R(\alpha_0)$ by the minimization of the objective function $F(\alpha_0) = -R(\alpha_0)$ subject to $0° \leq \alpha_0 \leq 90°$.

(e) Find the initial velocity v_0 that results in a peak altitude of $H = 1200$ ft when the projectile is fired at an angle of $45°$. Formulate this as an optimization problem and then find the optimum solution.

(f) Let $\overline{H} = 1500$ ft and $\overline{R} = 2000$ ft be the design values of peak altitude and down range distance. The objective is to find combinations of initial firing angles and initial velocities resulting in $H = \overline{H}$ and $R = \overline{R}$. Choose the objective function to be minimized as

$$F(\alpha_0, v_0) = \left(e_H^2 + e_R^2\right)^{1/2} = [(H - \overline{H})^2 + (R - \overline{R})^2]^{1/2}$$

and plot the surface $F(\alpha_0, v_0)$ as well as several equally spaced (in numerical value) contours for $0° \leq \alpha_0 \leq 90°$, $0 \leq v_0 \leq 600$ ft/s.

(g) Write an optimization program that starts from an initial point (α_0, v_0) in parameter space and locates an optimum point $\{(\alpha_0)^{opt}, (v_0)^{opt}\}$ where the objective function $F(\alpha_0, v_0)$ is a minimum. Fill in the following table:

Initial $\alpha_0(°)$	Initial v_0 (fps)	$(\alpha_0)^{opt}$	$(v_0)^{opt}$	Max H	R	$F[(\alpha_0)^{opt}, (v_0)^{opt}]$	Number of Iterations
20	400						
65	100						
45	300						
80	500						
5	200						

7.7 An alternative method for controlling the heading of a ship is to use rate feedback as shown in Figure E7.7. The saturation block is omitted. Using the same parameter values as in the text,

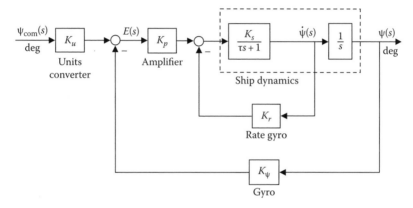

FIGURE E7.7

(a) Find the amplifier gain K_p and rate gyro gain K_r, which minimizes the ISE in response to a step command heading of 5°.
(b) Repeat part (a) for an IAE objective function.

7.8 Repeat the steps in estimating the parameters k and n if the observed chemical concentrations at the end of 5 min intervals are as given in the following table. Draw a graph similar to the one in Figure 7.30 showing simulated and observed values after 1 min intervals for the first hour.

t (min)	0	5	10	15	20	25	30
$\hat{x}$ (mol/L)	2.0000	1.3333	1.0000	0.8000	0.6667	0.5714	0.5000
t (min)		35	40	45	50	55	60
$\hat{x}$ (mol/L)		0.4444	0.4000	0.3636	0.3333	0.3077	0.2857

7.9 Suppose the hemispherical tank shown in Figure 7.31 is turned upside down.
(a) How does the mathematical model of the system change?
(b) Modify the Simulink diagram to reflect the new configuration.

    (c) Show that the surface plot in Figure 7.33 and zero contour plot in Figure 7.36 remain unchanged.

    (d) Repeat parts (a) and (b) and generate new surface and zero contour plots if the tank is cylindrical with radius $R = 5$ ft.

7.10 Compare the execution time required to draw the surface in Figure 7.33 where the objective function is evaluated over a 40-by-40 grid of points in the F_1–c plane when the numerical integrator is a fixed RK-4 integrator with step size 0.01 s and the default variable-step ode45 (Dormand Prince).

Hint: Insert the MATLAB commands "`tic`" and "`toc`" at the beginning and end of the MATLAB statements. The execution time will be returned by "`toc`."

7.11 The objective function defined in Equation 7.77, and shown in Figure 7.32, is modified to

$$F(t_{\text{fill}}) = \begin{cases} A, & 0 \leq t_{\text{fill}} < T_{\text{min}} \\ A\left(\dfrac{t_{\text{fill}}}{T_L} - 1\right)^2, & T_{\text{min}} \leq t_{\text{fill}} < T_L \\ 0, & T_L \leq t_{\text{fill}} \leq T_H \\ B\left(\dfrac{t_{\text{fill}} - T_H}{T_{\text{max}} - T_H}\right)^2, & T_H \leq t_{\text{fill}} < T_{\text{max}} \\ B, & T_{\text{max}} < t_{\text{fill}} \end{cases}$$

where T_{min} is the shortest time possible for a hemispherical tank with radius $R = 7.5$ ft to fill. The controllable parameters are confined to the ranges $0 \leq c \leq 4$ ft^3/min/ft$^{1/2}$ and $0 \leq F_1 \leq 20$ ft^3/min. The end points where the objective function is zero are $T_L = 190$ min and $T_H = 210$ min. Finally, $T_{\text{max}} = 500$ min. The numerical values of A and B, the limiting values of the objective function, are $A = 80$ and $B = 40$.

    (a) Find T_{min}.

    (b) Generate a new surface plot similar to the one shown in Figure 7.33 for the region $0 \leq F_1 \leq 20$, $0 \leq c \leq 4$.

    (c) Modify the objective function definition in "*Chap7_globe_contours.m*," and plot the contours corresponding to objective function values $0, 10, 20, \ldots, (F_0)_{\text{max}}$ where $(F_0)_{\text{max}}$ is the objective function value corresponding to a fill time of T_{min}.

    (d) Find several optimum points $\left(F_1^{\text{opt}}, c^{\text{opt}}\right)$ on the $F(F_1, c) = 0$ contour.

    (e) Run a simulation of the globe filling with $F_1 = F_1^{\text{opt}}$ and $c = c^{\text{opt}}$ from part (d) and verify that the fill time falls between T_L and T_H.

7.12 Write a program to implement a gradient-based search algorithm to find a point (F_1, c) where the objective function is zero. Test the algorithm starting from

    (a) $F_1 = 4$ ft^3/min, $c = 1$ ft^3/min/ft$^{1/2}$

    (b) $F_1 = 7.5$ ft^3/min, $c = 2$ ft^3/min/ft$^{1/2}$

    (c) $F_1 = 1$ ft^3/min, $c = 0$ ft^3/min/ft$^{1/2}$

7.13 For the hospital occupancy model, do a Monte Carlo simulation and plot the objective function $F(\mu_S)$ for $\mu_S = 15, 16, \ldots, 30$ with weights $c_1 = 1$, $c_2 = 5$. Use baseline values given in the text for the system parameters. Assume the hospital is initially operating at full occupancy.

7.14 Suppose the hospital has a holding facility where new patients wait for a bed to become available. Let the state variables in the discrete-time model be $x_B(i)$, the number of patients in rooms with beds at the end of the ith day, and $x_H(i)$, the number waiting for an assigned bed in the holding area at the end of the ith day. Patients are transferred from the holding area to a room with a bed on days when the number of emergency and scheduled arrivals is less than the number of discharged and dying patients. The number of beds is L_B; the holding area can accommodate L_H patients.

Repeat the Monte Carlo simulation described in the text using the baseline values of the system parameters and $L_H = 15$. Plot graphs similar to the ones in Figure 7.46. The weights are $c_1 = c_2 = 1$. Note that the occupancy rate is based on the number of patients with beds, that is, with $L_B = 200$ and $L_H = 15$, the occupancy rate is 90% if $x_B = 180$, $x_H = 0$, and 100% if $x_B = 200$, $x_H = 5$.

7.15 Investigate the effect of variability in the number of scheduled arrivals on the hospital's occupancy rate. Choose the mean $\mu_S = 21$ scheduled patients per day and simulate the percent occupancy as a function of the standard deviation σ_S where σ_S ranges from zero to three scheduled patients per day.

7.16 Use Monte Carlo simulation to obtain an empirical probability density function for random variable Y, the hospital's percent occupancy. Use the following values for the system parameters:

$$\mu_S = 24, \quad \sigma_S^2 = 9, \quad \mu_E = 6, \quad \sigma_E^2 = 4, \quad \mu_D = 28, \quad \sigma_D^2 = 9,$$
$$\mu_M = 2, \quad \sigma_M^2 = 0.25, \quad L = 200, \quad x_0 = 200$$

Hint: Simulate 100 records of sufficient length (in days) to obtain 100 observations y_1, y_2, $y_3, \ldots, y_{100}$ where y_i, $i = 1, 2, 3, \ldots, 100$ is the percent occupancy corresponding to the ith record. Plot the results

7.17 Consider a loan in the amount of P dollars to be repaid in n equal monthly installments of A dollars with interest at i per month. The unpaid balance P_k made after the kth payment is given by

$$P_k = P_{k-1} + iP_{k-1} - A = P_{k-1}(1 + i) - A, \quad k = 1, 2, \ldots, n$$

A Simulink block diagram is shown in Figure E7.17. Note that the loan amount P is the initial condition of the "Unit Delay" block and the simulation stop time is set to n. Also, be sure to set the "Solver options Type" to "Fixed-step," "Fixed-step size" to 1, and the integrator to "discrete no continuous states."

The terms of a car loan are $P = \$30{,}000$, $n = 48$ months, and $i = 0.005$ (0.5% per month). For a fixed value of monthly payment A, the unpaid balance at the end of the loan period is P_{48}. Positive values of P_{48} means A is too low and the loan has not been paid off in its entirety. A negative value of P_{48} implies A is too much and overpayment of the loan has occurred. The correct amount of the monthly payment A to retire the loan after the last (48th) payment is the value of A for which the unpaid balance at the end of the loan period is zero.

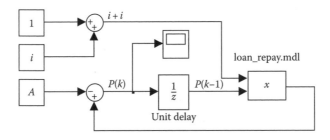

FIGURE E7.17

(a) Prepare a graph of P_{48} vs. A, for $A = \$600, \$625, \$650, \ldots, \$800$. Estimate the correct value of A to repay the loan.

(b) Write your own or use MATLAB's optimization toolbox to determine the correct A by finding the value of A, which minimizes the objective function P_{48}.

(c) Plot P_k vs. k, $k = 0, 1, 2, 3, \ldots, 48$ using the value of A found in part (b). Compare your answer for A with the correct value of A, which can be obtained from the formula

$$A = P\left[\frac{i(1+i)^n}{(1+i)^n - 1}\right]$$

7.4 LINEARIZATION

Chapter 4 introduced a number of important concepts instrumental in analyzing the behavior of linear systems. By linear systems, we are referring to actual systems modeled by linear algebraic and differential equations. Real-world systems are inherently nonlinear. However, in certain regions, they may respond in a way that a linear model provides an acceptable representation of the system's dynamics. Whenever we employ linear models to describe nonlinear systems, it must be with the understanding that the system remains within its so-called linear-operating region.

Consider the simple mechanical spring shown in Figure 7.47. Its deflection x from equilibrium depends on the magnitude and direction of the applied force F.

Measurements of deflection and force over a range of forces resulting in fracture from excessive compression or elongation produce a graph like the one shown in Figure 7.48.

The linear region of the spring is the section of the operating characteristic where x is proportional to F. Known as Hooke's law, the familiar form is

$$F = kx \tag{7.87}$$

where k is the spring constant, a measure of its stiffness. The linear model in Equation 7.87 is a valid model of the spring provided the applied force is confined to $F_1 \leq F \leq F_2$.

Numerous components behave in a similar fashion. The current in an electrical resistor is assumed proportional to the voltage across its terminals over a range of currents. Conductive heat flow due to a temperature difference between two points and fluid flow caused by pressure differences at different locations are additional examples of cause-and-effect relationships assumed to be linear over a range of operating conditions.

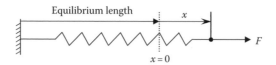

FIGURE 7.47 Deflection of a mechanical spring subjected to an applied force.

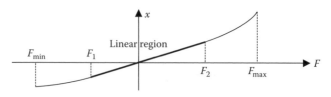

FIGURE 7.48 Operating characteristic of spring showing its linear region.

In the example of the spring, the static operating curve shown in Figure 7.48 can be divided into three distinct regions, that is, points $\{F, x(F)\}$ where

1. $F_{min} < F < F_1$
2. $F_1 \leq F \leq F_2$ (linear region)
3. $F_2 < F < F_{max}$

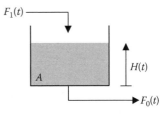

FIGURE 7.49 A liquid tank with input $F_1(t)$ and dependent variables $F_0(t)$ and $H(t)$.

The relation $x = x(F)$ between force and displacement in each region is based on empirical observation as opposed to an analytical model or equation based on scientific principles or natural laws. In contrast, the liquid tank with incompressible fluid shown in Figure 7.49 is modeled by the linear first-order differential equation based on conservation of volume,

$$A\frac{dH}{dt} + F_0 = F_1 \tag{7.88}$$

along with the operating characteristic of the tank, that is, the relationship between the out flow and the liquid level, which applies in both the steady state and otherwise.

$$F_0 = F_0(H) = cH^{1/2}, \quad H \geq 0 \tag{7.89}$$

Equation 7.89 is based on Bernoulli's principle from Physics. The constant c depends on the physical properties of the fluid, tank, and the discharge line.

Equation 7.88 was derived in Section 1.2. The discharge F_0 was assumed proportional to H, that is, $F_0(H) = cH$, resulting in a linear system model of the tank. A "real" tank is nonlinear by virtue of Equation 7.89.

7.4.1 DEVIATION VARIABLES

A linearized tank model can be obtained to approximate the nonlinear tank dynamics. The technique relies upon the concept of an operating point and deviation variables. To illustrate, let us suppose a linearized tank model is required, which provides a reasonable approximation to the nonlinear system provided the inflow, level, and outflow vary only slightly from the steady-state values $\overline{F}_1, \overline{H}, \overline{F}_0$ shown in Figure 7.50.

The operating point, for purposes of linearization, is characterized by an inflow $\overline{F}_1$ and the point $(\overline{H}, \overline{F}_0)$ where

$$\overline{F}_0 = c\overline{H}^{1/2} \tag{7.90}$$

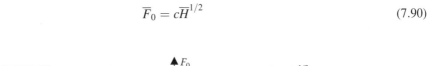
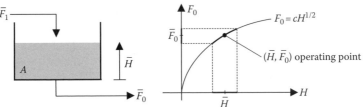

FIGURE 7.50 Operating point $(\overline{H}, \overline{F}_0)$ for tank linearization.

With steady-state conditions at the operating point, $\overline{F}_0 = \overline{F}_1$. From Equation 7.90,

$$\overline{H} = \frac{\overline{F}_0^2}{c^2} = \frac{\overline{F}_1^2}{c^2} \tag{7.91}$$

When the inflow $F_1(t)$ and outputs $H(t)$ and $F_0(t)$ differ from their operating point values, deviation variables $\Delta F_1(t)$, $\Delta H(t)$, and $\Delta F_0(t)$ are introduced according to

$$F_1(t) = \overline{F}_1 + \Delta F_1(t), \quad H(t) = \overline{H} + \Delta H(t), \quad F_0(t) = \overline{F}_0 + \Delta F_0(t) \tag{7.92}$$

Deviation variables relate the differences between actual values of the system variables and their operating point levels, that is,

$$\Delta F_1(t) = F_1(t) - \overline{F}_1, \quad \Delta H(t) = H(t) - \overline{H}, \quad \Delta F_0(t) = F_0(t) - \overline{F}_0 \tag{7.93}$$

Expanding F_0 in Equation 7.89 in a Taylor Series about the operating point $(\overline{H}, \overline{F}_0)$,

$$F_0 = \overline{F}_0 + \frac{\mathrm{d}}{\mathrm{d}H} F_0(H) \bigg|_{H=\overline{H}} (H - \overline{H}) + \frac{\mathrm{d}^2}{\mathrm{d}H^2} F_0(H) \bigg|_{H=\overline{H}} (H - \overline{H})^2 + \cdots \tag{7.94}$$

$$\Rightarrow F_0 - \overline{F}_0 = \Delta F_0 = \frac{\mathrm{d}}{\mathrm{d}H} F_0(H) \bigg|_{H=\overline{H}} \Delta H + \frac{\mathrm{d}^2}{\mathrm{d}H^2} F_0(H) \bigg|_{H=\overline{H}} \Delta H^2 + \cdots \tag{7.95}$$

If ΔH is small in absolute value, then the ΔH^2 term and all succeeding terms are higher order terms that can be ignored (to a first-order approximation). The result is a first-order Taylor Series approximation for the deviation flow ΔF_0, namely,

$$\Delta F_0 \approx \frac{\mathrm{d}}{\mathrm{d}H} F_0(H) \bigg|_{H=\overline{H}} \Delta H = F_0'(\overline{H}) \Delta H \tag{7.96}$$

Differentiating Equation 7.89 to find the first derivative $F_0'(H)$ and evaluating the result at $H = \overline{H}$ lead to

$$\Delta F_0 \approx \frac{1}{2} c \overline{H}^{-1/2} \Delta H \tag{7.97}$$

where the accuracy depends on the magnitude of ΔH (more about this point later).

Substituting expressions in Equation 7.92 for $F_1(t)$, $H(t)$, and $F_0(t)$ into Equation 7.88 gives

$$A \frac{\mathrm{d}}{\mathrm{d}t} [\overline{H} + \Delta H(t)] + \overline{F}_0 + \Delta F_0(t) = \overline{F}_1 + \Delta F_1(t) \tag{7.98}$$

$$\Rightarrow A \frac{\mathrm{d}}{\mathrm{d}t} \overline{H} + A \frac{\mathrm{d}}{\mathrm{d}t} \Delta H(t) + \overline{F}_0 + \Delta F_0(t) = \overline{F}_1 + \Delta F_1(t) \tag{7.99}$$

Knowing $\overline{F}_0 = \overline{F}_1$ and the fact that $A(\mathrm{d}/\mathrm{d}t)\overline{H} = 0$ leads to

$$A \frac{\mathrm{d}}{\mathrm{d}t} \Delta H(t) + \Delta F_0(t) = \Delta F_1(t) \tag{7.100}$$

Substituting the approximation in Equation 7.96 for $\Delta F_0(t)$ into Equation 7.100 results in the first-order linearized differential equation model

$$A \frac{d}{dt} \Delta H(t) + F_0'(\overline{H}) \Delta H(t) = \Delta F_1(t) \tag{7.101}$$

The nonlinear-operating characteristic for the tank, Equation 7.89, has been approximated by the linear relationship of Equation 7.96, which can be written as

$$F_0 = \overline{F}_0 + F_0'(\overline{H})(H - \overline{H}) \tag{7.102}$$

Equation 7.102 is the equation of the line tangent to the curve $F_0 = F_0(H)$ at the operating point $(\overline{H}, \overline{F}_0)$. Figure 7.51 illustrates the case when the tank constant $c = 0.5 \text{ ft}^3/\text{min}/\text{ft}^{1/2}$ and the operating point $(\overline{H}, \overline{F}_0) = (9 \text{ ft}, 1.5 \text{ ft}^3/\text{min})$. Note that $(\overline{H}, \overline{F}_0)$ is the origin in a new coordinate system with ΔH in the horizontal direction and ΔF_0 in the vertical direction.

Before we generalize the procedure for linearization of certain types of nonlinearities, we illustrate, through the next example, a case where the nonlinear term in the system model is a product of dependent variables.

Consider the well-stirred tank in Figure 7.52. The temperature of the liquid $T(t)$ as well as its level $H(t)$ is of interest. Accordingly, a second equation is required, one that introduces the additional dependent variable $T(t)$.

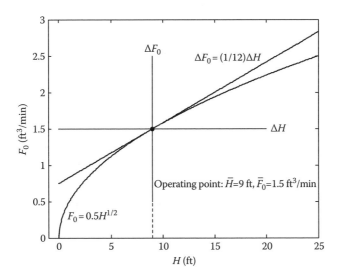

FIGURE 7.51   Nonlinear tank-operating curve and linearized approximation.

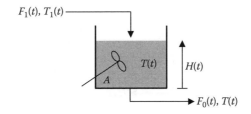

FIGURE 7.52   Stirred tank with inputs $F_1(t)$, $T_1(t)$ and dependent variables $H(t)$, $F_0(t)$, $T(t)$.

The rate at which energy is stored in the liquid holdup is equal to the difference in the rate of energy flowing in and out of the tank. If we substitute the word "mass" for "energy," we have the principle of conservation of mass, which led to the differential equation for the tank level in Equation 7.88. Applying the principle of conservation of energy in equation form gives

$$\frac{d}{dt}(c_p \gamma V T) = c_p \gamma F_1 T_1 - c_p \gamma F_0 T \tag{7.103}$$

where

    $T(t)$ is the uniform liquid temperature in tank, °F
    $F_1(t)$ is the input flow rate, ft^3/min
    $T_1(t)$ is the liquid temperature entering tank, °F
    $F_0(t)$ is the output flow rate, ft^3/min
    V is the volume of liquid in tank, ft^3
    c_p is the specific heat of liquid (Btu/lb-°F)
    γ is the specific weight of liquid (lb/ft^3)

The left-hand side accounts for the energy accumulation, and the right-hand side represents the difference in energy flows in the two streams. Replacing the tank volume V with the product AH in Equation 7.103 results in

$$A\frac{d}{dt}(HT) + F_0 T = F_1 T_1 \tag{7.104}$$

$$\Rightarrow AH\frac{dT}{dt} + AT\frac{dH}{dt} + F_0 T = F_1 T_1 \tag{7.105}$$

Equations 7.88, 7.89, and 7.105 comprise the nonlinear mathematical model of the system. Figure 7.53 illustrates the presence of two inputs (independent variables) and three dependent variables. The state variables are $T(t)$ and either $H(t)$ or $F_0(t)$, but not both since they are related algebraically according to Equation 7.89.

A steady-state operating point is established where $F_1(t) = \overline{F}_1$ and $T_1(t) = \overline{T}_1$ with dependent variables $H(t) = \overline{H}, F_0(t) = \overline{F}_0$, and $T(t) = \overline{T}$. Introducing deviation variables

$$\Delta T = T(t) - \overline{T}, \quad \Delta T_1 = T_1(t) - \overline{T}_1 \tag{7.106}$$

Equation 7.105 becomes

$$A(\overline{H} + \Delta H)\frac{d}{dt}(\overline{T} + \Delta T) + A(\overline{T} + \Delta T)\frac{d}{dt}(\overline{H} + \Delta H) + (\overline{F}_0 + \Delta F_0)(\overline{T} + \Delta T)$$

$$= (\overline{F}_1 + \Delta F_1)(\overline{T}_1 + \Delta T_1) \tag{7.107}$$

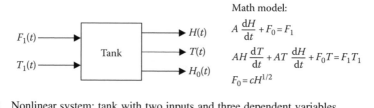

Math model:

$$A\frac{dH}{dt} + F_0 = F_1$$

$$AH\frac{dT}{dt} + AT\frac{dH}{dt} + F_0 T = F_1 T_1$$

$$F_0 = cH^{1/2}$$

FIGURE 7.53 Nonlinear system: tank with two inputs and three dependent variables.

Deviation variables are assumed to be small in magnitude, and, therefore, the products $\Delta H(d/dt)\Delta T$, $\Delta T(d/dt)\Delta H$, $\Delta F_0 \Delta T$, and $\Delta F_1 \Delta T_1$ are negligible by comparison. Equation 7.107 simplifies to

$$A\overline{H}\frac{d}{dt}\Delta T + A\overline{T}\frac{d}{dt}\Delta H + \overline{F}_0\overline{T} + \overline{F}_0\Delta T + \overline{T}\Delta F_0 = \overline{F}_1\overline{T}_1 + \overline{F}_1\Delta T_1 + \overline{T}_1\Delta F_1 \qquad (7.108)$$

Substituting ΔF_0 from Equation 7.96 into Equation 7.108 and rearranging terms give

$$A\overline{H}\frac{d\Delta T}{dt} + A\overline{T}\frac{d\Delta H}{dt} + \overline{F}_0\Delta T + \overline{T}F_0'(\overline{H})\Delta H = \overline{F}_1\overline{T}_1 - \overline{F}_0\overline{T} + \overline{F}_1\Delta T_1 + \overline{T}_1\Delta F_1 \qquad (7.109)$$

Recognizing that $\overline{F}_0 = \overline{F}_1$ and $\overline{T} = \overline{T}_1$ at the steady-state operating point, Equation 7.109 reduces to

$$A\overline{H}\frac{d\Delta T}{dt} + A\overline{T}\frac{d\Delta H}{dt} + \overline{F}_0\Delta T + \overline{T}F_0'(\overline{H})\Delta H = \overline{F}_1\Delta T_1 + \overline{T}_1\Delta F_1 \qquad (7.110)$$

Equations 7.101 and 7.110 are coupled linearized differential equations of the tank. It is left as an exercise problem to show that the state derivatives are expressible as

$$\begin{bmatrix} \dfrac{d}{dt}\Delta H \\[2ex] \dfrac{d}{dt}\Delta T \end{bmatrix} = \begin{bmatrix} -\dfrac{F_0'(\overline{H})}{A} & 0 \\[2ex] 0 & -\dfrac{\overline{F}_0}{A\overline{H}} \end{bmatrix} \begin{bmatrix} \Delta H \\[2ex] \Delta T \end{bmatrix} + \begin{bmatrix} \dfrac{1}{A} & 0 \\[2ex] 0 & \dfrac{\overline{F}_0}{A\overline{H}} \end{bmatrix} \begin{bmatrix} \Delta F_1 \\[2ex] \Delta T_1 \end{bmatrix} \qquad (7.111)$$

Simulation is an effective way to appreciate the limitations of a linearized model. The following example illustrates the point.

Example 7.2

The tank shown in Figure 7.52 with cross-sectional area $A = 100\ \text{ft}^2$ is initially in equilibrium with $\overline{F}_1 = \overline{F}_0 = 25\ \text{ft}^3/\text{min}$, $\overline{H} = 9\ \text{ft}$, and $\overline{T}_1 = \overline{T} = 150°\text{F}$. The input flow and temperature profiles are shown in Figure 7.54.

(a) Simulate the transient response of the nonlinear model when $\alpha = \beta = 0.1$.
(b) Repeat part (a) using the linear state model in Equation 7.111.
(c) Compare the nonlinear and linearized responses and comment on the results.
(d) Find the time constants τ_H and τ_T of the linearized system and show that $\tau_H = 2\tau_T$.
(e) Find expressions for $H(\infty)$ and $T(\infty)$ in response to constant inputs $F_1(t) = \hat{F}_1, t \geq 0$ and $T_1(t) = \hat{T}_1, t \geq 0$ based on the nonlinear model. Compute the numerical values for $H(\infty)$ and $T(\infty)$ when $\hat{F}_1 = (1 + \alpha)\overline{F}_1$ and $\hat{T}_1 = (1 - \beta)\overline{T}_1$.
(f) Repeat part (e) using the linearized model and compare the results.

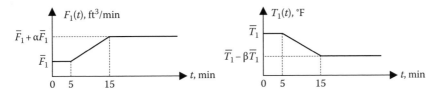

FIGURE 7.54   Tank system input profiles.

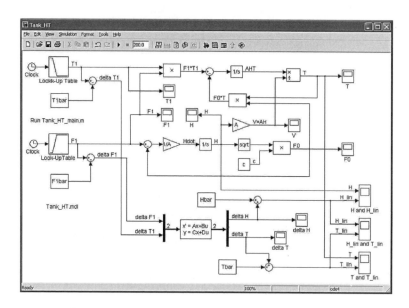

FIGURE 7.55   Simulink® diagram for simulation of nonlinear and linearized system.

(a) Figure 7.55 is a Simulink diagram for simulating the dynamic response of the nonlinear and linearized system models.

    The level H and temperature T of the nonlinear system are plotted in Figure 7.56. The transient period is on the order of 300 min for level and 200 min for temperature.

(b) The linearized system outputs are shown in Figure 7.57.

(c) The nonlinear and linearized transient responses for level and temperature are compared in Figure 7.58. For a 10% increase in inlet flow rate above $\bar{F}_1$ and a 10% decrease in inlet temperature below $\bar{T}_1$, the nonlinear system and linearized system approximation exhibit nearly identical transient responses.

(d) Referring to Equation 7.101, the time constant τ_H of the linearized system is

$$\tau_H = \frac{A}{F_0'(\bar{H})} = \frac{A}{(1/2)c\bar{H}^{-1/2}} = \frac{2A\bar{H}^{1/2}}{c} \tag{7.112}$$

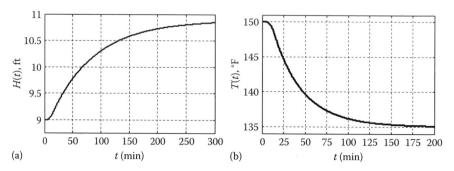

FIGURE 7.56   Nonlinear system response of (a) level and (b) temperature.

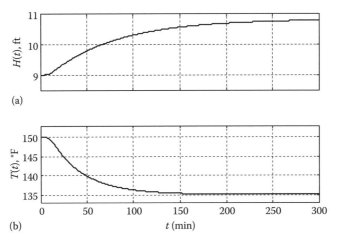

(a)

(b)

FIGURE 7.57 Linearized system (a) level and (b) temperature transient response ($\alpha = \beta = 0.1$).

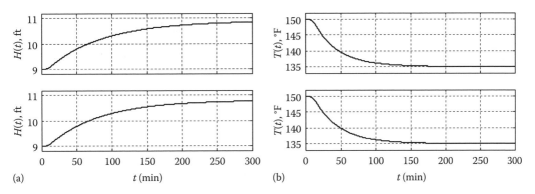

(a) t (min)

(b) t (min)

FIGURE 7.58 Comparison of (a) nonlinear and (b) linearized system transient responses.

Solving for c in Equation 7.90 and substituting the result into Equation 7.112 give

$$\tau_H = \frac{2A\overline{H}^{1/2}}{\overline{F_0}\overline{H}^{-1/2}} = \frac{2A\overline{H}}{\overline{F_0}} = \frac{2(100)(9)}{25} = 72 \text{ min} \tag{7.113}$$

From the second of the two state equations in Equation 7.111, the time constant τ_T is

$$\tau_T = \frac{A\overline{H}}{\overline{F_0}} = \frac{100(9)}{25} = 36 \text{ min} \tag{7.114}$$

$\tau_H = 2\tau_T$ follows directly from Equations 7.113 and 7.114.

(e) At steady state ($t = \infty$), dH/dt and dT/dt are zero. Setting dH/dt equal to zero in Equation 7.88 gives

$$F_0(\infty) = F_1(\infty) = \hat{F}_1 \tag{7.115}$$

According to the tank-operating characteristic (Equation 7.89),

$$F_0(\infty) = c[H(\infty)]^{1/2} \tag{7.116}$$

Solving for $H(\infty)$ in Equation 7.116,

$$H(\infty) = \left[\frac{F_0(\infty)}{c}\right]^2 = \left[\frac{\hat{F}_1}{c}\right]^2 \tag{7.117}$$

Setting $(dH/dt) = (dT/dt) = 0$ in Equation 7.105 gives

$$F_0(\infty)T(\infty) = F_1(\infty)T_1(\infty) \Rightarrow T(\infty) = T_1(\infty) = \hat{T}_1 \tag{7.118}$$

The tank constant c is obtained from the given operating point conditions, that is,

$$c = \frac{\overline{F}_0}{\overline{H}^{1/2}} = \frac{25}{9^{1/2}} = \frac{25}{3} \text{ ft}^3/\text{min}/\text{ft}^{1/2} \tag{7.119}$$

The numerical values of $H(\infty)$ and $T(\infty)$ are

$$H(\infty) = \left[\frac{\hat{F}_1}{c}\right]^2 = \left[\frac{(1+\alpha)\overline{F}_1}{c}\right]^2 = \left[\frac{1.1(25)}{25/3}\right]^2 = 10.89 \text{ ft}$$

$$T(\infty) = \hat{T}_1 = (1-\beta)\overline{T}_1 = 0.9(150) = 135°F \tag{7.120}$$

(f) Setting $(d/dt)\Delta H = (d/dt)\Delta T = 0$ in Equation 7.111 and solving for $\Delta H(\infty)$ and $\Delta T(\infty)$ give

$$\begin{bmatrix} \Delta H(\infty) \\ \Delta T(\infty) \end{bmatrix} = -\begin{bmatrix} -\dfrac{F_0'(\overline{H})}{A} & 0 \\ 0 & -\dfrac{\overline{F}_0}{A\overline{H}} \end{bmatrix}^{-1} \begin{bmatrix} \dfrac{1}{A} & 0 \\ 0 & \dfrac{\overline{F}_0}{A\overline{H}} \end{bmatrix} \begin{bmatrix} \Delta F_1 \\ \Delta T_1 \end{bmatrix} \tag{7.121}$$

$$= -\begin{bmatrix} -\dfrac{A}{F_0'(\overline{H})} & 0 \\ 0 & -\dfrac{A\overline{H}}{\overline{F}_0} \end{bmatrix} \begin{bmatrix} \dfrac{1}{A} & 0 \\ 0 & \dfrac{\overline{F}_0}{A\overline{H}} \end{bmatrix} \begin{bmatrix} \Delta F_1 \\ \Delta T_1 \end{bmatrix} \tag{7.122}$$

$$= \begin{bmatrix} \dfrac{1}{F_0'(\overline{H})} & 0 \\ 0 & 1 \end{bmatrix} \begin{bmatrix} \Delta F_1 \\ \Delta T_1 \end{bmatrix} = \begin{bmatrix} \dfrac{1}{F_0'(\overline{H})}\Delta F \\ \Delta T_1 \end{bmatrix} \tag{7.123}$$

The slope $F_0'(\overline{H})$ is obtained by differentiation of Equation 7.89 followed by substitution of the values $c = 25/3 \text{ ft}^3/\text{min}/\text{ft}^{1/2}$ and $\overline{H} = 9 \text{ ft}$. The result is $F_0'(\overline{H}) = 25/18 \text{ ft}^3/\text{min}/\text{ft}$. The deviation variables at steady state are

$$\Delta H(\infty) = \frac{1}{F_0'(\overline{H})}\Delta F = \frac{1}{F_0'(\overline{H})}\alpha\overline{F}_1 = \frac{1}{25/18}(0.1)(25) = 1.8 \text{ ft} \tag{7.124}$$

$$\Delta T(\infty) = \Delta T_1 = -\beta\overline{T}_1 = -0.1(150) = -15° \tag{7.125}$$

The steady-state level and temperature from the linearized model are

$$H(\infty) = \overline{H} + \Delta H(\infty) = 10.8 \text{ ft} \tag{7.126}$$

$$T(\infty) = \overline{T} + \Delta T(\infty) = 135°F \tag{7.127}$$

The steady-state level based on the linearized system model differs by 0.09 ft from the value based on the nonlinear model. The steady-state temperatures are the same from both the nonlinear and linearized system models.

7.4.2 Linearization of Nonlinear Systems in State Variable Form

The starting point is a nonlinear system model

$$\underline{\dot{x}} = \underline{f}(t, \underline{x}, \underline{u}) \tag{7.128}$$

$$\underline{y} = \underline{g}(t, \underline{x}, \underline{u}) \tag{7.129}$$

where

$\underline{x} = [x_1 \quad x_2 \quad \ldots \quad x_n]^T$ is the $n \times 1$ state vector
$\underline{y} = [y_1 \quad y_2 \quad \ldots \quad y_p]^T$ is a $p \times 1$ vector of outputs
$\underline{u} = [u_1 \quad u_2 \quad \ldots \quad u_m]^T$ is the $m \times 1$ input vector
t is time

Equations 7.128 and 7.129 are short for

$$\begin{bmatrix} \dot{x}_1 \\ \dot{x}_2 \\ \vdots \\ \dot{x}_n \end{bmatrix} = \begin{bmatrix} f_1(t, \underline{x}, \underline{u}) \\ f_2(t, \underline{x}, \underline{u}) \\ \vdots \\ f_n(t, \underline{x}, \underline{u}) \end{bmatrix}, \quad \begin{bmatrix} y_1 \\ y_2 \\ \vdots \\ y_p \end{bmatrix} = \begin{bmatrix} g_1(t, \underline{x}, \underline{u}) \\ g_2(t, \underline{x}, \underline{u}) \\ \vdots \\ g_p(t, \underline{x}, \underline{u}) \end{bmatrix} \tag{7.130}$$

The objective is to linearize Equations 7.128 and 7.129 about a nominal operating point in the state space $\underline{x}^0 = [x_1^0 \quad x_2^0 \quad \ldots \quad x_n^0]^T$ for a given (usually constant) input vector $\underline{u}^0 = [u_1^0 \quad u_2^0 \quad \ldots \quad u_m^0]^T$. The first-order Taylor Series approximation of the function $f_1(t, \underline{x}, \underline{u})$ about the point $(\underline{x}^0, \underline{u}^0)$ is given by

$$\dot{x}_1 = f_1(t, \underline{x}^0, \underline{u}^0) + \frac{\partial}{\partial x_1} f_1(t, \underline{x}^0, \underline{u}^0)(x_1 - x_1^0) + \frac{\partial}{\partial x_2} f_1(t, \underline{x}^0, \underline{u}^0)(x_2 - x_2^0) + \cdots$$

$$+ \frac{\partial}{\partial x_n} f_1(t, \underline{x}^0, \underline{u}^0)(x_n - x_n^0) + \frac{\partial}{\partial u_1} f_1(t, \underline{x}^0, \underline{u}^0)(u_1 - u_1^0)$$

$$+ \frac{\partial}{\partial u_2} f_1(t, \underline{x}^0, \underline{u}^0)(u_2 - u_2^0) + \cdots + \frac{\partial}{\partial u_m} f_1(t, \underline{x}^0, \underline{u}^0)(u_m - u_m^0) \tag{7.131}$$

Similar relations hold for $\dot{x}_2, \ldots, \dot{x}_n$. Introducing deviation variables

$$\Delta x_1 = x_1 - x_1^0, \quad \Delta x_2 = x_2 - x_2^0, \ldots, \quad \Delta x_n = x_n - x_n^0$$

$$\Delta u_1 = u_1 - u_1^0, \quad \Delta u_2 = u_2 - u_2^0, \ldots, \quad \Delta u_m = u_m - u_m^0$$

leads to the linearized approximation of Equation 7.128 by

$$\Delta \underline{\dot{x}} = A \Delta \underline{x} + B \Delta \underline{u} \tag{7.132}$$

where

$$
A = \begin{bmatrix}
\dfrac{\partial f_1}{\partial x_1}(\underline{x}^0, \underline{u}^0) & \dfrac{\partial f_1}{\partial x_2}(\underline{x}^0, \underline{u}^0) & \cdots & \dfrac{\partial f_1}{\partial x_n}(\underline{x}^0, \underline{u}^0) \\[2ex]
\dfrac{\partial f_2}{\partial x_1}(\underline{x}^0, \underline{u}^0) & \dfrac{\partial f_2}{\partial x_2}(\underline{x}^0, \underline{u}^0) & \cdots & \dfrac{\partial f_2}{\partial x_n}(\underline{x}^0, \underline{u}^0) \\[2ex]
\vdots & \vdots & & \vdots \\[2ex]
\dfrac{\partial f_n}{\partial x_1}(\underline{x}^0, \underline{u}^0) & \dfrac{\partial f_2}{\partial x_2}(\underline{x}^0, \underline{u}^0) & \cdots & \dfrac{\partial f_n}{\partial x_n}(\underline{x}^0, \underline{u}^0)
\end{bmatrix}
\tag{7.133}
$$

$$
B = \begin{bmatrix}
\dfrac{\partial f_1}{\partial u_1}(\underline{x}^0, \underline{u}^0) & \dfrac{\partial f_1}{\partial u_2}(\underline{x}^0, \underline{u}^0) & \cdots & \dfrac{\partial f_1}{\partial u_m}(\underline{x}^0, \underline{u}^0) \\[2ex]
\dfrac{\partial f_2}{\partial u_1}(\underline{x}^0, \underline{u}^0) & \dfrac{\partial f_2}{\partial u_2}(\underline{x}^0, \underline{u}^0) & \cdots & \dfrac{\partial f_2}{\partial u_m}(\underline{x}^0, \underline{u}^0) \\[2ex]
\vdots & \vdots & & \vdots \\[2ex]
\dfrac{\partial f_n}{\partial u_1}(\underline{x}^0, \underline{u}^0) & \dfrac{\partial f_n}{\partial u_2}(\underline{x}^0, \underline{u}^0) & \cdots & \dfrac{\partial f_n}{\partial u_m}(\underline{x}^0, \underline{u}^0)
\end{bmatrix}
\tag{7.134}
$$

and

$$
\begin{aligned}
\Delta \underline{\dot{x}} &= [\Delta \dot{x}_1 \quad \Delta \dot{x}_2 \quad \cdots \quad \Delta \dot{x}_n]^T \\
\Delta \underline{x} &= [\Delta x_1 \quad \Delta x_2 \quad \cdots \quad \Delta x_n]^T \\
\Delta \underline{u} &= [\Delta u_1 \quad \Delta u_2 \quad \cdots \quad \Delta u_m]^T
\end{aligned}
$$

The combined matrix $[A|B]$ of all partials is called the Jacobian matrix of the vector function $\underline{f}(t, \underline{x}, \underline{u})$ defining the state derivatives. In similar fashion, the linearized approximation to Equation 7.129 is given by

$$
\Delta \underline{y} = C\Delta \underline{x} + D\Delta \underline{u}
\tag{7.135}
$$

where

$$
\Delta \underline{y} = [\Delta y_1 \quad \Delta y_2 \quad \cdots \quad \Delta y_p]^T = \left[y_1 - y_1^0 \quad y_2 - y_2^0 \quad \cdots \quad y_p - y_p^0 \right]^T
$$

and

$$
y_i^0 = g_i(\underline{x}^0, \underline{u}^0), \quad i = 1, 2, \ldots, p
\tag{7.136}
$$

C and D are matrix of partials with components

$$
c_{ij} = \frac{\partial g_i}{\partial x_j}(\underline{x}^0, \underline{u}^0), \quad i = 1, 2, \ldots, p, \quad j = 1, 2, \ldots, n
\tag{7.137}
$$

$$
d_{ij} = \frac{\partial g_i}{\partial u_j}(\underline{x}^0, \underline{u}^0), \quad i = 1, 2, \ldots, p, \quad j = 1, 2, \ldots, m
\tag{7.138}
$$

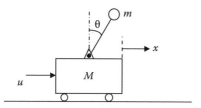

FIGURE 7.59 A nonlinear system: the inverted pendulum.

To illustrate the process of linearizing a nonlinear state variable model, consider the inverted pendulum previously introduced in Section 5.4, redrawn in Figure 7.59.

The coupled nonlinear differential equations describing the system (Equations 5.50 and 5.51) can be manipulated to read

$$\ddot{x} = \frac{ml\dot{\theta}^2 \sin\theta - (mg/2)\sin 2\theta + u}{M + m\sin^2\theta} \tag{7.139}$$

$$\ddot{\theta} = \frac{-(ml/2)\dot{\theta}^2 \sin 2\theta + (m + M)g\sin\theta - u\cos\theta}{l(M + m\sin^2\theta)} \tag{7.140}$$

State variables are x_1, x_2, x_3, x_4 where $x_1 = x$, $x_2 = \dot{x}$, $x_3 = \theta$, $x_4 = \dot{\theta}$. The state derivatives are given by

$$\dot{x}_1 = f_1(\underline{x}, \underline{u}) = x_2 \tag{7.141}$$

$$\dot{x}_2 = f_2(\underline{x}, \underline{u}) = \frac{mlx_4^2 \sin x_3 - (mg/2)\sin 2x_3 + u}{M + m\sin^2 x_3} \tag{7.142}$$

$$\dot{x}_3 = f_3(\underline{x}, \underline{u}) = x_4 \tag{7.143}$$

$$\dot{x}_4 = f_4(\underline{x}, \underline{u}) = \frac{-(ml/2)x_4^2 \sin 2x_3 + (m + M)g\sin x_3 - u\cos x_3}{l(M + m\sin^2 x_3)} \tag{7.144}$$

Choosing the outputs as x and θ,

$$y_1 = g_1(\underline{x}, \underline{u}) = x_1 \tag{7.145}$$

$$y_2 = g_2(\underline{x}, \underline{u}) = x_3 \tag{7.146}$$

Components of the linearized system matrices A, B, C, and D consist of the partials

$$a_{11} = \frac{\partial f_1}{\partial x_1}(\underline{x}^0, \underline{u}^0) = 0, \quad a_{12} = \frac{\partial f_1}{\partial x_2}(\underline{x}^0, \underline{u}^0) = 1,$$
$$a_{13} = \frac{\partial f_1}{\partial x_3}(\underline{x}^0, \underline{u}^0) = a_{14} = \frac{\partial f_1}{\partial x_4}(\underline{x}^0, \underline{u}^0) = 0 \tag{7.147}$$

$$a_{21} = \frac{\partial f_2}{\partial x_1}(\underline{x}^0, \underline{u}^0) = 0, \quad a_{22} = \frac{\partial f_2}{\partial x_2}(\underline{x}^0, \underline{u}^0) = 0, \quad a_{24} = \frac{\partial f_2}{\partial x_4}(\underline{x}^0, \underline{u}^0) = \frac{2mlx_4 \sin x_3}{M + m\sin^2 x_3} \tag{7.148}$$

The component a_{23} is equal to N_1/D_1 evaluated at the operating point $(\underline{x}^0, \underline{u}^0)$ where

$$N_1 = (M + m \sin^2 x_3) \left[mlx_4^2 \cos x_3 - mg(1 - 2\sin^2 x_3) \right]$$
$$- \left[mlx_4^2 \sin x_3 - \frac{mg}{2} \sin 2x_3 + u \right] (m \sin 2x_3) \tag{7.149}$$

$$D_1 = (M + m \sin^2 x_3)^2 \tag{7.150}$$

$$a_{31} = \frac{\partial f_3}{\partial x_1}(\underline{x}^0, \underline{u}^0) = a_{32} = \frac{\partial f_3}{\partial x_2}(\underline{x}^0, \underline{u}^0) = a_{33} = \frac{\partial f_3}{\partial x_3}(\underline{x}^0, \underline{u}^0) = 0, \quad a_{34} \frac{\partial f_3}{\partial x_4}(\underline{x}^0, \underline{u}^0) = 1 \tag{7.151}$$

$$a_{41} = \frac{\partial f_4}{\partial x_1}(\underline{x}^0, \underline{u}^0) = 0, \quad a_{42} = \frac{\partial f_4}{\partial x_2}(\underline{x}^0, \underline{u}^0) = 0, \quad a_{44} \frac{\partial f_4}{\partial x_4}(\underline{x}^0, \underline{u}^0) = \frac{-mlx_4 \sin 2x_3}{l(M + m \sin^2 x_3)} \tag{7.152}$$

The component a_{43} is equal to N_2/D_2 evaluated at the operating point $(\underline{x}^0, \underline{u}^0)$ where

$$N_2 = l(M + m \sin^2 x_3) \left[-mlx_4^2(1 - 2\sin^2 x_3) + (m + M)g \cos x_3 + u \sin x_3 \right]$$
$$+ \left[\frac{m}{2} lx_4^2 \sin^2 2x_3 - (m + M)g \sin x_3 + u \cos x_3 \right] lm \sin 2x_3 \tag{7.153}$$

$$D_2 = \left[l(M + m \sin^2 x_3) \right]^2 \tag{7.154}$$

$$b_{11} = \frac{\partial f_1}{\partial u}(\underline{x}^0, \underline{u}^0) = 0, \quad b_{21} = \frac{\partial f_2}{\partial u}(\underline{x}^0, \underline{u}^0) = \frac{1}{M + m \sin^2 x_3} \tag{7.155}$$

$$b_{31} = \frac{\partial f_3}{\partial u}(\underline{x}^0, \underline{u}^0) = 0, \quad b_{41} = \frac{\partial f_4}{\partial u}(\underline{x}^0, \underline{u}^0) = \frac{-\cos x_3}{l(M + m \sin^2 x_3)} \tag{7.156}$$

$$c_{11} = \frac{\partial g_1}{\partial x_1}(\underline{x}^0, \underline{u}^0) = 1, \quad c_{12} = \frac{\partial g_1}{\partial x_2}(\underline{x}^0, \underline{u}^0) = 0, \quad c_{13} = \frac{\partial g_1}{\partial x_3}(\underline{x}^0, \underline{u}^0) = c_{14} = \frac{\partial g_1}{\partial x_4}(\underline{x}^0, \underline{u}^0) = 0 \tag{7.157}$$

$$c_{21} = \frac{\partial g_2}{\partial x_1}(\underline{x}^0, \underline{u}^0) = c_{22} = \frac{\partial g_2}{\partial x_2}(\underline{x}^0, \underline{u}^0) = 0, \quad c_{23} = \frac{\partial g_2}{\partial x_3}(\underline{x}^0, \underline{u}^0) = 1, \quad c_{24} = \frac{\partial g_2}{\partial x_4}(\underline{x}^0, \underline{u}^0) = 0 \tag{7.158}$$

$$d_{11} = \frac{\partial g_1}{\partial u}(\underline{x}^0, \underline{u}^0) = d_{21} = \frac{\partial g_2}{\partial u}(\underline{x}^0, \underline{u}^0) = 0 \tag{7.159}$$

Suppose the steady-state operating point is $\underline{x}^0 = [0 \quad 0 \quad \pi \quad 0]^T$ and input $\underline{u}^0 = 0$. The nonzero elements of matrices A, B, C, and D are

$$a_{12} = 1, \quad a_{23} = -\frac{m}{M}g, \quad a_{34} = 1, \quad a_{43} = -\frac{g}{l}\frac{(m + M)}{M},$$

$$b_{21} = \frac{1}{M}, \quad b_{41} = \frac{1}{lM}, \quad c_{11} = c_{23} = 1 \tag{7.160}$$

For $M = 3$ kg, $m = 0.1$ kg, $l = 0.75$ m, $g = 9.8$ m/s², the system matrices are

$$A_1 = \begin{bmatrix} 0 & 1 & 0 & 0 \\ 0 & 0 & -0.3267 & 0 \\ 0 & 0 & 0 & 1 \\ 0 & 0 & -13.5022 & 0 \end{bmatrix}, \quad B_1 = \begin{bmatrix} 0 \\ 0.333 \\ 0 \\ 0.444 \end{bmatrix}, \quad C_1 = \begin{bmatrix} 1 & 0 & 0 & 0 \\ 0 & 0 & 1 & 0 \end{bmatrix}, \quad D_1 = \begin{bmatrix} 0 \\ 0 \end{bmatrix} \tag{7.161}$$

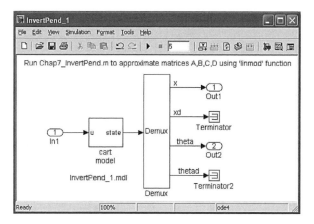

FIGURE 7.60   Simulink® model of inverted pendulum showing input and two outputs.

7.4.3   LINMOD FUNCTION

Simulink estimates the matrices A, B, C, and D in the linearized approximation by using small perturbations in the state and input(s) to numerically calculate the partial derivatives. The "linmod" and "linmod2" functions extract the linearized model coefficient matrices from a Simulink diagram of the nonlinear system.

The top level of a Simulink simulation of the inverted pendulum is shown in Figure 7.60.

The "cart" subsystem is shown in Figure 7.61.

The MATLAB statement

```
[sizes,X0,states] = InvertPend_1([],[],[],0);
```

in M-file "*Chap7_InvertPend.m*" returns the following information:

sizes = 4	x0 = 0
0	3.1416
2	0
1	0
0	
1	
1	

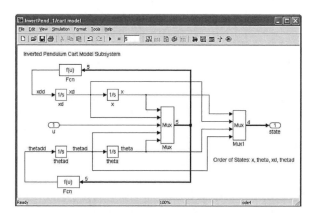

FIGURE 7.61   Cart subsystem showing internal states.

```
states = 'InvertPend_1/cartmodel/x Integrator'
        'InvertPend_1/cart model/theta Integrator'
        'InvertPend_1/cart model/xd Integrator'
        'InvertPend_1/cart model/thetad Integrator'
```

The first four components of the output 'sizes' reveal the number of continuous states (4), discrete states (0), outputs (2), and inputs (1) in the Simulink model. The ordering of the states is conveyed by the output vector "states," which in the present case is seen to be "x," "theta," "xd," and "thetad." "X0" reports the initial values of the state vector in the order defined by the output "states." It will soon become apparent why the ordering of the state vector is significant.

The same M-file "*Chap7_InvertPend.m*" contains the statement

```
[A2,B2,C2,D2] = linmod('InvertPend_1',x_operpt,u0)
```

which returns the linearized system matrices. The first argument "InvertPend_1" is the Simulink model file name, while "x_operpt" and "uo" are arrays with numerical values of the state and input at the operating point.

The "linmod" function returns the matrices

$$
A_2 = \begin{bmatrix} 0 & 0 & 1 & 0 \\ 0 & 0 & 0 & 1 \\ 0 & -0.3267 & 0 & 0 \\ 0 & -13.5022 & 0 & 0 \end{bmatrix}, \quad B_2 = \begin{bmatrix} 0 \\ 0 \\ 0.333 \\ 0.444 \end{bmatrix}, \quad C_2 = \begin{bmatrix} 1 & 0 & 0 & 0 \\ 0 & 0 & 1 & 0 \end{bmatrix}, \quad D_2 = \begin{bmatrix} 0 \\ 0 \end{bmatrix} \quad (7.162)
$$

The matrices in Equations 7.161 and 7.162 are different as a result of the difference in the ordering of the state vector in the two different linearized models of the system. That is, from Equation 7.161, when the state is $[\Delta x \quad \Delta \dot{x} \quad \Delta \theta \quad \Delta \dot{\theta}]^T$, we have

$$
\begin{bmatrix} \Delta \dot{x}_1 \\ \Delta \dot{x}_2 \\ \Delta \dot{x}_3 \\ \Delta \dot{x}_4 \end{bmatrix} = \begin{bmatrix} \Delta \dot{x} \\ \Delta \ddot{x} \\ \Delta \dot{\theta} \\ \Delta \ddot{\theta} \end{bmatrix} = \begin{bmatrix} 0 & 1 & 0 & 0 \\ 0 & 0 & -0.3267 & 0 \\ 0 & 0 & 0 & 1 \\ 0 & 0 & -13.5022 & 0 \end{bmatrix} \begin{bmatrix} \Delta x \\ \Delta \dot{x} \\ \Delta \theta \\ \Delta \dot{\theta} \end{bmatrix} + \begin{bmatrix} 0 \\ 0.333 \\ 0 \\ 0.444 \end{bmatrix} [\Delta u] \quad (7.163)
$$

On the other hand, when the state is $[\Delta x \quad \Delta \theta \quad \Delta \dot{x} \quad \Delta \dot{\theta}]^T$, Equation 7.162 implies

$$
\begin{bmatrix} \Delta \dot{x}_1 \\ \Delta \dot{x}_2 \\ \Delta \dot{x}_3 \\ \Delta \dot{x}_4 \end{bmatrix} = \begin{bmatrix} \Delta \dot{x} \\ \Delta \dot{\theta} \\ \Delta \ddot{x} \\ \Delta \ddot{\theta} \end{bmatrix} = \begin{bmatrix} 0 & 0 & 1 & 0 \\ 0 & 0 & 0 & 1 \\ 0 & -0.3267 & 0 & 0 \\ 0 & -13.5022 & 0 & 0 \end{bmatrix} \begin{bmatrix} \Delta x \\ \Delta \theta \\ \Delta \dot{x} \\ \Delta \dot{\theta} \end{bmatrix} + \begin{bmatrix} 0 \\ 0 \\ 0.333 \\ 0.444 \end{bmatrix} [\Delta u] \quad (7.164)
$$

Once the linearized system matrices A_1, B_1, C_1, and D_1 or A_2, B_2, C_2, and D_2 are known, the inverted pendulum dynamics can be approximated using either set, and the response should compare favorably with the nonlinear system response provided the state and input deviations from the operating point are kept small. Figure 7.62 is the Simulink diagram for comparing the nonlinear system model and the linearized model using the set of matrices A_2, B_2, C_2, and D_2 obtained from the "linmod" function.

Figure 7.63 shows the nonlinear and linearized response for $\theta(t)$ corresponding to a pulse input force of magnitude 2.5 N from 1 to 2 s. Agreement between the nonlinear and linearized responses is very good. Note the small deviation in $\theta(t)$ from $\theta^0 = \pi$ rad resulting from the particular input.

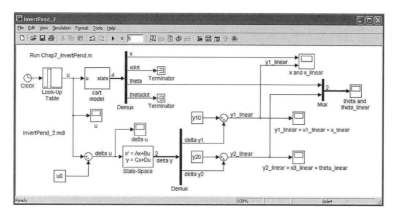

FIGURE 7.62 Simulink® diagram for comparing nonlinear and linearized models.

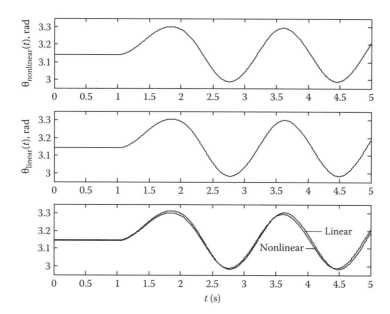

FIGURE 7.63 Nonlinear and linearized model response $\theta(t)$ for $u(t) = 2.5$ N, $1 \le t \le 2$.

Figure 7.64 exemplifies what happens when the state vector $\underline{x}(t)$ deviates by a significant amount from $\underline{x}^0$. The magnitude of the applied force pulse input is increased to 25 N. The nonlinear system model and linearized approximation no longer exhibit the same level of agreement as before.

Due to the absence of damping, the (nonlinear and linear) models predict sustained oscillations. Hence, the coefficient matrices A_1 and A_2 must possess a pair of purely imaginary characteristic roots, easily confirmed by checking the eigenvalues of each. The statements "`eig(A1)`" and "`eig(A2)`" both return two real eigenvalues 0,0, and two imaginary eigenvalues $\pm j3.674537$.

A closer look at Equations 7.163 and 7.164 reveals a simpler formulation of the governing equations. From Equation 7.164,

$$\Delta \ddot{x} = -0.3267\Delta\theta + 0.333\Delta u \tag{7.165}$$

$$\Delta \ddot{\theta} + 13.5022\Delta\theta = 0.444\Delta u \tag{7.166}$$

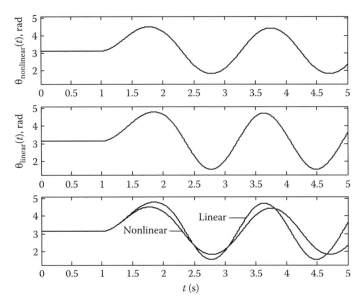

FIGURE 7.64 Nonlinear and linearized model response $\theta(t)$ for $u(t) = 25$ N, $1 \leq t \leq 2$.

The natural modes from Equation 7.166 are s_1, $s_2 = (-13.5022)^{1/2} = \pm j\,3.6745$, and the frequency of undamped oscillations is $\omega_n = 3.6745$ rad/s (see Figures 7.63 and 7.64). Furthermore, the two remaining characteristic roots are, from Equation 7.165, both zero. Laplace transforming Equations 7.165 and 7.166 and solving for $\Delta\theta(s)$ and $\Delta X(s)$ result in

$$\Delta\theta(s) = \frac{0.444}{s^2 + 13.5022}\Delta U(s) \tag{7.167}$$

$$\Delta X(s) = 0.333\left[\frac{s^2 + 13.0666}{s^2(s^2 + 13.5022)}\right]\Delta U(s) \tag{7.168}$$

The inverted pendulum is often used as an example of an inherently (open-loop) unstable system, and numerous linear controls texts demonstrate techniques for designing linear controllers to balance the pendulum in the upright position ($\theta = 0$). The steady-state operating point $[x^0, \dot{x}^0, \theta^0, \dot{\theta}^0; u^0] = (0, 0, 0, 0; 0)$ is unstable, easily verified by changing "x30" $= \theta^0$ to zero in M-file "*Chap7_InvertPend.m*" and observing the eigenvalues of the linearized system matrix A_1 or A_2. Of course, basic intuition suggests as much, that is, "What happens to the pendulum when it is displaced from the upright equilibrium position?" Exercise 7.24 looks at this case in more detail.

You can implement your own "linmod" function to numerically compute the linearized system matrices A, B, C, and D. To illustrate, suppose we wish to estimate a_{43} in Equation 7.163. The exact value of -13.5022 was computed from the analytical expression for the partial derivative $(\partial f_4/\partial x_3)(\underline{x}^0, \underline{u}^0)$ using Equations 7.153 and 7.154. A simple central difference formula to approximate $(\partial f_4/\partial x_3)(\underline{x}^0, \underline{u}^0)$ is

$$\frac{\partial f_4}{\partial x_3}(\underline{x}^0, \underline{u}^0) \approx \frac{f_4\left(x_1^0, x_2^0, x_3^0 + \Delta, x_4^0, u^0\right) - f_4\left(x_1^0, x_2^0, x_3^0 - \Delta, x_4^0, u^0\right)}{2\Delta} \tag{7.169}$$

From Equation 7.144, the numerator terms are

$$f_4\left(x_1^0, x_2^0, x_3^0 + \Delta, x_4^0, u^0\right) = \frac{-(ml/2)\left(x_4^0\right)^2 \sin 2\left(x_3^0 + \Delta\right) + (m+M)g \sin\left(x_3^0 + \Delta\right) - u \cos\left(x_3^0 + \Delta\right)}{l\left[M + m \sin^2\left(x_3^0 + \Delta\right)\right]}$$

(7.170)

$$f_4\left(x_1^0, x_2^0, x_3^0 - \Delta, x_4^0, u^0\right) = \frac{-(ml/2)\left(x_4^0\right)^2 \sin 2\left(x_3^0 - \Delta\right) + (m+M)g \sin\left(x_3^0 - \Delta\right) - u \cos\left(x_3^0 - \Delta\right)}{l\left[M + m \sin^2\left(x_3^0 - \Delta\right)\right]}$$

(7.171)

The operating point is $(\underline{x}^0; \underline{u}^0) = [0 \quad 0 \quad \pi \quad 0; 0]^T$. After substituting in the numerical values for m, M, g and choosing $\Delta = 0.01$, we have

$$\frac{\partial f_4}{\partial x_3}(\underline{x}^0, \underline{u}^0) \approx \frac{(0.1 + 3)(9.8)}{(0.75)(2)(0.01)} \left[\frac{\sin(\pi + 0.01)}{3 + 0.1 \sin^2(\pi + 0.01)} - \frac{\sin(\pi - 0.01)}{3 + 0.1 \sin^2(\pi - 0.01)}\right]$$
$$\approx -13.5020$$

(7.172)

which is very close to the analytically obtained value -13.5022. Another example of linearization involving nonlinear tanks is presented in Section 8.4.

7.4.4 Multiple Linearized Models for a Single System

When the inputs to a nonlinear system vary by a considerable amount, a single linearized model may no longer be sufficient to describe the excursions of the state vector about an individual operating point. It becomes necessary to linearize the system dynamics in terms of deviation variables about different operating points. The linearized models are applicable to specific regions in state space. While the initial state may have been at equilibrium, be mindful that the initial conditions of the deviation variables are no longer zero as the state transitions between different linearized regions in state space. The situation is illustrated in Figure 7.65 for an autonomous, second-order system with different linearized models in each of the four quadrants of state space.

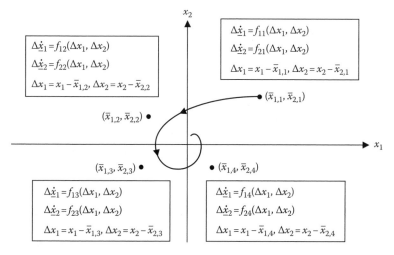

FIGURE 7.65 State trajectory of an autonomous, nonlinear, second-order system linearized about four different operating points.

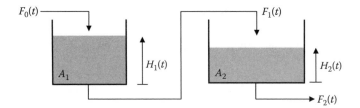

FIGURE 7.66   Second-order system consisting of two nonlinear first-order tanks.

An example of how to accommodate multiple operating points is illustrated using the nonlinear second-order system in Figure 7.66. The mathematical model describing the coupled dynamics of the two tanks is given in Equations 7.173 through 7.176.

$$A_1 \frac{dH_1}{dt} + F_1 = F_0 \tag{7.173}$$

$$F_1 = c_1 H_1^{1/2} \tag{7.174}$$

$$A_2 \frac{dH_2}{dt} + F_2 = F_1 \tag{7.175}$$

$$F_2 = c_2 H_2^{1/2} \tag{7.176}$$

Solving for the state derivative functions gives

$$\frac{dH_1}{dt} = f_1(H_1, H_2, F_0) = \frac{1}{A_1}(F_0 - c_1 H_1^{1/2}) \tag{7.177}$$

$$\frac{dH_2}{dt} = f_2(H_1, H_2, F_0) = \frac{1}{A_2}(c_1 H_1^{1/2} - c_2 H_2^{1/2}) \tag{7.178}$$

Suppose the flow into the first tank is constant, that is, $F_0(t) = \tilde{F}_0$, $t \geq 0$. Steady-state levels are obtained by setting both derivatives equal to zero. The result is

$$(H_1)_{ss} = \left(\frac{\tilde{F}_0}{c_1}\right)^2, \quad (H_2)_{ss} = \left(\frac{\tilde{F}_0}{c_2}\right)^2 \tag{7.179}$$

A typical state trajectory starting from $\{H_1(0), H_2(0)\}$ and ending at $\{(H_1)_{ss}, (H_2)_{ss}\}$ is shown in Figure 7.67. Four different operating points designated $(\overline{H}_1, \overline{H}_2)$ in the (H_1, H_2) state space are also shown.

Equations 7.177 and 7.178 are initially linearized about the operating point $(\overline{H}_1, \overline{H}_2)$ nearest to the initial state $\{H_1(0), H_2(0)\}$. The equations are relinearized as necessary, that is, when the state trajectory transitions from a neighborhood about one operating point to another region about a different operating point.

For nonlinear systems with inputs, linearization requires a nominal input for each operating point. The following example presents the results of using a single operating point compared with using multiple operating points for linearizing the two-tank system in Figure 7.66 with constant inflow.

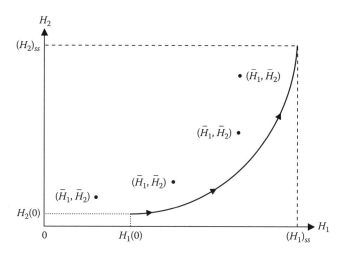

FIGURE 7.67 Several operating points and state trajectory for tanks subject to $F_0(t) = \tilde{F}_0, t \geq 0$.

Example 7.3

For the two-tank system shown in Figure 7.66, baseline parameter values are

$$A_1 = 25\,\text{ft}^2, \quad A_2 = 15\,\text{ft}^2, \quad c_1 = 3\,\text{ft}^3/\text{min}/\text{ft}^{1/2}, \quad c_2 = 4\,\text{ft}^3/\text{min}/\text{ft}^{1/2}$$

(a) Find the steady-state levels $(H_1)_{ss}$ and $(H_2)_{ss}$ when $\tilde{F}_0 = 12\,\text{ft}^3/\text{min}$.
(b) Choose a steady-state operating point $(\overline{H}_1, \overline{H}_2)$ where $\overline{H}_1 = 0.5(H_1)_{ss}$. Find the tank 2 level $\overline{H}_2$ at the operating point and the nominal inflow $\overline{F}_0$ at the operating point.
(c) Introduce deviation variables and linearize the model differential equations.
(d) Solve the linearized equations for the case where both tanks are initially empty and $\tilde{F}_0 = 12\,\text{ft}^3/\text{min}$. Compare the level responses of both tanks to the solutions obtained from the nonlinear equations.
(e) Plot state trajectories for the linearized and nonlinear systems.
(f) Establish four steady-state operating points corresponding to tank 1 levels of $0.125(H_1)_{ss}$, $0375(H_1)_{ss}$, $0.625(H_1)_{ss}$, and $0.875(H_1)_{ss}$.
(g) Repeat parts (d) and (e).

(a) From Equation 7.179, the steady-state levels are

$$(H_1)_{ss} = \left(\frac{\tilde{F}_0}{c_1}\right)^2 = \left(\frac{12}{3}\right)^2 = 16\,\text{ft}, \quad (H_2)_{ss} = \left(\frac{\tilde{F}_0}{c_2}\right)^2 = \left(\frac{12}{4}\right)^2 = 9\,\text{ft} \qquad (7.180)$$

(b) The required inflow $\overline{F}_0$ to maintain tank 1 level at $\overline{H}_1 = 0.5(H_1)_{ss} = 8$ ft is equal to the steady-state outflow from tank 1, that is,

$$\overline{F}_0 = \overline{F}_1 = c_1\overline{H}_1^{1/2} = 3(8)^{1/2} = 8.4853\,\text{ft}^3/\text{min} \qquad (7.181)$$

For steady-state conditions, tank 2 level must be

$$\overline{H}_2 = \left(\frac{\overline{F}_2}{c_2}\right)^2 = \left(\frac{\overline{F}_1}{c_2}\right)^2 = \left(\frac{3(8)^{1/2}}{4}\right)^2 = 4.5\,\text{ft} \qquad (7.182)$$

The steady-state operating point is $(\overline{H}_1, \overline{H}_2) = (8\,\text{ft}, 4.5\,\text{ft})$ and $\overline{F}_0 = 8.4853\,\text{ft}^3/\text{min}$.

(c) Introducing deviation variables $\Delta H_1 = H_1 - \overline{H}_1$, $\Delta H_2 = H_2 - \overline{H}_2$ for the tank levels and $\Delta F_0 = F_0 - \overline{F}_0 = \tilde{F}_0 - \overline{F}_0$ for the inflow to tank 1 produces a system of linearized differential equations

$$\Delta \dot{H}_1 = a_{11} \Delta H_1 + a_{12} \Delta H_2 + b_1 \Delta F_0 \tag{7.183}$$

$$\Delta \dot{H}_2 = a_{21} \Delta H_1 + a_{22} \Delta H_2 + b_2 \Delta F_0 \tag{7.184}$$

$$a_{11} = \left. \frac{\partial}{\partial H_1} f_1(H_1, H_2, F_0) \right|_{H_1 = \overline{H}_1, H_2 = \overline{H}_2, F_0 = \overline{F}_0} = \frac{-c_1}{2 A_1 \overline{H}_1^{1/2}} \tag{7.185}$$

$$a_{12} = \left. \frac{\partial}{\partial H_2} f_1(H_1, H_2, F_0) \right|_{H_1 = \overline{H}_1, H_2 = \overline{H}_2, F_0 = \overline{F}_0} = 0 \tag{7.186}$$

$$a_{21} = \left. \frac{\partial}{\partial H_1} f_2(H_1, H_2, F_0) \right|_{H_1 = \overline{H}_1, H_2 = \overline{H}_2, F_0 = \overline{F}_0} = \frac{c_1}{2 A_2 \overline{H}_1^{1/2}} \tag{7.187}$$

$$a_{22} = \left. \frac{\partial}{\partial H_2} f_2(H_1, H_2, F_0) \right|_{H_1 = \overline{H}_1, H_2 = \overline{H}_2, F_0 = \overline{F}_0} = \frac{-c_2}{2 A_2 \overline{H}_2^{1/2}} \tag{7.188}$$

$$b_1 = \left. \frac{\partial}{\partial F_0} f_2(H_1, H_2, F_0) \right|_{H_1 = \overline{H}_1, H_2 = \overline{H}_2, F_0 = \overline{F}_0} = \frac{1}{A_1} \tag{7.189}$$

$$b_2 = \left. \frac{\partial}{\partial F_0} f_2(H_1, H_2, F_0) \right|_{H_1 = \overline{H}_1, H_2 = \overline{H}_2, F_0 = \overline{F}_0} = 0 \tag{7.190}$$

Substituting values for A_1, A_2, c_1, c_2, $\overline{H}_1$, $\overline{H}_2$ in Equation 7.185 and Equations 7.187 through 7.189, the linearized tank model is

$$\Delta \dot{H}_1 = -0.0212 \Delta H_1 + 0.04 \Delta F_0 \tag{7.191}$$

$$\Delta \dot{H}_2 = 0.0354 \Delta H_1 - 0.0629 \Delta H_2 \tag{7.192}$$

(d) The general solution to Equations 7.183 and 7.184 with $a_{12} = b_2 = 0$, initial conditions $\Delta H_1(0) = H_1(0) - \overline{H}_1$, $\Delta H_2(0) = H_2(0) - \overline{H}_2$, and $\Delta F_0 = \tilde{F}_0 - \overline{F}_0$ is (see Exercise 7.25)

$$\Delta H_1(t) = -\frac{b_1 \Delta F_0}{a_{11}} + \left[\Delta H_1(0) + \frac{b_1 \Delta F_0}{a_{11}} \right] e^{a_{11} t} \tag{7.193}$$

$$\Delta H_2(t) = \frac{a_{21} b_1 \Delta F_0}{a_{11} a_{22}} + \frac{a_{21}}{(a_{11} - a_{22})} \left[\Delta H_1(0) + \frac{b_1 \Delta F_0}{a_{11}} \right] e^{a_{11} t}$$
$$+ \left[\Delta H_2(0) + \frac{a_{21}}{(a_{22} - a_{11})} \left\{ \Delta H_1(0) + \frac{b_1 \Delta F_0}{a_{22}} \right\} \right] e^{a_{22} t} \tag{7.194}$$

With $H_1(0) = H_2(0) = 0$, $\Delta F_0 = \tilde{F}_0 - \overline{F}_0 = 12 - 8.4853 = 3.5147 \, \text{ft}^3/\text{min}$, the linearized tank-level responses $H_1(t) = \overline{H}_1 + \Delta H_1(t)$ and $H_2(t) = \overline{H}_2 + \Delta H_2(t)$ become

$$H_1(t) = 14.6274(1 - e^{-0.0212t}), \quad t \geq 0 \tag{7.195}$$

$$H_2(t) = 8.2279 - 12.4195 e^{-0.212t} + 4.1916 e^{-0.0629t}, \quad t \geq 0 \tag{7.196}$$

The nonlinear system responses can be approximated by resorting to simulation with a fixed-step numerical integrator and small integration step. The Simulink diagram is shown in Figure 7.68.

An RK-4 integrator with step size of 0.01 s was used to approximate the tank 1 and tank 2 nonlinear system level responses. The linearized system responses in Equations 7.195 and 7.196 are plotted along with the nonlinear system responses in Figure 7.69.

Note that the nonlinear system step responses approach the correct steady-state levels $(H_1)_{ss} = 16$ ft and $(H_2)_{ss} = 9$ ft predicted in Equation 7.180. Can you verify whether the tank levels for the linearized system shown in Figure 7.69, namely, $H_1(350) = 14.62$ ft and $H_2(350) = 8.22$ ft, are correct?

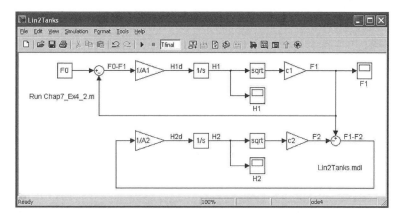

FIGURE 7.68 Simulink® diagram for nonlinear two-tank system.

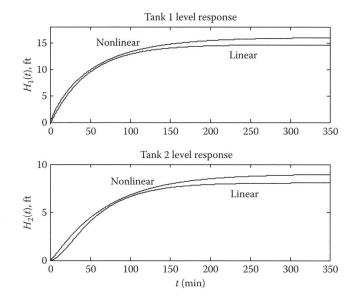

FIGURE 7.69 Comparison of linearized and nonlinear system tank-level responses.

(e) The state trajectories are shown in Figure 7.70.

(f) A similar procedure to the one used in parts (a) through (e) establishes four distinct steady-state operating points $(\overline{H}_1, \overline{H}_2)$ where $\overline{H}_1$ is one of the four values $0.125(H_1)_{ss} = 2$ ft, $0.375(H_1)_{ss} = 6$ ft, $0.675(H_1)_{ss} = 10$ ft, $0.875(H_1)_{ss} = 14$ ft. The corresponding values of $\overline{H}_2$ and $\overline{F}_0$ are shown in Table 7.7.

(g) The nonlinear and linearized system responses for both tanks are shown in Figures 7.71 and 7.72.

The static nonlinear operating curves for each tank are shown in Figure 7.73, and the state trajectories of the linearized and nonlinear systems are shown in Figure 7.74. The operating points listed in Table 7.7 are shown as well. Note the improved accuracy in the step response of the system linearized about multiple operating points compared to the case illustrated in Figure 7.70 where a single operating point $(\overline{H}_1, \overline{H}_2)$ was used.

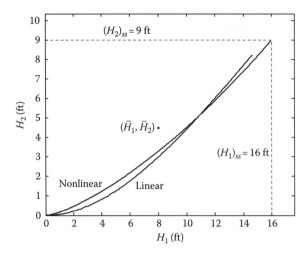

FIGURE 7.70 State trajectories of linearized ($\overline{H}_1 = 8$ ft, $\overline{H}_2 = 4.5$ ft) and nonlinear systems.

TABLE 7.7
Steady-State Operating Points $(\overline{H}_1, \overline{H}_2)$ and Corresponding $\overline{F}_0$

Region	$\overline{H}_1$ (ft)	$\overline{H}_2$ (ft)	$\overline{F}_0$ (ft³/min)
I	2	1.125	4.2426
II	6	3.375	7.3485
III	10	5.625	9.4868
IV	14	7.875	11.2250

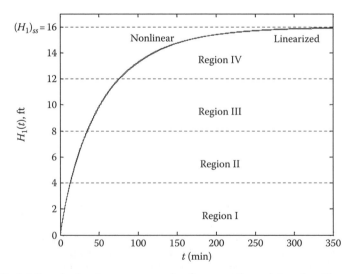

FIGURE 7.71 Tank 1 linearized system response using four operating points and nonlinear system response.

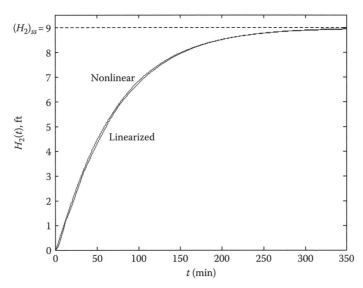

FIGURE 7.72 Tank 2 linearized system response using four operating points and nonlinear system response.

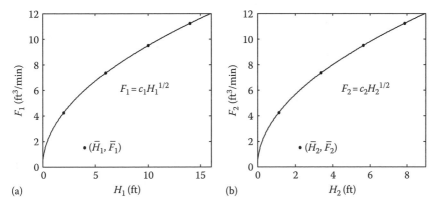

FIGURE 7.73 Static nonlinear operating curves for both tanks. (a) Tank 1 nonlinear system operating characteristic. (b) Tank 2 nonlinear system operating characteristic.

EXERCISES

7.18 The nonlinear tank model in which the outflow is based on Equation 7.89 can be thought of as exhibiting a variable fluid resistance, that is, $R = f(H)$. When the tank is linearized about an operating pt $(\overline{H}, \overline{F}_0)$, the resistance $\overline{R} = f(\overline{H}) = \Delta H / \Delta F_0$, which is the reciprocal of the slope of the tangent drawn to the function $F_0 = cH^{1/2}$ at the operating point (see Figure 7.51). Hence, for small variations about the operating point, the tank behaves similar to a linear tank with resistance $\overline{R}$.

(a) Show that the linearized resistance $\overline{R}$ about the point $(\overline{H}, \overline{F}_0)$ is equal to $2\overline{H}_0 / \overline{F}_0$.

(b) For the tank whose operating curve is shown in Figure 7.51, find the linearized resistance $\overline{R}$ when the tank level fluctuates by a small amount about
5 ft (ii) 10 ft (iii) 15 ft (iv) 20 ft.

(c) Comment on the apparent fluid resistance of a nonlinear tank as the level rises.

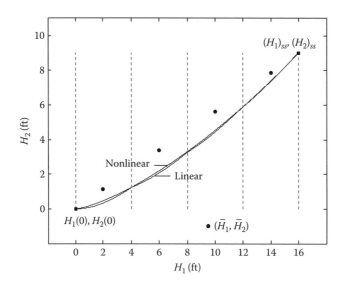

FIGURE 7.74   State trajectories for linearized and nonlinear systems.

7.19 A tank with nonlinear operating curve, shown in Figure 7.51, and cross-sectional area 25 ft² is initially filled to a height of 9 ft. There is no inflow.

(a) Employ Simulink (with suitable integrator and step size) to simulate the emptying of the tank. Graph $H_{\text{sim}}(t)$.

(b) Linearize the differential equation model about the initial point $H(0) = 9$ ft, $F_0(0) = 1.5$ ft³/min, that is, choose the operating pt as $(\overline{H}, \overline{F}_0) = (9, 1.5)$, and find the linear differential equation describing the deviation $\Delta H(t) = H(t) - \overline{H}$.

(c) Find the analytical solution for $\Delta H(t)$ and plot $H_{\text{lin}}(t) = \overline{H} + \Delta H(t)$ on the same graph as the simulated solution $H_{\text{sim}}(t)$. Comment on the results.

(d) Find the analytical solution for the level, $H_{\text{anal}}(t)$, and compare it with the simulated response $H_{\text{sim}}(t)$ in part (a) and linearized response $H_{\text{lin}}(t)$ in part (c).

7.20 Starting with Equations 7.101 and 7.110, obtain Equation 7.111 for the linearized state derivatives.

7.21 The nonlinear tank shown in Figure E7.21a has an adjustable valve in the discharge line. The valve opening is given by the normalized variable θ ($0 \le \theta \le 1$) where $\theta = 0$ is a closed valve and $\theta = 1$ represents a fully open valve. The outflow is obtained from $F_0 = F_0(\theta, H) = c(\theta)H^{1/2}$.

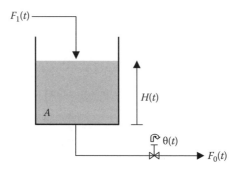

FIGURE E7.21a

An expression for $\Delta F_0(t)$ in the linearized differential equation model of the tank $A(\mathrm{d}/\mathrm{d}t)\Delta H(t) + \Delta F_0(t) = \Delta F_1(t)$ is obtained as follows:

$$F_0 = F_0(\bar{\theta}, \bar{H}_0) + \frac{\partial}{\partial\theta}F_0(\bar{\theta}, \bar{H}_0)\Delta\theta + \frac{\partial}{\partial H}F_0(\bar{\theta}, \bar{H}_0)\Delta H$$

$$\Rightarrow \Delta F_0 = \frac{\partial}{\partial\theta}[c(\theta)H^{1/2}]_{\theta=\bar{\theta}, H=\bar{H}}\Delta\theta + \frac{\partial}{\partial H}[c(\theta)H^{1/2}]_{\theta=\bar{\theta}, H=\bar{H}}\Delta H$$

$$= \bar{H}^{1/2}\frac{\mathrm{d}}{\mathrm{d}\theta}c(\theta)\bigg|_{\theta=\bar{\theta}}\Delta\theta + c(\bar{\theta})\frac{\mathrm{d}}{\mathrm{d}H}H^{1/2}\bigg|_{H=\bar{H}}\Delta H$$

$$= \bar{H}^{1/2}c'(\bar{\theta})\Delta\theta + c(\bar{\theta})\left[\frac{1}{2}\bar{H}^{-1/2}\right]\Delta H$$

Data points along the valve-operating characteristic $c(\theta)$ are shown in Figure E7.21b:

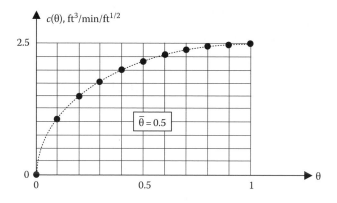

FIGURE E7.21b

(a) Find the linearized differential equation about the steady-state operating point where $\bar{\theta} = 0.5$, $\bar{H} = 9$ ft.
(b) Simulate the tank-level response when the inflow increases by 10% and valve opening decreases by 15% with respect to their operating point values. The initial conditions are $H(0) = H$, $\theta(0) = \theta$. Assume both changes are step inputs. The cross-sectional area of the tank is 50 ft^2.

7.22 In Example 7.2, let $\beta = 0$ and vary α from -0.5 to 0.5 in steps of 0.05.
(a) Plot the linearized level responses on the same graph.
(b) Plot the nonlinear level responses on the same graph.
(c) Repeat parts (a) and (b) for $\bar{H} = 4, 16$ and 25 ft.

7.23 The populations of two species coexisting in the same environment are governed by the predator–prey equations

$$\frac{\mathrm{d}x}{\mathrm{d}t} = x(a - bx - cy) + u_x$$

$$\frac{\mathrm{d}y}{\mathrm{d}t} = y(-k + \lambda x) + u_y$$

where $x = x(t)$ is the population of the prey at time "t," $y = y(t)$ is the population of predators at time "t," $u_x = u_x(t)$ is the net rate of new prey introduced at time "t," $u_y = u_y(t)$ is the net rate of new predators entering the environment at time "t," and a, b, c, λ, and k are parameters of the system.

(a) Find the nontrivial equilibrium points $(\bar{x}, \bar{y})$ in the x, y plane when the two inputs are $u_x = \bar{u}_x = 0$, $u_y = \bar{u}_y = 0$, $t \geq 0$. Leave your answers for $\bar{x}$ and $\bar{y}$ in terms of the system parameters.

(b) Introduce deviation variables Δx, Δy, Δu_x, and Δu_y and choose the outputs as $\Delta y_1 = \Delta x$ and $\Delta y_2 = \Delta y$. Linearize the state equations about the operating point where $x = \bar{x}$, $y = \bar{y}$, $u_x = \bar{u}_x$, and $u_y = \bar{u}_y$ and find the linearized system matrices A, B, C, and D in terms of the system parameters.

(c) Find the transfer functions $\Delta x(s)/\Delta u_x(s)$, $\Delta x(s)/\Delta u_y(s)$, $\Delta y(s)/\Delta u_x(s)$, and $\Delta y(s)/\Delta u_y(s)$.

(d) Suppose the numerical values of the system parameters are $a = 12$, $b = 0$, $c = 2$, $k = 20$, and $\lambda = 4$. Further, let the inputs be $u_x = 1$, $t \geq 0$ and $u_y = 0$, $t \geq 0$. Simulate the nonlinear and linearized system responses starting from the point $x(0) = 0$, $y(0) = 0$ and compare results using plots of

(i) $x(t)$ vs. t and $x_{lin}(t)$ vs. t on the same graph

(ii) $y(t)$ vs. t and $y_{lin}(t)$ vs. t on the same graph

(iii) $y(t)$ vs. $x(t)$ and $y_{lin}(t)$ vs. $x_{lin}(t)$ on the same graph

Comment on the results.

(e) Repeat part (d) with $u_x = 0$, $t \geq 0$, and $u_y = -1$, $t \geq 0$.

(f) Repeat parts (d) and (e) with $x(0) = 1$ and $y(0) = 1$.

7.24 The dynamics of an inverted pendulum with physical parameters $M = 4$ kg, $m = 0.15$ kg, and $l = 0.8$ m is to be linearized about a steady-state operating point $(\underline{x}^0; u^0) = [\underline{x}^0, \dot{x}^0, \theta^0, \dot{\theta}^0; u^0] = [0 \quad 0 \quad 0 \quad 0; 0]^T$.

(a) Find the linearized system matrices A, B, C, and D

(i) Analytically

(ii) Using "linmod"

(iii) By numerical approximation using a central difference approximation formula with suitably small Δ

(b) Find the eigenvalues of the coefficient matrix A for the three methods in part (a). Comment on the results.

(c) Use the A, B, C, and D matrices resulting from the analytical approach and simulate $\theta(t)$, $t \geq 0$ in response to the pulse input $u(t) = 0.01$ N, $1 \leq t \leq 2$.

(d) Simulate the nonlinear system response for $\theta(t)$, $t \geq 0$ due to the same input in part (c). Compare the linearized and nonlinear responses.

7.25 For the two-tank system in Figure 7.66,

(a) Show that the solution of the linearized differential equations in Equations 7.183 and 7.184 is given in Equations 7.193 and 7.194.

(b) Check the solution at $t = 0$ and $t = \infty$

(c) The system in Example 7.3 is linearized about a steady-state operating point where $\bar{H}_1 = 1$ ft. Find expressions for $\bar{H}_2$ and $\bar{F}_0$ in terms of $\bar{H}_1$ and the system parameters c_1, c_2, A_1, and A_2 and then evaluate them numerically.

(d) Plot the linearized system response for constant inputs of $\tilde{F}_0 = 2, 4, 8 \, \text{ft}^3/\text{min}$ and both tanks initially empty.

(e) Simulate the nonlinear system dynamics for the same constant input values, and plot the responses on the same graph used for the linearized system responses. Comment on the results.

7.26 Repeat Example 7.3 for the case where the tanks interact, that is, the flow out of the first tank F_1 enters tank 2 at the bottom and is modeled by

$$F_1 = c_{12}(H_1 - H_2)^{1/2}$$

where $c_{12} = 2 \, \text{ft}^3/\text{min/ft}^{1/2}$.

7.27 The nonlinear pendulum in Figure E7.27 is modeled by $J\ddot{\theta} + c\dot{\theta} + mgr\sin\theta = 0$.

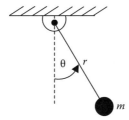

FIGURE E7.27

(a) The state components are $x_1 = \theta$ and $x_2 = \dot{\theta}$. Find the state derivative functions $f_1(x_1, x_2)$ and $f_2(x_1, x_2)$ in the state equations $\dot{x}_1 = f_1(x_1, x_2)$ and $\dot{x}_2 = f_2(x_1, x_2)$.
(b) Linearize the state equations about the equilibrium point $x_1 = 0$ rad, $x_2 = 0$ rad/s.
(c) The initial conditions are $x_1(0) = 0$ rad, $x_2(0) = 0.1$ rad/s. Find expressions for the linearized responses $x_1(t)$, $x_2(t)$ when the system parameters are

$$m = 0.25 \text{ slugs}, \quad r = 2 \text{ ft}, \quad c = 0.1 \text{ ft lb/rad/s}, \quad J = mr^2 = 1 \text{ ft lb s}^2$$

(d) Obtain the nonlinear system response by simulation, and plot the linearized and nonlinear system responses on the same graph.
(e) Repeat parts (c) and (d) when the initial conditions are $x_1(0) = 0.25$ rad, $x_2(0) = 0$ rad/s.

7.5 ADDING BLOCKS TO THE SIMULINK® LIBRARY BROWSER

7.5.1 INTRODUCTION

In order to keep development costs down, previously verified and validated models are reused in the development of new simulations. A verified model means the model was built right, whereas a validated model means the right model was built. As an example, in Section 5.12, various Kalman filters (continuous, discrete, and steady-state continuous) were developed in Simulink. These models were verified and validated by comparing them to known results (outputs and plots) from MATLAB scripts. It would be beneficial if these models were made available for use by other members of the simulation development team. What follows is the process by which models are added to a library and made available for modeling through the Simulink Library Browser.

Recall from Section 5.12 the case study of Kalman filtering led to the development of three different models: the continuous-time Kalman filter (CTKF), the discrete-time Kalman filter (DTKF), and the steady-state continuous-time Kalman filter (SSCTKF) whose top-level blocks are repeated in Figures 7.75 through 7.77 for convenience.

In each of Figures 7.75 through 7.77, the Kalman filter algorithms labeled CTKF Estimates, DTKF Estimates, and SSCTKF Estimates, respectively, have been selected to identify which particular blocks will be added to the library. In order to make these filters available to developers as individual drag and drop blocks within the Simulink Library browser, follow the procedure outlined next.

In Simulink, click on File → New → Library as shown in Figure 7.78.

This action opens an untitled library window shown in Figure 7.79.

Simply drag and drop the CTKF estimates block into the untitled library window. The result of this action is shown in Figure 7.80.

Repeating this procedure for the DTKF Estimates block and the SSCTKF block results in Figures 7.81 and 7.82.

From the library window, click File → Save to save the blocks into the library as shown in Figure 7.83.

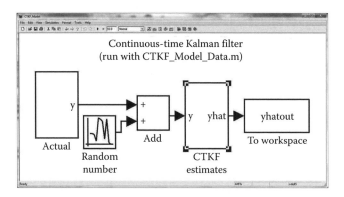

FIGURE 7.75 Continuous-time Kalman filter.

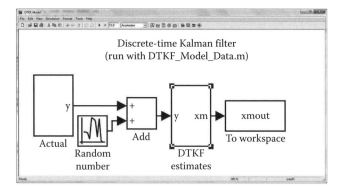

FIGURE 7.76 Discrete-time Kalman filter.

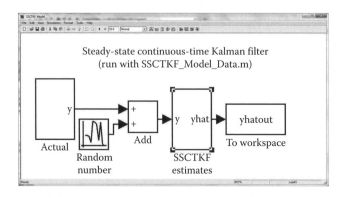

FIGURE 7.77 Steady-state continuous-time Kalman filter.

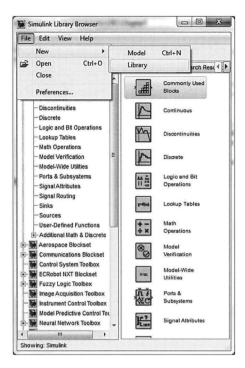

FIGURE 7.78 Creating a library.

FIGURE 7.79 Library: untitled.

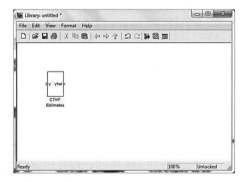

FIGURE 7.80 Drag and drop of CTKF estimates.

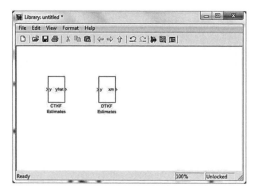

FIGURE 7.81   Drag and drop of DTKF estimates.

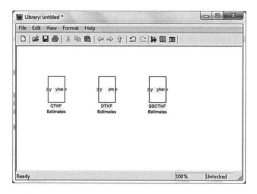

FIGURE 7.82   Drag and drop of SSCTKF estimates.

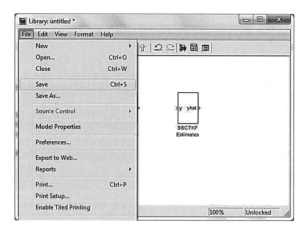

FIGURE 7.83   Saving the blocks into the library.

In the Save As dialog box, enter the name of the library, chosen here as "kflib" (to represent Kalman filter library) in Figure 7.84.

Once the library is saved, the name will change from "Library: untitled*" to "Library: kflib" as shown in Figure 7.85.

The next step in the process is the creation of the S-block M-file to load the library when Simulink is started. To view a template, type "edit slblocks" in the MATLAB command window. The M-file template is given as follows where executable lines are identified in bold.

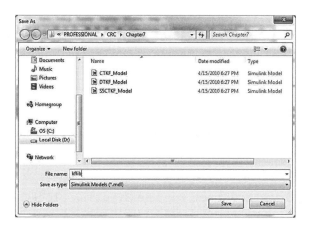

FIGURE 7.84 Saving the Kalman filter library, kflib.

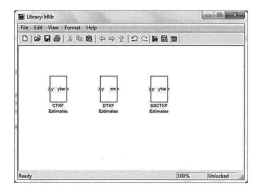

FIGURE 7.85 Library: kflib.

```
function blkStruct = slblocks
%SLBLOCKS Defines the block library for a specific Toolbox or Block-
   set. SLBLOCKS returns information about a Blockset to
% Simulink. The information returned is in the form of a Blockset-
   Struct with the following fields:
%
% Name     Name of the Blockset in the Simulink block library Block-
   sets & Toolboxes subsystem.
% OpenFcn   MATLAB expression (function) to call when you double-
   click on the block in the Blocksets & Toolboxes
% subsystem.
% MaskDisplay Optional field that specifies the Mask Display com-
   mands to use for the block in the Blocksets & Toolboxes
% subsystem.
% Browser    Array   of   Simulink   Library   Browser   structures,
   described below.
%
% The Simulink Library Browser needs to know which libraries in your
   Blockset it should show, and what names to give them. To
```

```
% provide this information, define an array of Browser data struc-
  tures with one array element for each library to display in the
% Simulink Library Browser. Each array element has two fields:
%
% Library    File name of the library (mdl-file) to include in the
  Library Browser.
% Name    Name displayed for the library in the Library Browser
  window. Note that the Name is not required to be the
% same as the % mdl-file name.
%
% Example:
% %Define the BlocksetStruct for the Simulink block libraries
% %Only simulink_extras shows up in Blocksets & Toolboxes
% %
% blkStruct.Name = ['Simulink' sprintf('\n') 'Extras'];
% blkStruct.OpenFcn = 'simulink_extras';
% blkStruct.MaskDisplay = sprintf('Simulink\nExtras');
%
% %
% % Both simulink and simulink_extras show up in the Library Browser.
% %
% blkStruct.Browser(1).Library = 'simulink';
% blkStruct.Browser(1).Name = 'Simulink';
% blkStruct.Browser(2).Library = 'simulink_extras';
% blkStruct.Browser(2).Name = 'Simulink Extras';
%
% Copyright 1990-2006 The MathWorks, Inc.
% $Revision: 1.20.2.10 $


% Name of the subsystem which will show up in the Simulink Blocksets
  and Toolboxes subsystem.
blkStruct.Name = ['Simulink' sprintf('\n') 'Extras'];


% The function that will be called when the user double-clicks on this
  icon.
blkStruct.OpenFcn = 'simulink_extras';


% The argument to be set as the Mask Display for the subsystem. You may
  comment this line out if no specific mask is desired.
%       Example:       blkStruct.MaskDisplay = 'plot([0:2*pi],sin
  ([0:2*pi]));'; No display for Simulink Extras.
blkStruct.MaskDisplay = '';


% Define the Browser structure array, the first element contains the
  information for the Simulink block library and the second for the %
  Simulink Extras block library.
Browser(1).Library = 'simulink';
Browser(1).Name = 'Simulink';
Browser(1).IsFlat = 0;% Is this library ''flat'' (i.e. no subsystems)?
Browser(2).Library = 'simulink_extras';
Browser(2).Name = 'Simulink Extras';
```

```
Browser(2).IsFlat = 0;% Is this library ''flat'' (i.e. no subsystems)?
blkStruct.Browser = Browser;
clear Browser;

% Define information about Signal Viewers
Viewer(1).Library = 'simviewers';
Viewer(1).Name = 'Simulink';
blkStruct.Viewer = Viewer;
clear Viewer;

% Define information about Signal Generators
Generator(1).Library = 'simgens';
Generator(1).Name = 'Simulink';

blkStruct.Generator = Generator;
clear Generator;

% Define information for model updater
blkStruct.ModelUpdaterMethods.fhSeparated-
  Checks = @UpdateSimulinkBlocksHelper;

% End of slblocks
```

For the Kalman filter library, the simplified S-block M-file was edited as shown in Figure 7.86. The primary changes are blkStruct.Name, Browser.Library, and Browser.Name on lines 4, 11, and 12, respectively. This file must be saved as "slblocks.m" in the same folder as the library file in order for Simulink to acknowledge existence of the "kflib" library at startup.

When Simulink is started, Figure 7.87 shows the Simulink Library Browser with an exploding directory named "My Kalman Filters" containing the three Kalman filters "CTKF Estimates," "DTKF Estimates," and "SSCTKF Estimates," which are now available to drag and drop into a Simulink model.

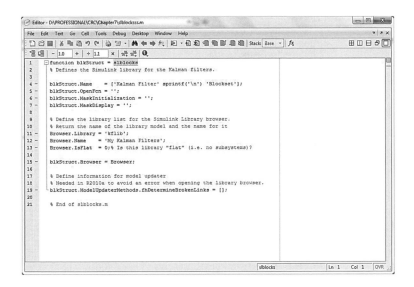

FIGURE 7.86 S-block M-file "slblocks.m."

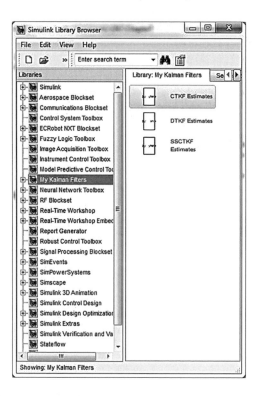

FIGURE 7.87 Simulink® Library Browser with my Kalman filters.

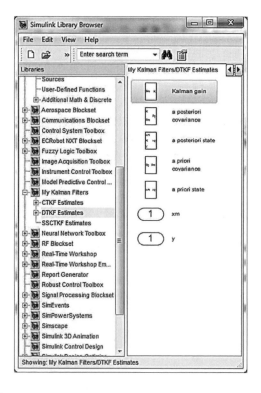

FIGURE 7.88 DTKF estimates subsystems.

By double-clicking on "DTKF Estimates" in the right window of the Simulink Library Browser, the subsystems of the algorithm (Kalman gain, a posteriori covariance, a posteriori state, a priori covariance, and a priori state) are displayed in the window as shown in Figure 7.88. These are also available for dragging and dropping for developing Simulink models.

7.5.2 SUMMARY

This section demonstrated how to create a Simulink library and add it to the Simulink Library Browser, thereby making custom models available to other members of a development team.

EXERCISE

7.28 In Simulink, create a simple model for the equation of a line $y = mx + b$ where x is the input signal, m is a gain block on the input signal, b is a constant block added to the output of the gain block, and y is the output signal. Once the model is built, create a library named "linelib" and add it to the Simulink Library Browser by editing the slblocks.m file accordingly.

7.6 SIMULATION ACCELERATION

7.6.1 INTRODUCTION

The default simulation option in Simulink is Normal mode. It is set by clicking Simulation → Normal in Simulink as shown in Figure 7.89 for the discrete-time Kalman filter model from Section 5.12. In this mode, the simulation is executed as a single (interpreted) process within the MATLAB/Simulink environment. Normal mode supports debugging, M-files, scopes/viewers, run-time diagnostics, parameter tuning, and algebraic loops. However, depending on the level of fidelity built into the Simulink model, or depending on how many replications of the Simulink model are run, the simulation could consume a lot of the user's time.

Simulink offers two *compiled* options: Accelerator mode (Figure 7.90) and Rapid Acceleration mode (Figure 7.91).

In Accelerator mode, the simulation executes as a single (compiled) process within the MATLAB/Simulink environment. This mode supports debugging, M-files, and scopes/viewers,

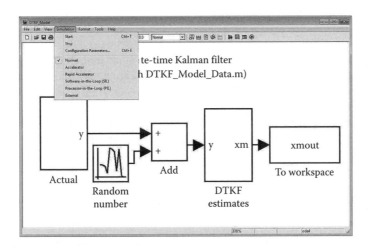

FIGURE 7.89 Normal simulation.

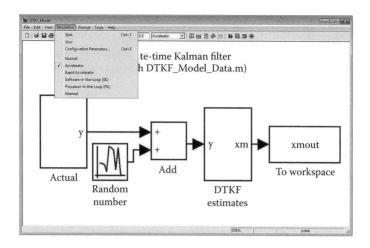

FIGURE 7.90   Accelerator mode simulation.

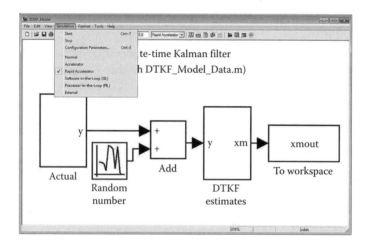

FIGURE 7.91   Rapidly accelerated simulation.

and allows the user to tune parameters. In order to run the simulation in Accelerator mode, simply select Simulation → Accelerator in Simulink as shown in Figure 7.90 and execute the model.

In Rapid Accelerator mode, the simulation executes as two separate processes: MATLAB/Simulink running as one process while another compiled process runs in parallel. This mode only supports scopes/viewers and parameter tuning. In order to run the simulation in Rapid Accelerator mode, simply select Simulation → Rapid Accelerator in Simulink as shown in Figure 7.91 and execute the model.

After selecting Accelerator mode for the discrete-time Kalman filter model, the MATLAB Command Window displays the following message just before running the simulation.

Building the *Accelerator* target for model: DTKF_Model
Successfully built the *Accelerator* target for model: DTKF_Model

This message (with italics added) indicates that Accelerator mode was selected and that a compiled version was created.

After selecting Rapid Accelerator mode for the discrete-time Kalman filter model, the MATLAB Command Window displays the following message just before running the simulation.

Building the *rapid accelerator* target for model: DTKF_Model
Successfully built the *rapid accelerator* target model: DTKF_Model

This message (with italics added) indicates that Rapid Accelerator mode was selected and that a compiled version was created. Note that while both Accelerator mode and Rapid Accelerator mode use aspects of MATLAB's Real-Time Workshop, the user does not need Real-Time Workshop to accelerate simulations. However, the user does need Real-Time Workshop to generate source code for other purposes.

One final comment is that the Rapid Accelerator mode lends itself toward running Monte Carlo simulations. Please see Section 5.10 for more information.

7.6.2 PROFILER

Sometimes, the user would like to know which sections of the simulation are consuming the most time. Simulink provides a tool called the Profiler for such analysis. To turn on the Profiler, click Tools → Profiler in Simulink (Figure 7.92) and rerun the simulation.

Output from the Profiler for the discrete-time Kalman filter simulation is shown in Figure 7.93. By examining this information, the user can take action (e.g., change an algorithm) to help reduce the amount of time the simulation is spending in any one area.

7.6.3 SUMMARY

This section demonstrated how to accelerate simulations using the Accelerator mode and the Rapid Accelerator mode. While both modes are compiled and are generally faster than the Normal (interpreted) mode, which one to use depends on the type of tool support needed. Both compiled modes support scopes/viewers and parameter tuning, but only Accelerator mode adds debugging and M-file support.

This section also briefly mentioned the Profiler—a tool to assist users in examining which areas of the simulation require the most amount of execution time. With this knowledge, the user can augment algorithms to increase simulation performance. This can pay dividends, particularly if the user is running many replications of a Monte Carlo simulation with a high level of fidelity.

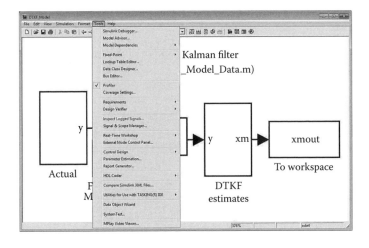

FIGURE 7.92 Simulink's profiler.

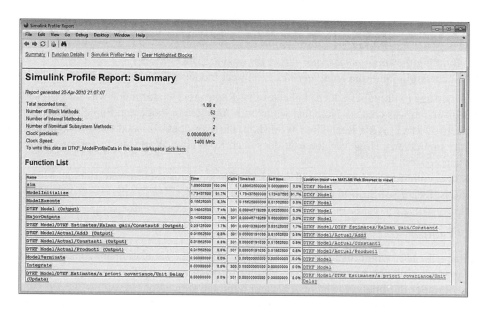

FIGURE 7.93   Profiler output (partial).

EXERCISE

7.29 Open the discrete-time Kalman filter from Section 5.12.

(a) Run the model in Normal mode.

(b) Run the model in Accelerator mode.

(c) Run the model in Rapid Accelerator mode.

(d) Turn on the Profiler and run the model to see where the model spends most of its time.

8 Advanced Numerical Integration

8.1 INTRODUCTION

Dynamic errors, an important aspect in digital simulation of dynamic systems, are introduced. Instead of focusing on truncation errors, the simulationist may be more concerned with errors in dynamic response, a yardstick of simulation accuracy involving comparisons of transient and sinusoidal responses of continuous-time and discrete-time models.

The subject of dynamic errors has been covered in great detail by Howe (1986). The commonly used numerical integrators are analyzed by considering the characteristic roots, magnitude, and phase properties of the "equivalent continuous-time system," that is, the continuous-time system whose sampled values coincide with the discrete-time (simulated) system outputs. The connection between digital simulation and discrete-time systems is further illustrated by exploring the subject of stability in both arenas.

Stiff systems, initially introduced in Chapter 6, are once again considered. Multirate integration schemes are presented as an alternative to the use of stiff integrators for the case where the overall system can be decomposed into several interconnected subsystems operating at different speeds.

Real-time simulation is a specialized application involving interactions between a digital simulation and real-time inputs from physical components or a human operator. The necessity of synchronizing with signals to and from external components places additional constraints on the simulation environment and numerical integrators. Real-time compatible numerical integrators are discussed along with numerical integrators not suitable for real-time implementation and an explanation of why they are not.

The chapter concludes with a look at some additional techniques for developing discrete-time models intended to approximate the dynamic behavior of linear time-invariant (LTI) continuous-time models.

8.2 DYNAMIC ERRORS (CHARACTERISTIC ROOTS, TRANSFER FUNCTION)

The use of numerical integrators to simulate the behavior of continuous-time systems introduces errors, that is, the transient and steady-state behavior of the discrete-time responses differs from that of the continuous-time outputs at the times where the simulated response is computed. Some insight with respect to the differences is possible by considering expressions for the truncation errors inherent in the various types of numerical integrators. We know that the local and global truncation errors are sensitive to the integration step size and the state derivative functions which define the continuous-time system model.

The differences in transient and steady-state sinusoidal responses are termed dynamic errors. Truncation errors, on the other hand, relate numerical solutions of differential equation models to various-order Taylor Series expansions of the continuous-time solutions. A mathematical framework for comparing dynamic errors resulting from numerical integration of linear continuous-time models is possible. Given that real-world system models are invariably nonlinear, the first step is therefore to linearize the system of nonlinear differential and algebraic equations about a steady-state operating point, similar to the procedures discussed in Section 7.4.

The dynamic errors associated with the use of fixed-step numerical integrators applied to linear system models fall in one of two categories (Howe 1986). One type of error focuses on differences between characteristic roots of the continuous-time system model and the apparent or equivalent continuous-time system. By equivalent continuous-time system, we mean the continuous-time system that generates sampled values identical with the discrete-time (simulated) system.

The second type of error relates to differences between the frequency response function of the continuous-time system and the discrete-time system used to approximate its behavior. Only linear first- and second-order systems will be considered because higher-order systems can be represented as linear combinations of these lower-order subsystems.

8.2.1 Discrete-Time Systems and the Equivalent Continuous-Time Systems

Consider a first-order linear system modeled by

$$\frac{dx}{dt} = f(x, u) = \lambda x + u \tag{8.1}$$

The characteristic root is λ, the pole of the system transfer function

$$H(s) = \frac{X(s)}{U(s)} = \frac{1}{s - \lambda} \tag{8.2}$$

Digital simulation of the system requires solution of a difference equation obtained by numerical integration of the state derivative function $f(x, u)$. For explicit Euler integration, the z-domain transfer function of the resulting discrete-time system can be obtained by z-transforming the difference equation

$$x_A(n + 1) = x_A(n) + T[\lambda x_A(n) + u(n)] \tag{8.3}$$

or equivalently from (see Section 4.7)

$$H(z) = H(s)\big|_{s \leftarrow \frac{z-1}{T}} = \frac{1}{s - \lambda}\bigg|_{s \leftarrow \frac{z-1}{T}} = \frac{1}{(z - 1/T) - \lambda} = \frac{T}{z - (1 + \lambda T)} \tag{8.4}$$

The discrete-time system pole is located at $z_1 = 1 + \lambda T$.

The equivalent continuous-time system is the system whose output $x(t)$, $t \geq 0$ is identical to the discrete-time output $x_A(nT)$ at times $t_n = nT$, $n = 0, 1, 2, \ldots$. To illustrate, suppose the input to the system in Equation 8.1 is $u(t) = 1$, $t \geq 0$. The response is

$$x(t) = \frac{1}{\lambda}[e^{\lambda t} - 1], \quad t \geq 0, \tag{8.5}$$

The use of explicit Euler integration with step size T to approximate the continuous-time step response produces the discrete-time approximation $x_A(n)$, short for $x_A(nT)$, $n = 0, 1, 2, \ldots$ obtained from

$$X_A(z) = H(z)U(z) \tag{8.6}$$

$$= \left[\frac{T}{z - (1 + \lambda T)}\right]\frac{z}{z - 1} \tag{8.7}$$

Partial fraction expansion of Equation 8.7 followed by inverse z-transformation of the resulting terms gives

$$x_A(n) = \frac{1}{\lambda}[(1 + \lambda T)^n - 1], \quad n = 0, 1, 2, \ldots \tag{8.8}$$

Let the equivalent continuous-time system be described by

$$\frac{dx}{dt} = f(x, u) = \lambda^* x + Ku \tag{8.9}$$

where λ^* and K are the characteristic root and gain parameter of the equivalent first-order continuous-time system, respectively. The step response is

$$x^*(t) = \frac{K}{\lambda^*}[e^{\lambda^* t} - 1], \quad t \geq 0 \tag{8.10}$$

Sampling the equivalent continuous-time system response every $T(s)$ gives

$$x^*(nT) = \frac{K}{\lambda^*}[e^{\lambda^* nt} - 1], \quad n = 0, 1, 2, \ldots \tag{8.11}$$

Equating the discrete-time responses in Equations 8.8 and 8.11,

$$\frac{1}{\lambda}[(1 + \lambda T)^n - 1] = \frac{K}{\lambda^*}[e^{\lambda^* nT} - 1], \quad n = 0, 1, 2, \ldots \tag{8.12}$$

Solving for K and λ^*,

$$e^{\lambda^* nT} = (1 + \lambda T)^n \Rightarrow \lambda^* = \frac{1}{T}\ln(1 + \lambda T) \tag{8.13}$$

$$\frac{K}{\lambda^*} = \frac{1}{\lambda} \Rightarrow K = \frac{\lambda^*}{\lambda} = \frac{\ln(1 + \lambda T)}{\lambda T} \tag{8.14}$$

The step response of the first-order continuous-time system in Equation 8.1 with characteristic root $\lambda = -2$ is shown in Figure 8.1. Also shown is the step response of the discrete-time system in Equation 8.3 corresponding to explicit Euler integration of the derivative function with step size $T = 0.05$ s. The step response of the equivalent continuous-time system in Equation 8.9 with λ^* and K computed from Equations 8.13 and 8.14 is also shown.

From Equation 8.4, the pole of the discrete-time system is $z_1 = 1 + \lambda T$. Replacing $1 + \lambda T$ in Equation 8.13 with z_1 leads to an expression relating the characteristic root of the equivalent continuous-time system and the pole of the discrete-time system. That is,

$$\lambda^* = \frac{1}{T}\ln z_1 \tag{8.15}$$

Solving Equation 8.15 for the discrete-time system pole leads to

$$z_1 = e^{\lambda^* T} \tag{8.16}$$

Uniform sampling of the equivalent continuous-time system response $x^*(t)$ every T s generates the discrete-time system signal $x_A(n)$ with pole z_1 given in Equation 8.16. In the general case, sampling

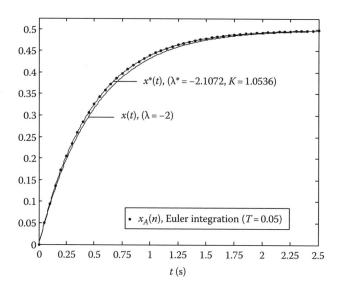

FIGURE 8.1   Step response of continuous-time, discrete-time, and equivalent continuous-time systems.

continuous-time signals with real and complex poles produces discrete-time system signals with z-plane poles given by

$$z_1 = e^{Ts_i}, \quad i = 1, 2, \ldots, n \tag{8.17}$$

where s_i are the poles of the continuous-time system (Jacquot).

Equation 8.17 applies to LTI systems and their characteristic roots as well. The sampled output of a continuous-time system with characteristic root (s-plane pole) s_1 is identical to the output from a discrete-time system with characteristic root (z-plane pole) located at $z_1 = e^{Ts_i}$. Looking at it from the opposite direction, the continuous-time system equivalent to a discrete-time system with a pole z_1 has an s-plane pole at $s_1 = 1/T \times \ln z_1$.

According to Equation 8.17, a continuous-time integrator with a pole at $s = 0$ in the s-plane is the continuous-time system equivalent to a discrete-time system with a pole at $z = 1$. Figure 8.2 illustrates the point by showing that a pure integrator generates a continuous-time signal $x(t)$ in response to the input $u(t)$, which matches the response of the discrete-time system with z-domain transfer function $H(z) = K/(z - 1)$ at the discrete times $t_n = nT$, $n = 0, 1, 2, \ldots$

Suppose $u(t) = e^{-at}$, $t \geq 0$ is the input to the integrator. The output $x(t)$ is

$$x(t) = \int_0^t u(t)dt = \frac{1}{a}(1 - e^{-at}), \quad t \geq 0 \tag{8.18}$$

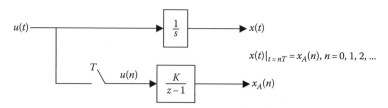

FIGURE 8.2   An integrator as the equivalent continuous-time system to a discrete-time system with pole at $z = 1$.

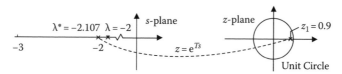

FIGURE 8.3 Mapping $z = e^{Ts}$ for finding the equivalent continuous-time system characteristic root $\lambda^* = -2.107$ when $z_1 = 0.9$, $T = 0.05$.

The discrete-time response is found from inverse z-transformation of

$$X(z) = \left(\frac{K}{z-1}\right)\frac{z}{z - e^{-aT}} = \frac{K}{1 - e^{-aT}}\left[\frac{z}{z-1} - \frac{z}{z - e^{-aT}}\right] \tag{8.19}$$

$$\Rightarrow x_A(n) = \frac{K}{1 - e^{-aT}}(1 - e^{-anT}), \quad n = 0, 1, 2, \ldots \tag{8.20}$$

and it follows that $x(nT) = x_A(n)$, $n = 0, 1, 2, \ldots$ provided

$$K = \frac{1 - e^{-aT}}{a} \tag{8.21}$$

Figure 8.3 shows the characteristic root of the continuous-time system in Equation 8.2 for the case when $\lambda = -2$. The pole of the discrete-time system resulting from explicit Euler integration with step size $T = 0.05$ is located at $z_1 = (1 + \lambda T) = 1 + (-2)(0.05) = 0.9$ in the z-plane. The characteristic root of the equivalent continuous-time system is $\lambda^* = 1/T \times \ln z_1 = 1/0.05 \times \ln 0.9 = -2.107$ in the s-plane.

8.2.2 Characteristic Root Errors

The fractional error in characteristic root incurred using numerical integration for digital simulation of a first-order continuous-time system with characteristic root λ is defined as (Howe 1986)

$$e_\lambda = \frac{\lambda^* - \lambda}{\lambda} \tag{8.22}$$

For an underdamped second-order system with complex poles $\lambda_{1,2} = -\zeta\omega_n \pm j\omega_d$ where ζ and ω_n are the damping ratio and natural frequency, respectively, and $\omega_d = \sqrt{1 - \zeta^2}\omega_n$ is the damped natural frequency, the characteristic root errors are

$$e_\zeta = \zeta^* - \zeta, \quad e_{\omega_n} = \frac{\omega_n^* - \omega_n}{\omega_n}, \quad e_{\omega_d} = \frac{\omega_d^* - \omega_d}{\omega_d} \tag{8.23}$$

ζ^*, ω_n^*, and ω_d^* are the damping ratio, natural frequency, and damped natural frequency of the equivalent continuous-time second-order system, respectively. The characteristic roots of the equivalent continuous-time system are

$$\lambda_{1,2}^* = -\zeta^*\omega_n^* \pm j\omega_d^* = -\zeta^*\omega_n^* \pm j\sqrt{1 - (\zeta^*)^2}\omega_n^* \tag{8.24}$$

High-order linear continuous-time systems can be represented as the sum of first- and second-order continuous-time systems. Hence, the characteristic root errors introduced in Equations 8.22 and 8.23 are sufficient to analyze transient response dynamic errors of higher-order systems comprising first- and second-order subsystems.

Example 8.1

The first-order system in Equation 8.1 is simulated using trapezoidal integration.

(a) Find an expression for e_λ, the fractional error in characteristic root.
(b) Find an asymptotic formula for e_λ valid for $|\lambda T| \ll 1$.
(c) Over what range of values for λT is the asymptotic formula for e_λ accurate?

(a) The difference equation for trapezoidal integration is based on

$$x_A(n+1) = x_A(n) + \frac{T}{2}\{f[x_A(n), u(n)] + f[x_A(n+1), u(n+1)]\} \tag{8.25}$$

where $f[x_A(n), u(n)]$ and $f[x_A(n+1), u(n+1)]$ refer to the derivative function in Equation 8.1. Z-transforming the difference equation and then solving for the ratio $X(z)/U(z)$ results in the z-domain transfer function

$$H(z) = \frac{X(z)}{U(z)} = T\left[\frac{z+1}{(2-\lambda T)z - (2+\lambda T)}\right] \tag{8.26}$$

The z-plane pole is

$$z_1 = \frac{2+\lambda T}{2-\lambda T} = \frac{1+\lambda T/2}{1-\lambda T/2} \tag{8.27}$$

From Equation 8.15, the characteristic root of the equivalent continuous-time system is

$$\lambda^* = \frac{1}{T}\ln z_1 = \frac{1}{T}\ln\left(\frac{1+\lambda T/2}{1-\lambda T/2}\right) \tag{8.28}$$

and the fractional error in characteristic root is

$$e_\lambda = \frac{\lambda^*}{\lambda} - 1 = \frac{1}{\lambda T}\ln\left(\frac{1+\lambda T/2}{1-\lambda T/2}\right) - 1 \tag{8.29}$$

(b) Equation 8.29 is expressed in the form

$$e_\lambda = \frac{1}{\lambda T}\left[\ln\left(1+\frac{\lambda T}{2}\right) - \ln\left(1-\frac{\lambda T}{2}\right)\right] - 1 \tag{8.30}$$

The asymptotic formula for e_λ is obtained by truncating the Taylor Series expansion

$$\ln(1+a) = a - \frac{a^2}{2} + \frac{a^3}{3} - \frac{a^4}{4} + \cdots \tag{8.31}$$

after the cubic term where $a = \lambda T/2$ and $a = -\lambda T/2$ in Equation 8.30. After simplification, the result is

$$e_\lambda \approx \frac{1}{12}(\lambda T)^2, \quad |\lambda T| \ll 1 \tag{8.32}$$

(c) A plot of the exact and asymptotic formulas for e_λ is shown in Figure 8.4. From the graph, it appears that the exact and asymptotic formulas for e_λ are nearly identical for $-0.5 \le \lambda T < 0$.

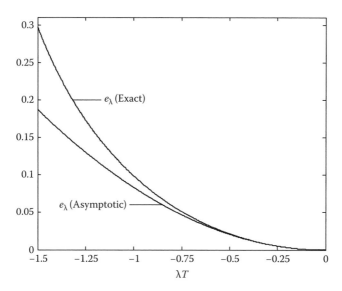

FIGURE 8.4 Exact and asymptotic fractional characteristic root errors for trapezoidal integration (with step size T) of first-order system $\dot{x} = \lambda x + u$.

The first-order continuous-time system in Equation 8.1 is asymptotically stable provided $\lambda < 0$. The graphs of exact and asymptotic error in Figure 8.4 are for $\lambda T < 0$; hence, they apply strictly to asymptotically stable, first-order systems. Equations 8.29 and 8.32 are not valid for $\lambda T = 0$, that is, when the continuous-time system reduces to a marginally stable integrator with characteristic root $\lambda = 0$.

Consider the use of trapezoidal integration with step size T to simulate the autonomous first-order system $\dot{x} = f(x) = \lambda x$ with initial condition $x(0)$. The discrete-time signal $x_A(n)$ satisfies the difference equation

$$x_A(n+1) = \left(\frac{1 + \lambda T/2}{1 - \lambda T/2}\right) x_A(n), \quad n = 0, 1, 2, 3, \ldots \tag{8.33}$$

with solution given by

$$x_A(n) = \left(\frac{1 + \lambda T/2}{1 - \lambda T/2}\right)^n x(0), \quad n = 0, 1, 2, 3, \ldots \tag{8.34}$$

Table 8.1 summarizes the results for a first-order system with characteristic root $\lambda = -0.5$ simulated using trapezoidal integration with four different step sizes. The results are consistent with the graphs in Figure 8.4.

TABLE 8.1
Effect of Parameter λT on Equivalent Characteristic Root and Fractional Characteristic Root Errors with Trapezoidal Integration

λ	T	λT	λ^*	e_λ (Exact)	e_λ (Asymptotic)
−0.5	0.015	−0.0075	−0.500002	4.68754×10^{-6}	4.68750×10^{-6}
−0.5	0.15	−0.075	−0.500234	4.69146×10^{-4}	4.68750×10^{-4}
−0.5	1.5	−0.75	−0.526538	5.12764×10^{-4}	1.30208×10^{-1}

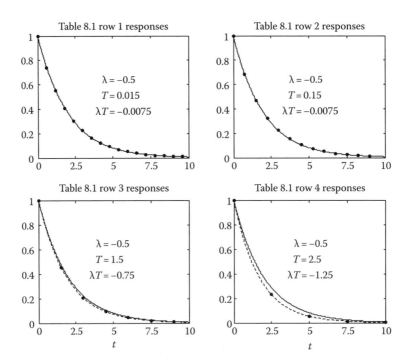

FIGURE 8.5 Responses of first-order continuous-time, discrete-time, dashed line, and equivalent continuous-time systems for conditions in Table 8.1.

Several different responses are shown in Figure 8.5. The top two plots show the response of the continuous-time system and the discrete-time response corresponding to the top two rows in Table 8.1. Due to the close agreement between λ and λ^*, the response of the equivalent continuous-time system is indistinguishable from the response of the actual system. Additionally, the discrete-time output (not all points shown) is in close agreement with the continuous-time response at times $0, T, 2T, \ldots$.

In the last two cases ($\lambda T = -0.75$ and $\lambda T = -1.25$), the difference between λ and λ^* is significant, and the response of the equivalent continuous-time system is noticeably different from the actual system response, particularly for the case where $\lambda T = -1.25$. The simulated (discrete-time) response is off as well.

Characteristic root errors resulting from simulation of second-order systems using specific numerical integrators are obtained in a straightforward manner. To illustrate, consider an underdamped second-order continuous-time system with characteristic roots $\lambda_{1,2} = -\zeta\omega_n \pm j\sqrt{1 - \zeta^2}\omega_n$. Similar to the approach used in Equation 8.4, replacing the Laplace variable s in the continuous-system transfer function with the reciprocal of the z-domain transfer function for Euler integration leads to the z-domain transfer function of the discrete-time system. This gives

$$H(z) = \frac{K\omega_n^2}{s^2 + 2\zeta\omega_n s + \omega_n^2}\bigg|_{s \leftarrow \frac{z-1}{T}} \tag{8.35}$$

$$= \frac{K(\omega_n T)^2}{z^2 - 2(1 - \zeta\omega_n T)z + 1 - 2\zeta\omega_n T + (\omega_n T)^2} \tag{8.36}$$

Setting the denominator to zero and solving for the poles of $H(z)$ give

$$z_{1,2} = 1 - \zeta\omega_n T \pm j\sqrt{(1 - \zeta^2)}\omega_n T \tag{8.37}$$

From Equation 8.15, the characteristic roots of the equivalent continuous-time system are

$$s_{1,2}^* = \frac{1}{T} \ln z_{1,2} \qquad (8.38)$$

Finding an expression for s_1^* is easier when the corresponding z-plane pole z_1 is written in polar form.

$$s_1^* = \frac{1}{T} \ln (Re^{j\theta}) = \frac{1}{T} \ln R + j\frac{\theta}{T} \qquad (8.39)$$

where R and θ are obtained from Equation 8.37 (after simplification) as

$$R = \sqrt{1 - 2\zeta\omega_n T + (\omega_n T)^2} \qquad (8.40)$$

$$\theta = \tan^{-1}\left(\frac{\sqrt{1 - \zeta^2}\,\omega_n T}{1 - \zeta\omega_n T}\right) \qquad (8.41)$$

Substituting Equations 8.40 and 8.41 into Equation 8.39 gives $s_1^* = a^* + jb^*$ where the real and imaginary components a^* and b^* are given by

$$a^* = \frac{1}{T} \ln\left(\sqrt{1 - 2\zeta\omega_n T + (\omega_n T)^2}\right) \qquad (8.42)$$

$$b^* = \frac{1}{T} \tan^{-1}\left(\frac{\sqrt{1 - \zeta^2}\,\omega_n T}{1 - \zeta\omega_n T}\right) \qquad (8.43)$$

The continuous-time pole s_1, the z-plane pole z_1, and the equivalent continuous-time system pole s_1^* are shown in Figure 8.6.

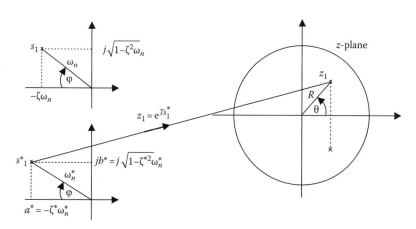

FIGURE 8.6 Relationship between second-order continuous-time system, discrete-time, and equivalent continuous-time system complex pole.

From Figure 8.6, it follows that

$$\omega_n^* = (a^{*2} + b^{*2})^{1/2} = \frac{1}{T} \left\{ \left[\ln\left(\sqrt{1 - 2\zeta\omega_n T + (\omega_n T)^2} \right) \right]^2 \right.$$
$$\left. + \left[\tan^{-1}\left(\frac{\sqrt{1 - \zeta^2}\,\omega_n T}{1 - \zeta\omega_n T} \right) \right]^2 \right\}^{1/2} \tag{8.44}$$

$$\zeta^* = \cos\varphi = \frac{-a^*}{\omega_n^*} = \frac{-\ln\left(\sqrt{1 - 2\zeta\omega_n T + (\omega_n T)^2} \right)}{\omega_n^* T} \tag{8.45}$$

Asymptotic formulas for ω_n^* and ζ^* are given in (Howe 1986) as

$$\omega_n^* \approx \left[1 + \frac{\zeta\omega_n T}{2} \right] \omega_n, \quad \omega_n T \ll 1 \tag{8.46}$$

$$\zeta^* \approx \zeta - \left(\frac{1 - \zeta^2}{2} \right) \omega_n T, \quad \omega_n T \ll 1 \tag{8.47}$$

Exact and approximate (asymptotic) expressions for the fractional error in natural frequency of the equivalent continuous-time system

$$e_{\omega_n} = \left(\frac{\omega_n^*}{\omega_n} \right) - 1 \tag{8.48}$$

are obtained from Equation 8.44 for the exact result and Equation 8.46 for the asymptotic one. Figure 8.7 shows exact and asymptotic fractional errors for several second-order continuous-time system damping ratios using explicit Euler integration.

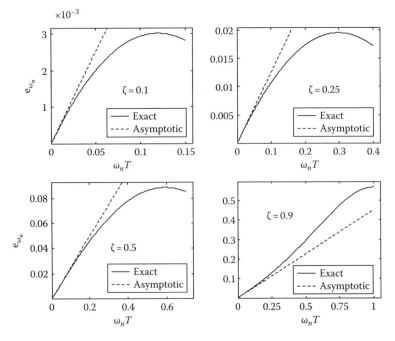

FIGURE 8.7   Exact and asymptotic fractional errors in natural frequency with explicit Euler integration.

Substituting Equation 8.46 into Equation 8.48 results in the asymptotic fractional error as a linear function of $\omega_n T$, that is,

$$e_{\omega_n} \approx 0.5\zeta(\omega_n T), \quad \omega_n T \ll 1 \tag{8.49}$$

From Equation 8.45, the damping ratio error e_ζ is expressible as

$$e_\zeta = \zeta^* - \zeta = \frac{-\ln\left(\sqrt{1 - 2\zeta\omega_n T + (\omega_n T)^2}\right)}{\omega_n^* T} - \zeta \tag{8.50}$$

where ω_n^* is given in Equation 8.44. From Equation 8.47, the asymptotic approximation for e_ζ is

$$e_\zeta \approx 0.5(\zeta^2 - 1)\omega_n T, \quad \omega_n T \ll 1 \tag{8.51}$$

The asymptotic expressions for e_{ω_n} in Equation 8.49 and e_ζ in Equation 8.51 are of order $O(\omega_n T)$ when using Euler integration to simulate an underdamped second-order system.

A plot of the exact and asymptotic formulas for the equivalent system damping ratio ζ^* as a function of $\omega_n T$ when $\zeta = 0.1$ is shown in the top half of Figure 8.8. Agreement between the two plots is excellent over the interval $0 \le \omega_n T \le 0.5$.

The equivalent continuous-time system is marginally stable when its two characteristic roots (transfer function poles) are purely imaginary, that is, $\zeta^* = 0$ (see Figure 8.6). From Equation 8.51 with $\zeta = 0.1$ and $\zeta^* = 0$, the dimensionless parameter $\omega_n T$ is computed as

$$\zeta^* - \zeta = 0 - 0.1 \approx 0.5[(0.1)^2 - 1]\omega_n T \tag{8.52}$$

$$\Rightarrow \omega_n T = \frac{-0.1}{0.5(-0.99)} = 0.202$$

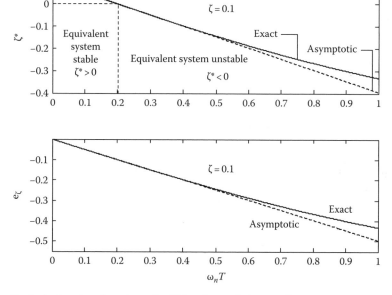

FIGURE 8.8 Equivalent system damping ratio (ζ^*) and error (e_ζ) vs. $\omega_n T$.

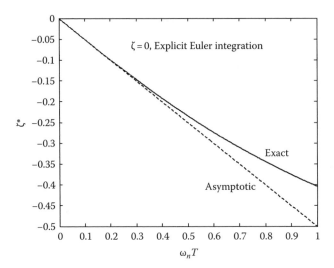

FIGURE 8.9   Equivalent system damping ratio using explicit Euler integration.

The damping ratio ζ^* of the equivalent continuous-time system is negative whenever $\omega_n T > 0.202$. The implication of $\zeta^* < 0$ is obvious from Figure 8.6, namely, the characteristic roots are located in the right half of the complex plane, and the equivalent system is unstable despite the fact that the actual continuous-time system is asymptotically stable with positive damping ratio $\zeta = 0.1$. The lower half of Figure 8.8 shows plots of the error $e_\zeta = \zeta^* - \zeta = \zeta^* - 0.1$ vs. $\omega_n T$ based on the exact and asymptotic formulas in Equations 8.50 and 8.51.

Figure 8.9 points out a serious shortcoming of using explicit Euler integration to simulate the response of a marginally stable ($\zeta = 0$) second-order system. The equivalent continuous-time system is unstable because $\zeta^* < 0$ regardless of how small $\omega_n T$ is chosen. The natural modes of the equivalent continuous-time system are oscillatory with increasing amplitude. The discrete-time system based on the use of explicit Euler integration is likewise unstable with a pair of complex poles outside the Unit Circle. This problem can be fixed by using trapezoidal integration instead of Euler integration (see Exercise 8.2).

Figures 8.7 through 8.9 are generated in M-file "Chap8_Fig2_7throughFig2_9.m."

Example 8.2

A second-order system with damping ratio $\zeta = 0.1$, natural frequency $\omega_n = 50$ rad/s, and steady-state gain $K = 1$ is initially in equilibrium. A unit step input is applied at $t = 0$. The step response is simulated using explicit Euler integration with step size T.

(a) Find the step response $x(t)$, $t \geq 0$.
(b) Find the equivalent system natural frequency ω_n^* and damping ratio ζ^* for $T = 0.001, 0.002,$ 0.004, 0.005 s.
(c) Plot the continuous-time system response $x(t)$ and the discrete-time system response $x_A(n)$, $n = 0, 1, 2, \ldots$ corresponding to the values of T in part (b).

(a) The unit step response of an underdamped second-order system is (see Chapter 2)

$$x(t) = K\left[1 - e^{-\zeta\omega_n t}\left(\cos\omega_d t + \frac{\zeta\omega_n}{\omega_d}\sin\omega_d t\right)\right], \quad t \geq 0 \tag{8.53}$$

TABLE 8.2
Comparison of Actual System and Equivalent System Parameters

$\omega_n T$	ω_n	ω_n^* Exact	ω_n^* Approximate	ζ	ζ^* Exact	ζ^* Approx.	ω_d	ω_d^* Exact	ω_d^* Approximate
0.05	50	50.099	50.125	0.1	0.0751	0.0753	49.749	49.958	49.983
0.10	50	50.147	50.250	0.1	0.0501	0.0505	49.749	50.084	50.186
0.20	50	50.084	50.500	0.1	0.0000	0.0010	49.749	50.084	50.500
0.25	50	49.975	50.625	0.1	−0.0249	−0.0237	49.749	49.959	50.611

Substituting the given values for the system parameters ζ, ω_n, and K and evaluating the damped natural frequency $\omega_d = \sqrt{1 - \zeta^2}\omega_n$ give

$$x(t) = 1 - e^{-5t}\left[\cos\left(50\sqrt{0.99}t\right) + \frac{1}{10\sqrt{0.99}}\sin\left(50\sqrt{0.99}t\right)\right], \quad t \geq 0 \qquad (8.54)$$

(b) Using Equations 8.44 and 8.46 for the exact and asymptotic equivalent system natural frequencies along with Equations 8.45 and 8.47 for the exact and asymptotic equivalent system damping ratios, the results are tabulated in Table 8.2. The damped natural frequency of the continuous-time system ω_d and the exact and asymptotic damped natural frequency approximation of the equivalent system are also shown.

Note that the equivalent continuous-time (as well as the discrete-time) system is on the verge of instability at $\omega_n T = 0.2$ in agreement with the top graph shown in Figure 8.8.

(c) The continuous-time response is plotted on the same graph as the simulated response for the four distinct values of T in Figure 8.10. Every third point of the discrete-time response is plotted in the top left graph. Every point is shown in the remaining plots.

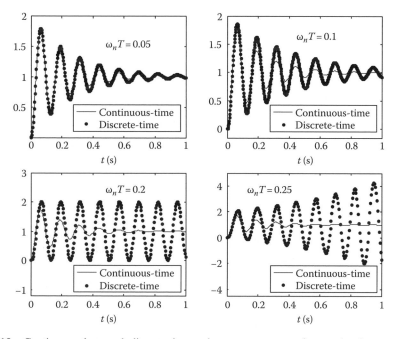

FIGURE 8.10 Continuous-time and discrete-time unit step responses of second-order system ($\zeta = 0.1$, $\omega_n = 50$ rad/s) using explicit Euler integration.

The damped natural frequency of the four simulated step responses corresponding to $\omega_n T = 0.05, 0.01, 0.2, 0.25$ appears to be in close agreement with the continuous-time system response. However, even the discrete-time response in the top left graph where the integration step size is $T = 0.001$ s deviates considerably from the continuous-time response in the neighborhood of the peaks and low points. The oscillatory discrete-time response in the lower left graph in Figure 8.10 is consistent with Figure 8.8, which shows the equivalent continuous-time system damping ratio is zero when $\omega_n T \approx 0.2$.

It is clear from this example that the use of explicit Euler integration to approximate the dynamics of an underdamped, stable, second-order system ($0 < \zeta < 1$) may result in an equivalent continuous-time system that is asymptotically stable ($0 < \zeta^* < 1$), marginally stable ($\zeta^* = 0$), or unstable ($\zeta^* < 0$). From Figure 8.6, the equivalent continuous-time system is marginally stable when $a^* = 0$. Setting the argument of the natural log term in the expression for a^* in Equation 8.42 to 1 and solving for $\omega_n T$ gives

$$(\omega_n T)_{max} = 2\zeta \Rightarrow T_{max} = \frac{2\zeta}{\omega_n} \quad (0 < \zeta < 1) \tag{8.55}$$

where $(\omega_n T)_{max}$ and T_{max} are the values of $(\omega_n T)$ and T, which result in marginally stable, discrete-time, and equivalent continuous-time systems. A plot of $(\omega_n T)_{max} = 2\zeta$ is shown in Figure 8.11 along with $\omega_n T$ ranging from 0.01 up to $(\omega_n T)_{max}$ when $\zeta = 0.707$.

Consider the second-order system

$$\ddot{x} + 2\zeta\omega_n\dot{x} + \omega_n^2 x = K\omega_n^2 u \tag{8.56}$$

with parameters $\zeta = 0.707$, $\omega_n = 10$ rad/s, and $K = 1$. Differentiating the unit step response in Equation 8.53 gives the unit impulse response (Ogata 1998). Alternatively, the impulse response can be obtained by inverse Laplace transformation of the system transfer function $H(s) = X(s)/U(s)$. Either way, the result is

$$h(t) = K\frac{\omega_n}{\sqrt{1 - \zeta^2}}e^{-\zeta\omega_n t}\sin \omega_d t, \quad \omega_d = \sqrt{1 - \zeta^2}\,\omega_n \tag{8.57}$$

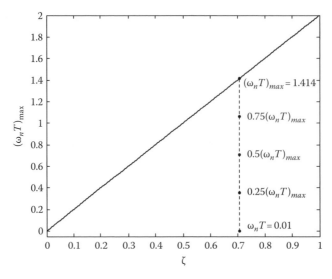

FIGURE 8.11 Plot of $(\omega_n T)$ vs. ζ resulting in marginally stable, second-order equivalent continuous-time system using explicit Euler integration.

Suppose we attempt to simulate the impulse response of the system in Equation 8.56 using explicit Euler integration. The difference equation for explicit Euler integration of the second-order system in Equation 8.56 was developed in Section 4.7 and is repeated in Equation 8.58.

$$x_{k+2} - 2(1 - \zeta\omega_n T)x_{k+1} + \left[1 - 2\zeta\omega_n T + (\omega_n T)^2\right]x_k = K(\omega_n T)^2 u_k, \quad k = -1, 0, 1, 2, \ldots$$

(8.58)

The unit impulse response of the second-order system in Equation 8.56 is identical to the response of the unforced system with initial conditions $x(0) = 0$, $\dot{x}(0) = \omega_n^2$ (see Exercise 8.6). Therefore, the impulse response can be simulated by solving the difference equation in Equation 8.58 with $u_k = -1$, 0, 1, 2, ... along with the appropriate initial conditions, namely, $x(0) = 0$ and $x(-1) = -\omega_n^2 T$. The simulated impulse responses for the values of $\omega_n T$ in Figure 8.11 are shown in Figures 8.12 and 8.13. Not all the data points for the discrete-time response when $\omega_n T = 0.01$ are shown.

Figure 8.12 illustrates the necessity of choosing the time step to achieve an accurate transient response. Indeed, all four simulated responses in Figure 8.12 are stable and converge to the correct steady state, but only one is reasonably accurate. Figure 8.13 represents the case where the discrete-time system (and the equivalent continuous-time system) are marginally stable with oscillatory natural modes.

It may have occurred to you that the impulse response of the second-order system in Equation 8.56 could be simulated by finding the impulse response h_k, $k = 0$, 1, 2, ... of the discrete-time system described by Equation 8.58, either analytically or by recursive solution of the difference equation with $u_k = \delta_k = 1$, $k = 0$, 1, 2, ... Think twice before doing so because $h_k \neq h(t)|t = kT$, $k = 0$, 1, 2, ...

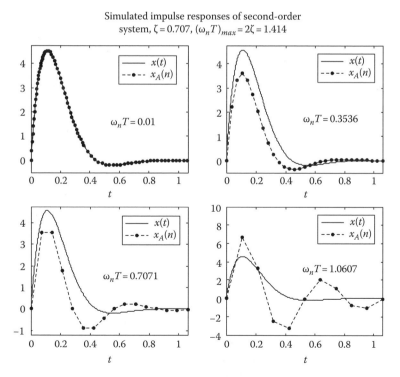

Simulated impulse responses of second-order system, $\zeta = 0.707$, $(\omega_n T)_{max} = 2\zeta = 1.414$

FIGURE 8.12 Continuous-time and simulated second-order system impulse responses using explicit Euler integration with different values for parameter $\omega_n T$.

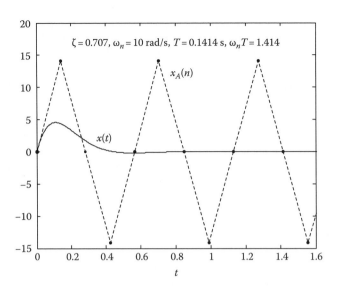

FIGURE 8.13   Simulated impulse response of second order system with marginally stable Euler integrator.

8.2.3 TRANSFER FUNCTION ERRORS

A second class of dynamic error involves the frequency response functions of the continuous-time system and the discrete-time system used to simulate it. The fractional error in the (discrete-time system) transfer function is

$$e_H = \frac{H(e^{j\omega T}) - H(j\omega)}{H(j\omega)} \tag{8.59}$$

where it is important to remember that

$$H(j\omega) = H(s)|_{s \leftarrow j\omega}, \quad H(z) = H(s)|_{s \leftarrow 1/H_I(z)}, \quad H(e^{j\omega T}) = H(z)|_{z \leftarrow e^{j\omega T}} \tag{8.60}$$

The fractional error in transfer function is a complex-valued, frequency-dependent function, which can be expressed in terms of a real and imaginary component, that is,

$$e_H = \frac{H(e^{j\omega T})}{H(j\omega)} - 1 = e_M + je_A \tag{8.61}$$

In polar form, the frequency response functions are expressed as

$$H(j\omega) = |H(j\omega)|e^{j\phi}, \quad \text{where } \phi = \text{Arg}[H(j\omega)] \tag{8.62}$$

$$H(e^{j\omega T}) = |H(e^{j\omega T})|e^{j\phi^*}, \quad \text{where } \phi^* = \text{Arg}[H(e^{j\omega T})] \tag{8.63}$$

Substitution of Equations 8.62 and 8.63 into Equation 8.61 yields

$$e_H = \frac{|H(e^{j\omega T})|e^{j\phi^*}}{|H(j\omega)|e^{j\phi}} - 1 \tag{8.64}$$

$$= \frac{|H(e^{j\omega T})|}{|H(j\omega)|}e^{j(\phi^* - \phi)} - 1 \tag{8.65}$$

Approximating $e^{j(\phi^* - \phi)}$ in a first-order Taylor Series expansion, that is,

$$e^{j(\phi^* - \phi)} \approx 1 + j(\phi^* - \phi) \tag{8.66}$$

$$\Rightarrow e_H \approx \frac{|H(e^{j\omega T})|}{|H(j\omega)|}[1 + j(\phi^* - \phi)] - 1 \tag{8.67}$$

$$\Rightarrow e_H \approx \frac{|H(e^{j\omega T})|}{|H(j\omega)|} + \frac{|H(e^{j\omega T})|}{|H(j\omega)|}j(\phi^* - \phi) - 1 \tag{8.68}$$

When the simulation is reasonably accurate, $H(e^{j\omega T}) \approx H(j\omega)$ over a range of frequencies and the term $(|H(e^{j\omega T})|)/(|H(j\omega)|)j(\varphi^* - \varphi)$ can be approximated by $j(\varphi^* - \varphi)$ (Howe 1986).

The final expression for e_H is therefore

$$e_H \approx \frac{|H(e^{j\omega T})|}{|H(j\omega)|} - 1 + j(\phi^* - \phi) \tag{8.69}$$

Comparison of Equations 8.61 and 8.69 reveals

$$e_M = \mathrm{Re}(e_H) = \mathrm{Re}\left\{ \left[\frac{H(e^{j\omega T})}{H(j\omega)} - 1 \right] \right\} \approx \frac{|H(e^{j\omega T})|}{|H(j\omega)|} - 1 \tag{8.70}$$

$$e_A = \mathrm{Im}(e_H) = \mathrm{Im}\left\{ \left[\frac{H(e^{j\omega T})}{H(j\omega)} - 1 \right] \right\} \approx \phi^* - \phi \tag{8.71}$$

From Equation 8.70, e_M, the real part of e_H (the fractional error in discrete-time transfer function), is approximately equal to the fractional error in the discrete-time transfer function gain. Furthermore, e_A, the imaginary part of e_H, is approximately equal to the phase error of $H(e^{j\omega T})$.

Consider the case of a continuous-time integrator approximated by explicit Euler integration with step size T. Setting $\lambda = 0$ in Equation 8.4 or referring to Equation 4.465, the z-domain transfer function is

$$H(z) = \frac{T}{z - 1} \tag{8.72}$$

Substituting expressions for $H(e^{j\omega T})$ and $H(j\omega)$ in the definition of e_H gives

$$e_H = \frac{T/(e^{j\omega T} - 1)}{1/j\omega} - 1 \tag{8.73}$$

$$= \frac{j\omega T - e^{j\omega T} + 1}{e^{j\omega T} - 1} \tag{8.74}$$

$$= \frac{1 - \cos \omega T + j(\omega T - \sin \omega T)}{\cos \omega T - 1 + j \sin \omega T} \tag{8.75}$$

Rationalizing Equation 8.75, that is, multiplying numerator and denominator by $\cos \omega T - 1 - j \sin \omega T$, and simplifying lead to

$$e_H = e_M + je_A = \frac{\omega T \sin \omega T}{2(1 - \cos \omega T)} - 1 + j\left(\frac{-\omega T}{2} \right) \tag{8.76}$$

From Equation 8.70, an approximation for the fractional gain error in $H(e^{j\omega T})$ is

$$\frac{|H(e^{j\omega T})|}{|H(j\omega)|} - 1 \approx e_M = \frac{\omega T \sin \omega T}{2(1 - \cos \omega T)} - 1 \tag{8.77}$$

and from Equation 8.71, the approximation for the phase error in $H(e^{j\omega T})$ is

$$\text{Arg}[H(e^{j\omega T})] - \text{Arg}[H(j\omega)] \approx e_A = -\frac{\omega T}{2} \tag{8.78}$$

Exact expressions for the fractional gain error and phase error for the explicit Euler integrator are (see Exercise 8.7)

$$\text{Fractional gain error} = \frac{|H(e^{j\omega T})|}{|H(j\omega)|} - 1 = \frac{\omega T}{[2(1 - \cos \omega T)]^{1/2}} - 1 \tag{8.79}$$

$$\text{Phase error} = \text{Arg}[H(e^{j\omega T})] - \text{Arg}[H(j\omega)] = -\tan^{-1}\left(\frac{\sin \omega T}{\cos \omega T - 1}\right) - \left(-\frac{\pi}{2}\right) \tag{8.80}$$

Figure 8.14 contains graphs of the exact and approximate expressions for the fractional error in gain for $0 \leq \omega T \leq 1$ rad. Note that e_M is a good approximation to the fractional gain error in $H(e^{j\omega T})$ provided $\omega T \ll 1$. An asymptotic approximation for e_M, which holds for $\omega T \ll 1$, can be obtained by replacing $\sin \omega T$ and $\cos \omega T$ in Equation 8.77 with the first two nonzero terms in the Taylor Series expansions,

$$\sin \omega T \approx \omega T - \frac{(\omega T)^3}{3!}, \quad \cos \omega T \approx 1 - \frac{(\omega T)^2}{2!} \tag{8.81}$$

eventually leading to $e_M \approx 0$, $\omega T \ll 1$ confirmed by the graph of e_M in Figure 8.14.

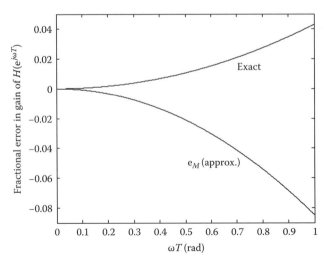

FIGURE 8.14 Approximate and exact fractional error in discrete-time transfer function gain using explicit Euler integration.

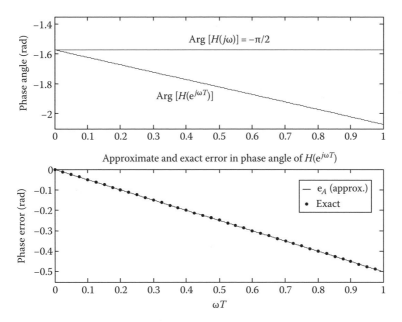

FIGURE 8.15   Phase angle plots for continuous-time and explicit Euler integrator.

The phase angle plots for the continuous-time integrator and explicit Euler integrator are shown in Figure 8.15. The top graph shows the constant phase angle $-\pi/2$ rad for the continuous-time integrator along with the phase angle of the discrete-time transfer function given in Equation 8.72. The lower graph shows e_A in Equation 8.78 and equally spaced points computed from the exact expression for the phase error in Equation 8.80. The linear approximation e_A is virtually identical to the exact expression for the phase error.

An asymptotic expression for $H(e^{j\omega T})$ can be derived starting with Equation 8.72.

$$H(e^{j\omega T}) = \frac{T}{e^{j\omega T} - 1} = \frac{T}{[1 + j\omega T + ((j\omega T)^2/2!) + ((j\omega T)^3/3!) + \cdots] - 1} \tag{8.82}$$

Truncating the power series for $e^{j\omega T}$ after the quadratic term gives

$$H(e^{j\omega T}) \approx \frac{T}{j\omega T + (j\omega T)^2/2}, \quad \omega T \ll 1 \tag{8.83}$$

$$\approx \frac{1}{j\omega} \frac{1}{(1 + j\omega T/2)}, \quad \omega T \ll 1 \tag{8.84}$$

The frequency response function of the continuous-time integrator is $H(j\omega) = 1/j\omega$. The second term

$$\frac{1}{1 + j\omega T/2} = \frac{1}{[1 - (\omega T/2)^2]^{1/2} e^{j\tan^{-1}(\omega T/2)}} \tag{8.85}$$

$$\approx e^{-j\omega T/2}, \quad \omega T \ll 1 \tag{8.86}$$

$$\Rightarrow H(e^{j\omega T}) \approx H(j\omega) e^{-j\omega T/2}, \quad \omega T \ll 1 \tag{8.87}$$

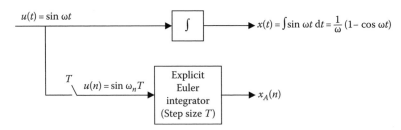

FIGURE 8.16 Continuous- and discrete-time integration of a sinusoidal input.

implying that the asymptotic behavior ($\omega T \ll 1$) of the explicit Euler integrator is that of a pure continuous-time integrator with an additional delay of $-\omega T/2$ rad.

To illustrate Equation 8.87, a sine wave $u(t)$ at a frequency of ω rad/s is input to an integrator shown in Figure 8.16. The signal $u(t)$ is sampled every T s, and the resulting discrete-time signal is input to an explicit Euler integrator updating at $1/T$ Hz.

The top half of Figure 8.17 shows the sinusoidal input $u(t) = \sin 2t$, the explicit Euler output $x_A(n)$, $n = 0, 5, 10, \ldots$ when the step size $T = 0.025$ s, and the continuous-time output $x(t)$. The parameter $\omega T = 2(0.025) = 0.05$ rad is small enough for the asymptotic formula in Equation 8.87 to accurately predict the characteristics of the discrete-time output $x_A(n)$. According to Equation 8.87, the steady-state amplitudes of $x_A(n)$ and $x(t)$ are equal for all input frequencies provided $\omega T \ll 1$. The amplitude is

$$|x_A(n)| = |x(t)| = |H(j\omega)| \cdot |u(t)| = \left|\frac{1}{j\omega}\right| \cdot 1 = \frac{1}{\omega} = \frac{1}{2} \tag{8.88}$$

easily verified by looking at the plots of $x_A(n)$ and $x(t)$ in the top half of Figure 8.17.

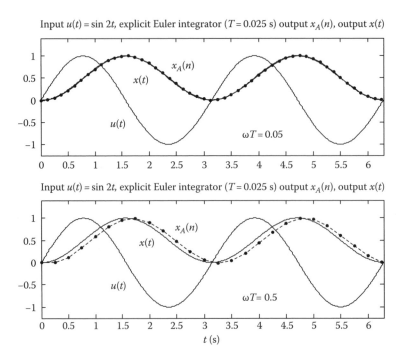

FIGURE 8.17 Explicit Euler and continuous-time integrator outputs ($\omega T = 0.05$ rad).

The asymptotic approximation for $H(e^{j\omega T})$ in Equation 8.87 also predicts a delay of $\omega T/2$ rad in $x_A(n)$ relative to the continuous-time output $x(t)$. The time delay can be estimated by zooming in on the responses in the top half of Figure 8.17.

Alternatively, the time delay can be determined by finding the time between occurrences of equal values of $x_A(n)$ and $x(t)$. For example, at $t = 4$ s, the discrete-time variable $n = 4/0.025 = 160$ and from M-file "*Chap8_Fig2_17.m*," $x_A(160) = 0.05603$. Setting $x(t_0) = x_A(160) = 0.5603$ and solving for t_0.

$$x(t_0) = \frac{1}{\omega}(1 - \cos \omega t_0) = x_A(160) = 0.5603 \tag{8.89}$$

$$t_0 = \frac{1}{2}\cos^{-1}[1 - 2(0.5603)] + \frac{2\pi}{\omega} = 3.9874\text{ s} \tag{8.90}$$

The simulated response $x_A(n)$ is lagging the output $x(t)$ by $4 - 3.9874 = 0.0126$ s, in close agreement with the predicted value of $T/2 = 0.0125$ s.

The lower half of Figure 8.17 illustrates the case where $\omega = 2$ rad/s, $T = 0.25$ s, and the asymptotic approximation in Equation 8.87 based on $\omega T \ll 1$ no longer applies. From Equation 8.79, the fractional gain error in $H(e^{j\omega T})$ when $\omega T = 0.5$ rad is 0.010493 (see Figure 8.14). Solving for $|H(e^{j\omega T})|$ in Equation 8.79,

$$|H(e^{j\omega T})| = (1 + \text{fractional gain error})|H(j\omega)|$$

$$= (1 + 0.010493)\left|\frac{1}{j2}\right| = 0.505247 \tag{8.91}$$

Since $|u(t)| = 1$, the predicted peak-to-peak swing in $x_A(n)$ is $2 \times 0.505247 = 1.010494$. The discrete-time response $x_A(n)$ was generated for different lengths of time (instead of two periods as in Figure 8.17) to capture the peak-to-peak swing in $x_A(n)$ using the MATLAB® statement "max (xA) −min (xA)" in "*Chap8_Fig2_17.m*." The results are tabulated in Table 8.3.

The time delay was computed in the same manner used for the case when $\omega T = 0.05$ rad. The results are $n = 4/0.25 = 16$, $x_A(16) = 0.4371$, $x(3.8639) = 0.4371$ and the time delay is equal to $4 - 3.8639 = 0.1361$ s. The asymptotic formula in Equation 8.87 for $H(e^{j\omega T})$ underestimates the time delay, that is, $T/2 = 0.125$ s.

8.2.4 Asymptotic Formulas for Multistep Integration Methods

The same steps used to obtain the asymptotic formula in Equation 8.84 for the explicit Euler integrator are applicable to the multistep integration formulas introduced in Section 6.4.

TABLE 8.3
Measured Peak-to-Peak Swing in $x_A(n)$
for Different Time Periods

Duration of Simulation (s)	Max(x_A) − Min(x_A)
$P = 2\pi/\omega = \pi$	1.010481
$25P = 25\pi$	1.010491
$50P = 50\pi$	1.010491
$100P = 100\pi$	1.010492

For example, simulating the response $x(t)$ of a continuous-time integrator subject to input $u(t)$, using a second-order explicit Adams–Bashforth (AB-2) numerical integrator, reduces to solve the difference equation

$$x_A(n+1) = x_A(n) + \frac{T}{2}[3u(n) - u(n-1)] \tag{8.92}$$

z-Transforming Equation 8.92 and solving for the z-domain transfer function give

$$\frac{X(z)}{U(z)} = H_I(z) = \frac{T}{2}\left[\frac{3 - z^{-1}}{z - 1}\right] \tag{8.93}$$

where the subscript I in $H_I(z)$ reminds us that we are dealing with the z-domain transfer function approximation of a continuous-time integrator. Replacing z by $e^{j\omega t}$ in Equation 8.93 produces the discrete-time system frequency response function

$$H_I(e^{j\omega T}) = \frac{T}{2}\left[\frac{3 - e^{-j\omega T}}{e^{j\omega T} - 1}\right] \tag{8.94}$$

Approximating the complex exponentials $e^{j\omega t}$ and $e^{-j\omega t}$ by power series up to the third-order term generates the asymptotic formula (see Exercise 8.10)

$$H_I(e^{j\omega T}) \approx \frac{1}{j\omega}\left[\frac{1}{1 - (5/12)(\omega T)^2}\right], \quad \omega T \ll 1 \tag{8.95}$$

According to the asymptotic approximation in Equation 8.95, the frequency response of an AB-2 integrator is identical in phase to that of an ideal continuous-time integrator while the gain is off by the factor in parenthesis in Equation 8.95. Hence, the phase error in the asymptotic approximation of $H_I(e^{j\omega t})$ is zero and the fractional gain error is

$$\frac{|H_I(e^{j\omega T})|}{|H(j\omega)|} - 1 \approx \left[\frac{1}{1 - (5/12)(\omega T)^2}\right] - 1 \tag{8.96}$$

$$\approx \frac{(5/12)(\omega T)^2}{1 - (5/12)(\omega T)^2} \tag{8.97}$$

$$\approx (5/12)(\omega T)^2, \quad |\omega T| \ll 1 \tag{8.98}$$

Equations 8.84 and 8.95 are special cases of a general formula (Howe 1986, 1995)

$$H_I(e^{j\omega T}) \approx \frac{1}{j\omega}\left[\frac{1}{1 + e_I(j\omega T)^k}\right], \quad \omega T \ll 1 \tag{8.99}$$

which holds for the multistep integrators in Section 6.4, namely, explicit Adams–Bashforth, implicit Adams–Moulton, and predictor–correctors. Numerical values for the error coefficient e_I depend on the order k and type of integrator. A table of values for low-order numerical integrators is given in Table 8.4. The error coefficients are identical to the constants in the local truncation error term for each integrator (see Table 6.9).

Frequency responses for a continuous-time integrator and AB-1 (explicit Euler) through AB-4 numerical integrators are shown in Figure 8.18a through d. Also shown are the frequency responses

TABLE 8.4
Error Coefficients in Asymptotic Formula in Equation 8.99 for kth Order, z-Domain Frequency Response Functions of Numerical Integrators

Numerical Integrator	Equation	Order k	Error Coefficient e_I
AB-1 (explicit Euler)		1	1/2
AB-2	6.180	2	5/12
AB-3	6.186	3	3/8
AB-4	6.187	4	251/720
AB-5	6.188	5	475/1440
AM-2 (trapezoidal)	6.191	2	−1/12
AM-3	6.192	3	−1/24
AM-4	6.193	4	−19/720
AM-5	6.194	5	−27/1440
AB-2 predictor	6.204	2[a]	−1/12[a]
AM-2 corrector	6.205		
AB-3 predictor	6.206	3[b]	−1/24[b]
AM-3 corrector	6.207		
AB-4 predictor	6.208	4[c]	−19/720[c]
AM-4 corrector	6.209		

[a] AB-2/AM-2.
[b] AB-3/AM-3.
[c] AB-4/AM-4.

for the same AB integrators based on the asymptotic formula in Equation 8.99 where e_I, $k = 1, 2, 3, 4$ are given in Table 8.4. The plots are generated in M-file "*Chap8_Fig2_18abcd.m*."

Figure 8.18a shows close agreement between the exact and asymptotic Euler magnitude functions up to $\omega T = \omega(1) \approx 0.5$ rad. Beyond that, the two plots begin to deviate from the continuous-time integrator magnitude function with the exact Euler the better approximation. Hence, for $\omega T \ll 1$, the Euler integrator introduces essentially zero gain error. The exact and asymptotic Euler phase plots also agree up to approximately $\omega T = 0.5$ rad. However, Figure 8.18 shows the Euler integrator introducing phase error with respect to the continuous-time integrator beginning around $\omega T = 0.04$ rad. Significant phase error in the neighborhood of $30°$ is present for $\omega T = 1$ rad.

The AB-2 integrator and its asymptotic approximation are both quite accurate in the range of frequencies for which $\omega T < 0.4$ rad. Beyond that, the asymptotic approximation of the magnitude begins to deviate from both the continuous-time and exact AB-2 magnitude functions. From Equation 8.99 with $k = 2$, the asymptotic curve approaches infinity at the point where

$$1 + e_I(j\omega T)^k = 1 + e_I(j)^2(\omega T)^2 = 1 - \frac{5}{12}\omega^2 = 0 \qquad (8.100)$$

$$\Rightarrow \omega = 1.5492 \, \text{rad/s}$$

The asymptotic phase plot is exact up to $\omega = 1.5492$ rad/s where it increases from $-90°$ to $90°$ due to the change in sign of the denominator. It follows from Equation 8.99 that the asymptotic phase plots for $k = 2, 6, 10, \ldots$ are similar to the one in Figure 8.18b and the asymptotic plots are exact, that is, $\arg\{H(e^{j\omega T})\} = -180$ deg for $k = 4, 8, 12, \ldots$.

8.2.5   Simulation of Linear System with Transfer Function $H(s)$

In addition to a simple continuous-time integrator, it is possible to approximate the discrete-time frequency response of higher-order linear systems simulated using numerical integrators like the ones represented in Table 8.4. We learned in Section 4.7 that $H(z)$ resulting from digital simulation of a continuous-time system with transfer function $H(s)$ is obtained by substituting $1/H_I(z)$ for s, where $H_I(z)$ is the z-domain transfer function of the numerical integrator.

Consider the first-order system governed by

$$\frac{dx}{dt} = \lambda x + Ku \tag{8.101}$$

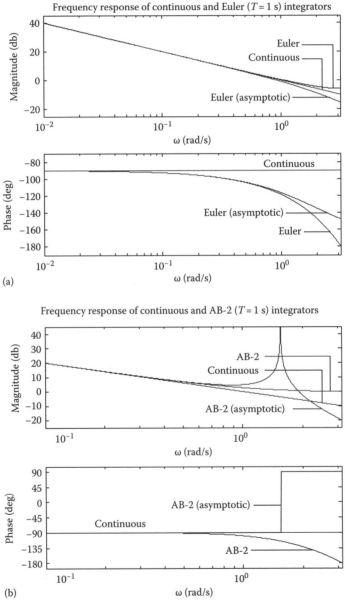

FIGURE 8.18   Exact and asymptotic frequency response of (a) AB-1 (Euler) integrator, (b) AB-2 integrator,

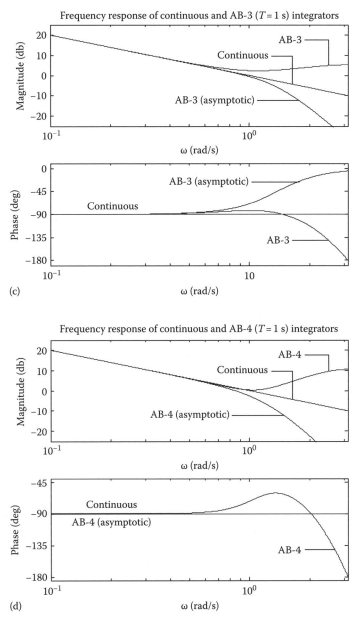

FIGURE 8.18 (continued) (c) AB-3 integrator, and (d) AB-4 integrator.

similar to Equation 8.1 except for the gain K on the right-hand side of the equation. Simulation of the system using a numerical integrator with transfer function $H_I(z)$ results in a discrete-time system with frequency response function

$$H(e^{j\omega T}) = H(s)\bigg|_{s \leftarrow 1/H_I(e^{j\omega T})} = \frac{K}{s - \lambda}\bigg|_{s \leftarrow 1/H_I(e^{j\omega T})} \tag{8.102}$$

$$= \frac{K H_I(e^{j\omega T})}{1 - \lambda H_I(e^{j\omega T})} \tag{8.103}$$

An approximate expression for $H(e^{j\omega T})$ is obtained using the asymptotic approximation for $H(e^{j\omega T})$ in Equation 8.99.

$$H(e^{j\omega T}) \approx \frac{K\left(1 + j\omega\left[1 + e_I(j\omega T)^k\right]\right)}{1 - \lambda\left(1/j\omega\left[1 + e_I(j\omega T)^k\right]\right)}, \quad \omega T \ll 1 \tag{8.104}$$

$$\approx \frac{K}{j\omega\left[1 + e_I(j\omega T)^k\right] - \lambda}, \quad \omega T \ll 1 \tag{8.105}$$

Using trapezoidal integration, $k = 2, e_I = -1/12$ from Table 8.4,

$$H(e^{j\omega T}) \approx \frac{K}{j\omega[1 + (1/12)(\omega T)^2] - \lambda}, \quad \omega T \ll 1 \tag{8.106}$$

The exact expression for $H(e^{j\omega T})$ is obtained from

$$H(z) = H(s)\Big|_{s \leftarrow 1/H_I(z)} = \frac{K}{s - \lambda}\Big|_{s \leftarrow \frac{2}{T}\left(\frac{z-1}{z+1}\right)} \tag{8.107}$$

$$\Rightarrow H(e^{j\omega T}) = \frac{K}{(2/T)((z-1)/(z+1)) - \lambda}\Big|_{z \leftarrow e^{j\omega T}} \tag{8.108}$$

$$= \frac{KT(e^{j\omega T} + 1)}{(2 - \lambda T)e^{j\omega T} - (2 + \lambda T)} \tag{8.109}$$

The use of trapezoidal integration for digital simulation of linear continuous-time systems is referred to as Tustin's method.

Example 8.3

The capacitor in the circuit shown in Figure 8.19 is initially uncharged when the switch closes.

(a) Find the discrete-time transfer function $H(z) = V_C(z)/E_0(z)$ using Tustin's method.
(b) Find the asymptotic form of the discrete-time frequency response function.
(c) Find the exact expression for the discrete-time frequency response function.
(d) Graph the frequency response function of the continuous-time system and the discrete-time frequency response functions obtained in parts (b) and (c) if the time constant of the circuit is 25 μs and the integration step size is 1 μs.
(e) Compute the gain and phase errors based on the asymptotic expression for $H(e^{j\omega T})$ when the input is a sinusoidal input at 1×105 Hz.

(a) The differential equation of the circuit is

$$\tau \frac{d}{dt} v_C(t) + v_C(t) = e_0(t), \quad (\tau = RC) \tag{8.110}$$

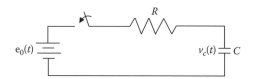

FIGURE 8.19 *RC* circuit for digital simulation.

Laplace transforming Equation 8.110 leads to the transfer function

$$H(s) = \frac{V_C(s)}{E_0(s)} = \frac{1}{\tau s + 1} = \frac{1/\tau}{s + 1/\tau} \tag{8.111}$$

$$\Rightarrow H(z) = \frac{V_C(z)}{E_0(z)} = \left. \frac{1}{\tau s + 1} \right|_{s \leftarrow \frac{2}{T}\left(\frac{z-1}{z+1}\right)} \tag{8.112}$$

$$= \frac{z + 1}{[1 + 2(\tau/T)]z + (1 - 2\tau/T)} \tag{8.113}$$

(b) The transfer function for the first-order system in Equation 8.101 is identical to $H(s)$ in Equation 8.111 when $\lambda = -1/\tau$ and $k = -1/\tau$. Making those substitutions in Equation 8.106 gives the asymptotic formula for $H(e^{j\omega t})$.

$$H(e^{j\omega T}) \approx \frac{1/\tau}{j\omega[1 + (1/12)(\omega T)^2] - (-1/\tau)}, \quad \omega T \ll 1 \tag{8.114}$$

$$\approx \frac{1}{j\omega\tau[1 + (1/12)(\omega T)^2] + 1}, \quad \omega T \ll 1 \tag{8.115}$$

(c) The exact expression for $H(e^{j\omega T})$ is from Equation 8.109,

$$H(e^{j\omega T}) = \left. \frac{(T/\tau)(e^{j\omega T} + 1)}{(2 - \lambda T)e^{j\omega T} - (2 + \lambda T)} \right|_{\lambda = -1/\tau} \tag{8.116}$$

$$= \frac{(T/\tau)(e^{j\omega T} + 1)}{[2 + (T/\tau)]e^{j\omega T} - [2 - (T/\tau)]} \tag{8.117}$$

(d) The MATLAB M-file "Chap8_Ex2_3.m" computes the magnitude and phase of

$$H(j\omega) = \left. \frac{1}{\tau s + 1} \right|_{s \leftarrow j\omega} = \frac{1}{1 + j\omega\tau} \tag{8.118}$$

The magnitude and phase of $H(e^{j\omega T})$ using the asymptotic and exact formulas in Equations 8.115 and 8.117 are computed. The gain and phase plots are shown in Figure 8.20.

(e) At $\omega = 1 \times 10^5$ Hz $\times 2\pi$ rad/cycle $= 2\pi \times 10^5$ rad/s,

$$H(j2\pi \times 10^5) = \frac{1}{1 + j2\pi \times 10^5(25 \times 10^{-6})} = \frac{1}{1 + j5\pi} = 0.0635e^{j(-1.5072)} \tag{8.119}$$

$$H(e^{j2\pi \times 10^5 \times 10^{-6}}) \approx \frac{1}{j2\pi \times 10^5(25 \times 10^{-6})[1 + (1/12)(2\pi \times 10^5 \times 10^{-6})^2] + 1} \tag{8.120}$$

$$\approx 0.0615e^{j(-1.5092)} \tag{8.121}$$

The fractional gain error and phase errors in the discrete-time frequency response function based on the asymptotic approximation for $H(e^{j\omega T})$ in Equation 8.115 at $\omega = 1 \times 10^5$ Hz are

$$\text{Fractional gain error} \approx \frac{|H(e^{j2\pi \times 10^5 \times 10^{-6}})|}{|H(j2\pi \times 10^5)|} - 1$$

$$\approx \frac{0.0635}{0.0615} - 1 = -0.0317 \tag{8.122}$$

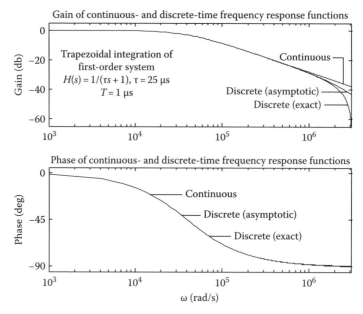

FIGURE 8.20 Continuous- and discrete-time (exact and asymptotic) bode plots for first-order system using trapezoidal integration.

$$\text{Phase error} \approx \text{Arg}\left[H\left(e^{j2\pi \times 10^5 \times 10^{-6}}\right)\right] - \text{Arg}[H(j2\pi \times 10^5)]$$

$$\approx -1.5092 - (-1.5072)$$

$$\approx -0.002 \text{ rad } (-0.1157 \text{ deg}) \tag{8.123}$$

Figure 8.20 shows the asymptotic formula for approximating $H(e^{j\omega T})$ is accurate up to approximately $\omega T = 10^6$ rad/s $\times 10^{-6}$ s $= 1$ rad. The exact and asymptotic discrete-time frequency response functions are close to the continuous-time frequency response function up until frequencies approaching the Nyquist frequency $\pi/T = 10^6\pi$ rad/s.

For an underdamped second-order system with damping ratio ζ and natural frequency ω_n, the asymptotic approximation of the discrete-time frequency response function using a kth-order numerical integrator with error coefficient e_I is obtained in the same manner employed for the first-order system, namely,

$$H(e^{j\omega T}) = \frac{\omega_n^2}{s^2 + 2\zeta\omega_n s + \omega_n^2}\bigg|_{s \leftarrow 1/H_I(e^{j\omega T})} \tag{8.124}$$

Substituting the asymptotic expression for $H(e^{j\omega T})$ in Equation 8.99 into Equation 8.124 results in (after simplification) (Howe 1986)

$$H(e^{j\omega T}) \approx \frac{1}{\omega_n^2 - \omega^2\left[1 + 2e_I(j\omega T)^k\right] + j2\zeta\omega_n\omega\left[1 + e_I(j\omega T)^k\right]}, \quad \omega T \ll 1 \tag{8.125}$$

and the fractional error in $H(e^{j\omega T})$ is approximated by the asymptotic formula

$$e_H = \frac{H(e^{j\omega T})}{H(j\omega)} - 1 \approx \frac{2e_I(j\omega T)^k\left[(\omega/\omega_n)^2 - j\zeta(\omega/\omega_n)\right]}{1 - (\omega/\omega_n)^2 + j2\zeta(\omega/\omega_n)}, \quad \omega T \ll 1 \tag{8.126}$$

Rationalizing Equation 8.126 leads to expressions for e_M and e_A, the real and imaginary components of e_H, which provide suitable approximations for the fractional gain error and phase error of $H(e^{j\omega T})$, respectively. The expressions are of the form

$$e_M = f_M(\zeta, \omega/\omega_n, k, e_l)(\omega T)^k, \quad e_A = f_A(\zeta, \omega/\omega_n, k, e_l)(\omega T)^k, \quad \omega T \ll 1 \qquad (8.127)$$

The functions $f_M(\zeta, \omega/\omega_n, k, e_l)$ and $f_A(\zeta, \omega/\omega_n, k, e_l)$ are further addressed in Exercise 8.13. The notable feature in Equation 8.127 is the dependence of both error measures on the term $(\omega T)^k$, emphasizing the importance of choosing the step size T and the integrator order k.

EXERCISES

8.1 Repeat Example 8.1 for the case where explicit Euler is used in place of trapezoidal integration.

8.2 A second-order system with damping ratio $\zeta = 0$ and natural frequency ω_n is simulated using trapezoidal integration with step size T.

  (a) Plot the equivalent continuous-time system damping ratio ζ^* as a function of the parameter $\omega_n T$ for $0 \le \omega_n T \le 1$.

  (b) Plot the equivalent continuous-time system natural frequency ω_n^* as a function of the continuous-time system natural frequency ω_n for $0 \le \omega_n \le 10$ rad/s when $T = 0.1$ s.

  (c) Repeat parts (a) and (b) for $\zeta = 0.1$ and 1.

  (d) What effect does changing the value of T have on the equivalent continuous-time system damping ratio and natural frequency?

8.3 Consider the overdamped continuous-time second-order system with transfer function

$$H(s) = \frac{1}{(\tau_1 s + 1)(\tau_2 s + 1)}$$

shown in Figure E8.3.

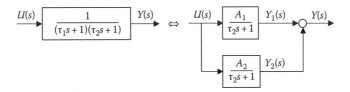

FIGURE E8.3

  (a) Decompose $H(s)$ into the sum of two first-order transfer functions $H(s) = H_1(s) + H_2(s)$ where $H_1(s) = A_1/(\tau_1 s + 1)$ and $H_2(s) = A_2/(\tau_2 s + 1)$ (see Figure E8.3) and express the constants A_1 and A_2 in terms of the time constants τ_1 and τ_2.

  (b) The equivalent realizations of the same second-order system are simulated using explicit Euler integration with step size T. Find the fractional error in the frequency response functions $H(e^{j\omega T})$, $H_1(e^{j\omega T})$, and $H_2(e^{j\omega T})$. Leave your answers in terms of τ_1, τ_2, and T.

  (c) Resolve the fractional errors into real and imaginary components, that is,

$$e_H = \frac{H(e^{j\omega T})}{H(j\omega)} - 1 = e_M + je_A, \quad e_{H_1} = \frac{H_1(e^{j\omega T})}{H_1(j\omega)} - 1 = e_{M_1} + je_{A_1}$$

$$e_{H_2} = \frac{H_2(e^{j\omega T})}{H_2(j\omega)} - 1 = e_{M_2} + je_{A_2}$$

For $\tau_1 = 1$ s, $\tau_2 = 10$ s, and $T = 0.05$ s, plot e_M, e_{M_1}, e_{M_2} vs. ωT on a single graph and e_A, e_{A_1}, e_{A_2} vs. ωT on a different graph. Comment on the results.

(d) Find exact expressions for the fractional gain error in $H(e^{j\omega T})$, $H_1(e^{j\omega T})$, and $H_2(e^{j\omega T})$. Plot the fractional gain error in $H(e^{j\omega T})$ vs. ωT and e_M vs. ωT on the same graph. Repeat for $H_1(e^{j\omega T})$ and e_{M_1} and then for $H_2(e^{j\omega T})$ and e_{M_2}.

(e) Find exact expressions for the phase error in $H(e^{j\omega T})$, $H_1(e^{j\omega T})$, and $H_2(e^{j\omega T})$. Plot the phase error in $H(e^{j\omega T})$ vs. ωT and e_A vs. ωT on the same graph. Repeat for $H_1(e^{j\omega T})$ and e_{A_1} and then for $H_2(e^{j\omega T})$ and e_{A_2}.

(f) Simulate the two configurations shown in Figure E8.3b when $u(t) = \sin 50t$, $t \geq 0$ using an explicit Euler integrator with step size $T = 0.01$ s. Plot the continuous-time input and output and the simulated response on the same graph for each configuration. Do the results agree with the graphs obtained in parts (d) and (e)?

8.4 For AB-2 integration,

(a) Find the discrete-time frequency response function $H_1(e^{j\omega T})$.

(b) Find expressions for the exact and asymptotic fractional gain and phase errors.

(c) Plot the results over a suitable range of values for ωT.

8.5 Generate a new table and figure similar to Table 8.2 and Figure 8.9 where a second-order system with damping ratio ζ and natural frequency ω_n is simulated using numerical integration for the following cases:

Z	ω_n (rad/s)	Numerical Integrator
0	50	Explicit Euler
0	50	Implicit Euler
0	50	Trapezoidal
0.1	50	Implicit Euler
0.1	50	Trapezoidal
0.707	1	Explicit Euler
0.707	1	Implicit Euler
0.707	1	Explicit Euler
2	0.01	Explicit Euler
2	0.01	Implicit Euler
2	0.01	Trapezoidal

8.6 For a second-order system described by

$$\ddot{x} + 2\zeta\omega_n\dot{x} + \omega_n^2 x = K\omega_n^2 u$$

(a) Show that the unit impulse response is identical to the response of the autonomous system ($u = 0$, $t \geq 0$) with initial conditions $x(0) = 0$, $\dot{x}(0) = K\omega_n^2$.

(b) Show that the initial conditions for the difference equation of the discrete-time system resulting from the use of explicit Euler integration are $x(0) = 0$ and $x(-1) = -K\omega_n^2 T$.

(c) Suppose the parameter values are $\zeta = 0.5$, $\omega_n = 10$, and $K = 1$. Simulate the continuous-time step and impulse responses using explicit Euler integration with $\omega_n T = 0.05$. Compare the simulated and analytical solutions.

8.7 Derive the exact expressions for the fractional gain error and phase error in the discrete-time transfer function $H(e^{j\omega T})$ using explicit Euler integration given in Equations 8.79 and 8.80.

8.8 Show that the asymptotic expression for the fractional error in the discrete-time transfer function $H(e^{j\omega T})$ resulting from explicit Euler integration of the first-order system $\dot{x} = \lambda x + u$ is given by

$$e_H = \frac{H(e^{j\omega T})}{H(j\omega)} - 1 \approx \frac{\omega\lambda}{2(\omega^2 + \lambda^2)}\omega T - j\frac{\omega^2}{2(\omega^2 + \lambda^2)}\omega T, \quad \omega T \ll 1$$

What does the system reduce to when $\lambda = 0$? Comment on what happens to the real and imaginary components.

8.9 Verify the curves plotted in Figures 8.14 and 8.15 for the fractional gain and phase errors based on explicit Euler integration by using the MATLAB functions "`real`," "`imag`," "`abs`," and "`angle`," that is,

$$\text{Fractional gain error} \approx e_M = \text{Re}(e_H) = \text{Re}\left\{\frac{H(e^{j\omega T})}{H(j\omega)} - 1\right\} = \text{Re}\left\{\frac{j\omega T}{e^{j\omega T} - 1} - 1\right\}$$

$$\text{Fractional gain error} = \frac{|H(e^{j\omega T})|}{|H(j\omega)|} - 1 = \left|\frac{\omega T}{e^{j\omega T} - 1}\right| - 1$$

$$\text{Phase error} \approx e_A = \text{Im}(e_H) = \text{Im}\left\{\frac{H(e^{j\omega T})}{H(j\omega)} - 1\right\} = \text{Im}\left\{\frac{j\omega T}{e^{j\omega T} - 1} - 1\right\}$$

$$\text{Phase error} = \text{Arg}[H(e^{j\omega T})] - \text{Arg}[H(j\omega)] = -\text{Arg}[e^{j\omega T} - 1] - \left(-\frac{\pi}{2}\right)$$

8.10 Derive the asymptotic formula for $H_1(e^{j\omega T})$ in Equation 8.95 starting with the exact expression for the discrete-time frequency response function in Equation 8.94.

8.11 Using trapezoidal integration to simulate the first-order system in Equation 8.1,
(a) Find the fractional error in transfer function e_H.

Hint: Start with Equation 8.109.

(b) Find the real and imaginary parts of e_H, that is, $e_H = e_M + je_A$.
(c) Compare e_M and e_A with the exact expressions for the fractional gain and phase errors.

8.12 For simulation of the first-order system $\dot{x} = \lambda x + u$ using a kth-order numerical integrator with error coefficient e_I,
(a) Show that the asymptotic expression for the fractional error in transfer function is given by

$$e_H = \frac{H(e^{j\omega T})}{H(j\omega)} - 1 = \frac{j\omega e_I(j\omega T)^k}{j\omega - \lambda}, \quad \omega T \ll 1$$

(b) Derive expressions for e_M and e_A when the order k is odd and different expressions when k is even.

8.13 Derive the asymptotic expressions for $H(e^{j\omega T})$ in Equation 8.125 and e_H in Equation 8.126. Find the functions $f_M(\zeta, \omega/\omega_n, k, e_I)$ and $f_A(\zeta, \omega/\omega_n, k, e_I)$ in Equation 8.127 when the numerical integrator order k is odd and even.

8.14 Show that the characteristic root error resulting from simulation of a first-order continuous-time system with characteristic root λ is approximated by

$$e_\lambda = \frac{\lambda^*}{\lambda} - 1 \approx -e_I(\lambda T)^k, \quad |\lambda T| \ll 1$$

where e_I and k are the error coefficient and order of the numerical integrator, respectively.

8.3 STABILITY OF NUMERICAL INTEGRATORS

We have seen a number of examples where digital simulation of a stable continuous-time system with a bounded input (or even no input with nonzero initial conditions) produced a sequence of numbers that grow without bound as time increases. The unstable conditions can be attributed to a combination of the numerical integrator and integration step size (for fixed-step integrators). Stability of fixed-step numerical integrators is reflected in the natural dynamics of the discrete-time system used to approximate the continuous-time system. The family of explicit multistep Adams–Bashforth integrators introduced in Section 6.4 is now examined in some detail.

8.3.1 ADAMS–BASHFORTH NUMERICAL INTEGRATORS

Difference equations resulting from the application of second-order and higher Adams–Bashforth integration are higher-order than the LTI continuous-time systems being simulated. For example, a first-order continuous-time system with a pole at $s = \lambda$ simulated using AB-2 integration produces a second-order discrete-time system with discrete-time input $u(n)$ and output $x(n)$, previously referred to as $x_A(n)$. The z-domain transfer function is

$$H(z) = \frac{X(z)}{U(z)} = \frac{1}{s - \lambda}\bigg|_{s \leftarrow \frac{1}{H_1(z)}} \tag{8.128}$$

$$= \frac{1}{s - \lambda}\bigg|_{s \leftarrow \frac{1}{T(3z-1)/2z(z-1)}} \tag{8.129}$$

$$= \frac{(T/2)(3z - 1)}{z^2 - (1 + (3/2)\lambda T)z + (1/2)\lambda T} \tag{8.130}$$

Note that $H_I(z)$ for AB-2 integration is given in Equation 8.93 of the previous section. Multiplying numerator and denominator in Equation 8.130 by z^{-1} followed by inverse z-transformation leads to the second-order difference equation

$$x(n + 1) - \left(1 + \frac{3}{2}\lambda T\right)x(n) + \frac{1}{2}\lambda Tx(n - 1) = \frac{T}{2}[3u(n) - u(n - 1)] \tag{8.131}$$

The states $x(n)$ and $x(n-1)$ are needed to compute the updated state $x(n+1)$. This is easily explained by referring to Figure 6.16. $P_1(t)$; the linear interpolating polynomial integrated to generate $x(n+1)$ depends on current and previous derivative functions, which in turn are functions of the current and previous discrete-time states $x(n)$ and $x(n-1)$.

The resulting z-domain transfer function in Equation 8.130 has two poles that are the roots of the characteristic polynomial in the denominator. The dominant pole for the case when $\lambda T \ll 1$ corresponds to an equivalent continuous-time system characteristic root λ^*, which can be estimated from the characteristic root error formula (Howe 1986)

$$e_\lambda = \frac{\lambda^* - \lambda}{\lambda} \approx -e_I(\lambda T)^k \tag{8.132}$$

where e_I and k are the integrator error coefficient and order, respectively. For AB-2 integration, $e_I = 5/12$ and $k = 2$. Hence,

$$\lambda^* \approx \lambda[1 - e_I(\lambda T)^k] \approx \lambda\left[1 - \frac{5}{12}(\lambda T)^2\right], \quad \lambda T \ll 1 \tag{8.133}$$

Suppose the continuous-time system pole is $\lambda = -100$ and AB-2 integration is used with a step size $T = 0.0001$ s. From Equation 8.133,

$$\lambda^* \approx -100 \left[1 - \frac{5}{12} \{(-100)(0.0001)\}^2 \right], \quad \lambda T \ll 1$$

$$\approx -99.99583 \tag{8.134}$$

The exact value of λ^* is obtained from

$$\lambda^* = \frac{1}{T} \ln z_1 \tag{8.135}$$

where z_1 is the dominant pole, that is, larger (in magnitude) root of the characteristic equation

$$z^2 - \left(1 + \frac{3}{2} \lambda T \right) z + \frac{1}{2} \lambda T = z^2 - 0.985z - 0.005 = 0 \tag{8.136}$$

The poles are located at $z_1 = 0.99005$, $z_2 = -0.00505$, and the equivalent characteristic root is from Equation 8.135

$$\lambda^* = \frac{1}{0.0001} \ln (0.99005) = -99.99581$$

There is no real equivalent system characteristic root for the extraneous pole z_2; however, $x(n)$, $n = 0, 1, 2, \ldots$ does include a transient component $c_2 z_2^k$, which rapidly vanishes to zero leaving the dominant component $c_1 z_1^k$ and input mode (if present) terms to accurately track the continuous-time system response $x(t)$, $t \geq 0$.

Numerical stability of the simulation becomes an issue when the AB-2 integration step size produces z-plane poles in proximity of the Unit Circle. For a given first-order continuous-time system with characteristic root $\lambda < 0$, the discrete-time system resulting from AB-2 integration is marginally stable when the dominant pole is located at 1 or -1. From Equation 8.136,

$$z = 1: (1)^2 - \left(1 + \frac{3}{2} \lambda T \right)(1) + \frac{1}{2} \lambda T = 0 \quad \Rightarrow \quad \lambda T = 0 \tag{8.137}$$

$$z = -1: (-1)^2 - \left(1 + \frac{3}{2} \lambda T \right)(-1) + \frac{1}{2} \lambda T = 0 \quad \Rightarrow \quad \lambda T = -1 \tag{8.138}$$

Combining the above two results imposes the condition for stability, namely,

$$-1 < \lambda T < 0 \quad \Rightarrow \quad 1 > -\lambda T > 0 \quad \Rightarrow \quad T < \frac{1}{-\lambda} \tag{8.139}$$

In other words, the AB-2 integration step size T is limited by the time constant $\tau = -(1/\lambda)$ of the first-order continuous-time system. Where is the second z-plane pole when $\lambda T = 0$ and $\lambda T = -1$?

Second-order systems can be analyzed in the same way by allowing λ to be complex in the case of an underdamped second-order system or a pair of distinct real values for an overdamped second-order system. For example, a stable, second-order system with complex poles located at $-7.5 \pm j5$ simulated with AB-2 integration using a step size $T = 0.1$ s generates a stable discrete-time system if the two z-plane poles (principal and extraneous) are located inside the Unit Circle. This is easily checked by substituting $\lambda T = (-7.5 + j5)(0.1) = -0.75 + j0.5$ into the characteristic equation,

$$z^2 - \left(1 + \frac{3}{2}\lambda T\right)z + \frac{1}{2}\lambda T\bigg|_{\lambda T = -0.75 + j0.5} = z^2 + (0.125 - j0.75)z - 0.375 + j0.25 = 0 \quad (8.140)$$

Solution of Equation 8.140 reveals that the poles are located inside the Unit Circle at

$$z_1 = -0.6188 + j0.6418 = 0.8916e^{j2.3379}$$

$$z_2 = 0.4938 + j0.1082 = 0.2157e^{j0.2157}$$

and the discrete-time system is therefore stable. Increasing T eventually causes one of the z-plane poles to be on the Unit Circle where the system becomes marginally stable.

A closed locus of λT points can be identified in the complex plane with the property that all interior points produce stable discrete-time systems using AB-2 integration. The locus of points is called a stability boundary and the interior points comprise the stability region. There is a different stability boundary for each AB integrator.

The starting point for locating the stability boundary is finding $H(z)$, the z-domain transfer function of the discrete-time system resulting from numerical integration of the stable, continuous-time system

$$\frac{dx}{dt} + \lambda x = u, \quad \text{Re}(\lambda) < 0 \quad (8.141)$$

A similar approach to the one used for finding $H(z)$ for AB-2 integration of the continuous-time system in Equation 8.141 is employed to find $H(z)$ for different-order AB integrators. For AB-1 (Euler), AB-3, and AB-4 integration, $H_1(z)$ in Equation 8.128 is

$$\text{AB-1: } H_1(z) = \frac{T}{z - 1} \quad (8.142)$$

$$\text{AB-3: } H_1(z) = \frac{T}{12}\left[\frac{23z^2 - 16z + 5}{z^2(z - 1)}\right] \quad (8.143)$$

$$\text{AB-4: } H_1(z) = \frac{T}{24}\left[\frac{55z^3 - 59z^2 + 37z - 9}{z^3(z - 1)}\right] \quad (8.144)$$

Replacing s by $1/H_1(z)$ in $H(s) = 1/(s - \lambda)$ leads to the z-domain transfer function $H(z)$. For AB-1 through AB-4 integration, the results are

$$\text{AB-1: } H(z) = \frac{T}{z - (1 + \lambda T)} \quad (8.145)$$

$$\text{AB-2: } H(z) = \frac{T(3z - 1)}{2z^2 - (2 + 3\lambda T)z + \lambda T} \quad (8.146)$$

$$\text{AB-3: } H(z) = \frac{T(23z^2 - 16z + 5)}{12z^3 - (12 + 23\lambda T)z^2 + 16\lambda Tz - 5\lambda T} \quad (8.147)$$

$$\text{AB-4: } H(z) = \frac{T(55z^3 - 59z^2 + 37z - 9)}{24z^4 - (24 + 55\lambda T)z^3 + 59\lambda Tz^2 - 37\lambda Tz + 9\lambda T} \quad (8.148)$$

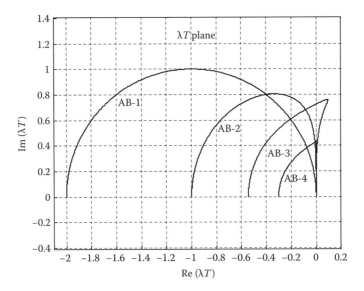

FIGURE 8.21 Stability boundaries for AB-1 through AB-4 integration.

Note the existence of one, two, and three extraneous z-plane poles in Equations 8.146 through 8.148. The stability boundaries are obtained by setting $z = e^{j\theta}$ in the denominators of Equations 8.145 through 8.148 and solving for λT. For example, with AB-3 integration, λT is given by

$$\lambda T = 12 \left(\frac{e^{j3\theta} - e^{j2\theta}}{5 - 16e^{j\theta} + 23e^{j2\theta}} \right) \tag{8.149}$$

Results for AB-1, AB-2, AB-3, and AB-4 integrators are obtained in the MATLAB M-file "*Chap8_AB_Stability_Boundaries.m*" and shown in Figure 8.21.

Only the top half of each stability boundary is shown since they are symmetric with respect to the real axis. Points along the top half of a stability boundary are computed by varying θ from 0 to π rad causing $e^{j\theta}$ to traverse the top half of the Unit Circle. The lower half is generated by sweeping θ from 0 to $-\pi$ rad.

A note of caution in finding the stability boundaries. The pole moving along the Unit Circle must be the largest in magnitude. For example, in the case of AB-4, the additional three poles must lie inside the Unit Circle. Values of λT for which this is not the case are ignored, that is, they are not points on the stability boundary (see Exercise 8.17).

Figure 8.21 confirms the result in Equation 8.139, namely, $\lambda T < -1$ for AB-2 integration of a stable, first-order system with real characteristic root λ. AB-1 integration is explicit Euler, and it is clear from Equation 8.145 that the lone z-plane pole of $H(z)$ migrates to $z = -1 = 1e^{j\pi}$ when $\lambda T = -2$, also confirmed by observing the leftmost point on the AB-1 stability boundary.

The equation of the stability boundary for AB-1 integration in the λT plane is easily derived. Figure 8.22 shows the z-plane pole of $H(z)$ in Equation 8.145 varying from 1 to -1 along the Unit Circle as θ increases from zero to π.

From Equation 8.145, the parameter λT on the AB-1 stability boundary is

$$\lambda T = e^{j\theta} - 1 = \cos\theta + j\sin\theta - 1 = (\cos\theta - 1) + j\sin\theta \tag{8.150}$$

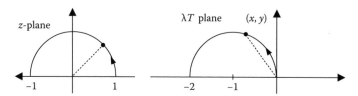

FIGURE 8.22 Variation of z-plane pole for determining AB-1 stability boundary.

If we let x and y be the real and imaginary parts of λT, respectively, that is, $\lambda T = x + jy$, then

$$x = \cos\theta - 1, \quad y = \sin\theta \tag{8.151}$$

$$\Rightarrow \ (x+1)^2 + y^2 = \cos^2\theta + \sin^2\theta = 1 \tag{8.152}$$

and the AB-1 stability boundary is therefore a circle with center at $(-1, 0)$ and radius 1 in the λT or x–y plane.

AB-1 integration is inappropriate for simulation of an undamped second-order system, a result we observed earlier in Section 3.6. The characteristic roots of a second-order system with $\zeta = 0$ are $\lambda = \pm j\omega_n$, and hence $\lambda T = \pm j\omega_n T$, which corresponds to the imaginary axis in the λT plane. From Figure 8.21, it is clear that the imaginary axis lies outside the AB-1 stability region (except for the origin).

A more general approach to locating stability boundaries is to view them as locus of points in the λT plane resulting from a mapping of the Unit Circle in the z-plane. Figure 8.23 illustrates the AB-2 stability boundary resulting from mapping points $z = re^{j\theta} = 1e^{j\theta}$ ($0 \le \theta < 2\pi$) along the Unit Circle in the z-plane according to the transformation

$$\lambda T = 2\left(\frac{e^{j\theta} - e^{j2\theta}}{1 - 3e^{j\theta}}\right) \tag{8.153}$$

obtained by solving for λT in the denominator of Equation 8.146 with z replaced by $e^{j\theta}$. The stability boundary (polar form $\lambda T = Me^{j\psi}$) is shown in Figure 8.23.

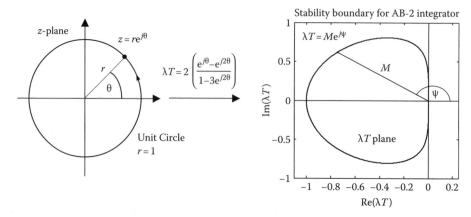

FIGURE 8.23 Mapping the Unit Circle to the AB-2 stability boundary.

Polar coordinates of the points along the AB-2 stability boundary are

$$M = 2 \left| \frac{e^{j\theta} - e^{j2\theta}}{1 - 3e^{j\theta}} \right| \tag{8.154}$$

$$= 2 \left[\frac{(\cos\theta - \cos 2\theta)^2 + (\sin\theta - \sin 2\theta)^2}{(1 - 3\cos\theta)^2 + (-3\sin\theta)^2} \right]^{1/2} \tag{8.155}$$

$$= 2 \left(\frac{1 - \cos\theta}{5 - 3\cos\theta} \right)^{1/2} \tag{8.156}$$

$$\psi = \mathrm{Arg}\left(\frac{e^{j\theta} - e^{j2\theta}}{1 - 3e^{j\theta}} \right) \tag{8.157}$$

$$= \tan^{-1}\left(\frac{4\sin\theta - \sin 2\theta}{4\cos\theta - \cos 2\theta - 3} \right) \tag{8.158}$$

Rectangular coordinates of λT on the AB-2 stability boundary are given in Exercise 8.30.

Example 8.4

For AB-2 integration,

(a) Find the image points on the AB-2 stability boundary in the λT plane of the following points:
$$z = 1, \left(\frac{\sqrt{2}}{2} \right)(1 + j), j, -1, e^{j4\pi/3}, -j, \left(\frac{\sqrt{2}}{2} \right)(1 - j).$$
(b) Is it possible for an AB-2 simulation of an undamped second-order system to be stable? Verify the result.

(a) The image points are computed using Equations 8.156 and 8.158 in the M-file "Chap8_Ex3_1. m." They are tabulated in Table 8.5.

(b) An undamped second-order system is governed by

$$\frac{d^2}{dt^2} x(t) + \omega_n^2 x(t) = u(t) \tag{8.159}$$

The characteristic roots are located on the imaginary axis at $\lambda = \pm j\omega_n$. Close-ups of the AB-2 stability boundary near the imaginary axis are shown in Figure 8.24. Observation of the left graph

---

TABLE 8.5
Points on Unit Circle and Image Points on AB-2 Stability Boundary

$z = re^{j\theta}$ r, θ	$z = a + jb$ a, b	$\lambda T = Me^{j\psi}$ M, ψ	$\lambda T = c + jd$ c, d
1, 0	1, 0	0, 0	0, 0
1, $\pi/4$	$\sqrt{2}/2$, $\sqrt{2}/2$	0.6380, $\pi/4$	−0.0596, 0.6352
1, $\pi/2$	0, 1	0.8944, $\pi/2$	−0.4, 0.8
1, π	−1, 0	1, π	−1, 0
1, $4\pi/3$	$-1/2, -\sqrt{3}/2$	0.9608, −2.0944	−0.6923, −0.6662
1, $3\pi/2$	0, −1	0.8944, $-\pi/2$	−0.4, −0.8
1, $7\pi/8$	$\sqrt{2}/2, -\sqrt{2}/2$	0.6380, $-\pi/4$	−0.0596, −0.6352

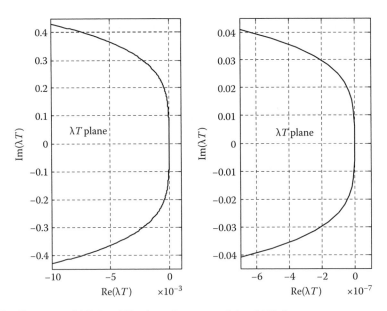

FIGURE 8.24 Close-ups of AB-2 stability boundary near origin of λT plane.

in Figure 8.24 implies $\lambda T = j\omega_n T$ is limited to approximately $j0.12$ for the AB-2 integrator to result in a stable discrete-time system. However, a closer look at the AB-2 stability region in the right graph indicates that the limit is considerably smaller. Further enlargement of the AB-2 stability region in the vicinity of the origin will show that the imaginary axis is exterior to the AB-2 stability region (with the exception of the origin, $\lambda T = 0$).

The instability of an AB-2 integrator for simulation of an undamped second-order system can be established by investigating the characteristic polynomial of the z-domain transfer function $H(z)$ given by

$$H(z) = \left.\frac{1}{s^2 + \omega_n^2}\right|_{s \leftarrow (2/T)[z(z-1)/(3z-1)]} \tag{8.160}$$

$$\Rightarrow \quad \frac{X(z)}{U(z)} = \frac{0.25T^2(9z^2 - 6z + 1)}{z^4 - 2z^3 + [1 + 2.25(\omega_n T)^2]z^2 - 1.5(\omega_n T)^2 z + 0.25(\omega_n T)^2} \tag{8.161}$$

Figure 8.25 is a plot of the loci of the four poles of $H(z)$ corresponding to numerical values of $\omega_n T = 0.05, 0.01, 0.15, \ldots, 0.95, 1$. The AB-2 integrator generates a discrete-time output $x(n)$ from the fourth-order system governed by

$$x(n + 4) - 2x(n + 3) + [1 + 2.25(\omega_n T)^2]x(n + 2) - 1.5(\omega_n T)^2 x(n + 1)$$
$$+ 0.25(\omega_n T)^2 x(n) = 0.25T^2[9u(n + 2) - 6u(n + 1) + u(n)] \tag{8.162}$$

Up until $\omega_n T \approx 0.3$, there is a pair of equivalent complex roots that die out rapidly due to the close proximity to the origin of the extraneous poles z_3 and z_4. At the same time, z_1 and z_2 appear to lie on the Unit Circle implying that the other pair of equivalent, continuous-time poles (corresponding to poles z_1 and z_2) lie on the imaginary axis in the s-plane since.

If z_1 and z_2 were actually on the Unit Circle, the damping ratio of the equivalent second-order continuous-time system would be zero and the discrete-time output would reflect an undamped second-order system response once the fast transient component vanishes. In reality, all the points along the two loci shown in Figure 8.25 are outside the Unit Circle, and the equivalent continuous-time second-order system damping ratio is slightly negative (see Exercise 8.20).

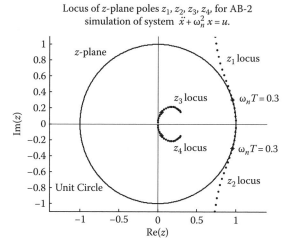

Locus of z-plane poles z_1, z_2, z_3, z_4, for AB-2
simulation of system $\ddot{x} + \omega_n^2 x = u$.

FIGURE 8.25 Locus of poles of $H(z)$ for AB-2 simulation of undamped second-order system $\ddot{x} + \omega_n^2 x = u$.

Howe (1986) includes asymptotic formulas for natural frequency and damping ratio errors incurred when using low-order multistep Adams–Bashforth integration methods. The formulas for $k = 1, 2, 3, 4$ are

$$k = 1: \quad e\omega_n \approx \zeta e_I \omega_n T, \quad e_\zeta \approx (\zeta^2 - 1)e_I \omega_n T, \quad \omega_n T \ll 1 \tag{8.163}$$

$$k = 2: \quad e\omega_n \approx (1 - 2\zeta^2)e_I(\omega_n T)^2, \quad e_\zeta \approx 2(\zeta - \zeta^3)e_I(\omega_n T)^2, \quad \omega_n T \ll 1 \tag{8.164}$$

$$k = 3: \quad e\omega_n \approx -(3\zeta - 4\zeta^3)e_I(\omega_n T)^3, \quad e_\zeta \approx (1 - 5\zeta^2 + 4\zeta^4)e_I(\omega_n T)^3, \quad \omega_n T \ll 1 \tag{8.165}$$

$$k = 4: \quad e\omega_n \approx -(1 - 8\zeta^2 + 8\zeta^4)e_I(\omega_n T)^4, \quad e_\zeta \approx -4(\zeta - 3\zeta^3 + 2\zeta^5)e_I(\omega_n T)^4, \quad \omega_n T \ll 1 \tag{8.166}$$

where e_I are the integration error coefficients given in Section 8.2. For accurate ($\omega_n T \ll 1$) simulations of undamped ($\zeta = 0$) second-order continuous-time systems, Equations 8.163 and 8.165 imply damping ratio errors of $-1/2(\omega_n T)$ with AB-1 and $3/8(\omega_n T)^3$ with AB-3 integration, respectively. Equations 8.164 and 8.166 imply the damping ratio error is zero for AB-2 and AB-4 integration to order $O(\omega_n T)^2$ and $O(\omega T)^4$, respectively. The actual damping ratio error is of order $O(\omega_n T)^3$ for AB-2 and $O(\omega_n T)^5$ for AB-4 integration.

8.3.2 Implicit Integrators

The Adams–Moulton implicit integrators were introduced in Section 6.4. The z-domain transfer functions for AM-2, AM-3, and AM-4 integrators are obtained in a similar fashion to the Adams–Bashforth integrators, that is, the difference equation approximation of a pure continuous-time integrator is developed and then z-transformed to produce $H_I(z)$. The results for AM-2 through AM-4 integrators are (see Exercise 8.22)

$$\text{AM-2:} \ H_I(z) = \frac{T}{2}\left(\frac{z+1}{z-1}\right) \tag{8.167}$$

$$\text{AM-3:} \ H_I(z) = \frac{T}{12}\left[\frac{5z^2 + 8z - 1}{z(z-1)}\right] \tag{8.168}$$

$$\text{AM-4:} \ H_I(z) = \frac{T}{24}\left[\frac{9z^3 + 19z^2 - 5z + 1}{z^2(z-1)}\right] \tag{8.169}$$

Replacing s by $1/H_I(z)$ in the first-order system transfer function $H(s) = 1/(s - \lambda)$ leads to the following expressions for the z-domain transfer functions using AM-2 through AM-4 integration,

$$\text{AM-2: } H(z) = \frac{T(z + 1)}{(2 - \lambda T)z - (2 + \lambda T)} \tag{8.170}$$

$$\text{AM-3: } H(z) = \frac{T(5z^2 + 8z - 1)}{(12 - 5\lambda T)z^2 - (12 + 8\lambda T)z + \lambda T} \tag{8.171}$$

$$\text{AM-4: } H(z) = \frac{T(9z^3 + 19z^2 - 5z + 1)}{(24 - 9\lambda T)z^3 - (24 + 19\lambda T)z^2 + 5\lambda Tz - \lambda T} \tag{8.172}$$

We may conclude from Equations 8.170 through 8.172 that AM-2 integration does not introduce extraneous roots (system poles), whereas AM-3 and AM-4 introduce one and two extraneous roots for each state. Stability boundaries for AM-2, AM-3, and AM-4 integration are obtained using the same method for the Adams–Bashforth integrators. Starting with AM-2, the characteristic polynomial for $H(z)$ in Equation 8.170 is

$$(2 - \lambda T)z - (2 + \lambda T) = 0 \tag{8.173}$$

Setting $z = e^{j\theta}$ and solving for λT yield

$$\lambda T = 2\left(\frac{z - 1}{z + 1}\right)\Bigg|_{z \leftarrow e^{j\theta}} = 2\left(\frac{e^{j\theta} - 1}{e^{j\theta} + 1}\right) \cdot \left(\frac{e^{-j\theta/2}}{e^{-j\theta/2}}\right) \tag{8.174}$$

$$= 2\left(\frac{e^{j\theta/2} - e^{-j\theta/2}}{e^{j\theta/2} - e^{-j\theta/2}}\right) \tag{8.175}$$

$$= j2\left[\frac{\sin(\theta/2)}{\cos(\theta/2)}\right] \tag{8.176}$$

$$= j2\tan(\theta/2) \tag{8.177}$$

From Equation 8.177, the top half of the Unit Circle, that is, $z = e^{j\theta}$, $0 \leq \theta < \pi$, is mapped into the imaginary axis from $\lambda T = 0$ to $\lambda T = j\infty$. The entire Unit Circle is mapped into the imaginary axis in the λT plane, which is the stability boundary for AM-2 or trapezoidal integration. In other words, the entire left-half plane is the stability region assuring that any stable continuous-time system (Re $\lambda < 0$) simulated by AM-2 integration leads to a stable discrete-time system regardless of the integration step size.

The stability regions for AM-3 and AM-4 integration are obtained from mapping the Unit Circle according to

$$\text{AM-3: } \lambda T = 12\left(\frac{e^{j2\theta} - e^{j\theta}}{5e^{j2\theta} + 8e^{j\theta} - 1}\right) \tag{8.178}$$

$$\text{AM-4: } \lambda T = 24\left(\frac{e^{j3\theta} - e^{j2\theta}}{9e^{j3\theta} + 19e^{j2\theta} - 5e^{j\theta} + 1}\right) \tag{8.179}$$

The stability boundaries for AM-3 and AM-4 integration are computed in "*Chap8_AM_Stability_Boundaries.m*" and shown along with the AM-2 stability boundary in Figure 8.26.

Note the restrictions imposed on AM-3 and AM-4 simulation of a stable, first-order system ($\lambda < 0$). The integration step size T is limited to less than $-6/\lambda$ and $-3/\lambda$, respectively. Equivalently, the step size $T < 6\tau$ for AM-3 and $T < 3\tau$ for AM-4 integration, where $\tau = -1/\lambda$ is the system time constant.

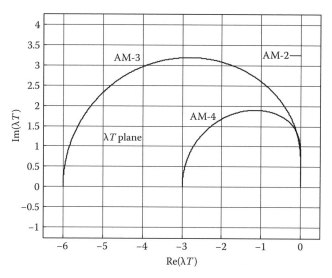

FIGURE 8.26   Stability boundaries for AM-2 through AM-4 integrators.

Example 8.5

The concentration of a chemical in the vessel shown in Figure 8.27 is determined by the differential equation

$$\frac{V}{Q}\frac{dc}{dt} + c = c_1 \tag{8.180}$$

where
  V is the constant volume of liquid in the vessel
  Q is the constant flow rate of liquid in and out of the vessel
  c_1 is the concentration of chemical in the liquid flowing in
  c is the concentration of the chemical in the well-stirred vessel

(a) Find the analytical solution for the concentration $c(t)$, $t \geq 0$ when the input $c_1(t) = \bar{c}_1, t \geq 0$.
(b) Find the difference equation for simulating the concentration response using AM-2 integration with step size T.
(c) Find an expression for the steady-state value $c(n)|_{n \to \infty}$ and compare it with $c(t)|_{t \to \infty}$.
(d) Repeat parts (b) and (c) for AM-3 integration.
(e) Numerical values of the system parameters are $Q = 25$ m³/min and $V = 150$ m³, and the initial concentration of chemical in the tank is $c(0) = 5$ mg/m³. The input $\bar{c}_1 = 60$ mg/m³. Simulate the concentration response using AM-2 and AM-3 integration with step size $T = 1.5\tau$, 1.25τ, τ, 0.75τ where $\tau = V/Q$ is the system time constant. Plot and compare the simulated responses and the analytical solution.

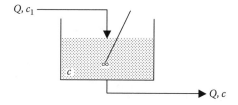

FIGURE 8.27   Chemical flowing in and out of a vessel with constant liquid volume.

(a) The analytical solution is obtained by Laplace transforming the differential equation with input c_1 constant along with the given initial condition. Alternatively, the step response of a first-order system is given in Equation 2.6 and repeated as follows using the current notation for the stirred tank.

$$c(t) = c(0)e^{-(Q/V)t} + c_1[1 - e^{-(Q/V)t}], \quad t \geq 0 \tag{8.181}$$

(b) Rewriting the differential equation as

$$\frac{dc}{dt} = -\frac{Q}{V}c + \frac{Q}{V}c_1 = -\frac{1}{\tau}c + \frac{1}{\tau}c_1 \tag{8.182}$$

and comparing it with $dx/dt = \lambda x + u$, the z-domain transfer function of the discrete-time system is obtained by replacing λ with $-1/\tau$ in Equation 8.170 and inserting $1/\tau$ in the numerator to give

$$H(z) = \frac{C(z)}{C_1(z)} = \frac{(T/\tau)(z + 1)}{[2 + (T/\tau)]z - [2 - (T/\tau)]} \tag{8.183}$$

$$= \frac{(T/2\tau)(z + 1)}{[1 + (T/2\tau)]z - [1 - (T/2\tau)]} \tag{8.184}$$

Inverting Equation 8.184 leads to the difference equation

$$\left(1 + \frac{T}{2\tau}\right)c(n + 1) - \left(1 - \frac{T}{2\tau}\right)c(n) = \frac{T}{2\tau}[c_1(n + 1) + c_1(n)], \quad n = 0, 1, 2, \ldots \tag{8.185}$$

which is used to update the state according to

$$c(n + 1) = \left[\frac{1 - (T/2\tau)}{1 + (T/2\tau)}\right]c(n) + \left[\frac{T/\tau}{1 + (T/2\tau)}\right]\bar{c}_1, \quad n = 0, 1, 2, \ldots \tag{8.186}$$

(c) The steady-state value $c(n)|_{n\to\infty}$ is obtained from Equation 8.186 after replacing $c(n)$ and $c(n + 1)$ with $c(n)|_{n\to\infty}$. Solving for $c(n)|_{n\to\infty}$ yields

$$c(n)|_{n\to\infty} = \bar{c}_1 \tag{8.187}$$

Hence, $c(n)|_{n\to\infty} = \bar{c}_1 = c(t)|_{t\to\infty}$, the final concentration of the continuous-time system.

(d) Replacing λ with $-1/\tau$, inserting $1/\tau$ in the numerator of Equation 8.171, and simplifying the result give

$$H(z) = \frac{C(z)}{C_1(z)} = \frac{5z^2 + 8z - 1}{[12(\tau/T) + 5]z^2 - [12(\tau/T) - 8]z - 1} \tag{8.188}$$

and the difference equation is

$$\left(12\frac{\tau}{T} + 5\right)c(n + 2) - \left(12\frac{\tau}{T} - 8\right)c(n + 1) - c(n) = 5c_1(n + 2) + 8c_1(n + 1) - c_1(n) \tag{8.189}$$

$$\Rightarrow c(n + 2) = \frac{1}{[12(\tau/T) + 5]}\left[\left(12\frac{\tau}{T} - 8\right)c(n + 1) + c(n) + 12\bar{c}_1\right], \quad n = 0, 1, 2, \ldots \tag{8.190}$$

Letting $c(n+2) = c(n+1) = c(n) = c(n)|_{n \to \infty}$ in Equation 8.190 and solving for $c(n)|_{n \to \infty}$ give the same result as Equation 8.187, that is, the AM-3 integrator also converges to $c(t)|_{t \to \infty} = \bar{c}_1$.

(e) The simulated responses using AM-2 and AM-3 integration along with the analytical solution are computed in "Chap8_Ex3_2.m" and shown in Figures 8.28 and 8.29. An RK-3 integrator would normally be used to generate $c(1)$, which is required to compute $c(2)$ in Equation 8.190. However, the exact value $c(T)$ was used instead.

Note the improvement in the AM-3 integrator compared with the AM-2 integrator.

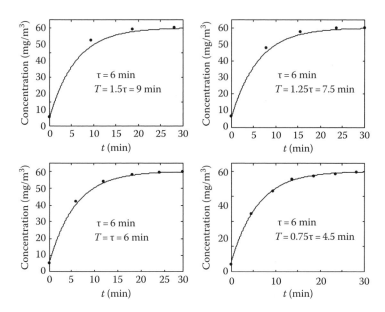

FIGURE 8.28 Analytical and simulated AM-2 concentration response.

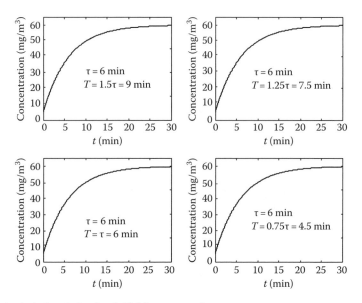

FIGURE 8.29 Analytical and simulated AM-3 concentration response.

8.3.3 RUNGA–KUTTA (RK) INTEGRATION

RKnumerical integration was introduced in Section 6.2. Unlike the multistep methods, RK integration algorithms are referred to as single pass or one step in nature. Depending on the order of the RK integrator, one or more state derivative function evaluations are required per step in order to advance the discrete-time state approximation to the next step. Fixed-step and variable-step RK formulas are popular in continuous-time system simulation.

Numerical stability with fixed-step RK integrators is important because of the limitations imposed on the integration step size. A similar approach to the one used for multistep methods is employed to obtain the stability boundary corresponding to a particular RK integrator. To illustrate, consider the second-order RK-2 integrator first introduced in Section 3.6 known as improved Euler or Heun's method. A continuous-time first-order system modeled by $dx/dt = f(x, u) = \lambda x + u$ is simulated using improved Euler integration by first predicting the updated state as

$$\hat{x}(n + 1) = x(n) + Tf[x(n), u(n)] \tag{8.191}$$

$$= x(n) + T[\lambda x(n) + u(n)] \tag{8.192}$$

$$= (1 + \lambda T)x(n) + Tu(n) \tag{8.193}$$

followed by correction to

$$x(n + 1) = x(n) + \frac{T}{2}\{f[x(n), u(n)] + f[\hat{x}(n + 1), u(n + 1)]\} \tag{8.194}$$

$$= x(n) + \frac{T}{2}\{f[x(n), u(n)] + f[(1 + \lambda T)x(n) + Tu(n), u(n + 1)]\} \tag{8.195}$$

$$= x(n) + \frac{T}{2}\{\lambda x(n) + u(n)] + \lambda[1 + \lambda T)x(n) + Tu(n)] + u(n + 1)\} \tag{8.196}$$

$$= \left[1 + \lambda T + \frac{(\lambda T)^2}{2}\right]x(n) + \frac{T}{2}[(1 + \lambda T)u(n) + u(n + 1)] \tag{8.197}$$

Taking the z-transform of Equation 8.197 and solving for the ratio $X(z)/U(z)$ give

$$H(z) = \frac{X(z)}{U(z)} = \frac{(T/2)(z + \lambda T + 1)}{z - [1 + \lambda T + (\lambda T)^2/2]} \tag{8.198}$$

Another popular RK-2 integrator, first introduced in Section 3.6, is the modified Euler integrator. The difference equation for modified Euler integration with a step size T can be obtained by reference to Figure 8.30. Note that the intervals of width $\hat{T} = T/2$ correspond to one-half the basic simulation frame rate $(1/T)$ to accommodate the input sampling rate of two samples per integration step T.

The first step in advancing the state using modified Euler integration with step size T is to compute the value $\hat{x}(n + 1)$ halfway through the integration interval, that is,

$$\hat{x}(n + 1) = x(n) + \hat{T}f[x(n), u(n)] \tag{8.199}$$

$$= x(n) + \hat{T}[\lambda x(n) + u(n)] \tag{8.200}$$

$$= (1 + \lambda\hat{T})x(n) + \hat{T}u(n) \tag{8.201}$$

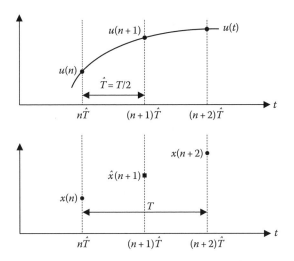

FIGURE 8.30   Modified Euler integration running at state update rate $(1/T)$.

The derivative function at $t = (n+1)\hat{T}$ is calculated using the predicted value $\hat{x}(n+1)$ in Equation 8.201. The updated state $x(n+2)$ is computed by taking a step of length $T = 2\hat{T}$ in the direction based on the midpoint derivative. Thus,

$$x(n+2) = x(n) + 2\hat{T}f[\hat{x}(n+1), u(n+1)] \tag{8.202}$$

$$= x(n) + 2\hat{T}[\lambda\hat{x}(n+1) + u(n+1)] \tag{8.203}$$

$$= x(n) + 2\lambda\hat{T}\hat{x}(n+1) + 2\hat{T}u(n+1) \tag{8.204}$$

$$= x(n) + 2\lambda\hat{T}[(1+\lambda\hat{T})x(n) + \hat{T}u(n)] + 2\hat{T}u(n+1) \tag{8.205}$$

$$= [1 + 2\lambda\hat{T}(1+\lambda\hat{T})]x(n) + 2\hat{T}[\lambda\hat{T}u(n) + u(n+1)] \tag{8.206}$$

In terms of the modified RK-2 integration step size $T = 2\hat{T}$, the difference equation for updating the discrete-time state $x(n)$ is

$$x(n+2) = \left[1 + \lambda T + \frac{(\lambda T)^2}{2}\right]x(n) + T\left[\frac{\lambda T}{2}u(n) + u(n+1)\right], \quad n = 0, 1, 2, 3, 4, \ldots \tag{8.207}$$

Note that $n = 0, 1, 2, 3, 4, \ldots$ in Equation 8.207 corresponds to times $0, T/2, T, 3T/2, 2T, \ldots$, and, therefore, $x(n)$, $n = 0, 2, 4, \ldots$ are the modified RK-2 states updated every T (s). The z-domain transfer function for modified RK-2 integration with step T is obtained by z-transforming Equation 8.207,

$$H(z) = \frac{T[z + (\lambda T/2)]}{z^2 - [1 + \lambda T + (\lambda T)^2/2]} \tag{8.208}$$

Difference equations and z-domain transfer functions for higher-order RK integrators are obtained in a similar fashion to the procedure outlined in Equations 8.199 through 8.208 for modified RK-2 integration. An RK-3 integrator with step size T requiring input samples at the beginning, one-third and two-thirds into the interval, is described by

$$k_1 = f[x(n), u(n)] \tag{8.209}$$

$$k_2 = f\left[x(n) + \frac{T}{3}k_1, u\left(n + \frac{1}{3}\right)\right] \tag{8.210}$$

$$k_3 = f\left[x(n) + \frac{2T}{3}k_2, u\left(n + \frac{2}{3}\right)\right] \tag{8.211}$$

$$x(n + 1) = x(n) + \frac{T}{4}(k_1 + 3k_3) \tag{8.212}$$

Using this RK-3 integrator with a sampling interval $\hat{T} = T/3$ to simulate the first-order continuous-time system $dx/dt = \lambda x + u$ results in the third-order difference equation (see Exercise 8.27)

$$x(n + 3) = \left[1 + \lambda T + \frac{(\lambda T)^2}{2} + \frac{(\lambda T)^3}{6}\right]x(n) + \left[\frac{T}{4} + \frac{\lambda^2 T^3}{6}\right]u(n)$$

$$+ \frac{\lambda T^2}{2}u(n + 1) + \frac{3T}{4}u(n + 2), \quad n = 0, 1, 2, 3, \ldots \tag{8.213}$$

where $x(n)$, $n = 0, 3, 6, 9, \ldots$ are the RK-3 states updated once every T(s).

z-Transforming Equation 8.213 leads to the z-domain transfer function

$$H(z) = \frac{(3T/4)z^2 + (\lambda T^2/2)z + (T/4) + (\lambda^2 T^3/6)}{z^3 - [1 + \lambda T + (\lambda T)^2/2 + (\lambda T)^3/6} \tag{8.214}$$

Consider the RK-4 integrator presented in Section 6.2, Equations 8.60 through 8.64 with integration step size T and input sampled at the beginning and midpoint of each interval. The z-domain transfer function is (Howe 1986, 1995)

$$H(z) = \frac{(T/6)\{z^2 + [4 + 2\lambda T + (\lambda T)^2/2]z + 1 + \lambda T + (\lambda T)^2/2 + (\lambda T)^3/4\}}{z^2 - [1 + \lambda T + (\lambda T)^2/2 + (\lambda T)^3/6 + (\lambda T)^4/24]} \tag{8.215}$$

The characteristic polynomials for the one-step RK integrators with z-domain transfer functions given in Equations 8.198, 8.208, 8.214, and 8.215 are summarized as follows.

$$\text{RK-2 (Improved Euler): } z - \left[1 + \lambda T + \frac{(\lambda T)^2}{2}\right] \tag{8.216}$$

$$\text{RK-2 (Modified Euler): } z^2 - \left[1 + \lambda T + \frac{(\lambda T)^2}{2}\right] \tag{8.217}$$

$$\text{RK-3 (Input sampling at } 3/T): z^3 - \left[1 + \lambda T + \frac{(\lambda T)^2}{2} + \frac{(\lambda T)^3}{6}\right] \tag{8.218}$$

$$\text{RK-4 (Input sampling at } 2/T): z^2 - \left[1 + \lambda T + \frac{(\lambda T)^2}{2} + \frac{(\lambda T)^3}{6} + \frac{(\lambda T)^4}{24}\right] \tag{8.219}$$

For an mth-order RK integrator requiring k_s input samples per integration step T, the characteristic polynomial is given by

$$\text{RK-}m\left(\text{Input sampling at } \frac{k_s}{T}\right): z^{k_s} - \left[1 + \lambda T + \frac{(\lambda T)^2}{2!} + \frac{(\lambda T)^3}{3!} + \cdots + \frac{(\lambda T)^m}{m!}\right] \tag{8.220}$$

Note that the bracketed expression in Equation 8.220 is the truncated Taylor Series approximation for $e^{\lambda T}$. Let us explore this point further. λ^*, the characteristic root of the equivalent continuous-time system, is related to the z-plane pole by

$$z = e^{\lambda^*(T/k_s)} \tag{8.221}$$

The z-plane pole for the RK-4 integrator is from Equation 8.219

$$z = \left[1 + \lambda T + \frac{(\lambda T)^2}{2} + \frac{(\lambda T)^3}{6} + \frac{(\lambda T)^4}{24}\right]^{1/2} \tag{8.222}$$

Substituting this z into Equation 8.221 with $k_s = 2$ and squaring both sides lead to

$$\left[1 + \lambda T + \frac{(\lambda T)^2}{2} + \frac{(\lambda T)^3}{6} + \frac{(\lambda T)^4}{24}\right] = e^{\lambda^* T} \tag{8.223}$$

Expanding the exponential term in Equation 8.223 in a fifth-order truncated power series eventually leads to the asymptotic formula for the fractional characteristic root error, that is,

$$\text{RK-4: } e_\lambda = \frac{\lambda^*}{\lambda} - 1 \approx -\frac{1}{120}(\lambda T)^4, \quad |\lambda T| \ll 1 \tag{8.224}$$

which implies the integrator error coefficient e_I for RK-4 is $-1/120$.

Example 8.6

Find the equivalent continuous-time system characteristic root for the system in Example 8.5 using RK-2, RK-3, and RK-4 integration with step size $T = 0.25$ s.
   From Equation 8.223 and similar expressions for RK-2 and RK-3,

$$\text{RK-2: } \lambda^* = \frac{1}{T} \ln\left[1 + \lambda T + \frac{(\lambda T)^2}{2}\right] \tag{8.225}$$

$$\text{RK-3: } \lambda^* = \frac{1}{T} \ln\left[1 + \lambda T + \frac{(\lambda T)^2}{2} + \frac{(\lambda T)^3}{6}\right] \tag{8.226}$$

$$\text{RK-4: } \lambda^* = \frac{1}{T} \ln\left[1 + \lambda T + \frac{(\lambda T)^2}{2} + \frac{(\lambda T)^3}{6} + \frac{(\lambda T)^4}{24}\right] \tag{8.227}$$

The characteristic root for the system in Example 8.5 is $\lambda = -1/6$. Substituting $\lambda T = (-1/6)(1/4) = -1/24$ in Equations 8.225 through 8.227 results in

$$\lambda^* = \begin{cases} -0.16661691, & \text{(RK-2)} \\ -0.16666719, & \text{(RK-3)} \\ -0.16666666, & \text{(RK-4)} \end{cases}$$

The stability boundaries for the RK integrators are obtained as before by mapping the Unit Circle in the z-plane into the λT plane using the denominator of $H(z)$ to define the mapping. The MATLAB M-file "Chap8_RK_Stability_Boundaries.m" finds and plots the top half of the RK-2, RK-3, and RK-4 stability boundaries shown in Figure 8.31.

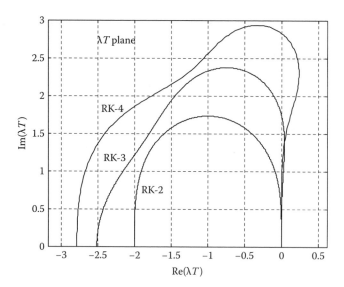

FIGURE 8.31 Stability boundaries for RK-2 through RK-4 integrators.

Unlike the Adams–Bashforth and Adams–Moulton integrators, the stability regions become larger for the higher-order RK integrators. There is a single stability boundary for all mth-order RK integrators, independent of the number of input samples required during each integration step. This is logical since stability of the discrete-time system associated with RK integration is an inherent system property unrelated to the possible existence of inputs.

The fractional characteristic root error e_λ for the mth-order numerical integrators discussed in this section and the previous section is related to the integrator error coefficient e_I according to (Howe 1995)

$$e_\lambda = \frac{\lambda^*}{\lambda} - 1 \approx -e_I(\lambda T)^m, \quad |\lambda T| \ll 1 \tag{8.228}$$

A comparison of characteristic root errors for comparable order, Adams–Bashforth, Adams–Moulton, and RK integrators, is shown in the middle three columns of Table 8.6.

Keep in mind the RK-m integrator requires m derivative function evaluations per step. The RK-4 integrator, for example, would take roughly four times longer than either AB-4 or AM-4 integrators to execute a single step. In order to keep the computational effort between the multistep AB-m and AM-m integrators comparable to the one-step RK-m integrators, the step size should be m times larger with RK-m integration.

TABLE 8.6
Characteristic Root Errors for AB, AM, and RK Integrators

m	e_λ, AB-m Step Size T	e_λ, AM-m Step Size T	e_λ, RK-m Step Size T	$\tilde{e}_\lambda$, RK-m Step Size T
2	$-\dfrac{5}{12}(\lambda T)^2$	$\dfrac{1}{12}(\lambda T)^2$	$-\dfrac{1}{6}(\lambda T)^2$	$-\dfrac{1}{6}(\lambda 2T)^2 = -\dfrac{4}{6}(\lambda T)^2$
3	$-\dfrac{3}{8}(\lambda T)^3$	$\dfrac{1}{24}(\lambda T)^3$	$-\dfrac{1}{24}(\lambda T)^3$	$-\dfrac{1}{24}(\lambda 3T)^3 = -\dfrac{27}{24}(\lambda T)^3$
4	$-\dfrac{251}{720}(\lambda T)^4$	$\dfrac{19}{720}(\lambda T)^4$	$-\dfrac{1}{120}(\lambda T)^4$	$-\dfrac{1}{120}(\lambda 4T)^4 = -\dfrac{256}{120}(\lambda T)^4$

The last column in Table 8.6 reflects the effect of increasing the step size with RK integration to make the computational effort approximately the same as the comparable order AB and AM integrators. In the case of RK-4, the effective characteristic root error $\tilde{e}_\lambda$ is proportional to $-(256/120)(\lambda T)^4$, and the ratio of e_λ for AB-4 integration to $\tilde{e}_\lambda$ for RK-4 integration is

$$\frac{e_\lambda}{\tilde{e}_\lambda} = \frac{-(251/720/\lambda T)^4}{-(256/120)(\lambda T)^4} = 0.1634 \tag{8.229}$$

making AB-4 integration roughly six times more accurate than RK-4 integration when execution time is taken into consideration.

Example 8.7

A simplified block diagram for the forward speed control of a ground vehicle is shown in Figure 8.32. The system parameters are the open-loop system gain K and poles located at $s = -a$ and $s = -b$.

(a) Find expressions for the natural frequency, damping ratio, and steady-state gain of the second-order closed-loop system in terms of the system parameters.
(b) Find the analytical solution for the unit step response.
(c) An RK-2 (improved Euler) simulation is performed for the cases where $K = 100, 250$ using step sizes of $T = 0.025$ and 0.1 s. The open-loop poles are located at $s = -a = -2$ s^{-1} and $s = -b = -5$ s^{-1}. Plot the analytical and simulated step responses for each case on separate graphs. Comment on the accuracy and numerical stability of the RK-2 integrator. Repeat using RK-4 integration with step sizes of 0.1 and 0.2 s.
(d) For the case where $K = 250$, $a = 2$, and $b = 5$, find the maximum value of T that can be used to implement RK-3 simulation. Verify the result.

(a) The closed-loop system transfer function is

$$\frac{V(s)}{V_{com}(s)} = \frac{K}{s^2 + (a+b)s + ab + K} \tag{8.230}$$

Comparing Equation 8.230 to the standard form of a second-order system transfer function

$$\frac{K}{s^2 + (a+b)s + ab + K} = \frac{K_{ss}\omega_n^2}{s^2 + 2\zeta\omega_n s + \omega_n^2} \tag{8.231}$$

and solving for the second-order system parameters K_{ss}, ω_n, and ζ results in

$$K_{ss} = \frac{K}{ab+K}, \quad \omega_n = (ab+K)^{1/2}, \quad \zeta = \frac{a+b}{2(ab+K)^{1/2}} \tag{8.232}$$

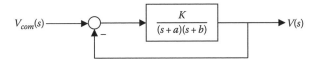

$$V_{com}(s) \longrightarrow \bigcirc \xrightarrow{} \boxed{\frac{K}{(s+a)(s+b)}} \longrightarrow V(s)$$

FIGURE 8.32 Block diagram of speed control system.

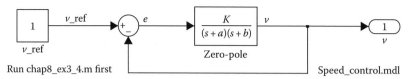

Run chap8_ex3_4.m first   Speed_control.mdl

FIGURE 8.33 Simulink® diagram for RK simulation of speed control system.

(b) The analytical solution for the step response is (see Equation 2.24)

$$y(t) = K_{ss}\left[1 = \frac{\omega_n}{\omega_d}e^{-\zeta\omega_n t}\sin(\omega_d t + \varphi)\right], \quad t \geq 0 \tag{8.233}$$

$$\omega_d = (1 - \zeta^2)^{1/2}\omega_n, \quad \varphi = \tan^{-1}\left(\frac{\omega_d}{\zeta\omega_n}\right) \tag{8.234}$$

(c) RK integrators "ode1" through "ode5" of order one through five are available in MATLAB and Simulink®. The Simulink diagram is shown in Figure 8.33.

The Simulink model file "speed_control.mdl" is called from the MATLAB M-file "Chap8_Ex3_4.m," which sets the system parameters, selects the numerical integrator as either RK-2 or RK-4, sets the timing parameters (step size and simulation duration), and plots the analytical and simulated responses. The results are shown in Figures 8.34 and 8.35.

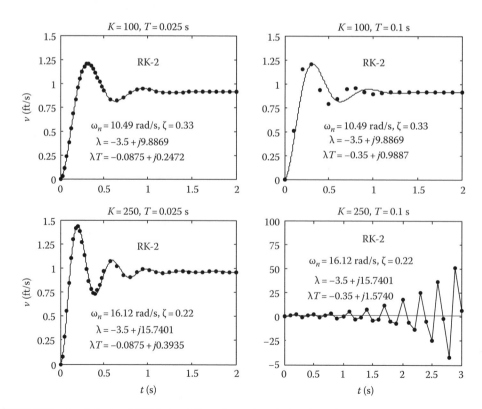

FIGURE 8.34 Analytical and RK-2 simulation of speed control system step response.

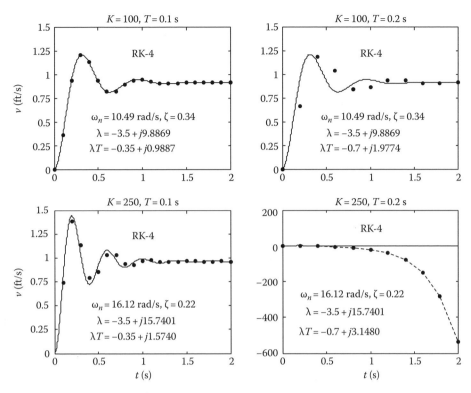

FIGURE 8.35   Analytical and RK-4 simulation of speed control system step response.

Some of the data points at the end of the simulated responses when $T = 0.025$ s in Figure 8.34 are omitted to make it easier to visualize the discrete-time nature of the response. Several of the simulated transient responses are quite accurate, whereas others deviate by a significant amount from the analytical solution. The RK-2 integrator is unstable when $K = 250$ and $T = 0.1$ s, and the RK-4 integrator exhibits instability for the case when $K = 250$ and $T = 0.2$ s. The reader should confirm that $\lambda T = -0.35 + j1.5740$ and $\lambda T = -0.7 + j3.1480$ fall outside the stability regions for RK-2 and RK-4, respectively.

(d) For the case when $K = 250$, the continuous-time system characteristic roots are $\lambda = -3.5 \pm j15.7401$. The limiting value of T for numerical stability is found by locating the intersection of the ray $\lambda T = (-3.5 + j15.7401)T$, $T > 0$ and the RK-3 stability boundary as shown in Figure 8.36. The M-file "*Chap8_Ex3_4.m*" contains MATLAB code, which tracks the values of λT along the RK-3 boundary as the point z rotates around the Unit Circle in the z-plane. The common point on the ray and stability boundary is located where the angle of λT on the stability boundary is equal to the constant angle of the ray (see Figure 8.36). It occurs at $\lambda T = -0.5198 + j2.3386$. The limiting step size is found from

$$\lambda T_{max} = (-3.5 + j15.7401)T_{max} = -0.5198 + j2.3386 \qquad (8.235)$$

Solving for T_{max} in Equation 8.235,

$$-3.5T_{max} = -0.5198 \quad \Rightarrow \quad T_{max} = 0.1485 \text{ s}$$

The Simulink model was run using the "ode3" RK-3 integrator with a step size of T_{max}. The marginally stable simulated response is shown in Figure 8.37.

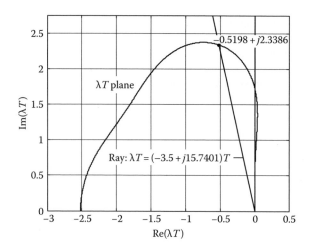

FIGURE 8.36 Finding λT_{max} point for RK-3 simulation of system.

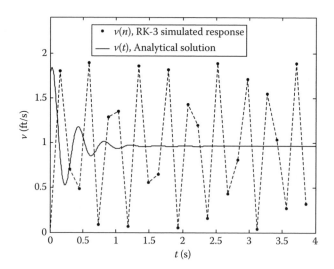

FIGURE 8.37 Analytical step response and marginally stable RK-3 simulated response.

EXERCISES

8.15 Show that the extraneous z-plane pole z_2 resulting from AB-2 integration of the first-order system $dx/dt = \lambda x + u$ is approximately equal to $0.5\lambda T$ when $\lambda T \ll 1$.

8.16 Simulate the unit step response of the first-order system $dx/dt = \lambda x + u$ using AB-2 integration, and plot both $x(t)$, $t \geq 0$ and $x(n)$, $n = 0, 1, 2, \ldots$ for the following cases:

λ	T
-0.1	1, 2, 5, 9, 10, 11
-2	0.1, 0.2, 0.3, 0.4, 0.5, 0.6
-50	0.002, 0.005, 0.01, 0.019, 0.02, 0.021

8.17 Show that the procedure for finding the AB-4 stability region must be modified to account for the existence of extraneous poles outside the Unit Circle, that is, for certain values of λT, the principal pole may lie on the Unit Circle; however, there may be other poles of $H(z)$ larger in magnitude.

8.18 The stability boundary for AB integration in polar form is $\lambda T = Me^{j\psi}$ where $M = |\lambda T|$ and $\psi = \text{Arg}(\lambda T)$ are both functions of the angle θ as shown in Figure 8.23 for AB-2 integration.

(a) Show that $M = (2 - 2\cos\theta)^{1/2}$ and $\psi = \tan^{-1}(\sin\theta/(\cos\theta - 1))$ for AB-1 integration.

(b) Derive Equations 8.156 and 8.158.

(c) Find M and ψ for AB-3 and AB-4 integration.

8.19 Investigate the stability of AB-3 and AB-4 integration for undamped continuous-time second-order systems. Specifically,

(a) Find the z-domain transfer functions of the system, and plot the loci of the poles as the parameter $\omega_n T$ varies, similar to Figure 8.25 for AB-2 integration.

(b) Include close-ups of the stability boundaries near the imaginary axis of the λT plane.

8.20 In Example 8.4,

(a) Find all the z-plane poles when $\omega_n T = 0.05, 0.1, \ldots, 0.45, 0.5$. Comment on how the results affect the stability of AB-2 integration of undamped second-order systems.

(b) Show that the difference equation for implementing AB-2 integration of the system $\ddot{x} + \omega_n^2 x = u$ is

$$x(n + 4) - 2x(n + 3) + \left[1 + 2.25(\omega_n T)^2\right]x(n + 2) - 1.5(\omega_n T)^2 x(n + 1)$$
$$+ 0.25(\omega_n T)^2 x_A(n) = 0.25T^2[9u(n + 2) - 6u(n + 1) + u(n)], \quad n = 0, 1, 2, 3, \ldots$$

(c) Find the difference equation for explicit Euler integration of the undamped second-order system.

(d) Write a MATLAB M-file that accepts values for ω_n and T and implements AB-2 integration to simulate the unit step response of the system. Use the explicit Euler integrator to compute the starting values $x(2)$ and $x(3)$.

(e) Plot the exact and simulated step responses for the following cases:

    (i) $\omega_n = 1$ rad/s, $T = 0.01$ s

    (ii) $\omega_n = 100$ rad/s, $T = 0.002$ s

    (iii) $\omega_n = 0.02$ rad/s, $T = 15$ s

    (iv) $\omega_n = 10$ rad/s, $T = 0.5$ s

8.21 Discuss the implications of the AB-3 stability boundary extending into the first quadrant of the λT plane. Illustrate by simulating the step response of the continuous-time second-order system

$$\frac{d^2}{dt^2}x(t) - \frac{d}{dt}x(t) + 49.25x(t) = u(t)$$

using AB-3 integration with step size $T = 0.1$ s. Plot the exact and simulated response on the same graph. What is the damping ratio and natural frequency of the continuous-time system?

8.22 Derive expressions for

(a) $H_I(z)$ for AM-2, AM-3, and AM-4 integrators given in Equations 8.167 through 8.169.

(b) $H(z)$ for AM-2, AM-3, and AM-4 integration of $dx/dt = \lambda x + u$, (Re $\lambda < 0$) given in Equations 8.170 through 8.172.

(c) λT in Equations 8.178 and 8.179.

8.23 Use the final value theorem (see Table 4.5) to obtain the final value for $c(n)|_{n \to \infty}$ given in Equation 8.187.

8.24 Find an expression for the equivalent continuous-time system characteristic root λ^* corresponding to the z-plane pole resulting from AM-2 simulation of the system $dx/dt = \lambda x + u$. Compute λ^* and e_λ (the characteristic root error) for the values of λ and T used in Example 8.5. Are your answers consistent with the responses in Figures 8.28 and 8.29?

8.25 For RK-2 integration of $dx/dt = \lambda x + u$ resulting in Equation 8.198 for $H(z)$,
  (a) Find the z-plane pole of the discrete-time system.
  (b) Find the equivalent continuous-time system characteristic root λ^*.
  (c) Find asymptotic formulas for λ^* and the fractional error in λ^*, that is, $e_\lambda = \lambda^*/\lambda - 1$.

8.26 Find the difference equation for the RK-4 integrator in Section 6.2.

8.27 Derive the result in Equation 8.214 for the z-domain transfer function of the RK-3 integrator in Equations 8.209 through 8.212.

8.28 Derive the expression in Equation 8.224 for the fractional characteristic root error incurred using RK-4 integration.

8.29 Consider an unstable, second-order system with DC gain $k_{SS} = 1$, natural frequency $\omega_n = 50$ rad/s, and damping ratio $\zeta = -0.02$. The initial conditions are $x(0) = 1$ and $\dot{x}(0) = 0$.
  (a) Use Simulink to simulate the transient response of the autonomous system using RK-2 and RK-4 integration with a step size of $T = 0.05$ s.
  (b) Find the analytical solution and plot it along with the RK-2 and RK-4 simulated responses on the same graph.
  (c) Comment on the results. Does λT lie inside the RK-2 and RK-4 stability regions?

8.30 Polar coordinates of the AB-2 stability boundary are expressed parametrically in Equations 8.156 and 8.158. Show that a parametric representation for the rectangular coordinates of the AB-2 stability boundary is given by

$$x = \text{Re}(\lambda T) = \frac{4\cos\theta - \cos 2\theta - 3}{5 - 3\cos\theta}$$

$$y = \text{Im}(\lambda T) = \frac{4\sin\theta - \sin 2\theta}{5 - 3\cos\theta}$$

for $0 \le \theta \le 2\pi$.

8.4 MULTIRATE INTEGRATION

The topic of stiff systems was introduced in Section 6.5. Recall that the stiffness property is a measure of the variation in magnitude between the smallest and largest characteristic roots (eigenvalues of the coefficient matrix A in state variable model) of a linear or linearized system. When the characteristic roots of a stiff system are as portrayed in Figure 8.38a, variable-step stiff integrators like MATLAB's "ode15s," "ode23s," "ode23tb" are more computationally efficient in simulating the system dynamics than fixed-step numerical integrators owing to the excessively small time steps necessary with fixed-step integrators to assure numerical stability.

When the system poles are clustered in distinct regions of the s-plane as shown in Figure 8.38b, the overall continuous-time system is composed of two or more subsystems that effectively operate at

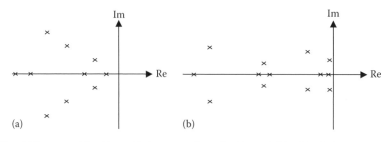

FIGURE 8.38 Stiff system (a) without distinct grouping of poles and (b) with distinct grouping of poles.

different speeds. Different time scales are required to view the time histories of the individual subsystem state variables. The pole locations in Figure 8.38b implies the existence of three subsystems, a relatively slow sixth-order subsystem associated with the six dominant poles nearest to the origin and imaginary axis, an intermediate speed fourth-order subsystem corresponding to the middle four poles, and a third-order fast subsystem arising from the three poles furthest from the origin.

Multirate integration methods are often effective in simulating continuous-time systems with identifiable subsystems like the one shown in Figure 8.38b. As the name suggests, numerical integrators running at different frame rates (step sizes) are tailored to the individual subsystems. The explanation and example that follow are geared toward a two-time scale system, that is, a system with characteristic roots in two distinct regions located an order of magnitude apart from the origin of the s-plane. By implication, a subset of the system's state variables are predominantly characterized by fast dynamics, that is, short time constants, high natural frequencies, and bandwidth, and the remaining states are just the opposite, namely, those associated with slow natural modes and longer transient responses.

Electromechanical control systems are frequently composed of fast and slow subsystems. Components in electronic controllers and sensors are much faster than the mechanical systems being controlled. The result is an overall system with fast and slow dynamics. Figure 8.39 is the block diagram of an aircraft pitch control system similar to one in Howe (1995). The airframe is modeled as a linear second-order system to account for the short-period longitudinal dynamics. The actuating signal for the controller is the difference between the commanded elevator deflection δ_i coming from the autopilot and the actual elevator deflection δ_e. The control surface actuator (lumped with the controller) moves the elevator. The pitch θ and pitch rate $\dot{\theta}$ are fed back to the autopilot, which receives the pitch angle command θ_{com} from the pilot.

The airframe dynamics and subsequent integrator constitute the slow subsystem, and the fast subsystem is composed of the remaining components. Since the slow and fast states are to be integrated at different rates, it is necessary to define the slow and fast states and express the state derivatives in terms of the states and the command input. We begin with the slow subsystem blocks and perform the steps necessary to generate an equivalent simulation diagram. The transfer function of the airframe dynamics is

$$\frac{\dot{\theta}(s)}{\delta_e(s)} = \frac{K_A \omega_n^2 (\tau_A s + 1)}{s^2 + 2\xi \omega_n s + \omega_n^2} \tag{8.236}$$

leading to the differential equation

$$\frac{d^2 \dot{\theta}}{dt^2} + 2\zeta \omega_n \frac{d\dot{\theta}}{dt} + \omega_n^2 \dot{\theta} = K_A \omega_n^2 \tau A \frac{d\delta_e}{dt} + K_A \omega_n^2 \delta_e \tag{8.237}$$

The simulation diagram is shown in Figure 8.40. The integrator outputs are chosen as the slow system states x_1, x_2, and x_3.

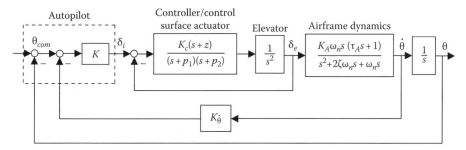

FIGURE 8.39 Block diagram of aircraft pitch control system.

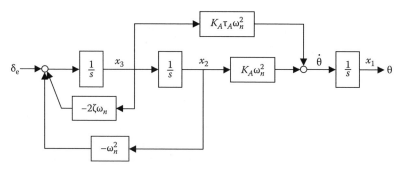

FIGURE 8.40   Simulation diagram of airframe dynamics with states x_1, x_2, and x_3.

FIGURE 8.41   Controller/control surface actuator.

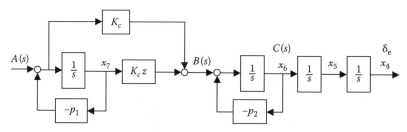

FIGURE 8.42   Simulation diagram of fast subsystem with states x_4, x_5, x_6, and x_7.

The fast systems states are obtained by breaking the controller/control surface actuator transfer function into serial first-order blocks as shown in Figure 8.41. The simulation diagrams for the first-order blocks are drawn using the techniques introduced in Section 2.4.

The simulation diagram with the fast states x_4, x_5, x_6, and x_7 is shown in Figure 8.42.

From Figures 8.39, 8.40, and 8.42, we are able to write algebraic and state derivative equation, which eventually lead to the following state model (see Exercise 8.31).

$$\underline{\dot{x}} = A\underline{x} + B\theta_{com} \tag{8.238}$$

$$\underline{y} = C\underline{x} + D\theta_{com} \tag{8.239}$$

where the matrices A, B, C, and D are given by

$$A = \begin{bmatrix} 0 & K_A\omega_n^2 & K_A\omega_n^2\tau_A & 0 & 0 & 0 & 0 \\ 0 & 0 & 1 & 0 & 0 & 0 & 0 \\ 0 & -\omega_n^2 & -2\zeta\omega_n & 1 & 0 & 0 & 0 \\ 0 & 0 & 0 & 0 & 1 & 0 & 0 \\ 0 & 0 & 0 & 0 & 0 & 1 & 0 \\ -K_CK & -K_CKK_{\dot{\theta}}K_A\omega_n^2 & -K_CKK_{\dot{\theta}}K_A\omega_n^2\tau_A & -K_C & 0 & -p_2 & K_C(z-p_1) \\ -K & -KK_{\dot{\theta}}K_A\omega_n^2 & -KK_{\dot{\theta}}K_A\omega_n^2\tau_A & -1 & 0 & 0 & -p_1 \end{bmatrix} \tag{8.240}$$

$$B = \begin{bmatrix} 0 \\ 0 \\ 0 \\ 0 \\ 0 \\ K_C K \\ K \end{bmatrix}, \quad C = \begin{bmatrix} 1 & 0 & 0 & 0 & 0 & 0 & 0 \\ 0 & 1 & 0 & 0 & 0 & 0 & 0 \\ 0 & 0 & 1 & 0 & 0 & 0 & 0 \\ 0 & 0 & 0 & 1 & 0 & 0 & 0 \\ 0 & 0 & 0 & 0 & 1 & 0 & 0 \\ 0 & 0 & 0 & 0 & 0 & 1 & 0 \\ 0 & 0 & 0 & 0 & 0 & 0 & 1 \end{bmatrix}, \quad D = \begin{bmatrix} 0 \\ 0 \\ 0 \\ 0 \\ 0 \\ 0 \\ 0 \end{bmatrix} \qquad (8.241)$$

Note that the output vector $\underline{y} = [y_1 \quad y_2 \quad y_3 \quad y_4 \quad y_5 \quad y_6 \quad y_7]^T$ is chosen to be identical to the state vector $\underline{x}$. Decomposing the state vector $\underline{x}$ into a vector of slow states $\underline{u} = [u_1 \quad u_2 \quad u_3]^T = [x_1 \quad x_2 \quad x_3]^T$ and fast states $\underline{w} = [w_1 \quad w_2 \quad w_3 \quad w_4]^T = [x_4 \quad x_5 \quad x_6 \quad x_7]^T$ leads to a definition of the slow state derivatives $\underline{\dot{u}} = \underline{f}(\underline{u}, \underline{w})$ as

$$\dot{u}_1 = \dot{x}_1 = f_1(\underline{u}, \underline{w}) = A_{1,2}u_2 + A_{1,3}u_3 \qquad (8.242)$$

$$\dot{u}_2 = \dot{x}_2 = f_2(\underline{u}, \underline{w}) = A_{2,3}u_3 \qquad (8.243)$$

$$\dot{u}_3 = \dot{x}_3 = f_3(\underline{u}, \underline{w}) = A_{3,2}u_2 + A_{3,3}u_3 + A_{3,4}w_1 \qquad (8.244)$$

and the fast state derivative vector $\underline{\dot{w}} = \underline{g}(\underline{u}, \underline{w}, \theta_{com})$ is

$$\dot{w}_1 = \dot{x}_4 = g_1(\underline{u}, \underline{w}, \theta_{com}) = A_{4,5}w_2 \qquad (8.245)$$

$$\dot{w}_2 = \dot{x}_5 = g_2(\underline{u}, \underline{w}, \theta_{com}) = A_{5,6}w_3 \qquad (8.246)$$

$$\dot{w}_3 = \dot{x}_6 = g_3(\underline{u}, \underline{w}, \theta_{com})$$
$$= A_{6,1}u_1 + A_{6,2}u_2 + A_{6,3}u_3 + A_{6,4}w_1 + A_{6,6}w_3 + A_{6,7}w_4 + B_6\theta_{com} \qquad (8.247)$$

$$\dot{w}_4 = \dot{x}_7 = g_4(\underline{u}, \underline{w}, \theta_{com}) = A_{7,1}u_1 + A_{7,2}u_2 + A_{7,3}u_3 - w_1 + A_{7,7}w_4 + B_7\theta_{com} \qquad (8.248)$$

where the coefficients $A_{i,j}$ are the elements in the coefficient matrix A in Equation 8.240.

Figure 8.43 portrays the slow and fast subsystems and the coupling between them.

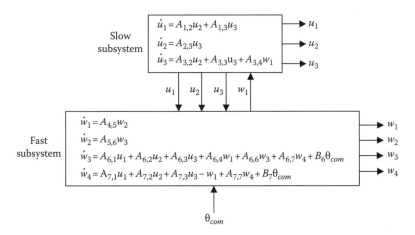

FIGURE 8.43 Slow and fast subsystem interaction.

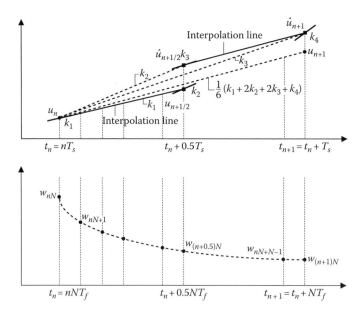

FIGURE 8.44 Multirate integration for one frame of slow state and N frames of fast state.

Once the system is decomposed into a slow and fast subsystem, the numerical integration routine and frame times (step size T_s for the slow subsystem and T_f for the fast one) must be selected. The numerical integrator to update the slow states is referred to as the "master" routine, and the integration method for advancing the fast states is called the "slave" routine (Palusinski 1986). The situation is illustrated in Figure 8.44 for the case where both "master" and "slave" are the classic RK-4 integrator (see Equations 6.60 through 6.64) with step sizes T_s and T_f, respectively. The quotient $N = T_s/T_f$ is called the frame ratio. A single slow and fast state is shown for simplicity.

There are several choices when it comes to scheduling the order of execution for slow and fast frames. Referring to Figure 8.44, starting at time t_n, we can take a half step through the slow frame starting from u_n in the direction defined by slope k_1. The endpoints (t_n, u_n) and $(t_n + 0.5T_s, u_{n+1/2})$ determine the equation of a line that is interpolated to provide the value of the slow state at the beginning of each fast frame. The fast state is then advanced using RK-4 integration up until the time $t_n + 0.5NT_f$, generating values for $w_{nN+1}, w_{nN+2}, \ldots, w_{nN+0.5N}$. Next, the slow state derivative function k_2 is evaluated at $t_n + 0.5T_s$ using the predicted slow state $u_{n+1/2}$ along with the previously computed fast state $w_{(n+0.5)N}$.

The step-by-step process for updating the slow state vector $\underline{u}$ and fast state vector $\underline{w}$ from t_n to t_{n+1} is outlined in the following.

8.4.1 Procedure for Updating Slow and Fast States: Master/Slave = RK-4/RK-4

1. Compute $\underline{k}_1 = \underline{f}(\underline{u}_n, \underline{w}_{nN})$
2. Compute $\underline{u}_{n+1/2} = \underline{u}_n + 0.5T_s\underline{k}_1$
3. Determine equation of lines connecting $(t_n, \underline{u}_n)$ and $(t_{n+1/2}, \underline{u}_{n+1/2})$
4. Use "slave" RK-4 to integrate fast state from t_n to $t_n + 0.5NT_f$ based on interpolated values for slow state at beginning of fast frame times.
5. Compute $\underline{k}_2 = \underline{f}(\underline{u}_{n+1/2}, \underline{w}_{(n+0.5)N})$
6. Compute $\hat{\underline{u}}_{n+1/2} = \underline{u}_n + 0.5T_s\underline{k}_2$
7. Compute $\underline{k}_3 = \underline{f}(\hat{\underline{u}}_{n+1/2}, \underline{w}_{(n+0.5)N})$
8. Compute $\hat{\underline{u}}_{n+1} = \underline{u}_n + T_s\underline{k}_3$
9. Determine equation of line connecting and $(t_n + 0.5T_s, \hat{\underline{u}}_{n+1/2})$ and $(t_n + 1, \hat{\underline{u}}_{n+1})$

10. Use "slave" RK-4 to integrate fast state from $t_n + 0.5NT_f$ to $t_{n+1} = t_n + NT_f$ based on interpolated values for slow state at beginning of fast frame times.
11. Compute $\underline{k}_4 = \underline{f}(\hat{\underline{u}}_{n+1}, \underline{w}_{(n+1)N})$
12. Compute updated slow state $\underline{u}_{n+1} = \underline{u}_n + \frac{1}{6}(\underline{k}_1 + 2\underline{k}_2 + 2\underline{k}_3 + \underline{k}_4)$

The choice of frame times T_f and T_s depends on the integrators used for the "master" and "slave" routines as well as the dynamics of the slow and fast subsystems. Baseline values of the system parameters for the following discussion are (see Figure 8.39)

Airframe dynamics: $K_A = 10$, $\tau_A = 0.8$ s, $\omega_n = 5$ rad/s, $\zeta = 0.2$
Controller/control surface actuator: $K_c = 4 \times 10^5, z = 12.5, p_1 = p_2 = 100$
Autopilot gain: $K = 0.1625$, pitch rate feedback sensor gain: $K_{\dot\theta} = 0.2$

Substituting the parameter values into Equation 8.240 gives the coefficient matrix A with eigenvalues (characteristic roots) equal to the closed-loop system poles. The result is

$$\lambda_1 = -0.67, \quad \lambda_{2,3} = -4.31 \pm j6.48, \quad \lambda_4 = -21.97, \quad \lambda_{5,6} = -11.07 \pm j37.46, \quad \lambda_7 = -148.59$$

Magnitudes of the system poles range from a low of $|\lambda_1| = 0.67$ to a high of $|\lambda_7| = 148.59$, demonstrating the stiffness of the system. The magnitude of the remaining poles suggests the existence of a slow subsystem characterized by the first three poles λ_1, λ_2, and λ_3, and a fast subsystem corresponding to the remaining four poles λ_4, λ_5, λ_6, and λ_7 located further from the origin than λ_1, λ_2, and λ_3.

8.4.2 Selection of Step Size Based on Stability

Simulation of the seventh-order control system with classic RK-4 integration and step size T is stable provided the points $\lambda_i T$, $i = 1, 2, \ldots, 7$ fall within the RK-4 stability region. Figure 8.45a shows the location of $\lambda_i T$, $i = 1, 2, \ldots, 7$ when the integration step size T is 0.01 s. Since all 7 $\lambda_i T$ points are inside the stability boundary, the RK-4 simulation is stable. The RK-4 simulation is marginally stable when the leftmost $\lambda_i T$ point is located on the stability boundary at -2.785. The step T_{max} is obtained from

$$-\max |\lambda_i| \cdot T_{max} = -148.59 T_{max} = -2.785 \Rightarrow T_{max} = 0.0187 \text{ s} \qquad (8.249)$$

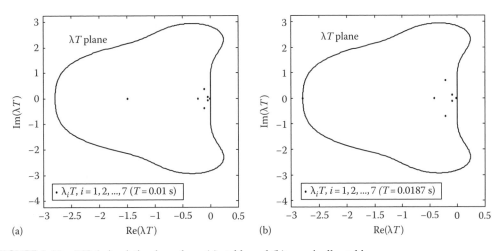

FIGURE 8.45 RK-4 simulation boundary: (a) stable and (b) marginally stable.

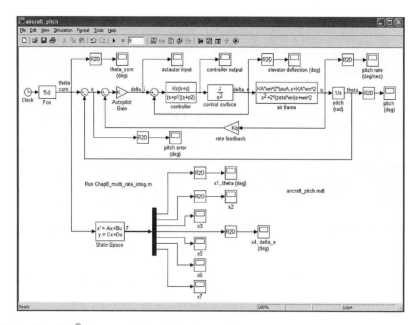

FIGURE 8.46 Simulink® diagram of aircraft pitch control system in Figure 8.39.

Figure 8.45b illustrates the case where $T = T_{max} = 0.0187$ s. The leftmost value of $\lambda_i T$ is on the RK-4 stability boundary at−2.785, leading to a z-plane pole of the discrete-time system located on the Unit Circle.

A Simulink diagram of the pitch control system is shown in Figure 8.46.

The diagram includes a "State-Space" block to implement the state equations in Equations 8.238 through 8.241. The pitch input command is given by the exponential rise

$$\theta_{com}(t) = \overline{\theta}_{com}(1 - e^{-t/\tau_{com}}), \quad t \geq 0 \tag{8.250}$$

which represents a real-world approximation to a step input provided τ_{com} is chosen appropriately, that is, $5\tau_{com}$ is set equal to 10–20 times the fixed integration step size.

Figure 8.47 shows the pitch responses from the scopes labeled "pitch (deg)" and "x1 theta (deg)" for the integration step sizes of 0.018 and 0.019 s. The numerical instability when $T = 0.019$ s is predicted by Equation 8.249.

The reader can run the M-file "*Chap8_Multi_Rate_Integ.m*" with different system parameter values to compare outputs $x_1(t)$, $x_4(t)$, and $x_6(t)$ from the state variable model and the equivalent signals $\theta(t)$ and $\delta_e(t)$ and the output of the controller block.

8.4.3 Selection of Step Size Based on Dynamic Accuracy

The transfer function of the closed-loop system in Figure 8.39 can be obtained using block diagram reduction or other graphical techniques such as Mason's gain formula for signal flow graphs. The result is

$$G_{\theta_{com} \to \theta}(s) = \frac{\theta(s)}{\theta_{com}(s)} = \frac{\beta_2 s^2 + \beta_1 s + \beta_0}{s^7 + \alpha_6 s^6 + \alpha_5 s^5 + \alpha_4 s^4 + \alpha_3 s^3 + \alpha_2 s^2 + \alpha_1 s + \alpha_0} \tag{8.251}$$

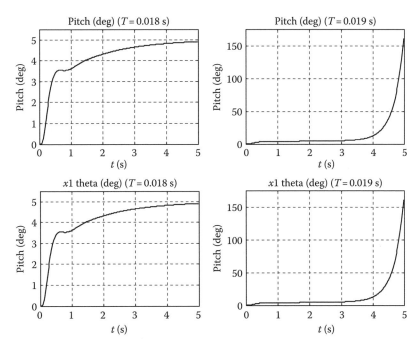

FIGURE 8.47 Simulink scope outputs showing stable ($T = 0.018$ s) and unstable ($T = 0.019$ s) pitch responses with RK-4 integration.

where

$$\beta_0 = KK_cK_A\omega_n^2z, \quad \beta_1 = KK_cK_A(1 + \tau_Az)\omega_n^2, \quad \beta_2 = KK_cK_A\omega_n^2\tau_A \tag{8.252}$$

$$
\left.
\begin{aligned}
\alpha_0 &= KK_cK_A\omega_n^2z, \\
\alpha_1 &= KK_cK_A\omega_n^2z + K_c\omega_n^2z(1 + KK_AK_{\dot\theta}), \\
\alpha_2 &= KK_cK_A\omega_n^2\tau_A + K_c\left[2\zeta\omega_nz + \omega_n^2 + KK_AK_{\dot\theta}\omega_n^2(1 + \tau_Az)\right] \\
\alpha_3 &= K_cz + 2\zeta\omega_nK_c + \omega_n^2p_1p_2 + KK_cK_AK_{\dot\theta}\tau_A\omega_n^2 \\
\alpha_4 &= K_c + 2\zeta\omega_np_1p_2 + \omega_n^2(p_1 + p_2) \\
\alpha_5 &= p_1p_2 + 2\zeta\omega_n(p_1 + p_2) + \omega_n^2 \\
\alpha_6 &= p_1 + p_2 + 2\zeta\omega_n
\end{aligned}
\right\} \tag{8.253}
$$

An equivalent implementation of the system with transfer function in Equation 8.251 consists of the input $\theta_{com}(t)$ feeding parallel first- and second-order components with the outputs of each block summed to generate the pitch response $\theta(t)$. Partial fraction expansion of Equation 8.251 using numerical values for $\beta_0, \beta_1, \beta_2$, and $\alpha_0, \alpha_1, \dots, \alpha_6$ based on the given system parameter values leads to the configuration shown in Figure 8.48.

Dashed lines identify the slow and fast subsystem components. Despite the fact that $\theta = x_1$ is classified as one of the slow states, it is clear from Figure 8.48 that $\theta(t)$ comprises both fast and slow components. However, it will be shown later that the fast component is negligible compared with the slow component.

The slow subsystem constants are

$$A_1 = 0.3373, \quad \lambda_1 = -0.67, \quad B_1 = 0.7930, \quad B_0 = 37.0325, \quad b_1 = 8.6254, \quad b_0 = 60.6057$$

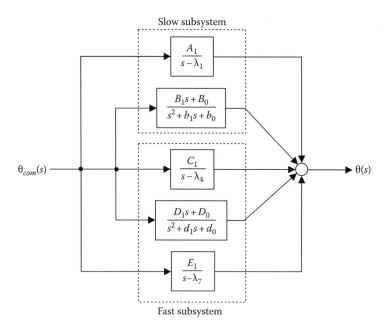

FIGURE 8.48 Parallel implementation of pitch control system transfer function.

and the fast subsystem constants are

$$C_1 = -1.7568, \quad \lambda_4 = -21.97, \quad D_1 = 0.6477, \quad D_0 = -49.8169, \quad d_1 = 22.1383, \quad d_0 = 1525.7,$$
$$E_1 = 0.0328, \quad \lambda_7 = -148.59$$

The poles of the slow subsystem second-order component (roots of $s^2 + b_1 s + b_0$) are $\lambda_{2,3} = -4.31 \pm j6.48$. The fast subsystem second-order component poles are roots of $s^2 + d_1 s + d_0$, namely, $\lambda_{5,6} = -11.07 \pm j37.46$.

Table 8.6 lists asymptotic formulas for the characteristic root errors resulting from the use of certain low-order numerical integrators. In particular, for RK-4 with step size T and integrator error coefficient $e_I = 1/120$,

$$\text{RK-4: } e_\lambda = \frac{\lambda^*}{\lambda} - 1 \approx -e_I (\lambda T)^4 \approx -\frac{1}{120}(\lambda T)^4, \quad |\lambda T| \ll 1 \tag{8.254}$$

where λ^* is the characteristic root of the equivalent continuous-time system.

For second-order systems Howe (1986) presents formulas for dynamic errors in damping ratio ζ, natural frequency ω_n, and damped natural frequency ω_d using first-order through fourth-order integration methods. For RK-4, the asymptotic expressions are

$$e_\zeta = \frac{\zeta^* - \zeta}{\zeta} \approx -4(\zeta - 3\zeta^3 + 2\zeta^5)e_I(\omega_n T)^4 \tag{8.255}$$

$$\approx -\frac{1}{30}(\zeta - 3\zeta^3 + 2\zeta^5)(\omega_n T)^4, \quad \omega_n T \ll 1 \tag{8.256}$$

$$e_{\omega_n} = \frac{\omega_n^* - \omega_n}{\omega_n} \approx -(1 - 8\zeta^2 + 8\zeta^4)e_I(\omega_n T)^4 \tag{8.257}$$

$$\approx -\frac{1}{120}(1 - 8\zeta^2 + 8\zeta^4)(\omega_n T)^4, \quad \omega_n T \ll 1 \tag{8.258}$$

$$e_{\omega_d} = \frac{\omega_d^* - \omega_d}{\omega_d} \approx -(1 - 12\zeta^2 + 16\zeta^4)e_I(\omega_n T)^4 \tag{8.259}$$

$$\approx -\frac{1}{120}(1 - 12\zeta^2 + 16\zeta^4)(\omega_n T)^4, \quad \omega_n T \ll 1 \tag{8.260}$$

For the first-order component in the slow subsystem in Figure 8.48,

$$e_\lambda \approx -\frac{1}{120}(\lambda_1 T)^4 \approx -\frac{1}{120}(-0.673)^4 T^4 \approx -0.00171 T^4 \tag{8.261}$$

The damping ratio, natural frequency, and damped natural frequency of the slow subsystem second-order component are found by equating the term $s^2 + b_1 s + b_0$ and the standard form of a quadratic characteristic polynomial $s^2 + 2\zeta\omega_n s + \omega_n^2$. The results are $\zeta_{slow} = 0.554$, $(\omega_n)_{slow} = 7.785$ rad/s, and $(\omega_d)_{slow.} = 6.481$ rad/s. Substituting the values of ζ_{slow} and $(\omega_n)_{slow}$ into Equations 8.256, 8.258, and 8.560 gives

$$e_\zeta \approx -\frac{1}{30}\left(\zeta_{slow} - 3\zeta_{slow}^3 + 2\zeta_{slow}^5\right)[(\omega_n)_{slow} T]^4 \tag{8.262}$$

$$\approx -\frac{1}{30}[0.554 - 3(0.554)^3 + 2(0.554)^5](7.785T)^4 \tag{8.263}$$

$$\approx -18.1562 T^4 \tag{8.264}$$

$$e_{\omega_n} \approx -\frac{1}{120}\left(1 - 8\zeta_{slow}^2 + 8\zeta_{slow}^4\right)[(\omega_n)_{slow} T]^4 \tag{8.265}$$

$$\approx -\frac{1}{120}[1 - 8(0.554)^2 + 8(0.554)^4](7.785T)^4 \tag{8.266}$$

$$\approx 21.4777 T^4 \tag{8.267}$$

$$e_{\omega_d} \approx -\frac{1}{120}\left(1 - 12\zeta_{slow}^2 + 16\zeta_{slow}^4\right)[(\omega_n)_{slow} T]^4 \tag{8.268}$$

$$\approx -\frac{1}{120}[1 - 12(0.554)^2 + 16(0.554)^4](7.785T)^4 \tag{8.269}$$

$$\approx 35.9894 T^4 \tag{8.270}$$

Choosing the RK-4 step size to limit the characteristic error in damped natural frequency to 0.025%,

$$T_{slow} = \left[\frac{(e_{\omega_d})_{des}}{35.9894}\right]^{1/4} = \left[\frac{0.00025}{35.9894}\right]^{1/4} = 0.0513 \text{ s} \tag{8.271}$$

The actual characteristic error in damped natural frequency will be slightly different from $(e_{\omega_d})_{des} = 0.025\%$ because $(\omega_n)_{slow} T_{slow} = 0.3997$, which is not an order of magnitude less than 1, a requirement for the asymptotic formula in Equation 8.260.

A similar procedure can be performed to determine an appropriate step size for RK-4 simulation of the fast subsystem. Suppose the fast subsystem step size is selected to limit the sum of the characteristic root errors associated with the fast poles $\lambda_4 = -21.97$ and $\lambda_7 = -148.59$. From Equation 8.261,

$$e_{\lambda_4} + e_{\lambda_7} \approx -\frac{1}{120}\left[\lambda_4^4 + \lambda_7^4\right]T^4 \leq E_{des} \tag{8.272}$$

$$\Rightarrow -\frac{1}{120}[(-21.97)^4 + (-148.59)^4]T^4 \leq E_{des} \tag{8.273}$$

Choosing $E_{des} = -0.02\%$,

$$T^4 \leq \frac{-0.0002}{-(1/120)[(-21.97)^4 + (-148.59)^4]}$$

$$\Rightarrow T_{fast} \leq 0.0026\,\text{s} \tag{8.274}$$

Once again, there will be a slight difference between $e_{\lambda_4} + e_{\lambda_7}$ and E_{des} when $T_{fast} = 0.0026$ s, because the product $|\lambda_7 T_{fast}| = |(-148.589)0.0026| = 0.3936$ is not significantly less than 1 as required in the asymptotic formula of Equation 8.254.

Henceforth, multirate integration using RK-4 for both slow and fast systems will be performed with $T_f = 0.0025$ s and $T_s = 0.05$ s resulting in a frame ratio $N = 20$.

8.4.4 ANALYTICAL SOLUTION FOR STATE VARIABLES

In most cases, analytical solutions for the state variables are not available with the possible exception of linear (or linearized) system models and elementary input signals. An advantage of knowing the analytical solution for the state variables is that it can serve as a benchmark for comparing results obtained by different simulation-based approaches. Consequently, the analytical solution for a subset of the state variables in the pitch control system will be determined with this purpose in mind.

Laplace transforming the pitch command signal given in Equation 8.250 gives

$$\theta(s) = \left[\frac{\beta_2 s^2 + \beta_1 s + \beta_0}{s^7 + \alpha_6 s^6 + \alpha_5 s^5 + \alpha_4 s^4 + \alpha_3 s^3 + \alpha_2 s^2 + \alpha_1 s + \alpha_0}\right]\frac{\overline{\theta}_{com}}{s(\tau_{com}s + 1)} \tag{8.275}$$

Choosing $\overline{\theta}_{com} = 5°, \tau_{com} = 0.01$ s and substituting the baseline parameter values into Equations 8.252 and 8.253 determine $\theta(s)$. Using MATLAB's "conv" function to expand the denominator into a ninth-order polynomial and then the "residue" function results in the partial fraction expansion of $\theta(s)$. Converting pairs of terms with complex poles and coefficients into real terms results in the analytical pitch response

$$\theta(t) = \theta_{slow}(t) + \theta_{fast}(t) + \theta_{forced}(t) \tag{8.276}$$

where $\theta_{slow}(t)$ comprises the slow subsystem natural mode terms,

$$\theta_{slow}(t) = 5\{-0.5047e^{-0.673t} + e^{-4.313t}[-0.6150\cos(6.4812t)$$
$$-0.3474\sin(6.4812t)]\} \tag{8.277}$$

$\theta_{fast}(t)$ is made up of fast subsystem natural mode terms,

$$\theta_{fast}(t) = 5\{0.0004548e^{-148.589t} + 0.1025e^{-21.975t}$$
$$+ e^{-11.069t}[0.0204\cos(37.4583t) + 0.0388\sin(37.4583t)]\} \tag{8.278}$$

and $\theta_{forced}(t)$ includes the input mode terms

$$\theta_{forced}(t) = 5(1 - 0.0035e^{-100t}) \tag{8.279}$$

The exponential decay in the forced component results from the exponential term in the command input (see Equation 8.250).

Plots of the slow component $\theta_{slow}(t)$ and fast component $\theta_{fast}(t)$ shown in the top half of Figure 8.49 suggest that the fast component contributes a negligible amount to the overall response. Hence, $\theta(t)$ is appropriately classified as a slow subsystem state variable.

The bottom half of Figure 8.49 shows the forced component given in Equation 8.279 and the total pitch response comprising the slow, fast, and forced components. Note that on the time scale used in Figure 8.49, the forced component appears to be a step input. In reality, it contains an exponential rise term with time constant $\tau_{com} = 0.01$ s.

A similar approach can be used to find the analytical solution for the fast state variable $x_4 = \delta_e$. The transfer function from $\theta_{com}(s)$ to $\delta_e(s)$ is

$$G_{\theta_{com} \to \delta_e(s)} = \frac{\delta_e(s)}{\theta_{com}(s)} = \frac{\gamma_4 s^4 + \gamma_3 s^3 + \gamma_2 s^2 + \gamma_1 s + \gamma_0}{s^7 + \alpha_6 s^6 + \alpha_5 s^5 + \alpha_4 s^4 + \alpha_3 s^3 + \alpha_2 s^2 + \alpha_1 s + \alpha_0} \tag{8.280}$$

$$\gamma_0 = 0, \quad \gamma_1 = KK_c \omega_n^2 z, \quad \gamma_2 = KK_c(\omega_n^2 + 2\zeta\omega_n z), \quad \gamma_3 = KK_c(2\zeta\omega_n + z), \quad \gamma_4 = KK_c \tag{8.281}$$

and the analytical solution for the elevator deflection $\delta_e(t)$ is

$$\delta_e(t) = (\delta_e)_{slow}(t) + (\delta_e)_{fast}(t) + (\delta_e)_{forced}(t) \tag{8.282}$$

$$(E_e)_{slow}(t) = 5\{0.0709e^{-0.673t} + e^{-4.313t}[0.0844\cos(6.4812t) - 0.1440\sin(6.4812t)]\} \tag{8.283}$$

$$(\delta_e)_{fast}(t) = 5\{0.050e^{-148.589t} + 0.252e^{-21.975t}$$
$$+ e^{-11.069t}[-0.2842\cos(37.4583t) - 0.1644\sin(37.4583t)]\} \tag{8.284}$$

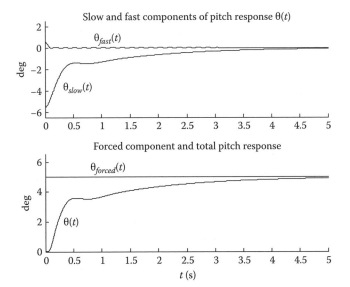

FIGURE 8.49 Total pitch response and its components.

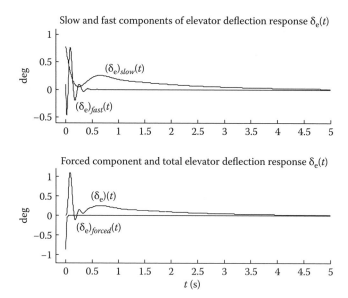

FIGURE 8.50　Total elevator deflection and its components.

$$(\delta_e)_{forced}(t) = 5(-0.1733e^{-100t}) \tag{8.285}$$

The total response $\delta_e(t)$ and its components $(\delta_e)_{slow}(t)$, $(\delta_e)_{fast}(t)$, and $(\delta_e)_{forced}(t)$ are shown in Figure 8.50.

From the top graph, it is clear that both fast and slow components are present in $\delta_e(t)$. Despite the existence of an appreciable slow component, $\delta_e(t)$ is nonetheless identified as a fast state variable. Multirate simulation of the overall system must integrate $\delta_e(t)$ at the fast frame rate due to the significant high-frequency component $(\delta_e)_{fast}(t)$.

8.4.5　Multirate Integration of Aircraft Pitch Control System

The MATLAB M-file "*Chap8_multi_rate_integ.m*" includes code for implementing multirate integration of Equations 8.242 through 8.248 with RK-4 as the "master" and "slave" integration routines.

Figure 8.51 shows Simulink and multirate integration results for the slow states $u_1 = x_1$, $u_2 = x_2$, and $u_3 = x_3$. The Simulink model was integrated using RK-4 with integration step size identical to the fast frame time $T_f = 0.0025$ s. The slow frame time was $T_s = 0.05$ making the frame ratio $N = 20$.

The Simulink and multirate simulation responses are in general agreement; however, the accuracy of each can only be established by comparison with the analytical solutions. Accordingly, Figures 8.52 and 8.53 show the analytical solutions for $x_1(t)$ and $x_2(t)$ on the same graph with the RK-4 and multirate simulation results. For purposes of clarity, not all simulated points are shown in the graph. The analytical solution for $x_1(t) = \theta(t)$ is given by Equations 8.276 through 8.279, and the one for $x_2(t)$ is obtained in the MATLAB M-file "*Chap8_multi_rate_integ.m*."

From Figures 8.52 and 8.53, it is clear that the simulated responses using Simulink with RK-4 and step size $T_f = 0.0025$ s are virtually identical with the analytical solutions. As expected, the responses obtained using multirate integration with RK-4/RK-4 and $T_f = 0.0025$ s, $T_f = 0.05$ s are not as accurate.

Simulated responses of two of the fast states, namely, $w_1 = x_4$ and $w_2 = x_5$, obtained using Simulink and multirate integration are shown in Figure 8.54. The M-file "*Chap8_multi_rate_integ.m*" plots the remaining fast states $w_3 = x_6$ and $w_4 = x_7$.

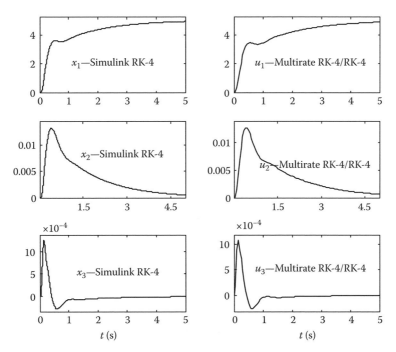

FIGURE 8.51 Simulation of slow states using Simulink® and multirate integration.

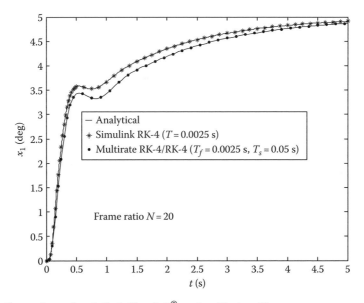

FIGURE 8.52 Comparison of analytical, Simulink®, and multirate $x_1(t)$ responses.

The analytical solution for $x_4 = \delta_e$, the elevator deflection, is shown in Figure 8.55 along with the responses from Simulink and multirate integration. Once again, the Simulink RK-4 and analytical responses are indistinguishable from each other, while the multirate solution deviates from both during the transient response period.

Multirate integration introduced errors in the transient response of each state variable in the aircraft pitch control system. The errors can be reduced by decreasing the frame ratio; however, the benefits from using multirate integration are lessened. An acceptable trade-off is generally possible.

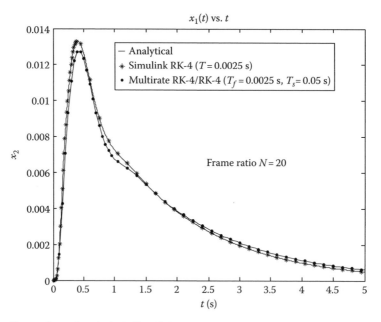

FIGURE 8.53 Comparison of analytical, Simulink, and multirate $x_2(t)$ responses.

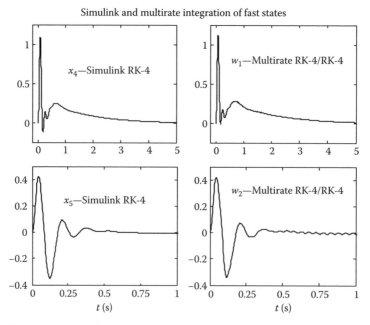

FIGURE 8.54 Simulation of fast states using Simulink and multirate integration.

Significant increases in performance are achieved using multirate integration for multiple time scale systems where the predominant number of states are associated with the slow subsystem(s). Moreover, the computational savings can be substantial when the times required to compute the slow subsystem state derivatives are appreciable due to the complex nature of the derivative functions or possibly due to the use of table lookups involved in the computation process.

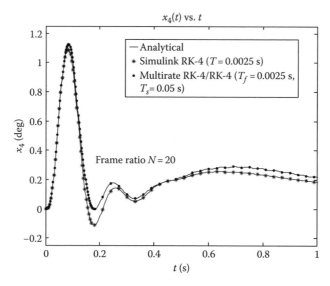

FIGURE 8.55 Comparison of analytical, Simulink, and multirate $x_4(t)$ responses.

8.4.6 NONLINEAR DUAL SPEED SECOND-ORDER SYSTEM

We now turn our attention to a second-order stiff system with a fast and a slow state. Furthermore, the system dynamics are nonlinear and the stiffness varies with the operating point of the linearized system. The system consists of two cylindrical tanks in series as shown in Figure 8.56.

Flow $F_0(t)$ into the first tank (open at the top) is completely controlled by a regulating valve in the inflow line. The outflow $F_2(t)$ from the second tank (sealed at the top) is a function of valve opening in the outflow line along with the liquid pressure at the bottom of the tank.

The system is modeled by differential and algebraic equations. Dynamics of the first tank are governed by

$$A_1 \frac{d}{dt} H_1(t) + F_{12}(t) = F_0(t), \quad H_1(t) \le L_1 \tag{8.286}$$

where
  A_1 is the cross-sectional area
  L_1 is the height of the first tank

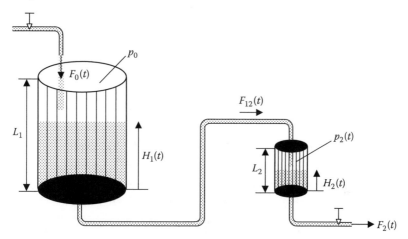

FIGURE 8.56 System of two different capacity tanks in series.

The flow from the first tank into the second tank $F_{12}(t)$ depends on the pressure differential between the bottom of the first tank and the top of the second tank.

$$F_{12}(t) = \begin{cases} c_1[p_0 + \gamma H_1(t) - \gamma L_2 - p_2(t)]^{1/2}, & p_0 + \gamma H_1(t) - \gamma L_2 - p_2(t) > 0 \\ 0, & \text{otherwise} \end{cases} \quad (8.287)$$

where
  c_1 is a constant related to the fluid resistance in the line connecting the tanks
  p_0 is atmospheric pressure (14.7 psi)
  γ is the specific weight of water (62.4 lb/ft³)

The air pressure above the liquid in the sealed tank $p_2(t)$ is related to the liquid level $H_2(t)$ according to

$$p_2 = \left(\frac{L_2}{L_2 - H_2}\right)p_0 \quad (8.288)$$

Equation 8.288 assumes that the air pressure in the sealed tank obeys the relationship $p_2 V_2 = \text{constant}$ and that $p_2 = p_0$ when $H_2 = 0$. Hence,

$$p_0 A_2 L_2 = p_2 A_2 (L_2 - H_2) \quad (8.289)$$

The differential equation for the second tank is

$$A_2 \frac{d}{dt} H_2(t) + F_2(t) = F_{12}(t), \quad H_2(t) \leq L_2 \quad (8.290)$$

The flow out of the second tank is governed by the algebraic relation

$$F_2(t) = c_2[\{p_2(t) + \gamma H_2(t)\} - p_0]^{1/2} \quad (8.291)$$

where c_2 is a constant related primarily to the physical construction of the valve in the discharge line and its percent opening.

Stiffness is a property of linear systems relating the magnitudes of the fast poles (eigenvalues) to the slower poles. We can linearize the nonlinear system modeled by Equations 8.286 through 8.291 about a steady-state operating point $(\overline{H}_1, \overline{H}_2)$ corresponding to a constant input flow $F_0(t) = \overline{F}_0, t \geq 0$.

At steady state, $\overline{F}_2 = F_2(\infty) = \overline{F}_0$ and it follows from Equation 8.291

$$\overline{F}_2 = c_2[(\overline{p}_2 + \gamma \overline{H}_2) - p_0]^{1/2} = \overline{F}_0 \quad (8.292)$$

$$\Rightarrow c_2\left[\left(\frac{L_2}{L_2 - \overline{H}_2}\right)p_0 + \gamma \overline{H}_2 - p_0\right]^{1/2} = \overline{F}_0 \quad (8.293)$$

Rearranging Equation 8.293 leads to a quadratic equation in $\overline{H}_2$,

$$\gamma \overline{H}_2^2 - \left[p_0 + \gamma L_2 + \left(\frac{\overline{F}_0}{c_2}\right)^2\right]\overline{H}_2 + \left(\frac{\overline{F}_0}{c_2}\right)^2 L_2 = 0 \quad (8.294)$$

It is left as an exercise to show that the steady-state operating level in the first tank is

$$\overline{H}_1 = L_2 + \frac{1}{\gamma}\left[\left(\frac{\overline{F}_0}{c_1}\right)^2 + \left(\frac{\overline{H}_2}{L_2 - \overline{H}_2}\right)p_0\right] \quad (8.295)$$

Given $\overline{F}_0$, Equations 8.294 and 8.295 can be solved in that order to find the operating point levels $\overline{H}_2$, $\overline{H}_1$ and ultimately the remaining dependent variable operating point values, namely, $\overline{F}_{12}$, $\overline{F}_2$, and $\overline{p}_2$.

The nonlinear system model in Equations 8.286 through 8.291 can be reduced to

$$\frac{dH_1}{dt} = f_1(H_1, H_2, F_0) \tag{8.296}$$

$$= \frac{1}{A_1}\left[F_0 - c_1\left\{ \gamma(H_1 - L_2) - \left(\frac{H_2}{L_2 - H_2}\right)p_0 \right\}^{1/2}\right], \quad H_1 \leq L_1 \tag{8.297}$$

$$\frac{dH_2}{dt} = f_2(H_1, H_2, F_0) \tag{8.298}$$

$$= \frac{1}{A_2}\left[c_1\left\{ \gamma(H_1 - L_2) - \left(\frac{H_2}{L_2 - H_2}\right)p_0 \right\}^{1/2} - c_2\left\{ \left(\frac{H_2}{L_2 - H_2}\right)p_0 + \gamma H_2 \right\}^{1/2}\right] \tag{8.299}$$

The linearized state model is

$$\frac{d}{dt}\Delta\underline{H}(t) = A\Delta\underline{H}(t) + B\Delta F_0(t) \tag{8.300}$$

$$\Delta\underline{y}(t) = C\Delta\underline{H}(t) + D\Delta F_0(t) \tag{8.301}$$

where

$$\Delta\underline{H}(t) = \begin{bmatrix} \Delta H_1(t) \\ \Delta H_2(t) \end{bmatrix} = \begin{bmatrix} H_1(t) - \overline{H}_1 \\ H_2(t) - \overline{H}_2 \end{bmatrix} \tag{8.302}$$

$$\Delta\underline{y}(t) = \begin{bmatrix} \Delta H_1(t) \\ \Delta H_2(t) \\ \Delta F_{12}(t) \\ \Delta F_2(t) \\ \Delta p_2(t) \end{bmatrix} = \begin{bmatrix} H_1(t) - \overline{H}_1 \\ H_2(t) - \overline{H}_2 \\ F_{12}(t) - \overline{F}_{12} \\ F_2(t) - \overline{F}_2 \\ p_2(t) - \overline{p}_2 \end{bmatrix} \tag{8.303}$$

The coefficient matrix A comprises the first partial derivatives

$$A_{11} = \frac{\partial}{\partial H_1}f_1(\overline{H}_1, \overline{H}_2, \overline{F}_0) \tag{8.304}$$

$$= \frac{-\gamma c_1}{2A_1}\left[\gamma(\overline{H}_1 - L_2) - \left(\frac{\overline{H}_2}{L_2 - \overline{H}_2}\right)p_0 \right]^{-1/2} \tag{8.305}$$

$$A_{12} = \frac{\partial}{\partial H_2}f_1(\overline{H}_1, \overline{H}_2, \overline{F}_0) \tag{8.306}$$

$$= \frac{p_0 c_1}{2A_1}\left[\gamma(\overline{H}_1 - L_2) - \left(\frac{\overline{H}_2}{L_2 - \overline{H}_2}\right)p_0 \right]^{-1/2}\left\{ \frac{L_2}{(L_2 - \overline{H}_2)^2} \right\} \tag{8.307}$$

$$A_{21} = \frac{\partial}{\partial H_1}f_2(\overline{H}_1, \overline{H}_2, \overline{F}_0) \tag{8.308}$$

$$= \frac{\gamma c_1}{2A_2}\left[\gamma(\overline{H}_1 - L_2) - \left(\frac{\overline{H}_2}{L_2 - \overline{H}_2}\right)p_0 \right]^{-1/2} \tag{8.309}$$

$$A_{22} = \frac{\partial}{\partial H_2} f_2(\overline{H}_1, \overline{H}_2, \overline{F}_0) \tag{8.310}$$

$$= \frac{-p_0 L_2 c_1}{2A_2} \left[\gamma(\overline{H}_1 - L_2) - \left(\frac{\overline{H}_2}{L_2 - \overline{H}_2} \right) p_0 \right]^{-1/2} \left\{ \frac{1}{(L_2 - \overline{H}_2)^2} \right\}$$

$$- \frac{c_2}{2A_2} \left[\left(\frac{\overline{H}_2}{L_2 - \overline{H}_2} \right) p_0 + \gamma \overline{H}_2 \right]^{-1/2} \left\{ \frac{p_0 L_2}{(L_2 - \overline{H}_2)^2} + \gamma \right\} \tag{8.311}$$

The components of the input matrix B and output matrix C are obtained from partial derivatives as well (see Exercise 8.41).

Example 8.8

The baseline numerical values of the system parameters are

$$R_1 = 15 \text{ ft}, \quad L_1 = 50 \text{ ft}, \quad c_1 = 4 \text{ ft}^3/\text{min}/(\text{lb}/\text{ft}^2)^{1/2}, \quad A_1 = \pi R_1^2 = 225\pi \text{ ft}^2$$

$$R_2 = 5 \text{ ft}, \quad L_2 = 7.5 \text{ ft}, \quad c_2 = 2 \text{ ft}^3/\text{min}/(\text{lb}/\text{ft}^2)^{1/2}, \quad A_2 = \pi R_2^2 = 56.25\pi \text{ ft}^2$$

and the baseline inflow under steady-state operating conditions is $\overline{F}_0 = 60 \text{ ft}^3/\text{min}$.
For the given baseline conditions,

(a) Find the steady-state operating point values $\overline{H}_1, \overline{H}_2, \overline{F}_{12}, \overline{F}_2$, and $\overline{p}_2$.
(b) Compute the numerical values of the components of matrix A.
(c) Find the eigenvalues of A and compute the stiffness ratio.
(d) Draw a Simulink diagram for simulating the system dynamics.
(e) Use the MATLAB "linmod" function to approximate the matrices A, B, C, and D.
(f) Compare the linearized system response and the simulated response of the nonlinear system for the case where the system is initially in steady state and the inflow to the first tank is given by

$$F_0(t) = \begin{cases} \overline{F}_0, & t \le 50 \\ \overline{F}_0 - 5, & t > 50 \text{ min} \end{cases} \tag{8.312}$$

(a) Determine the new steady-state levels in both tanks predicted by the nonlinear model and the linearized model.

(b) The steady-state operating levels are obtained from Equations 8.294 and 8.295 in the M-file "Chap8_Ex4_1.m." The results are

$$\overline{H}_1 = 23.52 \text{ ft}, \quad \overline{H}_2 = 2.01 \text{ ft}$$

$$\overline{F}_{12} = \overline{F}_2 = 60 \text{ ft}^3/\text{min}$$

$$\overline{p}_2 = 2891.4 \text{ lb}/\text{ft}^2$$

(c) The same M-file contains code for evaluating the components of matrix A using Equations 8.304 through 8.311. The result is

$$A = \begin{bmatrix} -0.0118 & 0.0993 \\ 0.1059 & -1.1440 \end{bmatrix}$$

(d) The eigenvalues of A are $\lambda_1 = -0.00255151$ and $\lambda_2 = -1.15318422$, and the stiffness ratio of the system linearized about the given steady-state operating point is

$$\frac{\lambda_2}{\lambda_1} = \frac{-1.15318422}{-0.00255151} = 451.96$$

The time constants of the linearized system are

$$\tau_1 = \frac{-1}{\lambda_1} \frac{-1}{-0.00255151} = 391.92 \, \text{min}$$

$$\tau_2 = \frac{-1}{\lambda_2} = \frac{-1}{-1.15318422} = 0.867 \, \text{min}$$

demonstrating the dual time scales involved.

(e) A Simulink diagram is shown in Figure 8.57.

(f) MATLAB statements in "*Chap8_Ex4_1.m*" for employing the "`linmod`" function are

```
[sizes,X0,states] = TwoTanks([],[],[],0)
H_opert = [H1_ss;H2_ss];
u0 = F0;
[A, B, C, D] = linmod('TwoTanks_linmod', H_opert, u0)
```

The first line returns the variable "`states`," which identifies the limited integrator outputs "H1" and "H2" as the first and second states, respectively. The last line refers to a Simulink model file "`TwoTanks_linmod.mdl`," which is similar to "`TwoTanks.mdl`" shown in Figure 8.57 except an input port block replaces the "`Constant`" block with parameter "`F0`" and the addition of five output port blocks to identify the system outputs. The last line produces the linearized system matrices

$$A = \begin{bmatrix} -0.0118 & 0.0993 \\ 0.1059 & -1.1440 \end{bmatrix}, \quad B = \begin{bmatrix} 0.0014 \\ 0 \end{bmatrix}, \quad C = \begin{bmatrix} 1 & 0 \\ 0 & 1 \\ 8.3200 & -70.2135 \\ 0 & 19.6334 \\ 0 & 526.6010 \end{bmatrix}, \quad D = \begin{bmatrix} 0 \\ 0 \\ 0 \\ 0 \\ 0 \end{bmatrix}$$

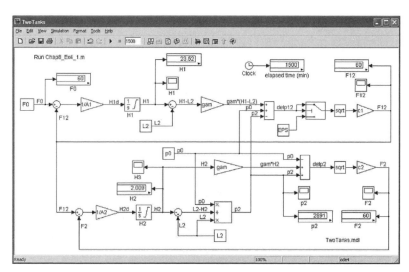

FIGURE 8.57 Simulink® diagram for simulation of two-tank system with stiff dynamics.

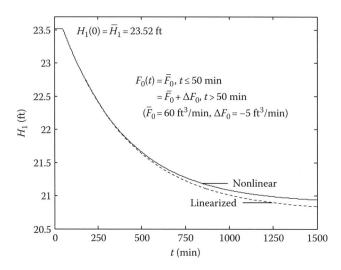

FIGURE 8.58 Tank 1 nonlinear and linearized system level responses.

Note that the coefficient matrix A using "linmod" is identical (to at least four places after the decimal point) to the previous result based on the analytical expressions for the partial derivatives in Equations 8.304 through 8.311.

(g) The Simulink diagram in Figure 8.57 is supplemented with additional blocks to generate the deviation input variable $\Delta F_0(t) = F_0(t) - \bar{F}_0$ into a "State-Space" block with output $\Delta \underline{y}(t) = [\Delta H_1(t) \; \Delta H_2(t) \; \Delta F_{12}(t) \; \Delta F_2(t) \; \Delta p_2(t)]^{\mathrm{T}}$. The linearized system out puts $H_1(t) = \bar{H}_1 + \Delta \bar{H}_1(t)$ and $H_2(t) = \bar{H}_2 + \Delta H_2(t)$ are compared with the simulated nonlinear system responses in Figures 8.58 and 8.59.

RK-4 simulation with a short time step $T = 0.1$ s was used to generate accurate approximations of the nonlinear responses. The Simulink model file is "*TwoTanks_NL_and_ L.mdl*." The linearized responses are approaching steady state after 1500 min in agreement with the larger time constant of 391.92 min.

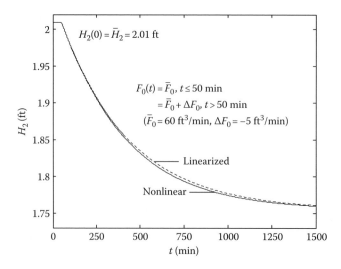

FIGURE 8.59 Tank 2 nonlinear and linearized system level responses.

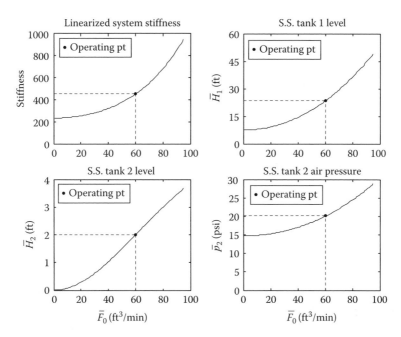

FIGURE 8.60 Graph of linearized system stiffness and nonlinear system steady-state character-istics.

(h) The new steady-state levels established when the inflow is held constant at 55 ft³/min are obtained from Equations 8.294 and 8.295. The result is $(H_1)_{ss} = 20.89$ ft and $(H_2)_{ss} = 1.76$ ft. From Equation 8.300 at steady state,

$$\Delta \underline{H}_{ss} = -A^{-1}B\Delta F_0 \tag{8.313}$$

$$= -\begin{bmatrix} -0.0118 & 0.0993 \\ 0.1059 & -1.1440 \end{bmatrix}^{-1} \begin{bmatrix} 0.0014 \\ 0 \end{bmatrix} [-5] = \begin{bmatrix} -2.7501 \\ -0.2547 \end{bmatrix}$$

$$\Rightarrow (H_1)_{ss} = \overline{H}_1 + (\Delta H_1)_{ss} = 23.52 + (-2.75) = 20.77$$

$$\Rightarrow (H_2)_{ss} = \overline{H}_2 + (\Delta H_2)_{ss} = 2.01 + (-0.25) = 1.76 \text{ ft}$$

It is clear from Figures 8.58 and 8.59 that the linearized system approximation to the nonlinear system about the given steady-state operating point is more than adequate when the perturbation in $F_0(t)$ about $\overline{F}_0$ is limited to -5 ft³/min.

Figure 8.60 contains graphs showing how the steady-state operating point levels $\overline{H}_1, \overline{H}_2, \overline{p}_2$ vary with changes in the flow $\overline{F}_0$. The baseline operating point in Example 8.8 with $\overline{F}_0 = 60$ ft³/min is also shown.

The stiffness of the linearized system is shown in the top left corner. Note how the stiffness increases from a little over 200 to around 950 before the first tank starts to overflow when the level reaches $L_1 = 50$ ft. At that point, the linearized system eigenvalues are -0.0015 and -1.4724, resulting in natural modes with time constants of approximately 0.679 and 645.8 min. The large difference in time constants of the linearized system results primarily from the significant disparity in the capacities of the two tanks.

8.4.7 MULTIRATE SIMULATION OF TWO-TANK SYSTEM

In view of the large difference between the linearized system time constants, multirate integration offers the possibility of reducing simulation execution time without significant loss of accuracy.

The first step is to choose the "master" and "slave" integration routines and determine the slow and fast frame times. The aircraft pitch example used the one-step RK-4 for "master" and "slave." For this example, the multistep AB-2 integrator will be used to integrate both the slow and fast states.

AB-2 integration is a popular numerical integrator, particularly in applications involving ground vehicle, aircraft, missile, ship, power plant, and chemical process simulators where a real-time solution of the model equations is required. Real-time numerical integration is the subject of the following section.

Looking at the AB stability regions in Figure 8.21 of the previous section, the simulation step size T is limited by the condition $\lambda T = -1$ for AB-2 integration of a linear first-order continuous-time system with characteristic root λ. Consequently, for small changes from the baseline operating point of the two-tank system, that is, ($\overline{F}_0 = 60\,\text{ft}^3/\text{min}$; $\overline{H}_1 = 23.52\,\text{ft}, \overline{H}_2 = 2.01\,\text{ft}$), AB-2 simulation will be stable provided

$$\lambda T = (-1.1532)T < -1 \quad \Rightarrow \quad T < 0.8672\,\text{min}$$

Figure 8.61 shows the results of AB-2 simulation of the system when the inflow $F_0(t) = 55\,\text{ft}^3/\text{min}, t \geq 0$.

The initial tank levels are the steady-state values when $\overline{F}_0 = 60\,\text{ft}^3/\text{min}$. The two plots on the left are the result of selecting the step size $T = 0.1$ min while the graphs on the right correspond to an integration step of $T = 0.87$ min, just slightly larger than the upper limit for AB-2 stability. The unstable nature of both tank level responses when $T = 0.87$ min is apparent. The unstable responses are similar to the stable transient responses up to a point. All graphs were generated in M-file "Chap8_TwoTanks_AB2.m."

The next set of graphs in Figure 8.62 illustrate AB-2 simulation of tank level responses when both tanks are initially empty and the inflow is a step input described by $F_0(t) = \overline{F}_0 = 60\,\text{ft}^3/\text{min}, t \geq 0$. The fluid level $H_2(t)$ remains at zero until the first tank level

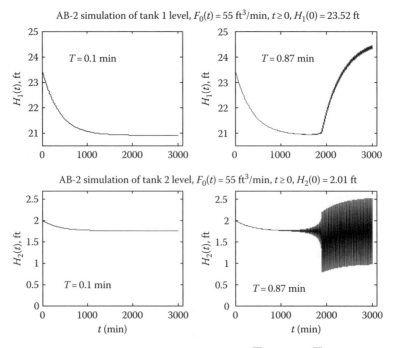

FIGURE 8.61 Stable and unstable AB-2 simulation for $H_1(0) = \overline{H}_1, H_2(0) = \overline{H}_2$.

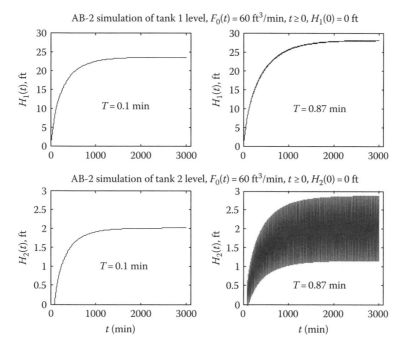

FIGURE 8.62 Stable and unstable AB-2 simulation for $H_1(0) = H_2(0) = 0$.

$H_1(t)$ reaches a height of $L_2 = 7.5$ ft, that is, high enough to push fluid from the bottom of the first tank up to the top of the second tank. Thus, $F_{12}(t) = 0$ as long as $H_1(t) < L_2$.

Due to the magnitude of the step, the system variables $H_1(t)$ and $H_2(t)$ are not confined to a small region about the initial steady-state operating point $(\overline{H}_1, \overline{H}_2) = (0, 0)$. Consequently, a single linearized model to accurately predict deviations in both levels is not valid, and the discussion in Section 7.4 dealing with multiple linearized models is applicable.

Multirate integration of the nonlinear two-tank system with AB-2 integration as the "master" routine and AB_2 integration as the "slave" routine is straightforward to implement (see "*Chap8_TwoTanks_Multirate_AB2.m*"). Time histories of each tank level when the fast frame time $T_f = 0.25$ min and the slow frame time $T_s = 25$ min (frame ratio $N = T_s/T_f = 100$) are shown in Figure 8.63. Note that every 200th point in the $H_2(t)$ response is plotted in the lower graph.

Results from another multirate simulation run are shown in Figure 8.64. The fast state $H_2(t)$ was updated using AB-2 integration with frame time $T_f = 0.1$ min while the slow state $H_1(t)$ was advanced by AB-2 integration every $T_s = 100$ min. The frame ratio was 1000. Every 200th point in the fast subsystem is plotted.

The solid lines in Figures 8.63 and 8.64 were obtained from Simulink RK-4 integration of the state equations with integration step size $T = 0.1$ min. Due to the small step size, they serve as accurate approximations to the exact solutions of the nonlinear state equations. Note that the AB-2/AB-2 response for $H_1(t)$ is quite accurate in both cases; however, the level response $H_2(t)$ is superior in the first case where the frame ratio is less. Exercise 8.45 explores the effect of frame ratio on the overall accuracy of the multirate simulation results.

8.4.8 SIMULATION TRADE-OFFS WITH MULTIRATE INTEGRATION

The case for multirate integration is based on the reduced number of derivative evaluations of the slow state required compared with the number of evaluations required when both slow and fast states are integrated at the same frame rate. The savings in execution time can be dramatic for

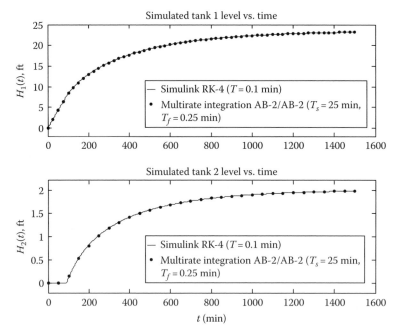

FIGURE 8.63   Multirate simulation (AB-2/AB-2) of nonlinear two-tank system ($T_s = 25$ min, $T_f = 0.25$ min, $N = 100$).

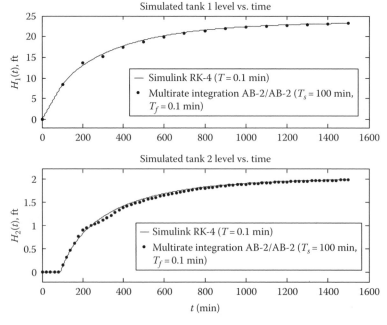

FIGURE 8.64   Multirate simulation (AB-2/AB-2) of nonlinear two-tank system ($T_s = 100$ min, $T_f = 0.1$ min, $N = 1000$).

high-order systems in which the majority of the state variables are associated with the slow subsystem. Even low-order systems experience significant reduction in simulation time when the slow derivatives are computationally more intensive. Be aware that real-world derivative functions often involve more than a few simple calculations. Logical branching, multidimensional lookup tables along with the sheer number of model equations to be evaluated contribute to the duration as well as uncertainty in the cpu time required to compute the state derivatives.

Without multirate integration, the total number of frames (integration steps) is given by t_{final}/T, where t_{final} is the simulation time and T is the integration step size. In the simplest case with only two states, fixed execution times of each derivative function and single-pass integration routines for "master" and "slave," the reduction in execution time from implementing multirate integration is straightforward. Suppose the cpu times required to execute the slow and fast derivative functions are Δ_s and Δ_f, respectively.

Case I: Without multirate integration ($T_s = T_f = T$)
The derivatives are numerically integrated at the simulation frame rate ($1/T$). The total execution time for fast derivative evaluations is

$$\Gamma_f = \left(\frac{t_{final}}{T_f}\right)\Delta_f = \left(\frac{t_{final}}{T}\right)\Delta_f \tag{8.314}$$

with a similar expression for the time required to perform slow derivative calculations,

$$\Gamma_s = \left(\frac{t_{final}}{T_s}\right)\Delta_f = \left(\frac{t_{final}}{T}\right)\Delta_s \tag{8.315}$$

The total time to compute both fast and slow derivatives is therefore

$$\Gamma_{w/o} = \Gamma_f + \Gamma_s = \left(\frac{t_{final}}{T}\right)\Delta_f + \left(\frac{t_{final}}{T}\right)\Delta_s = \frac{t_{final}}{T}(\Delta_f + \Delta_s) \tag{8.316}$$

Case II: With multirate integration ($T_s = NT_f = NT$)
The total time required for both fast and slow derivatives is

$$\Gamma_w = \Gamma_f + \Gamma_s = \left(\frac{t_{final}}{T_f}\right)\Delta_f + \left(\frac{t_{final}}{T_s}\right)\Delta_s \tag{8.317}$$

$$= \left(\frac{t_{final}}{T}\right)\Delta_f + \left(\frac{t_{final}}{NT}\right)\Delta_s \tag{8.318}$$

$$= \frac{t_{final}}{T}\left(\Delta_f + \frac{\Delta_s}{N}\right) \tag{8.319}$$

Assuming cpu times to execute fast and slow derivative functions are related by

$$\Delta_s = \alpha\Delta_f, \quad \alpha > 0 \tag{8.320}$$

From Equations 8.319 and 8.320,

$$\Gamma_w = \frac{t_{final}}{T}\Delta_f\left(1 + \frac{\alpha}{N}\right) \tag{8.321}$$

The cpu time (in seconds) required to evaluate two state derivatives using single-pass, multirate integration is illustrated in Figure 8.65 for the case where the transient response requires

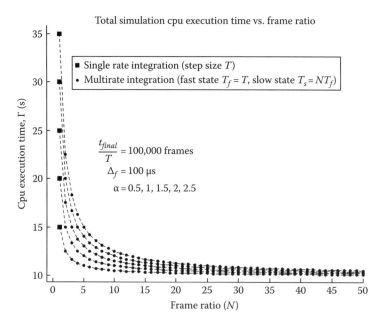

FIGURE 8.65 Total cpu time required to simulate transient response of system with single rate ($N = 1$) and multirate ($N > 1$) integration.

$t_{final}/T = 100{,}000$ simulation frames. This number of integration steps would be required, for example, if the step size needed to satisfy numerical stability and dynamic accuracy requirements was $T = 0.01$ s and the transient response lasted for 1000 s. The cpu time to execute the fast state derivative function Δ_f was fixed at 100 μs, and the slow state derivative requires $\alpha \Delta_f$ μs where α ranges from 0.5 to 2.5.

Observe from Figure 8.65 that the total cpu time without multirate integration varies from a low of 15 s when $\Delta_s = 0.5\Delta_f$ to a high of 35 s when $\Delta_s = 2.5\Delta_f$. The reduction in cpu time is more pronounced for lower values of frame ratio, that is, $N \leq 10$. Also, note that when the fast and slow state derivatives require the same amount of cpu time to execute, that is, $\alpha = 1$, the savings in overall cpu time is reduced from 20 min down to the limiting value of 10 min as expected.

The reduction in cpu time for the conditions illustrated in Figure 8.65 may seem trivial. The largest reduction in cpu time only approaches 25 s for the case where $\alpha = 2.5$ and the frame ratio is large. Simulation studies often entail multiple simulation runs with one or more system parameters varying from run to run. A two-parameter sensitivity study where each parameter assumes 10 numerical values requires 100 simulation runs. In this scenario, the use of multirate integration can achieve significant savings in overall computational time at the slight expense of reduced accuracy in the simulated responses.

EXERCISES

8.31 Derive the state equation matrices A, B, C, and D given in Equations 8.240 and 8.241.

8.32 In the aircraft pitch control system, find the maximum step size allowable for stable RK-4 simulation of the slow subsystem second-order component with poles located at $-4.3127 \pm j6.4812$.

8.33 In the aircraft pitch control system, use MATLAB to find

    (a) The analytical solution for state variable $x_3(t)$ and compare with the simulated results obtained with Simulink RK-4 and multirate RK-4/RK-4

    (b) The analytical solution for the pitch rate $\dot{\theta}(t)$ and compare it with the simulated response obtained from Simulink using RK-4

8.34 For the aircraft pitch control system represented by the block diagram shown in Figure 8.48,
  (a) Draw a simulation diagram and label the states $x_1, x_2, x_3, \ldots, x_7$.
  (b) Write the state equations and find the matrices A, B, C, and D in $\dot{\underline{x}} = A\underline{x} + B\theta_{com}, y = C\underline{x} + D\theta_{com}$. The output is $y(t) = \theta(t)$. Leave your answer in terms of parameters $A_1, B_1, B_0, b_1, b_0, \ldots, E_1$.
  (c) Using the given baseline values for the control system parameters $K, K_c, \ldots, K_{\dot{\theta}}$, evaluate the matrices A, B, C, and D.
  (d) Use MATLAB to verify that the eigenvalues of the coefficient matrix A are identical to those of the matrix A in Equation 8.240. Compare the eigenvalues to the roots of the characteristic polynomial in Equation 8.251.
  (e) Supplement the diagram shown in Figure 8.46 with additional Simulink blocks to simulate the pitch response based on the block diagram in Figure 8.48. Plot the three pitch responses on the same graph.

8.35 Label the five inputs to the summer in Figure 8.48 as $\theta_1, \theta_2, \ldots, \theta_5$. Find and plot the analytical solutions for $\theta_1(t), \theta_2(t), \ldots, \theta_5(t)$, on the same graph in response to the command pitch input in Equation 8.250. Comment on the results.

8.36 Simulate the aircraft control system pitch response to the input given in Equation 8.250 using multirate integration with $T_f = 0.001$ s and $T_s = 0.02$ s. Choose RK-1 for the "slave" routine and RK-4 for the "master" integration. Plot the response along with the analytical solution.

8.37 Consider the aircraft pitch control system operating in regulator mode, that is, zero input and initial condition $\theta(0) = \theta_0$.
  (a) Find analytical solutions for the pitch response $\theta(t)$ and the elevator deflection $\delta_e(t)$ when $\theta_0 = 10°$.
  (b) Find T_{max}, the maximum integration step for a stable simulation using RK-2 integration.
  (c) Simulate the pitch and elevator responses of the regulator control system ($\theta_0 = 10°$) using Simulink with RK-2 integration. Choose the step size $T = 0.1T_{max}$.
  (d) Simulate the pitch and elevator responses of the regulator control system using RK-2/RK-2 multirate integration. Choose the fast frame time T_f, so that the characteristic error in damping ratio of the fast subsystem second-order component in Figure 8.48 is 0.1%. Round T_f to three places after the decimal point. Choose the slow frame time T_s to make the frame ratio $N = T_s/T_f = 10$.
  (e) Plot the three pitch responses (analytical and two simulated) on the same graph. Repeat for the three elevator responses.

8.38 In the aircraft pitch control system, find the analytical solution for the fast state variables $x_5(t)$, $x_6(t)$, and $x_7(t)$ and plot the analytical, Simulink RK-4 and multirate RK-4/RK-4 solutions on the same graph similar to Figure 8.55.

8.39 Run the multirate integration of the aircraft pitch control system in the M-file "*Chap8_multi_rate_integ.m*" for the cases where the frame ratio $N = 20, 10, 5, 1$, and plot the simulated and analytical responses for $x_1(t) = \theta(t)$, $\dot{\theta}(t)$, and $x_4(t) = \delta_e(t)$.

8.40 Derive Equation 8.295 for the steady-state operating level in the first tank.

8.41 Find analytical expressions in terms of the system parameters A_1, c_1, A_2, L_2, and c_2 and the steady-state levels $\overline{H}_1, \overline{H}_2$ for the components of matrices B and C in Equations 8.300 and 8.301. Evaluate B and C for the given baseline values of the system parameters when $\overline{F}_0 = 60$ ft^3/min, and compare your results with those given in the text.

8.42 Generate responses similar to those in Figure 8.58 for the case where the initial conditions correspond to an input flow $F_0(t) = \overline{F}_0 = 20\,\text{ft}^3/\text{min} = 20$ ft^3/min. The inflow suddenly increases by $\Delta F_0(t) = 2.5$ ft^3/min at $t = 50$ min.

8.43 Plot an $\overline{H}_1$ vs. $\overline{H}_2$ operating characteristic for the two-tank system.

  Hint: Vary $\overline{F}_0$ from zero until the first tank begins to overflow. Find the steady-state values for $\overline{H}_1$ and $\overline{H}_2$.

8.44 Use AB-2 integration with step size T to simulate and plot the fluid level responses of both tanks like the ones shown in Figures 8.58 and 8.59 for $T = 0.05, 0.1, \ldots, 1.0$. Comment on the results.

8.45 For the baseline nonlinear two-tank system with tanks initially empty and tank one inflow given by $F_0(t) = 75$ ft³/min, $t \geq 0$.

(a) Run the Simulink model "*TwoTanks.mdl*" using RK-4 integration for a simulated time of 1500 min with decreasing step sizes T until there is negligible change in output for consecutive runs. Save the simulated tank levels at the end of each minute and denote them $H_{1,A}(n)$, $H_{2,A}(n)$, $n = 0, 1, 2, \ldots, 1500$. Assume that the simulated values are exact, that is, $H_{1,A}(n) \approx H_1(nT)$, $H_{2,A}(n) \approx H_2(nT)$, $n = 0, 1, 2, \ldots, 1500$.

(b) Run the MATLAB M-file "*Chap8_TwoTanks_Multirate_AB2_AB2.m*" or write your own to implement multirate AB-2/AB-2 integration for a simulated time of 1500 min with fixed frame time $T_f = 0.1$ min. Let the frame ratio $N = T_s/T_f$ vary according to 1, 5, 10, 15, 20, 25, 50, 75, 100, 500, 1000 and denote tank levels at the end of each minute by $\hat{H}_{1,A}(n), \hat{H}_{2,A}(n)$, $n = 0, 1, 2, \ldots, 1500$. Compute the mean squared errors for each value of N as

$$E_{H_1}(N) = \frac{1}{1500} \sum_{n=0}^{1500} \left[\left\{ \hat{H}_{1,A}(n) - H_{1,A}(n) \right\}^2 \right]$$

$$E_{H_2}(N) = \frac{1}{1500} \sum_{n=0}^{1500} \left[\left\{ \hat{H}_{2,A}(n) - H_{2,A}(n) \right\}^2 \right]$$

(c) Plot $E_{H_1}(N)$ and $E_{H_1}(N)$ vs. N and comment on the results.

8.46 Eight of ten natural modes of a linear 10th-order system are slow in comparison with the remaining two natural modes. The average cpu time required to compute the slow and fast state derivatives is 12 and 0.3 μs, respectively. A multirate integration scheme is proposed to simulate the transient response using RK-4 to integrate the slow states and RK-2 for the fast states. The fast states are updated at a rate of 250 Hz to assure numerical stability and reasonable dynamic accuracy. The dominant mode of the system corresponds to a real pole at $s = -0.05$.

A simulation study to investigate the effect of three parameters calls for $10 \times 10 \times 10$ simulation runs. Generate a graph like the one shown in Figure 8.65 relating the total simulation study cpu time vs. the multirate integration frame ratio.

8.5 REAL-TIME SIMULATION

Until now, the simulation execution time required by whatever computer resources might be available to update the state and algebraic variables of the system received minimal attention. A simulation study could "run long" for a number of reasons such as model complexity, dynamic accuracy and numerical stability requirements, limited cpu processing capabilities, and so forth; however, the consequences of waiting on the simulation to complete were not a critical concern. Simulations of this nature fall in the category of "off-line," "batch," or, more generally, nonreal-time simulation.

In some real-time simulations, a component that may have been simulated in the past has been replaced by the actual hardware. Alternatively, the component of interest may be physically integrated into the simulation from the beginning, making it unnecessary to simulate it beforehand. The component could be a gyroscopic sensor, a control surface actuator, an autopilot, or a

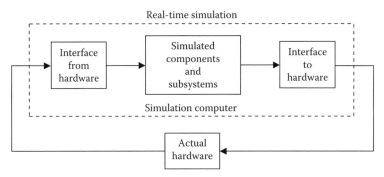

FIGURE 8.66   Hardware-in-the-loop real-time simulation.

combination of various sensors, actuators, and controllers in a particular system. The hardware must communicate with the simulation computer at precise intervals of time. The situation, illustrated in Figure 8.66, is referred to as "hardware-in-the-loop" simulation or HIL simulation for short.

HIL is used extensively in the development and testing of missile systems. Missile sensors are stimulated with input signals generated from real-time control computers representing the motion of targets during an engagement. Guidance and control hardware respond by providing inputs to the missile flight dynamics model, which is simulated in real-time to determine the missile's trajectory and calculate target intercept conditions (Eguchi 1998; Canova 1999).

The automotive industry incorporates real-time HIL simulation to design and test electronic control units (ECUs) for efficient operation of key systems such as power train control, the antilock braking system (ABS), and traction and cruise control. Classical simulation was performed off-line using simulation models of the vehicle's dynamics, sensors, and ECUs. While it was beneficial to demonstrate interaction of the various components and subsystems, it was still necessary to evaluate an ECU design using expensive prototype vehicles on a test track. Reproducing test track conditions to investigate unexpected results posed additional challenges.

One solution was to use HIL simulation composed of a real-time computer that runs a model of the vehicle to be controlled and the input/output (I/O) interfaces required to electrically connect to the controller. Benefits include a reduction in control system development and testing, no need for expensive prototype vehicles, elimination of risk that improper control software could lead to a hazardous failure during a test track run, and no concern about test track interactions with a prototype vehicle (Green 1997).

Figure 8.67 shows the main components of an HIL implementation for testing an ABS controller used by the German automaker Audi (Hanselmann). A digital-to-analog (D2A) converter generates wheel speed signals, sinusoidal voltages proportional in both frequency and amplitude to wheel speed, that replaces those from magnetic sensors in the actual vehicle. This accounts for the "interface-to-hardware" component in Figure 8.66. The "interface from hardware" consists of an analog-to-digital (A2D) converter for generating pressure sensor signals (in digital form) required by the vehicle dynamics model in the simulation computer to simulate the vehicle's response. Steering angle and other signals shown in Figure 8.67 are used for testing advanced levels of vehicle dynamics control such as automated braking on individual wheels at different intensity levels to stabilize vehicle motion in extreme situations.

It is imperative that the simulation computer be able to integrate the state variables in synchronization with real time. The beginning of each integration step must be properly aligned with the corresponding point in real time. In other words, the simulation must be capable of running fast enough on the digital computer that the computed outputs, in response to real-time inputs, occur at the exact time these outputs would take place in the real world.

A realistic vehicle dynamics model consists of coupled algebraic and differential equations with lookup tables for evaluating certain vehicle parameters that vary as driving conditions change.

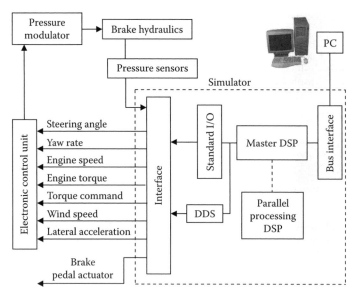

FIGURE 8.67 HIL simulation of vehicle ABS system. (From Hanselman, H. and Smith, K., *Test Meas. World Manage.*, 35, 1996.)

The equations are available as commercial C-language modules or in block diagram form (Simulink or other continuous simulation modeling program) with blocks representing transfer functions and system nonlinearities. Code is generated automatically from the block diagrams for real-time execution on the target DSP hardware.

Pushing the vehicle dynamics envelope to model evasive driving maneuvers adds to the complexity while imposing even more stringent timing requirements for numerically stable simulation of the differential equations. Additional mechanical degrees of freedom are present in the more detailed models used by Audi to account for slight movements in the vehicle's axles and suspension.

The time required to read input devices, perform simulation computations, and write to output devices determines the required frame rate for the simulation. In Audi's case, the simulation frame rate is less than 1 ms. The simulator generates signals and communicates them to the ECU in a matter of microseconds.

The simulated portion of the system in an HIL simulation may be all continuous-time, all discrete-time, or a combination of both. Furthermore, other types of signals, other than analog, are frequently encountered in HIL simulation. It is not uncommon for actual hardware to communicate with the simulation computer via I/O devices involving discrete digital (TTL), serial (RS-232, RS-422), instrumentation bus (IEEE-488) or network (Ethernet) signals (Ledin 2001).

Sometimes, the hardware in Figure 8.66 is actually a human such as a pilot in a flight simulator or an operator in a power plant simulator. A "human-in-the-loop" simulation can be used to evaluate the dynamic response of the system, the effectiveness of instrumentation displays and controls, or as a trainer to instruct the human in routine and emergency operation of the system. In the case of real-time interactive simulators (vehicle, aircraft, train, ship, plant, and so forth), several channels of output from the simulation computer may be used to drive motion systems as well as audio and visual displays to provide additional cues designed to enhance the overall sense of being physically immersed in a realistic, high-fidelity simulation environment. Figure 8.68 is a picture of a high-fidelity-driving simulator used for conducting research in traffic engineering, human factors, and design of new vehicle systems.

FIGURE 8.68   The National Advanced Driving Simulator used for Traffic Engineering Research and Vehicle System Design. (Courtesy of NHTSA, Washington, DC.)

8.5.1   Numerical Integration Methods Compatible with Real-Time Operation

The timing issues inherent in real-time simulation preclude the use of variable-step methods, which adaptively regulate the integration step size. The iterative nature of implicit methods makes the solution times unpredictable and, therefore, unsuitable for real-time applications as well. We will begin by looking at several one-step RK integrators (see Section 6.2) and determine whether they are compatible with real-time simulation. A continuous-time dynamic system is assumed to be modeled by the scalar, possibly nonlinear differential equation

$$\frac{dx}{dt} = f(x, u) \tag{8.322}$$

where
  $x = x(t)$ is the state
  $u = u(t)$ is the single input

For simplicity, the derivative function is assumed not to be an explicit function of "t." If the system is time-varying, the derivative function should be expressed as $f(t, x, u)$.

Figure 8.69 illustrates the sequence of operations for a real-time simulation running at a basic frame rate of $1/T$ using an integrator requiring two passes, that is, two derivative function

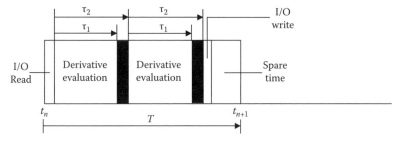

FIGURE 8.69   Real-time simulation with two-pass numerical integration method.

evaluations per frame. The initial operation is an I/O read of the input $u_n = u(t_n)$. The next process is the two derivative function evaluations including the calculations to advance the state from $x_A(n)$ to $x_A(n+1)$. Note that the time to evaluate the derivative function may be random due to the necessity of searching through tables of empirical data or as a result of branching when the code is executed. Lower and upper limits to compute the derivative functions and perform the calculations necessary for updating the state are τ_1 and τ_2 s. The final operation is an I/O write to the hardware interface as shown in Figure 8.66. The residual time before the frame ends is spare time to minimize the chances of a frame overrun and allow for expansion of the derivative function evaluation time should the model increase in complexity.

The following analysis of compatibility of real-time, single frame rate (as opposed to multirate) integration is based on the following assumptions:

1. The time required to complete the I/O read and write operations is negligible in comparison with the time to evaluate the derivative function and update the state.
2. The execution time to compute the derivative function is deterministic.
3. The spare time per frame is zero.

The net effect of these assumptions is that the frame time T is subdivided into m equal subframes where m is the number of passes through the derivative function. Several RK-m integrators will now be considered. In each instance, the test for compatibility with real-time simulation is whether or not the input $u(t)$ is needed at a point in time within the frame prior to it being available in real time.

8.5.2   RK-1 (Explicit Euler)

The simplest of all the numerical integrators, explicit Euler, is compatible with real-time simulation because the input $u(t)$ is needed only at the beginning of the frame. Thus, updating the discrete-time state from x_n to x_{n+1} with RK-1 requires u_n be available at t_n, the start time of the nth frame, which is certainly true (see Figure 8.70). Note that x_n is short for $x_A(n)$, the discrete-time approximation to $x(t_n)$.

8.5.3   RK-2 (Improved Euler)

Improved Euler RK-2 integration was introduced in Section 3.6 and again in Section 6.2. Figure 8.71 helps to explain why this one-step, two-pass numerical integrator is not suitable for real-time simulation under the previously assumed conditions. Specifically, the calculation of x_{n+1} commencing at $t_{n+(1/2)}$ requires knowledge of u_{n+1}, which is not available until $T/2$ s later at the end of the frame.

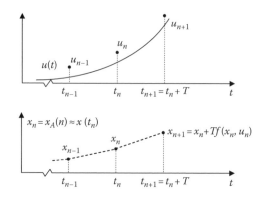

FIGURE 8.70   RK-1 (Euler) integration compatibility with real-time simulation.

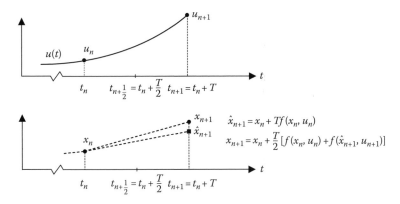

FIGURE 8.71 RK-2 (improved Euler) incompatibility with real-time simulation.

8.5.4 RK-2 (Modified Euler)

This version of RK-2 integration was first introduced in Section 3.6. The equations for updating the discrete-time state from x_n to x_{n+1} are

$$\hat{x}_{n+1/2} = x_n + \frac{T}{2}f(x_n, u_n) \tag{8.323}$$

$$x_{n+1} = x_n + Tf(\hat{x}_{n+1/2}, u_{n+1/2}) \tag{8.324}$$

The initial derivative evaluation starts at t_n and requires u_n. The second pass at evaluating the derivative function begins at $t_{n+1/2}$, precisely the time at which $u_{n+1/2}$ becomes available. Hence, both inputs are synchronized in real time with requirements of Equations 8.323 and 8.324. Howe (1995) refers to modified RK-2 integration as RTRK-2 to designate its suitability for real-time simulation.

Next, we look at two versions of RK-3 integration, one that is compatible with real-time simulation and the other that is not compatible.

8.5.5 RK-3 (Real-Time Incompatible)

The equations for the first RK-3 integrator are as follows:

$$\text{Starting at } t_n: k_1 = f(x_n, u_n), \quad \hat{x}_{n+1/2} = x_n + \frac{T}{2}k_1 \tag{8.325}$$

$$\text{Starting at } t_{n+1/3}: k_2 = f(\hat{x}_{n+1/2}, u_{n+1/2}), \quad \hat{x}_n = x_n + T(-k_1 + 2k_2) \tag{8.326}$$

$$\text{Starting at } t_{n+2/3}: k_2 = f(\hat{x}_n, u_{n+1}), \quad x_{n+1} = x_n + \frac{T}{6}[k_1 + 4K_2 + k_3] \tag{8.327}$$

The unsuitability for real-time implementation of Equations 8.325 through 8.327 stems from the requirement of needing $u_{n+1/2}$ at $t_{n+1/3}$, which is before it is available (see Equation 8.326) and a similar dilemma at time $t_{n+2/3}$ where u_{n+1} is required according to Equation 8.327.

8.5.6 RK-3 (Real-Time Compatible)

A real-time compatible RK-3 integrator is described by

$$\text{Starting at } t_n\text{: } k_1 = f(x_n, u_n), \quad \hat{x}_{n+1/3} = x_n + \frac{T}{3}k_1 \tag{8.328}$$

$$\text{Starting at } t_{n+1/3}\text{: } k_2 = f(\hat{x}_{n+1/3}, u_{n+1/3}), \quad \hat{x}_{n+2/3} = x_n + \frac{2T}{3}k_2 \tag{8.329}$$

$$\text{Starting at } t_{n+2/3}\text{: } k_3 = f(\hat{x}_{n+2/3}, u_{n+2/3}), \quad x_{n+1} = x_n + \frac{T}{4}[k_1 + 3k_3] \tag{8.330}$$

8.5.7 RK-4 (Real-Time Incompatible)

Fourth-order RK integration is widely used in applications not requiring real-time simulation. It can be shown that all RK-4 integrators require the input u_{n+1} for evaluation of the state derivative on the fourth pass at a time prior to the end of the current frame. Hence, none is compatible with real-time; however, a five-pass RK integrator with fourth-order accuracy suitable for real-time exists.

8.5.8 Multistep Integration Methods

The entire family of Adams–Bashforth numerical integrators presented in Section 6.4 is compatible with real time. The lower-order formulas are commonly used in real-time simulation applications. They are preferable to similar order real-time compatible RK integrators because they are single pass in nature and, hence, require less time to execute. For example, AB-m integration requires approximately $1/m$ as much time as any of the RK-m integrators. In HIL applications, AB-m simulation can run at frame rates roughly m times greater than any real-time compatible RK-m integrator. Dynamic errors are less for RK-m than AB-m integration with identical step size T; however, the advantage goes to AB-m integration running at $m \times (1/T)$ frames per second (fps) compared with RK-m integration at $1/T$ fps.

AB-1 integration is identical to explicit Euler. It is used sparingly in real-time simulation for the same reason it is used infrequently in nonreal-time simulation mode, namely, it is a first-order method, and even moderately accurate results require excessively small integration time steps. AB-2 through AB-4 are the most popular choices for real-time simulation. The stability regions of AB integrators higher than fourth order are quite small and become smaller as the order increases (see Figure 8.21). As a result, numerical stability constraints imposed by high-order AB integrators require the magnitude of λT (λ is the largest magnitude characteristic root of the stable linear or linearized system) be excessively small, thus requiring higher frame rates.

The predictor–corrector multistep methods (referred to by some as Adams–Moulton predictor–correctors) are not real-time compatible. They are two-pass integration algorithms, which combine an explicit Adams–Bashforth integrator to predict the new state followed by an implicit formula based on the predicted state to correct it. Equations 6.204 through 6.209 in represent second-through fourth-order methods. The dynamic error properties of predictor–corrector methods (see Table 8.4) are comparable to the single-pass implicit integrators (which are referred to in this text as Adams–Moulton integrators).

A real-time compatible predictor–corrector formula is possible. An example from Howe (1995) of a second-order method is now presented. The scalar state equation is the same as Equation 8.322. The first step is to generate an estimate of the state $\hat{x}_{n+1/2}$ at the midpoint of the current interval (see Figure 8.72).

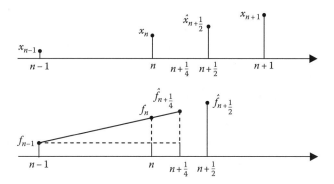

FIGURE 8.72 Diagram for illustrating real-time predictor–corrector method.

This is accomplished by using a form of modified Euler integration, that is, $\hat{x}_{n+1/2}$ is computed based on a step size of $T/2$ according to

$$\hat{x}_{n+1/2} = x_n + \frac{T}{2}\hat{f}_{n+1/4} \tag{8.331}$$

The derivative estimate $\hat{f}_{n+1/4}$ is obtained by linear extrapolation through (t_{n-1}, f_{n-1}) and (t_n, f_n) as shown in Figure 8.72. From the principle of similar triangles,

$$\frac{f_n - f_{n-1}}{t_n - t_{n-1}} = \frac{\hat{f}_{n+1/4} - f_{n-1}}{t_{n+1/4} - t_{n-1}} \tag{8.332}$$

Setting $t_n - t_{n-1} = T, t_{n+1/4} - t_{n-1} = 5T/4$ and solving for $\hat{f}_{n+1/4}$ give

$$\hat{f}_{n+1/4} = f_{n-1} + \frac{5}{4}(f_n - f_{n-1}) \tag{8.333}$$

Substituting $\hat{f}_{n+1/4}$ into Equation 8.331 results in the second-order predictor

$$\hat{x}_{n+1/2} = x_n + \frac{T}{8}(5f_n - f_{n-1}) \tag{8.334}$$

The derivative estimate $\hat{f}_{n+1/2}$ is calculated from

$$\hat{f}_{n+1/2} = f(\hat{x}_{n+1/2}, u_{n+1/2}) \tag{8.335}$$

Finally, the new state x_{n+1} is obtained from modified Euler integration,

$$x_{n+1} = x_n + T\hat{f}_{n+1/2} \tag{8.336}$$

Equations 8.334 through 8.336 describe a real-time, predictor–corrector algorithm (which Howe refers to as RTAM-2). It is a two-pass method since it requires two derivative function evaluations per step—the dynamic error coefficient $e_I = 1/24$ making it twice as accurate as the implicit (trapezoidal) and the AB-2/AM-2 predictor–corrector, since both have error coefficients of

$e_I = -1/12$ (see Table 8.4), and neither is compatible with real time. It is 10 times more accurate than AB-2 integration, which has an error coefficient $e_I = 5/12$.

For execution times comparable to single-pass formulas, this method would utilize a step size twice as large and generate state updates at half the frequency. After compensating for different step sizes, it still exhibits two and half times the dynamic accuracy of the single-pass AB-2 based on the approximate asymptotic formulas for small step sizes. The estimate $\hat{x}_{n+1/2}$ is available for real-time output; hence, the real-time predictor–corrector can output the state at the same frequency as the single-pass integrators. The integrator requires inputs at the beginning and midpoint of the frame making the sampling frequency twice that of a single-pass integrator. Higher-order real-time compatible predictor–correctors are possible (Howe 1995).

8.5.9 STABILITY OF REAL-TIME PREDICTOR–CORRECTOR METHOD

The stability region for the real-time predictor–corrector given in Equations 8.334 through 8.336 is obtained in the same way as for the explicit Adams–Bashforth, implicit Adams–Moulton, and RK integrators (see Section 8.3). Both the nonreal-time and real-time compatible predictor–correctors introduce extraneous roots in the z-domain and, therefore, are subject to stability limitations on step size. The characteristic polynomials for each integrator are

$$\text{Nonreal-time predictor–corrector: } \Delta(z) = z^2 - \left[1 + \lambda T + \frac{3}{4}(\lambda T)^2\right]z + \frac{1}{4}(\lambda T)^2 \qquad (8.337)$$

$$\text{Real-time predictor–corrector: } \Delta(z) = z^4 - \left[1 + \lambda T + \frac{5}{8}(\lambda T)^2\right]z^2 + \frac{1}{8}(\lambda T)^2 \qquad (8.338)$$

The stability regions are shown in Figure 8.73.

The real-time compatible predictor–corrector has a somewhat larger region, making it preferable from a stability standpoint.

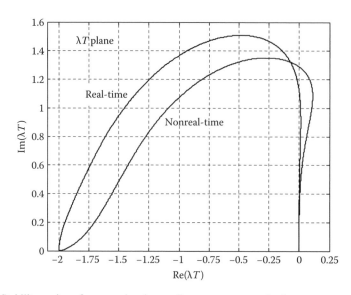

FIGURE 8.73 Stability regions for second-order predictor–corrector methods.

Example 8.9

Obtain difference equations for simulating the unit step response of the system

$$\frac{dx}{dt} = \lambda x + u, \quad \lambda = -0.25 \tag{8.339}$$

using the real-time compatible integrators

- (a) Modified Euler.
- (b) AB-2.
- (c) Real-time predictor–corrector.
- (d) Choose the step size T, so that $\lambda T = -0.25$, -1 for the modified Euler and real-time predictor–corrector and $\lambda T = -0.125$, -0.5 for the AB-2 integrator. Graph the step responses along with the exact solution and comment on the results.

(a) From Equation 8.323 for modified Euler, the estimated state at the halfway point is

$$\hat{x}_{n+1/2} = x_n + 0.5Tf_n \tag{8.340}$$

$$= x_n + 0.5T(\lambda x_n + u_n) \tag{8.341}$$

$$= (1 + 0.5\lambda T)x_n + 0.5Tu_n \tag{8.342}$$

and the second pass produces the updated state from Equation 8.324 as

$$x_{n+1} = x_n + T\hat{f}_{n+1/2} \tag{8.343}$$

$$= x_n + T(\lambda \hat{x}_{n+1/2} + u_{n+1/2}) \tag{8.344}$$

$$= x_n + [T\lambda\{(1 + 0.5\lambda T)x_n + 0.5Tu_n\} + u_{n+1/2}] \tag{8.345}$$

$$= [1 + \lambda T(1 + 0.5\lambda T)]x_n + T(0.5\lambda Tu_n + u_{n+1/2}) \tag{8.346}$$

(b) The AB-2 difference equation for computing the state is

$$x_{n+1} = x_n + 0.5T(3f_n - f_{n-1}) \tag{8.347}$$

$$= x_n + 0.5T[3(\lambda x_n + u_n) - (\lambda x_{n-1} + u_{n-1})] \tag{8.348}$$

$$= (1 + 1.5\lambda T)x_n - 0.5\lambda Tx_{n-1} + 1.5Tu_n - 0.5Tu_{n-1} \tag{8.349}$$

(c) The real-time predictor–corrector first step is from Equation 8.334,

$$\hat{x}_{n+1/2} = x_n + 0.125T(5f_n - f_{n-1}) \tag{8.350}$$

$$= x_n + 0.125T[5(\lambda x_n + u_n) - (\lambda x_{n-1} + u_{n-1})] \tag{8.351}$$

$$= (1 + 0.625\lambda T)x_n - 0.125\lambda Tx_{n-1} + 0.625Tu_n - 0.125Tu_{n-1} \tag{8.352}$$

The new state is obtained from Equations 8.335 and 8.336,

$$x_{n+1} = x_n + T\hat{f}_{n+1/2} \tag{8.353}$$

$$= x_n + T(\lambda \hat{x}_{n+1/2} + u_{n+1/2}) \tag{8.354}$$

$$= x_n + T[\lambda\{1 + 0.625\lambda T)x_n - 0.125\lambda Tx_{n-1} + 0.625Tu_n - 0.125Tu_{n-1}\} + u_{n+1/2}] \tag{8.355}$$

$$= [1 + \lambda T(1 + 0.625\lambda T)]x_n - 0.125(\lambda T)^2 x_{n-1} + \lambda T(0.625Tu_n - 0.125Tu_{n-1}) + Tu_{n+1/2}$$

$$\tag{8.356}$$

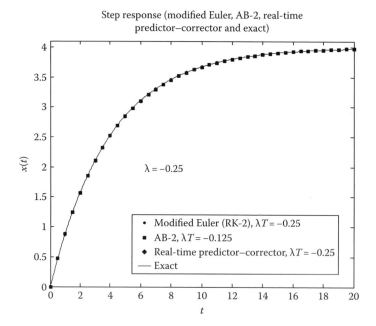

FIGURE 8.74 Unit step response of first-order system using three real-time compatible numerical integrators with $\lambda T = -0.25$, -0.125 and the exact solution.

(d) The difference Equations 8.346, 8.349, and 8.356 were solved recursively in the M-file "*Chap8_Ex5_1.m.*" The AB-2 integrator and real-time predictor–corrector were started with a single step of the improved Euler integrator. The results are shown in Figures 8.74 and 8.75 along with the exact solution for the unit step response,

$$x(t) = \frac{1}{\lambda}[e^{\lambda t} - 1], \quad t \geq 0 \tag{8.357}$$

The single-pass AB-2 integrator was running at twice the frame rate of the two-pass modified Euler and real-time predictor–corrector to keep the execution times comparable. In Figure 8.74, the simulated responses using the numerical integrators are in close agreement with the exact solution. In Figure 8.75, the accuracy of the numerical integrators has deteriorated as a result of the increased values of the parameter λT. The M-file "*Chap8_Ex5_1.m*" includes runs for intermediate values of λT as well.

8.5.10 EXTRAPOLATION OF REAL-TIME INPUTS

A solution to the problem of numerical integrators being incompatible with real-time simulation is to employ extrapolated input data. Consider the improved Euler integrator illustrated in Figure 8.71. The evaluation of $f_n = f(x_n, u_n)$ lasts from t_n to approximately $t_{n+1/2}$. After calculating $\hat{x}_{n+1}$, the evaluation of $\hat{f}_{n+1} = f(\hat{x}_{n+1}, u_{n+1})$ is scheduled to begin at $t_{n+1/2}$. The input u_{n+1} is required a half frame before it is available, thus explaining why improved Euler is incompatible with real-time simulation.

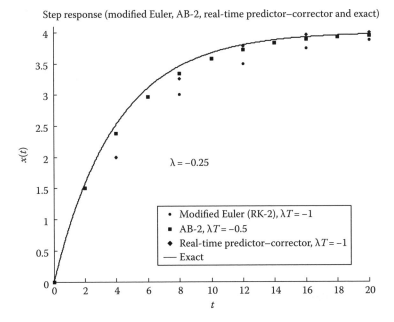

Step response (modified Euler, AB-2, real-time predictor–corrector and exact)

$\lambda = -0.25$

- •   Modified Euler (RK-2), $\lambda T = -1$
- ■   AB-2, $\lambda T = -0.5$
- ◆   Real-time predictor–corrector, $\lambda T = -1$
- — Exact

FIGURE 8.75 Unit step response of first-order system using three real-time compatible numerical integrators with $\lambda T = -1$, -0.5 and the exact solution.

A possible remedy is to use first-order (linear) extrapolation based on u_{n-1} and u_n to predict u_{n+1}. An alternative approach is to sample the input at $t_{n+1/2}$ and predict u_{n+1} based on linear extrapolation of u_n and $u_{n+1/2}$. The predicted values for each approach is denoted $\hat{u}_{n+1}$ in Figure 8.76.

Adopting the first approach leads to

$$\hat{u}_{n+1} = u_n + (u_n - u_{n-1}) \tag{8.358}$$

If we think of Equation 8.358 as the difference equation for a discrete-time system with input u_n and output $y_n = \hat{u}_{n+1}, n = 0, 1, 2, \ldots,$ the first several values of the output are

$$n = 0: \; y_0 = \hat{u}_1 = u_0 + (u_0 - u_{-1}) = 2u_0 \tag{8.359}$$

$$n = 1: \; y_1 = \hat{u}_2 = u_1 + (u_1 - u_0) = 2u_1 - u_0 \tag{8.360}$$

$$n = 2: \; y_2 = \hat{u}_3 = u_2 + (u_2 - u_1) = 2u_2 - u_1 \tag{8.361}$$

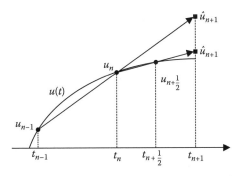

FIGURE 8.76 Use of extrapolation to make improved Euler compatible with real-time.

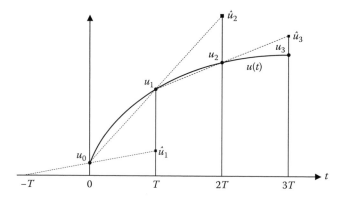

FIGURE 8.77   Linear extrapolation of input $u(t)$.

Figure 8.77 shows a continuous-time function $u(t)$, the first four sampled values u_0, u_1, u_2, and u_3, and the first three extrapolated values $\hat{u}_1, \hat{u}_2, \hat{u}_3$.

The z-transform of the output sequence y_n, $n = 0, 1, 2, 3, \ldots$ is by definition

$$Y(z) = y_0 + y_1 z^{-1} + y_2 z^{-2} + y_3 z^{-3} \cdots \tag{8.362}$$

$$= 2u_0 + (2u_1 - u_0)z^{-1} + (2u_2 - u_1)z^{-2} + (2u_3 - u_2) + \cdots \tag{8.363}$$

Rearranging the terms in Equation 8.363 gives

$$Y(z) = 2(u_0 + u_1 z^{-1} + u_2 z^{-2} + \cdots) - z^{-1}(u_0 + u_1 z^{-1} + u_2 z^{-2} + \cdots) \tag{8.364}$$

$$= (2 - z^{-1})U(z) \tag{8.365}$$

The same result follows directly from Equation 8.358 with $\hat{u}_{n+1}$ replaced by y_n. The z-domain transfer function of the linear extrapolator is therefore

$$G(z) = \frac{Y(z)}{U(z)} = 2 - z^{-1} \tag{8.366}$$

Before we discuss the dynamic errors incurred from the use of extrapolation, it is necessary to define the characteristics of an ideal extrapolator. Figure 8.78 illustrates the point for an arbitrary signal $u(t)$ sampled at regular intervals of T units of time.

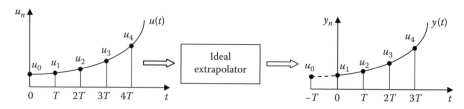

FIGURE 8.78   Illustration of an ideal extrapolator.

At time $t = nT$, if the input to an ideal extrapolator is $u_n = u(nT)$, the output $y_n = u_{n+1} = u[(n+1)T]$. Hence, an ideal extrapolator advances the input $u(t)$ by an amount T to the left along the t-axis. In contrast, a pure delay of the same duration shifts the input $u(t)$ by the same amount to the right along the t-axis.

The Laplace transform of the ideal extrapolator $G_I(s)$ can be obtained by replacing T in the transform for a pure delay of length T with $-T$ leading to

$$G_I(s) = \frac{Y(s)}{U(s)} = e^{-(-T)s} = e^{Ts} \tag{8.367}$$

The frequency response functions of the real and ideal extrapolators are

$$G(z)|_{z=e^{j\omega T}} = G(e^{j\omega T}) = 2 - e^{-j\omega T} \tag{8.368}$$

$$G_I(s)|_{s=j\omega} = G_I(j\omega) = e^{j\omega T} \tag{8.369}$$

The fractional error in $G(e^{j\omega T})$, the extrapolator frequency response function, is

$$e_G = \frac{G(e^{j\omega T}) - G_I(j\omega)}{G_I(j\omega)} = \frac{2 - e^{-j\omega T} - e^{j\omega T}}{e^{j\omega T}} = 2e^{-j\omega T} - e^{-2j\omega T} - 1 \tag{8.370}$$

The fractional error in extrapolator frequency response gain is

$$e_{|G|} = \frac{|G(e^{j\omega T})| - |G_I(j\omega)|}{|G_I(j\omega)|} = |2 - e^{-j\omega T}| - 1 \tag{8.371}$$

Replacing $e^{-j\omega t}$ with $\cos \omega T - j \sin \omega T$, Equation 8.371 reduces to

$$e_{|G|} = (5 - 4\cos \omega T)^{1/2} - 1 \tag{8.372}$$

An asymptotic formula for $e_{|G|}$ is (see Exercise 8.50)

$$e_{|G|} \approx (\omega T)^2, \quad \omega T \ll 1 \tag{8.373}$$

The phase error in extrapolator frequency response is

$$e_{\angle G} = \text{Arg}\{G(e^{j\omega T})\} - \text{Arg}\{G_I(j\omega)\} \tag{8.374}$$

$$= \text{Arg}\{2 - e^{-j\omega T}\} - \omega T \tag{8.375}$$

An asymptotic formula for $e_{\angle G}$ is (Howe 1995)

$$e_{\angle G} \approx -(\omega T)^3, \quad \omega T \ll 1 \tag{8.376}$$

Magnitude and phase angle plots of a real and ideal extrapolator are shown in Figure 8.79 for $0 \leq \omega T \leq 0.5$ rad. The graphs are in agreement with Equations 8.373 and 8.376, which imply that the magnitude error is more significant than the phase angle error.

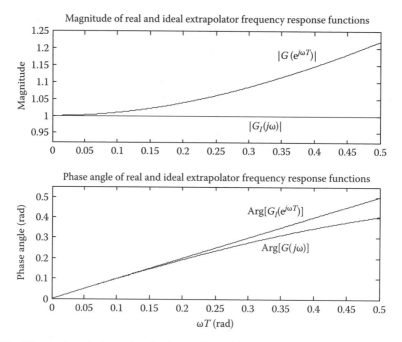

FIGURE 8.79 Magnitude and phase plots for first-order and ideal extrapolator.

Example 8.10

An input signal $u(t) = \sin \omega t$, $t \geq 0$ is sampled every $T = 0.1$ s, and the resulting discrete-time signal u_n, $n = 0, 1, 2, \ldots$ is input to an extrapolator governed by Equation 8.358.

(a) Graph the continuous-time signal $u(t)$, discrete-time signal u_n, and the extrapolator output $\hat{u}_{n+1}$, $n = 0, 1, 2, 3, \ldots$ for the following cases:
   (i) $\omega T = 0.1$ rad
   (ii) $\omega T = 0.25$ rad
   (iii) $\omega T = 0.5$ rad
   (iv) $\omega T = 1$ rad
(b) An improved Euler integrator with step size $T = 0.1$ s is used to simulate the response of the first-order system in Equation 8.339 to the sinusoidal input $u(t) = \sin \omega t$, $t \geq 0$. In order to simulate the real-time response, the input is extrapolated as shown in Figure 8.80 before being numerically integrated. Find the exact and simulated responses for the four cases in part (a) and plot the results.

(a) The signals $u(t)$, u_n, and $\hat{u}_{n+1}$ are generated in the script file "Chap8_Ex5_2.m" and the results are shown in Figures 8.81 and 8.82. The extrapolator gain error is first noticeable at $\omega T = 0.25$ rad, becoming progressively worse at $\omega T = 0.5$ rad and $\omega T = 1$ rad, respectively.

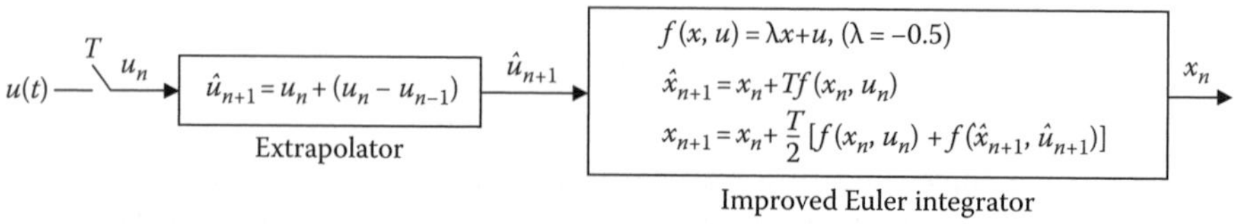

$$u(t) \xrightarrow{\quad T \quad u_n \quad} \boxed{\hat{u}_{n+1} = u_n + (u_n - u_{n-1})} \xrightarrow{\quad \hat{u}_{n+1} \quad} \boxed{\begin{array}{l} f(x, u) = \lambda x + u, \ (\lambda = -0.5) \\[4pt] \hat{x}_{n+1} = x_n + Tf(x_n, u_n) \\[4pt] x_{n+1} = x_n + \dfrac{T}{2}[f(x_n, u_n) + f(\hat{x}_{n+1}, \hat{u}_{n+1})] \end{array}} \xrightarrow{\quad x_n \quad}$$

Extrapolator   Improved Euler integrator

FIGURE 8.80 Real-time simulation of first-order system dynamic response.

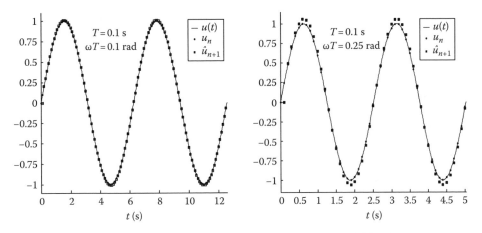

FIGURE 8.81   Continuous-time, sampled and extrapolated inputs ($\omega T = 0.1$, 0.25 rad).

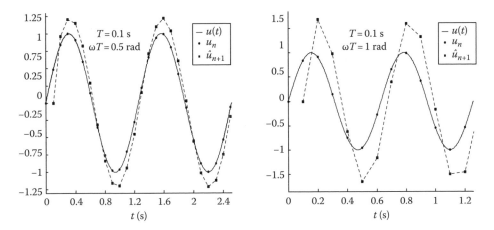

FIGURE 8.82   Continuous-time, sampled and extrapolated inputs ($\omega T = 0.5$, 1 rad).

(b) The analytical solution for the response is obtained by Laplace transformation of the differential equation $\dot{x} = \lambda x + u$ with sinusoidal input $u = \sin \omega t$. The Laplace transform of $x(t)$ is

$$X(s) = \frac{\omega}{(s - \lambda)(s^2 + \omega^2)} \tag{8.377}$$

which is easily inverted by partial fractions to give

$$x(t) = \frac{\omega}{\lambda^2 + \omega^2} \left(e^{\lambda t} - \cos \omega t - \frac{\lambda}{\omega} \sin \omega t \right) \tag{8.378}$$

$$= \frac{\omega}{(\lambda^2 + \omega^2)} e^{\lambda t} - \frac{1}{(\lambda^2 + \omega^2)^{1/2}} \sin (\omega t + \varphi), \quad \varphi = \pi + \tan^{-1} \left(\frac{\omega}{\lambda} \right) \tag{8.379}$$

The exact response $x(t)$ and the simulated responses x_n for the cases when $\omega T = 0.1$ rad and $\omega T = 0.25$ rad are plotted in Figure 8.83. Results for the remaining two cases, $\omega T = 0.5$ rad and $\omega T = 1$ rad, are shown in Figure 8.84. Error in the simulated response due to extrapolator gain error is significant at input frequencies $\omega T = 0.5$ rad and $\omega T = 1$ rad where the asymptotic approximations in Equations 8.373 and 8.376 are no longer valid.

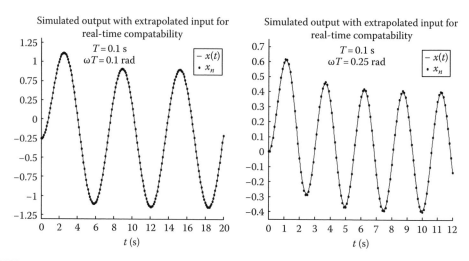

FIGURE 8.83   Exact and simulated (improved Euler) responses ($\omega T = 0.1$, 0.25 rad).

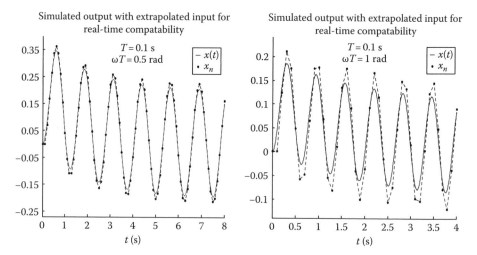

FIGURE 8.84   Exact and simulated (improved Euler) responses ($\omega T = 0.5$, 1 rad).

8.5.11   ALTERNATE APPROACH TO REAL-TIME COMPATIBILITY: INPUT DELAY

When numerical integrators are not compatible with real-time simulation, it is because the input(s) are required at points in time prior to their occurrence. One solution to this dilemma is to use input values previously sampled in place of the input data required by the formula in the numerical integration algorithm. Refer to Figure 8.85, which shows an input $u(t)$ and delayed versions $u(t - T/2), u(t - T)$.

Let us assume once again that improved Euler, a second-order, two-pass RK integrator incompatible with real-time simulation, is to be used. Starting at time t_n, the first stage is an Euler prediction of the state at t_{n+1}. However, instead of using the current input u_n, suppose the input from one-half a time step in the past is used, namely, $u_{n-1/2}$. That is, $\hat{x}_{n+1}$ is computed from

$$\hat{x}_{n+1} = x_n + Tf(x_n, u_{n-1/2}) \tag{8.380}$$

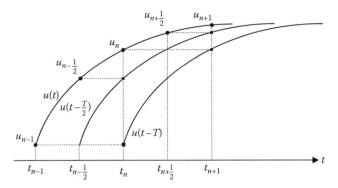

FIGURE 8.85 Use of delayed input to make numerical integrator real-time compatible.

Starting at time $t_{n+1/2}$, the second pass to compute the new state is

$$x_{n+1} = x_n + \frac{T}{2}[f(x_n, u_{n-1/2}) + f(\hat{x}_{n+1}, u_{n+1/2})] \tag{8.381}$$

Assuming Equation 8.381 requires approximately $T/2$ units of time to execute, the updated state x_{n+1} is available at time t_{n+1}. Hence, by using $u_{n-1/2}$ in place of u_n and $u_{n-1/2}$ instead of u_{n+1}, the improved Euler integrator is running in real time. Equations 8.380 and 8.381 applied to the first-order system $dx/dt = \lambda x + u$ lead to the difference equation

$$x_{n+1} = \left[1 + \lambda T + \frac{(\lambda T)^2}{2}\right]x_n + \frac{T}{2}(1 + \lambda T)u_{n-1/2} + \frac{T}{2}u_{n+1/2} \tag{8.382}$$

Equation 8.382 is similar to the difference equation for simulation of the first-order system using classical improved Euler integration except for the presence of the delayed input, that is, u_n is replaced by $u_{n-1/2}$ and u_{n+1} is replaced by $u_{n+1/2}$.

There is of course a penalty incurred as a result of using "old" values from the delayed input $u(t)$. To illustrate, consider the case where sampled values are obtained from the input delayed a full time step T as shown in Figure 8.86.

Simulation of the system $dx/dt = \lambda x + u$ with real-time, improved Euler integration leads to a discrete-time system with z-domain transfer function

$$G_{R/T}(z) = \frac{X(z)}{U(z)} = z^{-1}G(z) = \frac{b_1 z + b_0}{z(z - a_0)} \tag{8.383}$$

where

$$a_0 = 1 + \lambda T + \frac{(\lambda T)^2}{2}, \quad b_0 = \frac{T}{2}(1 + \lambda T), \quad b_1 = \frac{T}{2} \tag{8.384}$$

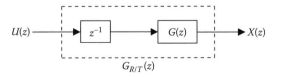

FIGURE 8.86 z-Domain transfer function for real-time implementation.

The continuous-time system transfer function is

$$G(s) = \frac{1}{s - \lambda} \tag{8.385}$$

The dynamic errors in the discrete-time frequency response functions are

$$e_G = \frac{G(z)|_{z \leftarrow e^{j\omega T}} - G(s)|_{s \leftarrow j\omega}}{G(s)|_{s \leftarrow j\omega}} \tag{8.386}$$

$$= \frac{(b_1 e^{j\omega T} + b_0)/(e^{j\omega T} - a_0)}{1/(j\omega - \lambda)} - 1 \tag{8.387}$$

$$= \frac{(j\omega - \lambda)(b_1 e^{j\omega T} + b_0)}{e^{j\omega T} - a_0} - 1 \tag{8.388}$$

$$5e_{G_{R/T}} = \frac{G_{R/T}(z)|_{z \leftarrow e^{j\omega T}} - G(s)|_{s \leftarrow j\omega}}{G(s)|_{s \leftarrow j\omega}} \tag{8.389}$$

$$= \frac{(b_1 e^{j\omega T} + b_0)/(e^{j\omega T}(e^{j\omega T} - a_0))}{1/(j\omega - \lambda)} - 1 \tag{8.390}$$

$$= \frac{(j\omega - \lambda)(b_1 e^{j\omega T} + b_0)}{e^{j\omega T}(e^{j\omega T} - a_0)} - 1 \tag{8.391}$$

The fraction gain errors are

$$e_{|G|} = \frac{|G(e^{j\omega T})| - |G(j\omega)|}{|G(j\omega)|} \tag{8.392}$$

$$= \frac{|(b_1 e^{j\omega T} + b_0)/(e^{j\omega T} - a_0)|}{|1/(j\omega - \lambda)|} - 1 \tag{8.393}$$

$$e_{|G_{R/T}|} = \frac{|G_{R/T}(e^{j\omega T})| - |G(j\omega)|}{|G(j\omega)|} \tag{8.394}$$

$$= \frac{|(b_1 e^{j\omega T} + b_0)/[e^{j\omega T}(e^{j\omega T} - a_0)]|}{|1/(j\omega - \lambda)|} - 1 \tag{8.395}$$

$$= \frac{|(b_1 e^{j\omega T} + b_0)/[(e^{j\omega T} - a_0)]|}{|1/(j\omega - \lambda)|} - 1 \tag{8.396}$$

$$= e_{|G|} \tag{8.397}$$

The phase error are

$$e_{\angle G} = \angle G(e^{j\omega T}) - \angle G(j\omega) \tag{8.398}$$

$$= \text{Arg}\left(\frac{b_1 e^{j\omega T} + b_0}{e^{j\omega T} - a_0}\right) - \text{Arg}\left(\frac{1}{j\omega - \lambda}\right) \tag{8.399}$$

$$e\angle G_{R/T} = \angle G_{R/T}(e^{j\omega T}) - \angle G(j\omega) \tag{8.400}$$

$$= \mathrm{Arg}\left(\frac{b_1 e^{j\omega T} + b_0}{e^{j\omega T}(e^{j\omega T} - a_0)}\right) - \mathrm{Arg}\left(\frac{1}{j\omega - \lambda}\right) \tag{8.401}$$

$$= \mathrm{Arg}\left(\frac{b_1 e^{j\omega T} + b_0}{(e^{j\omega T} - a_0)}\right) - \omega T - \mathrm{Arg}\left(\frac{1}{j\omega - \lambda}\right) \tag{8.402}$$

$$= e_{\angle G} - \omega T \tag{8.403}$$

The fractional gain and phase errors for the classical and real-time, improved Euler integrators are graphed in Figure 8.87 for the case when $\lambda = -0.5$. As expected from Equation 8.397, the fractional gain errors are equal and from Equation 8.403, the real-time, improved Euler integrator introduces an additional phase lag of ωT rad. Note that the fractional gain error varies from zero to approximately -2% over the interval $0 \le \omega T \le 0.5$ rad.

Also, note that $e_{\angle G} \approx 0$ for $0 \ v \ \omega T \le 0.5$ rad. Hence, the classical improved Euler integrator contributes essentially zero phase shift with respect to the continuous-time frequency response.

The phase angles (in deg) of the two discrete-time and the continuous-time frequency response functions are shown in Figure 8.88. As expected from Equation 8.403, the separation between the top two plots $\angle G(e^{j\omega T})$ and $\angle G(j\omega)$ and the bottom plot $\angle G_{R/T}(e^{j\omega T})$ is ωT rad. For example, at $\omega T = 0.3$ rad, $\angle G(e^{j\omega T}) = \angle G(j\omega) = -1.4031$ rad (-80.3914 deg) and $\angle G_{R/T}(e^{j\omega T}) = -1.7031$ rad (-97.5801 deg).

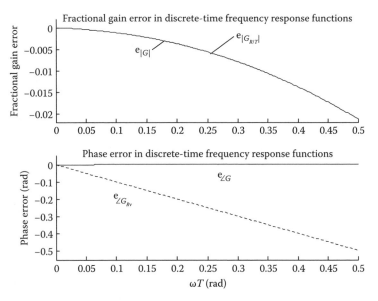

FIGURE 8.87 Dynamic errors from simulation of $dx/dt = -0.5x + u$ with classical and real-time, improved Euler integration ($T = 0.1$ s).

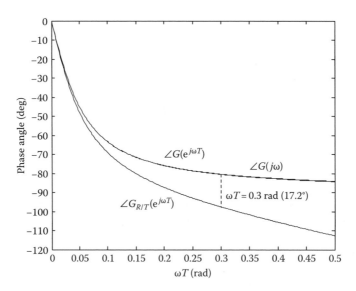

FIGURE 8.88 Phase of discrete-time and continuous frequency response functions.

Example 8.11

An object with thermal capacitance C and thermal resistance R shown in Figure 8.89 is exposed to a surrounding temperature that varies according to $T_0(t) = \overline{T}_0 + \Delta T_0 \sin 2\pi f_0 t$, $t \geq 0$. The mathematical model governing $\hat{T}(t)$, the temperature of the object, consists of Equations 8.404 and 8.405.

Baseline system parameter values are

$$C = 200\,\text{Btu}/°\text{F}, \quad R = 0.005°\text{F}/\text{Btu/h},$$

$$\overline{T}_0 = 50°\text{F}, \quad \Delta T_0 = 20°\text{F}, \quad f_0 = 1 \text{ cycle every } 24\,\text{h}, \quad T(0) = 50°\text{F}$$

Find the difference equations for simulating the temperature response using

(a) Improved Euler integration
(b) Real-time, improved Euler integration using a one-step delayed version of the input
(c) Find the analytical solution for $\hat{T}(t)$.
(d) Simulate the temperature response over two cycles in $T_0(t)$ by recursive solution of the difference equations in parts (a) and (b). Choose the time step T, so that $T/RC = 0.25$. Plot the analytical and numerical solutions on the same graph.
(e) Repeat part (d) for $f_0 = 1$ cycle every 3 h.

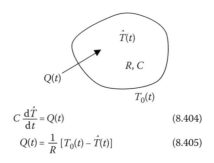

$$C \frac{d\hat{T}}{dt} = Q(t) \tag{8.404}$$

$$Q(t) = \frac{1}{R}[T_0(t) - \hat{T}(t)] \tag{8.405}$$

FIGURE 8.89 Thermal system for Example 8.11.

(a) Combining Equations 8.404 and 8.405 leads to the differential equation of the system.

$$\tau \frac{d\hat{T}}{dt} + \hat{T}(t) = T_0(t), \quad \tau = RC \tag{8.406}$$

The state derivative function is

$$f(\hat{T}, T_0) = \frac{d\hat{T}}{dt} = \frac{1}{\tau}(T_0 - \hat{T}) \tag{8.407}$$

and the difference equation for implementing standard improved Euler integration is

$$\hat{T}_{n+1} = \left[1 - \frac{T}{\tau} + \frac{1}{2}\left(\frac{T}{\tau}\right)^2\right]\hat{T}_n + \frac{T}{2\tau}\left(1 - \frac{T}{\tau}\right)T_{0,n} + \left(\frac{T}{2\tau}\right)T_{0,n+1,n=0,1,2,\ldots} \tag{8.408}$$

where

$$T_{0,n} = \overline{T} + \Delta T_0 \sin \omega nT, \quad n = 0, 1, 2, \ldots \quad (\omega = 2\pi f_0) \tag{8.409}$$

(b) Delaying the input $T_0(t)$ by T h before sampling leads to the difference equation

$$\hat{T}_{n+1} = \left[1 - \frac{T}{\tau} + \left(\frac{T}{\tau}\right)^2\right]\hat{T}_n + \frac{T}{2\tau}\left(1 - \frac{T}{\tau}\right)T_{0,n-1} + \left(\frac{T}{2\tau}\right)T_{0,n}, \quad n = 0, 1, 2, \ldots \tag{8.410}$$

(c) The analytical solution for $\hat{T}(t)$ is obtained by Laplace transforming Equation 8.406 followed by inverse Laplace transformation of the expression for $\hat{T}(s)$. The steps are left for an exercise. The result is

$$\hat{T}(t) = \overline{T}_0\left[\overline{T}(0) - \hat{T}(0) + \frac{\tau\omega\Delta T_0}{1 + (\tau\omega)^2}\right]e^{-t/\tau} + \frac{\Delta T_0}{1 + (\tau\omega)^2}[\sin \omega t - (\tau\omega)\cos \omega t] \tag{8.411}$$

(d) The simulated responses are determined by recursive solution of the appropriate difference equation in "*Chap8_Ex5_3.m.*" The step size is determined from

$$T = 0.25RC = 0.25(0.005°F/Btu/h)(200\,Btu/°F) = 0.25\,h$$

The continuous-time input $T_0(t)$ is shown in the top half of Figure 8.90. The discrete-time input $T_{0,n}$, $n = 0, 1, 2, 3, \ldots$ are the sampled values at 0.25 h intervals; however, only the sampled values at the end of each hour are shown in Figure 8.90. The lower half of Figure 8.90 shows the continuous-time output $\hat{T}(t)$ and the discrete-time outputs at the end of each hour, that is, every fourth value.

The continuous-time response $\hat{T}(t)$ and the simulated response $\hat{T}_n$ generated by improved Euler integration are indistinguishable from each other at the end of the integration steps. The discrete-time response $\hat{T}_{R/T,n}$ is simply $\hat{T}_n$ delayed by $T = 0.25$ h. There is close agreement between the simulated and analytical responses because the dynamic errors are very small when $\omega T = 0.065$ (see Figure 8.87).

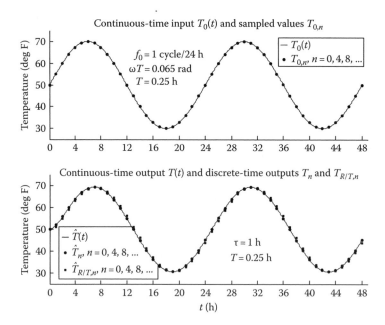

FIGURE 8.90   Continuous- and discrete-time inputs and outputs ($f_0 = 1$ cycle/24 h).

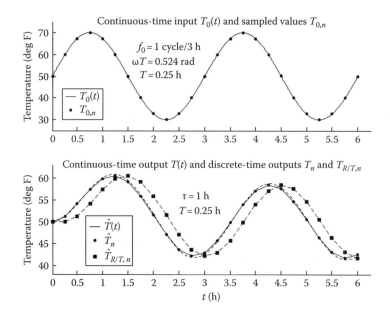

FIGURE 8.91   Continuous- and discrete-time inputs and outputs ($f_0 = 1$ cycle/3 h).

(e) The period of input temperature fluctuations is reduced from 24 to 3 h. The new radian frequency is $\omega = 2\pi f_0 = 2\pi(1/3) = 2.094$ rad/h and $\omega T = 0.524$ rad. A slight difference between the simulated response $\hat{T}_n$ and the continuous-time response is now evident as shown in Figure 8.91. According to Figure 8.87, the two are in phase and the fractional gain error is approximately -0.02 (-2%).

The real-time, iImproved Euler temperature response $\hat{T}_{R/T,n}$ is once again a delayed version of $\hat{T}_n$, the delay being $T = 0.25$ h. There is a significant difference between the analytical solution and the real-time, improved Euler response.

EXERCISES

8.47 Rework Example 8.9 using two half steps of RK-2 to generate the starting value for the real-time predictor–corrector.

8.48 Use the real-time predictor–corrector to simulate the response of the first-order system in Example 8.9 for $\lambda T = 0.1$ and a sinusoidal input $u(t) = \sin \omega t$, $t \geq 0$. Write a MATLAB script file that accepts values for the radian frequency in the range $0.1\, \omega_{BW} \leq \omega \leq 10\omega_{BW}$, where ω_{BW} is the system bandwidth and plots the simulated response and exact solution.

8.49 Suppose a zero-order extrapolator $y_n = \hat{u}_{n+1} = u_n$, $n = 0, 1, 2, \ldots$ is used instead of the first-order extrapolator in Equation 8.358.
(a) Find the z-transform $G(z) = Y(z/U(z))$ of this extrapolator.
(b) Find an expression for the fractional error in the frequency response function e_G.
(c) Find expressions for the fraction error in gain $e_{|G|}$ and the error in phase $e_{\angle G}$.
(d) Find asymptotic formulas for the errors in part (c).
(e) Plot the magnitude and phase of the zero order and ideal extrapolator.

8.50 Derive the asymptotic expression in Equation 8.373 for the fractional error in extrapolator frequency response gain.

8.51 Estimate the fractional gain and phase errors from the graph in Example 8.10 (Figure 8.79) for the case when $\omega T = 0.5$ rad. Compare the results with the exact values given in Equations 8.372 and 8.375.

Hint: Run "*Chap8_Ex5_2.m*" and enlarge the plots to facilitate the measurements needed to estimate the respective errors.

8.52 Repeat Example 8.10 part (b) using the second-order system

$$\frac{d^2x}{dt^2} + 2\xi\omega_n \frac{dx}{dt} + \omega_n^2 x = K\omega_n^2 u \quad (K = 2, \omega_n = 10\,\text{rad/s})$$

in place of the first-order system. Plot the exact and simulated responses for $\zeta = 0.1$, $\zeta = 0.707$, and $\zeta = 2$ when $\omega T = 0.1$ rad, $\omega T = 0.25$ rad, $\omega T = 0.5$ rad, and $\omega T = 1$ rad. Note that there are a total of 12 distinct combinations of ζ and ωT.

8.53 Derive the analytical expression for $\hat{T}(t)$ in Equation 8.411.

8.54 Run the M-file "Chap8_Ex5_3.m."
(a) Zoom in the bottom graph in Figure 8.91 in order to accurately measure the peak amplitudes (with respect to $T_0 = 50°$ F) after the transient response has died out. Calculate the fractional error in $|G(e^{j\omega t})|$ and compare to the value estimated from Figure 8.87.
(b) Measure the time phase shift in $\hat{T}(t)$ and $\hat{T}_n$ with respect to the input $T_0(t)$, and convert the value to degrees. Compare your answer with the phase angle estimated from Figure 8.87.

8.55 Rework Example 8.11 and include the real-time, modified Euler integrator given in Equations 8.323 and 8.324.

8.56 The classic RK-4 integrator introduced in Section 6.2 is incompatible with real-time simulation. Choose a step size of $T = 0.01$ s for the RK-4 integrator given by

$$k_1 = f(x_n, u_n), \quad x_{n+1/2} = x_n + 0.5Tk_1$$

$$k_2 = f(x_{n+1/2}, \hat{u}_{n+1/2}), \quad \hat{x}_{n+1/2} = x_n + 0.5Tk_2$$

$$k_3 = f(\hat{x}_{n+1/2}, u_{n+1/2}), \quad \hat{x}_{n+1/2} = x_n + 0.5Tk_3$$

$$k_4 = f(\hat{x}_{n+1}, \hat{u}_{n+1})$$

$$x_{n+1} = x_n + \frac{T}{6}(k_1 + 2k_2 + 2k_3 + k_4)$$

to simulate the response of the system $dx/dt = x + u$ when the input is given by $u = u(t) = \sin 25t$, $t \geq 0$. The initial condition $x(0) = 0$. Compute $\hat{u}_{n+1/2}$ and $\hat{u}_{n+1}$ based on linear extrapolation through the points (t_{n-1}, u_{n-1}) and (t_n, u_n). Plot the exact solution and the simulated response on the same graph.

8.6 ADDITIONAL METHODS OF APPROXIMATING CONTINUOUS-TIME SYSTEM MODELS

Several additional methods for simulating the dynamics of continuous-time systems are presented in this section. Explanations of each are followed by the application of the methods to a linear continuous-time system to produce the z-domain transfer function, difference equations, and frequency response functions of the resulting discrete-time systems.

8.6.1 SAMPLING AND SIGNAL RECONSTRUCTION

A special case of this method was introduced briefly in Exercise 4.74. A discrete-time system to approximate an LTI continuous-time system can be synthesized by sampling the continuous-time input and then reconstituting the input using a reconstruction process. The reconstructed signal is applied to the LTI continuous-time system. Finally, the output is sampled to produce a discrete-time signal. The process is illustrated in Figure 8.92.

The sampled values u_k, $k = 0, 1, 2, \ldots$ can be used to reconstruct a piecewise continuous approximation to $u(t)$ in different ways. The simplest approach is to use a zero-order hold (ZOH) circuit, which generates a zero-order polynomial fit through the sampled values to produce the piecewise constant staircase function $\tilde{u}(t)$ shown in Figure 8.93. A single value of u_k, $k = 0, 1, 2, \ldots$ in each interval is all that is required to reconstruct the continuous-time signal approximation for that interval.

The piecewise constant function $\tilde{u}(t)$ can be decomposed into a series of rectangular pulses as shown in Figure 8.94.

Expressing $\tilde{u}(t)$ in terms of the unit step function $\hat{u}(t)$,

$$\tilde{u}(t) = u_0[1 - \hat{u}(t - T)] + u_1[\hat{u}(t - T) - \hat{u}(t - 2T)] + u_2[\hat{u}(t - 2T) - \hat{u}(t - 3T)] + \cdots \quad (8.412)$$

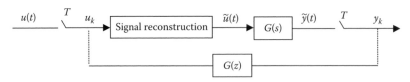

FIGURE 8.92 Sampling and signal reconstruction to approximate a linear time-invariant continuous-time system.

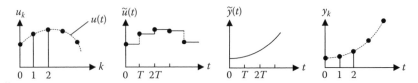

FIGURE 8.93 Representative signals in Figure 8.92 using a ZOH reconstruction device.

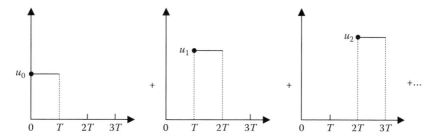

FIGURE 8.94 ZOH output $\tilde{u}(t)$ shown as a sum of rectangular pulses.

Laplace transforming Equation 8.412 gives

$$\tilde{U}(s) = u_0 \left[\frac{1}{s} - \frac{e^{-Ts}}{s} \right] + u_1 \left[\frac{e^{-Ts}}{s} - \frac{e^{-2Ts}}{s} \right] + u_2 \left[\frac{e^{-2Ts}}{s} - \frac{e^{-3Ts}}{s} \right] + \cdots \qquad (8.413)$$

The output of the continuous-time system with transfer function $G(s)$ is

$$\tilde{y}(t) = \mathcal{L}^{-1}\{\tilde{Y}(s)\} = \mathcal{L}^{-1}\{G(s)\tilde{U}(s)\} \qquad (8.414)$$

$$= \mathcal{L}^{-1}\left\{ G(s)u_0 \left[(1 - e^{-Ts}) + u_1 e^{-Ts}(1 - e^{-Ts}) + u_2 e^{-2Ts}(1 - e^{-Ts}) + \cdots \right] \frac{1}{s} \right\} \qquad (8.415)$$

The discrete-time output y_k consists of the sampled values $\tilde{y}(t)|_{t=kT}$, $k = 0, 1, 2, \ldots$ $Y(z)$ is obtained by z-transforming Equation 8.415 after setting $z = e^{Ts}$, resulting in

$$Y(z) = \mathfrak{z}\left\{ \left[u_0(1 - z^{-1}) + u_1 z^{-1}(1 - z^1) + u_2 z^{-2}(1 - z^{-1}) + \cdots \right] \mathcal{L}^{-1}\left\{ \frac{G(s)}{s} \right\} \right\} \qquad (8.416)$$

$$= \mathfrak{z}\left\{ \left[u_0 + u_1 z^{-1} + u_2 z^{-2} + \cdots \right](1 - z^{-1}) \mathcal{L}^{-1}\left\{ \frac{G(s)}{s} \right\} \right\} \qquad (8.417)$$

The inside bracketed expression is recognized as $U(z) = \mathfrak{z}\{u_k\}$. Hence,

$$Y(z) = U(z)(1 - z^{-1})\mathfrak{z}\left\{ \mathcal{L}^{-1}\left\{ \frac{G(s)}{s} \right\} \right\} \qquad (8.418)$$

where $\mathfrak{z}\{\mathcal{L}^{-1}\{G(s)/s\}\}$ represents the z-transform of the discrete-time signal obtained from uniform sampling of the continuous-time signal $\mathcal{L}^{-1}\{G(s)/s\}$. The z-domain transfer function resulting from the sampling and ZOH reconstruction method illustrated in Figure 8.92 is given by

$$G(z) = \frac{Y(z)}{U(z)} = (1 - z^{-1})\mathfrak{z}\left\{ \mathcal{L}^{-1}\left\{ \frac{G(s)}{s} \right\} \right\} \qquad (8.419)$$

The errors resulting from the use of Equation 8.419 are related to the signal reconstruction process. As you might expect, properties of the input $u(t)$, sampling interval T, and the method of reconstructing the input from the sampled values u_k, $k = 0, 1, 2, \ldots$ play a central role in the process.

We now illustrate the application of Equation 8.419 in finding a discrete-time system approximation of a second-order continuous-time system.

Example 8.12

Consider an underdamped second-order system with damping ratio $\zeta = 1/\sqrt{10}$, natural frequency $\omega_n = \sqrt{10}$ rad/s, and steady-state gain of unity.

(a) Find the z-domain transfer function and difference equation of the discrete-time system approximation. Leave your answer in terms of the sampling period T.
(b) Input to the continuous-time system is $u(t) = 5(1 - e^{-2t})$, $t \geq 0$. Find the continuous-time system response $y(t)$, $t \geq 0$.
(c) Plot the continuous-time system response $y(t)$ and the discrete-time approximation y_k, $k = 0$, 1, 2, ... for $T = 0.05$, 0.1, 0.25, 0.5 s.

(a) The transfer function of the continuous-time system is

$$G(s) = \frac{k\omega_n^2}{s^2 + 2\zeta\omega_n s + \omega_n^2} = \frac{10}{s^2 + 2s + 10} \tag{8.420}$$

$$\frac{G(s)}{s} = \frac{10}{s(s^2 + 2s + 10)} = \frac{1}{s} - \frac{s + 2}{s^2 + 2s + 10} \tag{8.421}$$

$$\mathcal{L}^{-1}\left\{\frac{G(s)}{s}\right\} = 1 - e^{-t}\left(\cos 3t + \frac{1}{3}\sin 3t\right) \tag{8.422}$$

From Table 4.4,

$$\mathfrak{z}\{1\} = \frac{z}{z - 1} \tag{8.423}$$

$$\mathfrak{z}\{e^{-kT}\cos 3kT\} = \frac{z^2 - (e^{-T}\cos 3T)z}{z^2 - (2e^{-T}\cos 3T)z + e^{-2T}} \tag{8.424}$$

$$\mathfrak{z}\{e^{-kT}\sin 3kT\} = \frac{(e^{-T}\sin 3T)z}{z^2 - (2e^{-T}\cos 3T)z + e^{-2T}} \tag{8.425}$$

Using Equations 8.423 through 8.425 in Equation 8.419 for $G(z)$ results in (after simplification)

$$G(z) = \frac{b_1 z + b_2}{z^2 + a_1 z + a_2} \tag{8.426}$$

$$b_1 = 1 - e^{-T}\left(\cos 3T + \frac{1}{3}\sin 3T\right), \quad b_2 = e^{-2T} - e^{-T}\left(\cos 3T - \frac{1}{3}\sin 3T\right) \tag{8.427}$$

$$a_1 = -2e^{-T}\cos 3T, \quad a_2 = e^{-2T} \tag{8.428}$$

Equation 8.426 leads to the difference equation of the discrete-time system

$$y_k + a_1 y_{k-1} + a_2 y_{k-2} = b_1 u_k + b_2 u_{k-1} \tag{8.429}$$

(b) The continuous-time system response to the input $u(t)$ is obtained from

$$y(t) = \mathcal{L}^{-1}\{G(s)U(s)\} = \mathcal{L}^{-1}\left\{\frac{10}{s^2 + 2s + 10} \cdot 5\left(\frac{1}{s} - \frac{1}{s + 2}\right)\right\} \tag{8.430}$$

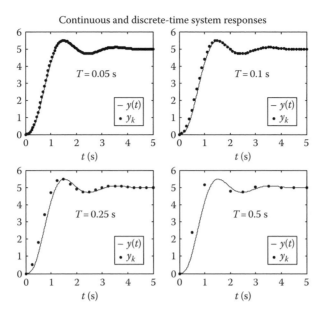

FIGURE 8.95 Illustration of "sample and ZOH reconstruction" method.

Partial fraction expansion of the terms in brackets followed by inverse Laplace transformation leads to

$$y(t) = 5 - 5e^{-2t} - \frac{10}{3}e^{-t}\sin 3t, \quad t \geq 0 \tag{8.431}$$

(c) The MATLAB M-file "Chap8_Ex6_1.m" includes statements to solve Equation 8.429 in recursive fashion. Figure 8.95 shows the continuous-time system response and the discrete-time response when $T = 0.05, 0.1, 0.25, 0.5$ s.

For signals that are not band limited such as the input $u(t) = 5(1 - e^{-2t})$, a good rule of thumb is to sample 10 times faster than the shortest time constant ($\tau = 0.5$ s in this case). The top left graph in Figure 8.95 corresponds to $T = \tau/10 = 0.05$ s, and the agreement between the continuous-time and discrete-time responses is excellent.

The outputs of the ZOH for the two extremes ($T = 0.05$ and 0.5 s) are shown in Figure 8.96, illustrating the importance of the sampling process.

The ZOH has characteristics similar to a low-pass filter. To see this, suppose the first sampler in Figure 8.92 produces a train of impulses of strength $u(kT)$ at the sampling instants kT, $k = 0, 1, 2,\ldots$ instead of the discrete-time signal $u_k = u(kT)$, $k = 0, 1, 2,\ldots$. Knowing the output of the ZOH is u_k, $kT \leq t < (k+1)T$ implies that the ZOH is effectively integrating the kth impulse for $kT \leq t < (k+1)T$. The situation is portrayed in Figure 8.97. The transfer function of the ZOH is therefore

$$G_{ZOH}(s) = \frac{1 - e^{-Ts}}{s} \tag{8.432}$$

Keep in mind that the impulse sampler is a mathematical fiction that allows the zero-order hold to be modeled by the continuous-time transfer function in Equation 8.432.

The frequency response function is obtained by replacing s with $j\omega$ in Equation 8.432.

$$G_{ZOH}(j\omega) = \frac{1 - e^{-j\omega T}}{j\omega} \tag{8.433}$$

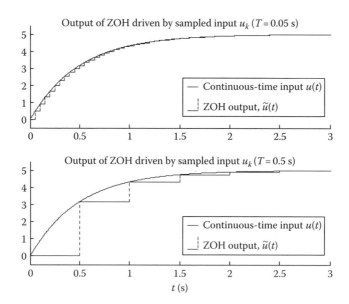

FIGURE 8.96 Effect of sampling rate on ZOH reconstruction of input $u(t)$.

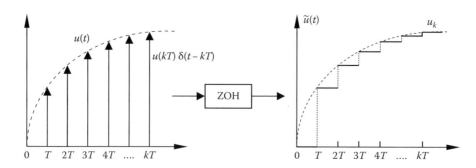

FIGURE 8.97 Impulse sampler feeding ZOH device.

Equation 8.432 can be manipulated into the form (Kuo 1980)

$$G_{ZOH}(j\omega) = T\frac{\sin(\omega T/2)}{\omega T/2}e^{-j(\omega T/2)} \tag{8.434}$$

$$= \left(\frac{2\pi}{\omega_s}\right)\frac{\sin \pi(\omega/\omega_s)}{\pi(\omega/\omega_s)}e^{-j\pi(\omega/\omega_s)} \tag{8.435}$$

where $\omega_s = 2\pi/T$ is the sampling frequency. Equation 8.434 reveals that the ZOH introduces a half sample period $(T/2)$ delay, which explains the need for choosing T small when the input contains significant high-frequency components.

The magnitude and phase of $G_{ZOH}(j\omega)$ are shown in Figure 8.98 for the case where $T = 0.05$ s and $\omega_s = 2\pi/T = 125.67$ rad/s. Note the DC gain $|G_{ZOH}(j0)| = T$.

For band-limited inputs with cut-off frequency ω_0, the minimum sampling frequency is $\omega_s = 2\omega_0$. The actual sampling period should be chosen to minimize the attenuation of $G_{ZOH}(j\omega)$

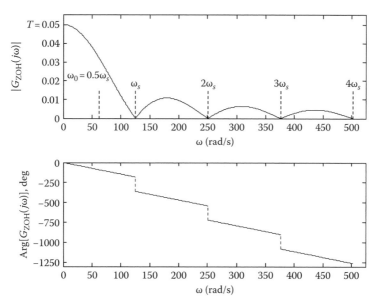

FIGURE 8.98 Frequency response of $G_{ZOH}(j\omega)$.

over the information band $(0, \omega_0)$. Furthermore, additive noise components above the cutoff frequency will also be passed, since there is no sharp drop in attenuation at ω_0.

The "c2d" function in the MATLAB control system toolbox introduced in Section 4.10 supports sampling and ZOH signal reconstruction to find the z-domain transfer function given in Equation 8.419. The syntax for calling the "c2d" function using ZOH approximation is `sysd = c2d(sysc,T,'zoh')` where ``sysc`` is created using the control system toolbox command "tf" to represent the continuous-time transfer function.

8.6.2 First-Order Hold Signal Reconstruction

More accurate signal reconstruction methods are possible using polynomial fits through several data points, resulting in different expressions for the z-domain transfer function $G(z)$. The output of a first-order hold circuit that approximates the sampled continuous-time signal by a sequence of linear functions is shown in Figure 8.99.

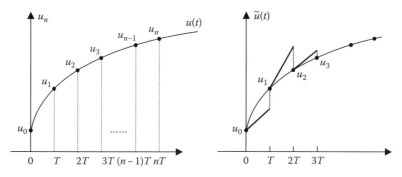

FIGURE 8.99 First-order hold reconstruction of a sampled continuous-time signal.

The analytical expression for the piecewise continuous output of the first-order hold is given by

$$\tilde{u}(t) = u_n + \frac{u_n - u_{n-1}}{T}(t - nT), \quad nT \le t < (n+1)T \quad (n = 0, 1, 2, \dots) \tag{8.436}$$

where u_{-1} is assumed to be zero. A derivation of $G(z)$ based on a first-order hold approximation is possible using a similar approach to the derivation leading to the z-domain transfer function in Equation 8.419 using the zero-order hold approximation. However, it is quite laborious and unnecessary, since the "c2d" function includes the first-order hold approximation method. The approximation is invoked by issuing also the command sysd = c2d(sysc, T, 'foh').

8.6.3 MATCHED POLE-ZERO METHOD

Another approach to developing a discrete-time approximation to a continuous-time system is by the process of matching the z-plane poles and zeros to their s-plane counterparts. This method can be applied to any asymptotically stable, LTI system with nonzero steady-state gain.

Consider an nth-order, stable, LTI system with transfer function $G(s)$. Uniform sampling every T s of the system's impulse response produces a discrete-time signal from an equivalent nth-order discrete-time system with z-domain transfer function $G(z)$. The n poles of $G(z)$ are obtained by a mapping of the s-plane poles according to

$$z_i = e^{s_i T}, \quad i = 1, 2, \dots, n \tag{8.437}$$

Two examples of this are

$$G(s) = \frac{1}{s+a} \Rightarrow g(t) = \mathcal{L}^{-1}\{G(s)\} = e^{-at} \tag{8.438}$$

$$gk = g(kT) = e^{-akT} \Rightarrow G(z) = \mathfrak{z}\{gk\} = \frac{z}{z - e^{-aT}} \tag{8.439}$$

$$G(s) = \frac{s+\alpha}{(s+\alpha)^2 + \beta^2} = \frac{s+\alpha}{[s - (\alpha + j\beta)][s - (\alpha - j\beta)]} \tag{8.440}$$

$$g(t) = \mathcal{L}^{-1}\{G(s)\} = e^{-\alpha T}\cos \beta T \tag{8.441}$$

$$g_k = g(kT) = e^{-akT}\cos \beta kT \tag{8.442}$$

$$G(z) = \frac{z - e^{-\alpha T}\cos \beta T}{z^2 - (e^{-aT}\cos \beta T)z + e^{-2\alpha T}} \tag{8.443}$$

$$= \frac{z - e^{-\alpha T}\cos \beta T}{[z - e^{-\alpha + j\beta)T}][z - e^{-(\alpha - j\beta)T}]} \tag{8.444}$$

When zeros of $G(s)$ are present as in Equation 8.440, they are not mapped into zeros of $G(z)$ according to Equation 8.437. However, in the matched pole-zero method, a discrete-time transfer function is created with the poles and zeros of $G(z)$ determined from Equation 8.437.

Two additional steps complete the process. First, the term z^{n-m}, where m is the order of the numerator polynomial of $G(s)$, is inserted in the numerator of $G(z)$ (Smith 1987). An alternative approach inserts the term $(z+1)^{n-m}$ in the numerator of $G(z)$. Second, the gains of the two transfer functions are matched at some frequency by appropriate choice of a gain term in $G(z)$.

The matched pole-zero method is illustrated for the second-order system in Example 8.12. The poles of $G(s)$ in Equation 8.420 are $s_{1,2} = \alpha \pm j\beta$, $(\alpha = -1, \beta = 3)$. Since there are no zeros of $G(s)$, $m = 0$ and the z-domain transfer function $G(z)$ is of the form

$$G(z) = K' \frac{z^2}{(z - e^{s_1 T})(z - e^{e^{s_2} T})} \tag{8.445}$$

$$= K' \frac{z^2}{[z - e^{(\alpha+j\beta)T}][z - e^{(\alpha-j\beta)T}]} \tag{8.446}$$

$$= K' \frac{z^2}{z^2 - 2(e^{\alpha T} \cos \beta T)z + e^{2\alpha t}} \tag{8.447}$$

Substituting the given values of α and β into Equation 8.447 results in

$$G(z) = K' \frac{z^2}{z^2 - 2(e^{-T} \cos 3T)z + e^{-2T}} \tag{8.448}$$

The DC gains of $G(s)$ and $G(s)$ are

$$G(s)|_{s=0} = \frac{10}{s^2 + 2s + 10}\bigg|_{=0} = 1 \tag{8.449}$$

$$G(z)|_{z=1} = K' \frac{z^2}{z^2 - 2(e^{-T} \cos 3T)z + e^{-2T}}\bigg|_{z=1} \tag{8.450}$$

$$= K' \frac{1}{1 - 2e^{-T} \cos 3T + e^{-2T}} \tag{8.451}$$

Equating the DC gains gives

$$K' = 1 - 2e^{-T} \cos 3T + e^{-2T} \tag{8.452}$$

Substituting K' in Equation 8.452 into Equation 8.448 gives

$$G(z) = \frac{(1 - 2e^{-T} \cos 3T + e^{-2T})z^2}{z^2 - 2(e^{-T} \cos 3T)z + e^{-2T}} \tag{8.453}$$

A frequency response plot of the continuous-time system transfer function $G(s)|_{s=j\omega}$ and the approximating discrete-time system transfer functions $G(z)|_{z=e^{j\omega T}}$ based on the two methods are shown in Figure 8.100 for sampling times of $T = 0.05$ s and $T = 0.25$ s, respectively. $G_1(e^{j\omega T})$ refers to the discrete-time transfer function in Equation 8.426 arrived at by using the ZOH method, and $G_2(e^{j\omega T})$ corresponds to the one in Equation 8.453 obtained using the matched pole-zero method.

The plots extend from zero (DC) to the Nyquist frequency (π/T), which is 62.83 rad/s for $T = 0.05$ s and 12.57 rad/s when $T = 0.25$ s. An accurate (magnitude and phase) approximation of the continuous-time system frequency response characteristics is possible using the ZOH approximation method or the matched pole-zero technique with $T = 0.05$ s for frequencies up to around 5 rad/s. The magnitude functions for both discrete-time systems and the continuous-time system are nearly identical over the entire range of frequencies shown for $T = 0.05$ s.

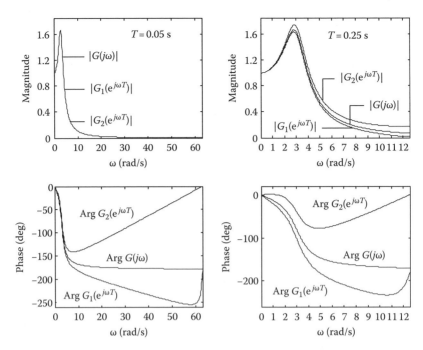

FIGURE 8.100 Frequency response of continuous-time and approximate discrete-time systems.

The "c2d" function in the MATLAB control system toolbox implements a "modified matched pole-zero" approximation. A $(z + 1)^{(n-m)-1}$ term is inserted in the numerator where m and n are the orders of the numerator and denominator of $G(s)$. The resulting $G(z)$ will contain an $(n - 1)$st-order polynomial in the numerator. The current output of the nth-order discrete-time system y_k depends on outputs $y_{k-1}, y_{k-2}, \ldots, y_{k-n}$ and most importantly only on the past inputs $u_{k-1}, u_{k-2}, \ldots, u_{k-n}$. With an nth-order term in the numerator of $G(z)$, y_k will depend on the current input u_k as well. In real-time applications, the current output would have to wait for an A/D read, implementation of the difference equation followed by a D/A write to hardware, all performed in theoretically zero time. The problem is mitigated to a large extent when these operations consume a small fraction of the sample time T.

The matched pole-zero and modified matched pole-zero methods are applied to the continuous-time transfer function in Equation 8.420 in "*Chap8_matched_pole.m*" with a sampling time of $T = 0.05$ s. Results are as follows:

$$\text{Matched pole-zero: } G(z) = \frac{0.0237\, z^2}{z^2 - 1.8811\, z - 0.9048} \tag{8.454}$$

$$\text{Modified matched pole-zero: } G(z) = \frac{0.01187(z + 1)}{z^2 - 1.8811\, z + 0.9048} \tag{8.455}$$

An important property of the ZOH approximation and matched pole-zero methods is related to the stability of the resulting discrete-time systems. Note that the characteristic polynomials of the transfer functions $G(z)$ in Equations 8.426, 8.454, and 8.455 are identical, namely, $z^2 - 2(e^{-T} \cos 3T)$ $z + e^{-2T}$. The continuous-time system poles are mapped to the z-plane according to Equation 8.437 in each case. Consequently, continuous-time system poles in the left-hand plane are mapped to the interior of the Unit Circle in the z-plane and, therefore, produce stable discrete-time modes as well.

8.6.4 BILINEAR TRANSFORM WITH PREWARPING

The use of trapezoidal integration to discretize a continuous-time system with transfer function $G(s)$ was discussed in Section 4.7. The z-domain transfer function of the discrete-time system approximation was shown to be

$$G(z) = G(s)\Big|_{s \leftarrow \frac{2}{T}\left(\frac{z-1}{z+1}\right)} \tag{8.456}$$

An alternate derivation of Equation 8.456 is based on the transformation $z = e^{Ts}$, which can be written in terms of a pair of infinite series expansions according to

$$z = \frac{e^{(T/2)s}}{e^{(-T/2)s}} = \frac{1 + (Ts/2) + (1/2!)(Ts/2)^2 + (1/3!)(Ts/2)^3 + \cdots}{1 - (Ts/2) + (1/2!)(Ts/2)^2 - (1/3!)(Ts/2)^3 + \cdots} \tag{8.457}$$

Truncating both series after the linear term gives

$$z = \frac{1 + (T/2)s}{1 - (T/2)s} \tag{8.458}$$

Solving for s in Equation 8.458 gives

$$s = \frac{2}{T}\left(\frac{z-1}{z+1}\right) \tag{8.459}$$

Equation 8.459 is known as the bilinear transform, and the process for obtaining the discrete-time approximation is commonly referred to as Tustin's method. The left half of the s-plane consisting of points $s = \sigma + j\omega$, $-\infty < \sigma < 0$ is mapped into the interior of the Unit Circle, $|z| < 1$. Consequently, the method produces stable discrete-time systems regardless of the step size T provided the continuous-time system is stable. For this reason, it is among the most popular methods for simulation of continuous-time systems.

The frequency response of discretized systems obtained using the bilinear transform in Equation 8.459 is examined by considering the image of points along the $j\omega$ axis, that is, $s = j\omega$, $-\infty < \omega < \infty$. From Equation 8.458 with $s = j\omega$,

$$z = \frac{1 + (T/2)j\omega}{1 - (T/2)j\omega} = 1e^{j\theta} \tag{8.460}$$

where

$$\theta = 2\tan^{-1}\left(\frac{\omega T}{2}\right), \quad -\infty < \omega < \infty \tag{8.461}$$

The entire length of the $j\omega$ axis from $-j\infty$ (pt A) to $j\infty$ (pt C) is mapped one-to-one into the Unit Circle starting at $\theta = -\pi$ (pt A') to $\theta = \pi$ (pt C') (see Figure 8.101).

Compressing the $j\omega$ axis into the Unit Circle according to Equation 8.461 results in a warping of the frequency response. This can be overcome by prewarping a critical frequency, say ω_0, in the s-plane before applying the bilinear transform to the continuous-time transfer function $H(s)$.

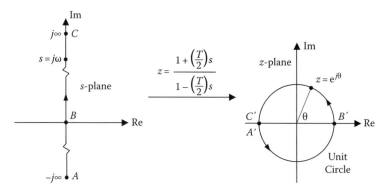

FIGURE 8.101 Bilinear transform mapping of the imaginary axis in the s-plane.

Frequency response of the resulting z-domain transfer function $\hat{H}(z)$ and the continuous-time transfer function $H(s)$ will agree at the selected critical frequency, that is,

$$H(s)\big|_{s=j\omega_0} = \hat{H}(z)\big|_{z=e^{j\omega_0 T}} \tag{8.462}$$

The prewarped critical frequency $\hat{\omega}_0$ is obtained from (Jacquot 1981)

$$\hat{\omega}_0 = \frac{2}{T}\tan\left(\frac{\omega_0 T}{2}\right) \tag{8.463}$$

The following example illustrates the process of prewarping a second-order continuous-time filter transfer function to force agreement in the frequency response functions at the natural frequency of the filter.

Example 8.13

An analog filter is described by

$$H(s) = \frac{s + \omega_n^2}{s^2 + 2\zeta\omega_n s + \omega_n^2} \quad (\zeta = 0.25, \omega_n = 1000\,\text{rad/s}) \tag{8.464}$$

(a) Find $H(z)$ using the bilinear transform with a sampling time of $T = 0.001$ s.
(b) Find the transfer function $\hat{H}(s)$ resulting from prewarping the natural frequency ω_n.
(c) Find $\hat{H}(z)$ using the bilinear transform on the prewarped transfer function $\hat{H}(s)$.
(d) Plot the magnitude and phase of $H(s)$, $H(z)$, and $\hat{H}(z)$ on the same graph and comment on the results.

(a) Substituting the filter parameter values ζ and ω_n into Equation 8.464 gives

$$H(s) = \frac{s + 10^6}{s^2 + 500s + 10^6} \tag{8.465}$$

$H(z)$ is obtained by replacing s with the right-hand side of Equation 8.459. The MATLAB control system toolbox functions "BILINEAR" and "c2d" are both designed to facilitate implementation of the bilinear transform. One form of the function "BILINEAR," which is applicable in this case, is

$$[\text{NUMd, DENd}] = \text{BILINEAR}(\text{NUM, DEN, FS})$$

where the parameters "NUM" and "DEN" are row vectors describing the numerator and denominator of $H(s)$ in descending powers of s, and "FS" is the sampling frequency in Hz. The numerator and denominator of $H(z)$ are specified in the output arrays "NUMd" and "DENd."

The M-file "*Chap8_Ex6_2.m*" contains the call to the "BILINEAR" function and the result is

```
NUMd = 0.1670 0.333 0.1663
DENd = 1.000 -1.0000 0.6667
```

Invoking the "c2d" function with "sysd = c2d(sysc, T, 'tustin')" results in

```
Transfer function:

0.167z^2 + 0.3333z + 0.1663
----------------------
z^2 - z + 0.6667
sampling time = 0.001 sec
```

in agreement with the results obtained using the "BILINEAR" command.

(b) Prewarping the natural frequency using Equation 8.463 yields

$$\hat{\omega}_n = \frac{2}{T} \tan\left(\frac{\omega_n T}{2}\right) = \frac{2}{0.001} \tan\left(\frac{1000(0.001)}{2}\right) = 1092.6 \, \text{rad/s} \tag{8.466}$$

The prewarped transfer function is therefore

$$\hat{H}(s) = \frac{s + \hat{\omega}_n^2}{s^2 + 2\zeta\hat{\omega}_n s + \hat{\omega}_n^2} = \frac{s + 1.1938 \times 10^6}{s^2 + 546.3s + 1.1938 \times 10^6} \tag{8.467}$$

(c) $\hat{H}(z)$ results from the bilinear transformation applied to the transfer function in Equation 8.467. Equivalently, the MATLAB statement

```
sysd_prewarp = c2d(sysc, T, 'prewarp', w_crit)
```

can be found in "*Chap8_Ex6_2.m*" with "w_crit" set equal to the natural frequency $\omega_n = 1000$ rad/s. The resulting z-domain transfer function appears as

```
0.1902 z^2 + 0.3798 z + 0.1896
------------------------------
z^2 - 0.8928z + 0.6524
```

(d) The magnitude and phase plots for the continuous-time system frequency response $H(j\omega)$ and the two discrete-time systems (with and without prewarping the natural frequency $\omega_n = 1000$ rad/s) are shown in Figure 8.102. As expected, Equation 8.462 is verified at the critical frequency of 1000 rad/s.

There is one additional method included in the "c2d" function for converting continuous-time models to discrete-time models. It is called the impulse invariant method. It is predicated on making the discrete-time system impulse response proportional to the sampled values of the continuous-time system impulse response function. The syntax for implementing this method is "sysd = c2d(sysc, T, 'imp')."

EXERCISES

8.57 Implement the "c2d" function using the zero-order hold approximation in Example 8.12 and show that the results are consistent with Equations 8.426 through 8.428 when $T = 0.05$ s.

8.58 Derive the expression for $G_{FOH}(s)$, the transfer function of a first-order hold driven by an impulse sampler. Plot the frequency response for the case when $T = 0.05$ s, and compare the result with the frequency response plot of a ZOH with $T = 0.05$ s shown in Figure 8.98.

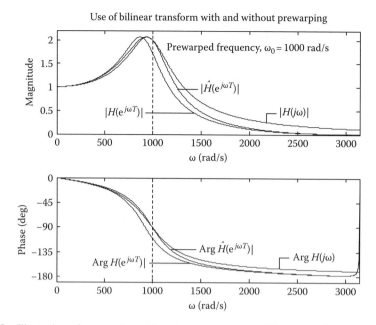

FIGURE 8.102   Illustration of prewarping critical frequency prior to bilinear transform.

8.59 Redo Example 8.12 using
    (a) The first-order hold approximation and compare the results with the ZOH approximation method
    (b) The bilinear transform method and compare the results with the ZOH approximation method

8.60 Apply the bilinear transform to the prewarped continuous-time transfer function $\hat{H}(s)$ in Equation 8.46 and compare the result with z-domain transfer function $\hat{H}(z)$ given in part (c) of Example 8.13.

8.61 The circuits in Figure E8.61 are low- and high-pass filters.

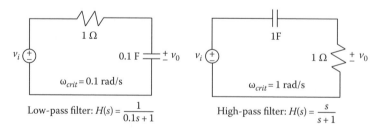

FIGURE E8.61

    (a) Find the z-domain transfer function of the approximating discrete-time filters based on the bilinear transform with sample time $T = 0.005$ s for the low-pass filter and $T = 0.05$ s for the high-pass filter.
    (b) Repeat part (a) after first prewarping the critical frequencies of each filter.
    (c) Compare the frequency responses of the continuous-time and discrete-time filters.
    (d) Find and plot the unit step responses of the low-pass filter and its discrete-time approximations.
    (e) Repeat part (d) for the high-pass filter.

8.62 For the transfer function $G(s) = 10/(s^2 + 2s + 10)$ in Example 8.12,
  (a) Convert to state-space form $\dot{x} = A\underline{x} + Bu, y = C\underline{x} + Du$ using the MATLAB function "tf2ss."
  (b) Convert the continuous-time state-space model to discrete-time form using the MATLAB function "c2dm." Choose the sample time $T = 0.05$ s and specify "zoh" as the method of approximation.
  (c) Obtain the unit step response by solving the discrete-time state model equations recursively and plot the results.

8.63 For the filters in Exercise 8.62,
  (a) Obtain discrete-time filter approximations to each using the impulse invariant method.
  (b) Compare the frequency response functions of the continuous- and discrete-time filters.
  (c) Compare the impulse response functions of the continuous- and discrete-time filters.

8.7 CASE STUDY: LEGO MINDSTORMS™ NXT

8.7.1 INTRODUCTION

In the November 2008 issue of *Mechanical Engineering*, the American Society of Mechanical Engineers surveyed its members for the trend they thought would have the most significant impact 10 years hence. In second place, with 26% of the responses, was Mechatronics—the integration of mechanical and electronic design. Incidentally, first place (28%) went to nanotechnology and microelectromechanical systems—devices that are demanded by mechatronicians.

The discipline of Mechatronics is a nexus of four technical sub-disciplines: mechanical systems, electronics systems, control systems, and computers. The intersection of mechanical and electronic systems is electromechanics; electronic and control systems intersect at control electronics; control systems and computers combine to form digital control systems; and finally, computers and mechanical systems form mechanical computer-aided design (CAD). Figure 8.103 displays these relationships graphically.

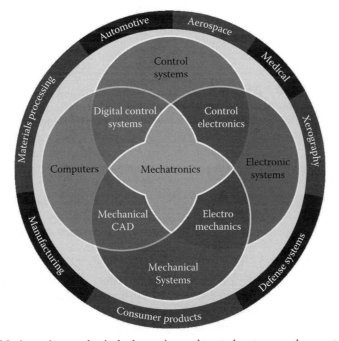

FIGURE 8.103 Mechatronics: mechanical, electronics, and control systems, and computers.

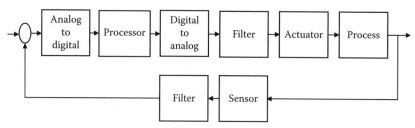

FIGURE 8.104   Feedback control system block diagram.

From a feedback control systems perspective, mechatronics is the implementation or realization of a controller design. A feedback control system involves either tracking a changing input or regulating a constant reference input. The block diagram for a generic feedback control system is given in Figure 8.104.

The feedback control system diagram begins with the reference input, which is fed into a comparator. The comparator measures the difference between the reference input and the feedback signal, generating an error signal. In order for the computer to process the error signal, it must first be converted by an A2D converter. Once the control signal is processed, it is converted by a D2A converter. It is this signal, fed into the actuator, that controls the process. Note that this signal may need to be amplified in order to actuate the controlling hardware. A sensor is connected to the process in order to measure the performance of the system. The sensor provides the feedback signal, which is used in the comparator. This loop is continually repeated for the process to be controlled. Simply stated, control design synthesis involves mathematically modeling and analyzing a physical process (e.g., missile airframe) and then designing a controller (e.g., acceleration autopilot). Simulink's graphical environment facilitates this "model-based design" approach. These steps of controller design synthesis are usually performed on the same host development platform, that is, a personal computer.

Beyond modeling, analysis, and design, mechatronics adds the implementation step. In this step, the controller design is realized on the actual (e.g., flight) hardware. Generated source code is compiled and assembled for a particular microprocessor that is executed for the digital controller design. Source code interfaces for actuators and sensors, called device drivers, are typically provided by vendors that manufacture these various pieces of hardware. An application that facilitates this extended development process is called an Integrated Development Environment (IDE). By adding MATLAB's Real-Time Workshop to the host's suite of tools, code generation, compilation, and assembly for a specific microprocessor are enabled. The repetitive process of making changes to the controller design in the Simulink model, generating, compiling, assembling, downloading, and running code on the microprocessor is known as rapid prototyping. This allows the engineer to build a little and test a little, thereby rooting out errors early in the development process and potentially avoiding the larger costs associated with redesigning the system late in the development process.

By combining the popular Lego Mindstorms™ NXT (henceforth referred to NXT) robotics platform with MATLAB's Simulink and Real-Time Worshop tools, this development process can be demonstrated end to end. Therefore, the remainder of this section is devoted to

- Product requirements, software download, and installation
- Creating a Simulink model that provides a noisy input signal and then running the "unfiltered" model on the NXT to observe how the noisy signal affects the physical motor
- Modifying the Simulink model by adding a discrete-time Kalman filter (DTKF) (Section 5.12), which filters the noisy input signal and then running the "filtered" model on the NXT observing the effect of the filtered signal on the physical motor

8.7.2 REQUIREMENTS AND INSTALLATION

The software and hardware requirements are

- MATLAB, Simulink, Real-Time Workshop, and Real-Time Workshop's Embedded Coder available from The Mathworks
- Cygwin™ and the GNU ARM™ compiler available as a download
- NXT hardware and corresponding device drivers available from Lego

Once The Mathworks software is installed and functioning properly, the third-party software download and installation (Cygwin™ and the GNU ARM™ compiler) is facilitated by a MATLAB m-file script. Cygwin™ is a UNIX shell environment that runs on Windows and the GNU ARM™ compiler compiles C source code for the ARM processor that runs on the NXT. The script is available from The Mathworks' Web site at http://www.mathworks.com by searching for "ECRobot Installer." (ECRobot is an abbreviation for Embedded Controller Robot.) Locate the hyperlink "Download the ECRobot installer" to download (ecrobot_installer_v1_2.zip at the time). Extract the files and then follow the README.pdf instructions. The ECRobotInstaller contains three MATLAB m-file scripts:

- A script "download_ecrobot_tools" to download all the necessary software. *Note:* Before running the script in Step 1: Automated Download of the README file, it may need to be edited to accommodate the current version of nxtOSEK.
- A script "install_ecrobot_tools" to configure and install the necessary software.
- A script "update_nxt_firmware" that updates the firmware on the NXT to run ARM binary files

Note: At the time of this writing, sg.exe had been removed from nxtOSEK. Therefore, sg.exe is obtained by downloading and extracting osek_os-1.1.lzh for nxtOSEK from the Web site http://lejos-osek.sourceforge.net/download.htm. Copy /toppers_osek/sg/sg.exe to the nxtOSEK/toppers_osek/sg directory.

At this point, follow Step 5: Verify that everything works as outlined in the README file. Note that the README file contains answers to commonly asked questions. If you have additional questions, please e-mail mindstorms@mathworks.com.

8.7.3 NOISY MODEL

In this section, a Simulink model is created that generates a noisy input signal, which drives an NXT motor. First, the model is built and simulated to view the noisy signal. Then, C source code is generated, compiled, assembled, downloaded, and run on the NXT to observe how the noisy signal affects the physical motor.

Knowing ahead of time that C source code will be generated with MATLAB's Real-Time Workshop, the model is architected such that the portion of the model that is generated into C source code exists within a function—where the function is driven by a scheduler. Upon starting Simulink, a new block set has been installed and added to the Simulink Library Browser called "ECRobot NXT Blockset." In this blockset, there is a block called "ExpFcnCalls Scheduler." This block generates function-call events according to the rate specified within the block parameters. For the model shown in Figure 8.105, this block generates function calls at the rate of 100 ms. The function-call scheduler expects to be connected to a demux block in case there are multiple functions being called by the scheduler. Even though there is only one function, a demux block is still necessary. The output of the demux block is the input into a subsystem block—which contains the function. However, at this top-level, a servo motor interface block (from the ECRobot NXT

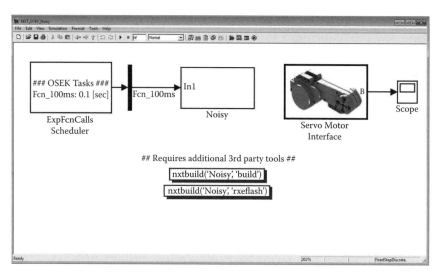

FIGURE 8.105 Top-level block diagram of the noisy model.

Blockset) is connected to a scope, so the (noisy) output can be viewed. The blocks nxtbuild('Noisy', 'build') and nxtbuild('Noisy','rxeflash') are annotation blocks with call-back functions enabled to execute the corresponding command in MATLAB.

By double clicking on the subsystem block named "Noisy," the function-call subsystem (from the standard Simulink library blockset) is seen as in Figure 8.106.

By double clicking on the function-call subsystem, the model that generates the noisy signal is seen as in Figure 8.107.

The creation of the function-call sSubsystem automatically places the f() block in this subsystem to indicate that the included elemental blocks are part of the function. A random number block with a mean of 32 and a variance of 32^2 generates the random signal. The saturation block limits possible signals to ± 100 as these are the limits of the NXT motor signals. The data type conversion is set to int8 to represent the signed 8-bit integer, that is, -128 to 127. (The andom number, saturation, and data type conversion blocks are all part of the standard Simulink library blockset.) Finally, the Servo Motor Write block (from the ECRobot NXT blockset) is connected to port B. Port B is the second output (top/left) from the Lego "brick" as seen in Figure 8.108

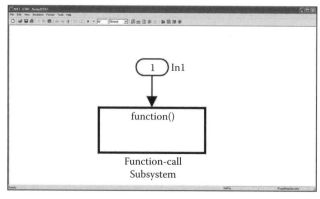

FIGURE 8.106 Function-call subsystem.

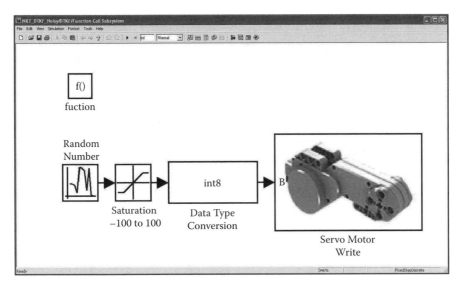

FIGURE 8.107 Noisy signal model.

FIGURE 8.108 Lego Mindstorms™ NXT "brick." (LEGO® and LEGO® Mindstorms® NXT™ are trademarks of the LEGO® Group, which does not sponsor nor endorse this book. This photo of the LEGO® Mindstorms® NXT™ brick is used here with permission. © 2010 The LEGO® Group.)

Upon running a Simulink simulation, the noisy data may be viewed from the scope block as seen in Figure 8.109.

Alternatively, the noisy data are available in the MATLAB Workspace as a variable "structure with time" named "noisy." A plot of this noisy data is shown in Figure 8.110.

The next part of this exercise is to generate, compile, assemble, download, and run C source code for the "noisy" function on the NXT to observe how the noisy signal affects the physical motor.

By clicking on the annotation block nxtbuild ('Noisy', 'build'), the Real-Time Workshop code generator is invoked, which creates C source code and corresponding header files from the Simulink function. The following text appears in MATLAB's Command Window.

Starting Real-Time Workshop build procedure for model: Noisy
Successful completion of Real-Time Workshop build procedure for model: Noisy
Generating ECRobot NXT scheduler file(s) for model: Noisy
Successful completion of ECRobot NXT scheduler file (s) generation for model: Noisy
Executing GNU-ARM toolchain for building executable . . .

Successful C source code and header file generation result in the Real-Time Workshop Report appearing as shown in Figure 8.111.

On the left side of the Real-Time Workshop Report window, hyperlinks indicate the various sections of the C source code related to the function from the Simulink model. In particular, by clicking on "Noisy.c," one can view portions of the code that correspond directly with the elemental blocks that constitute the function of the Simulink model. The rest of the messages in MATLAB's Command Window correspond to the build portion of compiling and assembling the binary image file named "Noisy.rxe."

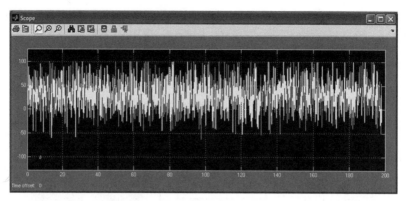

FIGURE 8.109 Noisy output viewed from the Scope Block.

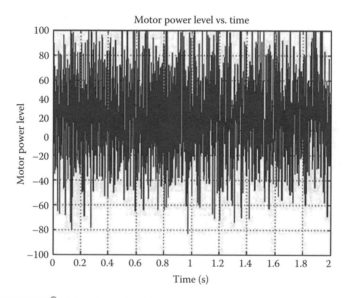

FIGURE 8.110 MATLAB® plot of the noisy data.

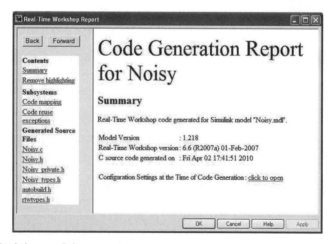

FIGURE 8.111 Real-time workshop report.

.
.
.

(many messages corresponding to the build process, i.e., compiling and assembling)

.
.
.

Generating binary image file: Noisy_rom.bin
Generating binary image file: Noisy_ram.bin
Generating binary image file: Noisy.rxe

Once the binary image file "Noisy.rxe" has been created, click on the annotation block nxtbuild ("Noisy," "rxeflash") to load the binary image into the flash memory of the NXT. For this part of the procedure, MATLAB's Command Window shows the following:

Execute NeXTTool for uploading a program to the enhanced NXT standard firmware:./ nxtprj/Noisy.rxe
Executing NeXTTool to upload Noisy.rxe...
Noisy.rxe = 26144
NeXTTool is terminated.
Note: NeXTTool is a utility that transfers files from the PC to the NXT.

At this time, the NXT is ready to run the noisy motor program. Be certain there is a motor connected to Port B on the brick. Upon running this program, the motor indeed runs erratically, exhibiting its response to the noisy input.

8.7.4 FILTERED MODEL

In this section, the noisy Simulink model is modified by adding a DTKF (Section 5.12), which filters the noisy input signal. The model is simulated in Simulink to view the filtered signal. Then, as before, C source code is generated, compiled, assembled, downloaded, and run on the NXT to observe how the physical motor responds to the filtered signal.

As shown in Figure 8.112, the top-level block for the filtered model is similar to that of the noisy model, except the name of the subsystem block has been changed to "Filtered" and the annotation blocks have been updated as well.

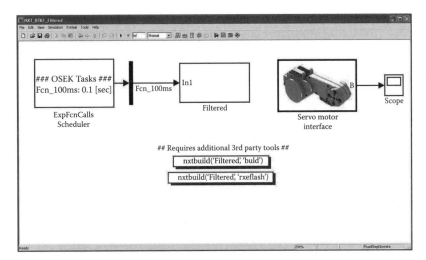

FIGURE 8.112 Top-level block diagram for the filtered model.

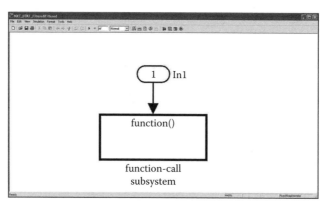

FIGURE 8.113 Function-call subsystem.

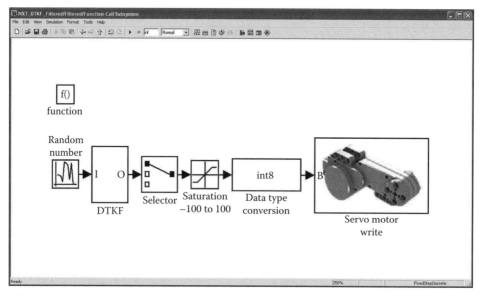

FIGURE 8.114 Filtered signal model.

By double clicking on the subsystem block named "Filtered," the function-call subsystem (from the standard Simulink library blockset) is seen as in Figure 8.113.

By double clicking on the function-call subsystem, the model that generates the filtered signal is seen as in Figure 8.114.

A subsystem block named "DTKF" has been added to the model in order to filter the noisy signal. This is the same DTKF that was developed in Section 5.12 for the meteorite. However, rather than setting the variables in MATLAB (as in Section 5.12), the parameters are set directly in the Simulink blocks. Also, while the DTKF had three states: position, velocity, and acceleration, the position of the meteorite was of primary interest in that example. Therefore, the first output of the DTKF (corresponding to position) is selected as the input to the saturation block. The major subsystems of the DTKF are the a priori and a posteriori calculations of the state and covariance matrix updates as seen in Figure 8.115.

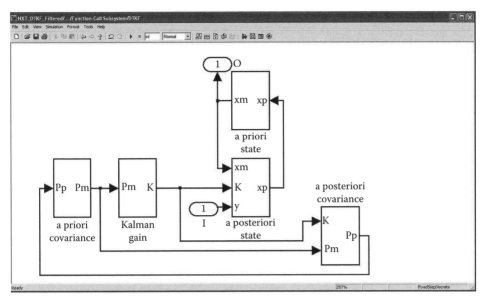

FIGURE 8.115   Discrete-time Kalman filter subsystems.

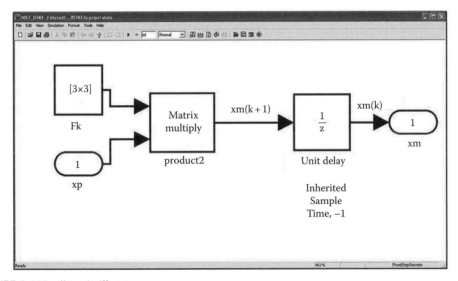

FIGURE 8.116   "A priori" state.

The details of the DTKF are shown in Figures 8.116 through 8.120. Notice in Figure 8.116 (a priori state) and 8.117 (a priori covariance) that the unit delay block inherits the sample time, that is, the function-call scheduler time, by setting sample time equal to −1 in the block properties.

Upon running a Simulink simulation, the filtered data may be viewed from the Scope Block as seen in Figure 8.121. Notice that the filtered value appears to be approximately 32, which was the mean of the random number block.

Alternatively, the filtered data are available in the MATLAB Workspace as a variable "structure with time" named "filtered." A plot of the filtered data is shown in 122.

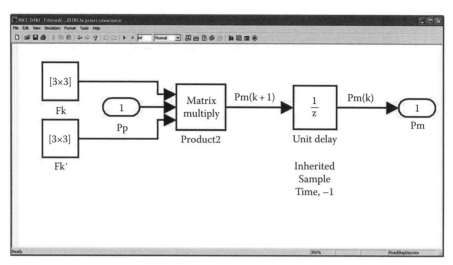

FIGURE 8.117 "A priori" covariance.

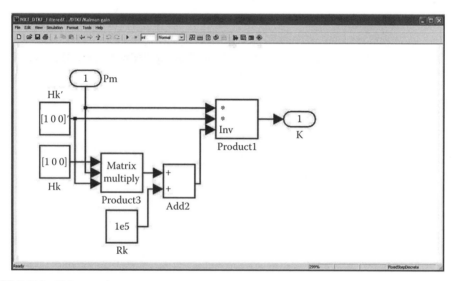

FIGURE 8.118 Kalman gain.

By clicking on the annotation block nxtbuild("Filtered," "build"), the Real-Time Workshop code generator is invoked, which creates C source code and corresponding header files from the Simulink function. The following text appears in MATLAB's Command Window.

Starting Real-Time Workshop build procedure for model: Filtered
Successful completion of Real-Time Workshop build procedure for model: Filtered
Generating ECRobot NXT scheduler file(s) for model: Filtered
Successful completion of ECRobot NXT scheduler file(s) generation for model: Filtered
Executing GNU-ARM toolchain for building executable . . .

Successful C source code and header file generation result in the Real-Time Workshop Report appearing as shown in Figure 8.123.

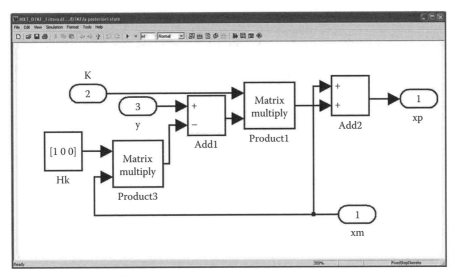

FIGURE 8.119 "A posteriori" state.

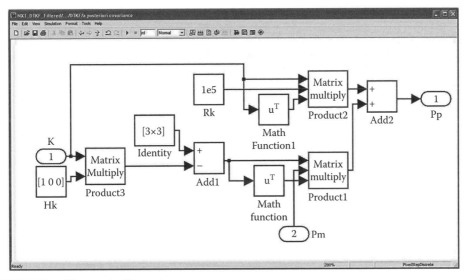

FIGURE 8.120 "A posteriori" covariance.

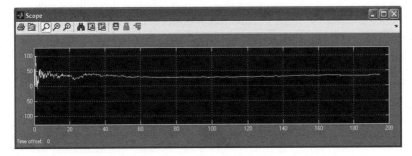

FIGURE 8.121 Filtered output viewed from the Scope Block.

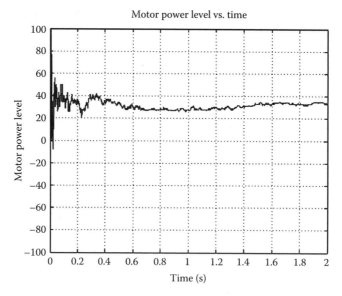

FIGURE 8.122　MATLAB plot of the filtered data.

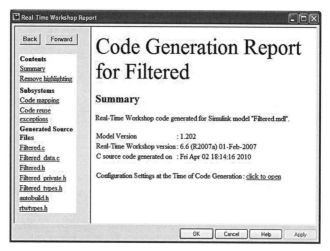

FIGURE 8.123　Real-time workshop report.

On the left side of the Real-Time Workshop Report window, hyperlinks indicate the various sections of the C source code related to the function from the Simulink model. In particular, by clicking on "Filtered.c," one can view portions of the code that correspond directly with the elemental blocks and the DTKF blocks that constitute the function of the Simulink model. The rest of the messages in MATLAB's Command Window correspond to the build portion of compiling and assembling the binary image file named "Filtered.rxe."

.
.
.

(many messages corresponding to the build process, i.e., compiling and assembling)

.

.

.

Generating binary image file: Filtered_rom.bin
Generating binary image file: Filtered_ram.bin
Generating binary image file: Filtered.rxe

Once the binary image file "Filtered.rxe" has been created, click on the annotation block nxtbuild ("Filtered,", "rxeflash") to load the binary image into the flash memory of the NXT. For this part of the procedure, MATLAB's Command Window shows the following:

Execute NeXTTool for uploading a program to the enhanced NXT standard firmware:./
    nxtprj/Filtered.rxe
Executing NeXTTool to upload Filtered.rxe . . .
Filtered.rxe = 28720
NeXTTool is terminated.

At this time, the NXT is ready to run the filtered motor program. As before, be sure there is a motor connected to Port B on the NXT brick. Upon running this program the motor indeed runs smoothly, exhibiting its response to the filtered input.

8.7.5 SUMMARY

In this Case Study, the build-a-little/test-a-little rapid prototyping development process was facilitated by an IDE. The technology (software and hardware) that enabled the IDE was made available by tools from The Mathworks (MATLAB, Simulink, Real-Time Workshop, and RTW's Embeddeed Coder), Cygwin™, GNU ARM™, and Lego (Mindstorms™ NXT).

Once the IDE is enabled with the technology, the intent was to demonstrate how easy it is to rapidly change the model from within Simulink, and then with two mouse clicks: generate, compile, assemble, download, and run the model on the NXT.

It is left as an exercise for the student to now unleash his/her creativity in developing various applications on this platform.

EXERCISE

8.64 While a noisy input signal was generated from within Simulink, modify the model such that a sensor provides the input. Examine the various sensors that are available physically, as well as from the ECRobot NXT Blockset in the Simulink Library Browser. For additional assistance, see the samples that were part of the software installation, for example, TestUltrasonicSensor. mdl.

References

Akai, T. J., *Applied Numerical Methods*, John Wiley & Sons, New York, 1994.

Allen, R. W. and T. Rosenthal, Systems technology/requirements for vehicle dynamics simulation models, Society of Automotive Engineers, SAE 941075, 1994.

Aycin, M. and R. Benekohal, Stability and performance of car-following models in congested traffic, *Journal of Transportation Engineering*, 127, 2–12, 2001.

Banks, J., J. S. Carson II et al., *Discrete-Event System Simulation*, 4th edn., Pearson/Prentice-Hall, Upper Saddle River, NJ, 2005.

Baruh, H., *Analytical Dynamics*, WCB/McGraw-Hill, Boston, MA, 1999.

Beltrami, E., *Mathematical Models in the Social and Biological Sciences*, Jones and Bartlett, Boston, MA, 1993.

Bender, J. G. and R. E. Fenton, A study of automatic car following, *IEEE Transactions on Vehicular Technology*, VT-18, 134–140, 1966.

Borse, G. J., *Numerical Methods with MATLAB*, PWS Publishing, Boston, MA, 1997.

Bracewell, R., *The Fourier Transform and Its Applications*, McGraw-Hill, New York, 1986.

Braun, M., *Differential Equations and Their Applications*, Springer-Verlag, New York, 1978.

Brown, D. and P. Rothery, *Models in Biology: Mathematics, Statistics and Computing*, John Wiley & Sons, West Sussex, U.K., 1993.

Bryson, A. E., *Dynamic Optimization*, Addison-Wesley, Menlo Park, CA, 1999.

Buckley, P., *Techniques of Process Control*, John Wiley & Sons, New York, 1964.

Burns, R. S., *Advanced Control Engineering*, Butterworth Heinemann, Oxford, U.K., 2001.

Cadzow, J. A., *Discrete-Time Systems—An Introduction with Interdisciplinary Applications*, Prentice-Hall, Englewood Cliffs, NJ, 1973.

Canova, B. S., P. H. Christensen, M. D. Lee, B. R. Tripp, M. H. Pack, and D. L. Pack, Simulation to support operational testing: A practical approach, in *Proceedings of the 1999 Winter Simulation Conference*, pp. 1071–1078, 1999.

Chapra, S. and R. Canalel, *Numerical Methods for Engineers with Software Programming Applications*, 4th edn., McGraw-Hill, New York, 2002.

Close, C. M., *Modeling and Analysis of Dynamic Systems*, 3rd edn., John Wiley & Sons, New York, 2002.

Converse, A. O., *Optimization*, Holt, Rinehart & Winston, New York, 1970.

Coutinho, F. A. B., L. F. Lopez, M. N. Burattini, and E. Massad, Modeling the natural history of HIV infection in individuals and its epidemiological implications, *Bulletin of Mathematical Biology*, 63, 1041–1062, 2001.

Culshaw, R. V. and S. Ruan, A delay differential equation model of HIV infection of CD4+ T cells, *Mathematical Biosciences*, 165, 27–39, 2000.

Dabney, J. B. and T. L. Harman, *Mastering Simulink 4*, Prentice Hall, Upper Saddle River, NJ, 2001.

Daniels, R. W., *An Introduction to Numerical Methods and Optimization Techniques,* Elsevier/North Holland, New York, 1978.

D'Azzo, J. J. and C. H. Houpis, *Linear Control System Analysis and Design*, 4th edn., McGraw-Hill, New York, 1995.

Dorf, R. C. and R. H. Bishop, *Modern Control Systems*, 10th edn., Pearson/Prentice-Hall, Upper Saddle River, NJ, 2005.

Edelstein-Keshet, L., *Mathematical Models in Biology*, McGraw-Hill, New York, 1988.

Eguchi, H., K. Obana, and M. Kamiya, Hardware-in-the-loop missile simulation facility, *Proceedings of SPIE*, 3368, 2–9, 1998.

Etkin, B., *Dynamics of Flight*, John Wiley & Sons, New York, 1982.

Farlow, S. J., *An Introduction to Differential Equations and Their Applications*, McGraw-Hill, New York, 1994.

Fausett, L. V., *Numerical Methods—Algorithms and Applications*, Prentice-Hall, Upper Saddle River, NJ, 2003.

Fishwick, P. A., *Simulation Model Design and Execution—Building Digital Worlds*, Prentice-Hall, Upper Saddle River, NJ, 1995.

Franklin, G. F., J. D. Powell, and A. Emami-Naeini, *Feedback Control of Dynamic Systems*, 4th edn., Prentice-Hall, Upper Saddle River, NJ, 2002.

Gawthrop, P. and L. Smith, *METAMODELLING: Bond Graphs and Dynamic Systems*, Prentice-Hall, London, U.K., 1996.

Gear, W. C., *Numerical Initial Value Problems in Ordinary Differential Equations*, Prentice-Hall, Englewood Cliffs, NJ, 1971.

Gordon, G., *System Simulation*, Prentice-Hall, Englewood Cliffs, NJ, 1978.

Green, R. and K. Jackson, The design drive—Advanced HITL simulation systems for automotive controllers, *Modern Simulation and Training Journal*, 56–58, 1997.

Haberman, R., *Mathematical Models—Mechanical Vibrations, Population Dynamics and Traffic Flow*, Prentice-Hall, Englewood Cliffs, NJ, 1977.

Hannon, B. and R. Matthias, *Dynamic Modeling with STELLA II*, Springer-Verlag, New York, 1994.

Hanselmann, H. and K. Smith, Real-time simulation replaces test drives, *Test & Measurement World Magazine*, 35–40, February 15, 1996.

Haraldsdottir, A. and R. Howe, Multiple frame rate integration, *Flight Simulation Technologies Conference*, Atlanta, GA, September 7–9, 1988, Technical Paper (A88–53626 23–09), 1988.

Hartley, T. T., *Digital Simulation of Dynamic Systems—A Control Theory Approach*, Prentice-Hall, Englewood Cliffs, NJ, 1994.

Hasdorff, L., *Gradient Optimization and Nonlinear Control*, John Wiley & Sons, New York, 1976.

Hay, J. L., R. E. Crosbie, and R. I. Chaplin, Integration routines for systems with discontinuities, *The Computer Journal*, 17, 275–279, 1973.

Hethcote, H., Qualitative analyses of communicable disease models, *Mathematical Biosciences*, 28, 335–356, 1976.

Hoffman, J. D., *Numerical Methods for Engineers and Scientists*, Marcel Dekker, New York, 1992.

Hostetter, G. H., M. S. Santina, and Paul D'Carpio-Montalvo, *Analytical, Numerical and Computational Methods for Science and Engineering*, Prentice-Hall, Englewood Cliffs, NJ, 1991.

Howe, R., Transfer function and characteristic root errors for fixed-step integration algorithms, *Transactions of SCS*, 2, 293–320, 1986.

Howe, R., *Dynamics of Real-Time Digital Simulation,* Applied Dynamics International, Ann Arbor, MI, 1995.

Hultquist, P. F., *Numerical Methods for Engineers and Computer Scientists*, Benjamin Cummings, Menlo Park, CA, 1988.

Huntsinger, R., Personal notes.

Hutton, D. V., *Fundamentals of Finite Element Analysis*, McGraw-Hill, New York, 2004.

Isham, V., Mathematical modeling of the transmission dynamics of HIV infection and AIDS, *Journal of the Royal Statistical Society*, 151, 5–30, 1988.

Jackson, L. B., *Signals, Systems and Transforms*, Addison-Wesley, Reading, MA, 1991.

Kailath, T., *Linear Systems*, Prentice-Hall, Englewood Cliffs, NJ, 1980.

Karayanakis, N., *Computer-Assisted Simulation of Dynamic Systems with Block Diagram Languages*, CRC Press, Boca Raton, FL, 1993.

Karnopp, D. C., D. L. Margolis, and R. C. Rosenberg, *System Dynamics—Modeling and Simulation of Mechatronic Systems*, 4th edn., John Wiley & Sons, New York, 2000.

Keen, R. E. and J. D. Spain, *Computer Simulation in Biology—A Basic Introduction*, John Wiley & Sons, New York, 1992.

Kelton, W. D., R. P. Sadowski, and D. A. Sadowski, *Simulation with Arena*, McGraw-Hill, New York, 1997.

Kermack, W. D. and A. D. McKendrick, A contribution to the mathematical theory of epidemics, *Proceedings of the Royal Society of London*, 115, 700–721, 1927.

Korn, G. A. and J. V. Wait, *Digital Continuous-System Simulation*, Prentice-Hall, Englewood Cliffs, NJ, 1978.

Kraniauskas, P., *Transforms in Signals and Systems*, Addison-Wesley, Wokingham, U.K., 1992.

Kuo, B., *Digital Control Systems*, Holt, Rinehart & Winston, New York, 1980.

Ledin, J., *Simulation Engineering*, CMP Books, Lawrence, KS, 2001.

Linz, P. and R. L. C. Wang, *Exploring Numerical Methods—An Introduction to Scientific Computing Using MATLAB*, Jones and Bartlett, Boston, MA, 2003.

Mathews, J. H. and K. D. Fink, *Numerical Methods Using MATLAB*, 3rd edn., Prentice-Hall, Upper Saddle River, NJ, 1999.

McClamroch, N. H., *State Models of Dynamic Systems*, Springer-Verlag, New York, 1980.

McLeod, J., PHYSBE . . . A physiological simulation benchmark experiment, *SIMULATION*, 7, 324–329, 1966.

Meerschaert, M. M., *Mathematical Modeling*, 2nd edn., Academic Press, San Diego, CA, 1999.

Mesterton-Gibbons, M., *A Concrete Approach to Mathematical Modeling*, Addison-Wesley, Redwood City, CA, 1988.

Miller, K. S., *Partial Differential Equations in Engineering Problems*, Prentice-Hall, Englewood Cliffs, NJ, 1975.

Miller, R. E., *Optimization Foundations and Applications*, John Wiley & Sons, New York, 2000.

Mokhtari, M. and M. Marie, *Engineering Applications of MATLAB 5.3 and SIMULINK 3*, Springer-Verlag, London, U.K., 2000.

Natke, H. G., *Introduction to Multi-Disciplinary Model-Building*, WIT Press, Southampton, U.K., 2003.

Nekoogar, F. and G. Moriarty, *Digital Control Using Digital Signal Processing*, Prentice-Hall, Upper Saddle River, NJ, 1999.

Nise, N. S., *Control Systems Engineering*, 2nd edn., Benjamin Cummings, Redwood City, CA, 1995.

Ogata, K., *Discrete-Time Control Systems*, 2nd edn., Prentice-Hall, Englewood Cliffs, NJ, 1995.

Ogata, K., *System Dynamics*, 3rd edn., Prentice-Hall, Upper Saddle River, NJ, 1998.

Ogata, K., *Modern Control Engineering*, 4th edn., Prentice-Hall, 2002.

O'Neil, P. V., *Advanced Engineering Mathematics*, Wadsworth, Belmont, CA, 1983.

Oppenheim, A. V., R. W. Schafer, and R. J. Buck, *Discrete-Time Signal Processing*, 2nd edn., Prentice-Hall, Eaglewood Cliffs, NJ, 1999.

Orfanidis, S., *Introduction to Signal Processing*, Prentice-Hall, Upper Saddle River, NJ, 1996.

Palm, W. J., *Modeling, Analysis and Control of Dynamic Systems*, John Wiley & Sons, New York, 1983.

Palusinski, O. A., Simulation of dynamic systems using multirate integration techniques, *Transactions of the Society for Computer Simulation*, 2, 257–273, 1986.

Papoulis, A., *The Fourier Integral and Its Applications*, McGraw-Hill, New York, 1962.

Parks, T. W. and C. S. Burrus, *Digital Filter Design (Topics in Digital Signal Processing)*, John Wiley & Sons, New York, 1987.

Perelson, A., Dynamics of HIV infection of CD4+ T cells, *Mathematical Biosciences*, 114, 81–125, 1993.

Ralston, A. and H. S. Wilf, *Mathematical Methods for Digital Computers*, John Wiley & Sons, New York, 1965.

Rao, S. S., *Applied Numerical Methods for Engineers and Scientists*, Prentice-Hall, Upper Saddle River, NJ, 2002.

Recktenwald, G., *Numerical Methods with MATLAB—Implementation and Application*, Prentice-Hall, Upper Saddle River, NJ, 2000.

Reseck, J., *SCUBA, Safe and Simple*, Simon and Schuster, New York, 1990.

Richmond, B., *An Introduction to Systems Thinking: STELLA Software*, High Performance Systems Inc., Hanover, NH, 2001.

Riggs, D. S., *Control Theory and Physiological Feedback Mechanisms*, The Williams & Wilkins Co., Baltimore, MD, 1970.

Rohrs, C. E., J. L. Melsa, and D. G. Schultz, *Linear Control Systems*, McGraw-Hill, New York, 1993.

Schilling, R. J. and S. L. Harris, *Applied Numerical Methods for Engineers Using MATLAB and C*, Brooks/Cole, Pacific Grove, CA, 2000.

Shampine, L., *Numerical Solution of Ordinary Differential Equations*, Chapman & Hall, New York, 1994.

Shearer, J. L., *Dynamic Modeling and Control of Engineering Systems*, Prentice-Hall, Upper Saddle River, NJ, 1997.

Shevell, R. S., *Fundamentals of Flight*, 2nd edn., Prentice-Hall, Englewood Cliffs, NJ, 1989.

Shier, D. R., *Applied Mathematical Modeling*, CRC Press, Boca Raton, FL, 2000.

Smith, W. A., *Elementary Numerical Analysis*, Prentice-Hall, Englewood Cliffs, NJ, 1986.

Smith, J. M., *Mathematical Modeling and Digital Simulation for Engineers and Scientists*, 2nd edn., John Wiley & Sons, New York, 1987.

Speckhart, F. H., *A Guide to Using CSMP—The Continuous System Modeling Program*, Prentice-Hall, Englewood Cliffs, NJ, 1976.

Theusen, G. J. and W. J. Fabrycky, *Engineering Economy*, Prentice-Hall, 1971.

Theusen, G. J. and W. J. Fabrycky, *Engineering Economy*, 9th edn., Prentice-Hall, Upper Saddle River, NJ, 2001.

Tse, I. E., F. S. Hinkle, and R. T. Marse, *Mechanical Vibrations: Theory and Applications*, Allyn and Bacon, 1963.

Wellstead, P. E., *Introduction to Physical System Modelling*, Academic Press, London, U.K., 1979.

Wilde, D. J., *Optimum Seeking Methods*, Prentice-Hall, Englewood Cliffs, NJ, 1964.

Woods, R. L. and K. L. Lawrence, *Modeling and Simulation of Dynamic Systems*, Prentice-Hall, Upper Saddle River, NJ, 1997.

Index

M